Professional Practice in Earth Sciences

Series editor

James W. LaMoreaux, Tuscaloosa, AL, USA

More information about this series at http://www.springer.com/series/11926

Zoran Stevanović
Editor

Karst Aquifers – Characterization and Engineering

Springer

Editor
Zoran Stevanović
Centre for Karst Hydrogeology,
Department of Hydrogeology,
Faculty of Mining and Geology
University of Belgrade
Belgrade
Serbia

ISSN 2364-0073 ISSN 2364-0081 (electronic)
Professional Practice in Earth Sciences
ISBN 978-3-319-12849-8 ISBN 978-3-319-12850-4 (eBook)
DOI 10.1007/978-3-319-12850-4

Library of Congress Control Number: 2015932419

Springer Cham Heidelberg New York Dordrecht London

Printed on acid-free paper

Springer International Publishing AG Switzerland is part of Springer Science+Business Media (www.springer.com)

Foreword

Springer's *Professional Practice in Earth Sciences* (PPES) Book Series was developed to meet the need for present state-of-the-art guidelines to be applied in multiple disciplines of the earth system sciences. The series portfolio contains practical training guidebooks (PTG) and supporting material for academic courses, laboratory manuals, work procedures, and protocols for environmental sciences and engineering.

Items published in the series are directed at researchers, students, and anyone interested in the practical application of science. Books in the series cover the applied components of selected fields in the earth sciences and enable practitioners to better plan, optimize, and interpret their results.

The series is subdivided into the different fields of applied earth system sciences:

Laboratory Manuals and work procedures
Environmental methods and protocols
Training guidebooks

It is my pleasure as Editor of the PTGs to bring the expertise of recognized professionals in their fields to a broader audience. These professionals have presented training courses over many years utilizing the material in the PTGs. Not only will the reader benefit from the years of experience of the presenter, but they will also benefit from the participants' real-life experiences that have been discussed in classes and incorporated into the training materials over time. This makes these PTGs even more valuable because the information contained therein is based on tried and true applications in the field.

Professor Zoran Stevanović is internationally known for his expertise in karst terrains. He has developed a course, *Characterization and Engineering of Karst Aquifers*, with the collaboration of his peers and former students and was recognized by UNESCO for its excellence. A concise explanation of karst and exploration methods enables a better understanding of these unique environments. Systematic and step-by-step tutorials help the reader to learn rapidly how to facilitate sustainable use and protection of groundwater. Many practical examples and lessons learned allow knowledge transfer to similar cases. Numerous illustrations and figures help to better visualize this complicated scientific area of study.

The PTG based on this course is one of the first two PTGs in the series; the other one being Hydrogeological Report Planning and Review Guide by Dr. John Moore and myself. Being the first publications in the series puts added stress on the authors to present a world class product so as to establish the ongoing reputation of the series. It is the authors' hope that the reader will find this to be the case and spread the word about these excellent publications.

James W. LaMoreaux

Preface

When my distinguished colleague and friend Jim LaMoreaux sent me a letter at the beginning of 2013 asking if I could edit a practical guidebook on karst hydrogeology for the new Springer series *Professional Practice in Earth Sciences* I was very honored for several reasons.

First, I felt this as a kind of recognition of the importance and high quality of the school to which I belong, the University of Belgrade, which has a long tradition in hydrogeology and karst-related disciplines. The school was founded at the beginning of the twentieth century by the famous karstologist Jovan Cvijić. Thanks to him and his doctoral thesis *Das Karstphänomen*, karst explorers worldwide use the local Slavic terms doline, ponor, polje, uvala, and others. Since that time, many other Yugoslavian and Serbian karst hydrogeologists have become known and many successful projects have been implemented in the engineering regulation of karst surface and groundwater.

Secondly, I was personally very proud to get such an invitation because I felt it as an appreciation of my earlier efforts and work in karst hydrogeology. Together with colleagues, today I try to strengthen the capacity and ability of our Centre for Karst Hydrogeology at the University and the Department for Hydrogeology.

Thirdly, there is a great practical justification for such a book. In 2012, we started to prepare materials and the organizational setup for the international course and field seminar *Characterization and Engineering of the Karst Aquifers*, a course which will traditionally be held in the city of Trebinje, Bosnia and Herzegovina, the heart of classical Dinaric karst, and which is certified by the University of Belgrade as one of the regular courses of the M.S. Program of the Department of Hydrogeology. The course however is developed to be of interest also to other graduate students or students in their senior years of undergraduate studies in geology, environmental sciences, and engineering interested in the research of karst environments and in the development and engineering of karst water resources. The professionals and decision-makers involved in engineering and management of karst waters or environments should also benefit from the course by improving their understanding of karst processes and sensitivity.

The first course in 2014 supported by UNESCO was attended by more than 20 participants from 11 countries, and lectured by ten professors.

There are many excellent books written on different aspects of karst science. During the last decade the published books of Derek Ford and Paul Williams, and of my compatriots Petar Milanović and Neven Kresic, traced a new path and I am quite certain will soon be recognized as the new classical works on karst hydrogeology. For this reason writing something very new and fresh seemed to be no easy task, but the idea to edit a book which should be less theoretical and more a catalog of possible engineering interventions in karst and their implications sounded attractive and finally got support from the Springer authorities.

For this book I asked for support from several well-known karst scientists, who happen also to be good friends of mine from all over the world, but I invited a few younger and prospective researchers working in and studying classical karst of southeastern and Mediterranean Europe to participate as well. I am proud to have worked with and even to have been a mentor to four of them who have already obtained their Ph.Ds at the University of Belgrade. Their topics considered modern methods and innovative techniques in time series analyses of spring discharges, water reserves and recharge components assessment, and the mathematical and physical modeling of karst. I firmly believe they have properly exploited the chance to be co-authors of this book and have demonstrated a good capability for individual scientific work. However, the final judgment will be with the readers and future attendees of the international course.

The practical guidebook *Characterization and Engineering of the Karst Aquifers* consists of three parts. The first is more introductory. It includes a historical review of karst research as well as a presentation of basic karst terminology, a description of landscape and features, and a discussion of the distribution and importance of karst water resources. The chapter on the characterization of karst aquifer consists of several subchapters. It is a guide to the analysis and evaluation of karst aquifer from various angles essential for its optimal utilization and/or protection. Although presentation of research methodology is not the principal aim of this book an overview of existing and mostly applied methods is given, along with an assessment of their practicability and limitations.

The second part of the book comprises several contributions written by leading experts in related fields. The topics include a consideration of the brief theory and practice of some of the most important steps in karst aquifer engineering control and protection such as the relationship between surface waters and groundwater, budgeting and assessing groundwater resources, evaluating spring discharges regime, and estimating aquifer vulnerability to pollution. The chapters explaining how to model the karst environment physically and mathematically and how to organize monitoring of its water regime are also included in this section. The last chapter presents a catalog of possible engineering interventions aiming to control karstic flows. Plunging ponors and cavities, placing blankets over karstic surfaces, and constructing grout curtains are some such measures aiming to use the hydropower potential of karst waters or prevent them from being diminished.

The third part covers practical experiences in karst waters management. It consists of one introductory article and 15 case studies in different fields of karst hydrogeology. The introductory article aims to provide general guidance on how to deal with problems in karst, create a conception for research, evaluate obtained results, and optimize technical solutions. Discussion of typical conflicts between the water users and possible management resolutions as well as assessment of the environmental impacts of various measures and constructions applied in karst are also discussed. The following case studies describing the most common problems in karst water management are separated into three groups. The first group includes assessment of groundwater availability, and forecasts the effects of further intensive over-extraction as a permanent or temporary measure to overcome water shortage. The second group includes four presentations dealing with karst waters inflow in underground mines, leakage from reservoirs, and mixture of salt or lake waters and fresh groundwater. The third group consists of six contributions which discuss the quality of karst groundwater in terms of protection, contamination, remediation, heating, sharing.

Although the majority of the users of this book will be professionals with a geology/hydrogeology background we have tried to utilize language which is not purely technical but, rather, accessible to a wider audience. In this way presented methodology, case studies, and obtained experiences should also benefit water managers working in karst environments.

I gratefully acknowledge the work and contributions of all my colleagues who are co-authors of this book. I believe their expertise and experience will benefit a wide reading audience and will help toward more successful and sustainable projects to be implemented in karst environment in the future. I am especially grateful to Jim LaMoreaux for giving me the opportunity to edit this book, and to my friend Neven Kresic for his advices and some ideas brought from his infinite mind storage. My personal special thanks go to Beverly Lynch, who made most of the chapters of this book legible and to Branislav Petrović, who provided technical support in preparation of the illustrations for the chapters that I wrote.

Belgrade, September 2014 Zoran Stevanović

Contents

Contributors

Bartolomé Andreo European Topic Centre for Spatial Information and Analysis (ETC/SIA), University of Málaga, Málaga, Spain

Ognjen Bonacci Faculty of Civil Engineering, Architecture and Geodesy, University of Split, Split, Croatia

Veselin Dragišić Department of Hydrogeology, Faculty of Mining and Geology, University of Belgrade, Belgrade, Serbia

Francesco Fiorillo Department of Science and Technology, University of Sannio, Benevento, Italy

Nico Goldscheider Karlsruhe Institute of Technology (KIT), Institute for Applied Geosciences (AGW), Division of Hydrogeology, Karlsruhe, Germany

Igor Jemcov Department of Hydrogeology, Faculty of Mining and Geology, University of Belgrade, Belgrade, Serbia

Neven Kresic Amec Foster Wheeler, Kennesaw, VA, USA

Neno Kukurić IGRAC, International Groundwater Resources Assessment Centre, Delft, The Netherlands

James W. LaMoreaux PELA GeoEnvironmental, Tuscaloosa, AL, USA

Judit Mádl-Szőnyi Faculty of Science, Eötvös Loránd University, Budapest, Hungary

Peter Malík Geological Survey of Slovak Republic, Bratislava, Slovakia

Ana I. Marín Centre of Hydrogeology, Department of Geology, University of Málaga (CEHIUMA), Málaga, Spain

Alex Mikszewski Amec Foster Wheeler, Kennesaw, NC, USA

Petar Milanović National Committee of IAH for Serbia, Belgrade, Serbia

Saša Milanović Centre for Karst Hydrogeology, Department of Hydrogeology, Faculty of Mining and Geology, University of Belgrade, Belgrade, Serbia

Mario Parise National Research Council, IRPI, Bari, Italy

Milan M. Radulović Faculty of Civil Engineering, University of Montenegro, Podgorica, Montenegro

Nataša Ravbar Karst Research Institute, Research Centre of the Slovenian Academy of Sciences and Arts, Postojna and Urban Planning Institute of the Republic of Slovenia, Ljubljana, Slovenia

Vesna Ristić Vakanjac Centre for Karst Hydrogeology, Department of Hydrogeology, Faculty of Mining and Geology, University of Belgrade, Belgrade, Serbia

Zoran Stevanović Centre for Karst Hydrogeology, Department of Hydrogeology, Faculty of Mining and Geology, University of Belgrade, Belgrade, Serbia

Ljiljana Vasić Centre for Karst Hydrogeology, Department of Hydrogeology, Faculty of Mining and Geology, University of Belgrade, Belgrade, Serbia

Vladimir Živanović Department of Hydrogeology, Faculty of Mining and Geology, University of Belgrade, Belgrade, Serbia

Part I
Characterization of Karst Aquifer

Chapter 1
Historical Overview on Karst Research

James W. LaMoreaux and Zoran Stevanović

Most people understand general terms such as springs, groundwater, geology, and hydrology. Newer more familiar terms include bottled water, spring water, sparkling water, or mineral water. Vessels containing these waters come in different shapes, sizes, and colors with various labels as shown in Dr. Philip E. LaMoreaux's book, Springs and Bottled Waters of the World: Ancient History, Source, Occurrence, Quality and Use (Fig. 1.1).

Dr. LaMoreaux's book provides a good summary of the phenomenon sweeping the world of people changing to drinking bottled water instead of drinking water from municipal or other public supplies. Many of these people, however, do not realize that sources of these bottled waters come from springs with thousands of years of history and that these or similar sources have provided water to many different people over the course of civilization. More specifically related to this chapter is that the sources of many of these springs are located in karst or limestone terrains.

Historically, the use of springs for water dates back to earliest civilization. An example is cave drawings made 17,000 years ago in Lascaux Cave in Southern France (Fig. 1.2).

These paintings and engravings during Paleolithic times provide evidence of man using caves for water supply, habitation, and areas of protection.

J.W. LaMoreaux
PELA GeoEnvironmental, Tuscaloosa, AL, USA

Z. Stevanović (✉)
Centre for Karst Hydrogeology, Department of Hydrogeology,
Faculty of Mining and Geology, University of Belgrade, Belgrade, Serbia
e-mail: zstev_2000@yahoo.co.uk

Z. Stevanović (ed.), *Karst Aquifers – Characterization and Engineering*,
Professional Practice in Earth Sciences, DOI 10.1007/978-3-319-12850-4_1

Fig. 1.1 Springs and bottled waters of the world

Fig. 1.2 Documentation of paleolithic cave dwellings

Some of the major points that man throughout time has understood to varying degrees is that hydrogeology plays a major role in environmental planning, execution, and implementation. Much of the world's future water resources will be from groundwater, and life requires sustainable water supplies to exist. Man in karst or limestone terrains has perhaps understood this better than man in other terrains due to the unique nature of karst. In fact, the earliest hydrologic concepts of the hydrologic cycle, water source, occurrence, and quality were developed in relation to karst environments.

An average of about 20 % of the earth's surface is covered by karst or limestone terrains varying by as much as from 0 to 40 %. Yugoslavia where the term originated has one of the most karstic environments at 33 % (Milanović 1981). The geographic region from where the word karst comes is in northwest Yugoslavia near the Italian border. The Slavs used the word kar meaning rock, and the Italians used carso; through Germanization of the two words, this unique terrain became known as karst. Karst typically occurs in carbonate rock formations which are also the sources of many of the famous and not so famous springs in the world.

Springs from limestone, dolomite, and marble have been sites of earliest settlements and religious shrines. Many of the areas in which karst occurs are abundant in resources including water supplies, rock quarries, minerals, and oil. Dorothy Crouch wrote two comprehensive texts on interrelationships between water supplies, groundwater primarily from springs, and development of civilization. These areas have produced diverse topographic expression by weathering under varied geologic and climatic conditions (Figs. 1.3 and 1.4).

Owing to variability and solubility of limestone under different climatic and geologic conditions, man's inhabitation and development have been a challenge.

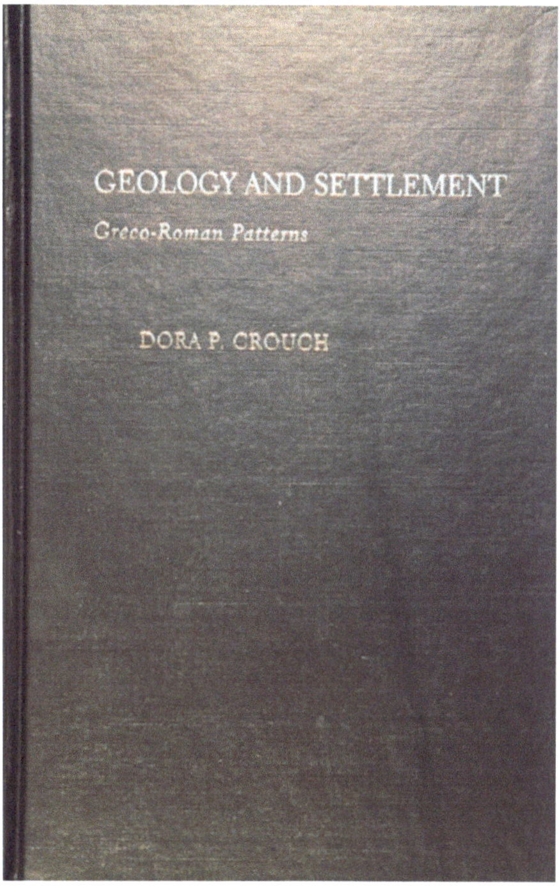

Fig. 1.3 Dorothy Crouch documented early environmental geology of karst

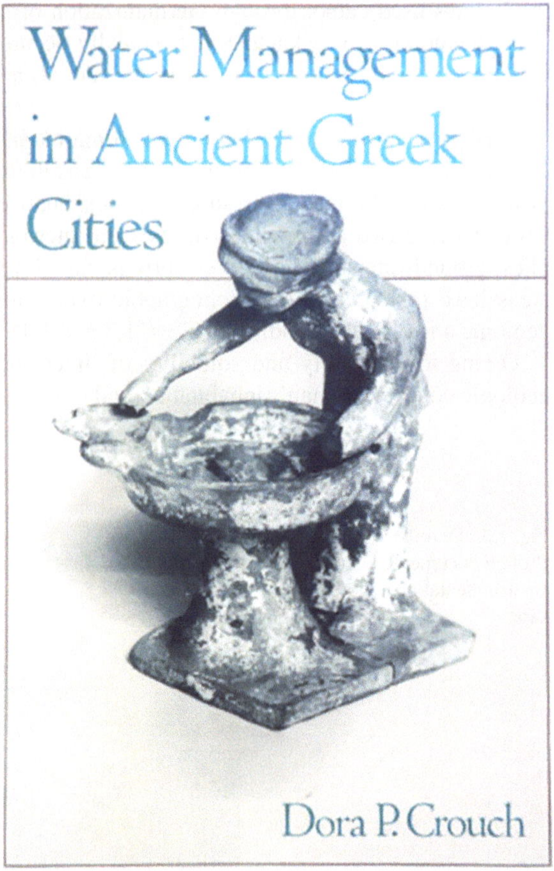

Fig. 1.4 Dorothy Crouch documented early environmental geology of karst

Famous examples of karst terrains and their differences include the barren white desert of Cretaceous chalk near Bahariya Oasis in Egypt, in contrast to large areas of limestone in Southern Europe that are very productive with rich soil.

To provide some historical perspective to the knowledge of karst terrains and their role in history, several examples follow. More extensive descriptions of and through historical development, old documents, and evidence of the importance of karst are presented in detail in the works of LaMoreaux (1991), LaMoreaux and LaMoreaux (2007), Ford and Williams (2007), and Krešić (2013).

Cuneiform tablets are the first written records of hydrological research. Assyrian King Salmanassar III's expedition in 852 BC to the headwaters of the Tigris River is described in the tablets shown in Fig. 1.5.

The headwaters discharge from a karst spring and inscriptions at the cave, from which the spring discharges, state that it is the source of water for the Tigris. It is also the first known reference to stalagmites.

Arabs, Greeks and Romans, and Chinese among others contributed to early knowledge. Some of the earliest known wells date from the Neolithic era about

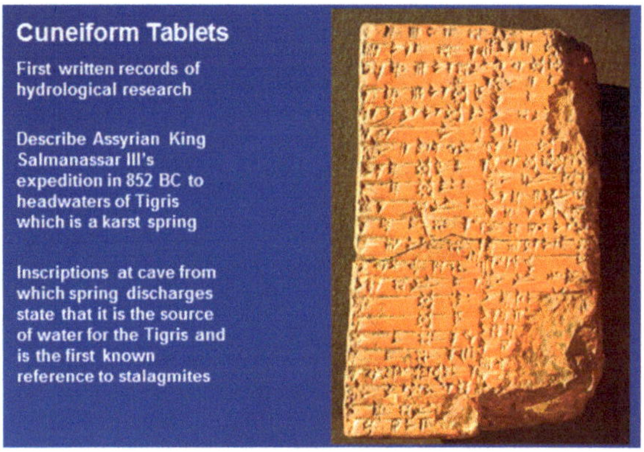

Fig. 1.5 Cuneiform hydrological records, 852 BC

Fig. 1.6 Early religious texts document basic hydrology of Eastern area

10,000 BC. Well construction correlates with common human behaviors including use of farming, animals, and metal tools.

Not only does water development correlate with human behaviors, but it also relates to the earliest evolution of laws and ethics. Some examples of this include writings of the early prophets and the Ten Commandments of the Bible. Descriptions of groundwater documented in the oldest Biblical records mention its use in the cities of Palmyra, Jericho, Sidron, and elsewhere. Dr. O.E Meinzer, former Chief of the Groundwater Branch of the US Geological Survey (USGS), used to comment that parts of the Bible read like a water supply paper (Fig. 1.6).

The spring in Palmyra is warm (33 °C), sulphurous, radioactive, and believed to have curative powers. At Wadi Garandal, at the base of Mount Sinai occur the Hot Springs of Pharoah (thermal 32 °C). These springs were used by Pharaoh's soldiers and by miners and are now a tourist attraction (Fig. 1.7).

Fig. 1.7 Discharge from the hot springs of Egypt

Other contributions from the Middle East include those of the Arabian Order of the Brothers of Purity. In 940 AD, they described caves, springs, and artesian water in An Encyclopedia of Knowledge.

During Greek and Roman times, Empedocles in 400 BC and Aristotle in 300 BC theorized on the hydrologic cycle. Eratosthenes in 100 BC described ponors and poljes. Strabo, at about the time of Jesus Christ (150 BC) devoted his 8th book of 17 volumes to wells and springs. Seneca, an important Roman philosopher in 50 AD, reported on springs, caves, and disappearing springs. Vitruvius in 40 AD was possibly the first hydrogeologist to describe how to search for groundwater.

Aqueducts as architectural master works were developed and designed by the Romans to enable the long distance transportation of high quality waters (Fig. 1.8).

Fig. 1.8 Roman aqueduct, the three-level type

Fig. 1.9 Famous Roman-era springs in Bath, England

At the zenith of the Roman Empire, several aqueducts delivered the incredible amount of around 13 m^3/s of water supply mainly tapped at karstic springs (issued from Simbruini Mountains and Aniene River basin and other catchments) to the center of Rome. Famous springs developed by the Romans include Bath in England (Figs. 1.9 and 1.10) and Baden Baden in Germany.

Hannibal after sacking Rome, camped at Les Bouillens, which is famous as the source of Perrier. Perrier's CO_2 content was rigidly controlled by an edict of Louis XV.

The Chinese also made many contributions to the knowledge of caves and karst, writing a book on caves in 221 BC that describes caves and hydrography. Hongshan Karst Spring in Shanxi Province has been in use since the Song Dynasty (1000 AD). In 1040 AD, Hongshan Spring discharge was separated into three channels to irrigate 100,000 acres. Karst scientist Fan Chengda in 1175 AD, also during the Song Dynasty, explained speleothems and creation of stalactites.

Xu Xiake, known as the "father of karst studies in China," lived during the years 1586–1641 AD in the Ming Dynasty. He studied geomorphology of caves and identified tropical karst features in and around Guilin. According to custodians of the Karst Museum in Guilin, he visited and described about 340 caves in China (Fig. 1.11).

Springs all over the world experience various problems. One of the recurring problems in karst terrains is contamination of the water source. Early on, some of the spring sources became contaminated. They represent some of the first recorded problems with pollution of spring waters. This problem continues today

Fig. 1.10 Bath, England

particularly in developing countries where people may not realize the relationship of water supplies in karst terrain to the disposal of garbage and/or sewerage. Better education is a critical part of communication between the scientific community and the lay public.

As such, problems related to springs and karst terrains are a continuing source of interest to scientists, engineers, planners, regulators, and stakeholders. Whereas research in geology, hydrogeology, and speleology has been utilized to address contamination and other problems that occur in karst terrains in the past, professionals working in the field have realized that solving these problems in the future requires a more multidiscipline approach to develop better solutions.

Before delving further into the multidisciplinary approach, it is necessary to bring the study of karst into more modern times. Karst research in the late seventieth and early eightieth century in Europe began with observation and description. An example is that of the German Melchior Goldast who described the Blautopf, one of Germany's largest karst springs. In addition, the large greatly variable flow of the spring of Vaucluse, France, that lends its name to this type of spring has been measured at regular intervals since 1854.

Balthazar de la Motte Hacquetin in the late 1700s described many karst phenomena in Slovenia and Austria. Kranjc (2006) states that Hacquet is "the father of hydrology" in Europe, as he anticipated problems that would be the subject of controversy in the future. There were many karst explorers of that time, one of whom, for example, the French Edouard Martel, is considered the "father of speleology". In 1907, Emil Racovita published "Essai Sur Les Problemes Biospeologiques". His work is considered to be the birth of biospeleology.

Fig. 1.11 Xu Xiake, Father of karst studies in China

The renowned Serbian scientist, Jovan Cvijić (Fig. 1.12) is often called the "father of karst geomorphology and hydrology" (Ford 2005). Even though in his works Cvijić used the term "subterranean hydrography" and not the exact term hydrogeology, he was one of the real founders of the new disciplines of karstology and karst hydrogeology (Stevanović 2012; Stevanović and Milanović 2013). During his life he served as Professor of Geography at the University of Belgrade, where he also did his undergraduate studies. He later became rector of the university and a member of and president of the Serbian Royal Academy. He has been honored by his likeness appearing on the national currency of Serbia.

Cvijić helped establish the basis for karst science in his doctoral dissertation "Das Karstphänomen" (1892). He followed his dissertation with Karst Poljes of Western Bosnia and Herzegovina in 1900 and Geomorphology in 1924 and 1926. In Geomorphology, Cvijić addressed water tables, karst springs, seepages, and estavellas. His research provided systematic treatment of karrens, dolines, karst rivers, karst valleys, poljes, and other types of karst phenomena (Fig. 1.13).

Fig. 1.12 Jovan Cvijić, famous karst researcher of the 19th and 20th centuries (by Uroš Predić, 1923)

Although Cvijić was the first to separate karst into three morphologic types of limestone terrains, holokarst, merokarst, and transition karst, the use of his nomenclature is relatively limited today. Monroe (1970) in US Geological Survey, Water Supply Paper 1899-K provides a dictionary of karst terms many of which are a derivation of Cvijić's original definitions and terminology.

Many other notable scientists helped develop systematic scientific thought on hydrogeology and karst in particular. Darcy (1856) discovered one of the basic laws of hydrogeology. It has been used in all aspects of hydrogeology, petroleum engineering, and soil science as well as other cases involving flow of liquid through porous media. In experiments during construction of the Dijon, France municipal water supply, Darcy determined basic laws related to water flow

Fig. 1.13 Sketches from Cvijić's early works. Karrens in Igrište uvala on Kučaj Mt. in Eastern Serbia (*left*) and Trebič pothole (abyss) near Trieste declared by Cvijić as representative of this group (vertical fissures and horizontal room-like expansions; right cross-section scheme) (reproduced by permission from Cvijić and karst, eds. Stevanović and Mijatović 2005)

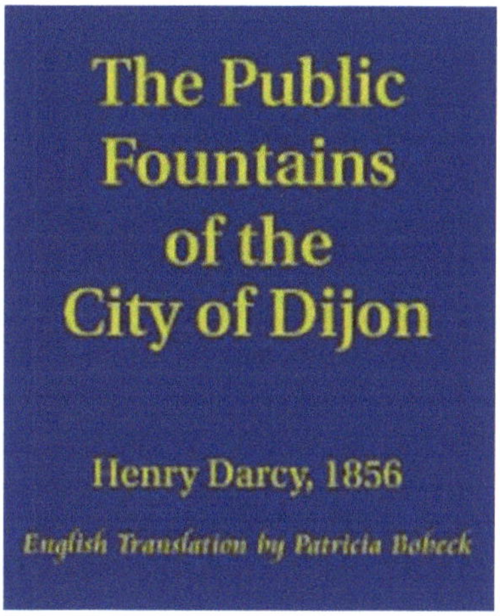

Fig. 1.14 Bobeck's English translation of Darcy's "Les Fontaines publiques de la ville de Dijon"

through sand and published the results. Darcy's historic work was recently translated from the French into English by Patricia Bobeck in 2004 (Fig. 1.14).

Darcy's law set the stage for quantifying occurrence, storage, and discharge of groundwater and became a basic requirement for all aspects of environmental planning and development of this resource.

With much qualitative work having been accomplished by the end of the nineteenth and beginning of the twentieth centuries, quantitative methods for determining ground water velocity and permeability were the next stage of development. These tools required better definition of the physical character and thickness of karst rocks to make them more reliable. During the period beginning in the 1940s, with the need for water during the war years, one of the most difficult problems facing the development and management of groundwater was application of quantitative methods to karst rocks.

In 1935, Theis published his equation to describe non-steady groundwater flow. This was a major breakthrough for modern quantitative investigations. In addition, many new techniques evolved during the latter part of the century to more precisely describe the physical character of geologic systems. Quantitative methods applied to fractured and soluble rocks included sequential satellite imagery, air photography, remote sensing, geochemical laboratory sophistication and computer recording, storage, evaluation and recovery of data, and computer graphic portrayal. More detailed knowledge of the geologic system allowed more meaningful results to be obtained from pumping tests on karst aquifers.

In addition to meeting demands for more water of acceptable quantity, it became necessary for protection against pollution to trace the movement of water from recharge to storage to discharge in karst systems. Tracing groundwater movement in karst required some innovative thinking. Petar Milanović describes the use of geobombs as a unique approach. Other tracing studies involved the injection of various tracers, including sodium fluorescein, potassium and lithium chlorides, pitchblende, plant spores, and even radioactive substances, when little was known about the dangers of radioactive materials. Selected examples include the first large quantitative tracer experiment in 1878 in which tracers were injected into sinkholes of the Danube. A later study in 1908 again injected large amounts of tracers in sinkholes in the Danube near Fridingen.

During the latter part of the twentieth century, a diversity of multidisciplinary talent has been brought to the forefront to study complex problems in karst settings around the world. Professionals involved in these studies include hydrologists, speleologists, chemists, geologists, engineers, biologists, botanists, and mathematicians. Numerous areas of research by professionals in these disciplines have provided results from collaboration on tracing techniques, isotope studies, geomorphology, geochemistry, speleology and sedimentation, and depositional environments for carbonate rocks.

Zötl in 1989 in his "Bibliography of Karst Research" published in the Annotated Bibliography of Karst Terranes as Volume IV identifies the work of selected professionals during this period. Volume IV is one of five volumes by LaMoreaux PE and co-authors; the first two were published by the Alabama Geological Survey (1970, 1976) and the last three through the International Association of Hydrogeologists (IAH) Karst Commission (1986, 1989, 1993). These documents helped establish the bibliographic reference set for future studies in karst. There are also several terminology and encyclopedia books such as the one edited by Gunn (2004). Some of the selected main references for karst hydrogeology are noted within the chapters of this book. Greatly expanded databases also provide a major advantage to contemporary researchers. Two examples are the International Atlas of Karst Phenomena edited by Karl-Heinz Pfeffer and a new digital karst map of the USA by David Weary and Daniel Doctor (2014).

White (2015) notes that various sources of information rapidly increased the knowledge of caves and karst, a schematic of which is shown in Fig. 1.15.

Use of X-ray diffraction and the scanning electron microscope revealed many minerals in cave deposits. Applications of standard engineering fluid mechanics to cave streams and water filled conduits replaced previous assumptions of Darcy's law behavior in karst aquifers. Another important advance was application of isotope geochemistry to provide an absolute chronology for both speleothems and clastic sediments in caves.

A flood of analyses and information from field surveys and laboratory works resulted in different, sometimes contradictory, theories on many aspects of karst hydrogeology, such as concepts of karst origin (for example the prevalence of meteoric or hypogenic factors in its creation), groundwater table and movements (separate concentrated or diffuse flows), the impact of epikarst on springs discharge and so on. Recently Ford and Williams (2007) presented and perfectly explained many of those theories and approaches. Bakalowicz (2005) noticed that "karst schizophrenia" (in reference to the term "hydroschizophrenia" imposed by Llamas 1975) may suggest the wrong idea that each karst unit is characterized by its own specific characteristics with no possibility of approaching it by general laws or any common approach. As well, in order to provide something very new, innovative, and if possible extraordinary, relatively simple things can sometimes become very complicated.

To help address these complications, it helps to develop a framework into which to incorporate this information. The first series of environmental atlases in

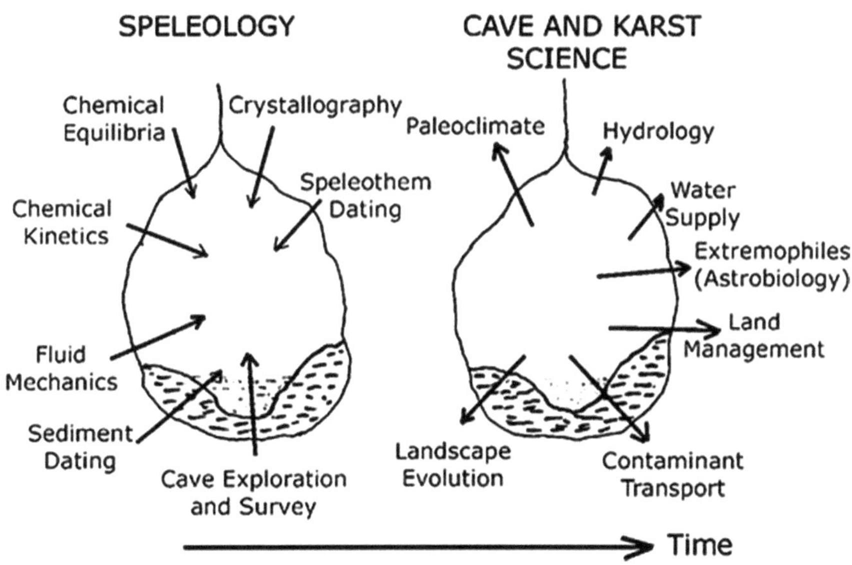

Fig. 1.15 William White (2015): Schematic of karst research.

the USA was published in the 1970s, 10 years before the environmental movement in the USA described the fragile character of karst terranes (LaMoreaux 1971; Moser and Hyde 1974). They were the culmination of a series of studies: well and spring inventories, test drilling, remote sensing, pumping tests, and chemical analysis of karst waters.

These atlases have become the basis for water supply development, urban planning, municipal rules and regulations, and resolution of damage litigation. They provide guidance for the location and development of wells, pumpage from springs, protection of groundwater supplies, identification of possible areas of subsidence, and construction of all types. Information from these atlases and from later maps has been gathered together under the title "Engineering Aspects of Karst," The National Atlas, US Geological Survey (Fig. 1.16).

http://www.nationalatlas.gov/articles/geology/a_karst.html.

The development of the study of karst terrains and its increasing importance has been recognized by the scientific and engineering communities. Political leaders, regulators, and the general public have realized that karst terrains are major sources of water supplies and that their yield, reliability, and protection from contamination must be managed accordingly (White 2015). Not only has the value of karst for water supplies taken on greater significance, but also, as more vulnerable karst areas are subject to development, the protection of the public from hazards that occur in karst areas has come under increased scrutiny as well.

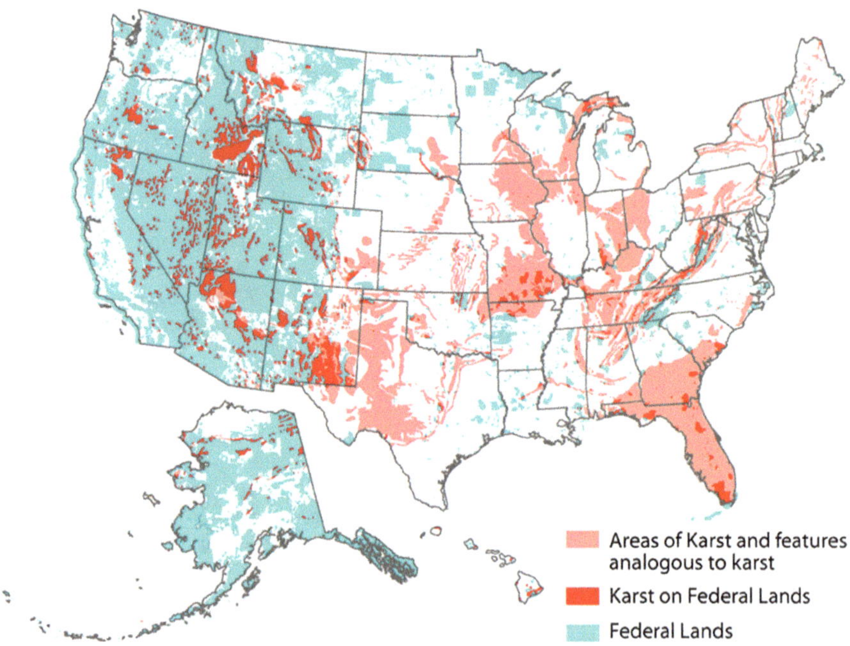

Fig. 1.16 Example of karst atlas mapping program: 16 % of federally owned land in the USA is underlain by karst

Symposia and conferences which were originally organized on a regional basis to address karst science have now become worldwide meetings. Many proceedings, special issues, and book series have been published from presentations at these conferences and international collaborations that originated as a result of them. Selected journals that have an emphasis on research in karst include Environmental Earth Sciences (formerly Environmental Geology), Carbonates and Evaporites, and Hydrogeology Journal. The Cave and Karst Systems of the World Book Series is one that has developed from the worldwide interest in karst.

Entities organizing these conferences and projects include IAH, IASH, FAO, and UNESCO within the International Hydrologic Decade (IHD) and International Hydrologic Programmes (IHP) and university groups. For example, the IHD included a Commission for the Study of Carbonate Rocks in Mediterranean Countries, and since 1970, a permanent Commission for Karst Hydrogeology exists within IAH. The IAH promotes cooperation between scientists who are working on hydrogeologic problems and is affiliated with the International Union of Geological Sciences (IUGS), among others.

This chapter is the beginning of a Practical Training Guide Book Series which will facilitate training of practitioners and the lay public alike in this burgeoning area of scientific study.

References

Bakalowicz M (2005) Karst groundwater: a challenge for new resources. Hydrogeol J 13:148–160

Darcy H (1856) Les Fontaines publiques de la ville de Dijon. Dalmont, Paris

Ford D (2005) Jovan Cvijić and the founding of karst geomorphology. In: Stevanović Z, Mijatović B (eds) Cvijić and karst/Cvijić et karst. Board of Karst and Speleology, Serbian Academy of Science and Arts, Belgrade (special edition), pp 305–321

Ford D, Williams P (2007) Karst hydrogeology and geomorphology. Wiley, England

Gunn J (ed) (2004) Encyclopedia of caves and karst science. Fitzroy Dearborn, New York, p 902

Kranjc A (2006) Baltazar Hacquet (1739/40-1815), the pioneer of karst geomorphologists. Acta Carsologica 35(2):163–168

Krešić N (2013) Water in karst: management, vulnerability and restoration. McGraw Hill, New York

LaMoreaux PE (1971) Environmental Hydrogeology of Karst, Geological Survey of Alabama, Tuscaloosa, AL USA

LaMoreaux PE (1991) History of karst hydrogeological studies. In: Proceedings of the international conference on environmental changes in karst areas. IGU – UIS, Quadreni del Dipartimento di geografia, No. 13. Universita di Padova, Padua, 15–27 Sept 1991, pp 215–229

LaMoreaux PE, Tanner JT (eds.) (2001) Springs and bottled waters of the World: ancient history, source, occurrence, quality and use, Springer

LaMoreaux PE, LaMoreaux J (2007) Karst: the foundation for concepts in hydrogeology. Environ Geol 51:685–688

Llamas MR (1975) Noneconomic motivations in ground-water use: hydroschizophrenia. Ground Water 13(3):296–300

Milanović P (1981) Karst hydrogeology. Water Resources Publications, Littleton

Monroe WH (1970) A glossary of karst terminology: U.S. Geological Survey, Water-Supply Paper 1899-K

Moser PH, Hyde LW (1974) Environmental geology as an aid to growth and development in Lauderdale, Colbert and Franklin counties, Alabama. Alabama Geological Survey Atlas, series 6, p 45

Stevanović Z (2012) History of hydrogeology in Serbia. In: Howden N, Mather J (eds) History of hydrogeology. International contribution to hydrogeology. CRC Press and Balkema, Boca Raton, pp 255–274

Stevanović Z, Mijatović B (eds) (2005) Cvijić and karst/Cvijić et karst. Board on Karst and Speleology Serbian Academy of Science and Arts, Belgrade (special edition)

Stevanović Z, Milanović S (2013) Karst in Serbian hydrogeology: a tradition in research and education. Eur Geol 35:41–45

Weary D, Doctor D (2014) Digital karst map of the United States. Draft, US Geological Survey, Reston, VA

White WB (2015) Introduction, In: Andreo B, Carrasco F, Duran JJ, Jimenez P, LaMoreaux JW (eds) Hydrogeological and environmental investigations in karst systems, Springer, Heidelberg, pp xxi–xxv

Chapter 2
Karst Environment and Phenomena

Zoran Stevanović

In saxis ac speluncis permanant aquarum liquidus umor et uberibus flent omnia guttis.

Titus Lucretius Carus

(in: De rerum natura; around 70 BC)
(Water flows through rocks and caves and everything drips plentiful tears)

2.1 Past Karst as a Human Shelter and Mythic Understanding of Karst

A karst environment is something extraordinary. Other landscapes or rocks may be equally beautiful or remarkable, but it is only karst that often provides the opportunity to delve into its inner secrets to enable confirmation or revision of our ideas, theories, and engineering solutions.

How did this relationship with karst come to be? Caves were the first shelters of big animals, and then of Neanderthals and humans whose movement and settlements were chosen in accordance with the presence of caves as safe havens. Most archaeological evidence is found in the caves, giving us the chance to reconstruct human evolution and the kinds of life in ancient times. This is also how we know that caves served as sanctuaries or burial places as well.

For example, Shanadar cave in northern Iraq (Fig. 2.1), close to edge of the Mesopotamian basin built into massive Lower Cretaceous limestones, is one

Z. Stevanović (✉)
Centre for Karst Hydrogeology, Department of Hydrogeology,
Faculty of Mining and Geology, University of Belgrade, Belgrade, Serbia
e-mail: zstev_2000@yahoo.co.uk

Z. Stevanović (ed.), *Karst Aquifers – Characterization and Engineering*,
Professional Practice in Earth Sciences, DOI 10.1007/978-3-319-12850-4_2

Fig. 2.1 Shanadar Cave (northern Iraq)

of the world's archaeological treasuries well known for Neanderthal skeletons (c. 60,000 years ago). The Shanadar site was discovered in 1951 (Solecki 1955, 1975) when 14 m of stratified cultural layers were excavated from the cave along with stone tools associated with Neanderthals. But more importantly, analyses of pollen taken from soil samples and from around the skeleton of Shanadar IV revealed different plant species probably chosen for their medicinal purposes and possibly also providing evidence of Neanderthal funeral rituals (Maran and Stevanović 2009).

The famous *Petralonian Archantropus* of Petralona, Greece, discovered in 1960 is assumed to date back some 700,000 years (http://www.petralona-cave.gr). Findings as well as replicas and reconstructions are exhibited at the Anthropological Museum of Petralona along with animals' bones covered by $CaCO_3$ (Fig. 2.2).

Remnants of *Homo sapiens* are found in numerous caves. In fact, caves were their principal settlements. Caves such as Altamira in Spain, Chauvet and Lascaux in France, Zhoukoudian and Zengpiyan in China, and many others are famous for Palaeolithic paintings, skeletons, bones of mammals, or used tools. Later on, in the Middle Ages, when the caves became associated with devils and hells, people avoided entering them (Cigna 2005).

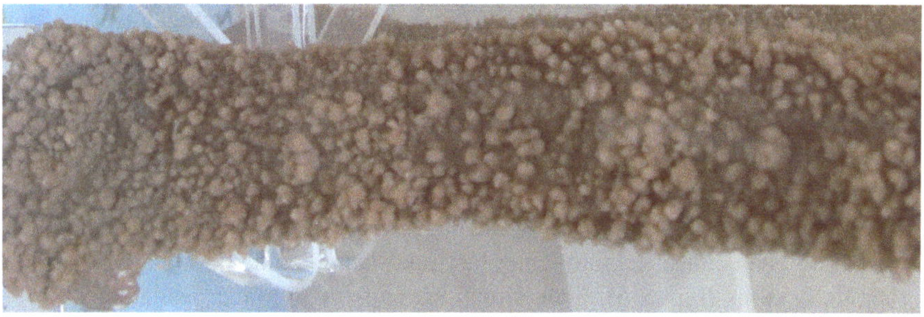

Fig. 2.2 Animal (panther) bone with stalagmitic cover. Details from exhibition in the Anthropological Museum of Petralona

2.2 Present—Man and Karst

Specific karst landscape and hydrographic networks as well as the high permeability of karstified rocks greatly influence both the distribution of flora and fauna and human life. Karstified rocks of different lithology cover more than 10–15 % of the continental ice-free surface of our planet (Ford and Williams 2007). In many places worldwide, where is karst, there are also limited natural resources. Of course, this depends greatly on climate conditions, and karst in arid areas is very different from karst in humid ones, or in zones permanently covered by ice (permafrost). Secondly, great differences for karst properties are powered by the altitude: Karstic terrains in high mountains often function as recharge zones with a shortage of water on their surface, while in littoral karst saltwater intrusion may disturb the freshwater supply (Fig. 2.3). One way or another, the karst environment is not always friendly and various kinds of intervention are sometimes needed to adapt the ambience to human needs.

Water shortage at high plateaus above erosional bases and drainage areas is well known in many places in the world. An extreme example is the mountain

Fig. 2.3 Glacial sinking lake at the Durmitor high mountains (Montenegro *left*) and littoral karst in direct contact with seawater (Zakinthos Island, Greece)

area above the Boka Kotorska Bay in Montenegro. Here, the annual average rainfall sum reaches 5,000 mm, but on account of fast infiltration, only collected rain water or no water at all is available at the surface, with the exception of small springs which drain rare perched aquifers.

The "fight for water" in such extreme conditions may result in migration of local inhabitants, reduction in the number of cattle and small ruminants, and limited cropping (Fig. 2.4). On the other hand, during periods of floods, many karstic poljes have been converted into temporary lakes, and people have had very limited cropping seasons (Fig. 2.5). Such situations and very limited available resources for normal life are well known in the Mediterranean region and in Dinaric karst in particular, and several anthropo-geographic studies describe the impact of current life conditions on human personality and psychology types. Although every generalization is relative, Cvijić (1914) highlighted the following as general

Fig. 2.4 Cisterns—specific reservoirs and intake wells for collecting rain water in high karstic mountains of Montenegro (Vilusi site *left,* and Grahovo village *right*)

Fig. 2.5 Channeled Trebišnjica River in mid of the Popovo polje, eastern Herzegovina. Previously the largest sinking stream in all of Europe, but also a cause of floods

Fig. 2.6 Villager in high Kurdistan mountains (Qandil Mts. northern Iraq, photograph courtesy of Alec Holm)

characteristics of Dinaric people: intellectual sensitivity, instinct to feel danger and to survive, great energy, and highly imaginative, often leading to mysticism. In addition, cheerfulness and a love of humor can be evident, but so too can vanity, pride, and, not rarely, stubbornness.

There are many karstic regions with rugged reliefs which result in the isolation or limited movement of the local population (Fig. 2.6). Although Papua New Guinea is not a purely karstic country, it is an excellent example of diversity and a variety of islanders. This country, with its glacier-capped mountains that reach 4,500 m asl and with the deepest caves in the Southern Hemisphere, is the most linguistically diverse region in the world. The isolation of local inhabitants has resulted in some 800 languages on the same island.

The karst terrains and their underground world are also an ideal place for tourism and extreme sports. Despite opposition from the "green" society which objects to tourist (show) caves as destructive of nature, they are ideal sites for learning about karst processes and the variety of nature. It is estimated that 150 million people visit show caves each year. Cigna (2005) assumed that with this figure the total amount spent to visit caves is around 3.2 billion USD annually and that around 100 million people worldwide earn salaries from show cave business. Beautiful canyons carved in limestones or evaporitic rocks also attract many visitors, while the wild rivers which created them are good places for rafting or canoeing.

There are many geoparks or reservation areas linked with karst (Fig. 2.7). The Nahanni National Park in Canada, the national park Plitvice in Croatia, the Lascaux cave in France, and Grand Canyon National Park, USA, were the first four sites containing karst features inscribed onto the UNESCO World Heritage list in 1978/1979. Today, the list contains 45 world heritage sites which "contain karst of outstanding universal value" (Williams 2008). The largest number (five) is in China and in Australia.

UNESCO is also an umbrella for geoparks as another cluster of protected outstanding geoheritage sites. As of January 2012, there were 90 Global Geoparks spread across 26 countries but mainly concentrated in Europe and in China. Many of them are pure karstic or include karst features. Some of the typical karstic are

Fig. 2.7 Tower in deltaic quartz conglomerate, fluvio karst in Meteora (World Heritage Site in Greece) *left*; Brenta Dolomites in Adamello-Brenta Geopark (northern Italy) *right*

the Carnic Alps (Austria), the Réserve Géologique de Haute Provence (France), the Swabian Alps (Germany), the Sierras Subeticas (Spain), Dong Van karst geopark (Vietnam), and Shilin geopark (China).

Speleology is a science, but it is also one of the most attractive disciplines open for wider purposes. When one declares himself a speleologist, the first impression is that he is caver or cave lover who visits and explores caves and potholes but who is not necessarily going to explain their origin or functioning or collect paleoenvironmental records. Many Web sites today are polygons for the race for the longest cave or deepest speleoobjects or for photographs showing the most beautiful speleothems. The Mammoth Cave system with the connected Flint Ridge Cave system in Kentucky, USA, is the world's longest natural cave network with more than 590 km of explored passages. Another world record for the deepest shaft in the world is Voronja (Krubera) Cave in the western Caucasus Mountains, Georgia, which is 2,190 m deep (Williams 2008).

Furthermore, diving in caves is a discipline which is becoming more and more prevalent in aquatic sport and may provide excellent information for engineering projects aiming to control karstic groundwater flows (Fig. 2.8).

Diving may be supported by the use of remote (autonomous) underwater vehicles for smaller passages. In contrast, for large underwater cave systems, it is common even to ride dive scooters as in the case of karst of Yacatan, Mexico, or Florida, USA.

2.3 Water and Karstic Rocks

No karst landscape or small karstic feature could be created without water. Water is the main agent for the destruction and dissolution of rocks. But when we consider that despite their voids or cavities, volcanic rocks such as the basaltic type do

Fig. 2.8 Diving in large karstic channels of Buna Spring (Bosnia and Herzegovina) (*photograph courtesy* of Claude Touloumdijan)

not belong to the karst group, we see that the sedimentary rocks are practically an exclusive media in which karstic processes take place. Thus, karstic rocks always deal with water: They were produced from water; later, again under the influence of water, they were modified at the surface; and today, they are "drinking" rain or stream water and store this water in the ground.

Sedimentary karstic rocks can be generally classified into the two major groups:

- carbonate rocks and
- evaporite rocks.

The carbonate rocks are formed from calcium and magnesium minerals—calcite, dolomite aragonite, and magnesite—and include the two major groups:

- limestones ($CaCO_3$) and
- dolomites ($CaCO_3$ x $MgCO_3$).

With a wide range of varieties, it is conventionally recognized that pure limestones contain 90–100 % of calcite (Fig. 2.9), pure dolomites (often called as dolostones) 90–100 % of dolomite mineral, while dolomitic limestones have 50–90 % of calcite and 50–10 % of dolomite.

The classification of carbonate sediments and facies is given by Dunham (1962), Folk (1965) and Wilson (1975).

Fig. 2.9 Calcite minerals (*left*) and banks of marly limestones (*right*) (*photographs courtesy* of Aleksandra Maran)

Discussing the origin of dolomites, Benson (1984) stated that some dolomite may primary precipitate, but mostly, it is formed through the replacement of original aragonite or calcite. This process of *dolomitization* has several driving mechanisms, and no consensus among specialists as to which is the primary one has yet been reached. The pure limestones are usually younger than dolomites, while dolomitic complexes prevail among Paleozoic and older, Pre-Cambrian formations.

Most carbonates are of organic origin. The constituent components of carbonate rocks are the following:

- allochemicals,
- micrite,
- cement, and
- non-carbonates (impurities).

There are six allochemical constituents: skeletal grains, ooids, oncoids, peloids, intraclasts, and aggregate grains (Benson 1984). The term micrite is proposed by Folk (1965) and explains microcrystalline lime mud which is common in slowly transported carbonate masses. This is an equivalent of detrital matrix in terrigenous deposits. The cement (connection material) resulted from saturated solutions and can be calcitic, aragonitic, or dolomitic. Non-carbonate material is terrigenous detritus, and its content in carbonates is usually minor.

Due to the dominant presence of the same basic minerals, cherts, conglomerates, breccias (Fig. 2.10), and finally marbles as metamorphosed sedimentary complex also belong to this carbonate group.

To the *evaporite group* belong rocks and minerals which contain SO_4 or Cl anions:

- anhydrite ($CaSO_4$),
- gypsum ($CaSO_4 \times 2H_2O$)
- halite (HCl), (Fig. 2.11), and
- sylvite (KCl).

Fig. 2.10 Carbonate breccia (*photograph courtesy* of Aleksandra Maran)

Fig. 2.11 Halite deposits, Konarsiah salt diapir (central Iran *left*), and laminated impure salty sequences, Ocnele Mari, Romania (*right*)

The quartzites and quartzitic rocks (SiO_2) can be distinguished as an additional group, but due to the structure of these rocks and their texture, karstified parts can be totally absent, or only surficial forms such as karrens may be developed.

There are the two main karstification forces: *mechanical* and *chemical*.

Mechanical erosion usually starts with surface water flows. The primary porosity of most carbonate sedimentary rocks does not allow for fast decomposition if it is created only by the pressure of water flows. As illustrated on Figs. 2.12 and

Fig. 2.12 Karstification over horizontal bedding plane of Hamanlei Fm. (limestones of Doggerian age) in eastern Ethiopia (a small coin is to a scale)

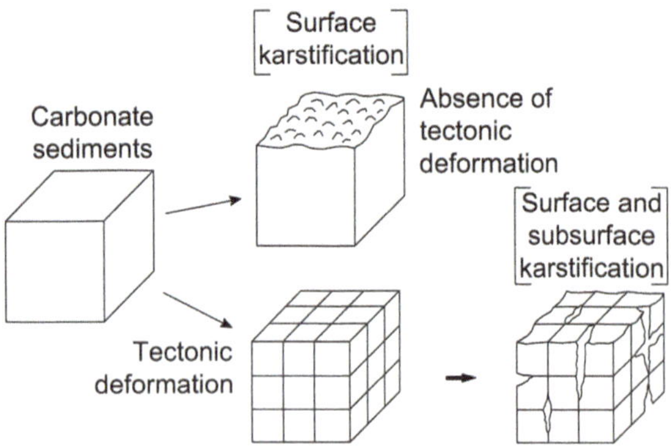

Fig. 2.13 Karstification mechanism (modified from Drogue 1982): Degradation and limited surficial karstification take place in compact carbonate blocks without tectonic deformation, while mechanical and chemical erosions as geodynamical factors create secondary porosity in the rock interior when tectonic deformation exists

2.13, in the case of the absence of tectonic deformation, karstification stays on the rock surface. Thus, small fissures and joints at the surface of the rocks or rough bedding planes are needed to have the process initiated in the rock interior.

When deeper thermal flows exist, the situation is quite different: Hypogenic karstification inside the rock mass takes place, and the process is activated, but again it starts along the deep faults as privileged paths. Mechanical erosion causes the expansion of the small, even microfissures, but during geological times, larger voids, caverns, and finally caves can be created. However, the process of mechanical erosion cannot be separated from the chemical process, and as a rule, they are integrated and function mutually. Which prevails is a matter of local conditions and many other factors.

Chemical erosion or *corrosion* is the process of rock dissolution. But not all rocks are equally soluble. The ranking list established by Freeze and Cherry (1979) for some representative minerals shows that calcite is six times less soluble than gypsum, but even 1,000 times less than halite. However, carbonate minerals are among the most reactive on the earth's surface. According to Wollast (1990), the calcium and magnesium bicarbonates are the most abundant ions present in freshwaters and when exposed to weathering result in roughly 50 % of the chemical denudation of the continents.

The dissolution of calcite (2.1) and dolomite (2.2) is expressed by the following equations:

$$CaCO_3 + H_2O + CO_2 \leftrightarrow Ca^{2+} + 2HCO_3 \tag{2.1}$$

$$CaMg(CO_3)_2 + 2H_2O + 2CO_2 \leftrightarrow Ca^{2+} + Mg^{2+} + 4HCO_3 \tag{2.2}$$

Similarly, dissolution of gypsum (2.3) is as follows:

$$CaSO_4 \times 2H_2O \rightarrow Ca^{2+} + SO_4^{2-} + 2H_2O \tag{2.3}$$

Solubility also depends on temperature and pressure. When the dissolution process is endothermic (heat is absorbed), solubility increases with rising temperatures, but when the process is exothermic (heat is released), solubility decreases with rising temperatures.

The concentration of carbon dioxide enhances solubility. CO_2 is present in the air, in the soil where it is produced mostly by biological activity, and in deep confined parts where it comes ascendantly with basinal fluids and other gases such as H_2S through open, hydrogeologically active faults. As dissolved CO_2 is driven by the temperature and partial pressure of the atmosphere (Bakalowicz 2005), climate is therefore one of the main factors which drives karst processes.

In conclusion, the karstification process is very rapid with respect to geological time. Bakalowicz (2005) affirms the statement that a few thousand years, generally less than 50,000 years, are required to develop an integrated karst network.

2.4 Karst Classifications and Distribution

2.4.1 Classifications

There are many classifications and regionalizations of karst. The classification principles include karstification factors and driving mechanisms, but also lithology types, morphological forms, and genetic or climatic conditions.

Sweeting (1972), by accepting Cvijić's basic classification of *holokarst*, *mero-karst*, and *transitional karst,* distinguishes as follows:

1. Holokarst (fully developed),
2. Fluvio karst,
3. Glacial–nival karst,
4. Tropical karst, and
5. Arid and semi-arid karst.

Gvozdeckiy (1981) classifies several morphogenetical types of karst in the territory of the former USSR. Among them are paleorelief (buried) karst, covered karst, relict tropical karst, permafrost karst, and seashore karst. In addition, based on lithology, Gvozdeckiy makes the following divisions:

1. Limestone karst,
2. Dolomite karst,
3. Marble karst,
4. Chalky and marly karst,
5. Gypsum–anhydrite karst, and
6. Salty rocks karst.

Herak et al. (1981) categorize orogenic and epiorogenic karst based on tectogenetic factors. To the first group belong the following:

1. Lenticular karst (lens in orogenic structure),
2. Folded karst (system of anticlinoriums and synclinoriums formed before karstification, e.g., Alpine orogenic belt, Figs. 2.14 and 2.15),
3. Dissected karst (intensive tectonic disturbance, faulting, erosion, and dissolution, Fig. 2.16),
4. Accumulated karst (development of large forms as poljes, deep karstification often below erosional base, Fig. 2.17).

According to Herak et al. (1981), epiorogenic karst includes epicontinental rock sequences which overlie older platforms and orogens (e.g., Russian or Arabian Pre-Cambrian platforms, Paris basin, etc.). The group comprises tabular, homoclinal, folded, basinal, and deep types of epiorogenic karst.

In accordance with the main genetic factors of the creation of caves, i.e., speleogenesis, Klimchouk et al. (2000) differentiate the two main groups which may be identified as karst typology:

1. *Hypergene* or (*meteoric, unconfined, phreatic*) *karst.*
2. *Hypogene* (*deep, confined*) *karst.*

While in the first group karstic features are created by infiltrated meteoric waters, juvenile waters and gases are the main dissolution agents of hypogene karst. In addition to the two main groups, there are younger coastal caves developed in rocks which have newly appeared.

Several other classifications recognize the karst types in accordance with some specific developments or forms created by karstic or other mostly tectonic and

Fig. 2.14 Highly folded limestones at bank of Dokan Lake (north Iraq)

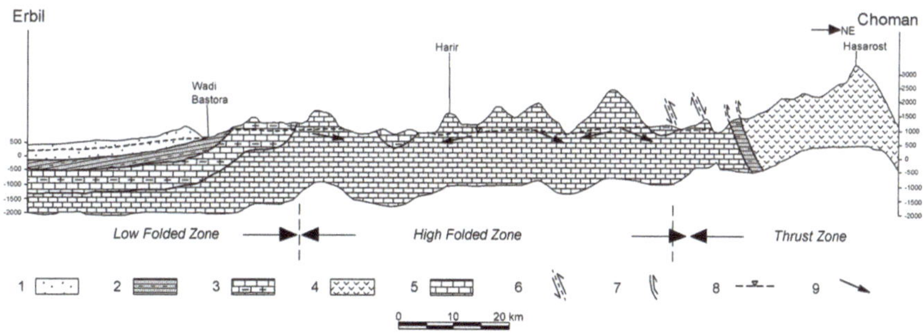

Fig. 2.15 Typical folded karst: The system of anticlines and synclines in Alpine orogenic belt: cross section from Zagros Mts. to Erbil plain and Mesopotamian basin, northern Iraq (after Stevanović and Iurkiewicz 2009)

erosional synergetic processes, as well as those formed under specific climate conditions. Below are explanations of some of these types.

Exokarst is a surficial type of karst, while *endokarst* considers internally developed forms inside unconfined karstic rocks. *Cryptokarst* refers to confined karst

Fig. 2.16 Dissected karst: Tectonically disturbed limestones and subvertical residual strata of Pila Spi Fm. (Darbandikhan area, northern Iraq)

Fig. 2.17 Accumulated karst: Small karstic polje Nava de Cabra, Sierra Subeticas Mt. Spain (*left*), and one of the submarine springs in the Boka Kotorska Bay, Montenegro (*right*)

and forms developed beneath the blanket of soil, till, residual clays, or similar sediments (Ford and Williams 2007).

Buried karst, paleokarst, and *fossil karts* are synonyms for covered karst, while *exhumed karst* is covered, but its actual status and exposure to the surface is different than in the creation phase.

Bare karst is a term commonly used for karst which either developed, or exists now, with a lack of vegetation. However, the same name is often applied to the karst which consists of big blocks with fast vertical karstification but without *epikarst*. The late term is used for the net of small fissures and a destroyed subsurficial part of karst. The synonym for bare karst is *naked karst* and is commonly used to explain either both a lack of vegetation and a lack of *epikarst*, or just one of them.

Barré karst or barrier karst is an isolated karst bordered by generally impervious rocks (Herak et al. 1981) or karst where the process has been interrupted due to the transgression of sea or lake waters. The latter may cause secondary plunging and dislocation of primary discharge outlets.

Pseudo karst represents landscape similar to karst and created karst-like forms but which has been created by other processes (Ford and Williams 2007).

A huge number of books and papers have been written on different aspects of karst hydrogeology, geomorphology, speleology and other karst-related disciplines over the last hundred-plus years. LaMoreaux PE and co-authors made a great effort to collect as many of them as possible and by 1993 he had edited five volumes of the Annotated Bibliography of karst terranes, three of them being published by the International Association of Hydrogeologists (1986, 1989, 1993).

2.4.2 General Distribution

Ford and Williams (2007) stated that surface and subsurface outcrops of potentially soluble karstic rocks occupy around 20 % of the planet's ice-free land, but as previously mentioned probably not more than 10–15 % is extensively karstified. The same authors found that probably more than 90 % of the evaporitic rocks anhydrite and gypsum do not crop out, while this percentage in the case of salty rocks is almost 99 %: Salt is remarkably soluble (360 g/l at 20 °C) and highly ductile, and the salt deposits as a rule quickly erode after precipitation (Zarei and Raeisi 2010, Fig. 2.18).

The sketch maps showing regional distribution of carbonate rocks are presented in Ford and Williams' book *Karst Hydrogeology and Geomorphology* (2007). The same map with a somewhat better resolution is available at the Web address of the University of Auckland, New Zealand http://www.sges.auckland.ac.nz/sges_research/karst.shtm.

Based on the experience of the creation of the Hydrogeological World Map (WHYMAP, BRG and UNESCO, http://www.whymap.org/whymap/EN/Home/whymap_node.html), the project World Karst Aquifer Map (WOKAM) was established in 2012. The ultimate goal of this project is to create a world map and database of karst aquifers. The new map will be based on the detailed Global

Fig. 2.18 Salt spring and deposits in the foothill of Konarsiah diapir (Iran)

Lithological Map (GLiM) but will show not only carbonate rock outcrops, but also deep and confined karst aquifers, large karst springs, including thermal and mineral springs, drinking water abstraction sites, and selected caves (Goldscheider et al. 2014).

The statistics concerning areas of carbonate rock outcrops are also available at the Web address of the University of Auckland, New Zealand. The total surface of carbonate outcrops on the continents, excluding Antarctica, Greenland, and Iceland, is around 17.7 million km^2 or 13.2 %. The largest portion is in Central and North America, with around 4 million km^2, but the largest contribution of carbonate rocks in the total surface coverage is in the Middle East and Central Asia, where they occupy around 23 % of the total land.

2.4.3 Regional Distribution

Previously presented classifications of karst also include regional and climatic factors. Although we have to agree that mixing classification criteria is not a scientifically correct approach, for practical reasons, we may distinguish several basic

Fig. 2.19 The littoral karst (Zakinthos, Greece *left*) and high alpine karst (St. Moritz, south Switzerland)

regional types of karst in accordance with their geographical position and dominant karstic occurrences:

1. Geosynclinal carbonate karst,
2. Platform carbonate karst,
3. Tropical karst,
4. Hypogenic and evaporitic karst,
5. Glacial karst.

Geosynclinal carbonate karst was formed in large sedimentary basins later exposed to epirogenic uplifting and to intensive orogenic folding. The typical example is the Mediterranean karst which was created in the Tethys sedimentary basin and includes the three subgroups: *littoral, hilly-mountains,* and *high alpine karst.* To the first group belong islands, shorelines, and adjacent coastal areas of northern African and Near East countries, Turkey, Greece, Albania, former Yugoslavia, Italy, France, and Spain as well as of some other smaller countries (Fig. 2.19). The second group encompasses mountain chains of the Atlas, Pyrenees, Provencale Mts., Apennines, Dinarides, Pindes, Hellenides, and Taurides, with hilly and mountain karstic relief dissected by numerous karstic poljes and wide valleys. High alpine karst extends over the central Alps (Austria, south Germany, south Switzerland, north Italy, and north Slovenia, Fig. 2.19) and is characterized by rough relief, steep slopes, and highly folded rocks. The climate conditions in the Mediterranean basin are quite diverse: from glacial karst at the tops of the Alps to the semi-arid and arid karst in North Africa and the Near East. The alpine systems extend further to the Caucasus and Zagros as well as to the Himalaya Mts., all specific by their altitudes and climate varieties which influenced karstification intensity and resulted in the creation of different karst landscape and forms.

The "classical" Mediterranean holokarst is mostly orogenic (folded) and hypergene, containing all forms typical of karstified carbonate rocks. The following text contains descriptions of main karstic forms typical of the Mediterranean, but also other similar, dominantly carbonate karstic environments worldwide.

Fig. 2.20 Karrenfeld on the shore of Skadar Lake (Montenegro) and in Ponoarele Mehedinti (SW Romania)

Fig. 2.21 Doline in karst of Kučaj (east Serbia): Cvijić photograph from 1895 and Stevanović photograph of the same area in 1985 (from "Cvijić and Karst," Stevanović and Mijatović (eds.) 2005, reprint with permission)

Karrens (German term, or *lapiés* in French) are small, dissolutional, and regularly linear forms of cm dimensions formed over rock surfaces (Fig. 2.20). When karrens are assembled, they create *karrenfeld* (Ford and Williams 2007). Many other similar small forms such as *solution channels* (*rillenkaren*) or *pans* (*kamenitze*), *pockets* (*tafoni*), or *microrills* are regularly present in holokarstic areas.

Dolines (*sinkholes*) are small to intermediate enclosed depressions in karst (Ford and Williams 2007). They are circular or elliptical and can be created by dissolution, subsidence, or collapse, but based on many in-field case studies, Cvijić (1893) recognized the solutional origin as the dominant one (Fig. 2.21). He also classified dolines as (1) bowl-shaped (Fig. 2.22), (2) funnel-shaped, and (3) shaft-like dolines.

Dry valley is the term employed for a completely dry riverbed which has been created by surface streams in the initial or previous karstification stage. It should be distinguished from a *sinking stream* which has a temporary flow throughout the year.

A blind valley is created by perennial or temporary streams and ends at some usually lithological (rocks of lesser permeability and solubility) barrier. It is

Fig. 2.22 Several very small and shallow bowl-shaped dolines at the karst plateau of Rax Mt. (Austria)

Fig. 2.23 Spectacular stone bridges carved in Jurassic and Cretaceous limestones in Carpathian karst of eastern Serbia (Velika Prerast *left*, and Valja Prerast *right*)

therefore a closed basin with an active or fossil ponor (sink point) at its end. *The stone bridge* may be opened after long "digging" of sinking water. Stone bridges are sometimes of spectacular dimensions because the process continues to deepen and erode the rock mass, causing the distance between the riverbed and the top of the bridge to expand quickly on a geological timescale (Fig. 2.23).

An *uvala* is elongated depression, a fossil valley (Fig. 2.24). The bottom can be rough with numerous dolines, but also filled with deluvial sediments.

Fig. 2.24 Uvala in Dinaric karst (Pluzine, Montenegro)

In many cases, the orientation of the uvala indicates fossils or the actual direction of groundwater flow underneath.

A polje is the largest landform on the karst surface. It is a tectonical–erosional depression filled by lacustrian, deluvial, or fluvial sediments or a combination thereof and is dissected by perennial or temporary streams. Commonly, on one side of the polje edge, there are drainage zones marked with springs and estavelles, while on the other margin there are ponors (Fig. 2.25) which transfer percolated surface waters to the other lower-positioned poljes or regional erosional bases (sea or basin or riverbed; see also Sect. 16.4). A small polje may have an extension of a few km^2, or even less, but the largest forms can extend over a hundred square kilometers.

The karstic plateau or karstic plain is a flat or subhorizontal surface consisting of karstified rocks either naked or with a thin soil cover, usually positioned at higher altitudes. Residual plateaus are often left after the paleostream incises a deep canyon (Fig. 2.26).

Canyons and gorges are deep-incised landforms very frequent in carbonate karst (Fig. 2.27). If the bottom is reached and cut in impervious bedrock, which is very common, this means that karstification was completed in the studied vertical section.

A shaft (*pit, aven, abyss, pothole, jama*) is a vertical or subvertical cavity, usually tectonically formed and additionally adapted by percolated water. The majority of shafts are fossil or still active ponors. Some of the deepest single vertical

Fig. 2.25 Alluvial ponors at the margin of Dasht Arjan karstic polje (Zagros Mt., south-central Iran)

Fig. 2.26 High abandoned plateaus and meanders on the *left side* of the Verdone canyon (Provençale karst, France)

shafts ("pitch") which may reach a depth of 500 m are explored in the Slovenian Alps (Vrtoglavica, Kanin Mt.) and in Croatia (Lukina jama). In the "Stone Sea" area in the Montenegrin karst above the Boka Kotorska Bay, there are more than 300 vertical shafts within an area of only 8 km^2 (Milanović 2005).

Caves are mostly explored karstic and speleological forms (Figs. 2.28 and 2.29). A cave is a subhorizontal or slightly inclined cavity or developed system of cavities and galleries, created by a combination of mechanicals and chemical erosional factors. In principle, the term is used when cavities are completely, or at least partly, passable for humans, although for smaller forms, the term *cavity* or *channel* or *cavern* could be adequate. The conventional hydro(geo)logical classification of caves includes still active (with perennial flows), temporary active,

Fig. 2.27 Lazarev Canyon (east Serbia)

Fig. 2.28 Gigantic entrance into the Potpeć cave (Dinaric karst, western Serbia)

and dry (with percolated water), while in hypergene (unconfined) karst, they are genetically *ponor cave*, *spring cave*, or a combination of both, *ponor–spring cave* (groundwater *passage* or *tunnel*). An additional group are hypogene caves created by thermal fluids and water containing H_2S and other specific microconstituents which stimulate corrosion of karstic rocks.

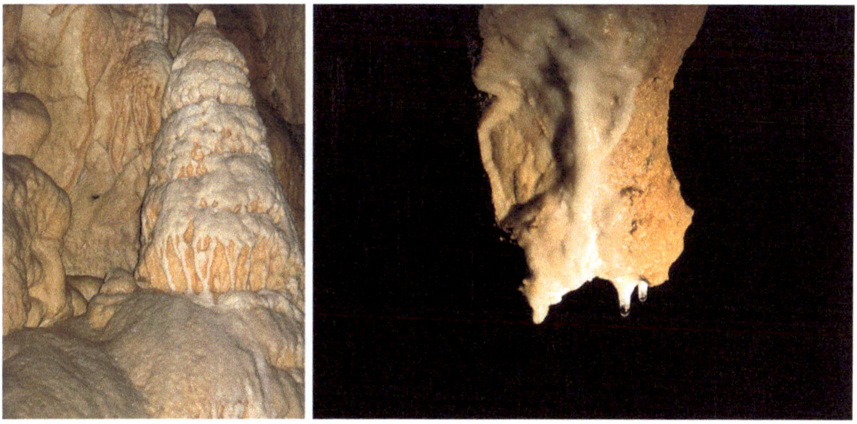

Fig. 2.29 The stalagmite (*left*) and stalactite with water drips (*right*). Speleothems from Cloşani cave (SW Romania)

The Glossary of karst hydrogeology (Paloc et al. 1975) includes definitions of the terms above and other terms from karstology, speleology, karst hydrogeology and geomorphology, and karst sciences in general. Another important glossary, *A Lexicon of Cave and Karst Terminology with Special to Environmental Karst Hydrolog*y (EPA/600/R-02/003, 2002, EPA: Washington, DC. Speleogenesis Glossary), is available at the Web site http://www.speleogenesis.info/directory/glossary/.

Platform carbonate karst is generally characterized by thick sedimentary complex formed in large platforms, and less folded in comparison with geosynclinal karst. It can also be exposed to the surface as *exhumed karst* or *paleokarst,* or still *buried karst* overlain by younger rocks.

The four typical representatives are the following: *Russian platform, Yucatan—Floridian karst, Edwards Aquifer,* and *Chalky aquifers* of Great Britain and France.

Russian platform karst is one of the largest world karst systems extending over the Russian Plain to the east and comprising also the Pre-Ural (Ural Mt. Foreland) and Ural. With the exception of a few basins, it developed in a platform depositional environment and has a very long geological history, from Pre-Cambrium until Neogene, with Paleozoic and Miocene sediments encompassing the largest portion. Both carbonate and evaporitic facies developed, and they are often covered by younger moraines or fluvio-glacial sediments. This type of karst Maksimovich (1963) called "Russian karst."

Yucatan—Floridian karst was created in younger Tertiary carbonate deposits under the strong influence of the Caribbean Sea water fluctuation during Pleistocene. The tectonic subsidence also caused an impermanent submergence of the carbonate rocks. These highly karstified rocks are widely exposed to the surface with large nets of underground channels and specific karstic features such as *cenotes* (vertical shafts with groundwater formed mostly by cave roof collapse), distributed all over the Yucatan Peninsula (Mexico) or large karstic springs such as the Wakulla discharge system in the Floridian karst which, along with the two neighboring systems, consists of 50 km of underwater caves.

Edwards aquifer is an arch-like zone around 250 km long extending through south-central Texas. The Edwards massive and thick-bedded limestones and dolomitic limestones are between 100 and 250 m thick (Eckhardt 2010). The hilly and slightly undulated terrain is separated from the low-lying coastal plain by long and large escarpments. The aquifer outcrops at a large area covering more than 3,000 km^2, but large springs issue from a few outlets in the marginal confined–artesian part.

Chalky limestones are pure $CaCO_3$ in chemical composition and characterized by specific *vuggy* porosity. The presence of chert or flint nodules in *chalk* is also common. The chalk primary porosity is generally larger than in other carbonates. The largest extension of these sediments of Cretaceous age is in Great Britain, NW France, Denmark, and Germany. Along the shorelines where most of the chalky rocks are exposed, very high cliffs can be formed.

There are also other specific regional karst developments. Ford and Williams (2007) described *Nullarbor Plain* in SW Australia covering a surface of some 200,000 km^2. The *Guateng Group* and other dolomitic formations are well developed in central and northern South Africa but are often overlain by a relatively thin cover of younger rocks.

Tropical karst includes different subtropical varieties characterized by undulated relief with specific rough hills and residuals in topography. Although *Yucatan—Floridian karst* is also tropical and subtropical in terms of climate factors, it is added to the previous group because of other different properties such as flat relief and the absence of specific residuals.

Cockpit karst is typical of Jamaica and several other Caribbean Islands. *Cockpit* is of similar origin to *doline*, but its creation has been more influenced by humid tropical conditions. Cockpits are usually extended between small hills and may be connected and elongated (Fig. 2.30).

Fig. 2.30 Sketch of cockpit forms (elongated valleys between the hills)

Tower karst is famous due to its very specific residual landforms (Figs. 2.31 and 2.32) present in south China and in north Vietnam, as well as in several other countries of SE Asia (e.g. Indonesia, Philippines). The size and shape of the hills can be very different, symmetrical, or irregular. A very similar environment is also called *Pinnacle karst* or *Cone karst*. The terms *cupola* and *hum* are also used to describe the residuals which are rounded and conical. If residuals are very rare, such karst is often called *Relict karst*. *Fenglin* is the Chinese term indicating isolated forms (*teeth*), while *fengcong* is a group of hills from the same bedrock. *Fenglin* means "peak forest," and there are numerous wonderful geoheritage sites and parks in China, Papua, Malaysia, and Belize consisting of huge fenglins and karrens making the surrounding area almost impassable for visitors if corridors are not artificially opened. *Feng* means valley, and *fengcong* may be developed as an accentuated cockpit karst (Ford and Williams 2007).

Fig. 2.31 Tower karst—Fengcong along Li Yiang River (*left*) and in Xiangqiao geopark (Guanxi province, south China)

Fig. 2.32 Tower karst—Ha Long Bay (Vietnam) UNESCO World (geo) heritage site (*photograph courtesy* of Boris Prokić)

Fig. 2.33 Specific karst combination: platform (structure) and evaporitic (lithology), in Horn of Africa (Karkar Fm. near Boohodle, Somalia)

Hypogenic and evaporitic karst is also developed under specific climate conditions, mostly arid or semi-arid. Low rainfall rates and the absence of large paleostreams as main generators of mechanical karstification in humid karst have been compensated for by the presence of highly soluble evaporitic rocks deposited in a warm, often desert environment (Fig. 2.33). The hypogene karst and process of carving caves in confined conditions is well explained in the book *Speleogenesis* of Klimchouk et al. (2000).

The very large caves and long system of tens of kilometers of underground channels and labyrinths are created in gypsum (called also *maze*) and halite rocks. Such caves are well explored and studied in Ukraine, Poland, Hungary, Italy, and the USA (Klimchouk et al. 2000; Ford and Williams 2007). In addition, evaporitic and salty rocks are widely present in Central Asia, the Middle East, and the Arabian Peninsula, where many sulfuric caves and springs are recorded.

Glacial karst is characterized by the presence of both recent forms created by glacier actions and "paleoforms" developed in bedrock before Quaternary glaciations (pre-glacial karst). In addition to Antarctica and Arctic poles, permanently ice-covered areas (permafrost) are widely distributed in Greenland, Russia, and Canada and in high glacial mountains such as Patagonia (Argentina) or the European Alps. An excellent explanation of the glacier mechanism and creation of landforms can be found in Ford and Williams' book *Karst hydrogeology and geomorphology* (2007).

References

Anthropological Museum of Petralona (http://www.petralona-cave.gr). Accessed 15 Jan 2014

Bakalowicz M (2005) Karst groundwater: a challenge for new resources. Hydrogeol J 13:148–160

Benson DJ (1984) Carbonate rocks and geological processes. Lithology. In: LaMoreaux PE, Wilson BM, Memon BA (eds) Guide to the hydrology of carbonate rocks. IHP studies and reports in hydrology, vol 41. UNESCO, Paris, pp 21–30

BRG and UNESCO. The world hydrogeological map. http://www.whymap.org/whymap/EN/Home/whymap_node.html. Accessed 12 Nov 2013

Cigna A (2005) Show caves. In: Culver DS, White WB (eds) Encyclopedia of caves. Elsevier, Academic Press, Amsterdam, pp 495–500

Cvijić J (1893) Das Karstphaenomen. Versuch einer morphologischen monographie, Geograph. Abhandlungen Band, V, Heft 3, Wien, p 114

Cvijić J (1914) Jedinstvo i psihički tipovi dinarskih i južnih slovena (Unity and psychology types of Dinaric and South Slaves). In: Lukić R (ed) Works of Jovan Cvijić, speeches and articles (1987) (Reprinted in Serbian). Serbian Academy Science and Arts, Belgrade, pp 237–294

Dunham RJ (1962) Classification of carbonate rocks according to depositional texture. In: Ham WE (ed) Classification of carbonate rocks. American Association of Petroleum Geologists, Memoires 1, pp 108–121

Eckhardt G (2010) Case study: protection of Edwards aquifer springs, the United States. In: Kresic N, Stevanović Z (eds) Groundwater hydrology of springs: engineering, theory, management and sustainability. Elsevier, Amsterdam, pp 526–542

EPA (2002) A lexicon of cave and karst terminology with special to environmental karst hydrology. EPA/600/R-02/003, Washington DC. Speleogenesis glossary. Also available at the web site http://www.speleogenesis.info/directory/glossary/. Accessed 12 Jan 2014

Freeze RA, Cherry JA (1979) Groundwater. Prentice-Hall, Englewood Cliffs

Folk RL (1965) Petrology of sedimentary rocks. Hemphill Publications, Cedar Hill

Ford D, Williams P (2007) Karst hydrogeology and geomorphology. Wiley, England

Goldscheider N, Chen Z, WOKAM Team (2014) The world karst aquifer mapping project—WOKAM. In: Proceedings of international conference Karst without boundaries. Kukurić N, Stevanović Z, Krešic N (eds) DIKTAS, Trebinje, 11–16 June 2014, p 391

Gvozdeckiy NA (1981) Karst. Izdatelstvo Misl, Moscow, p 214

Herak M, Magdalenic A, Bahun S. (1981) Karst hydrogeology. In: Halasi Kun GJ (ed) Pollution and water resources. Columbia University seminar series, vol XIV, part 1. Hydrogeology and other selected reports. Pergamon Press, New York, pp 163–178

Klimchouk AB, Ford DC, Palmer AN, Dreybrodt W (eds) (2000) Speleogenesis; evolution of karst aquifers. National Speleological Society of America, Huntsville

LaMoreaux PE, Tanner JM, ShoreDavis P (1986) Hydrology of limestone terranes; annotated bibliography of carbonate rocks, vol 3. International contributions to hydrogeology, vol 2. Verlag Heinz Heise, Hannover

LaMoreaux PE, Prohic E, Zötl J, Tanner JM, Roche BN (1989) Hydrology of limestone terranes; annotated bibliography of carbonate rocks, vol 4. International contributions to hydrogeology, vol 10. Verlag Heinz Heise, Hannover

LaMoreaux PE, Assaad FA, McCarley A (1993) Hydrology of limestone terranes; annotated bibliography of carbonate rocks, vol 5. International contributions to hydrogeology, vol 14. Verlag Heinz Heise, Hannover

LaMoreaux PE, LaMoreaux J (2007) Karst: the foundation for concepts in hydrogeology. Environ Geol 51:685–688

Maksimovich GA (1963) Osnovii karstovedenia, vol I and II. Perm

Maran A, Stevanović Z (2009) Iraqi Kurdistan environment—an invitation to discover. IK Consulting Engineers and ITSC Ltd., Belgrade

Milanović P (2005) Water potential of south-eastern Dinarides. In: Stevanović Z, Milanović P (eds) Water resources and environmental problems in karst. Proceedings of international conference KARST 2005, University of Belgrade, Institute of Hydrogeology, Belgrade, pp 249–257

Paloc H, Zötl JG, Emplaincourt J et al (1975) Glossaire d'hydrogeologie du karst. Choix de 49 termes specifiques en Allemand, Anglais, Espagnol, Français. Italien, Russe et Yougoslave (Glossary of karst hydrogeology. A selection of 49 specific terms) In: Burger A, Dubertet L (eds) Hydrogeology of karstic terrains with a multilingual glossary. Publication of IAH and International Union of Geology Science, Paris

Solecki Ra (1955) Shanidar cave: a palaeolithic site in Northern Iraq and its relationship to the Stone Age sequence of Iraq. Sumer 11:14–38

Solecki Ra (1975) Shanidar IV, a Neanderthal flower burial in Northern Iraq. Science 190:880–881
Stevanović Z (2010) Utilization and regulation of springs. In: Kresic N, Stevanović Z (eds) Groundwater hydrology of springs: engineering, theory, management and sustainability. Elsevier, Amsterdam, pp 339–388
Stevanović Z (2012) History of hydrogeology in Serbia. In: Howden N, Mather J (eds) Hystory of hydrogeology. CRC Press and Balkema, Boca Raton, pp 255–274 (International contribution to hydrogeology)
Stevanović Z, Mijatović B (eds) (2005) Cvijić and karst/Cvijić et karst. Special edition of Board on Karst and Speleology, Serbian Academy of Science and Arts, Belgrade
Stevanović Z, Iurkiewicz A (2009) Groundwater management in northern Iraq. Hydrogeol J 17(2):367–378
Sweeting MM (1972) karst landforms. Macmillan Press, London
University of Auckland School of Environment, New Zealand. World Map of carbonate rock outcrops v3.0. http://www.sges.auckland.ac.nz/sges_research/karst.shtm. Accessed 15 Dec 2013
Williams P (2008) World heritage caves and karst. IUCN, Gland, p 57
Wilson JL (1975) Carbonate facies in geologic history. Springer, New York
Wollast R (1990) Rate and mechanism of dissolution of carbonates in the system $CaCO_3$–$MgCO_3$. In: Sturmm W (ed) Aquatic chemical kinetics; reaction rates of processes in natural waters. Wiley, New York, pp 431–445
Zarei M, Raeisi E (2010) Karst development and hydrogeology of Konarsiah salt diapir. Carbonates Evaporites 25:217–229

Chapter 3
Characterization of Karst Aquifer

Zoran Stevanović

Proper knowledge of an *aquifer system* is prerequisite for its utilization, protection from pollution, and sustainable development. To recognize and understand a karstic system fully is a very long and difficult process, one which will probably never be completed. Why? Simply the reason is that the system is so complex and heterogeneous. However, during the last century, karst scientists took many important steps forward and today, we know much more about karst system properties than did our predecessors, the founders of karstology.

Could the recognition of a natural system such as karst be reasonably compared with the recognition of a human being? This question may well seem strange, but there is actually some sense behind it. Initially, information is usually collected "second-hand" and from remote sources, and this introductory information could be similar for both: name, address (location), and some general data. Then, the subject is met and its size and shape observed which might be sufficient to create a very general overall impression. Then, the most sensitive step, talking or surveying starts, and word by word, or applied method by applied method, information on personality, i.e., on system behavior, is collected. While the external shape or figure of the human being could be equated to the geometry of the aquifer, the behavior of the karstic aquifer system could be identified with the following properties of aquifer: permeability and storativity, flow pattern and direction, processes of recharge and discharge, natural water quality, and several others. And many of these properties are very variable in space and in time. Therefore, it is a very

Z. Stevanović (✉)
Centre for Karst Hydrogeology, Department of Hydrogeology,
Faculty of Mining and Geology, University of Belgrade, Belgrade, Serbia
e-mail: zstev_2000@yahoo.co.uk

Z. Stevanović (ed.), *Karst Aquifers – Characterization and Engineering*,
Professional Practice in Earth Sciences, DOI 10.1007/978-3-319-12850-4_3

delicate task first to collect all necessary information and even more so to have the *character* of each individual karstic aquifer system properly explained in order that it may be fully understood.

What is the *aquifer system*? First, aquifer is usually defined as a porous media (rock mass) that can store, transmit, and discharge an important amount of water. "Important" here does not necessarily mean "significant" or "huge," but refers rather to an amount sufficient to be observed and economically valuated to satisfy demands of the consumer(s). If groundwater quantity within the rock media is too small, instead of "aquifer," the terms "*aquitard*" and "*aquifuge*" are used. The latter means an almost total absence of water in rocks. In reference to karst, *aquifer* is almost always discussed even though there are many compact carbonate rocks with low permeability. Finally, the attribute "system" indicates the complexity and functionality of an aquifer.

How can karstic aquifers be distinguished? There are several criteria, of which just a few will be emphasized here.

In accordance with *discharge* and accordingly with a transmitted and stored amount of water, there are

- Karst aquifer of high productivity,
- Karst aquifer of moderate productivity,
- Karst aquifer of low-to-moderate productivity (Box 3.1).

When an aquifer produces just a small amount of water (low productive), the terms *fissured* or *fractured* are more appropriate than *karstic* because the system of voids and joints is probably not karstified, i.e., not expanded under the influence of mechanical and corrosional agents. It is also common to entitle aquifer of low-to-moderate productivity as combined *karstic-fissured aquifer*, or vice versa.

Box 3.1

On the Hydrogeological map of Dinaric karst prepared under the GEF regional project Protection and Sustainable Use of the Dinaric karst transboundary aquifer system (DIKTAS), implemented by UNDP and executed by UNESCO, the following karst and fissured aquifers are distinguished (Fig. 3.1):

1. Highly productive karst aquifer (KA1),
2. Moderately productive karst aquifer (KA2),
3. Fissured aquifers (FA).

In total, 15 lithostratigraphic units are grouped to KA1 in accordance with aquifer permeability, transmissivity, flow type, and discharge. Nineteen units are classified into the second group KA2, while seven are attributed to the FA group (Stevanović 2011).

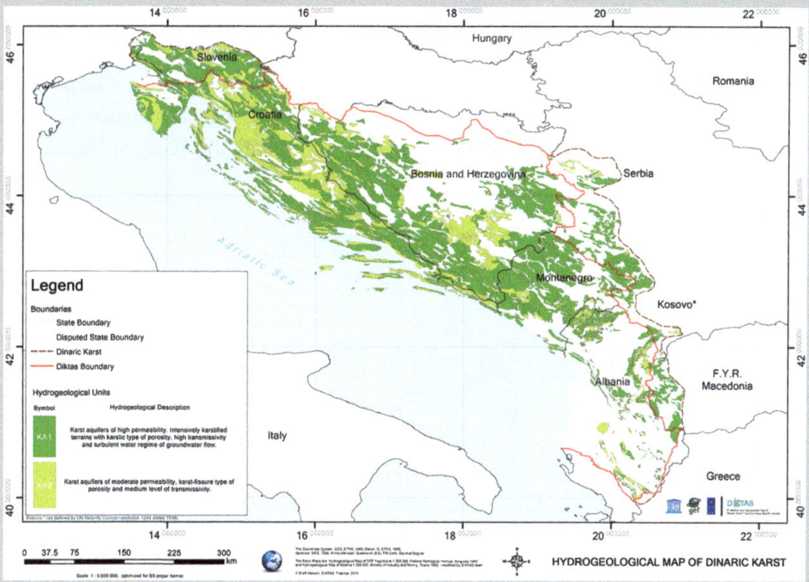

Fig. 3.1 Extension of highly productive (KA1) and moderately productive karst aquifers (KA2) in Dinaric karst (http://dinaric.iwlearn.org/, printed with permission)

According to the dominant type of rocks of which karstic aquifer consists, a classification similar to that presented in Sect. 2.5 can be made as follows:

- Carbonate karst aquifer,
- Dolomitic karst aquifer,
- Marble karst aquifer,
- Chalky karst aquifer,
- Anhydritic karst aquifer,
- Gypsum karst aquifer,
- Halitic karst aquifer.

Bakalowicz (2005) separated carbonate aquifers into the following major groups: the fracture, non-karstic aquifers, and the truly karst aquifers, with all intermediate stages of karst conduit development.

Finally, taking into consideration structures and hydrodynamic properties, there are

- Unconfined karstic aquifer,
- Confined karstic aquifer,
- Semi-confined karstic aquifer.

The first group is characterized by the absence of any cover, free *water table,* or *potentiometric pressure* in karstic voids equal to atmospheric at concerned depth. Water table as surface-connecting water table in voids represents the upper limit of saturation (the term *phreatic* is also commonly used, Fig. 3.2a). *Confined* karst aquifer lies below the impermeable cover, and its *potentiometric pressure* or *hydraulic head* are sufficient to raise the groundwater level over the base of the overlying

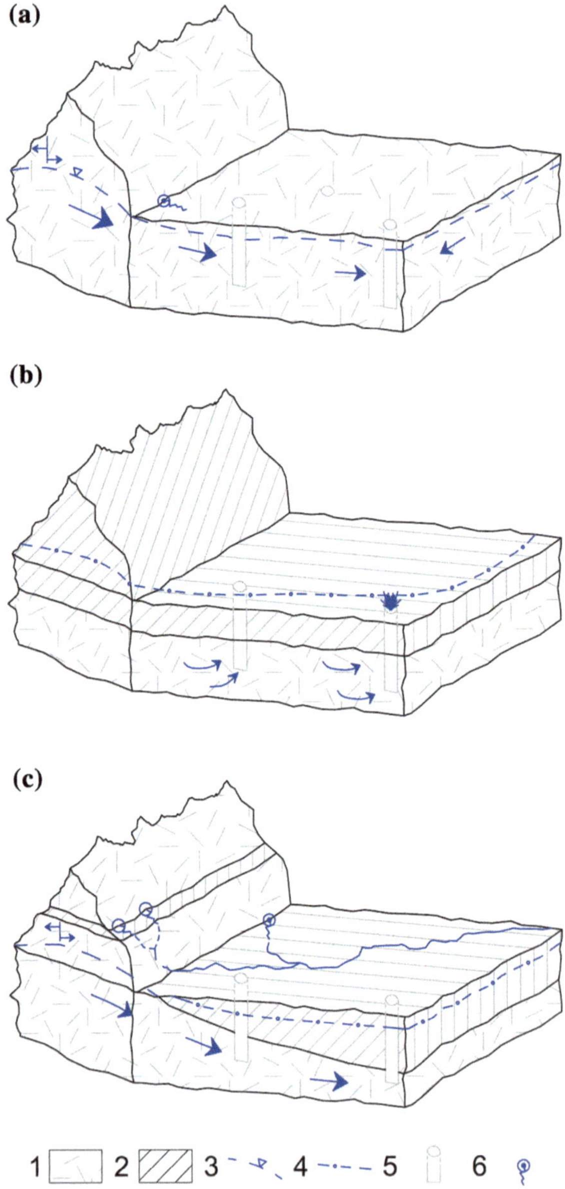

Fig. 3.2 **a** Unconfined karstic aquifer. **b** Confined karstic aquifer. **c** Semi-confined karstic aquifer, and perched aquifer within its upper part drains through small springs. *Legend 1* karstic aquifer, *2* impervious rocks, *3* groundwater table, *4* hydraulic head (piezometry), *5* borehole, and *6* spring

bed when it is penetrated by the borehole (Fig. 3.2b). *Semi-confined* (transient) karst aquifer includes confined and unconfined sections. *Perched aquifer* is the term usually applied to an aquifer lens localized within aquitard ("pockets"), or isolated unconfined beds of limited extension above the regional water table. In the latter case, perched aquifer is disconnected from deeper parts by some impervious layers or by non-karstified rocks (Fig. 3.2c). Younger (2007) stated that great care must be taken not to confuse perched aquifers with regional unconfined aquifer below.

3.1 Aquifer Geometry and Elements

The size of the aquifer is defined by its boundaries. They are *lateral or horizontal*, and *vertical*, while in accordance with watertightness, the boundaries are *permeable, semipermeable,* and *impermeable*. In the last case, there is a factual aquifer limit, while if boundaries are permeable or even semipermeable, there is just a lithological break, but no water barrier because circulation through the boundary is possible (*relative barrier*). The term *groundwater body,* widely introduced by the Water Framework Directive of the European Union (EU WFD 2000), considers the case of possible linkage of two or more vertically or laterally interconnected aquifers.

The subsurface catchment in most cases is different from the topographical (Fig. 3.3). Herak (1981) stated that the actual catchment area of the Cetina River in the Croatian part of Dinaric karst is 2.7 times greater than its topographical frame. This case and several other cases are discussed in Bonacci's reference book on karst hydrology, *Karst hydrology with special reference to the Dinaric karst* (1987) and are presented by the same author also in Chap. 5.

The flexibility and variability of *lateral boundaries* of a karst aquifer is also discussed in Chap. 6. It can be the result of variation in the water table during the high- and low-water periods or of a possible temporary reorientation of flow direction. Herak (1981) concluded that most of the catchment areas are asymmetrical and only approximately determinable even in regions where intensive survey has been conducted.

Concerning the determination of *vertical boundary,* the case of the top boundary is much simpler than the bottom one. The top is defined by land surface or

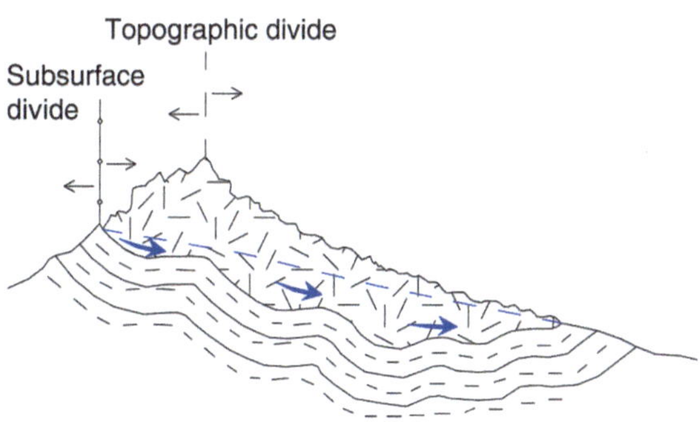

Fig. 3.3 One of the very typical and not rare cases of significant difference between topographical and underground catchment

in the case of confined aquifer by the base of the overlying bed. The bottom boundary is either a *karstification base* or the top of an underlying bed. According to Milanović (1981, 1984), a karstification base is a boundary which is not well-defined and can be considered a transitional zone between karstified and non-karstified rocks.

To locate an even approximate karstification base can be very difficult, but it is an important task of hydrogeological survey. As more information from drilled cores, down-the-hole camera recording and geophysical logging is collected, the chance for relatively correct approximation of this imaginary line increases.

The field hydrogeological survey and tracing tests are essential methods for assessment of karst surface geometry and delineation of the catchment. This topic is discussed in Chap. 4. In hydrogeological practice, it is also common to apply an inverse method to estimate the recharge area based on the results obtained from water budget calculation. Although uncertain in many respects, the method may nonetheless still provide a general view of the geometry of the studied aquifer, especially if similar karstic terrain and aquifer are properly explored in terms of permeability and storativity whereby the analogy between the two is somehow confirmed (Box 3.2).

Box 3.2

Petrič (2002) calculated the recharge area of the Vipava springs (Slovenia) based on observed rainfall data, estimated evapotranspiration rate from the relevant maps and average discharge of the springs covering the 30-year period from 1961–1990. From the average rainfall sum of 2,075 mm/year, and an evapotranspiration rate of 640 mm/year, the difference of 1,435 mm/year has been obtained. This value, which can be equalized with effective infiltration to karst aquifer due to minimal runoff, was compared with the average spring discharge of 6.78 m^3/s. Although there was almost no difference between the two values, Petrič found that the total catchment area is 149 km^2.

$$F = Q/I_{ef}$$
$$F = 6.78 \times 31.536 \times 10^6\,m^3/1.435\,m$$
$$F = 149\,km^2,$$

where

F catchment area of the springs,
Q total annual spring discharge,
I_{ef} effective infiltration on annual basis.

From the geological map, it is concluded that flysch impermeable rocks cover some 9 km^2 (allogenic recharge area), while carbonate karst aquifer covers the remaining 140 km^2.

The method of multiple nonlinear correlations applied to the karst spring hydrograph for the assessment of the catchment area is discussed in Sect. 15.2.

The elements of a typical karst aquifer are the following:

- Top surface,
- Epikarst,
- Vadose zone,
- Saturated zone, and
- Karstification base.

When might the three first elements and the last one be absent? The answer is in the case of confined karstic aquifer. It is thus only the presence of a *saturated zone* that is certain, but if saturation is also missing, then rocks are dry and in fact, we are not dealing with an aquifer.

The schematic representation of an unconfined karstic aquifer and its elements is shown on Fig. 3.4.

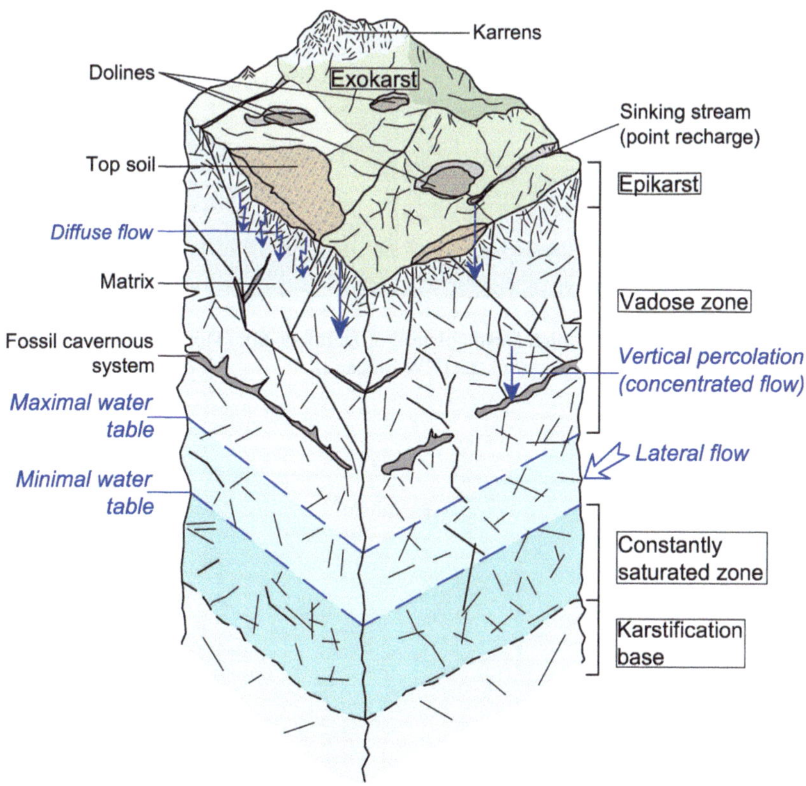

Fig. 3.4 Model of unconfined karst aquifer

The *top surface* or *exokarst* may contain variable features, some of which stimulate aquifer recharge, while others obstruct this process. For instance, dolines could have a dual function: As a depression in relief, they collect water which circulates to their bottoms, but if the bottom is filled with thick and impermeable deluvial soil, the water will remain in the soil or just refill the formed swamp until it finally evaporates (Fig. 3.5).

The top surface includes soil horizons resulting from the weathering process, but soil can also be brought about artificially. It is common in soil-poor karst terrains for local villagers from generation to generation to destroy rocks and expand soil surface and the thickness of cultivated land (Fig. 3.5).

An unsaturated zone usually consists of two parts: an upper zone or epikarst and an underlying vadose zone.

Epikarst is a "skin" of the karst system underlying the soil cover and opens to the surface (Figs. 3.6 and 3.7). Its role in the functioning of a karst system,

Fig. 3.5 Doline—swamp in high karstic plateau of Durmitor Mt. (Monetnegro) (*left*) and cultivated bottom of doline in Vrachanski karst (NW Bulgaria)

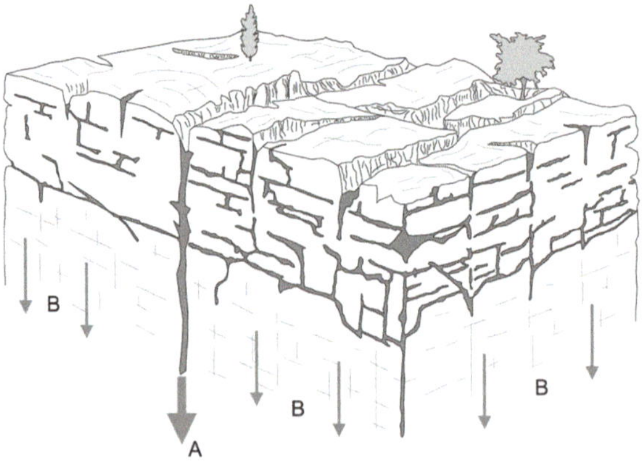

Fig. 3.6 Scheme of the epikarstic aquifer in the zone of alteration and superficial fracturation. *A* is vertical drainage through large fissure, *B* is seepage, i.e., effluent phenomenon (from Mangin 1975)

Fig. 3.7 Epikarst covered by thin soil layer (*left*) and small discharge spring draining perched aquifer (*right*)

in particular its functions of recharge and discharge, has been studied by many authors. Mangin (1974, 1975) explained the functions of a karstic system locating "epikarstic aquifer" above a real "karst aquifer" and also suggesting the term *eukarstic aquifer*. Atkinson (1977) was among the first who explained the epikarst drainage mechanism and its impact on the entire spring hydrograph. Williams (1983, 2008) called epikarst a *subcutaneous zone* and analyzed its role in karst hydrology and cave hydrogeology. He emphasized that porosity and permeability are in principle greater near the surface than down deep.

Kiraly et al. (1995) attributed the base of epikarst to *Faraday cage* with respect to lower aquifer parts. The hydrodynamic analyses on a 3-dimensional model of finite elements indicated that epikarst has a large impact on (1) The shape of karst hydrograph, (2) The base flow component, (3) The water table fluctuation in conduit network and surrounding rock blocks, and (4) The recharge of low permeable rocks.

Klimchouk (2000) highlighted the diffuse karstification of the underlying bulk rock mass and the higher degree of fissuring of the epikarstic zone. He stated that contrast in the hydraulic conductivity of these two zones allows some groundwater storage in the epikarstic zone and flow concentration in its base. Trček (2003) further discussed diffuse vertical recharge from epikarst which decreases with depth and converges into lateral flow. Perrin et al. (2003) analyzed the storage in epikarst based on isotopic data and developed a conceptual model. Bakalowicz (2005) emphasized production of carbon dioxide in a perched saturated zone. This CO_2 as an important karstification agent is then transported to deeper aquifer parts.

According to Ford and Williams (2007), the epikarst is typically 3–10 m deep. It consists usually of a particularly weathered zone of limestone which gradually gives way to the main body of the vadose zone that comprises largely unweathered bedrock.

Common among the explanations provided is the existence of perched aquifer in the uppermost part of the vadose zone resulting in water retention and a delay in water infiltration to the lower parts. This delay depends on the epikarst permeability and may take a few days or a few months.

Can epikarst be absent? Krešić and Mikszewski (2013) aptly stated that the concept of epikarst was applied indiscriminately to explain aquifer behavior and functioning. They stated that epikarst is mentioned in many studies as a factor influencing aquifer regimes even for terrains where it is completely absent and no perched aquifer is present (Figs. 3.8 and 3.9).

We may conclude that epikarst, when present, is located within the uppermost part of the vadose zone and is partly saturated, storing a certain amount of water and rerouting vertical infiltration to the deeper phreatic zone of the karst aquifer.

Fig. 3.8 Lack of epikarst. Almost unique degree of fissuration in vertical section in massive limestones of Triglav (Slovenia, *left*) and Verdone (France, *right*)

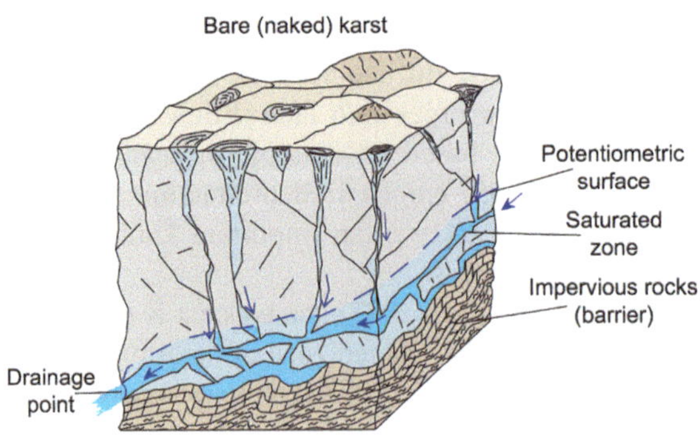

Fig. 3.9 A scheme of terrain with dominant bare (naked) karst. Very deep potholes and cavities very quickly convey water to the saturated bottom part. This situation is typical for Dinaric holokarst

In contrast, in many highly karstified terrains water from rainfall or sinking streams infiltrates so quickly to the saturated zone that we have completely dried caves and fossil channels and a total absence of water in the uppermost part, as discussed in Chap. 2. Therefore, the field hydrogeology survey and the assessment of the presence of perched aquifer and epikarst and its thickness and permeability must precede any hydrodynamic analysis or modeling of karst aquifer regime.

The vadose zone is a deeper, transient zone in karst aquifer. In the vertical direction, it ends with the groundwater table. The terms *unsaturated zone* or *zone of aeration* are also commonly used. The latter refers to the atmospheric pressure and to the fact that some of the voids may contain air. The circulation in the vadose zone is dominantly vertical, but along with the transfer of water which percolates from the top surface or epikarst to the water table, some lateral flows also may occur. This strongly depends on the orientation and angle of karst voids and channels. Therefore, preferential paths for water circulation are created in the matrix, fissures, and caverns. Parizek (1976) presented a strong relationship between the main water paths and large fault zones (Fig. 3.10).

The saturated zone represents the main aquifer layer and water storage space. Its thickness depends on several factors such as recharge, permeability and storativity, position and size of discharge points, hydraulic head, strike and dip of bedding planes, and location and utilization of artificially introduced intakes. Water also flows through the system of interconnected voids, joints, fissures, and cavities of various sizes.

Figure 3.11 is a visual interpretation after Drogue (1982) of the system of different elements for water circulation. The various shape and size voids and cavities are separated into blocks. Their density is in accordance with the vertical position: The maximal density but small fissures is in the first level blocks, closest to the surface. There is now an essential difference in the size of voids and cavities which belong to the blocks in the second and third deepest level, but the density decreases with the depth. However, the main drainage system at the deepest part can be enlarged by mechanical erosion of lateral flows which are concentrated near discharge points.

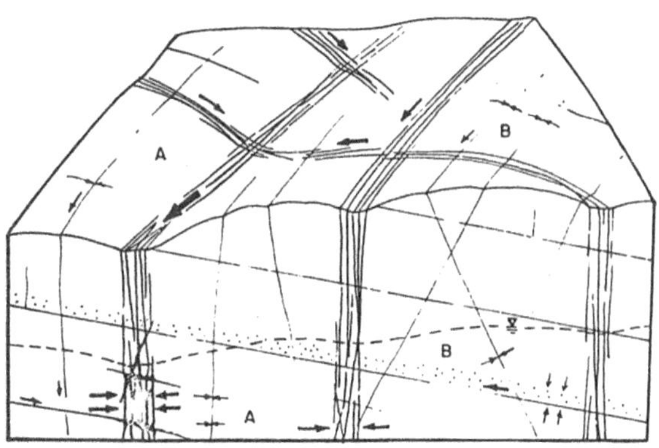

Fig. 3.10 Massive blocks in carbonate rocks dissected by large faults and fissures which enable water infiltration and circulation (from Parizek 1976)

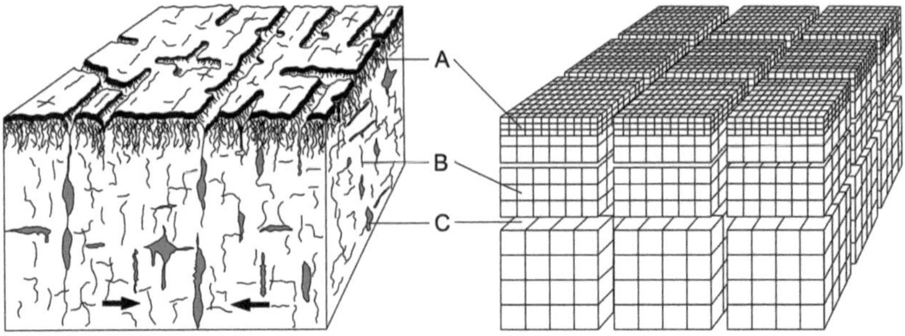

Fig. 3.11 Scheme of karstic blocks. *A* The uppermost blocks are altered, very permeable and allow rainfall infiltration. *B* Mid blocks, less permeable with fissures and small cavities as the dominant elements. *C* Deepest block, also fissures and cavities, and locally larger caverns which evacuate (drain) water from the system (modified from Drogue 1982)

As mentioned already the *karstification base* is an aquifer bottom, relatively located and non-fixed. The depth to the *karstification base* can be very large, sometime thousands of meters (Box 3.3). In the Dinaric karst, the exploratory borehole found cavities in Permian–Triassic clastic limestones at a depth of 2,236 m (Milanović 1981). However, local karst development should always be differentiated: For example, small paleo-cavitation which is disconnected from the rest of the voids in a rock mass should always be differentiated from the real karstification which results in interconnected cavities that enable active circulation (*effective porosity*).

Box 3.3

- Milanović (2006) evaluated results of pressure permeability tests in 140 boreholes drilled in eastern Herzegovina. The karstification exponentially decreased, and it is almost 30 times greater in the subsurface zone than at a depth of 300 m. He also stated that at greater depths, karstification is developed along major faults.
- The case of inverse karstification, greater at bigger depths and developed along restrictive pathways, is described by Stevanović (2010a). In the Bogovina area in the Carpathian karst of eastern Serbia, the local maximal karstification depth has been estimated to be at over 800 m, while karstified intervals with elements of paleo-karstification have been confirmed 471 m deep on a borehole SB-1. However, the major karstified zone was identified during the drilling at a depth of 60–90 m. This is a very active conduit system, which conveys most of the local groundwater. Some of the large caverns in that system are filled with secondary deposited clays and sands and are not currently active.
- Very similar variable porosity (cavernosity) with a maximum depth between 65 and 125 m has been recorded during the drilling of several exploratory boreholes in karst of Zagros Mt. in Iraq (Fig. 3.12).

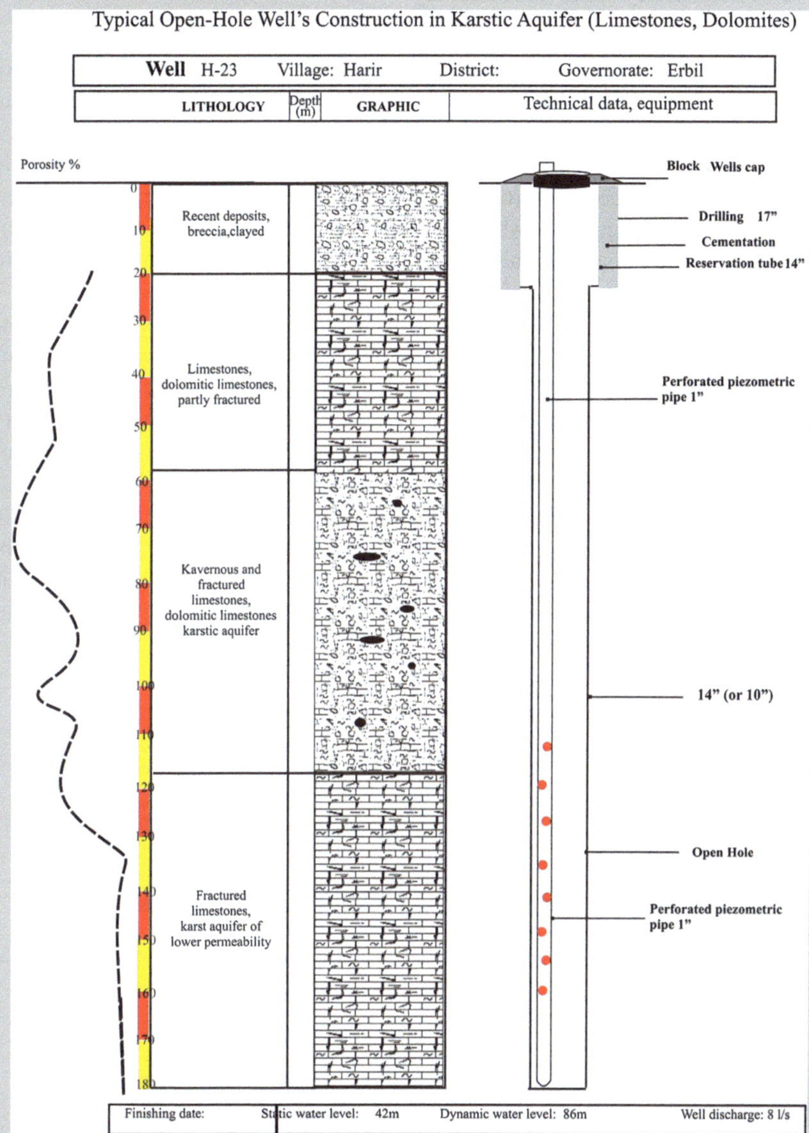

Fig. 3.12 Drilling log of the borehole drilled in the Harir area (Erbil Governorate, Iraq). Porosity curve indicates maximum cavernosity at the three intervals: 70, 100, 120 m

3.2 Permeability and Storativity

In Chap. 2, the type of rocks subjected to karstification was noted. In addition, it was highlighted that not all rocks are equally soluble and that mechanical and chemical agents are crucial for the karstification process as is the presence of water and pores. The *porosity* refers to the proportion of volume of rock occupied by pores. The two types of porosity exist: *primary* and *secondary*. The former is the result of rock diagenesis, while the latter results from tectonic fabrics, exogenetic factors, and karstification. Theoretically, the final result of the karstification process is the total conversion of primary into secondary porosity.

In principle, the matrix of the majority of karstic rocks has a small *primary porosity* commonly referred to as a *microscopic porosity*. This particularly concerns limestones and dolomites (Fig. 3.13). Castany (1984) noticed that chalk consisting of calcite particles has *interstitial porosity*. Synonymous with that term is the term *vuggy porosity* which relates to visible pores between some of the constituent components (Fig. 3.14).

The *secondary porosity* is also called *macroscopic porosity* which is a visible porosity that could be presented in poorly cemented or oolitic carbonates but which mainly refers to bedding planes, joints, fissures, fractures, and cavities formed by a dissolutional, and not diagenetic process (Fig. 3.15). The secondary

Fig. 3.13 Carbonate breccia: well cemented and compact with very small porosity (*left*) and poorly cemented (*right*)

Fig. 3.14 Vuggy porosity of Miocene marine limestones, the widely spread formation in the Alpine karst (named Asmari Fm. in Iran, Pila Spi Fm. in Iraq, Sarmathian Fm. in Balkan countries)

Fig. 3.15 Secondary porosity, but restricted: cavities partly filled with recrystallized calcite in limestone core samples

porosity is responsible for transferring gravitational water, but if voids are not connected, there would still be no free flow system created in the aquifer. For this reason, hydrogeologists must try to distinguish the *total porosity* from *effective porosity* (Box 3.4).

The total porosity is thus the ratio between the total volume of voids and the total volume of rock (sample, core). Voids are, therefore, not necessarily connected and enable water circulation. According to Castany (1984):

$$P = V_{\mathrm{v}} / V \tag{3.1}$$

where

P total porosity,
V_{v} volume of voids,
V total volume of rock.

In contrast, the *effective porosity* is the volume of interconnected voids against the total volume of rock.

$$P_{\mathrm{e}} = V_{\mathrm{e}} / V \tag{3.2}$$

where

P_{e} effective porosity,
V_{e} volume of interconnected voids,
V total volume of rock.

Some authors also recognize *tertiary porosity* which creates very large caverns. In accordance with explained differences, we may distinguish the following:

- matrix porosity, as mainly primary,
- fissure (fracture) porosity, as secondary, and
- cavernous porosity, as tertiary.

The presence of two or all types of porosity is termed *dual porosity*.

Box 3.4

Castany (1984) presented total porosity of different kinds of carbonate rocks taken from several locations. The values range widely from only 0.1 % for Carrara marbles in Italy to 45 % for chalk in France. Ford and Williams (2007) also found that matrix porosity of some samples of chalk from England is around 30 %, but its secondary porosity is very low, less than 0.1 %.

Krešić and Mikszewski (2013) presented the case of Biscayne aquifer from Florida, with very high primary porosity of 40–50 % which is increased further by karstification.

Milanović (1981) calculated effective porosity of littoral karstic aquifer in carbonate rocks to be in the range of 1.4–3.5 %, based on results of analyses of recession discharge curve (see Chap. 7 for method explanation) of the Ombla Spring near Dubrovnik and groundwater fluctuations. Milanović also cited data of Vlahović (1975) and Torbarov (1976) who estimated effective porosity in neighboring Dinaric karst aquifers in Nikšić karst polje and in the Trebišnjica springs basin. Values of around 6 % for the highly tectonized zone in Nikšić and 1.2–1.5 % for the entire Trebišnjica basin are obtained.

Secondary porosity does not mean that fissures and cavities, even interconnected ones, always remain open for water circulation. The karstic paleo-flows, and especially mineralized and thermal waters, have caused many previously active systems to become filled with sediments such as clay, sand, gravel, or calcite and aragonite (Fig. 3.15). If the plug is compact, solid and recrystallized, then permeability in this segment is lost, but if unconsolidated sediments are present, then long and forced pumping or big floods which pressurize water flow may clean out cavities and reactivate them. This explains why water turbidity at karstic outlets (springs) may significantly increase, and this is also the reason why hydrogeological parameters which define karst aquifer behavior change over time.

Effective porosity is often equated with *specific yield*. Castany (1984) highlighted their differences: Effective porosity is an index of available interconnected pore or cavity space, whereas specific yield is the ratio of the amount of water which can freely flow by gravity from the rock against the total volume of the rock. Younger (2007) defined specific yield as the "amount of water which drains freely from the unit of volume of initially saturated rock per unit decline in water table elevation." The point is that not all water will flow out from the pores or cavities, and specific yield is thus smaller than effective porosity for *specific retention*.

Analyzing the effective porosity and infiltration/filtration capacities of rocks, Younger (2007) introduced the terms "fillable effective porosity" and "drainable effective porosity." Considering that some adhered water always remains in an aquifer and that “new” water can only fill dry pores, voids, or cavities, he declared *fillable effective porosity* to be less than effective porosity. *Drainable effective porosity* is something similar to the above described specific yield: effective porosity minus specific retention.

The two main characteristics of karstic aquifers are *anisotropy* and *heterogeneity*. *Anisotropy* means that one physical property varies with direction. In the case of a karstic environment, anisotropy is very typical and has almost become the first association with the word “karst.” A great variation in fissuration or cavernosity in the vertical section has already been confirmed, but at the horizontal level compact blocks, small fissures and also large cavities may be jointly present. The latter refers more to *heterogeneity* as a variation of a property from site to site within the same formation (Box 3.5). While in an unkarstified rock heterogeneity within a common permeability area may be perhaps 1–50, in karstic rocks it may increase to perhaps 1–1 million (Ford and Williams 2007).

Box 3.5

Figures 3.16 and 3.17 show various properties of karstic and karstic-fissured aquifers recorded during the intensive field survey and tests conducted for the regional water supply system of Bogovina in the Carpathian karst of eastern Serbia (Stevanović 2010b).

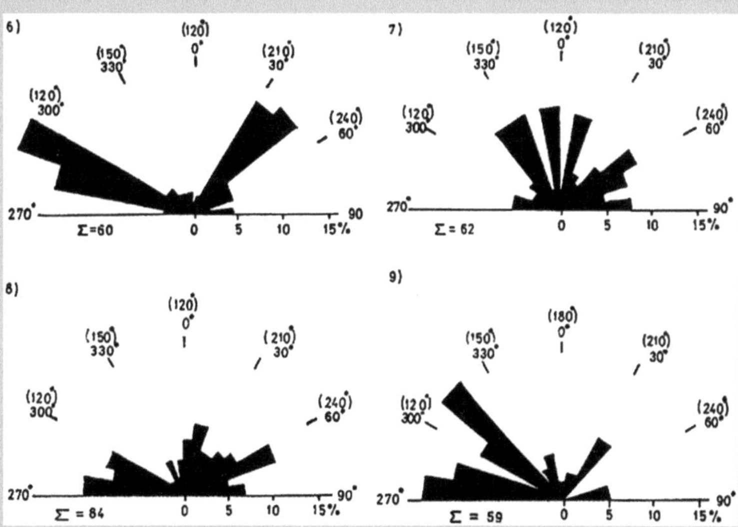

Fig. 3.16 Rose-chart diagrams of the orientation and frequency of measured fissures and cavities within four selected blocks in Aptian limestones

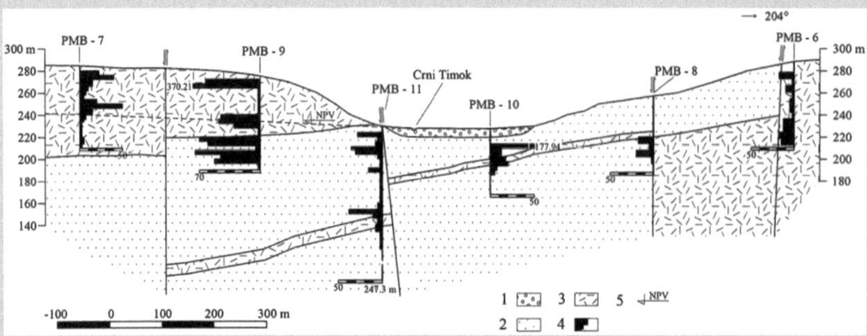

Fig. 3.17 Results of water injection tests conducted in Aptian limestones and sandstones at Bogovina dam site show considerable differences in permeability within the same blocks. *Legend 1* alluvium, *2* sandstones and sandy limestones, *3* karst, *4* volume of water injected per intervals, and *5* water table

Krešić (2013) noticed that the density, size, and geometry of individual conduits in the karst aquifer vary enormously depending on specific properties such as mineral composition, tectonic fabric, recharge mechanism, and position of erosional base.

Such differences are clearly evidenced in the case of seven piezometers situated very close together and located at short distances from and perpendicular to the Trebišnjica River (Bosnia and Herzegovina). This case is described by Milanović (2006). During the summer months, the riverbed is dry, while the groundwater table is below the bottoms of all piezometers. During the floods and river flow of 10–30 m^3/s, the groundwater table rises but not to over a depth of 23 m. Although in the same lithological unit, and in close proximity, the amplitude of the differences in water table of those piezometers is 17 m. This is the result of the permeability of the overlying sinking zone (ponors in riverbed) and aquifer heterogeneity (Fig. 3.18).

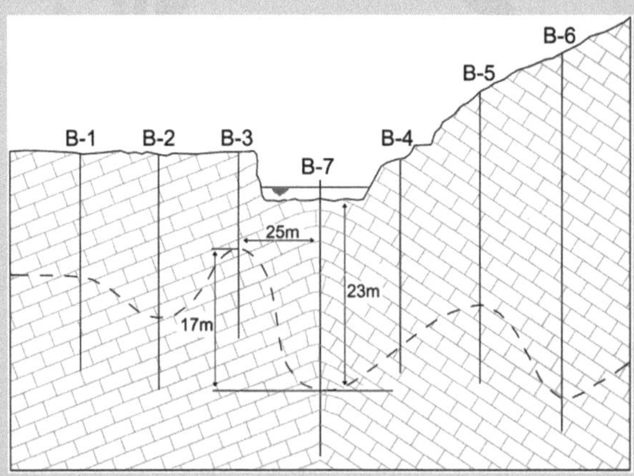

Fig. 3.18 Cross section of karst aquifer in the Trebišnjica River basin, Kočela locality. The position of the water table of an unconfined aquifer in a high water period is marked with a broken line (from Milanović 2006, printed with permission)

The last presented case confirms what hydrogeologists often realize in daily practice: A distance of just 1 m or even less is enough for completely different drilling results: One borehole in a completely compact limestone block without a single drop of water and another which attains a privileged underground water course with large discharge and small drawdown (see Box 3.9).

The *anisotropy* and *heterogeneity* influence other aquifer properties—*permeability* and *hydraulic conductivity*, *transmissivity*, and *storage coefficient* as standard hydrogeological parameters. The recommended references for these topics are those written by Meinzer (1923a), Castany (1984), Palmer et al. (1999), Kiraly (2002), Ford and Williams (2007), Worthington and Ford (2009), Krešić (2013). We shall refer just briefly to the terminology and provide short explanations necessary to understand some further topics.

Permeability is a property of aquifer that allows a fluid to flow under a pressure gradient. It can also be defined as the ability to convey water through the rocks. This aquifer property is quite different from effective porosity, a fact which surprises many non-hydrogeologists. Although effective porosity indicates the proportion of rock sample or mass occupied by interconnected pores, for permeability it is the size of the openings which connect pores and voids that is important. Therefore, limestones with a very small porosity, e.g., 1 % or even less, may be very permeable because their connected cavities as a preferential path may transmit a large amount of water. In contrast, chalk with a high effective porosity is much less permeable: Due to small pores, specific retention of chalk is high. Younger (2007) stated that the list of rocks with both high effective porosity and high permeability is very long, but in any case, effective porosity is the prerequisite for permeability.

The difference in *hydraulic head* and/or *potentiometric pressure* is a driving force for water flow. *Head* is the result of measured water pressure and atmospheric pressure at a given point. *Potentiometric pressure* is a theoretical and imaginary surface connecting all points to which water released from aquifer (confined) would rise. If this rise enables a free flow from the well, such a well is called an *artesian well*. The same term is applied to the potentiometric pressure at that exact point: *artesian pressure* (the term derives from an old self-flowing well in Artois, Belgium, but today is rarely used because a free flow from a well can also result from other factors).

The difference in potentiometric pressure to the distance over which the change occurs is called the *hydraulic gradient*. Its presence is prerequisite for water movement, but velocity still very much depends on the permeability and effective porosity of the aquifer (Box 3.6). If the size of openings is too small, for instance of a microscopic scale, water will move very slowly. In karst that could be the case with sediments which fill the conduits and in that case molecular diffusion could be more active than the impulse of the natural hydraulic gradient (Younger 2007).

Box 3.6

In the case of karst and high pressure (hydraulic gradient), the water can be propagated very quickly to great distances. Many tracing tests conducted in karst worldwide confirmed velocities of a few centimeters per second, and even up to tens of centimeters per second, which may be approximated to the existence of a waterfall in the karst ground.

Referring to 380 conducted experiments in Dinaric karst, Komatina (1983) concluded that the frequency of fictive groundwater velocities in Dinaric karst is as follows: in 70 % of cases from 0 to 5 cm/s; in 20 % of cases 5–10 cm/s; and in 10 % of cases more than 10 cm/s. Groundwater velocity directly depends on the hydrological period and the water table position (Fig. 3.19).

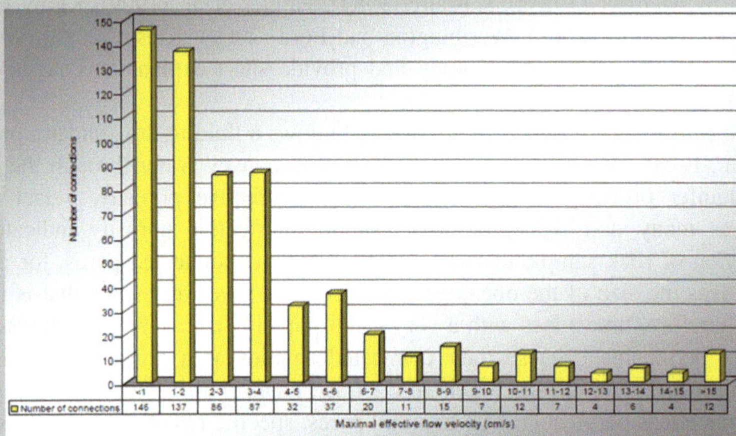

Fig. 3.19 Frequency of maximal flow velocities based on 623 confirmed ponor spring connections in Croatia (after Pekaš, DIKTAS database http://dinaric.iwlearn.org/)

During the dry season and low aquifer water table, water circulation in the karst system is characterized by a slow movement of aquifer waters. But the water waves labeled with dye take two- to five-fold less time to travel the same distance during a season of high hydrological activity (Milanović 2001). For instance, to cover the distance (34 km) from Gatačko Polje to the Trebišnjica Spring (Dinaric karst, Herzegovina), the underground flow takes 35 days when the water table is low and inflow is small. During the high water levels and large inflow, the well-distinguished water wave takes only 5 days to cover the same distance (Milanović 1981). Similarly, between Čaprazlije ponor and Mali Rumin Spring, also in Dinaric karst, the velocity was as follows: in the dry period 5.1 cm/s; at high water level 28.8 cm/s (Komatina 1983).

A large velocity value has been obtained by tracing tests conducted in Prespa Lake, connected with the Ohrid Lake (Macedonia/Albania, see details in Chap. 5). The maximum values in the test conducted in 2002 were between 19 and 80 cm/s.

In principle, Darcy law (1856) expressed as a quantity of water flowing through homogenous intergranular aquifer which is proportional to the difference in pressure (hydraulic gradient) is not applicable in karst. The reasons have already been mentioned: *anisotropy* and *heterogeneity.* An adequate and detailed explanation can be found in Krešić's works (2007, 2013), but perhaps the simplest way to understand why not to use Darcy law is provided by Krešić and Mikszewski (2013): The hydraulic head may go up and down along the same pipe as the cross-sectional area increases or decreases, respectively.

However, in some cases, an approximation of homogenous media can be applied with caution. Such exceptions include, for instance, the prevalence of matrix porosity, or a fissure–karstic system of limited permeability which allows only laminar, but not turbulent flows, or chalk with vuggy porosity with prevailing laminar and diffuse flow. In such limited circumstances, an approach called equivalent porous medium (EPM) could still be acceptable. But many mathematical EPM models were developed based on the inclusion of "problematic" privileged paths (cavities) and of course there have been many negative results when the models have been applied.

If *Darcy law* is not viable in karst, how then can parameters which define hydraulic properties of the aquifer be managed? *Hydraulic conductivity* (K) is probably the most problematic because it directly implies homogeneity (conductance) of intergranular porous media. Hydraulic conductivity (K) is equal to groundwater flow (Q) divided by the hydraulic gradient (i) and cross-sectional area of flow (A):

$$K = Q/i \cdot A \tag{3.3}$$

$$i = \Delta H/L \tag{3.4}$$

where

ΔH difference in hydraulic head between two observation points (loss of energy), and
L distance between observation points.

The unit for K is the same as the unit for velocity (m/s). The feasible application considers knowledge of conductivity in all three dimensions or directions (K_x, K_y, K_z), but this would never be an easy task in karst simply because all elements of the karstic system, such as the position of bedding planes or openings and the orientation of every fracture (dip and strike), will never be known, no matter how detailed a survey is conducted. Figures 3.20 and 3.21 show some examples of great differentiation in the development of privileged groundwater paths identified during a hydrogeological survey in northern Algeria.

Another parameter *transmissivity* (T) (or *transmissibility*) integrated hydraulic conductivity (K) and saturated thickness (h):

$$T = K \times h \quad (\mathrm{m}^2/\mathrm{s}) \tag{3.5}$$

The aquifer thus produces more water (is more transmissive) when hydraulic conductivity and saturated thickness are greater. In confined aquifer, thickness is stable except when over time a high pumping rate causes a gradual reduction

Fig. 3.20 Subhorizontal karstic channel in totally impermeable block of Upper Cretaceous limestones. It functions as a single conductive path for groundwater, but it is able to transfer enormous flow (foundation of the Hammam Grouz dam, Algeria)

Fig. 3.21 Core samples taken from the exploratory borehole at Ourkiss dam site (Oum el Bouaghi, Algeria). Great variation in lithology of Turonian limestones: fissuration degree, texture, and impurity. No sample taken from a section of around 1 m in length indicates the presence of a cavern

of pressure and the aquifers become unconfined. In contrast, the saturated thickness in unconfined aquifers is variable and depends on fluctuation of water table. Therefore, transmissivity is variable, too. Knowing that fast changes in the water table in karst are very common, it can be concluded that not only anisotropy is an obstacle to the use of transmissivity as an aquifer property, but also too is the regime of karst aquifer. However, considering that most laboratory and infield tests result in a calculation of K and T and that there are just a few other options for the assessment of hydraulic behavior, application of these parameters in hydrogeological practice is still inevitable (Boxes 3.7 and 3.8). This application must, though, always be done with caution and with comparison of the results obtained from one site to another.

The effective storage volume in karstic aquifer depends on its area, aquifer thickness, effective porosity, maximal water level, and karstification base (Issar 1984). *The storage coefficient* (S, ε) is equal to *specific yield* in unconfined aquifers. The specific yield is the change in the amount of water in storage per unit area of unconfined aquifer that occurs in response to a unit change in head (Castany 1984). Younger (2007) highlighted that storativity in confined aquifer is a very different physical parameter from specific yield although for both, the symbol S is applied. Ford and Williams (2007) and Krešić (2013) stated that in confined aquifer, *specific storage* (S_S) represents the stored or released volume of water by the unit volume of aquifer per unit surface, due to the change of hydraulic head. The unit is m^{-1} in length. *Storability* or *storage coefficient* for confined aquifer is the product of specific storage (S_S) and aquifer thickness (h):

$$S = S_S \times h \tag{3.6}$$

S is a non-dimensional number and is often presented in the form of % of the total rock mass.

Box 3.7

Figure 3.22 shows a hydrodynamic scheme where the application of standard procedures for calculating hydrogeological parameters supposedly will not result in significant errors and can be used just as an approximation or as "equivalent" parameters as indicated by Marsaud (1997). On the other hand, the pumping test remains the most convenient method to assess aquifer potential for groundwater exploitation.

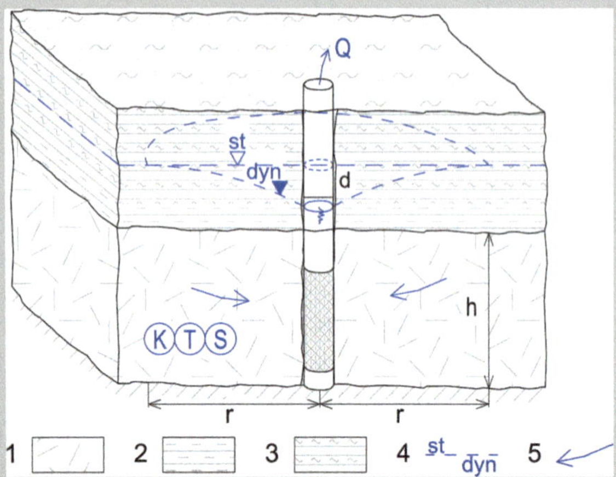

Fig. 3.22 Schematic presentation of a confined karstic aquifer with dominant fissure porosity and presence of small cavities. A fully penetrated well pumps a constant amount of water, and drawdown indicates a non-steady-state flow with slow decline over time. *Legend 1* karstic aquifer, *2* low permeable sand and clay, *3* impervious marls, *4* potentiometric surface, *st* static, *dyn* dynamic, *5* groundwater streamline. Symbols are explained further on in text

Theis (1935) formulated an equation for non-steady-state hydrodynamic conditions which calculates drawdown (*d*) as a result of pumping rate (*Q*), transmissivity (*T*), and pumping well function *W*(*u*). Non-steady-state flow means that pressure in confined aquifer declines with time under constant pumping rate.

$$d = (Q/4\pi \times T) \times W(u) \tag{3.7}$$

where

d drawdown in m
Q constant pumping rate in m^3/s
W(*u*) well function, values *W*(*u*) = *f*(*u*) including theoretical curve can be found in hydrogeology literature

When $u < 0.05$, the well function *W*(*u*) can be expressed with error less than 3 % as follows:

$$W(u) = \ln(2.25 \times T \times t)/r^2 \times S \tag{3.8}$$

where

t time since the beginning of pumping
r radius of well or distance to the point where drawdown is recorded (parameters *T*, *S*, and *K* are associated to sector defined by *r*)
S storage coefficient

Transmissivity (*T*) is calculated using Eq. (3.7) and values for *d* and *W*(*u*) which are obtained as matched cross-points of superimposed theoretical curve $W(u) = f(1/u)$, and curve from log-log graph $d = f(t)$ plotted based on results of pumping test:

$$T = Q/(4\pi \times d) \times W(u) \quad (\mathrm{m}^2/\mathrm{s}) \tag{3.9}$$

The storage coefficient (*S*) is calculated from transformed Eq. (3.8), whereas values for *t* and $1/u$ are obtained from matched cross-point of superimposed theoretical curve and pumping test curve $d = f(t)$.

$$S = 4T \times t \times u/r^2 \tag{3.10}$$

Finally, hydraulic conductivity (*K*) is calculated from transmissivity (*T*) and average thickness of confined karst aquifer (*h*):

$$K = T/h \quad (\mathrm{m}/\mathrm{s}) \tag{3.11}$$

Mijatović (1968) applies Maillet's method (1905) for analyzing spring hydrograph in recession periods by calculating discharge coefficient and volume of discharged waters from aquifer sections which are characterized by different fissuration and permeability. The storativity (*S*) could be estimated in accordance with the volume of discharged waters and decline of the water table. He also applies this equation for calculating *S*:

$$S = 2.25 \times T \times t/r^2 \tag{3.12}$$

t time for emptying water reserves from studied aquifer section; other symbols are as above

Storativity (*S*) could be estimated very roughly by using formula (3.2) for effective porosity (P_e) and calculating the relation of volume of discharged water for the certain period of time (V_t), and the total volume of the rock (*V*).

$$S = V_t/V \times 100\,\% \tag{3.13}$$

Case example:

The karstic spring discharge has declined from 43.5 to 40 l/s during 1 day. The water table, which has been measured in the well located 300 m from the spring inside the catchment area, depleted by 2 m during these 24 h (without any rainfall event). Previous pumping tests of this well resulted in a transmissivity value of 1×10^{-3} m²/s. Considering that the well is located at the edge of the catchment and without any inside–outside lateral flow, what is the storativity value for the aquifer section which becomes completely drained between the well and spring? (Fig. 3.23)

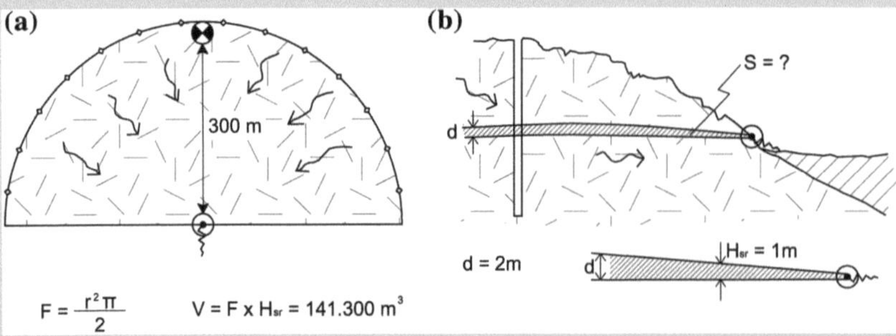

Fig. 3.23 Sketch map and cross section within the spring's catchment

By applying formula (3.12), we calculate storativity to be:

$$S = 2.25 \times T \times t/r^2;\ S = 2.25 \times 0.001 \times 86{,}400/300^2;\ S = 0.00216(0.22\,\%)$$

By applying formula (3.13), we obtain as follows:

$$S = V_t/V;\ S = 0.0035 \times 86{,}400/141{,}300 = 0.00214(0.22\,\%)$$

Obtained storativity rates specify an aquifer which is not highly karstified and with modest storage. The fissure porosity in the studied part probably prevails over the cavernous one. Secondly, although the results of the two equations are relatively similar, this is not always the case and these approximation formulae have to be compared with results of field surveys and several methods suitable for an assessment of aquifer storage (Stevanović et al. 2010).

Box 3.8

Castany (1984) presented results of calculated storativity and transmissivity obtained from infield surveys and tests undertaken in a variety of karstic aquifers in central Europe and in the Mediterranean region (Table 3.1).

Table 3.1 Storativity and transmissivity for some carbonate karstic aquifers (reproduced from Castany 1984)

Carbonate rock category	Age	Location	Storativity *S* (%)	Transmissivity *T* (m^2/s)
Fissured limestones	Upper Jurassic	Moutier (Switzerland)	1–1.5	
	Turonian–Cenomanian	Israel	1	$0.1 \times 10^{-2} - 1.3 \times 10^{-1}$
	Upper Cretaceous	Tunisia	0.5–1	
	Miocene	Murcia (Spain)	0.7–1	
	Jurassic	Lebanon	0.1–2.4	$0.1. \times 10^{-2} - 6 \times 10^{-2}$
Karstic-fissured limestones	Lias	Tunisia	4–5	
	Upper Jurassic	Tunisia	5–7	
	Urgonian (lower cretaceous)	Salon (France)	1–5	10^{-3}
	Jurassic	Parnassos (Greece)	5	$1–2 \times 10^{-3}$
	Jurassic	Vaucluse (France)	1–5	
	Jurassic	Grand Causses (France)		10^{-2}
Fissured dolomite	Jurassic	Grand Causses (France)		10^{-3}
	Lias	Morocco		$10^{-2}–10^{-4}$
	Jurassic	Parnassos (Greece)		3×10^{-5}
Fractured marble		Almeria (Spain)	10–12	
Fissured dolomite		Murcia (Spain)	7	
Marly limestone	Jurassic	Grand Causses (France)		10^{-3}

Referring to storativity, in many karst studies there is a statement of the two dominant types of karst systems (Atkinson 1977; Bonacci 1993; Padilla et al. 1994; Panagopoulos and Lambrakis 2006): (1) Poorly developed karst systems with large storage capacity (diffuse) and (2) well-developed karst aquifer with larger conduits but without significant storage of water. In contrasting these two systems, large phreatic reservoirs in highly karstified limestones, such as those of Yucatan, are neither evaluated nor classified often (Stevanović et al. 2010).

Therefore, there are many karstic aquifers with large effective porosity, permeability, and storativity where due to a large and generally full reservoir, the impulse of newly infiltrated water is slowly transferred to discharge point. This is typical for deep siphonal karstic aquifer drained by ascending springs. Although it is not the same as, it is somehow similar to the reaction of a large surface reservoir on an increased river flow upstream of a water tail (of course, not dramatically increased), impulse will slowly transmit until it reaches the last point—the dam. Advantages of such type of drainage and large karstic reservoirs for engineering regulation are discussed in Sect. 15.5.

3.3 Flow Types and Pattern

Ten years before Henry Darcy did his survey in Dijon and formulated his famous law, Poiseuille's (1846) studied water flow through small tubes and found that *specific discharge* as flow per unit of a cross-sectional area is directly proportional to the hydraulic head loss between one end of the tube and other (Ford and Williams 2007). Poiseuille imposed gravitational acceleration, fluid density, dynamic viscosity, and hydraulic gradient as additional impact factors influencing specific discharge rate.

Darcy's law assumes *laminar flow*. This approximates the constant diameter of the tube or pore system which the water is flowing through. Hypothetically, the water particles move in parallel threads in the flow direction (non-disturbed parallel streamlines). *The turbulent flow* appears when the velocity and diameter of tubes increase and particles start to fluctuate with transverse mixing (streamlines are disturbed). The Reynold's number (R_e) identifies the critical velocity. It depends on the diameter of the tube (d), fluid velocity (v), and the two factors imposed by Poiseuille: fluid density (ρ) and dynamic viscosity (μ):

$$R_e = \rho \times v \times d/\mu \tag{3.14}$$

Some other standard hydraulic formulae are also applicable for specific cases of groundwater flow in karst and many mathematical deterministic models include equations of Bernouli, Chézy, and Maning. Details can be found in reference literature.

Krešić (2013) noticed that the traditional hydraulic of tubes (pipes, channels) is based on the principle of flow continuity (Fig. 3.24), which assumes that there is no inflow or outflow through the tube's walls. The elementary flow tube (Q) is directly proportional to the elementary cross-sectional area (A) and the average flow velocity (V_{av})

$$Q = A \times V_{av} \quad (m^3/s) \tag{3.15}$$

Krešić (2013) also stated that ideal tubes are rare in karst aquifer and that an extensive exchange of water between main conduits ("tubes") and the rock matrix usually takes place. Influx or outflux in the same conduit depends on pressure, and if the conduit is fully saturated, it discharges into the matrix. During the recession periods, the process is reversed, and water from the matrix discharges back to the conduit with a free water surface (Fig. 3.25).

Krešić (2007) highlighted the four factors which complicate calculation of flow through natural karst conduits:

- Inconsistency of pressure in the same conduit (presence of segments under pressure and with free surface);
- Irregularity of conduit's walls which requires estimation of the coefficient of roughness (and has to be inserted into the flow equation);

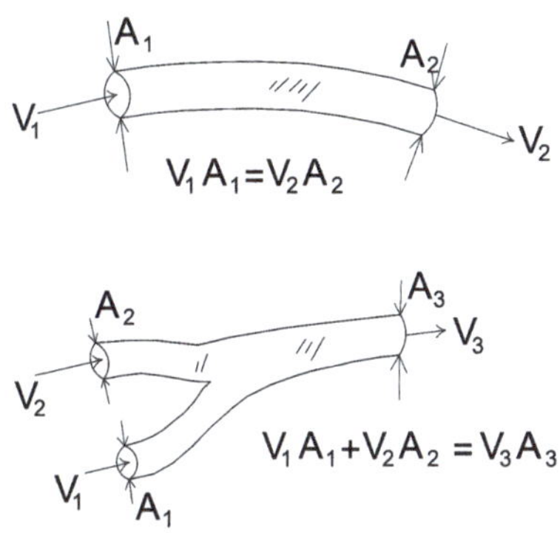

Fig. 3.24 Flow tubes and principle of flow continuity (after Krešić 2013, reprinted with permission of McGraw-Hill Education)

Fig. 3.25 Temporary dried cervical channel ("tube") in Devonian carbonate matrix (Zengpiyan, Guilin, China) (*left*) and active turbulent flow in Stopića cave (Zlatibor, Serbia) (*right*)

- Variability of cross sections along the conduit (even within short distances); and
- Changeability of flow types (laminar and turbulent) in the same conduit, caused by factors such as cross-sectional area, wall roughness, and flow velocity.

Finally, determination of the position of the main conduit (if there is just one preferential) is sometimes considered to be the main target of a survey (whatever the task is: water extraction or prevention of leakage) and a positive result heightens the possibility of having a successful project (Box 3.9). Bakalowicz (2005) emphasizes geophysics and field mapping of fracture distribution as generally efficient enough to locate drillings in productive zones. Goldscheider and Drew (2007) edited a book on different methods applied in karst hydrogeology which can, along with book of Petar Milanović's (1981), be used as a reference to the topic.

Box 3.9

The borehole HG-31 was drilled under the investigation program for the regional water system "Bogovina" in eastern Serbian karst (Stevanović 2010a). The location was chosen, as it was for all other exploratory boreholes, based on results of extensive infield survey. The logging and airlifting of HG-31 fully confirmed the choice: During the drilling at an interval of 77.6–98.2 m, a very small percentage of cores were removed and this part was drilled very fast. Analyzed core fragments were dark gray and round, which was supposed to be a possible indicator of active water circulation near an edge of an open cavernous zone. A continual yield of 2 l/s with a stable level in the small diameter piezometer HG-31 during the airlifting was also an indicator of promising well productivity. Considering these positive results of borehole HG-31, a good yield from well planned at that location was expected. The design of extraction well IE-5 was similar to the others drilled in the well field (Q_{av} = 40 l/s) and included an open-hole interval with a diameter of 295 mm in the deepest part of the well, at a depth of 59.5–110 m (Fig. 3.26).

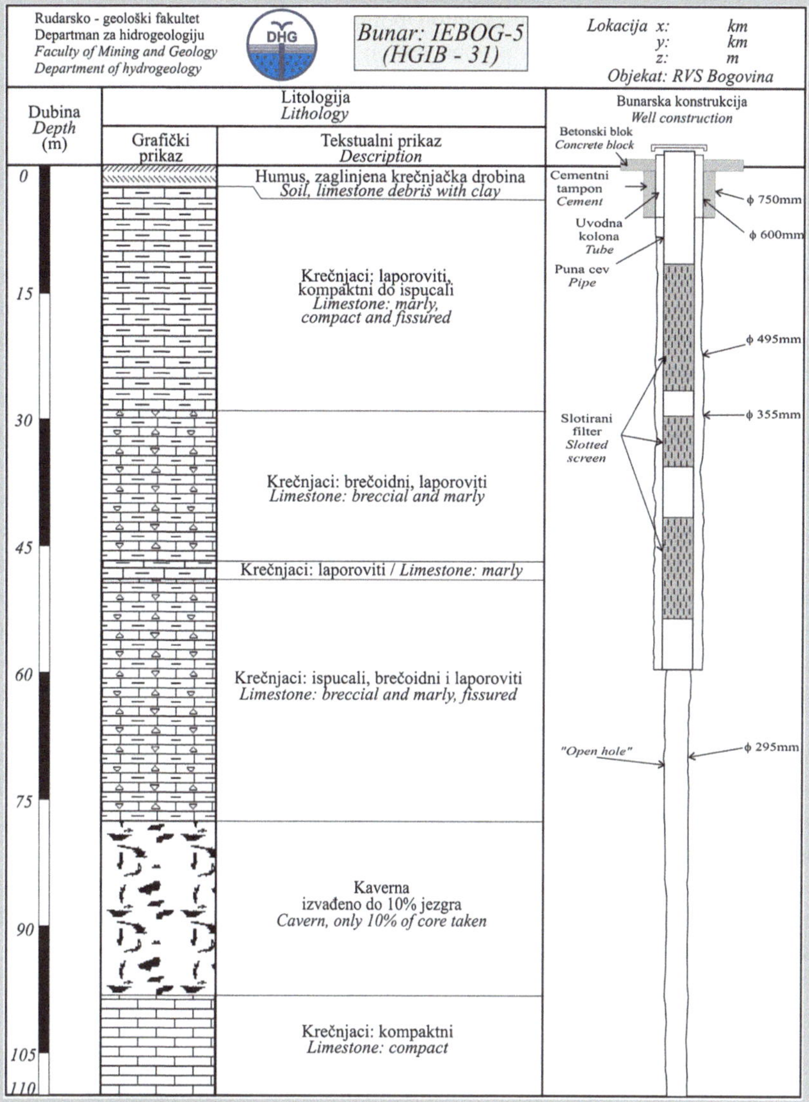

Fig. 3.26 Design and log of well IE-5 (Bogovina, Serbia; from Stevanović 2010a)

Unfortunately, this well is not as productive as expected. Its discharge is 15 l/s per drawdown of 21 m. The smaller capacity is most probably the result of having moved the well location by just 1 m from the HG-31 borehole. Although such situations have happened elsewhere in karst, previous positive experience with other wells and the wish to keep piezometer HG-31 as an important nearby observation point prevailed and brought about an undesired result. However, a good lesson was learned. Later, a remediation project including controlled explosions within the open-hole interval was prepared in order to open the route between the two holes, but after long discussion, this solution did not come into effect (mostly because of sufficient productivity of other wells).

White presented a classification (1969, and updated later) of flow types in carbonate aquifer system in regions of low-to-moderate relief: (1) *Diffuse flow* characterized shaley limestones and crystalline dolomites with high primary porosity or uniformly distributed fractures; (2) *Free flow* is linked to thick, massive soluble rocks, both perched and deep unconfined aquifers with conduits developed along bedding, joints, fractures, or fold axes; (3) *Confined flow* is diffuse or free flow and takes place in artesian and sandwiched aquifers overlain by beds of low permeability.

By analyzing the epikarst flow to the underlying vadose zone, the seepage flow in the upper parts has been differentiated from vertical drainage through a large fissure in the zone of lower aquifers (Fig. 3.6). Vertical drainage is the result of the connection of numerous small seepages which may reflect *diffuse flow* into one or a few vertically percolated flows. The latter can be termed *concentrated flow*. These conditions again certify the dual or multi-porosity of a karst aquifer. Finally, the part of lowest aquifer in the zone of saturation of unconfined aquifers is characterized by movement of water under described hydraulic laws but in principle with a water table (potentiometric pressure) slightly inclined toward discharge points or base of erosion. Such conditions were described by Cvijić in his theory of water circulation (1918). His coherent synthesis introduced three main superpositioned zones in specific dynamic coexistence. Vertical gravity circulation is dominant in the uppermost part, while the deeper saturated zones are characterized by (sub)horizontal and/or ascending circulation. Cvijić concluded that permanent lowering of the saturated part of the carbonate rocks is a logical consequence of the dynamic evolution of the karst (Fig. 3.27).

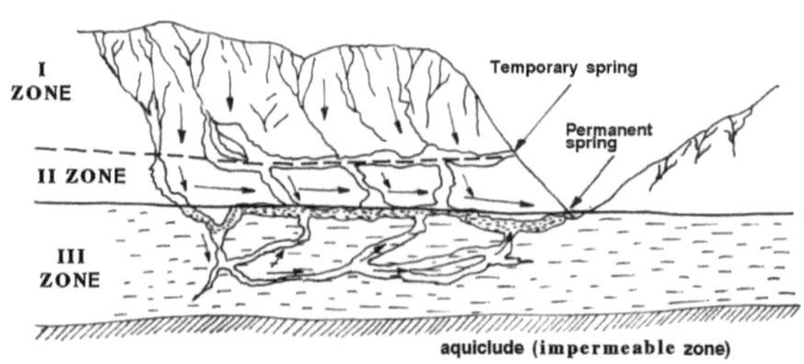

Fig. 3.27 Cvijić's concept of three "hydrographic" zones in karst: *I zone*—"dry" with percolation only; *II zone*—"transition" zone of water table fluctuation causing both vertical percolation but also gravity discharge; *III zone*—"stagnant" zone, fully saturated with siphons and conduits with ascending flow

Mangin (1975) suggested the existence of two types of flow through the vadose zone. Figure 3.28 shows that percolation of infiltrated rain or water of sinking streams might be *slow* or *fast* depending on aquifer properties.

This is not very different from the previous separation of *diffuse* and *concentrated* flow.

Another classification considers flow hierarchy as defined by Tóth (1999, 2009). Figure 3.29 shows the general concept of Tóth's theory and superpositions of gravity-driven flows: *regional flow, intermediate flow,* and *local flow*. They are characterized by the functionally different flow regimes superhydrostatic, hydrostatic, and subhydrostatic, respectively. Application of this theory in deep structures and confined aquifers is discussed in Sect. 17.5.

It is clear from the below scheme that the system of karstic conduits could also be superimposed even in the same hydraulic zone. Therefore, a network of karstic paths which enables water transfer may be vertically but also horizontally arranged. In many field experiments, the main or "chief" conduit has been identified, while others have a subordinate role (Milanović 1981). The main conduit is often termed *principal* (*dominant*), while the others are *peripheral*. Figure 3.30 shows the result of a tracing experiment and curve tracer concentration versus time. While the

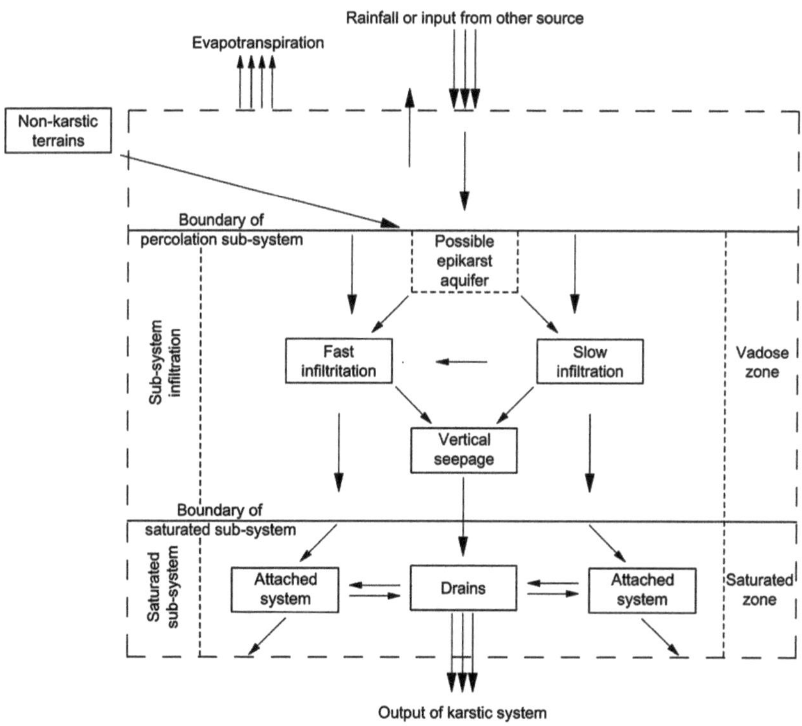

Fig. 3.28 Scheme of a karst aquifer function (from Mangin 1975)

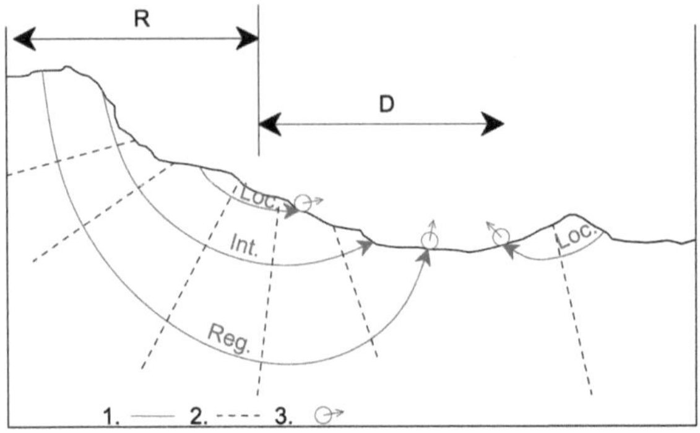

Fig. 3.29 Conceptual interpretation of Tóth's solution for gravity-driven flow in unconfined drainage basin. *Legend 1* streamline (flow line), *2* equipotential line, *3* spring, Reg/Int/Loc—regional, intermediate, and local flow, *R* recharge zone, *D* drainage zone

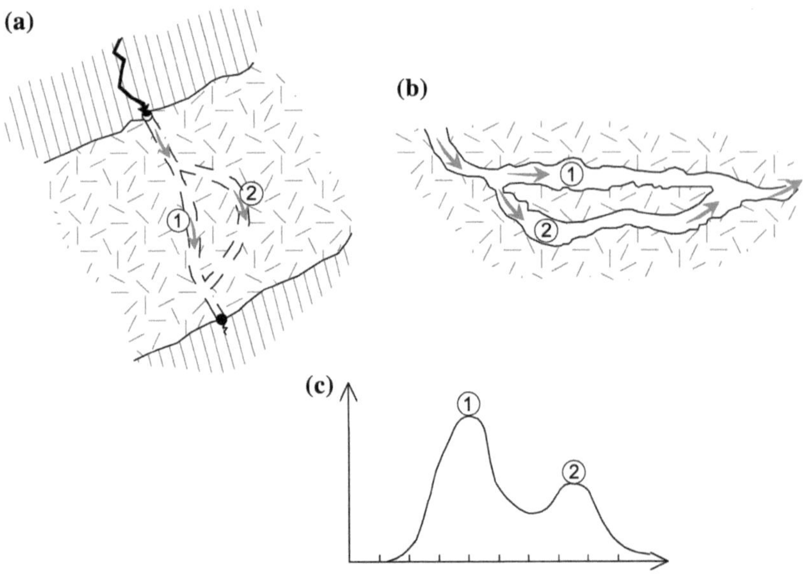

Fig. 3.30 Principal (*1*) and peripheral (*2*) channels and flows connecting ponor and spring. **a** Map—lateral difference. **b** Cross section—vertical difference. **c** Tracer concentration curve

first and higher peak indicates the time when the tracer was first recorded passing through the main conduit, the second peak is a result of peripheral flow.

The replacement of roles of superimposed conduits happens often in karst. When during the recession periods pressure in an aquifer decreases and water from pressurized cavities ("tubes") starts to flow freely instead, and the upper cavity has dried out, the peripheral bottom conduit assumes the main role in water transportation to the discharge point. Cases of groundwater piracy and altering of the flow to the "new" catchment, or changing directions of flow throughout the year, are similar (Box 3.10). The final stage of evolution of karst aquifer considers adaptation of water flow to the base of erosion. All transition occurrences may be considered as incidental indicators of the karstification process, which is ongoing. This is the case with the dual function system of estavelles (spring ponor), or temporary aquifer piracy, or conversion of a flow's direction during a limited period of time.

Box 3.10

Figure 3.31 shows a characteristic subterranean hydraulic mechanism in the Kučaj karst massif in eastern Serbian karst. At higher water periods, drainage takes place at two discharge points (Mrljiš and Fundonj springs), both at the edge of the karst massif but dislocated in two directions. Such a mechanism has been confirmed by the tracing of the Bogovina cave ponor (tracer simultaneously appeared in both springs). During the recession periods, the decline in the potentiometric pressure allows flow only to the regional base level of erosion in the area—to the Mrljiš spring and the Crni Timok River valley (Stevanović 2010a).

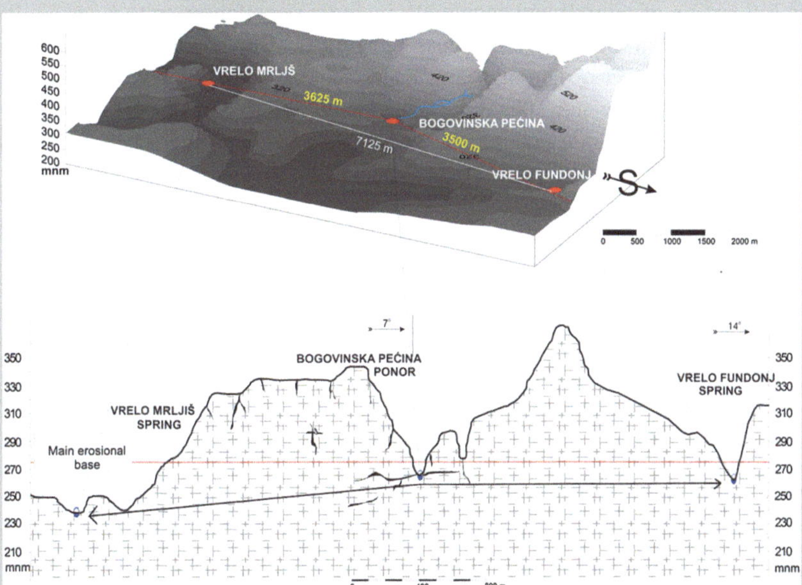

Fig. 3.31 Two-directional flow from ponor zone of Bogovina cave converts into one-directional flow during the recession periods (from Stevanović 2010a)

3.4 Aquifer Recharge

Two main types of recharge exist: *natural* and *artificial*. The main focus here will be on the natural recharge processes.

Lerner et al. (1990) describes natural recharge as the flow of the infiltrated water which reaches the groundwater table and results in an increase of water volume stored in aquifer.

The natural recharge sources are as follows:

- Precipitation,
- Surface waters: sinking streams, lakes, and sea water,
- Underground flows from adjacent catchment.

1. *Precipitation* includes rain water, snow melting, and condensate water. In most of the continents between 60°N and 60°S, except on high mountains rainfall is the dominant recharge component and its contribution is dispersed throughout a whole year, while *snow melting* is relatively short and regularly results in high peaks in the springs' hydrograph. *Condensation* also requires differences in air and soil temperature and humidity and usually contributes the least to replenishment of an aquifer's water reserves.

The infiltration of rainfall and snowmelt, including condensation to a certain level, all produce the earlier discussed *diffuse flow*. In fact, the character of flow is more dictated by soil cover or tectonic fabric and fissures' aperture, but, however, small the seepage of rainy water, it cannot be neglected as a factor. The exception is very high rainy storm episodes or intensive and fast snow melting when diffuse flow may convert into *concentrated flow* (Box 3.11).

Box 3.11

An example of the temporary functioning of a small sinking brook resulting from snow melting was described by Raeisi (2010). Thirty kilograms of sodium fluorescein were injected into this sinking brook ending in a nearby ponor (swallow hole) in the northern flank of Barm-Firooz anticline (80 km NW of Shiraz, Iran) (Fig. 3.32). The sampling sites were more than 30 springs of the two karstic formations Sarvak Fm. (Cretaceous) and Asmari-Jahrum Fm. (Paleogene), but a tracer only appeared at 18 km from the Sheshpeer spring.
Four experiments were performed on the catchment area of Sheshpeer Spring from December 1991 to April 1992 to determine the effect of external parameters such as flow rate, specific conductivity, temperature, and dissolved ions of recharged water on the physicochemical characteristics of the Sheshpeer spring. In the second experiment, snowmelt flow was measured every 2 h for 6 days. The ratio of average daily maximum flow rate to the average daily minimum flow rate varied from 5 to 10. However, no effect of

such a daily oscillatory recharge had been observed on the Sheshpeer spring hydrograph. A distance of 15 km between input and output points seems to be enough to suppress the effect of daily oscillations of the input flow rate.

Fig. 3.32 Karrenfeld in the core of Barm-Firooz anticline (Iran). Small and short brooks form only during the snow melt

2. *Concentrated recharge* also called *point recharge* is typical for the waters which infiltrate from *sinking streams* and concentric ponors. The ponor capacity is an equivalent of spring discharge but with inverse function. It is a volume of water in a unit of time which may be absorbed by the rocks. Some ponors are really huge and may swallow and transfer yields equivalent to thousands of m^3/s of water, but in principle, such yields are rare nowadays. Therefore, big cave ponors (Fig. 3.33) played a more evident role in the intensity of the karstification process in the past.

In contrast, many sinking streams are characterized by very small apertures in their riverbeds and the only means to identify the presence of sink streams (swallets) is to conduct simultaneous hydrometry at consecutive sections along the riverbed and to find evidence of the water losses (see Chap. 6).

As a result of aquifer's high permeability, recharge of karstic aquifer by lake water or undesired sea water is frequent in coastal aquifers (see Sect. 16.4).

3. *Underground flows from adjacent aquifers* and catchment are not often discussed in karst literature. Most authors consider this as simply a flow continuum, but for aquifer budget and assessment of water reserves, this is a very important issue. The lateral inflow or seepage from overlying aquifers (Fig. 3.34) may be continual, and then, revision of the basin geometry is probably required. But seepage may also take place for a limited period depending on potentiometric pressure and water table fluctuation in both receiving and emitting aquifers.

Fig. 3.33 Active cave ponors in the Carpathian and Dinaric karst (Sohodol, Runcu gorge, Romania, *left*) and (Rakov Škocjan, Slovenia, *right*)

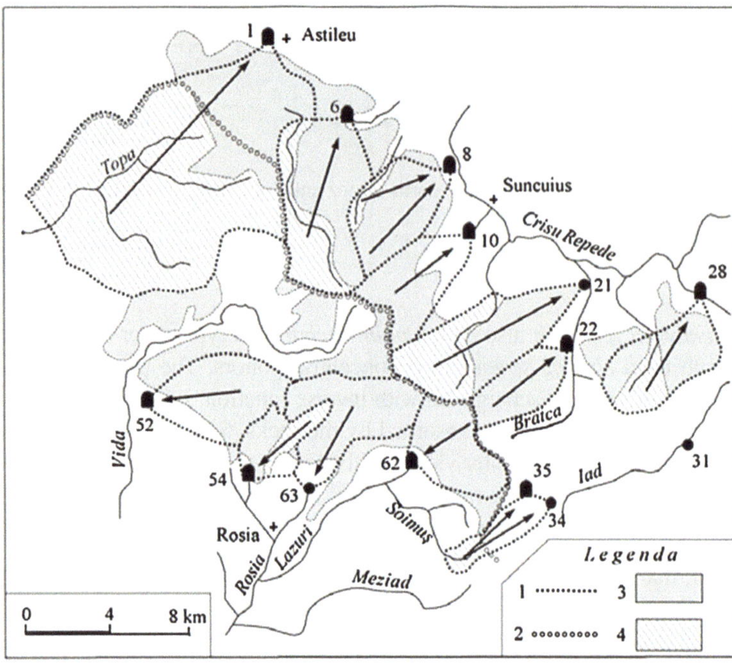

Fig. 3.34 The sketch map of Padurea Craiului Mt. (Romania): Karstic aquifers with a large extension of non-karstic terrains contributing to allogenic recharge. *Legend 1* groundwater divide, *2* watershed divide, *3* karstic terrains, *4* non-karstic rocks. Reproduced by permission from Oraşeanu and Iurkiewicz (2010)

When infiltration results from water formed inside a delineated karstic catchment, it is considered as *direct recharge* or *autogenic recharge*. If water for infiltration arrives from non-karstic terrains and contributing catchments, *indirect recharge* or *allogenic recharge* is spoken about. The system consists exclusively from karstic rocks with autogenic recharge is also called *unary karst system*, while *binary karst*

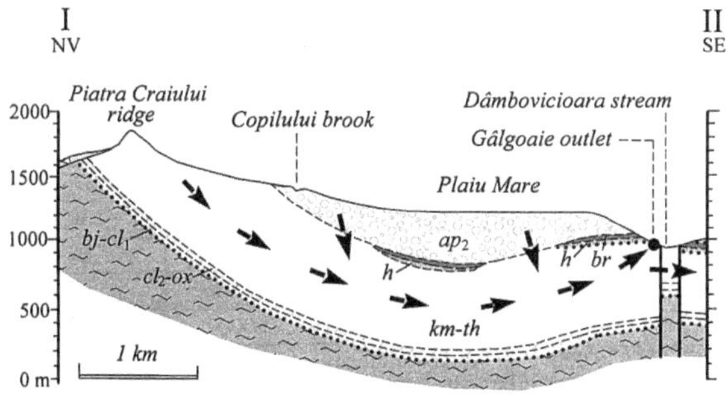

Fig. 3.35 Model of Dambovicioara passage (Padurea Craiului Mt. Romania). Vertical seepage is from overlying recent deposits (*gray*) into the main karstic Mesozoic carbonate karstic aquifer (*white*). Reproduced by permission from Oraşeanu and Iurkiewicz (2010)

system includes non-karstic rocks and their contributing catchment (Marsaud 1997; Bakalowicz 2005).

Many still active ponors are located directly at or near the contact of karst and impervious or low permeable rocks which convey perennial streams toward those aquifer recharge points (Fig. 3.35). There are also many blind dry valleys formed at such contacts providing evidence of karstification and enabling reconstruction of the paleo-hydrographic situation.

Effective infiltration as the amount of water which will efficiently reach the water table and storage (Box 3.12) depends on many factors, and in the case of water originating from the atmosphere involves climatic and water balance elements such as rainfall intensity, evapotranspiration, wind and other elements discussed in Chap. 6. However, along with these climatic factors there are some other factors which may be grouped as follows:

3.4.1 Non-geological Factors of Natural Recharge

Topography: Slope is one of the main factors, depressions such as uvala or dolines regularly favor water stabilization and infiltration, while higher slopes stimulate a runoff component. Therefore, terrains with smaller slopes are more suitable for groundwater recharge. Digital elevation models, topography maps, and remote sensing support slope analysis.

Soil cover: If this is missing, the infiltration will be more effective. If it is present and thick, then it regularly reduces efficiency of infiltration (Fig. 3.36). Pedology maps and digital Corine land cover maps in combination with remote sensing may facilitate this assessment.

Soil moisture: When soil cover is present then moisture balance becomes an important factor for infiltration rate. If soil is fully saturated from previous rains, melted snow or flooding then runoff and evapotranspiration will again be dominant in the water balance, and less water will infiltrate. But in arid karst or during

Fig. 3.36 The vertical section of composite carbonate—gypsum layers prosperous for active recharge from soil cover and moisture point of views: soil is very thin and layers are dry due to arid environment (Taalex Fm. in Laasqoray Xudun—Sool Plateau, Somalia)

the long and dry seasons when soil moisture deficit grows, macro pores and cracks are activated causing the infiltration capacity also to rise (Fig. 3.36).

Vegetation: Plant distribution and density also play an important role in the water cycle. The grass, cultivated crops, and trees are all extensive water consumers, and if those are extended and dense, the transpiration will increase causing a reduction in water available for infiltration. Moreover, some experiences show that during the vegetation seasons in old and dense forests, rains of little intensity with a sum of less than 3 mm/day will not reach the land and will not have a chance to infiltrate. *Interception* represents the difference between the total precipitation which falls and the precipitation that actually reaches the soil; thus, water is intercepted by the leaves and plants.

3.4.2 The Geological Factors Influencing Recharge

Lithology: Not all dissoluble rocks have the same infiltration capacity; those more soluble and with greater primary porosity such as the evaporitic group will infiltrate more water. Overlying layers such as thick overburdening impervious rocks and lithological impurities also have a great impact causing reduction of inflow.

Tectonic fabric: It is probably the central geological factor. If surficial fissuration is very small or missing and the bedding plane is horizontal, then infiltration will not take place or will be minimal. In fact only small intensity surficial karstification will then take place. Such a situation is presented on Figs. 2.12 and 2.13. In contrast, both epikarst and bare karst with developed vertical cavities are suitable environments for intense infiltration. Tectonically fabricated vertical or subvertical strata also support water infiltration between bedding planes (Fig. 3.37).

Aquifer saturation: If aquifer is completely full of water, and the water table reaches land surface, no new water can be added to the ground and then runoff prevails.

To conclude, Fig. 3.38 illustrates the explained components of natural recharge of a typical unconfined karst aquifer. The rainfalls produce mostly diffuse recharge

Fig. 3.37 Structures suitable for receiving and conveying waters: bedding plane aperture (Devonian karst of south China, *left*), widely open fractures (Eramosa dolomitic karst, Ontario, Canada, *center*), and highly inclined carbonate rocks (Sierra de la Nieves, south Spain, *right*)

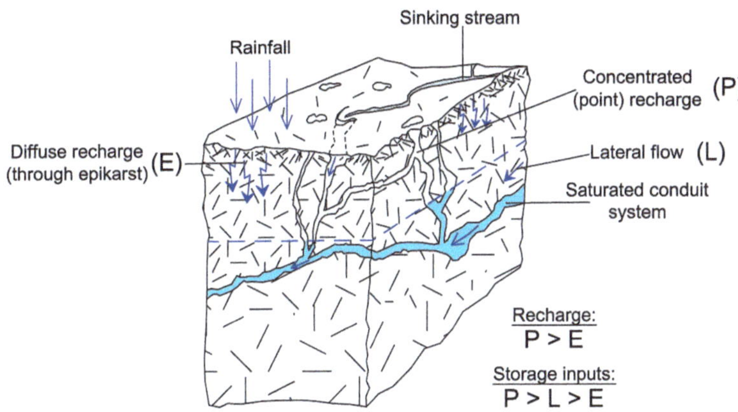

Fig. 3.38 Schematic model of natural recharge components in an unconfined karstic aquifer

and slow percolation through epikarst (*E*). Sinking streams lose all their water in a big concentric ponor which extends to a great depth in the form of a pothole enabling concentric water flow to the water table (*P*). Finally, lateral flow (*L*) conveys water from a remote part or from another aquifer or adjacent basin. When vertical cavities are sufficiently large to convey sunken water, the point recharge is higher than diffuse through epikarst: $P > E$, and similarly $P > L > E$.

The term *artificial recharge* is often improperly identified with *engineering regulation of aquifer*. Although in intergranular media this is justifiable because the majority of interventions consider precisely the artificial transport and induced inflow (whether with or without pressure) into the aquifer, this is not often a solution applied in karst. There is simply a general awareness that water in karst may "escape" through many unknown corridors and that effects of such interventions would be minimal. But similar interventions with regulating the riverbeds or redirecting surface streams to recharge ponors in other basins are known in engineering practice (see Chap. 13). However, the artificial recharge of aquifer in platform karst is more promising and recharge wells are being used in Florida to convey waters from shallow to the underlying Floridian karstic aquifer.

Finally, artificial recharge may take place unintentionally. Water use in irrigation can lead to substantial increases in recharge. This may be the indirect result of soil watering or leakage from irrigation conveying channels. Younger (2007) pointed out that not only seepage from water pipelines and septic tanks, but also intensive watering of parks and gardens increases recharge in Ryadh, Saudi Arabia. Such recharge is commonly termed as *irrigation return flow*.

Areas urbanized on karst are ideal places to have significant recharge of aquifers. Knowing that losses from water-conveying pipes presented by water utilities are as great as ca. 10 % to as much as 60–70 %, which is occurring in some undeveloped cities, it is clear that a large portion of lost water may re-infiltrate in the aquifer. Sharp and Garcia-Fresca (2004) estimated the drinking water leakage in Austin, Texas, to be around 7.7 % or 21 mm/year as water available to be re-infiltrated into the Edwards aquifer. The amount is deemed significant considering that natural recharge of local aquifer is in the range of 30–100 mm/year. An example of incidental recharge of karstic aquifer is also presented in Sect. 3.6, Box 3.18. Nevertheless, seepage from waste water reservoirs and channels, storm sewers, or septic tanks is also a common recharge source in urban areas.

From above, it is clear that effective infiltration is variable element. If we thus express it as % of precipitation as common in hydrology practice, we should consider the time component, i.e., period in which effective infiltration is taking place. For a rough and general assessment of groundwater recharge, an average annual value may be acceptable as presented in Table 3.2 (Box 3.12).

Box 3.12

Table 3.2 Estimated effective infiltration rates of some karstic aquifers

Type of aquifer rock category	Location	Effective infiltration in % of annual precipitation (%)	Author	Source (reference)
Cretaceous chalk	London basin, England	20–35	Anon, 1972	Lerner et al. (1990)
Paleozoic dolomites	Ghaap Plateau, South Africa	2–25	Smith, 1978	Lerner et al. (1990)
Taalex gypsum, Eocene	Ceerigaboo, Somalia	30	Stevanović et al. 2012a	–
Pliocene–Pleistocene limestones	Morroco	14–19.5	Bolelli, 1951	LaMoreaux et al. (1984)
Aptien-Albien limestones	Dj. Sidi Rheriss, Oum el Bouaghi, Algeria	38	Stevanović, 1985	–
Eocene limestones	Dyr el Kef, Tunisia	33.2–90.0	Schoeller, 1948	LaMoreaux et al. (1984)

(continued)

Table 3.2 (continued)

Type of aquifer rock category	Location	Effective infiltration in % of annual precipitation (%)	Author	Source (reference)
Senonian and Jurassic limestones	Dj. Chounata, Tunisia	33–53	Tixernot et al. 1951	LaMoreaux et al. (1984)
Middle Cretaceous limestones	Na'am spring basin, Israel	53	Mero, 1958	LaMoreaux et al. (1984)
Pila Spi Fm., Tertiary limestones	Dohuk, northern Iraq	35	Stevanović, 2003	–
Bekhme Fm., Cretaceous limestones	Harir, northern Iraq	40	Stevanović and Iurkiewicz, 2004	–
Mesozoic limestones	Lilaia springs, Greece	51.6	Aronis et al. 1961	LaMoreaux et al. (1984)
Mesozoic limestones	Monte Simbriuni, Italy	69	Boni and Bono, 1984	Burger and Dubretret (1984)
Mesozoic limestones	Monte Lepini, Italy	78	Boni and Bono, 1984	Burger and Dubretret (1984)
Jurassic and Cretaceous limestones	Kučaj-Beljanica Mts. east Serbia	31	Stevanović, 1991	–
Triassic limestones	Tara Mt. west Serbia	47	Stevanović, 1995	–
Mesozoic limestones	Malé Karpaty, Slovakia	24.4–52.4	Kullman, 1977	Burger and Dubretret (1984)

Numerous methods are applied to estimate recharge of a karstic aquifer; some of them are discussed in Chaps. 4–6 and other contributions to this book. Among them are groundwater budgeting, stochastic input–output modeling, isotopic methods, Cl^- ion balance, and GIS application methods.

3.5 Aquifer Discharge

Drainage of karst aquifer is a crucial process for most planned engineering works, whether we intend to tap and utilize karstic waters or defend ourselves from them. Therefore, the majority of research activities in karst hydrogeology usually concentrate on the discharge points as the places where direct access to issuing groundwater is possible. In many cases, concentration of works near discharge points is also dictated by the topography and accessibility, and while many spring sites are in reachable foothills or in flat areas, their catchments could be in high mountains not easily reached, glaciers or dense tropical forests.

Drainage of karst aquifer or aquifer's discharge takes place as (Stevanović 2010b) follows:

Natural, carried out through springs, submerged springs, estavelles, subsurface outflow and

Artificial, effectuated through groundwater extraction by wells, galleries, canals, or similar water intake structures.

The topic of artificial drainage and tapping karstic waters is discussed in Chap. 11. From the point of view of water budget, natural drainage of aquifer also includes a few other components such as runoff from karstic surface or evapotranspiration, both discussed in Chap. 6. The subsurface outflow as an important way of drainage, even essential for many engineering works in karst, is also discussed in Chap. 6 and Sect. 15.5. What follows is thus focused on natural discharge and interrelated factors.

The intensity and velocity of groundwater flow toward discharge points depend mostly on the hydraulic gradient. The greater the inclination of the water table and the larger the hydraulic head, the greater the energy that will push stored water to flow out from the aquifer (Fiorillo 2011). The limitations on effectual drainage can be the opening of discharge point(s), i.e., *orifice,* and its dimensions which may not allow quick discharging or even total emptying of an aquifer. When the water table and hydraulic head decline over time, the discharge rates also decline usually in dry seasons, but it sometimes takes a long time before the aquifer completely dries out (Fig. 3.39). The *coefficient of discharge,* also called the *recession coefficient* (α), which expresses the diversity of an aquifer's characteristics and the storage capacities which influence discharge intensity, is explained in Chap. 7.

Groundwater flow aims to reach the lowest discharge point of an aquifer which commonly corresponds to the local or regional erosional base. *Erosional base* or *base level of erosion* is the lowest level of the terrain where erosion by water is still possible. The erosional base is changeable in geological time, and the evolution of the karstic process follows and adapts to that descending level. Many dry karstic caves had previously functioned as springs and superpositioned big cave openings carved in karst at the cliffs in many canyons provide evidence of this process (Fig. 3.40).

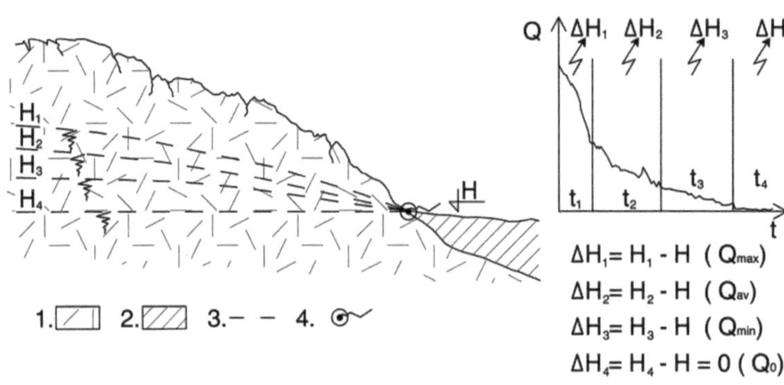

Fig. 3.39 Discharge intensity depends on hydraulic head. The maximum discharge is during the period t_1 when the hydraulic head has maximal value (ΔH_1), while during the period t_4, the spring is running dry ($\Delta H_4 = 0$, $Q = 0$). *Legend 1* karstic aquifer, *2* impervious rocks, *3* water table or hydraulic head, *4* spring, *Q* discharge, *t* time

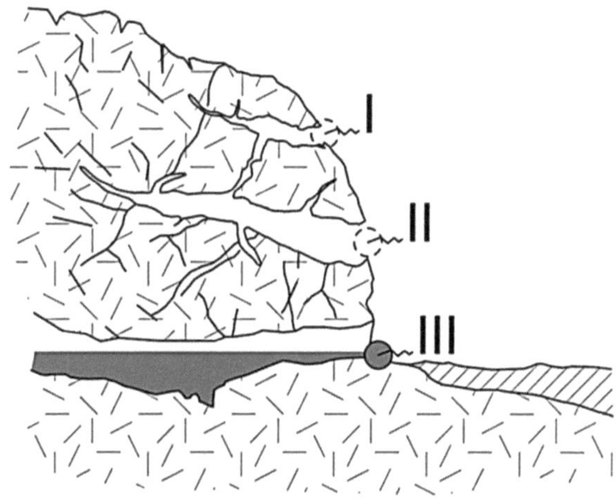

Fig. 3.40 Evolution of karst aquifer and adaptation of discharge points to the erosional base. *I*, *II* previous drainage points at the cliff; *III* actual deep ascending discharge point

The erosional base is a principal controlling factor of the development of karstic landforms (Ford and Williams 2007) and of groundwater deepening. However, the *erosional base* cannot be equaled to the *karstification base,* and in mature karst, the process of karstification and rock dissolution may take place deep below the actual erosional base.

The *regional base level* is equivalent to major rivers, lakes, or seas and zones to which the regional groundwater flow tries to accommodate. There are also *local erosional bases*, which may be represented by a lithological barrier or non-permeable layer within an aquifer, fault, anticline plunge, depression, or other geological and topographical elements, which influence and predispose groundwater drainage. One such local erosional base is shown in Fig. 3.2 (see C: case of perched aquifer).

A spring is a natural opening (*orifice*) of the land surface from which groundwater visibly flows out from an aquifer. Thanks to springs and their usually sufficient openings, karstic aquifers mostly discharge concentrated flows; diffuse flows are definitely present in karst to a lesser extent than in any other aquifer system.

Many authors afforded classifications of the springs as outlets of aquifers. Along with the description and explanation of the function of some major springs in the USA, Meinzer (1923b) distinguished categories in accordance with springs' discharge rates. Bögli (1980) provided a very comprehensive and well-explained classification which includes several groups of springs. Herak et al. (1981) stated that characteristics of springs are defined by the properties of aquifer, the position of connected aquicludes and aquitards, and by the climate and vegetative cover. Ford and Williams (2007) also explained the main spring types and distinguished the three major groups (free draining, dammed, and confined). In addition, they provided a list of the world's largest springs. Krešić (2010) wrote a chapter on types and classification of springs in the textbook "Groundwater hydrology of springs," while in the same book, Stevanović (2010b) focused on utilization and regulation of springs based on their characteristics.

There are several classifications of springs made in accordance with established criterion. Table 3.3 is an attempt to present some of those criteria, ten in total, and accordingly to provide common classifications and typology of springs.

Table 3.3 Classification of springs (criteria and typology)

Criterion	Type of springs	Synonyms/ other names	Short explanation	Sketch/section
1. Flow type and pressure in aquifer	1.1 Descending	Gravity	Free draining aquifer. Aquifer with water table (phreatic) inclined toward discharge point: gravity downward flow	
	1.2 Ascending	Artesian, confined, raising	Confined aquifer. Water driven by pressure: upward flow	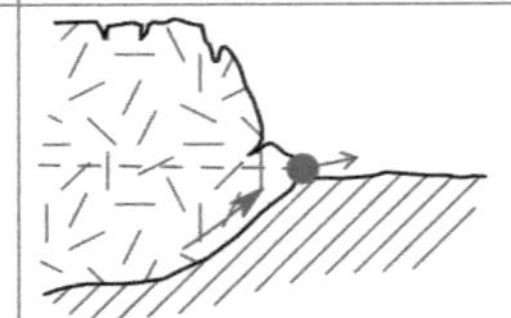
	1.3 Overflowing		Usually free draining aquifer. Often upper aquifer's outlet functioning periodically. Slightly inclined or subhorizontal water table: flow pours over barrier	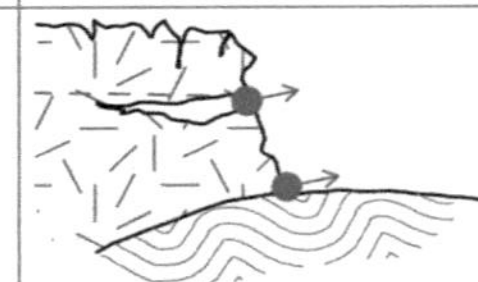
2. Barrier position and type	2.1 Contact	Dammed, bedding	Lithologically predisposed. Water issuing at the lithological contact with non-karstic rocks of lower permeability (total barrier)	
	2.2 Fracture	Fault	Tectonically guided flow. Water issuing from an open crashed zone, fracture, fault or other lineament	

(continued)

Table 3.3 (continued)

Criterion	Type of springs	Synonyms/ other names	Short explanation	Sketch/section
	2.3 Hanging		Lithologically or tectonically predisposed local barrier. Water issuing above main (regional) erosional base. Often drainage of perched aquifers	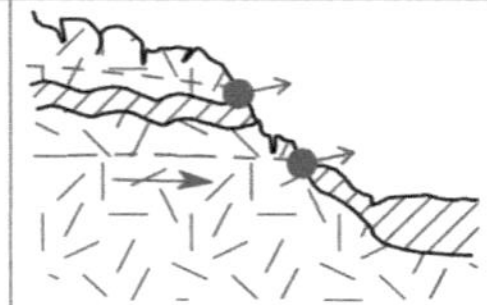
	2.4 Anticlinal		Tectonically predisposed. Water issuing from anticline plunge (margin of the structure)	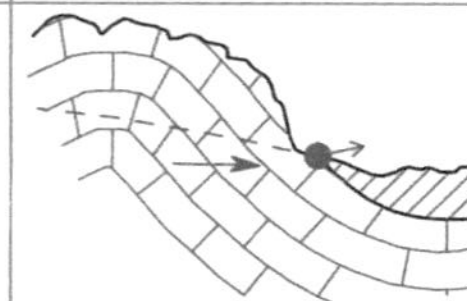
	2.5 Depressional		Predisposed by topography and more often by river incision. Depression in relief intersects water table enabling free flow from aquifer. The same is true with cutoff of confined aquifers	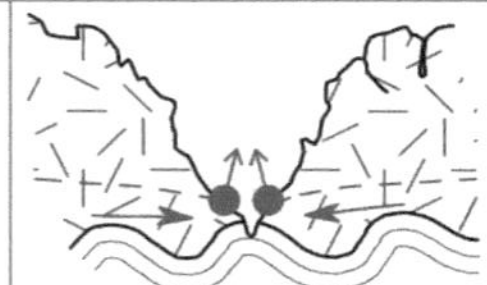
	2.6 Impounded	Subaqueous, submarine, sublacustrine, submerged, vrulja	Discharge point is impounded by fresh, lake, or river water, or is under the sea level (totally or intertidal). The karstification is thus below the erosional base and barrier may be absent	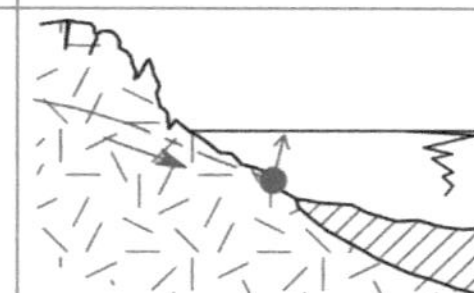 See also Box 3.13

(continued)

Table 3.3 (continued)

Criterion	Type of springs	Synonyms/ other names	Short explanation	Sketch/section
3. Spring–aquifer relation	3.1 Primary		Water issuing directly from original karstic aquifer	As in case 1.1–1.3
	3.2 Secondary	Masked	Water issuing from connected lateral aquifer or from rock debris away from original discharge point	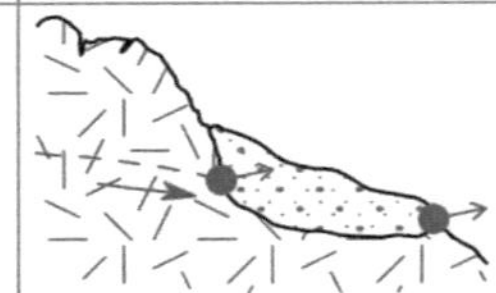
4. Discharge point feature	4.1 Cave		Water issuing from a hydrologically active cave	See Fig. 3.41i
	4.2 Siphonal		Water issuing from siphon or from siphonal reservoir, usually as an ascending flow	See Fig. 3.41j
	4.3 Lake	Vauclusian	Water issuing from the lake usually extends to depth as funnel-like vertical channel formed by erosion of a strong ascending flow. The name of the famous French spring Fontaine de Vaucluse (near Avignon) is the derivation of all similar types	Fig. 3.48
	4.4 Pond		As above, but water issuing from smaller and shallow basin or doline, often representing an aquifer "eye." Such a type is the cenotes at Yucatan, Mexico if they generate flow (not stagnant water)	See Fig. 3.41k

(continued)

Table 3.3 (continued)

Criterion	Type of springs	Synonyms/ other names	Short explanation	Sketch/section
5. Periodicity	5.1 Constant	Perennial, steady	Permanently discharging spring, despite flow variations (to prevent mixture of spring with stable discharge)	
	5.2 Intermittent	Episodic	Temporary activating spring, discharging usually during and immediately after intensive rains, floods and snow melting	See Fig. 3.411
	5.3 Periodic	Rhythmic, ebb and flow	Specific hydraulic mechanism. The pressure and water level in the connected aquifer's siphon vary causing alternate discharge and break events	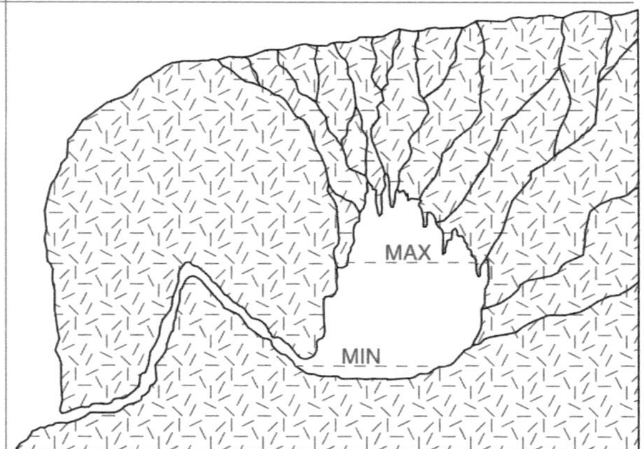 Mechanism of ebb-and-flow occurrence after Radovanović (1897). Discharge activates when water reaches maximal level in connected siphon

(continued)

Table 3.3 (continued)

Criterion	Type of springs	Synonyms/ other names	Short explanation	Sketch/section
	5.4 Inactive	Fossil, paleo	Dry spring. Not functional anymore because of actual karstification phase which lowers water table. The synonyms "fossil," "paleo-" relevant only for that case. Evidence of ex-spring site might be presence of travertine deposits. Drying can also be result of anthropogenic impacts (e.g., forced pumping of nearby wells) in which case spring may recover	
	5.5 Estavelle		Partly ponor (sink point), partly spring (discharge point). Characterized by reversed discharge	
6. Discharge rate	6.1 Large	Strong	Habitually, with minimal flows (Qmin) over 100 l/s	
	6.2 Medium	Moderate	Habitually, with Qmin in range of 10–100 l/s	
	6.3 Small	Weak	Habitually, with Qmin less than 10 l/s, or even less than 1 l/s	
	6.4 Of first, second, etc. category (magnitude)		In accordance with average discharge Meinzer (1923b) classified springs in eight groups: 1 > 10,000 l/s, 2.1,000–10,000 l/s, 3.100–1,000 l/s, 4. 10–100 l/s, 5. 1–10 l/s, 6. 0.1–1 l/s, 7. 0.01–0.1 l/s, and 8.0 < 0.01 l/s	
7. Discharge regime	7.1 Extremely variable		Habitually, with ratio Qmin:Qmax over 1:100	
	7.2 Variable		Habitually, with ratio Qmin:Qmax between 1:10 and 1:100	
	7.3 Stable		Habitually, with ratio Qmin:Qmax under 1:10	

(continued)

Table 3.3 (continued)

Criterion	Type of springs	Synonyms/ other names	Short explanation	Sketch/section
8. Intake	8.1 Tapped	Captured	Springs usually with simple structure for collecting and protecting water from pollution, but not allowing changes of natural flow. Water may be used in situ (tap) or delivered by pipes for further centralizing water supply (utility)	
	8.2 Non-tapped	Non-captured	Natural discharging point without any intervention around	
	8.3 Regulated flow	Artificially arranged discharge	Complex water tapping structure enables the maintenance of or increase in the flow (storage box, pumps, wells, etc.)	
9. Water quality	9.1 Freshwater	Cold	Habitually, low-mineralized water (TDS less than 1,000 ppm) and with temperature below 20 °C	
	9.2 Mineral	Acidic, sulfuric, gas-lift, etc.	Habitually, water with mineral substances (TDS more than 1,000 ppm). Derived synonyms depend on prevailing ions, specific components or discharge mechanism	
	9.3 Thermal	Lukewarm, subthermal	Habitually, with water temperature between 20–100 °C, but synonyms are relevant for temperatures between 20–30 °C	
	9.4 Hot (water)	Hyper-thermal	Habitually, with water temperature over 100 °C	

(continued)

Table 3.3 (continued)

Criterion	Type of springs	Synonyms/ other names	Short explanation	Sketch/section
10. Utilization	10.1 Drinking water supply		Satisfying potable water criteria and standards	
	10.2 Bottling water		Satisfying potable water criteria and standards. Not additionally purifying	
	10.3 Technical water supply		Non-potable springwater	
	10.4 Medicinal purpose		Used for balneotherapy or for recreation. In the former case, water contains active medicinal substances	
	10.5 Irrigation		Non-potable water but requires adequate quality to be low or slightly mineralized. The springwater often includes animals watering	
	10.6 Hydropower generation		Hydraulic head utilized for hydropower production (usually small plants)	
	10.7 Heating/ cooling		Water utilized directly (hot) or with support of heat pumps (low enthalpy resources)	

Legend for sketches/sections: *1* karstic aquifer massive rocks, *2* karstic aquifer stratified rocks, *3* impervious rocks, *4* impervious folded rocks, *5* sand, intergranular aquifer, *6* spring, *7* groundwater table, *8* potentiometric line (hydraulic head), *9* direction of flow, and *10* fault

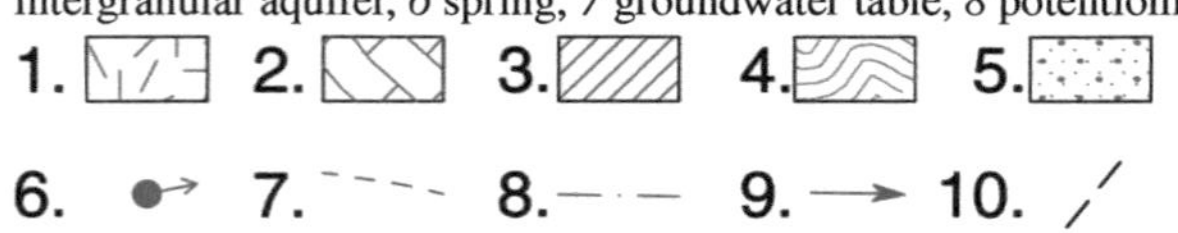

The unavoidable mixture of criteria makes it not easy task to group or classify the springs: The same spring can be characterized from several angles, and one spring may also change its behavior depending on the water season (high/low waters).

Krešić (2010) also considered *seep* as a kind of discharge which cannot be visually observed but could be indicated by the wet soil. *Seepage spring* is a term also often used for such drainage. Diffuse flows are relatively rare in karst since concentrated flows prevail. However, such occurrences exist in wetlands, impound karst, or follow the discharge of secondary springs.

Fig. 3.41 **a** Gravity spring Bekhal (north Iraq), **b** ascending spring Syri Kalter ("Blue eye," south Albania), **c** overflow outlet of Iskretz Spring (Svoge, Bulgaria), **d** contact Istok spring (Triassic limestone–Jurassic ophiolites) (Metohija, Serbia), **e** fault gravity spring of Sava River (Bohinj, Slovenia), **f** hanging Margoon waterfall spring (Shiraz, Iran), **g** anticline plunge—Soosan spring (Kazeroon, Iran), **h** impounded spring—sublacustrine (vrulja) Karuč spring (Skadar Lake, Montenegro), **i** cave spring—Le Loue (Jura Mt. France), **j** siphonal spring of Cetina River (south Croatia), **k** pond spring—"eye" (Laas Caanood, Somalia), **l** intermittent spring (Dokan, Iraq; courtesy A.Holm), **m** small, stable and locally tapped spring Stapari (west Serbia), **n** mineral sulfuric spring Awa Spi ("white water," Sangaw, Iraq), and **o** thermal lake spring Heviz (Hungary)

Fig. 3.41 continued

In accordance with their supposed origin Ford and Williams (2007) classify springs in terms of: (1) *emergence* (no evidence of origin), (2) *resurgence* (re-emergence of a known swallet stream, i.e., re-appearance of sinking river), and (3) *exsurgence* (autogenic seepage water).

In reference to their exact position against the shoreline, springs are *terrestrial* (*continental*) or *coastal*.

The following Fig. 3.41 illustrates some of the above-defined types of springs, but also springs which are well known, beautiful, or specific in terms of some of their properties.

Box 3.13

The complexity of the application of different classification criteria is demonstrated at Sopot Spring in Boka Kotorska Bay, Montenegro.

Sopot Spring is located at the shoreline, some 50 m from the sea, and 2 km from the city of Risan. The upper outlet is the cave 20 m above sea level which is hydrologically active only during the periods of intensive rains at the Orjen Mountain and "Stone Sea" area above Risan, which is characterized by the highest precipitation rate in all of Europe (3,000–5,000 mm/year in average). Rainy water quickly infiltrates into the highly karstified massive Cretaceous limestones where in an area of 8 km^2 more than 300 vertical shafts are registered (Milanović 2006). Depending on the saturation level, but commonly after 2–3 days, the cave as the outlet of an upper aquifer is activated, producing enormous discharges. This periodical discharge is highly impressive, and the spring cave functions with a discharge of over 150 m^3/s, one of the world's largest (Stevanović et al. 2010). Some estimates even indicate 200 m^3/s, but precise measurements are extremely difficult due to a very steep cliff. The water then flows a very short distance from the upper cave and then with a crashing noise falls to the sea over a cascade around 20 m high (Fig. 3.42).

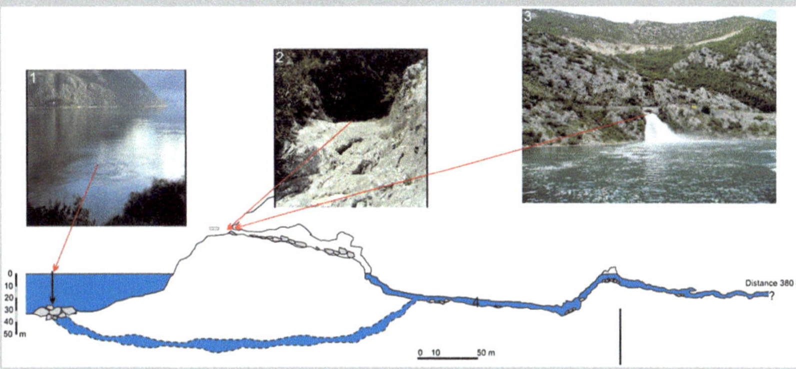

Fig. 3.42 Sopot Spring discharge mechanism (Boka Kotorska bay, Montenegro; photographs Milanović S, reprinted from Stevanović et al. 2010 with permission)

The discharge zone is predisposed by the contact of Cretaceous limestones with Eocene flysch sediments which belong to the regional Budva–Cukali tectonic zone. Direct contact is under the sea and not visible, and groundwater permanently discharges along that contact. The diving exploration located main discharge points at depths of 28 and 36 m (Milanović 2007). Research at Sopot was conducted as part of a Yugoslav-French project that lasted for 3 years. In total, 380 m of flooded canals were examined. The system is permanently discharging; during the summer season, only the two submarine springs (locally "vrulja") are active and they drain the Sopot aquifer system.

In accordance with the above site description and the proposed criteria, the Sopot Spring can be classified as follows:

- Dominantly gravity spring (but with lower ascending, and upper overflowing channels),
- Contact spring (partly impounded),
- Primary spring,
- Cave spring (with siphons at depth),
- Constant spring,
- Large spring (to moderate during the low-water seasons),
- Extremely variable spring,
- Non-tapped spring,
- Fresh water spring (with salty intrusion during the low-water seasons),
- Non-utilized spring.

It is clear that classification of such a complex discharge system may be problematic even under the same and clearly defined criterion.

The water table fluctuations and variation of pressure in an aquifer system depend on many factors such as the previously described recharge intensity, permeability, actual saturation (storage), and hydraulic head as the difference between input (recharge) and output (discharge) points. But the spring type and aquifer drainage regime are, almost by definition, closely related: The ascending springs have a more stable regime, while variation in the greater discharges characterizes gravity springs. The regime of springs which discharge from confined aquifers, such as depressional, fault, or thermal springs, is even more stable as a result of small pressure variations or limited orifices.

The spring hydrograph is thus the result of various processes which take place on the land surface or inside the aquifer system. When the main recharge is from sunken stream waters, the spring and stream hydrographs could be of similar shape. When the main recharge is from percolated rains and/or from snow melting, the differences between the maximal recharge episode and discharge peaks represent the *residence* or *travel time* necessary to estimate aquifer character. The stochastic analysis of spring hydrographs and correlative rainfall/spring flow diagram are the "books" for reading and understanding the karst aquifer behavior (Box 3.14). The method of time series, i.e., autocorrelation and cross-correlation analyses, was developed with the aim of characterizing a karst aquifer (Mangin 1984; Bonacci 1993; Krešić 2013). The correlation and cross-correlation analyses are discussed in Sects. 15.1 and 15.2.

Box 3.14

The different aquifer reaction in recharge events is presented in the case of Sarchinar Spring in north Iraq. It is one of the largest springs in the region, supplying municipal water for the Sulaimaniya city of more than 700,000 inhabitants. The outlet drains a large catchment area, some 200 km^2 of the Sarchinar–Chaq Chaq karstic system of High Folded tectonic zone (Piramagroon Mt. as a part of Zagros Mts. chain), with a dominant presence of carbonate and clastic rocks of Cretaceous age (Fig. 3.43).

Sarchinar Spring is an ascending spring issuing at the anticline plunge. The recharge of the spring is based on: (a) the diffuse infiltration of rainfall

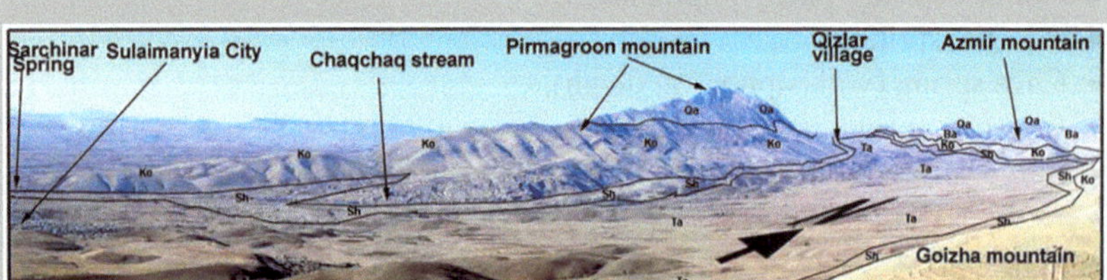

Fig. 3.43 Sarchinar–Chaq Chaq karstic system. Dominant karstic formations are Kometan (*Ko*) and Qamchuga (*Qa*), while low permeable rocks belong to Shiransh (*Sh*) and Tanjero (*Ta*) Fms

through the exposed outcrops of thickly bedded and highly fractured limestone layers; (b) the percolation of runoff and intermittent stream water of Chaq Chaq Valley through tectonically active zones (Ali et al. 2009). During the recession periods of the extremely dry years 1999 and 2000, the minimal discharge of the system was around 600 l/s. The recorded maximum during the period of 1999–2005 was 7,454 l/s (March, 2003).

The main factor which directly reflects the regime of karstic aquifers in the study area is related to the unequal distribution of recharge. The rainy season usually ends in April and no single rainfall event would occur until late September. During the rainy season, the system reacts to rainfall events with a delay of a few days representing the minimum travel time for the recharging inputs (Fig. 3.44). Meanwhile, a slow reaction could be observed during the low-water periods after approximately 1 month (rainfall at the beginning of May transferred as output at the beginning of June). Further on, during a recession period, the spring hydrograph displays a typical monotone depletion of the accumulated resources as a base flow (Stevanović and Iurkiewicz 2009).

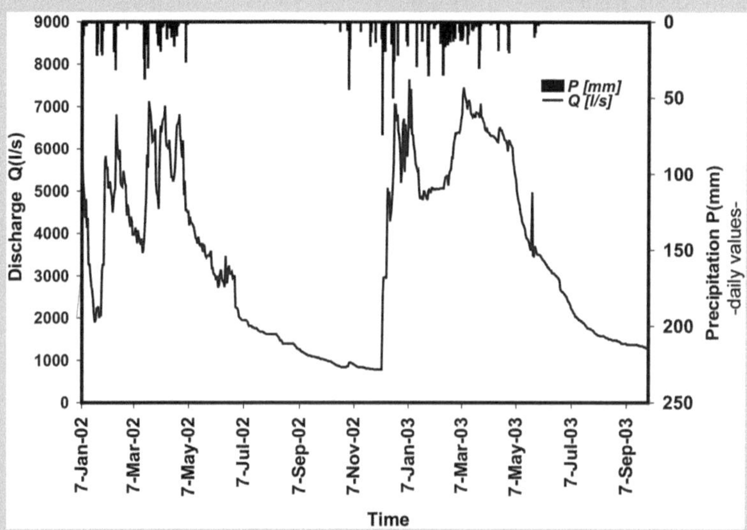

Fig. 3.44 Hydrograph of Sarchinar Spring (after Stevanović and Iurkiewicz 2009)

The regime of this spring is determined by several factors. The principal factors are arid climate and cyclic recharge variations (6 months without rainfall) on one side, and significant aquifer storage capacity and slow drainage on the other. Recession analysis and obtained coefficient α confirm a very slow drainage: It was inferred that to exhaust the dynamic resources of Sarchinar reservoir completely would theoretically require a period of several years of continuous discharge without any additional recharge (Ali et al. 2009)

Krešić (2013) presented the complex shape of hydrographs based on Jevdjević's (1956) explanations of the functioning of single hydrographs. The single hydrograph shows the transformation between the input function (precipitation) and the output function (discharge). The recharge waves may pass quickly through the system but may also simply accumulate if the deficit in stored reserves is large (after long recession). In fact, the common complex hydrograph is a result of the superpositioning of single hydrographs which correspond to separate rainfall episodes (Fig. 3.45).

Mangin (1984) and Padilla and Pulido-Bosch (1995) applied correlation and cross-spectral analysis on several karstic springs in France and in Spain and made an attempt to generalize results obtained from a single hydrograph of the spring. The parameters that can be assumed are the *response time*, the distinction between

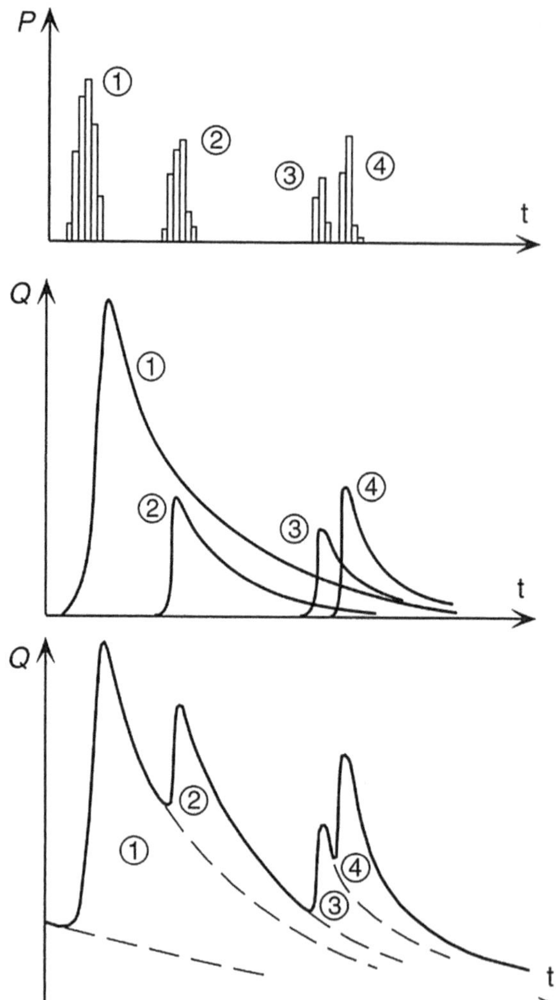

Fig. 3.45 Complex spring hydrograph resulted from several rainfall events (1–4). *P* precipitation, *Q* discharge (after Jevdjević (1956) and modified by Krešić, reprint with permission from Water in karst (2013), Mc Graw Hill Education)

quickflow, *intermediate flow* and *base flow*, and the *mean delay*. They stated that the "method offers quantifiable and objective criteria for differentiation and comparisons of karstic aquifers." Figure 3.46 shows four typical single hydrographs with various memory effects (prolongation of recharge on hydrograph shape) proposed by Mangin to be widely applied as etalons on similarly obtained hydrograph shapes.

Padilla and Pulido-Bosch (1995) additionally examined three out of four etalon springs and confirmed lag time, i.e., response of aquifer to rainfall events. The response at Aliou and Baget was immediate, while at Torcal was after 12–35 days.

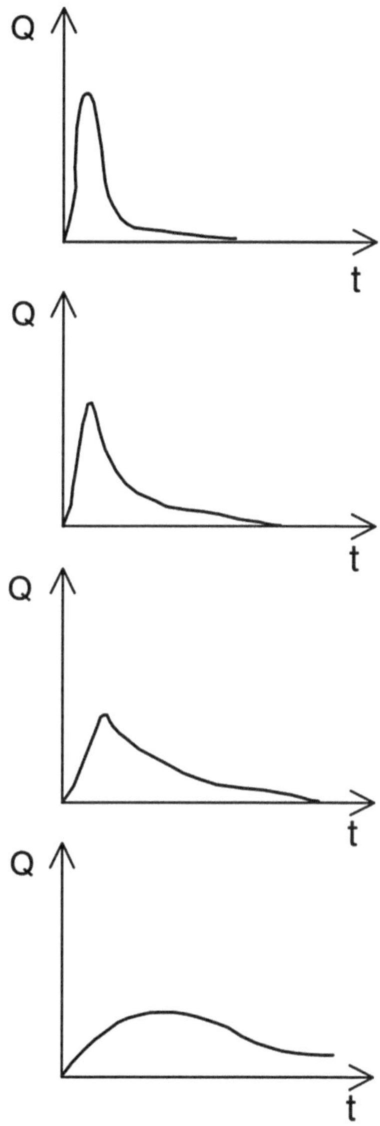

Fig. 3.46 Mangin's typical single hydrographs of four tested springs proposed as etalons. From the *top Aliou* (memory effect: reduced, up to 5 days); *Baget* (memo: moderate, 10–15 days); *Fontestorbes* (memo: large, 50–60 days); and *Torcal* (memo: significant, 70 days)

Iurkiewicz (2003) presented various single hydrographs as a unit step response function for the surveyed springs in the Banat Mts. in Romania. Figure 3.47 shows results obtained for the central compartment (Miniş–Nera zone).

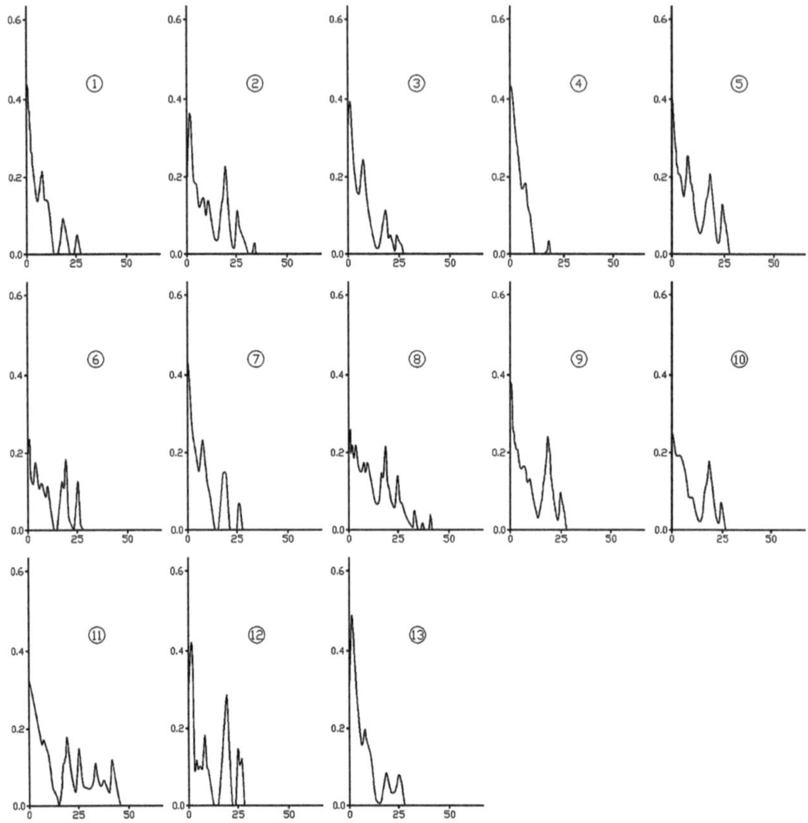

Fig. 3.47 Single hydrographs of karstic springs in Miniş—Nera zone (Romania) (after Iurkiewicz 2003)

Box 3.15

The Fontaine de Vaucluse is one of the most famous and best explored karstic springs in the world. It is located in Provance in southern France, about 30 km from the city of Avignon. The deep siphonal channels, enormous variation in discharges, and great minimal flows have always attracted researchers from all over the world. Blavoux et al. (1992) define the catchment area of the Fontaine de Vaucluse system as over 1,100 km^2. The Lower Cretaceous limestones of Urgonian facies are 1,500 m thick and highly karstified. The thickness of the unsaturated zone in the entire basin is over 800 m; four sinkholes at the plateau are more than 500 m deep, but their bottoms still do not reach the saturated zone (Blavoux et al. 1992).

The spring is a siphonal lake which is an upper outlet and not functional during the low-water seasons. The huge fallen blocks are masking the downstream part, and springwater is always discharging through this thick debris creating the Sorgue River, one of the tributaries of the Rhône. The water fluctuates in the discharge zone for about 25 m in an average hydrological year (Fig. 3.48).

Fig. 3.48 The Fontaine de Vaucluse in a low-water period in August, 2011. *Left photograph* deep siphonal channel and dry water gauges on the wall. *Right* Sorgue river flow some 300 m downstream from spring site (discharge from debris)

The Fontaine de Vaucluse is thus a lake spring with a deep siphonal channel used as *locus typicus* for all such springs worldwide. It is surveyed by many speleo-divers and robots (remotely operated underwater vehicles, sonors) that also were constructed for exploring deep channels. The Vaucluse's museum “Le Monde Souterrain” (The Underground World) provides information on the long history of these surveys starting with Ottonelli in 1878, but also undertaken by the famous Cousteau (in 1944, 1955), Touloumdijan (in 1980s), and many others. The deepest point of the siphon at −308 m was reached by the machine the “spelenaute.”

The average spring discharge is around 20 m^3/s, making that spring the largest in France. The spring discharge has been recorded since 1878 making this spring one of the best explored in the world in terms of drainage regime. Minimal discharge is 3.7 m^3/s, while the maximum exceeds 100 m^3/s (Blavoux et al. 1992).

Cognard-Plancq et al. (2006) confirmed quick system responses to rainfall in comparison with the large recharge area: The peak of the hydrograph occurred 24–72 h after the rainfall events. The springwater level and discharge depletion are slow, which can be explained by the existence of the large storage capacity of the aquifer. Bonacci (2007) also studied the discharge regime for 127 years (1878–2004) and identified a generally decreasing trend of aquifer discharge equal to 0.0468 m^3/s per year (Fig. 3.49).

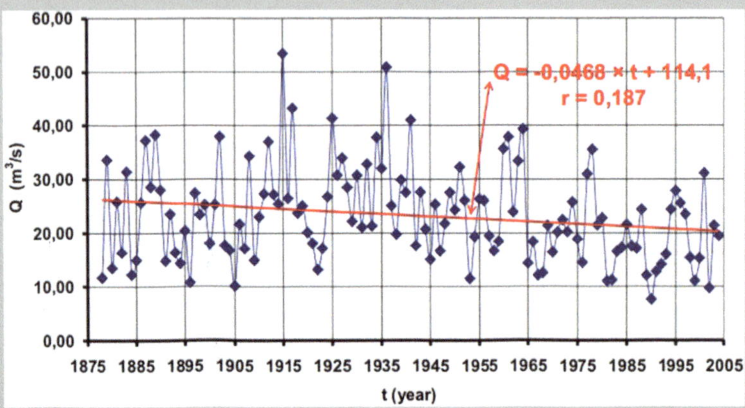

Fig. 3.49 Time data series of average annual discharges (Q) at the Fontaine de Vaucluse with trend line for the period 1878–2004 (after Bonacci 2007, printed with permission)

That trend, although not significant, is not easy to explain because during the same period, the annual rainfall in the catchment has an increasing trend. Along with a question on the accuracy of the measured spring flows and rainfalls, or problematic delineation of the catchment, Bonacci (2007) noticed that the weak relationship between runoff and rainfall might be the result of some others factors such as air temperature, groundwater level, interannual rainfall distribution, changes of catchment area during the time, preceding soil wetness, anthropological influences, and climate change.

Stochastic methods and an established relationship between input–output signals enable not only the time data series to be refilled by missing (unmeasured) data but also forecast discharges under different climatic scenarios. For instance, prior to a technical decision to tap or not tap a karstic spring, it is possible and even advisable to estimate karstic aquifer drainage behavior under long drought episodes (with limited or no recharge). Figure 3.50 shows hydrographs and results obtained from the created stochastic model of Veliko Vrelo Spring in the Carpathian karst of Serbia. The established rainfall–discharge function for

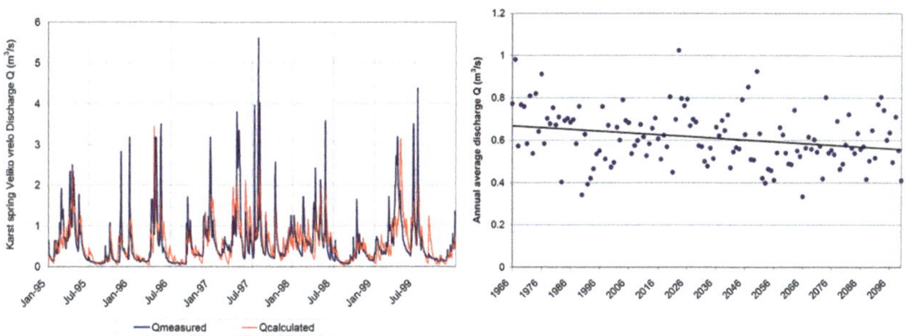

Fig. 3.50 Measured and simulated discharges (*Q*) of Veliko Vrelo Spring (*left*, after Stevanović et al. 2010, printed with permission) and forecasted annual mean discharges of the same spring including general trend up to end of 21 Ct (from Stevanović et al. 2012a)

daily values is used as one of the parameters for long-term prediction of spring discharges (see also Box 15.2.5 in Sect. 15.2). The time series of annual mean discharges of Veliko Vrelo are obtained by using the bias-corrected regional climatic model E-obs for precipitations and air temperatures for the period until the year 2100 (Stevanović et al. 2012a). The result shows a depletion of groundwater reserves by around 15–20 % for the predicted climatic scenario.

The karstic springs are widely utilized as a source of drinking water supply. It is known that some 20 % or even a few percent more of the global population largely depends on karstic groundwater (Ford and Williams 2007), but there are no statistics yet as to how much of this water is provided directly from the springs and how much from other structures. Discussion on actual and possible problems when natural spring flow is not regulated is provided in Sect. 15.5.

Many countries utilize karstic springs simply because there are no other alternatives, but in many other countries, awareness of their importance and the good water quality they provide is a principal factor for such a decision. The karstic aquifers have a significant proportion of the water supply in the following regions: southeastern Europe (Alps and Carpathians, Box 3.16), the Mediterranean basin, the Near East and Middle East, the Arabian Peninsula and Horn of Africa, southeastern Asia, North Africa, the Caribbean basin and Central America, and the southern part of the USA.

Box 3.16

Drinking water supply for about half of the population or four million citizens in Austria is effectuating from karstic aquifers (Zötl 1974). Vienna has the largest water supply system based on karst waters. The system consists of two major gravity pipelines, 130 and 200 km long (Drennig 1973). The first Viennese mountain spring pipeline was completed in 1873 by tapping water from the Kaiserbrunn (Fig. 3.51) and other springs issuing from Schneeberg, Rax, and Schneealpe Mountains, while the second pipeline was completed in 1910 by tapping the Kläffer and other smaller springs discharging from Hochschwab Mountain (Styria). The system was established as an alternative to Danube alluvial waters which often caused hydric epidemic. The total catchment is around 600 km^2. The long concrete tunnels and channels provide an average yield of 4.5 m^3/s, for some 1.7 million inhabitants. These two springwater supply pipelines which meet around 95–97 % of the amount of water required for the municipality of Vienna (on average 140×10^6 m^3/year). The water is of excellent quality. Generally, no treatment is applied; only chlorination is required, primarily for cleaning the distribution pipes.

Fig. 3.51 Kaiserbrunn's old design sketches (photographs made by courtesy of the Kaiserbrunn Museum, Vienna) and spring's intake (*inside view*)

Natural drainage of aquifers through springs can therefore cover water demands on a wide scale: from the supply to multi-millions of towns at the regional level, to the very local level where the supply to just one or several houses is concerned. Although the latter is not a big problem in terms of amount of water, for the big consumer a very large aquifer and spring discharge are required. The population

Fig. 3.52 Camels around one of the rare freshwater springs draining Taalex Fm. in Puntland province of Somalia

growth and increased demands have caused many large cities to substitute or enhance their primary water system based on springwater, with surface waters or groundwater from other aquifers (Stevanović 2010b).

Table 3.3 and criterion *Utilization* present a wide range of uses of karstic waters. In arid regions in the Near East and Middle East, it is, for instance, very common to tap karstic springs and to construct gravity channels for irrigating arable land. Use of pipes instead of channels significantly increases efficiency in water provision because losses from channels are usually very high due to seepage or evaporation resulting from systems' openings. Springwater is also widely used for watering animals, and fresh water of good quality provides security for animal health and growth. Thus, in a rural environment, it is common to see a large numbers of animals occupying the springs or the ponds and swamps formed nearby (Fig. 3.52).

The use of karstic waters in hydropower generation by utilizing high hydraulic head is limited mostly to the Alps (Austria, Switzerland), while thermal properties of karstic waters and springs are utilized elsewhere. Finally, the number of karstic springs utilized in the world's water bottling industry which runs an annual revenue of around $13 billion is very large and karstic aquifers probably lead the list of aquifers where such sources originate.

The conflict of interest in utilizing karstic waters is present especially in the undeveloped world. This is not very different from conflicts related to any other aquifer or surface water, but the recognized precious water quality of karstic sources may sometimes be an additional factor for disagreement.

3.6 Quality of Karst Groundwater

The dissolution of rocks and the duration of direct water–rock contact results in variable groundwater quality at discharge points. The mineral components of karst waters depend upon the composition of the rocks through which water is percolating: Hydrocarbonate (HCO_3)–calcium (Ca) type of waters is created from the dissolution of calcium carbonate which is a dominant type of water in limestone, (see Box 3.17), while the hydrocarbonate (HCO_3)–magnesium (Mg) type of groundwater is present to a lesser extent, and is regularly connected to dolomitic rocks.

Langmuir (1984) listed processes which control and influence the quality of groundwater before it reaches the spring site or well head. These processes are as follows:

1. The composition of the infiltrated atmospheric precipitation;
2. Evapotranspiration losses from groundwater recharge and shallow groundwaters;
3. The acidity and degree of undersaturation of groundwater recharge;
4. The availability and solubility of carbonate and associated rocks, including halite, gypsum, and anhydrite;
5. Rates of solution of the rocks and contact time;
6. Hydrological processes such as dilution by fresh water recharge and mixing of dissimilar groundwater;
7. Anthropogenic processes, including groundwater pollution by wastes and leachates from solid wastes.

Calcium carbonate is highly soluble when carbon dioxide is present, and its concentration in water ranges from 200 to 300 mg/l under normal conditions. The sodium chloride and calcium sulfate waters, which result from the dissolution of

Table 3.4 Chemical composition and general water quality of some karstic aquifers (after Stevanović and Papić 2008, modified)

Karstic aquifer	Lithology	Minerals	Chemical composition	Expected quality	Remarks (possible problems for drinking water quality)
Carbonate	Limestones	Calcite	Hydrocarbonates, calcium	Low mineralization (0.3 g/l), pH 7–7.5	Problem: turbidity, microbiology, nitrates
	Dolomites	Dolomite	Hydrocarbonates, calcium, magnesium	Medium mineralization (0.5 g/l), pH ~ 8	Problem: hardness, iron
	Marbles	Calcite (dolomite, quartz)	Hydrocarbonates, calcium, (magnesium, silica, iron)	Low mineralization, pH ~ 8, iron	Problem: hardness, iron turbidity, microbiology
Evaporite	Gypsum	Gyps	Sulfates, calcium	Higher mineralization (2–3 g/l), pH ~ 6	Problem: bitter odor
	Halite	Halite	Chloride, sodium (sulfates, calcium)	Very high mineralization, salty water	Problem: salty, bitter odor

halite and gypsum, are undesirable from a water supply perspective due to changes in the organoleptic properties of water (Stevanović and Papić 2008). The issue of rock dissolution is discussed in Sect. 2.3.

The typical groundwater quality of water issuing from different karstic aquifers is presented in Table 3.4.

Groundwater quality can also be the result of a cation exchange process, which is a frequent occurrence in limestones whose fissures may include clay minerals; as such, an initial hydrocarbonate–calcium type of water may be replaced by a hydrocarbonate–sodium composition. This is a typical example of a natural softening of "hard" calcium waters by "soft" sodium waters. Herak et al. (1981) stated that even small lenses or intercalations may strongly influence the chemical properties of the water; thus, it is almost impossible to give a common composition of waters in karst.

Box 3.17

The typical HCO_3–Ca or HCO_3–Ca/Mg type of water characterizes the carbonate karst of the Carpathian and Dinaric mountain arches in SE Europe. The chemical composition of these waters reflects entirely the conditions of their formation, the intensive water exchange, and the rapid filtration.

- Around 300 analyzed samples from the Carpathian karst of Serbia confirmed low mineralization of waters (total dissolved solids TDS of 0.2–0.4 g/l, only 10 % samples below, and 6 % above these values) with a prevailing content of HCO_3 in the anion and of Ca in the cation composition. The average content of HCO_3 is 87 % mval and of Ca is 75 % mval, whereas the ions Mg, Na, K, SO_4, Cl, NO_3, usually occur inferiorly. The ions NH_4, Fe, Mn, as well as microelements, are usually absent in results of conventional laboratory analyses, and if they do occur, it is always within the limits of the drinking water quality standards.
- Based on chemical analyses of around 180 samples taken from karstic aquifers in Dinaric karst, Petrik (1976) concluded that $CaCO_3$ hardness was between 100–250 mg/l for 50 % of all samples. In continental Dinaric karst, ion Cl is usually present in groundwater with a concentration less than 10 mg/l, but along the coast and in the islands, it may significantly increase (over 5,000 mg/l) due to sea water intrusion.
- Dynamic karstic water regime may be very problematic concerning low discharges in recession periods, but during the floods, we often witness another problem of high turbidity and considerable change in natural water quality. Water temperature in unconfined aquifers regularly does not exceed annual amplitude of 5–6 °C. For instance, those are usual values for Fontaine de Vaucluse (maxima in late summer and minima in late winter), while in Carpathian karst, the average amplitude of groundwater temperature regularly does not exceed 3 °C. The water mineralization throughout the hydrological year rarely varies higher than 0.1 g/l.

Evaporitic rocks bind more soluble characterize water with higher salinity and increased content of Cl and SO_4 anions.

- Figure 3.53 shows diagrams of low-mineralized springwaters issuing from carbonate Cretaceous and Paleogene aquifers in central part of northern Iraq (Qamchuga, Dokan Fms.) and mineralized springwaters from Neogene evaporitic aquifers (Fars Fm. Garmian area, N Iraq).

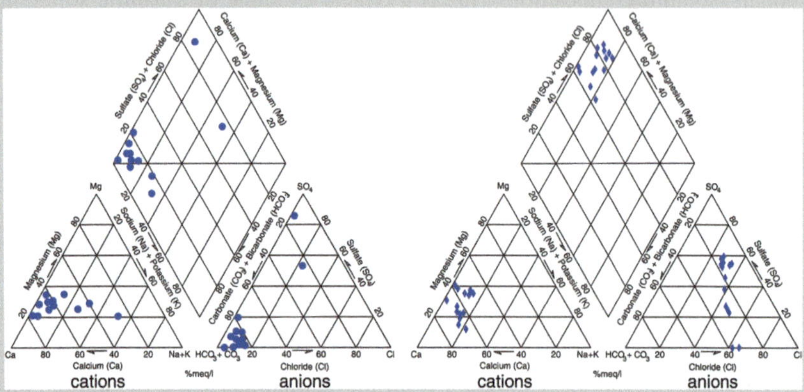

Fig. 3.53 Piper's diagrams of low-mineralized karstic springwater from carbonate Cretaceous and Paleogene aquifers (*left*) and mineralized springwater from Neogene evaporitic aquifers (*right*) in northern Iraq

- In the Somali provinces of Somaliland and Puntland (the Horn of Africa), both carbonate and evaporite karstic aquifers are present. In terms of geological formations, water from carbonate aquifer of the Jurassic age is the best quality (Stevanović et al. 2012b). Water mineralization is moderate and the electro conductivity (EC) is commonly around 600 μS/cm. The bicarbonate type of water prevails, but concentration of sulfate and chloride might be occasionally slightly higher in some samples. In contrast, in evaporite karst of Eocene Taalex Fm (Fig. 3.54), EC is very high, ranging from minimal 890–7,270 μs/cm. Evaporites are generated during the precipitation from an over-saturated brine solution which is usually highly soluble where concentrations rapidly increase, even if the portion of evaporitic material is relatively small. When recharging fresh water comes into contact with a thin layer of anhydrite or gypsum, the EC will rapidly increase (usually beyond the potable limit). Water is often highly sulfatic, with calcium as the predominant cation. SO_4 is in the range 125 up to 3,100 mg/l, with an average concentration of 1,300 mg/l.

Fig. 3.54 Fissured gypsum of Taalex Fm. (Somaliland)

Analyzing rock-water interactions, Younger (2007) concluded that due to low velocities of groundwater in saturated zones, there is normally sufficient time for even slow geochemical reactions to alter water chemistry considerably. Dissolution of calcite under laboratory conditions requires no more than 24 h to approach equilibrium; considering the years or even centuries of calcite–water contact in saturated zones that equilibrium is more standard than exceptional in limestone aquifers. Therefore, even brief contact between percolated water and karstic rocks may sometimes significantly change the water quality (Box 3.18).

Box 3.18

Investigating causes of an accidental water leakage which appeared in several houses located on the slope of a hill formed from Miocene vuggy limestone in the Belgrade suburb Rakovica (Serbia), water samples were taken from the waterworks' pipeline on the top of the hill, and from a small cave 40 m beneath the top of the hill (Fig. 3.55). The seepage sites followed a thin marly horizon interstratified as lower permeable layer between upper and lower limestone horizons. The chemical composition of the water from the pipeline and underneath the cave was almost equal, confirming the direct infiltration and fast percolation in Miocene aquifer of artificially treated chlorinated water. The electrical conductivity of 655 μg/l and trihalomethanes as a residual product of water chlorination of 7.2 μg/l indicate that seepage is probably from the crashed pipeline. Although not far from the cave, the quality of water which further filtrated through limestones of vuggy porosity quickly changed and adapted to become similar to natural springwaters

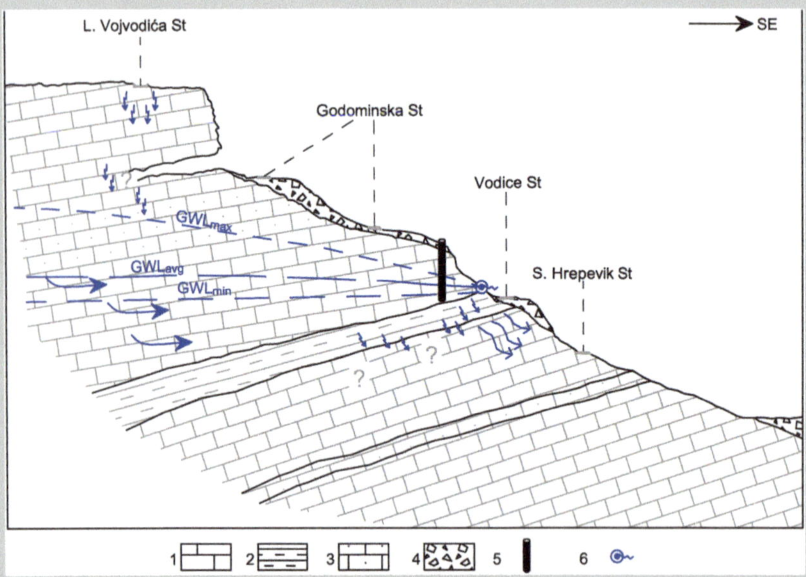

Fig. 3.55 Cross section of Miocene limestone aquifer in accident area in Rakovica, Belgrade. *Small arrows* indicate leakage and percolation toward houses in Vodice Street. *Legend 1* Miocene limestones, *2* miocene marls, impermeable barrier to groundwater flow, *3* miocene sandy limestones, *4* debris, *5* dug well, *6* seepage sites. *GWL* groundwater levels (max–min)

which drain Miocene aquifer. The seepage water sampled in the basements of houses had an increased salinity, almost double that in the pipeline (EC of 1,105 μg/l), while the content of trihalomethanes was reduced to 2.2 μg/l. Although the latter value was still greater than the natural one, the reason for this accidental leakage was clearly recognized as damage on the pipeline was found and then repaired. As an additional remedial measure, the groundwater table was depleted by forced over-pumping of a nearby dug well and leakage was soon stopped. In conclusion, even though the travel time through very porous media was very short, the water quality had significantly changed.

It is almost a rule that groundwater in open karst structures is low mineralized which is a result of the intensive water exchange and rapid filtration. In deeper parts of the aquifer, slower filtration results in an increase in mineralization. This variation is often minimal, but nevertheless indicates a certain differentiation that may be important under specific circumstances (e.g., when pollution is involved). Similarly, when the conventional classification into gravity and ascending springs is concerned, in principle the latter are easier to protect because they drain "lower" and slower circulation zones. When a specific type of "combined" springs with a drainage system consisting of gravity channels and ascending siphons in the same discharge zone is indicated, the water quality of various channels should be monitored because irrespective of the "unity" of a karst aquifer, local variations in water quality are very probable.

The best waters in terms of quality are generally those from the deepest parts of an aquifer, particularly water near or below the base level of erosion or discharging underground into laterally connected aquifers. In these parts of a karst aquifer, the self-purification of water is the highest, and the quality, regardless of possible slightly increased mineralization, is always the best. This of course excludes the very deep aquifer zones where the water exchange might be very slow, while the temperature might also increase as a result of a natural or possibly an anomalous geothermal grade.

The above facts led us to consider a hypothetical "bedding" character of karst groundwater and separation of hydrochemical zones:

1. a zone of fast water exchange and propagation which corresponds to the highest levels of respective karst channels, the least favorable hydrochemical and bacteriological conditions for protection;
2. a zone of slow water exchange with dominantly horizontal or siphonal circulation, including the subsurface outflow area (always of the best qualitative properties), and, finally,
3. a deep zone of retarded water exchange, often unsuitable for water supply because of the increased temperature, mineral content or presence of specific micro constituents.

In reference to the above, Langmuir (1984) noticed that waters extracted from wells often have higher TDS than adjacent springwater in the same aquifer. This is because springwaters generally discharge from zones of enlarged porosity, whereas randomly sited wells most often draw water from less permeable zones.

Isotopic analyses may significantly contribute to assessing groundwater origin, contact time, recharge conditions, and consequently even the catchment size. Most of the analyses concerns stable environmental isotopes such as ^{2}H (deuterium—D), ^{16}O, ^{18}O (oxygen 16 and oxygen 18), ^{13}C (carbon 13), and radioactive man-made isotopes (as a result of nuclear explosions) such as ^{3}H (tritium) or ^{14}C (carbon 14).

Due to their openings and fast groundwater circulation, karst aquifers have very low attenuation capacities and are extremely vulnerable to pollution. This issue is discussed in several contributions in Chap. 17. In addition to easy infiltration of the contaminants, the results of many tracing experiments indicate that under convenient conditions, the pollution may migrate very fast, as far as 1 km of rectilinear distance in as little as 24 h. Active hydraulic connections between the ground and surface waters, where the harmful components may be carried from great distances including non-karst terrains and infiltrated into the narrowest spring zone, have a particular significance in this regard.

Therefore, in the case of carbonate karst, the quality of natural karst waters is excellent almost by definition: It is confirmed in many places worldwide that water issuing from unpopulated catchment areas on mountain massifs is sanitary and pure while only exceptionally could there be a small amount of bacteria present, but if pollution sources are present in the catchment of an unconfined karstic aquifer, then severe hazards follow (Box 3.19).

Box 3.19

The water supply of the town Shiraz in central–south Iran was initially based on the tapping of karstic springs which drain the Asmari limestone formation (Oligocene–Lower Miocene age). The city is located in a flat area representing the core of syncline which consists of impervious marly sediments of Mishan Fm. (Middle Miocene). The Mishan's sediments are very thick, reaching some 800 m. The basin is surrounded by high hills consisting predominantly of Asmari limestones, and along the marginal parts and in contact with Mishan Fm. (and adjacent underlying Gachsaran Fm.), many large springs are located making this a suitable site for the establishment of this city which further grew and became the strategic center of the Persian Empire.

Figure 3.56 presents the three stages of evolution of karstic aquifer in accordance with the concrete intensity of water utilization. The normal

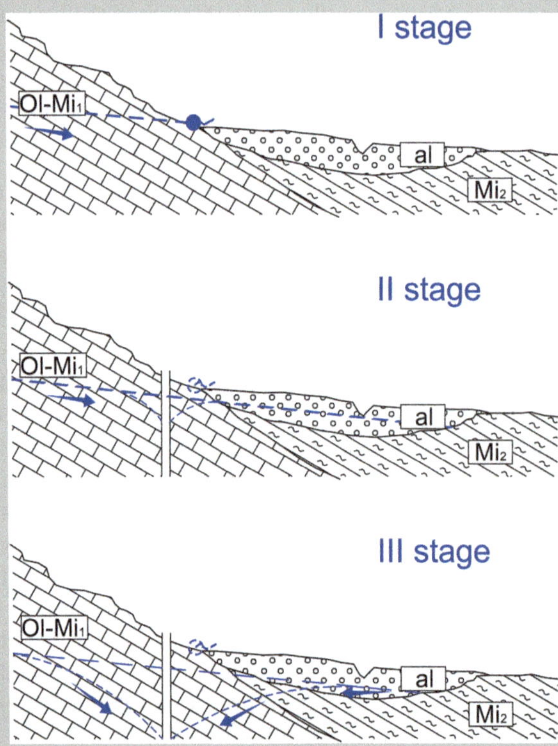

Fig. 3.56 Evolution of the water table during three stages of extraction of groundwater for water supply of the city of Shiraz in Iran. *Legend* Ol-Mi_1 Asmari limestones—karstic aquifer, Mi_2 Gachsaran and Mishan Fms.—impervious rocks, al—alluvial sands, gravels, clay—intergranular aquifer

functioning and natural discharging of the spring marks stage I. Along with city growth (today nearly 1.5 million inhabitants) and the pumping of drilled wells also located along the Asmari–Mishan contact, the springs started to dry periodically, and afterward, with further intensification of groundwater extraction, the springs completely dried out (stage II). The further aquifer development resulted in significant depletion of the groundwater table and the reverse infiltration of groundwater from the alluvium of a temporary stream which passes through the city center (stage III). Currently, a major problem is the pollution of alluvial and, consequently, of karstic aquifers by nitrates resulting from infiltrated sewage and irrigation waters.
(This situation is explained in personal communication with Dr Ezzat Raeisi from Shiraz University, Iran).

The direct impact of pollutants on the karstic aquifer might in some cases be reduced and even minimized. Figure 3.57 shows several case examples of the functioning of a single karstic channel or siphon and its role in purifying or amortizing the impurity of the water flow through it. The siphon could be pressurized with a potentiometric line above the channel (cases a, c, e) or with gravity-free flow (unconfined, cases b, d, f). Additionally, the siphon may be totally empty of any sediments (a, b), or partly filled with sediments (c, d). Finally, the end of a channel could be sealed with some fine sediment which belongs to laterally connected aquifer of intergranular porosity (e, f). Each of these options results in a specific water quality. There are also many factors which influence the duration of rock–water contact such as potentiometric pressure, channel length and inclination, permeability of sediments' plug, and water viscosity. However, the cases presented can be used as a starting point for estimating whether the attenuation effect on contaminated water could be smaller or larger.

Figure 3.57 contains diagrams showing the expected reduction of bacterial concentration in water samples for each of the presented cases. Experimental determinations of the life cycle of different microorganisms in limestones are very important. Depending on the conditions, the shortest life of *Escherichia coli* is 70–210 days, which proves how old pollution of this type is. At low temperatures (4–8 °C), life cycles vary from 40 days for *Salmonella* (at contamination rates to 10^5 microbes/l of water) to 120 days for enterococci (at contamination rates to 10^8 microbes/l of water) (Gavich 1985). Therefore, slower water movement through channels and longer rock–water contact proves the reduction of bacteria in the system. The worst scenario for water quality presented on Fig. 3.57 is first presented case indicated by (a) an empty and pressurized channel. From the point of view of water purification, every additional case presented is better, while significant improvement of water quality and even elimination of bacteria may be expected when an unconfined siphonal channel is laterally connected to fine porous sediments (case f). The last case can be similar to the functioning

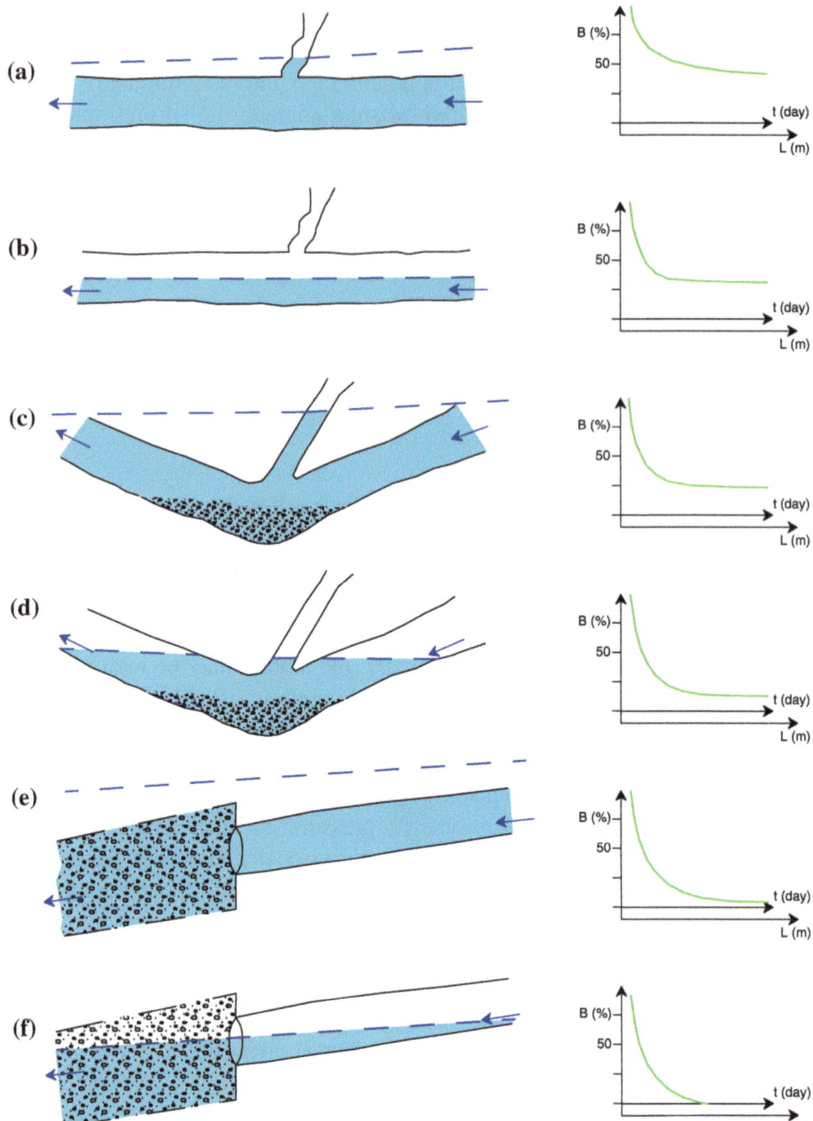

Fig. 3.57 The karstic drainage channel: **a** pressurized and empty, **b** unconfined, empty, **c** pressurized, knee shaped, partially filled with sediments, **d** unconfined, knee shaped, partially filled with sediments, **e** pressurized, with adjacent porous aquifer, **f** unconfined, with adjacent porous aquifer. The diagrams on the *right* show a reduction of bacteria (*B*) in the percentage based on residence time (*t*) and the length of the channel (*L*), i.e., distance between input and output points

of a secondary spring type. There are many such springs worldwide located in mountain foothills. Probably, the most famous and well explored are those issuing from Alps mountain chain. Some of them are located at the northern margin of Le Mans Lake in France, in proximity of Evian mineral water source. These springs discharge from fine lacustrine–glacial sediments to which water partly arrives from the adjacent high-Alpine karstic aquifer with catchment altitudes reaching 2,000 m.

Finally, after detailed discussion of karstic aquifer properties, we can repeat in conclusion that each karstic aquifer is an individual case. Nonetheless, we may still try to highlight some general outlines of the character of karst aquifer:

Accessible. Not everywhere and not always, but still reachable even in deep submerged siphons, if not by humans, then by instruments of navigation. It is just a matter of time before the new technology (scanners, acoustic log, floating sensors, nanotechnology or something else) will enable complete tracing of main and secondary channels and complete recognition of groundwater paths.

Variable. Karstic aquifer is a non-homogenous and anisotropic system changing its properties from place to place. It is a fact that for a minimal distance, one may find an extremely permeable cavity and a totally compact impermeable block. Similar to spatial variability, the karst aquifer system is also time variable. The springs which discharge enormous water during the flood periods often become dry during the long recession periods.

Unpredictable. The system is dynamic with a lot of changes during a hydrological cycle. Many engineering solutions and designs have failed because of karst aquifer variability but also due to improper or insufficient research. However, it also true that prior to construction of any type of structure in karst, properly conducted investigations and an appropriate project concept can significantly reduce the risks or minimize them to acceptable levels.

Precious. No other aquifers can provide water of such high natural quality as carbonate or dolomitic karst can (Fig. 3.58). Generally, low-mineralized water with dominant HCO_3 and Ca, Mg ions is an ideal arrangement for the human

Fig. 3.58 Pristine low-mineralized water of cave spring Genal in Sierra de las Nieves (south Spain), however, characterized by rapid chemical variations (personal communication B. Andreo)

organism. In many places and utilities worldwide where pollutants are absent, chlorination is the only required treatment of water.

Vulnerable. Due to the presence of large voids and cavities as preferential paths which also cause turbulent flows, karstic groundwater can be easily polluted and pollutants can be quickly transported a long distance. Therefore, the pollution risk is much higher than in the case of other aquifers, but more problematic is the very limited attenuation capacity of karst (see Chap. 8 and Sects. 17.1–17.4).

Beautiful. As mentioned repeatedly in Chap. 2, there are so many natural wonders created in a karstic environment. Some are already protected, while others await further evaluation and decisions before they are eligible for inclusion on the long list of internationally or locally protected objects.

References

Ali SS, Stevanović Z, Jemcov I (2009) The mechanism and influence on karstic spring flow—Sarchinar spring case example, Sulaimaniya, NE Iraq. Iraqi Bull Geol Min Baghdad 5(2):87–100

Atkinson TC (1977) Diffuse flow and conduit flow in limestones terrain in the Mendip Hills, Somerset (GB). J Hydrol 35:93–110

Bakalowicz M (2005) Karst groundwater: a challenge for new resources. Hydrogeol J 13:148–160

Bonacci O (1987) Karst hydrology with special reference to the Dinaric karst. Springer, Berlin

Bonacci O (1993) Karst spring hydrographs as indicators of karst aquifers. Hydrol Sci J 38(1):51–62

Bonacci O (2007) Analysis of long-term (1878–2004) mean annual discharges of the karst spring Fontaine de Vaucluse (France). Acta Carsologica 36(1):151–156

Bögli A (1980) Karst hydrology and physical speleology. Springer, Berlin

Blavoux B, Mudry J, Puig JM (1992) The karst system of the Fontaine de Vaucluse (Southeastern France). Environ Geol Water Sci 19(3):215–225

Burger A, Dubertret L (eds) (1984) Hydrogeology of karstic terrains. Case histoires. International contributions to hydrogeology, IAH, vol 1. Verlag Heinz Heise, Hannover

Castany G (1984) Hydrogeological features of carbonate rocks. In: LaMoreaux PE, Wilson BM, Memon BA (eds) Guide to the hydrology of carbonate rocks. IHP Studies and reports in hydrology, vol 41. UNESCO, Paris, pp 47–67

Cognard-Plancq AL, Gévaudan C, Emblanch C (2006) Historical monthly rainfall-runoff database on Fontaine de Vaucluse karst system: review and lessons. In: Duran JJ, Andreo B, Carrasco FY (eds) Karst, Cambio Climatico y Aguas Subterraneas. Proceedings of 3rd international symposium on karst "Groundwater in the Mediterranean Countries", Malaga, Spain, IGME Publications, vol 18, pp 465–475

Cvijić J (1918) Hydrographie souterraine et evolution morphologique du karst. Recueil Trav Inst Geogr Alpine, Grenoble 6(4):40

Darcy H (1856) Les fontaines publiques de la ville de Dijon. Dalmont, Paris

Drennig A (1973) Die I. Wiener Hochquellenwasserleitung. Magistrat der Stadt Wien, Abteilung 31—Wasserwerke, Wien, p 303

Drogue C (1982) L'aquifère karstique: un domain perméable original. Le Courier du CNRS, March 1982 44:18–23

Ford D, Williams P (2007) Karst hydrogeology and geomorphology. Wiley, Chichester

Fiorillo F (2011) Tank-reservoir emptying as a simulation of recession limb of karst spring hydrographs. Hydrogeol J 19:1009–1019

Gavich IK (ed) (1985) Metody ohrany podzemnyh vod od zagrjaznenija i istošćenija. Nedra, Moscow

Goldscheider N, Drew D (eds) (2007) Methods in karst hydrogeology. In: International contribution to hydrogeology, IAH, vol 26. Taylor & Francis/Balkema, London
Herak M, Magdalenic A, Bahun S (1981) Karst hydrogeology. In: Halasi Kun GJ (ed) Pollution and water resources. Columbia University seminar series, vol XIV, part 1, Hydrogeology and other selected reports. Pergamon Press, New York, pp 163–178
Issar A (1984) Storage volume in karstic aquifers. In: LaMoreaux PE, Wilson BM, Memon BA (eds) Guide to the hydrology of carbonate rocks. IHP studies and reports in hydrology, vol 41. UNESCO. Paris, pp 264–265
Iurkiewicz A (2003) Analiza sistemică în investigarea hidrodinamică a acviferelor carstice (Systemic analysis in hydrodynamic research of karstic aquifers). Ph.D. thesis, University of Bucharest, Bucharest
Jevdjević V (1956) Hydrology, part I (in Serbian). Institute for development of water research "J.Černi", Belgrade
Kiraly L (2002) Karstification and groundwater flow. In: Gabrovšek F (ed) Evolution of karst: from Prekarst to Cessation. Institut za raziskovanje krasa ZRC SAZU, Postojna—Ljubljana, pp 155–190
Kiraly L, Perrochet P, Rossier Y (1995) Effect of the epikarst on the hydrograph of karst springs: a numerical approach. Bull d'Hydrogéologie 14:199–220
Klimchouk A (2000) The formation of epikarst and its role in vadose speleogenesis. Speleogenesis: evolution of karst aquifers. National Speleological Society Inc., Huntsville, pp 91–99
Komatina M (1983) Hydrogeologic features of Dinaric karst. In: Mijatovic B (ed) Hydrogeology of the Dinaric karst, Field trip to the Dinaric karst, Yugoslavia, 15–28 May 1983. "Geozavod" and SITRGMJ, Belgrade, pp 45–58
Krešić N (2007) Hydrogeology and groundwater modeling. CRC Press, Taylor & Francis Group, Boca Raton
Krešić N (2010) Types and classification of springs. In: Kresic N, Stevanović Z (eds) Groundwater hydrology of springs. Engineering, theory, management and sustainability. Elsevier Inc., Amsterdam, pp 31–85
Krešić N (2013) Water in karst. Management, vulnerability and restoration. McGraw Hill, New York
Krešić N, Mikszewski A (2013) Hydrogeological conceptual site model: data analysis and visualization. CRC Press, Boca Raton
Langmuir D (1984) Physical and chemical characteristics of carbonate water. In: LaMoreaux PE, Wilson BM, Memon BA (eds) Guide to the hydrology of carbonate rocks. IHP studies and reports in hydrology, vol 41. UNESCO. Paris, pp 264–265
LaMoreaux PE, Wilson BM, Memon BA (eds) (1984) Guide to the hydrology of carbonate rocks. IHP studies and reports in hydrology, vol 41. UNESCO, Paris
Lerner DN, Issar AS, Simmers I (eds) (1990) Groundwater recharge. A guide to understanding and estimating natural recharge. International contributions to hydrogeology, IAH, vol 8. Verlag Heinz Heise, Hannover
Maillet E (ed) (1905) Essais d'hydraulique soutarraine et fluviale. Herman et Cie, Paris 1:218
Mangin A (1974) Contribution a l'étude hydrodynamique des aquifères karstiques, 2eme partie. Concept méthodologiques adoptés. Systèmes karstiques étudiés. Ann Spéléol 29(4):495–601
Mangin A (1975) Contribution a l'étude hydrodynamique des aquifères karstiques, 3eme partie. Constitution et functionnement des aquifères karstiques. Ann Spéléol 30(1):21–124
Mangin A (1984) Pour une meilleure connaissance des systèmes hydrologiques à partir des analyses corrélatoire et spectrale. J Hydrol 67:25–43
Marsaud B (1997) Structure et fonctionnement de la zone noyée des karsts à partir des résultats expérimentaux (Structure and functioning of the saturated zone of karsts from experimental results). Documents du BRGM 268, Editions de BRGM, Orleans
Meinzer OE (1923a) Outline of groundwater hydrology with definitions. USGS water supply paper, 494, p 71
Meinzer OE (1923b) The occurrence of ground water in the United States. USGS water supply paper, 489

Mijatović B (1968) Metoda ispitivanja hidrodinamičkog režima kraških izdani pomoću analize krive pražnjenja i fluktuacije nivoa izdani u recesionim uslovima (Method of hydrodynamic analyses of karst aquifer regime based on discharge and groundwater fluctuation curves from recession periods). Vesnik Geozavoda, ser. B, vol 8, Belgrade

Milanović P (1981) Karst hydrogeology. Water Resources Publications, Littleton

Milanović P (1984) Water resources engineering in karst. CRC Press, Boca Raton

Milanović P (2006) Karst of eastern Herzegovina and Dubrovnik littoral. ASOS, Belgrade

Milanović S (2007) Hydrogeological characteristics of some deep siphonal springs in Serbia and Montenegro karst. Environ Geol 51(5):755–760

Oraseanu I, Iurkiewicz A (2010) Karst hydrogeology of Romania. Belvedere Publ., Oradea

Padilla A, Pulido-Bosch A, Mangin A (1994) Relative importance of baseflow and quickflow from hydrographs of karst spring. Ground Water 32(2):267–277

Padilla A, Pulido-Bosch A (1995) Study of hydrographs of karstic aquifers by means of correlation and cross-spectral analysis. J Hydrol 168:73–89

Palmer AN, Palmer MV, Sasowsky ID (eds) (1999) Karst modeling. Special Publication 5, Karst Water Institute, Charles Town, WV

Panagopoulos G, Lambrakis N (2006) The contribution of time series analysis to the study of the hydrodynamic characteristics of the karst systems: application on two typical karst aquifers of Greece (Trifilia, Almyros Crete). J Hydrol 329:368–376

Parizek R (1976) O prirodi i značaju tragova i površinskih obilježja lomova u karbonatnim i drugim terenima (On nature and importance of traces and surficial records of fractures in carbonate and other terrains). In: Hydrology and water richness of karst. Proceedings of the Yugoslav—American symposium, Dubrovnik, June 1975. Zavod za hidrotehniku Gradjevinskog fakulteta, Sarajevo, pp 39–79

Pekaš Ž (unpublished) Hydrogeological overview. In: Transboundary diagnostic analysis country report, Croatia. http://dinaric.iwlearn.org/. Accessed 20 Dec 2013

Perrin J, Jeannin PY, Zwahlen F (2003) Epikarst storage in a karst aquifer: a conceptual model based on isotopic data, Milandre test site. Switz J Hydrol 279:106–124

Petrič M (2002) Characteristics of recharge—discharge relations in karst aquifer. Carsologica, Institut za raziskovanje krasa ZRC SAZU, Postojna, Ljubljana

Petrik M (1976) Karakteristike voda na Dinarskom kršu (Water characteristics in Dinaric karst) In: Hydrology and water richness of karst. Proceedings of the Yugoslav—American symposium, Dubrovnik, June 1975. Zavod za hidrotehniku Gradjevinskog fakulteta, Sarajevo, pp 30/1–30/9

Poiseuille JML (1846) Recherches expérimentales sur le mouvement des liquides dans les tubes de très petits diameters. Académie des Sci Paris Memoir Sav Etrang 9:433–545

Radovanović S (1897) Podzemne vode; Izdani, izvori, bunari, terme i mineralne vode (Ground waters; aquifers, springs, wells, thermal and mineral waters), vol 42. Srpska književna zadruga. Belgrade, p 152

Raeisi E (2010) Sheshpeer spring, Iran. In: Kresic N, Stevanović Z (eds) Groundwater hydrology of springs. Engineering, theory, management and sustainability. Elsevier Inc., Amsterdam, pp 516–525

Sharp JJ, Garcia-Fresca B (2004) Urban implications on groundwater recharge in Austin, Texas (USA). In: Proceedings of XXXIII IAH congress "Groundwater flow understanding from local to regional scale". Zacatecas, published on CD—T5-31

Stevanović Z (2010a) Regulacija karstne izdani u okviru regionalnog vodoprivrednog sistema "Bogovina" (Management of karstic aquifer of regional water system "Bogovina", Eastern Serbia). University of Belgrade, Faculty of Mining and Geology, Belgrade

Stevanović Z (2010b) Utilization and regulation of springs. In: Kresic N, Stevanović Z (eds) Groundwater hydrology of springs. Engineering, theory, management and sustainability. Elsevier Inc. BH, Amsterdam, pp 339–388

Stevanović Z (2011) Annual consultancy report for 2011. DIKTAS project documents. http://dinaric.iwlearn.org/. Accessed 16 Nov 2013

Stevanović Z, Papic P (2008) The origin of groundwater. In: Dimkić M, Brauch HJ, Kavanaugh M (eds) Groundwater management in large river basins. IWA Publishing, London, pp 218–246
Stevanović Z, Iurkiewicz A (2009) Groundwater management in Northern Iraq. Hydrogeol J 17(2):367–378
Stevanović Z, Milanović S, Ristić V (2010) Supportive methods for assessing effective porosity and regulating karst aquifers. Acta Carsologica 39(2):313–329
Stevanović Z, Ristić Vakanjac V, Milanović S (eds) (2012a) Climate changes and water supply. Monograph. SE Europe cooperation programme. University of Belgrade, Belgrade, p 552
Stevanović Z, Balint Z, Gadain H, Trivić B, Marobhe I, Milanović S et al (2012b) Hydrogeological survey and assessment of selected areas in Somaliland and Puntland. Technical report no. W-20, FAO-SWALIM (GCP/SOM/049/EC) Project, Nairobi. http://www.faoswalim.org/water_reports
Theis CV (1935) The relation between lowering of the piezometric surface and rate and duration of a discharge of a well using ground-water storage. Trans Am Geophys Union 16:519–524
Torbarov K (1976) Proračun provodljivosti i efektivne poroznosti u uslovima krša na bazi analize krive recesije (A calculation of permeability and effective porosity in karst on the basis of recession curve analysis). In: Hydrology and water richness of karst. Proceedings of the Yugoslav—American symposium, Dubrovnik, June 1975. Zavod za hidrotehniku Gradjevinskog fakulteta, Sarajevo, pp 97–106
Tóth J (1999) Groundwater as a geologic agent: an overview of the causes, processes, and manifestations. Hydrogeol J 7:1–14
Tóth J (2009) Gravitational systems of groundwater flow theory, evaluation, utilization. Cambridge University Press, Cambridge
Trček B (2003) Epikarst zone and the karst aquifer behaviour: a case study of the Hubelj catchment, Slovenia. Geološki zavod Slovenije, Ljubljana
Vlahović V (1975) Karst Nikšićkog Polja i njegova hidrogeologija (Hydrogeology of karst of the Nikšićko Polje). Društvo za nauku i umjetnost Crne Gore, Podgorica
Water Framework Directive WFD of the European Union (2000) Act 2000/60/EC. Official J EU, L 327/1, Brussels. http://ec.europa.eu/environment/water/water-framework/. Accessed 14 Jan 2014
White WB (1969) Conceptual models for carbonate aquifers. Ground Water 7(3):15–21
Williams PW (1983) The role of the subcutaneous zone in karst hydrology. J Hydrol 61:45–67
Williams PW (2008) The role of the epikarst in karst and cave hydrogeology: a review. Int J Speleol 37(1):1–10
Worthington SRH, Ford D (2009) Self-organized permeability in carbonate aquifers. Ground Water 47(3):319–320
Younger P (2007) Groundwater in the environment: an introduction. Blackwell Publishing, Malden, MA
Zötl JG (1974) Karsthydrogeologie. Springer, New York

Chapter 4
Overview of Methods Applied in Karst Hydrogeology

Nico Goldscheider

4.1 The Duality of Karst Aquifers and Investigation Methods

Karst aquifers have a different hydraulic structure and behavior than porous media and therefore require specific investigation methods (Goldscheider and Drew 2007). As discussed in Chap. 3, they are characterized by a high degree of heterogeneity and discontinuity, resulting in a duality of recharge, infiltration, porosity, flow, and storage (Bakalowicz 2005; Ford and Williams 2007). Recharge either originates from the karst area itself (autogenic) or from adjacent non-karst areas that drain toward the karst aquifer (allogenic). Infiltration occurs diffusely through soil and epikarst or concentrated via dolines or swallow holes. Karst aquifers show double or triple porosity, consisting of intergranular pores and fractures (often summarized as matrix porosity), and solutional conduits. Flow in the network of conduits and caves is often rapid and turbulent, while flow in the matrix is generally slower and laminar (Fig. 4.1). Storage occurs in the matrix and conduits, but residence times in the matrix are several orders of magnitude longer than those in the conduits (Kovacs et al. 2005).

The heterogeneity and duality of karst systems have to be considered while choosing appropriate investigation methods. Most methods used in karst are not fundamentally different than those used in porous media hydrogeology, but specific adaptations are required concerning the application and combination of adequate methods, as well as the interpretation of the results.

The duality of karst aquifers (matrix vs. conduits) also results in a duality of investigation methods: Artificial tracer tests are the ideal methodological tool to

N. Goldscheider (✉)
Karlsruhe Institute of Technology (KIT), Institute for Applied Geosciences (AGW), Division of Hydrogeology, Kaiserstr. 12, 76131, Karlsruhe, Germany
e-mail: goldscheider@kit.edu

Z. Stevanović (ed.), *Karst Aquifers – Characterization and Engineering*, Professional Practice in Earth Sciences, DOI 10.1007/978-3-319-12850-4_4

Fig. 4.1 View of a karst spring illustrating the duality of karst: Rapid and turbulent flow in the conduits, while porosity and permeability in the adjacent rock matrix are much lower. This duality also requires a duality of investigation methods. Karst spring at Cirque de Consolation, French Jura Mountains

study the rapid flow components in the conduit network (Goldscheider et al. 2008), whereas natural tracers, such as stable or radioactive isotopes, can be used to obtain information on the slow-flow and long-storage components in the matrix (Geyer et al. 2008; Maloszewski et al. 2002).

The hydraulic structure of karst systems results in a high degree of variability. Karst systems often react quickly and strongly on precipitation events or snowmelt. The water table or hydraulic head in conduits can vary by tens of meters or even more than 100 m. Flow velocities in conduits were reported to vary by a factor of 10 or more, and the discharge of karst springs is also highly variable. Discharge variations by several orders of magnitude within a few hours or days are common at karst springs. Along with water table and discharge, the physical, chemical, and microbial compositions of karst groundwater and spring water also show marked variations (Ravbar et al. 2011).

This variability represents a major challenge in the exploitation and management of karst water resources, but also requires methodological adaptations. Isolated measurements and observations are never adequate to characterize a karst system. Continuous monitoring of water levels, spring discharge, and water quality is crucial in karst research. The dynamics of karst can only be captured on the basis of continuous, long-term, and event-based observations (Pronk et al. 2006; Savoy et al. 2011). Therefore, adapted monitoring and sampling techniques are crucial in studying karst systems. For the same reasons, tracer tests, water balances, and other methods should ideally be applied repeatedly, during different flow conditions, to capture the variability of flow velocities, flow directions, and drainage divides (Göppert and Goldscheider 2008).

The following sections present an overview of methods used to study karst aquifer systems, their structure, hydraulic duality, and dynamics. Numerical models and other mathematical approaches are not described in this section, but presented in Chap. 10 as well as in several review articles (Hartmann et al. 2014) and book chapters (Kovacs and Sauter 2007). The focus of this chapter is on experimental and field-based methods, methodological adaptations, and combinations of methods that are most specific for karst. More general, hydrogeological methods are described in various other textbooks, such as Fetter (2001).

4.2 The Karst Hydrogeology Toolbox

Available methods in karst hydrogeology include geological, geophysical and speleological methods, hydrologic and hydraulic techniques, the use of natural tracers, such as isotopes and hydrochemical parameters, as well as the application of artificial tracer tests (Goldscheider and Drew 2007). The choice of appropriate methods and the sequence of methods applied depend on the practical or scientific research questions, but also on the level of previous knowledge and on the available time, money, staff, and instrumentation. The following sections briefly present the most important methods of the "karst hydrogeology toolbox" and discuss the advantages, fields of applications, and limitation of the individual methods (condensed in Table 4.1).

Table 4.1 Summary of methods available in the "karst hydrogeology toolbox" (modified after Goldscheider and Drew 2007)

Group of methods	Applications and advantages	Limitations and drawbacks
Geological methods	External and internal geometry of the aquifer system	Not always direct relations between geological data and groundwater flow
	Orientation and characteristics of potential flow paths	Limited data availability in deep and confined settings
	Potential hydraulic properties, e.g., karstifiability and porosity	
Geophysical methods	Determining geologic structures and overburden thickness	Generally, no direct and clear information on groundwater
	Locating fracture zones and other preferential flow paths	Non-uniqueness of all results
	Low costs compared to drilling	The greater the investigation depth, the lower the resolution
	Data can be obtained over wide areas (site coverage)	Noise problems and various technical limitations
Speleological methods	Locating and mapping of the past and active conduit network	In many cases, only a limited and unrepresentative part of the conduit system is accessible
	Direct observations and experiments inside the aquifer	
	Understanding the temporal evolution of the karst system	
Hydrological methods	Establishing dynamic water balances (input, output, storage)	Water budgets are often problematic, because of unknown and complex catchment boundaries
	Spring hydrographs to characterize system behavior and properties	
Hydraulic methods	Determination of hydraulic parameters and boundary conditions	Scale-dependency of hydraulic properties results in limited representativeness of data
	Determination of flow directions and water table variations	Conventional hydraulic methods assume laminar (Darcian) flow

(continued)

Table 4.1 (continued)

Group of methods	Applications and advantages	Limitations and drawbacks
Hydrochemical methods	Information on water quality and contamination problems	Temporal variability requires event-based sampling or continuous monitoring
	Natural tracers for the origin, movement and mixing of water and for water–rock interactions	
Isotopic methods	Natural tracers for the origin, movement and mixing of water and for water–rock interactions	Spatiotemporal input function often not known precisely
	Determination of residence times and water ages	Ambiguities (non-uniqueness) in interpreting data
Artificial tracer methods	Determination of connections and linear flow velocities	Ambiguities in interpreting negative tracer results
	Delineation of catchment areas	Limited applicability in regional systems with long transit times
	Information on contaminant transport	Visible coloring or toxicity concerns for some tracers
	Usually highly reliable, precise, and unambiguous information	

4.3 Geologic and Geophysical Methods

A detailed *understanding of the geologic setting* is always required for the characterization and management of karst water resources, because the stratigraphic sequence and geologic structure define the external boundaries and internal geometry of the karst aquifer system. Fractures and fold axes often predetermine cave patterns and karst water flow paths. Every conceptual and numerical hydrogeological model requires a geologic model as a basis. Therefore, geologic maps and sections are indispensable bases for karst hydrogeology. In most cases, existing geologic information has to be complemented by purpose-specific geologic fieldwork. Geologic methods in karst terrains are similar to conventional geologic methods and are not further described here. The reader is referred to Goldscheider and Andreo (2007) and geologic textbooks.

Box 4.1: KARSYS—Method for the conceptualization of karst systems

Jeannin et al. (2013) have proposed an approach (KARSYS) for the hydrogeological conceptualization of karst aquifer systems, based on geological information and other available data, such as data on springs and caves (Fig. 4.2). KARSYS is based on several generalized but defensible assumptions that make it possible to predict flow directions, delineate spring catchments, and determine the extensions and dimensions of the unsaturated and saturated zones. KARSYS is most suitable for mountainous and shallow karst systems, such as the Swiss Jura Mountains and Alps for which the

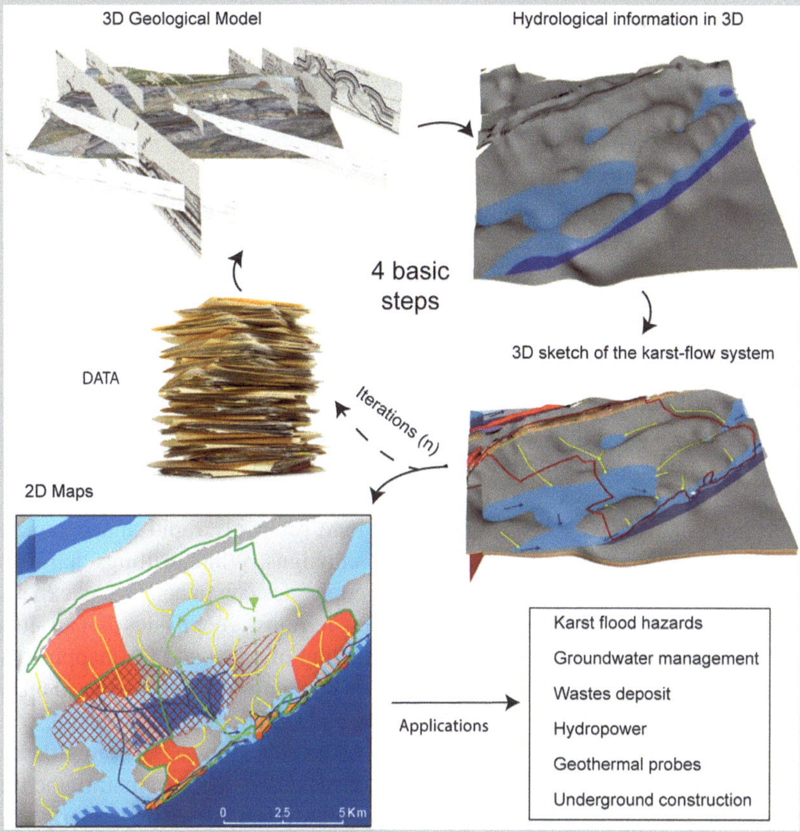

Fig. 4.2 Illustration of the KARSYS approach for the conceptualization of karst systems (Jeannin et al. 2013). The approach uses geological maps and other data as a basis to establish a 3D conceptual model of the karst flow system. The 3D model is then used to generate a set of maps that can be applied for a wide range of hydrological and engineering questions

approach was initially developed, but has also been applied for other karst systems, related to various applied and research questions, such as groundwater protection zoning, flooding, and engineering problems.

In deep and confined hydrogeological settings, but also in many lowland areas with insufficient natural rock outcrops, direct geological information has to be obtained from drillings. However, drillings are expensive and a large number of drillings would be necessary to localize linear elements, such as conduits or fracture zones. In such cases, geophysics comes into play.

Exploration geophysics is the science of seeing into the earth without digging or drilling (Bechtel et al. 2007). Geophysical methods can help to establish the geological setting of a karst system, but can also be used to identify hydraulically relevant structures, such as fracture zones or large conduits.

Different geophysical methods measure different physical properties, such as gravity, electrical resistivity, or acoustic wave velocities. Geophysical methods can be grouped into active and passive ones. Passive methods use existing geophysical fields, such as the natural gravity field of the earth, while active methods introduce a signal into the earth, as it is done in explosion seismic.

Observed geophysical anomalies can be interpreted as geological heterogeneities and structures. This so-called data inversion represents the major work step and challenge of all geophysical investigations. Non-uniqueness is a key problem of geophysical data inversion: Different geological heterogeneities can cause the same geophysical anomaly, which means in turn that any observed anomaly can be interpreted in many different ways. Non-uniqueness can be partly overcome by combining different geophysical methods (Fig. 4.3) and/or by checking geophysical interpretations by a limited number of well-selected drillings. Another general problem of geophysics is the trade-off between depth of investigation and resolution: The greater the depth, the lower the resolution.

The major advantages of geophysics include relatively rapid and inexpensive site coverage compared to detailed drilling campaigns. While drillings deliver precise but expensive data for selected points, geophysics can generate less expensive but less precise data for larger areas. In karst terrains, geophysics is particularly useful for identifying suitable locations for drilling of wells and for mapping overburden thickness, but also for geotechnical investigations, such as the identification of potential sinkhole hazards below building land (Bechtel et al. 2007).

A wide range of methods is available in karst geophysics, summarized by Bechtel et al. (2007). *Seismic methods* involve measuring and evaluating the travel time of different types of acoustic waves. Seismic refraction is particularly useful for mapping overburden thickness. Seismic reflection is commonly used in gas and oil exploration and delivers the most detailed picture of deep subsurface structures.

Gravity methods required precise location control and various corrections. Microgravity is suitable for karst applications, such as the detection of cavities in the subsurface, which generate negative anomalies (Fig. 4.3), or bedrock pinnacles covered by sediments, which generate positive anomalies.

Electrical and electromagnetic methods include a wide range of techniques that are based on very different operational principles but are all sensitive to the electrical and/or magnetic properties of the subsurface (Bechtel et al. 2007). In the context of karst and groundwater studies, these methods are particularly suitable to detect differences in porosity, water saturation, and water chemistry, which all generate anomalies of electrical resistivity. Application examples include the identification of the saltwater–freshwater interface in coastal aquifers or mapping of fracture zones or cavities (Fig. 4.3).

Box 4.2: Advantage of combining different geophysical methods

Figure 4.3 presents a combined depiction of two geophysical images of Port Kennedy Bone Cave in Pennsylvania, USA. The cave is a major site for Pleistocene fossils that was initially discovered in the nineteenth century. Later on, the cave was filled with waste material, and the cave location got lost. In 2005, the cave was rediscovered by an interdisciplinary research effort including the application of microgravity and electrical resistivity tomography. The combined application of different geophysical methods helps to overcome the problem of non-uniqueness that is inherent to all geophysical investigations.

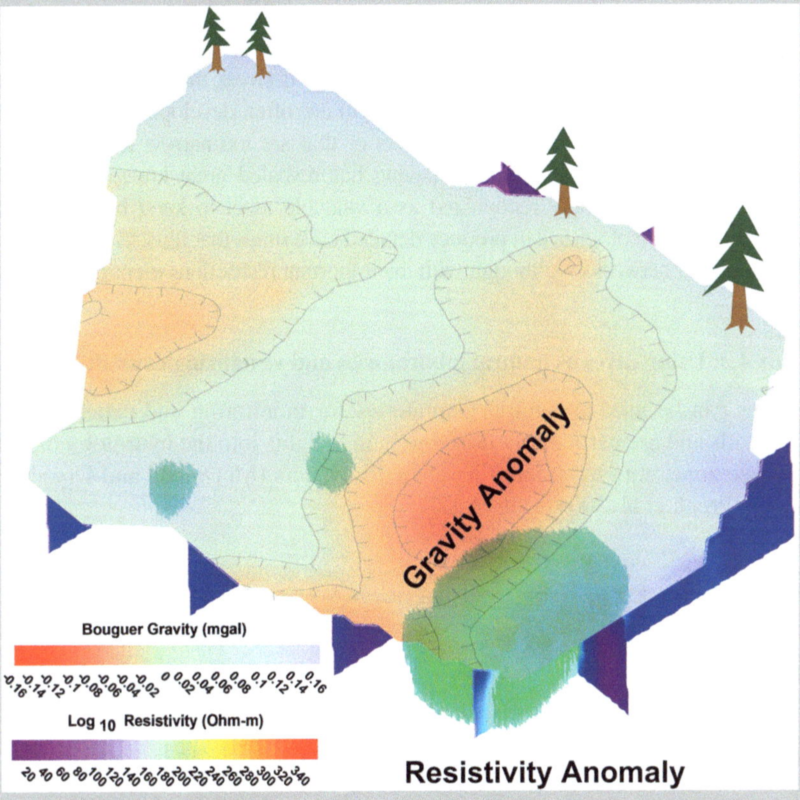

Fig. 4.3 Geophysical images of Port Kennedy Bone Cave (PA, USA) that has been rediscovered by microgravity and electrical resistivity tomography (courtesy of Tim Bechtel, Enviroscan Inc.)

Borehole geophysics (*Logging*) is an indispensable supplement to the analysis of drilling cores. Drilling cores allow detailed lithological, mineralogical, geochemical, and paleontological analyses, while borehole geophysics makes it possible to study a wide range of physical underground properties in situ. Temperature logs are a simple type of borehole geophysics and an example of a relevant underground property that cannot be obtained from drilling cores. Bechtel et al. (2007) summarize available borehole geophysical methods and their uses in karst and hydrogeology studies.

4.4 Speleological Methods

Caves make it possible to access the karst system and directly observe and study flow and transport in the unsaturated zone and in the aquifer (Jeannin et al. 2007). Many accessible caves represent older parts of the drainage network that are no longer active today, while large parts of the active flow system are often developed in deep phreatic conduits that can only be accessed by divers or that are too narrow to be entered by humans (Palmer 1991). Nevertheless, caving has revealed great knowledge on karst systems and is increasingly recognized as a valuable asset to karst hydrogeological investigations. Speleologists can produce detailed cave maps that illustrate the geometry of karst conduit networks—although with the important restrictions mentioned above.

Box 4.3: Using caves as natural laboratories and visualizing conduit systems

Caves can be used as natural laboratories for monitoring and experimental methods and are particularly useful to gain insights into the hydrology of the vadose zone, e.g., by monitoring cave drip waters (McDonald and Drysdale 2007; Pronk et al. 2009) (Fig. 4.4).

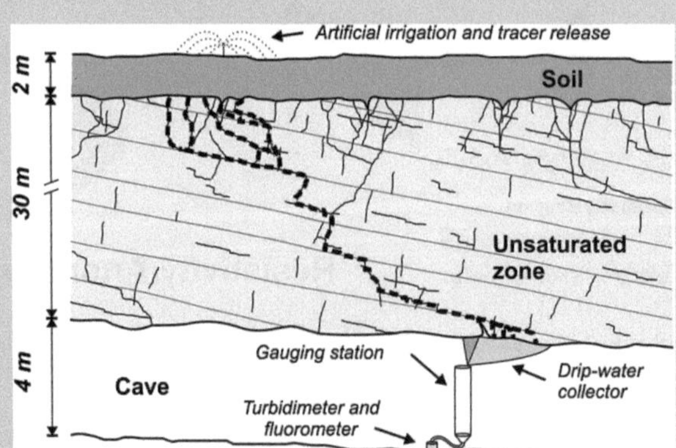

Fig. 4.4 Example illustrating the utilization of a shallow cave as a monitoring site for tracer tests to study flow and transport in the unsaturated zone (Pronk et al. 2009, slightly modified)

The combination of tracer tests, speleology, and hydrology makes it possible to obtain a more detailed picture of the underground drainage system than one of these methods alone (Lauber et al. 2014; Smart 1988). Figure 4.5 represents a classical example of this approach.

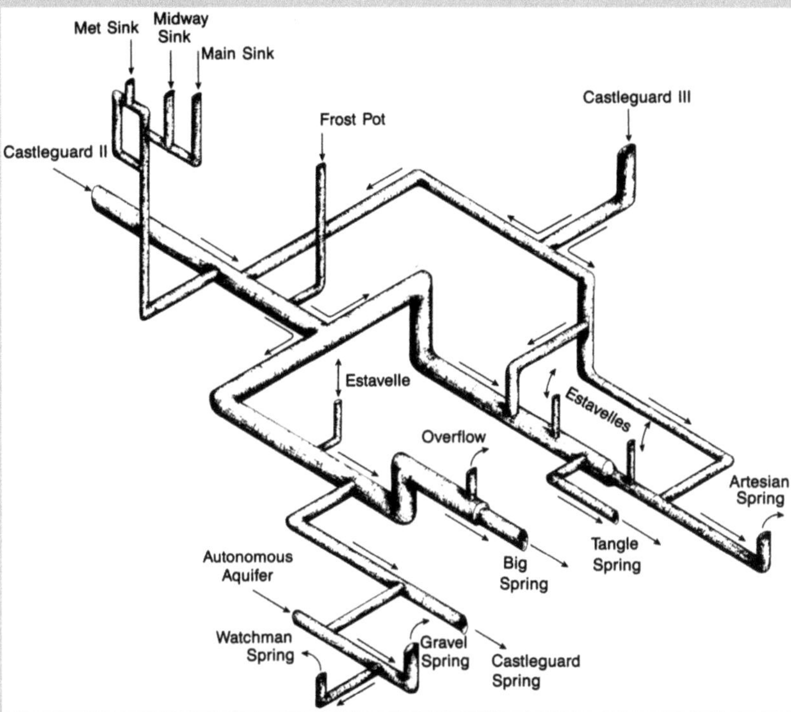

Fig. 4.5 Representation of a karst conduit drainage network that has been obtained by the combined application of speleological investigations (mapping of accessible caves), surface hydrology (mapping of springs, swallow holes, and estavelles), and artificial tracing technique (characterization of inaccessible conduits) (Smart 1983; Copyrights Arctic and Alpine Research journal, Alberta, Canada)

4.5 Hydrologic and Hydraulic Methods

Hydrologic methods often focus on the establishment and quantification of dynamic water balances (Groves 2007). The general form of any water balance is:

$$Input = Output \pm \Delta Storage. \tag{4.1}$$

Concerning the quantification of input into karst aquifer systems, the described duality of recharge requires monitoring of both diffuse recharge via the soil and concentrated infiltration via swallow holes. The former depends on rainfall and snowmelt, but intermediate storage in the soil or snowpack and evapotranspiration losses have to be considered. The latter can be determined by continuous flow monitoring at sinking streams. Quantification of the output requires continuous discharge monitoring at karst springs. However, in many case, not all groundwater flows to monitored springs but may drain toward adjacent surface waters or other aquifers. In these cases, water balances are more difficult and require sophisticated hydrogeological methods (as discussed in Chap. 6). Water storage can occur at many places in the hydrologic system, such as the soil, the unsaturated zone, or the aquifer, thus requiring a wide range of different methods. Differences in aquifer storage can be observed by water-level measurements at observation wells and piezometers (hydraulic methods, see below).

Continuous discharge monitoring at karst springs is not only required to establish dynamic water balances, but also delivers hydrographs that are invaluable tools in karst hydrogeology, particularly in combination with chemographs, i.e., time series of various physicochemical or chemical parameters (Grasso and Jeannin 2002; Hartmann et al. 2013). Spring hydrographs and chemographs can be used for the overall characterization of karst systems (Kovacs and Perrochet 2008) and to better understand the system behavior during floods (Winston and Criss 2004) or droughts (Fiorillo 2009).

Flow nets are two-dimensional representations of groundwater flow and consist of equipotential lines and flow lines. Real groundwater flow is always a three-dimensional phenomenon and can be described by equipotential surfaces and flow lines (Fetter 2001). Equipotential surfaces are locations of equal hydraulic head, which can be measured at piezometers. However, the construction of meaningful flow nets or flow fields in karst aquifers is complicated by their heterogeneity, hydraulic discontinuity, and anisotropy. Therefore, the interpretation of piezometer data and the construction of flow nets in karst aquifers need to be done with care and caution, and predicted flow directions must be checked with tracer tests and compared with geological and speleological information.

Hydraulic borehole methods, such as pumping tests and slug tests, are key methods to quantify the hydraulic properties of alluvial aquifers, such as transmissivity and storage capacity, and to identify the hydraulic boundary conditions. Specific methodological adaptations are required, concerning both the implementation and interpretation of hydraulic tests (Kresic 2007). For example, the time-drawdown curves of pumping tests in karst aquifers often display characteristic steps that result from the consecutive dewatering of discrete karstified fractures or bedding planes. Water-filled karst conduits near the pumped well can induce quasi-stationary conditions, similar to an equipotential boundary of an alluvial aquifer, such as a lake or a river (Fig. 4.6).

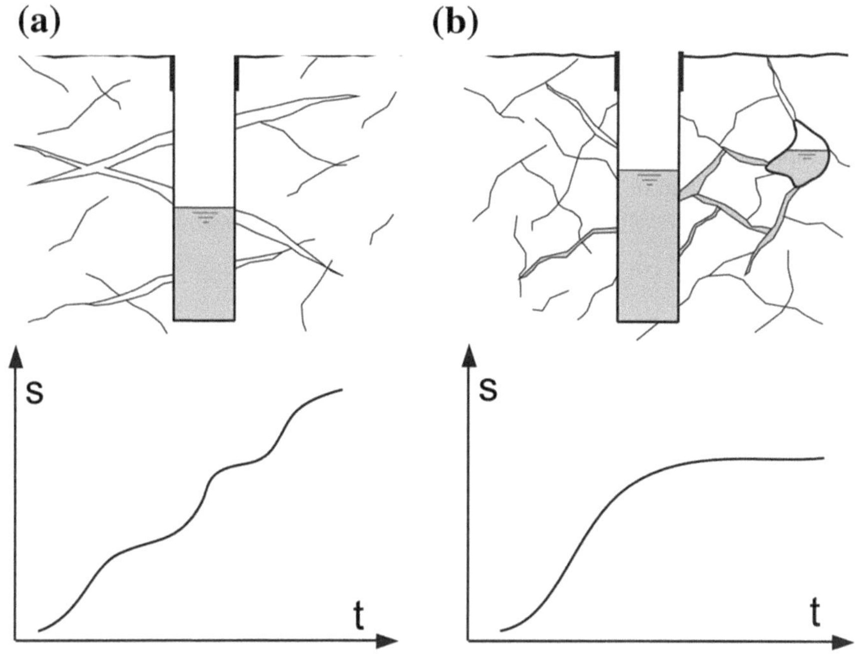

Fig. 4.6 Examples of characteristic time—drawdown curves obtained from pumping tests in karst aquifers (not to scale). **a** The consecutive drainage of three discrete karstified fractures or bedding planes creates a stepped curve. **b** A nearby water-filled karst conduit causes quasi-stationary conditions (modified after Kresic 2007, Copyrights Taylor and Francis/Balkema, London)

4.6 Hydrochemical and Isotopic Methods

Hydrochemical and microbiological methods are either applied to characterize water quality and contamination problems, or the obtained parameters are used as "natural tracers" to determine transit times, the origin and mixing of different water types as well as hydrologic or biogeochemical processes in the aquifer system. Due to the high degree of variability of karst systems, continuous monitoring or event-based high-frequency sampling is generally preferred. Hunkeler and Mudry (2007) group the available parameters into (i) precipitation-related parameters, (ii) soil-related parameters, (iii) carbonate-rock-related parameters, (iv) parameters related to other rock types, and (v) compounds of anthropogenic origin.

The primary result of detailed physicochemical and hydrochemical monitoring at a karst spring is a so-called chemograph. As described above, chemographs can be combined with hydrographs in order to gain insights into the behavior of the karst system, particularly in response to intense precipitation events or droughts (Grasso et al. 2003; Raeisi et al. 2007).

Box 4.4: Understanding the behavior of a karst system from hydrographs and chemographs

The primary increase of discharge at a karst spring, documented in a hydrograph, is often the result of a hydraulic pressure pulse in the conduit system, while the arrival of freshly infiltrated water occurs later and can be observed by means of physicochemical, hydrochemical, and/or microbiological parameters (Fig. 4.7).

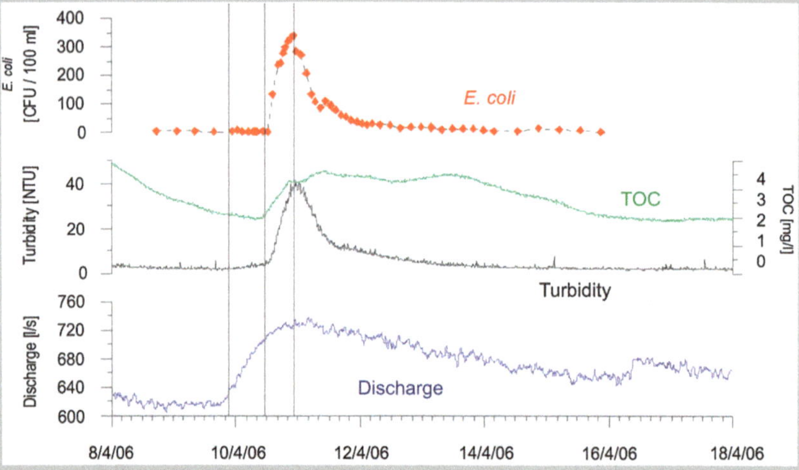

Fig. 4.7 Variability of discharge, fecal bacteria (*Escherichia coli*), total organic carbon (TOC), and turbidity observed at a karst spring following an intense precipitation event during a high-flow period. Discharge increases first, as a result of a hydraulic pressure pulse in the aquifer. The three "soil-related parameters" increase about 12 h later and indicate the arrival of freshly infiltrated water. This example also illustrates the extreme variability of contamination at karst springs and the potential usefulness of continuous monitoring of TOC and turbidity as "early warning system" for fecal bacteria contamination (*Data* Pronk et al. 2006, 2007)

Stable and radioactive isotopes are often used as natural tracers in karst hydrogeology (Criss et al. 2007). Deuterium (D or ^{2}H) and oxygen-18 (^{18}O) are the most important precipitation-related parameters. The comparison of the seasonal variation of D and ^{18}O in rainfall with the observed variability in karst spring waters allows to quantify mean transit times (Maloszewski et al. 2002). Tritium with its half-life time of 12.3 years can be used to determine groundwater ages. Other radioactive isotopes, such as Radon, can be used to quantify mixing processes in discharge areas (Eross et al. 2012) or for various other purposes, similarly as described above for the different hydrochemical parameters.

4.7 Artificial Tracer Methods

Tracer techniques using artificial tracers, such as fluorescent dyes, are a major technique for the investigation of karst aquifer systems. They can be used, among others, to identify underground connections, for example, between swallow holes and springs or within cave systems, to delineate karst spring catchments, to quantify linear flow velocities and other relevant parameters, and to obtain information on the processes of contaminant transport and attenuation (Benischke et al. 2007; Goldscheider et al. 2008).

The basic principle of a tracer test is simple: A specific substance (i.e., an artificial tracer) is injected into the hydrogeological systems at a particular place and time, while other sites, usually springs or pumping wells, are monitored for the arrival of this tracer. Unlike most other hydrogeological methods, tracer tests deliver not just hints and indications, but clear and quantitative evidence on underground connections. Tracer tests can be applied in all types of hydro(geo)logical environments, but they are particularly useful in karst aquifers, where underground flow paths are often long, complicated, and unexpected, whereas flow velocities in the conduit network are fast, thus limiting the required duration of the sampling period (Benischke et al. 2007; Goldscheider et al. 2008).

A wide range of substances can be used as artificial tracers (Käss 1998), but some specific fluorescent dyes are by far the most important hydrogeological tracers, because of their favorable properties: Extremely low detection limits and absence in natural waters, thus requiring limited injection quantities; highly water soluble; non-toxic to humans and the environment; and largely conservative behavior, i.e., microbiologically and chemically stable and mobile. The best and most important fluorescent dyes include Uranine (=Sodium-Fluorescein), Amidorhodamine G, Eosin and Sodium-Naphthionate; Sulforhodamine B is also a good water tracer, but with some ecotoxicological limitations. The detection limit of Uranine is as low as 0.005 μg/L. This means that injection quantities of 100 to 10 g are generally sufficient even for large-scale tracer tests with flow distances of many kilometers (Goldscheider et al. 2008).

Salts can also be used as artificial tracers. However, there is always some natural background in the water and the detection limits are much higher than those of fluorescent dyes, so that tracer tests over large distances would require enormous injection quantities (>100 kg). Different types of particles, colloids, and microbiological substances (e.g., bacteriophages) can also be used as water tracers (Goldscheider et al. 2007), but this section focuses on the application of fluorescent dyes.

Box 4.5: Where and how to inject the tracers?

Tracer tests are powerful tools, but the preparation and operation of tracer tests is rather expensive and laborious. Therefore, tracer tests usually require careful preliminary investigations, such as geological mapping, hydrological measurements and water balances, and the application of hydrochemical and isotopic methods to obtain first indications on aquifer properties, catchment areas, flow, and mixing processes. On this basis, tracer tests can be used to check research hypotheses and obtain clear evidence on underground connections and catchments.

Simple tracer tests require one single injection site, while multi-tracer tests involve the injection of several different tracers into different injection sites. Successful tracer injections require flowing water. Therefore, sinking surface waters (i.e., swallow holes), cave streams, or water wells are favorable injections sites. When no naturally flowing water is available, large quantities of flushing water need to be used, but this injection technique has a lower chance of success. Usually, the tracer powder is dissolved in an appropriate container (plastic bottle, canister, and barrel) and then slowly released into the injection site, along with naturally or artificially flowing water (Fig. 4.8).

Fig. 4.8 Tracer injection into Blauhöhle cave, Germany (photo courtesy of A. Kücha). The tracer powder was transported to this hard-to-reach injection site in a plastic canister, mixed with water from the cave stream, and directly released into flowing water to ensure optimal tracer transport. In-cave dye tracing makes it possible to obtain spatially resolved information on conduit flow (Lauber et al. 2014)

Box 4.6: Estimation of required trace injection quantities

Two groups of methods are available to determine appropriate injection quantities. Several authors have suggested purely empirical methods, based on experience and requiring a minimum of prior knowledge, such as the formula by Käss (1998):

$$M = L \times K \times B \quad (4.2)$$

where

M is required tracer mass (kg),
L is relevant distance (km),
K is a coefficient for the tracer type (e.g., one for Uranine, two for Amidorhodamine G, 15 for Naphthionate, and 20,000 for NaCl), and
B is a coefficient for the hydrogeological framework conditions (e.g., 0.1–1.0 for karst conduits)

According to this formula, a tracer test in karst over 10 km distance would require 1–10 kg of Uranine (or 2–20 tons of salt!).

Alternatively, the required tracer injection quantity can be determined on the basis of quantitative transport equations, such as the Advection–Dispersion Model (ADM) described further below. This approach requires more input data (that are usually not available prior to the tracer test) but also represents a more objective and defensible basis of estimation (Field 2003).

Three groups of methods are available for sampling and monitoring the arrival of the injected tracers at springs or other observation sites: integrative sampling, discrete sampling, or continuous monitoring. In the case of fluorescent dyes, integrative sampling can be done by means of charcoal adapters that are placed in the water, usually for a time period of several days. Fluorescent dyes accumulate in the charcoal and can later be eluted and analyzed in the laboratory. This technique is cheap and timesaving but delivers only qualitative results (basically Yes or No). It can thus be used for a large number of minor sampling sites, such as small and rather unimportant springs.

Discrete sampling means to take water samples at given time intervals, either manually (grab sampling) or by means of automatic samplers. The samples can then be analyzed quantitatively in the laboratory thus generating a series of time-concentration data. This sampling method is highly reliable and reproducible, but also requires sufficient staff and/or auto-samplers.

Continuous monitoring of fluorescent tracers involves the application of field fluorimeters. Different models are available, such as downhole or flow-through fluorimeters with up to four optical channels that make it possible to monitor several fluorescent tracers and turbidity at the same time (Schnegg 2002). The UV channel can also be used for organic carbon monitoring (Pronk et al. 2006).

The primary result of a tracer test is a breakthrough curve (BTC), where tracer concentrations are plotted versus time (Fig. 4.9). A well-documented BTC is a much more convincing evidence for an underground connection than an isolated result from a single charcoal bag or from scarce water samples. Furthermore, several relevant flow and transport parameters can be directly obtained from the BTC, such as the time of first arrival, the peak time, and the maximum concentration. Concentrations are either plotted as absolute values (e.g., in μg/L) or normalized by the respective injection quantity to allow better comparison of BTCs from multi-tracer tests.

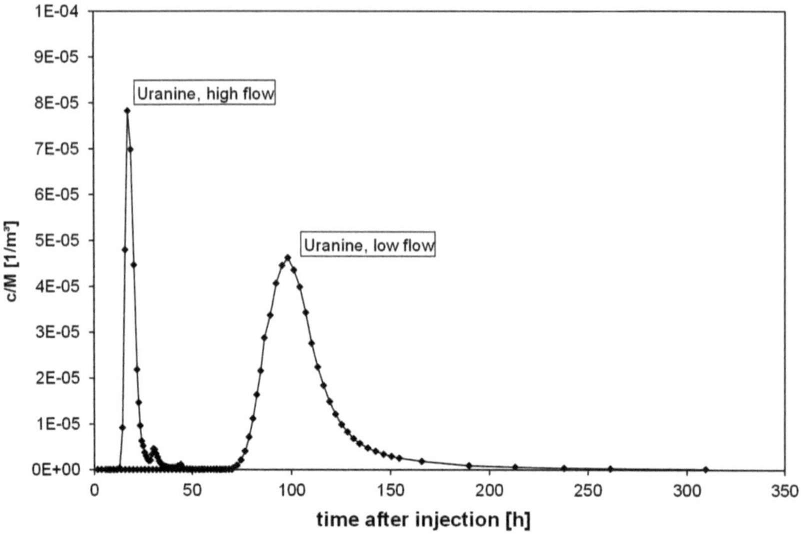

Fig. 4.9 Breakthrough curves of two tracer tests carried out in the same conduit system during low- and high-flow conditions, demonstrating the variability of flow velocities and dilutions in karst aquifers. Concentrations (c) are normalized by the injection mass (M) to allow unambiguous comparison [*Data* Göppert and Goldscheider (2008)]

Linear flow velocities can be calculated by dividing the linear distance between injection and sampling point by the corresponding transit time. Tracer recovery can be calculated as follows:

$$M_R = \int_{t=0}^{\infty} (Q \cdot c)dt \tag{4.3}$$

where

M_R is recovered tracer mass,
Q is discharge or pumping rate,
c is concentration, and
t is time

More advanced transport parameters can be obtained by different analytical, stochastic, or numerical modeling approaches. The usual procedure consists of fitting a theoretical BTC to the measured data (best-fit approach and inverse modeling). The most conventional approach is the one-dimensional ADM (Kreft and Zuber 1978). This model can also be used for the prediction of tracer transport (forward modeling) and thus for the determination of optimal injection quantities and sampling strategies (Field 2003). Several computer codes are available for direct and inverse tracer transport modeling, such as CXTFIT (Toride et al. 1999).

Box 4.7: Repeated tracer tests to capture hydrologic variability

Repeated tracer tests during low-flow and high-flow conditions are rarely done but are highly useful, as they make it possible to quantify the variability of flow velocities and contaminant transport processes (Göppert and Goldscheider 2008) as well as the shifting of catchment boundaries (Ravbar et al. 2011) as a function of the hydrologic condition. Flow velocities obtained from such repeated tracer tests can vary by a factor of 5 (Fig. 4.9; Göppert and Goldscheider 2008) or even 10 or more (Pronk et al. 2007). This type of information is also important for the delineation of protection zones.

References

Bakalowicz M (2005) Karst groundwater: a challenge for new resources. Hydrogeol J 13:148–160
Bechtel TD, Bosch FP, Gurk M (2007) Geophysical methods. In: Goldscheider N, Drew D (eds) Methods in karst hydrogeology. International Contribution to Hydrogeology, IAH, vol 26. Taylor and Francis/Balkema, London, pp 171–199
Benischke R, Goldscheider N, Smart CC (2007) Tracer techniques. In: Goldscheider N, Drew D (eds) Methods in karst hydrogeology. International Contribution to Hydrogeology, IAH, vol 26. Taylor and Francis/Balkema, London, pp 147–170
Criss R, Davisson L, Surbeck H, Winston W (2007) Isotopic methods, In: Goldscheider N, Drew D (eds.), Methods in karst hydrogeology. International Contribution to Hydrogeology, IAH, vol 26. Taylor and Francis/Balkema, London, pp 123–145
Eross A, Madl-Szonyi J, Surbeck H, Horvath A, Goldscheider N, Csoma AE (2012) Radionuclides as natural tracers for the characterization of fluids in regional discharge areas, Buda Thermal karst, Hungary. J Hydrol 426:124–137
Fetter CW (2001) Applied hydrogeology, 4th edn. Prentice Hall, Upper Saddle River
Field MS (2003) A review of some tracer-test design equations for tracer-mass estimation and sample-collection frequency. Environ Geol 43:867–881
Fiorillo F (2009) Spring hydrographs as indicators of droughts in a karst environment. J Hydrol 373:290–301
Ford D, Williams P (2007) Karst hydrogeology and geomorphology. Wiley, Chichester
Geyer T, Birk S, Liedl R, Sauter M (2008) Quantification of temporal distribution of recharge in karst systems from spring hydrographs. J Hydrol 348:452–463
Goldscheider N, Andreo B (2007) The geological and geomorphological framework, In: Goldscheider N, Drew D (eds.) Methods in karst hydrogeology. International Contribution to Hydrogeology, IAH, vol 26. Taylor and Francis/Balkema, London, pp 9–23
Goldscheider N, Drew D (eds) (2007) Methods in karst hydrogeology. International Contribution to Hydrogeology, IAH, vol 26. Taylor and Francis/Balkema, London
Goldscheider N, Haller L, Poté J, Wildi W, Zopfi J (2007) Characterizing water circulation and contaminant transport in Lake Geneva using bacteriophage tracer experiments and limnological methods. Environ Sci Technol 41:5252–5258
Goldscheider N, Meiman J, Pronk M, Smart C (2008) Tracer tests in karst hydrogeology and speleology. Int J Speleol 37:27–40
Göppert N, Goldscheider N (2008) Solute and colloid transport in karst conduits under low- and high-flow conditions. Ground Water 46:61–68
Grasso DA, Jeannin PY (2002) A global experimental system approach of karst springs' hydrographs and chemographs. Ground Water 40:608–617

Grasso DA, Jeannin PY, Zwahlen F (2003) A deterministic approach to the coupled analysis of karst springs' hydrographs and chemographs. J Hydrol 271:65–76
Groves C (2007) Hydrological methods, In: Goldscheider N, Drew D (eds) Methods in karst hydrogeology. International Contribution to Hydrogeology, IAH, vol 26. Taylor and Francis/Balkema, London, pp 45–64
Hartmann A, Goldscheider N, Wagener T, Lange J, Weiler M (2014) Karst water resources in a changing world: review of hydrological modeling approaches. Rev Geophys 52(3):218–242
Hartmann A, Weiler M, Wagener T, Lange J, Kralik M, Humer F, Mizyed N, Rimmer A, Barbera JA, Andreo B, Butscher C, Huggenberger P (2013) Process-based karst modelling to relate hydrodynamic and hydrochemical characteristics to system properties. Hydrol Earth Syst Sci 17:3305–3321
Hunkeler D, Mudry J (2007) Hydrochemical methods. In: Goldscheider N, Drew D (eds) Methods in karst hydrogeology. International Contribution to Hydrogeology, IAH, vol 26. Taylor and Francis/Balkema, London, pp 93–121
Jeannin PY, Eichenberger U, Sinreich M, Vouillamoz J, Malard A, Weber E (2013) KARSYS: a pragmatic approach to karst hydrogeological system conceptualisation. Assessment of groundwater reserves and resources in Switzerland. Environ Earth Sci 69:999–1013
Jeannin PY, Groves C, Häuselmann P (2007) Speleological investigations, In: Goldscheider N, Drew D (eds) Methods in karst hydrogeology. International Contribution to Hydrogeology, IAH, vol 26. Taylor and Francis/Balkema, London, pp 25–44
Käss W (1998) Tracing technique in geohydrology. Balkema, Brookfield, p 581
Kovacs A, Perrochet P (2008) A quantitative approach to spring hydrograph decomposition. J Hydrol 352:16–29
Kovacs A, Perrochet P, Kiraly L, Jeannin PY (2005) A quantitative method for the characterisation of karst aquifers based on spring hydrograph analysis. J Hydrol 303:152–164
Kovacs A, Sauter M (2007) Modelling karst hydrodynamics. In: Goldscheider N, Drew D (eds.) Methods in karst hydrogeology. International Contribution to Hydrogeology, IAH, vol 26. Taylor and Francis/Balkema, London, pp 201–222
Kreft A, Zuber A (1978) Physical meaning of dispersion equation and its solution for different initial and boundary conditions. Chem Eng Sci 33:1471–1480
Kresic N (2007) Hydraulic methods. In: Goldscheider N, Drew D (eds.) Methods in karst hydrogeology. International Contribution to Hydrogeology, IAH, vol 26. Taylor and Francis/Balkema, London, pp 65–91
Lauber U, Ufrecht W, Goldscheider N (2014) Spatially resolved information on karst conduit flow from in-cave dye tracing. Hydrol Earth Syst Sci 18:435–445
Maloszewski P, Stichler W, Zuber A, Rank D (2002) Identifying the flow systems in a karstic-fissured-porous aquifer, the Schneealpe, Austria, by modelling of environmental O-18 and H-3 isotopes. J Hydrol 256:48–59
McDonald J, Drysdale R (2007) Hydrology of cave drip waters at varying bedrock depths from a karst system in southeastern Australia. Hydrol Process 21:1737–1748
Palmer AN (1991) Origin and morphology of limestone caves. Geol Soc Am Bull 103:1–21
Pronk M, Goldscheider N, Zopfi J (2006) Dynamics and interaction of organic carbon, turbidity and bacteria in a karst aquifer system. Hydrogeol J 14:473–484
Pronk M, Goldscheider N, Zopfi J (2007) Particle-size distribution as indicator for fecal bacteria contamination of drinking water from karst springs. Environ Sci Technol 41:8400–8405
Pronk M, Goldscheider N, Zopfi J, Zwahlen F (2009) Percolation and particle transport in the unsaturated zone of a karst aquifer. Ground Water 47:361–369
Raeisi E, Groves C, Meiman J (2007) Effects of partial and full pipe flow on hydrochemographs of Logsdon river, Mammoth Cave Kentucky USA. J Hydrol 337:1–10
Ravbar N, Engelhardt I, Goldscheider N (2011) Anomalous behaviour of specific electrical conductivity at a karst spring induced by variable catchment boundaries: the case of the Podstenjsek spring, Slovenia. Hydrol Process 25:2130–2140
Savoy L, Surbeck H, Hunkeler D (2011) Radon and CO_2 as natural tracers to investigate the recharge dynamics of karst aquifers. J Hydrol 406:148–157

Schnegg PA (2002) An inexpensive field fluorometer for hydrogeological tracer tests with three tracers and turbidity measurement. In: XXXII IAH and ALHSUD congress Groundwater and human development, Oct 2002, Balkema, Rotterdam, Mar del Plata, Argentina, pp 1484–1488
Smart CC (1983) The hydrology of the Castleguard karst, Columbia Icefields, Alberta, Canada. Arct Alp Res 15:471–486
Smart CC (1988) Artificial tracer techniques for the determination of the structure of conduit aquifers. Ground Water 26:445–453
Toride N, Leij FJ, van Genuchten MT (1999) The CXTFIT code for estimating transport parameters from laboratory or field tracer experiments. US Salinity Laboratory, USDA, ARS, Riverside
Winston WE, Criss RE (2004) Dynamic hydrologic and geochemical response in a perennial karst spring. Water Resour Res 40(W05106):11

Part II
Engineering Aspects of Control and Protection of Karst Aquifer

Chapter 5
Surface Waters and Groundwater in Karst

Ognjen Bonacci

5.1 Introduction

Karst is defined as a terrain, generally underlain by limestone or dolomite, in which the topography is chiefly formed by the dissolving of rock, and which is characterised by sinkholes, sinking streams, closed depressions, subterranean drainage and caves (Field 2002). A wide range of closed surface depressions, a well-developed underground drainage system and a strong interaction between circulation of surface water and groundwater typify karst. Due to very high infiltration rates, especially in bare karst, overland and surface flow is rare in comparison with non-karst terrains.

Carbonate rocks are more soluble than many other rocks. They are subject to a number of geomorphological processes. The processes involved in the weathering and erosion of carbonate rocks are many and diverse. The varied and often spectacular surface landforms are merely a guide to the presence of unpredictable conduits, fissures and cavities beneath the ground. But at the same time, these subsurface features can occur even where surface karstic landforms are completely absent. Diversity is considered as the main feature of karstic systems. They are known to change very fast over time and in space, so that an investigation of each system on its own is needed. Karstification is a continuous process governed by natural and man-made interventions.

In karst terrains, groundwater and surface water constitute a single dynamic system. The groundwater and surface water are hydraulically connected through

O. Bonacci (✉)
Faculty of Civil Engineering, Architecture and Geodesy,
University of Split, Split, Croatia
e-mail: obonacci@gradst.hr

Z. Stevanović (ed.), *Karst Aquifers – Characterization and Engineering*,
Professional Practice in Earth Sciences, DOI 10.1007/978-3-319-12850-4_5

numerous karst forms which facilitate and govern the exchange of water between the surface and subsurface (Katz et al. 1997). A complex underground conduit system, as well as interplay of pervious and impervious layers of karst massif are an inherent characteristic of many (practically all) karst systems. Groundwater and surface water exchange with both adjacent and distant aquifers through underground routes or inflows from surface streams, natural lakes and in recent time artificial reservoirs. Due to this reason, one of the almost inevitable characteristics of open streams, creeks and rivers in karst regions is that they either have partial water loss along their course or completely sink into the underground (Bonacci 1987).

de Marsily (1986) states that the study of the water cycle or hydrology in its wider sense is usually divided into three separate disciplines: meteorology, surface hydrology and hydrogeology or groundwater hydrology. What is difference, and what is identical in karst hydrology and hydrogeology? Usual definitions of hydrogeology and hydrology are UNESCO and WMO (1992): (1) hydrogeology is branch of geology, which deals with groundwater and especially its occurrence, while (2) hydrology is science that deals with the processes governing the land areas of the Earth and treats various phases of the hydrological cycle. From these definitions is hardly possible to strictly distinguish between the two scientific disciplines. In engineering practice, the division is grounded in argument that hydrology deals with surface water and hydrogeology with groundwater. However, strictly enforcing such division could have harmful consequences on the development of both sciences, especially in case of investigations of the karst water circulation. Synthesis of hydrogeological and hydrological approach could expedite progress in karst surface water–groundwater system understanding.

Hydrogeology generally deals with groundwater occurrence and circulation in aquifers. Aquifers are in turn geological units involved in transmission of quantities of water under ordinary hydraulic gradient. At the same time, interest of hydrology is mainly focused on water balance, which is basically accounting of the inflow to, outflow from and storage within a hydraulic unit such as a drainage basin or aquifer. Very often it is impossible and harmful to separate two abovementioned approaches, but in practice it predominantly occurs.

Introductory part will be concluded with Atkinson's (1986) remark: "In soluble rock terrains, more than in most other terrains, the unexpected should always be expected".

5.2 Catchments in Karst

A catchment area (drainage basin or watershed) is the entire geographical area drained by any water body (spring, river, lake, aquifer, marsh, etc.). It is characterised by all run-off being conveyed to the same profile, outlet or the same water body zone.

Karst catchment represents complex water transport system in which heterogeneity of surface and underground karst forms, serving for flow circulation and storage, makes discovering and quantification of water through them difficult. Numerous and extremely different surface and underground karst forms make possible unexpected connections of water in karst medium space which changes in time. Changes of underground flow path during the time are caused by: (1) different recharges from different parts of surface area mainly caused by variable distribution of areal precipitation; (2) different groundwater levels (GWLs) in karst aquifers and their fast changes in time and space; (3) anthropogenic influences; and (4) exogenic and endogenic forces (Bonacci 2004).

The determination of the catchment boundaries and the catchment area is the starting point in all hydrological analyses and one of the essential data which serve as a basis for water resources protection, management, understanding and modelling of water circulation through karst massif. In karst landscape, the definition of catchment area and boundaries is a difficult and complex task, which very often remains unsolved (Bonacci 1987). The differences between the topographic and hydrological catchments in karst terrain are, as a rule, so large that data about the topographic catchment are useless in hydrological and hydrogeological analyses and water management practice. Determination of a karst catchment is an unreliable procedure due to unknown morphology of underground karst features (mainly karst conduits and characteristics of karst aquifers) and their connections with surface karst forms. The variability, in time and space, of karst aquifer as well as conduit parameters makes this process extremely sensible and complex.

Box 5.1

Herold et al. (2000) analysed the influence of tectonic structures on karst flow patterns in karstified limestones and aquitards in the Jura Mountains, Switzerland. In the early phase of karstification, fewer parts of the aquifer are oriented towards one spring, i.e. in this phase, there are a great number of springs with a small catchment area. As the hydrological activity increases, the respective catchment area of a spring becomes larger and deeper. Consequently, certain springs stop functioning, and the remaining active springs become larger and have a greater capacity (Bögli 1980). It is already evident in this phase that the catchment area of karst springs changes in time depending upon the water quantity and its altitude in the aquifer. This variability can be greater or smaller depending upon the local and regional geological and geomorphological conditions.

Figure 5.1 shows one relatively simple example of the GWL changes in karst aquifer. With number 1 is designated situation when the GWL is in minimum. Number 2 shows situation when the GWL is in average. Maximum GWL is designated with number 3. In this situation, the depressions in karst (mostly poljes in the karst) are flooded. Karst spring is designated with **A**. The swallow hole (ponor), **B**,

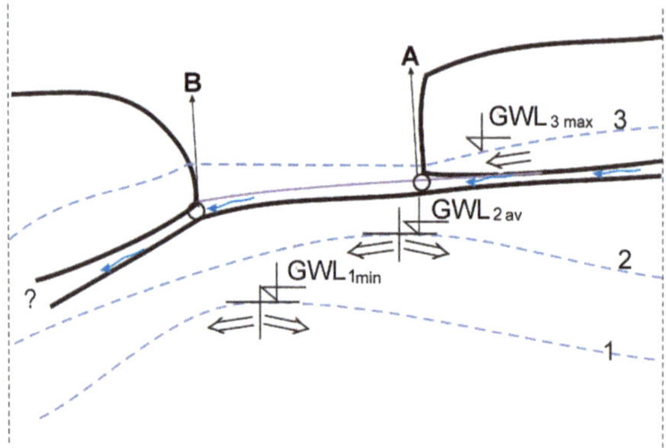

Fig. 5.1 An example of the GWL changes in karst aquifer

during the flood can act as spring. In this case, it is an estavelle. Extremely large heterogeneities in geometric and hydraulic parameters exist in the vadose and phreatic zone. Boundary between these two zones is very fast changeable in time and space due fast raising and falling of the GWL in karst aquifers.

The full characterisation of the conduit network is the way to find accurate catchment in karst and to protect its water resources. In some situations at very high GWL (caused by intensive rainfall), generally, fossil and inactive underground karst conduits are activated, causing the redistribution of the catchment areas, i.e. overflow from one to other catchment (Roje-Bonacci and Bonacci 2013). Limited maximum outflow capacity of karst springs (Bonacci 2001a) and limited inflow capacity of swallow holes (ponors) cause overflow from one catchment to the other in large karst space.

Earthquakes may cause collapses of surface (mostly doline) and subsurface features (caves and large conduits) in karst areas. This can be reason of instantaneous changes of water circulation and redistribution of catchments. Especially in recent time anthropogenic actions in karst (inter-basin water transfer, dams and reservoirs, groundwater pumping, motorway and railway construction, etc.) strongly, suddenly and generally unpredictably affect natural hydrological and hydrogeological regime at the local and even large regional scale (Bonacci and Andrić 2010).

Figure 5.2 is an attempt to present schematically all possible relationships of water circulation between two karst springs (**a** and **b**) and their topographic catchment areas (**A** and **B**). Water from the spring, **b**, with catchment, **B**, can flow by surface stream to the catchment, **A**, or to any other catchment. Water sinking in the swallow hole located in the catchment, **B**, can reappear in the same catchment or in the catchment, **A**, as well as in any other catchment. Water sinking in swallow holes located in the catchment, **A**, can reappear in the same catchment or in any other catchment excluding the catchment, **B**.

Figure 5.3 represents seven possible relationships of water circulation between two karst springs (**a** and **b**) and their topographic catchment areas (**A** and **B**). It should be stressed that in reality there are much more possible combinations. These examples are given with the goal to point out on changes of catchment areas

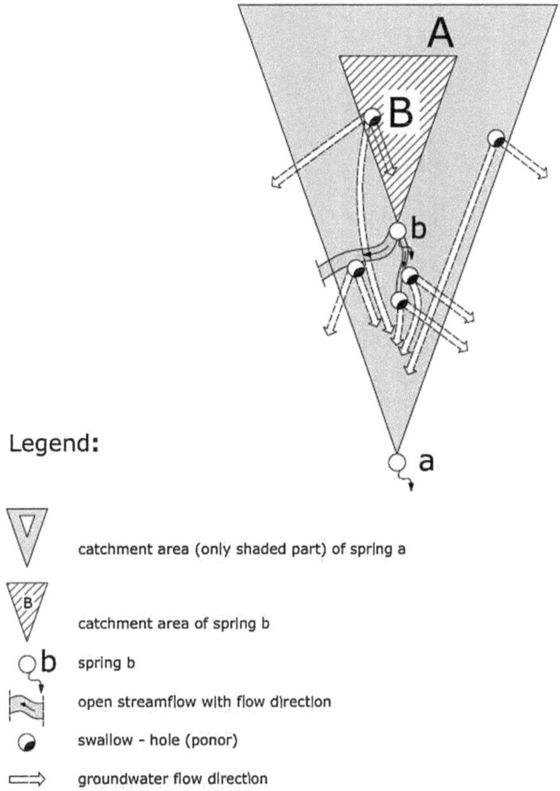

Fig. 5.2 Schematic presentation of all possible relationships of water circulation between two karst springs (*a* and *b*) and their topographic catchment areas (*A* and *B*)

depending on existence of different karst phenomena: (1) swallow hole; (2) open stream; and (3) losing stream. Hydrologic catchment areas are designated as, $\mathbf{A_a}$, and, $\mathbf{A_b}$, for springs, **a**, and, **b**, consequently.

Possible explanation of the GWL influence on the hydrological functioning of karst springs is given in Fig. 5.4. Explanation of the estavelle functioning and flooding of karst depression (mostly polje in karst) is given in Fig. 5.4a. If the spring **b** is at higher altitude than spring **a,** it can be intermittent, when the GWL in its topographic catchment is lower than spring **b** exit (Fig. 5.4a, b). Figure 5.4c presents how different geological setting (existence of low permeable rocks) influences on the hydrological functioning of the spring, **b**.

5.3 Karst Aquifers

Aquifer is rock formation that is able to retain large quantities of water (White 2002). The specific characteristic of karst aquifer is the existence of solutionally and by erosion generated and permanently enlarged karst voids of different

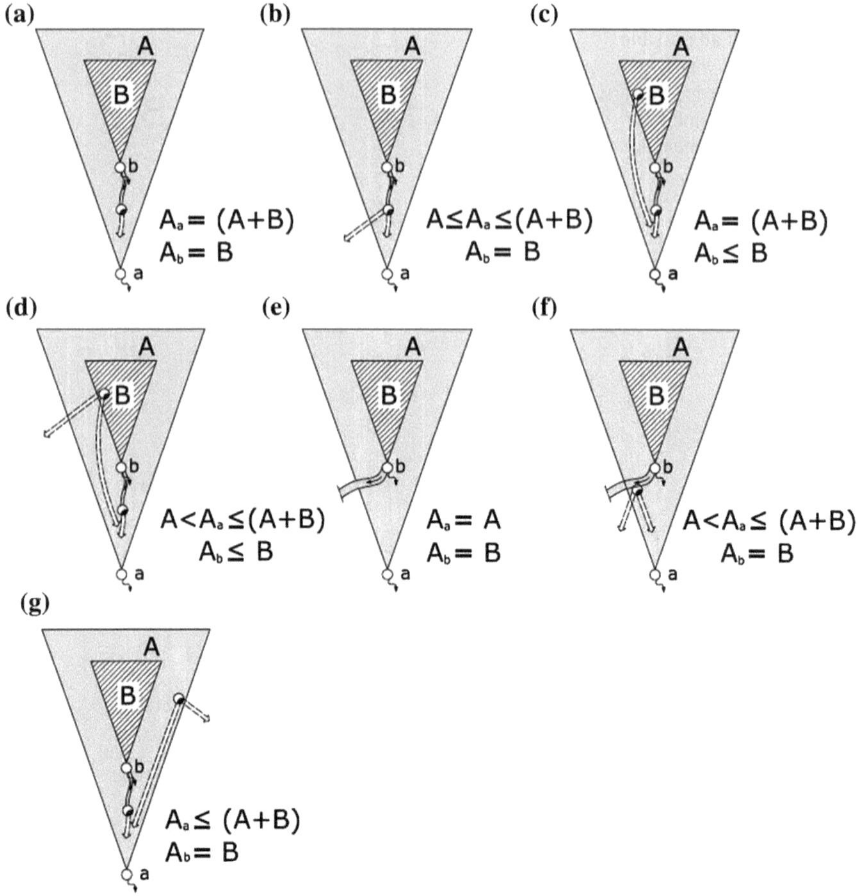

Fig. 5.3 Seven possible relationships of water circulation between two karst springs (*a* and *b*) and their topographic catchment areas (*A* and *B*) shoved in Fig. 5.2

dimensions. Circulation of groundwater in karst aquifers is quite different from water circulation in other non-karstic-type aquifers. In karst aquifers, water is being collected in networks of interconnected cracks, caverns and channels. The extremely enhanced heterogeneity of karst aquifers is caused by multiple porosity and anisotropy (Ford and Williams 2007).

Karst aquifer triple permeability (matrix, fissures and fractures and conduits) results in its heterogeneity and anisotropy. Water flowing through karst aquifer continuously dissolves surrounding rocks and spreads the dimensions of preferable voids. Process of karstification is temporal variable and relatively rapid in comparison with common geological processes. Each karst aquifer has specific hydrogeological, hydrological and hydraulic characteristics. There are three different types of karst aquifer: (1) only with large karst conduits; (2) only with narrow karst joints; and (3) system combined of highly developed and interconnected

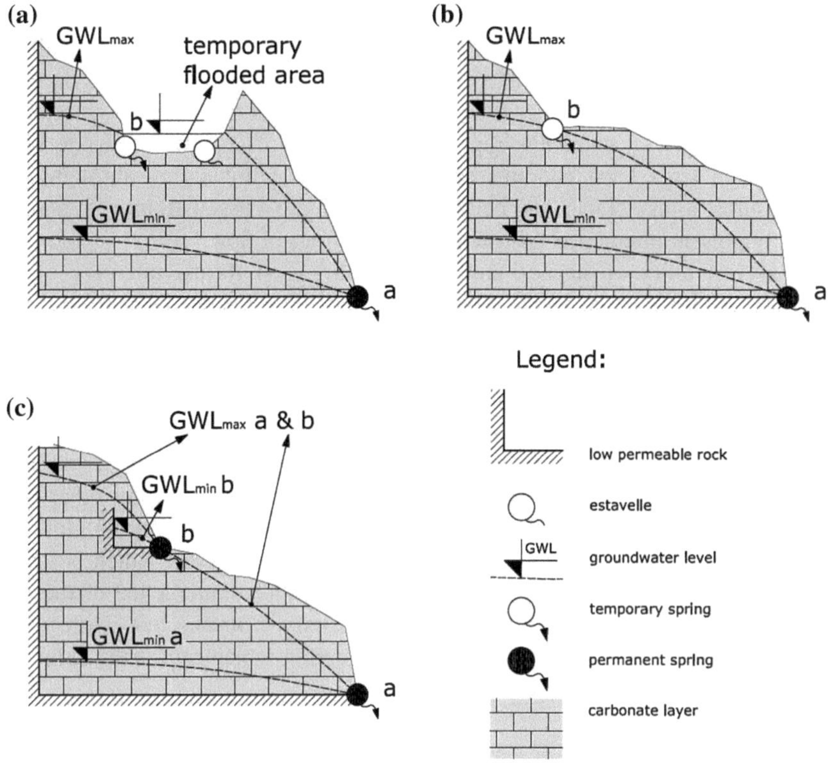

Fig. 5.4 Explanation of the GWL influence on the hydrological functioning of karst springs

large karst conduits and narrow karst fissures (Bonacci 1993). In karst aquifer, generally, it is not possible to define representative elementary volume, as it is case in other non-karstic aquifers.

Due to generally unknown spatial distribution of the karst conduit network and interplay between high and low permeable layers, the hydraulic conductivity of karst aquifers is extremely anisotropic and heterogeneous. The conduit porosity of karst aquifer ranges from solutionally widened joints and bedding planes with apertures of about 1 cm to large and irregularly shaped channels many metres in diameter.

Karst aquifers are generally continuous. However, numerous subsurface morphologic features in karst massive (caves, jamas, fractures, faults, impermeable layers, karst conduits, etc.) strongly influence the continuity of the aquifer, so that an aquifer commonly does not function as a continuum in a catchment especially during periods of abrupt groundwater rise. One of the most important characteristics of karst aquifers is the high degree of heterogeneity in their hydraulic properties. Karst aquifers can be very deep (hundreds of metres) with endless cracks, fractures, joints, bedding planes and conduits serving as groundwater pathways. In karst aquifer investigations, problem is that subsurface water is highly heterogeneous in terms of location of conduits, location of vertically moving water and flow velocities. Karst aquifers are some of the most complex and difficult systems to decipher. The highly heterogeneous nature of karst aquifers leads to the inability to predict groundwater flow direction and travel times. For karst aquifers' investigation, special challenge represents existence of concurrent fast turbulent flow through large karst conduits and slow, diffuse laminar flow through small karst fissures, joints, cracks and bedding plains. There exists a significant and permanently present interaction between these two types of flow.

Great variability of surface and underground karst forms, interplay of pervious and impervious layers as well as fast and large range of the GWL rising and decreasing in karst massif creates practically endless possibilities of contact between two or more karst aquifers which can belong and feed to different karst water bodies. In last about hundred years and especially in recent time, anthropogenic influences created new and very fast redistribution of surface water and groundwater in karst areas, which had caused changes of connections between aquifers of neighbouring (in some cases distant) karst springs and/or other water bodies (Bonacci 2004). De Waele (2008) explained the case of the Su Gologone karst springs influenced by the different water level in reservoir which is located downstream of them. Milanović (1986) found that the submergence of the karst spring zone of the Trebišnjica River (Dinaric karst of Bosnia and Herzegovina) affects the dynamic and emptying of the karst aquifer, which causes the redistribution of catchments of many karst springs.

Box 5.2

Figure 5.5 represents an attempt to schematically expose all possible relationships between the aquifers of two karst springs, **a** and **b,** showed in Figs. 5.2 and 5.3. In Fig. 5.5a, b, two aquifers are not connected. The spring acts as intermittent when the GWL_{min} is lower than spring exit. In Fig. 5.5c, d exists overflow from the aquifer of the spring **a** to the aquifer of the spring **b** and vice versa, respectively. Figure 5.5e presents overflow of groundwater from both analysed springs in any other catchments. Overflow appears only after heavy precipitations and generally last short time (maximum few days after the rainfall termination). Figure 5.5f represents existence of underground connection between two analysed springs. It can last different time during the year depending on relationship between the GWLs. Figure 5.5g shows case when groundwater from aquifer of the spring, **b,** emerges on the surface and by this way feeds aquifer of the spring, **a**. Example given in Fig. 5.5h shows possibilities of existence overflow as well as underground contacts between the aquifers of two analysed springs. It is obvious that in reality exist more different cases.

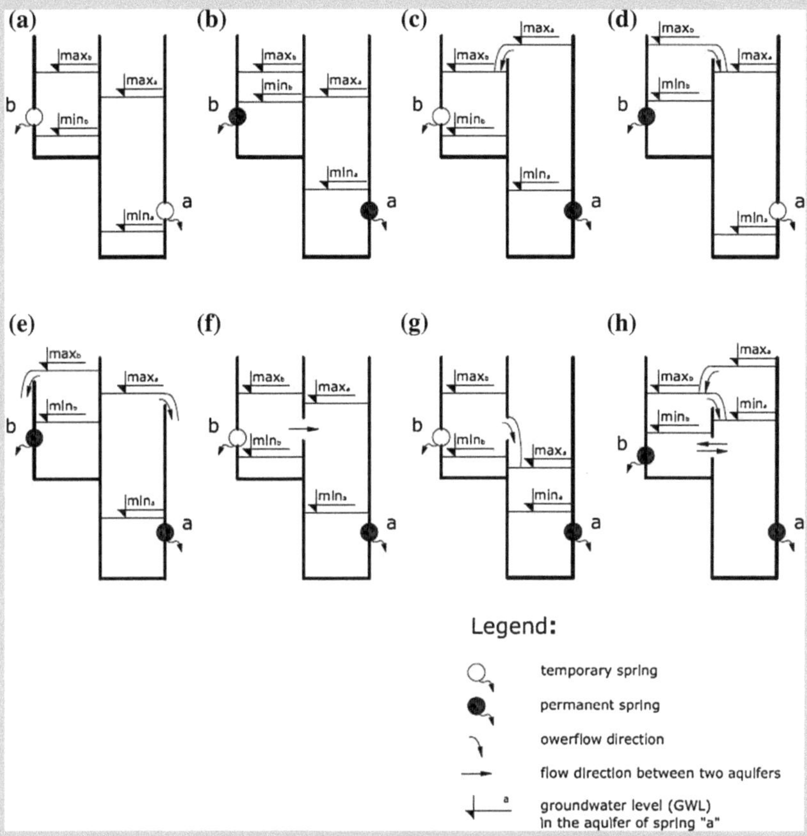

Fig. 5.5 Schematic presentation of all possible relationships between the aquifers of two karst springs, *a* and *b*

Underground flow direction is controlled by the GWL in neighbouring aquifers. As it is previously stressed, the GWLs in karst are extremely variable in time and space, as well as in different parts of them. Shape of the GWL in karst aquifer strongly depends on spatial distribution of intensive precipitation in analysed catchment. The catchment area of karst spring can be larger than 100 km^2, while cells of intensive precipitations (which cause fast GWL rising) are rarely larger of 5–10 km^2. It means that the GWL rising exists in only one part of the catchment area of the karst aquifer (Eagleson 1970; Dahlström 1986). For each case of intensive precipitation, it can be different part of the aquifer.

Figure 5.6 schematically presents connection between two neighbouring karst spring aquifers. Discharge from the spring **b** aquifer to the spring **a** aquifer, Q_{a-b}, depends on dimension of area through which groundwater flow, hydrogeological and hydraulic characteristics of this area and slope of the groundwater piezometric

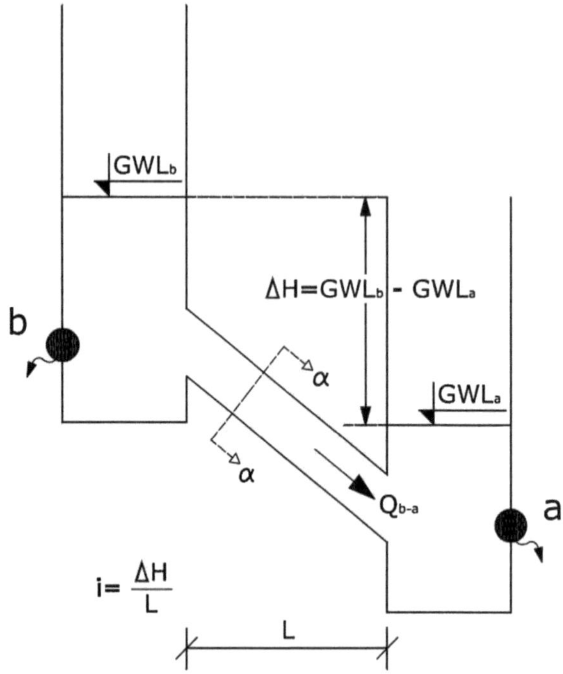

Fig. 5.6 Schematic presentation of connection between two neighbouring karst spring aquifers

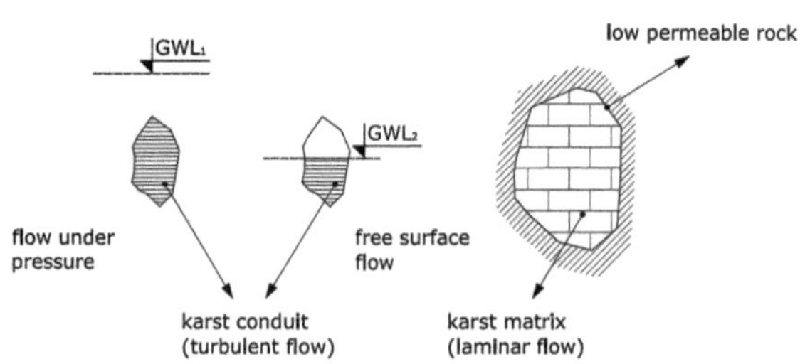

Fig. 5.7 Three possible type of flow in karst underground

line, **i**. Figure 5.7 presents three possible type of flow: (1) flow under pressure in karst conduit; (2) flow with free surface in karst conduit; and (3) flow through karst matrix. In most cases, all three types of flow exist at the same time what depends on structure of contact area and appearance of large karst underground features in it. Along a karst conduit, the shape and diameter of its cross section can vary significantly. Because of this flow in one part of the same conduit can be partly under pressure, while in the other part can be with the free surface. These characteristics strongly influenced on the groundwater travel time to the spring or residence time in the aquifer. They can vary from few hours (if a large conduit flow prevails) to decades (if there are only small karst joints).

Figure 5.8 is schematized cross section through karst massif which connected Prespa and Ohrid Lakes (Macedonia, Greece and Albania). The Prespa Lake does

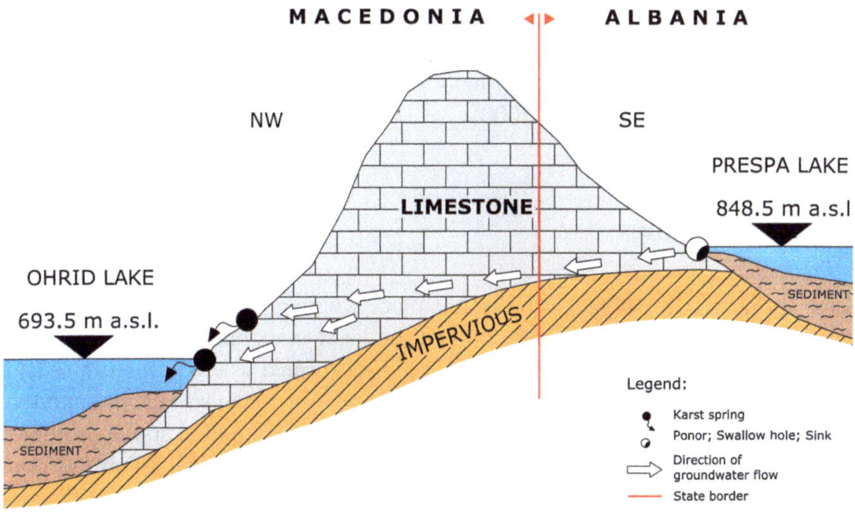

Fig. 5.8 Schematized cross section through karst massif which connected Prespa and Ohrid Lakes (Macedonia, Greece and Albania)

not have surface outflow. The waters from it outflow through karst underground massif into the Ohrid Lake (Popovska and Bonacci 2007).

For karstified rocks, the hydraulic conductivity depends on the density and aperture of the joints existing in karst matrix. Fractures may either seal with time or increase in aperture as a consequence of natural or anthropogenic actions. Hydraulic conductivity **K** of karst massive varies in range 10^{-4}–10^{-1} m/s and generally decrease with depth (Perrin et al. 2011).

Box 5.3

Using Dupuit's assumption, Bonacci (2001b) determined hydraulic conductivity **K** in the karst massif around the Ombla Spring (Croatia). In relatively small area of about 50 km^2, the values of the hydraulic conductivity ranged from 0.702×10^{-3} to 26.414×10^{-3} m s^{-1} during 262 h of measurement. Changes in values of **K** in both time and space can be attributed to differences in the position of investigated pairs of piezometers, their connection with main karst conduits, their GWLs and differences in development of karstification process in the analysed karst massif. These points to a significant influence of the time and space scale effect upon the results obtained by investigations and measurements in karst. Another analysis leads to conclusion that piezometric relations in karst aquifer are more uniform during the descent than during rising of the GWL. The rising phase is relatively short and lasts 5–10 % of the year. During this time all processes are rapid, flow is mostly turbulent and more non-homogenous than during descending phase (Bonacci and Roje-Bonacci 2000).

5.4 Karst Springs

Karst spring can be defined as a discrete place where groundwater flows naturally from the hydrologically active fissures of the karst mass onto the land surface or into a body of surface water. Generally, spring is formed where groundwater table intersects with the earth's surface or groundwater rises to the surface through rock faults, fractures or depressions. Lehmann (1932) mentioned the karst-hydrological contrast expressed by the presence of numerous places through which the water sinks into the karstified mass, whereas there are relatively few karst springs.

Hydrological characteristic of karst springs, especially their minimum and maximum values, can be very different in comparison with same characteristics of non-karst springs. Water surfacing from the karst springs can be considered as groundwater contribution in a large scale (Rimmer and Salingar 2006). It may not actually depend on the surface area on the catchment but is a function of total subsurface area contributing to groundwater flow.

Karst morphology and water pathways within the karst massif feeding the karst springs change with altitude and outflow capacity. Due to very special and complex underground and surface karst forms, which control surface water and groundwater behaviour, there are very different cases of karst springs. As discussed in Chap. 3 in literature, there are many classifications, systematisations and/or definitions of karst springs from different points of view (Bögli 1980; Bonacci 1987; Smart and Worthington 2004) in accordance with different scientific disciplines (hydrology, hydrogeology, geomorphology, geochemistry, geography, etc.). No one of them is so many-sided that would be able to clearly explain their complex functioning.

Karst springs can be perennial (permanent) or intermittent (temporary, ephemeral or seasonal). From perennial spring, stream flows above land throughout the whole year, while from the intermittent ones flows at irregular intervals related to seasonal variations in rainfall. The intermittent spring falls dry several times or most of the year. Seasonal ones act only during a certain (wet or rainy) seasons while ephemeral springs are active only for a short time as a consequence of intensive precipitation. Intermittent springs are very often in karst areas. They flowing at irregular intervals related to seasonal variations in rainfall, which control the GWL. In many cases, ephemeral karst springs are active for a short time (few days or even hours) after intensive precipitation. Rhythmic (ebb and flow) springs are special kind of intermittent springs, which appear exclusively in karstified terrains (Bonacci and Bojanić 1991).

Andreo et al. (2009) and Ravbar and Goldscheider (2009) distinguish the following three zones of contribution and connection between karst aquifer and karst springs: (1) inner; (2) intermediate; and (3) lower. In response to hydrogeological and hydrological settings, parts of an aquifer can either permanently or temporary contribute to the spring. The inner zone comprises parts of the system that always contribute and the connection is sure and direct to the spring. The outer zone comprises the morphologically uplifted part of the system that contributes only a small portion of the total amount (less than 1 %). It could comprise part of the aquifer system that temporarily contributes to the spring. The intermediate zone is located between nearer inner and further outer zone. It represents transition between them.

Due to fact that water discharging from the karst springs integrates the signal of geological and hydrological processes over large spatial areas and long periods of time, they are an indirect source of information (Manga 2001). By using a variety of techniques and approaches (for example: isotopic tracers, water chemistry, discharge, water temperature, electrical conductivity, etc.), it is possible to determine the mean-residence time of water, infer the spatial pattern and extent of groundwater flow or estimate basin-scale hydraulic properties. All previously mentioned in combination with dense and continuous monitoring of the GWL can help in determination of karst springs catchments.

The amount of water that flows from a spring depends on several factors but mostly from the rainfall on its catchment as well as size of the spring recharge catchment. Other factors are given as: (1) the size of caverns within rocks; (2) the hydraulic characteristics of aquifer soil; (3) the water pressure in the aquifer; and (4) the size and shape of spring exit.

Human intervention, especially construction of dams and reservoirs as well as interbasin water transfers can introduce instantaneous and distinct changes in the volume of discharge from a spring and especially in changes of its extreme discharges (minimum and maximum). For example, increased groundwater withdrawals can reduce the hydraulic pressure in an aquifer, causing water levels to decline and spring flows to decrease. The increasing in the natural GWL, caused by the reservoir's construction, can lead to formation of new springs or increasing of the spring characteristic discharges (minimum, mean or maximum) (Bonacci and Jelin 1988).

5.5 Karst Ponors

Ponor, swallow hole or sinkhole can be defined as a hole or opening in the bottom or side of a depression where a surface stream or lake flows either partly or completely underground into the karst groundwater system, and/or as a hole in the bottom or side of a closed depression, through which water passes to an underground channel (Field 2002). Ponors are situated commonly close to the terminus of a polje. The following classification of ponors from a morphological viewpoint is given by Milanović (1981): (1) large pits and caves; (2) large fissures and caverns; (3) system of narrow fissures; and (4) alluvial ponors. Practically, all underground phenomena (jamas, karst conduits, caves and even bedding planes) can take over the function of ponors. Jamas most commonly function as ponors and present a pathway for fast and direct contact of the surface water with the karst underground.

Estavelles belong to a special type of ponors and/or springs. They have a double hydrological function. In one period, they operate as ponors. This happens generally in the dry period of the year, when the GWL in the surrounding karst massif is situated under their surface openings. In the wet period of the year when the GWL are high, they functioning as springs (Bonacci 1987).

Figure 5.9 gives a schematic explanation of a ponor swallow capacity, $_{\mathbf{P}}\mathbf{Q}_{\mathbf{O}}$, (in m^3/s) as a function of a water level, **H** in a flooded karst area (mostly karst polje). When the flow in the main karst channel is not under pressure, **(H < H*)**, a ponor's discharge curve has a form indicated in Fig. 5.9 as, ($_PQ_O = f\,(H{-}H_1)$). When the flow in the main karst channel comes under pressure, **(H > H*)**, the discharge curve changes suddenly (point **H***, **Q*** in Fig. 5.9). Then the ponor swallow capacity depends exclusively upon the difference, $\mathbf{\Delta H_3}$, between the water level in the polje, **H**, and the level of the spring exit, H_3. In this case, the equation for the ponor discharge curve is given as:

$$_PQ_O = c \times A \times ((2 \times g \times (H{-}H_3))^{0.5} \qquad (5.1)$$

where

c	is the discharge coefficient (usually ranging between 0.7 and 0.9),
A	is the average cross-sectional area of the main channel in m^2,
g	is the acceleration of gravity in m s^{-2},
$(\Delta H_3 = H{-}H_3)$	the dimension of difference given in metres

If there is a large cave system in the karst massif that is never completely filled with water, that is, if flow under pressure does not exist all the way up to the spring level, then the difference between water levels, $\mathbf{\Delta H_2}$ is less than $\mathbf{\Delta H_3}$, and the discharge coefficient $\mathbf{c_2}$ is different from $\mathbf{c_3}$. In general, $\mathbf{c_3}$ exceeds $\mathbf{c_2}$.

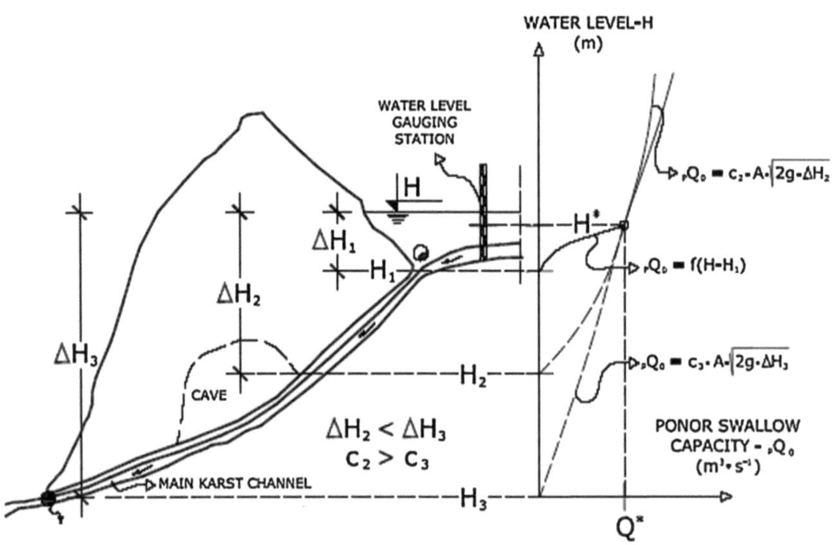

Fig. 5.9 Schematic explanation of a ponor swallow capacity

This explanation is valid for the case where the ponor swallow capacity is not under influence of the GWL in the adjacent karst massif. When the GWL ($\mathbf{H_{GWL}}$) is higher than the water level (**H**), the ponor acts as an estavelle. When the GWL, ($\mathbf{H_{GWL}}$), is lower than water level (**H**), the ponor swallows water and its capacity depends upon the difference ($\Delta \boldsymbol{H} = \boldsymbol{H} - \boldsymbol{H_{GWL}}$) (Bonacci 2013).

5.6 Karst Open Streamflows

Flow regime in open streamflows in karst depends mostly upon the interaction between the groundwater and the surface water (Bonacci 1987). The GWL in karst greatly depends upon the effective porosity of the matrix, while groundwater connections between different parts of karst massive as well as some open streamflows depend on the existence, features and dimensions of karst conduits. Some of these conduits recharge, while some drain off water from the rivers in karst. Generally, recharging or drainage depends on the GWL. The influence of karst differs from one streamflow to another, and therefore general conclusions should be carefully drawn.

Open streamflows in karst very often disappear underground a number of times and emerge again in different karst springs, usually under a different name. Sinking, losing and underground streams are frequent karst phenomena. Their occurrence in karst terrains is more the rule than the exception. Such streamflows are more typical, significant and relatively frequent karst phenomena than is reflected in their treatment in the karst literature (e.g. Hess et al. 1989; Yuan 1991; Bonacci 1999; Potié et al. 2005; Bonacci and Andrić 2008; Prelovšek et al. 2008; Cavalera and Gilli 2009; Bonacci and Andrić 2010; Bonacci et al. 2013). A synonym for a sinking and losing stream is an *influent stream*. Such streams have an integral function in karst hydrology and hydrogeology.

A losing streamflow can be defined as an open streamflow that loses water as it flows downstream. A losing streamflow is a surface stream that contributes water to the karst groundwater system in localised areas. It has cracks in its bed that allow water to seep into the groundwater. These losses can be massive in particular river sections, whereas in others they are small and difficult or even impossible to observe without performing especially precise measurements. Losing streams segments are important groundwater recharge zones for underlying karst aquifers. The water level in a losing stream is higher than the GWL, as opposed to the water level in a gaining stream which is lower than the GWL. Due to very rapid rise and fall of the GWLs in karst terrains, some losing rivers or their losing stretches can intermittently act as gaining streams. Figure 5.10 presents an attempt at the conceptualisation of losing streamflows. Water infiltrated from these sections can either flow in another catchment or can reappear in the downstream reaches of same river (at the spring, **B** in Fig. 5.10b).

Occasionally, permanent water courses flow beyond the GWL, even for 50 m or more. Bonacci (1987, 1999) called these river sections "suspended" or "perched".

A sinking surface streamflow can be defined as a surface river or stream flowing onto or over karst that then disappears completely underground through a swallow hole (ponor or sinkhole) and which may or may not rise again and flow as a resurgent surface river or stream. Infiltration from sinking streams into the karst groundwater

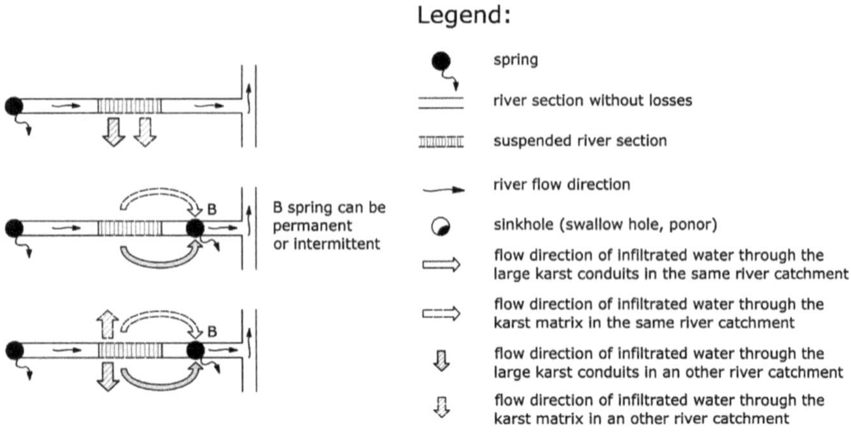

Fig. 5.10 An attempt for the conceptualisation of losing streamflows

system is the most rapid form of recharge for carbonate aquifers (Hess et al. 1989). Sinking streams represent the most direct access to the sensitive and highly vulnerable karst groundwater system. The unique nature of sinking rivers is their development and evolution of conduit flow routes and caves through soluble rocks. The evolution of most of the world's largest and most significant karst caves and springs is formed as a consequence of large volumes of concentrated recharge from sinking rivers (Ray 2005). Figure 5.11 presents an attempt at the conceptualisation of sinking streamflows. Sinking stream can reappear at the surface through a typically large karst spring (Fig. 5.11a) though there are some cases when it reappears through many permanent and intermittent karst springs dissipated over a large area.

Underground or subterranean streamflows are subsurface karst passages that have the main characteristics of open rivers or streams. In an underground streamflow, water flows through caves, caverns, karst conduits and large galleries in the karst underground. The karst underground system provides access to fragments of the abandoned conduit system, which have hydraulic geometries comparable, though not identical, to those of surface rivers or streams.

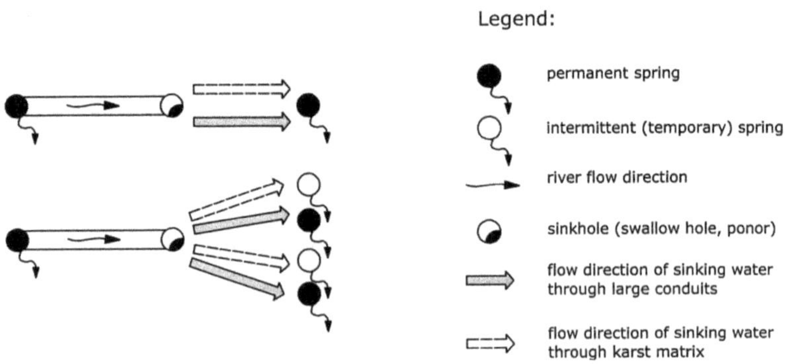

Fig. 5.11 An attempt for the conceptualisation of sinking streamflows

Box. 5.4

Figure 5.12 is the map of the Disu underground stream system in China, which has a catchment area of 1,004 km^2 (Yuan 1991). The system has a total length of 241.1 km and includes a main conduit that is 57.2 km long and 12 tributaries. The Disu underground system is the longest identified subterranean stream in China. In the upstream section, it is about 100 m in depth, with karst conduits usually in a simple fissure-shape, from several metres to 30 m wide, and ten to tens of metres high. The average hydraulic gradient is about 12 ‰. At the middle and lower reaches, it is 30–50 m below the bottom of the valleys. The cross section of the conduit here varies between 145 and 184 m^2, and the average hydraulic gradient is 1 ‰. Discharges at the exit of the Disu underground river vary from the minimum 4.03 $m^3\ s^{-1}$ in dry season to the maximum 544.9 $m^3\ s^{-1}$ (Yuan 1991).

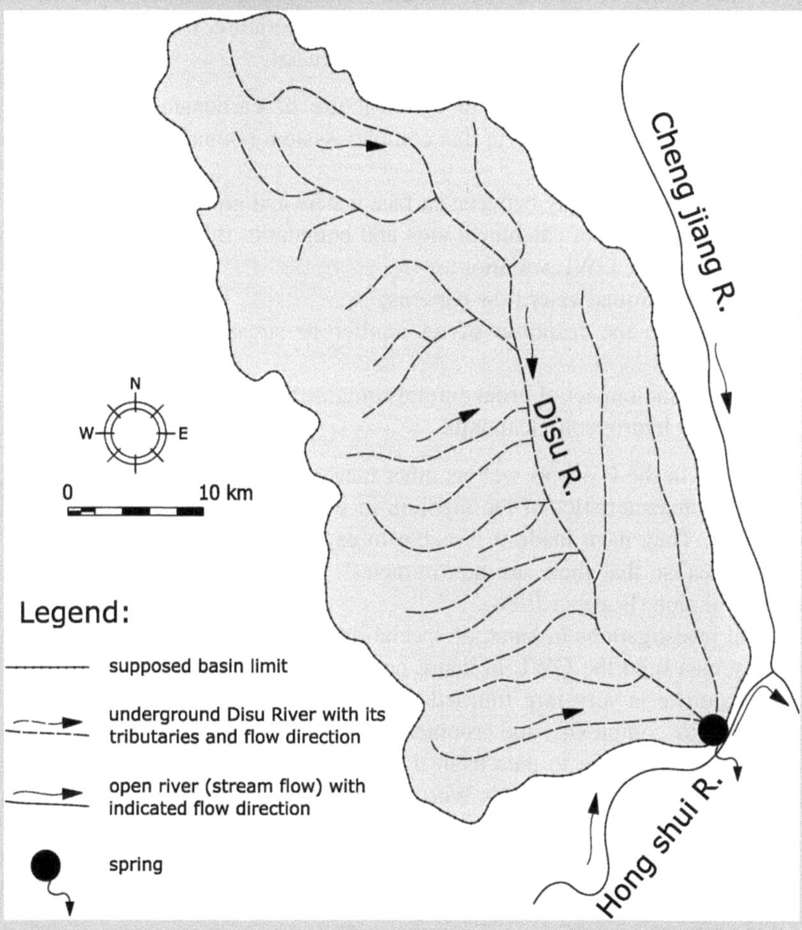

Fig. 5.12 Map of the Disu (China) underground stream system (Yuan 1991)

Through sinking and losing streamflows, contaminants can quickly enter the karst groundwater, with little or no filtration. This can cause the rapid and extensive pollution of a downstream water body and can have a negative impact on the surrounding environment.

5.7 Piezometers as a Crucial Source of Information in Karst

The piezometer boreholes represent an exceptionally important source of information of a wide range, necessary for all types of investigations related to the regime of water circulation in karst. One of the main goals of the GWL measurements in karst massif is determination of the hydraulic head, which underlies the interpretation of groundwater flow, the quantification of aquifer properties and the calibration of flow models (Post and von Asmuth 2013). The measurements of the GWL as well as many other parameters (e.g. water temperature, chemical composition and electrical conductivity) in piezometers are crucial:

(1) for investigating the permeability structure of carbonate aquifers and the behaviour of groundwater in this complex system (Bonacci and Roje-Bonacci 2012);
(2) to understand interplay between surface waters and groundwater;
(3) to define changes of catchment area and boundaries during the time as a consequence of the GWL variation;
(4) to establish groundwater flow patterns;
(5) to determine the response of an aquifer to stresses such as pumping or recharging;
(6) to understand impact of grout curtains on groundwater behaviour;
(7) to identify hydrogeological units.

The changes in the GWL as well as other measured water parameters can clearly explain the characteristics of the medium in which the process of the water flow takes place. They have made it possible to explain numerous "mysteries" in the karst and realise that there are no "miracles" in the karst regions related to the water circulation (Bonacci 1988).

For all investigations in karst, of special importance is to have at disposal continuously measured the GWL in dense network of deep piezometers. In practice, this prerequisite is very rare fulfilled, not only due to high price of project, but mostly due to complexity and troubles in determination of suitable position of piezometers. According to data from the literature (Drogue 1980; Bonacci 1999; Bonacci and Roje-Bonacci 2000; Worthington 2003), only every second piezometer provides the necessary information on the water circulation in karst and aquifer characteristics. At least half of them are drilled in impermeable, compact or less permeable parts of the karst massif in which all dynamic processes of water flow are very slow.

Box 5.5

- The probability of intersecting a major karst conduit during the drilling of one piezometer is small. Worthington (2003) estimated that it is between 0.37 and 7.5 %.
- Yuan (1986) reported the fact that from a wells had been excavated in the Malian valley, Waxian country, Gaungxi province (China), only two yield satisfactory results, giving a discharge of 190 and 33 l s^{-1} under 5 m drawdown. The discharges from the other seven are smaller than 0.5 l s^{-1} with a drawdown of more than 5 m. The distance between the wells was of the order of magnitude of 100 m.
- Studying geometry of the karst aquifer over 1,000 m^2 experimental area in the vicinity of Montpellier, France, Drogue (1980) made continual measurements of the GWL and water temperature at 19 piezometers. This study showed that the GWL in piezometers, which were close to one another (less than 10 m), responded differently to rainfall in the catchment, mainly because of varying connections with the main karst conduits, subsurface karst features and karst springs. The GWL in piezometers connected to small fissures reacts much more slowly than in piezometers connected to main karst conduit and/or spring.

A piezometer represents a window open onto the inner side of the karst massif. The question of whether this window will be really wide open or just slightly open depends upon ability and skill of the researcher more than on the application of modern technology (Bonacci and Roje-Bonacci 2012).

Reliable GWL measurements are fundamental to all hydrogeological investigations. Post and von Asmuth (2013) worn that the measurement of the hydraulic head is not as trivial as simply lowering a measurement tape down to the water level in a borehole. Their paper aims to provide quantitative guidance on the likely sources of error and when these can be expected to become important.

References

Andreo B, Ravbar N, Vias JM (2009) Source vulnerability mapping in carbonate (karst) aquifers by extension of the COP method: application to pilot sites. Hydrogeol J 17:749–758

Atkinson TC (1986) Soluble rock terrains. In: Fookes P, Vaughan PR (eds) Handbook of engineering geomorphology. Chapman and Hall, New York, pp 241–257

Bögli A (1980) Karst hydrology and physical speleology. Springer, Berlin

Bonacci O (1987) Karst hydrology with special references to the Dinaric karst. Springer, Berlin

Bonacci O (1988) Piezometer—the main source of hydrologic information in the karst. Vodoprivreda, Belgrade 20(115):265–278

Bonacci O (1993) Karst springs hydrograph as indicator of karst aquifer. Hydrol Sci J 38(1–2):51–62
Bonacci O (1999) Water circulation in karst and determination of catchment areas: example of the River Zrmanja. Hydrol Sci J 44(3):373–386
Bonacci O (2001a) Analysis of the maximum discharge of karst springs. Hydrogeol J 9:328–338
Bonacci O (2001b) Heterogeneity of hydrologic and hydrogeologic parameters in karst: example from Dinaric karst. IHP-V Techn Doc in Hydrol 49(II):393–399
Bonacci O (2004) Hazards caused by natural and anthropogenic changes of catchment area in karst. Nat Haz Earth Syst Sci 4(5/6):655–661
Bonacci O (2013) Poljes, ponors and their catchments. In: Shroder JH (ed) Treatise on geomorphology 6. Academic Press, San Diego, pp 112–120
Bonacci O, Andrić I (2008) Sinking karst rivers hydrology: case of the Lika and Gacka (Croatia). Acta Carsologica 37(2–3):185–196
Bonacci O, Andrić I (2010) Impact of inter-basin water transfer and reservoir operation on a karst open streamflow hydrological regime: an example from the Dinaric karst (Croatia). Hydrol Proc 24:3852–3863
Bonacci O, Bojanić D (1991) Rhytmic karst springs. Hydrol Sci J 36(1):35–47
Bonacci O, Jelin J (1988) Identification of a karst hydrological system in the Dinaric karst (Yugoslavia). Hydrol Sci J 33(5):483–497
Bonacci O, Roje-Bonacci T (2000) Heterogeneity of hydrological and hydrogeological parameters in karst: examples from Dinaric karst. Hydrol Proc 14:2423–2438
Bonacci O, Roje-Bonacci T (2012) Impact of grout curtain on karst groundwater behaviour: an example from the Dinaric karst. Hydrol Proc 26:2765–2772
Bonacci O, Željković I, Galić A (2013) Karst rivers particularity: an example from Dinaric karst (Croatia/Bosnia and Herzegovina). Envi Earth Sci 70:963–974
Cavalera T, Gilli E (2009) The submarine river of Port Miou (France), a karstic system inherited from the Messinian deep stage. Geoph Res Abs 11, EGU 2009–5591
de Marsily G (1986) Quantitative hydrogeology, groundwater hydrology for engineers. Academic Press, San Diego
Dahlström B (1986) Estimation of areal precipitation. Nordic hydrological programme report, No 18
De Waele J (2008) Interaction between a dam site and karst springs: The case of Supramonte (Central-East Sardinia, Italy). Engin Geol 99:128–137
Drogue C (1980) Essai d'identification d'une type de structure de magasine carbonates, fissures. Mém Hydrogéol Série Soc Géologique de France 11:101–108
Eagleson PS (1970) Dynamic hydrology. McGraw Hill, New York
Field MS (2002) A lexicon of cave and karst terminology with special reference to environmental karst hydrology. USEPA, Washington DC
Ford D, Williams P (2007) Karst hydrogeology and geomorphology. Wiley, Chichester
Herold T, Jordan P, Zwahlen F (2000) The influence of tectonic structures on karst flow patterns in karstified limestones and aquitards in the Jura Mountains, Switzerland. Eclog Geolog Helvet 93:349–362
Hess JW, Wells SG, Quinlan JF, White WB (1989) Hydrogeology of the South-Central Kentucky karst. In: White WB, White EL (eds) Karst hydrology concepts from the Mammoth Cave area. Van Nostrand Reinhold, New York, pp 15–63
Katz BG, DeHan RS, Hirten JJ, Catches JS (1997) Interactions between ground water and surface water in the Suwannee river basin. Florida. J Am Wat Res Assoc 33(6):1237–1254
Lehmann O (1932) Die Hydrographie des Karstes (Karst hydrography). Enzyklopädie Bd. 6b, Leipzig-Wien
Manga M (2001) Using springs to study groundwater flow and active geologic processes. Annual Rev Earth Plan Sci 29:201–228
Milanović P (1981) Karst hydrogeology. Water Resources Publication, Littelton
Milanović P (1986) Influence of the karst spring submergence of the karst aquifer regime. J Hydrol 84:141–156

Perrin J, Parker BL, Cherry JA (2011) Assessing the flow regime in a contaminated fractured and karstic dolostone aquifer supplying municipal water. J Hydrol 400(3–4):396–410
Potié L, Ricour J, Tardieu B (2005) Port-Mioux and Bestouan freshwater submarine springs (Cassis-France) investigations and works (1964-1978). In: Stevanović Z, Milanović P (eds) Water resources and environmental problems in karst, Proceedings of international conference KARST 2005. University of Belgrade, Institute of Hydrogeology, Belgrade, pp 266–274
Popovska C, Bonacci O (2007) Basic data on the hydrology of Lakes Ohrid and Prespa. Hydrol Proc 21:658–664
Post VEA, von Asmuth JR (2013) Review: hydraulic head measurements–new technologies, classic pitfalls. Hydrogeol J 21:737–750
Prelovšek M, Turk J, Gabrovšek F (2008) Hydrodynamic aspect of caves. Internat J Speleol 37(1):11–26
Ravbar N, Goldscheider N (2009) Comparative application of four methods of groundwater vulnerability mapping in a Slovene karst catchment. Hydrogeol J 17:725–733
Ray JA (2005) Sinking streams and losing streams. In: Culver DC, White WB (eds) Encyclopedia of caves. Elsevier, Amsterdam, pp 509–514
Rimmer A, Salingar Y (2006) Modelling precipitation-streamflow processes in karst basin: The case of the Jordan River sources, Israel. J Hydrol 331:524–542
Roje-Bonacci T, Bonacci O (2013) The possible negative consequences of underground dam and reservoir construction and operation in coastal karst areas: an example of the hydro-electric power plant (HEPP) Ombla near Dubrovnik (Croatia). Nat Haz Earth Sys Sci 13:2041–2052
Smart C, Worthington SRH (2004) Springs. In: Gunn J (ed) Encyclopedia of caves and karst science. Fitzroy Dearborn, New York, pp 699–703
UNESCO, WMO (1992) International glossary of hydrology. WMO, Geneve & UNESCO, Paris
White WB (2002) Karst hydrology: recent developments and open questions. Eng Geol 65(2–3):85–105
Worthington SRH (2003) A comprehensive strategy for understanding flow in carbonated aquifer. In: Palmer AN, Palmer MV, Sasowsky ID (eds) Karst modelling. Special Publ 5. the Karst Waters Institute, Charles town, pp 30–37
Yuan D (1986) On the heterogeneity of karst water. IAHS Publ 161:281–292
Yuan D (1991) Karst of China. Geological publishing house, Beijing

Chapter 6
Budget and General Assessment of Karst Groundwater Resources

Zoran Stevanović

There is a wide range of possible interventions in karstic aquifers. The majority of them concerns tapping structures for utilization of groundwater for drinking, heating, or irrigation purposes, but engineering works also include water drainage or dewatering of mines and construction sites. For each of these tasks, knowledge of the groundwater budget and more importantly of the groundwater reserves is required. Development of a conceptual aquifer model and assessment of water budget elements, such as groundwater recharge, flow directions, and discharge, are essential for the successful planning and implementation of engineering work in karst.

6.1 Budget Equation and Parameters

Groundwater budgeting or balancing is a process of definition or estimation of different input (recharge) and output (discharge) components. The process is not very different from the economic *input–output* law (Leontief 1986; Miller and Blair 2009). The term *balance* is from the Latin adjective *bilanx,* indicating a weighing machine with two equalizing pans.

The general budget equation can be simply presented as follows:

$$\text{Input} = \text{Output} \pm \text{Storage} \tag{6.1}$$

Z. Stevanović (✉)
Centre for Karst Hydrogeology, Department of Hydrogeology,
Faculty of Mining and Geology, University of Belgrade, Belgrade, Serbia
e-mail: zstev_2000@yahoo.co.uk

Z. Stevanović (ed.), *Karst Aquifers – Characterization and Engineering*,
Professional Practice in Earth Sciences, DOI 10.1007/978-3-319-12850-4_6

To ensure as much as possible proper budgeting in the economy, Ranković (1979) proposed a few principles:

1. Clearly defined period of budgeting;
2. Non-changeable methods of calculating input/output parameters;
3. Cautions and the confidence to eliminate subjective factors that may bias the results.

These principles can also be applied in hydrogeology budgeting.

The budget equation in case of a watershed or any aquifer system is as follows:

$$\text{Recharge} = \text{Discharge} \pm \text{Storage (Reserves)} \tag{6.2}$$

Chosen budgeting periods can vary from very short term, i.e., monthly, seasonal, or annual, to medium (multi-annual) and long term (historical). Since variation of input and output components and variation in stored water reserves is quite normal during shorter observation periods, budget equilibrium is better established over a longer period as more realistic results would be obtained given the absence of impact of external factors such as climate change or human activities.

Many authors have described groundwater budget and its components (Brown et al. 1969; Yevjevich 1981; Boni and Bono 1984; Boni et al. 1984; Stevanović 1984; Bonacci 1987; Scanlon et al. 2002; Healy et al. 2007; Kresic 2007). The typical scheme of an open (unconfined) karstic aquifer and main budget components is shown in Fig. 6.1.

For such cases, the general budget equation can be presented in the following mode:

$$P + I_s + I_g = R_f + E_t + E_g + Q_s + Q_{sb} + Q_a \pm R \pm E \tag{6.3}$$

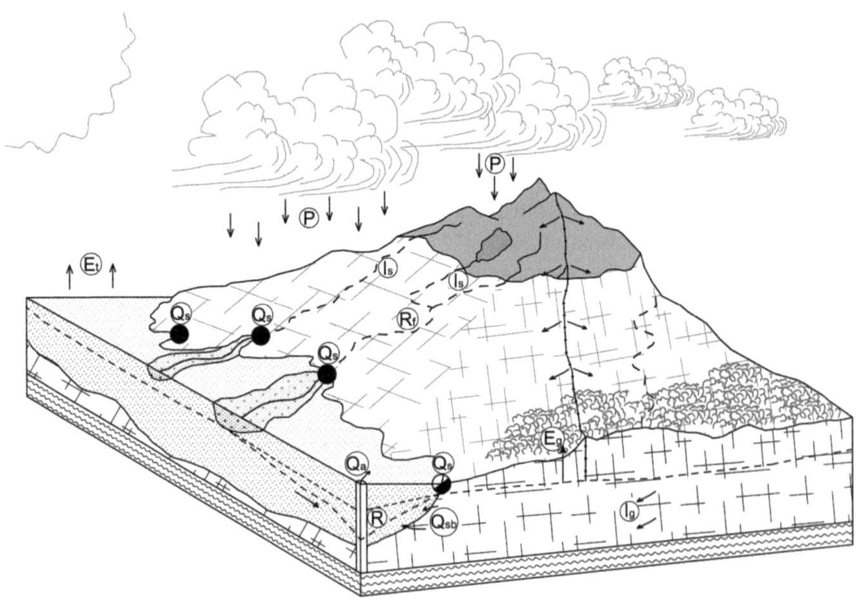

Fig. 6.1 Groundwater budget components in a karst aquifer in open and semi-confined structures

where

P precipitation,
I_s inflow (surficial),
I_g inflow (via subsurface, as groundwater inflow but may also include hypogenic flow component),
R_f runoff,
E_t evapotranspiration,
E_g evaporation from water table,
Q_s spring discharge,
Q_{sb} subsurface drainage (groundwater outflow),
Q_a artificial withdrawal (extraction),
R groundwater reserves variation,
E error.

In addition to the above, other input or output components could also influence water budget. For instance, leakage from reservoirs or artificial recharge could be important recharge constituents. Some components could be easily measured and quantified, while others are very difficult, even impossible, to define properly. Although not directly involved in budget equation, various climatic factors significantly alter recharge/discharge parameters. For instance, air temperature, humidity, winds, sun radiation, and latitude all directly influence the evapotranspiration rate.

P—precipitation. This recharge parameter is the easiest to determine. It includes rainfall, snow, and condensate water. Although measurements can be organized without difficulty in flat areas, some remote mountain karstic terrains are very difficult to access and launch monitoring stations from (Fig. 6.2). The difference in rainfall distribution in high and lowlands should always be taken into consideration. The use of isohyets maps (Fig. 6.3) or of the Thiessen polygons method could bridge data gaps.

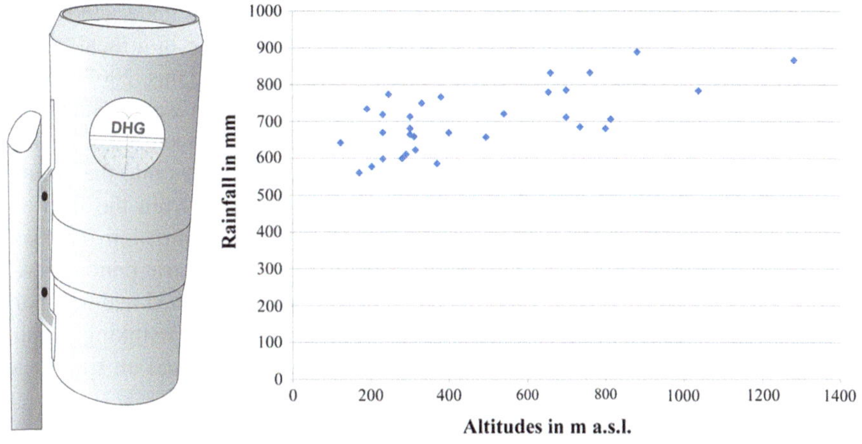

Fig. 6.2 Simple rain gauge (*left*) and an example of relation between minimal annual rainfall total (in mm) and altitudes (in m a.s.l)—data from 60 rainfall gauges for 30-year period in Carpathian karst of eastern Serbia

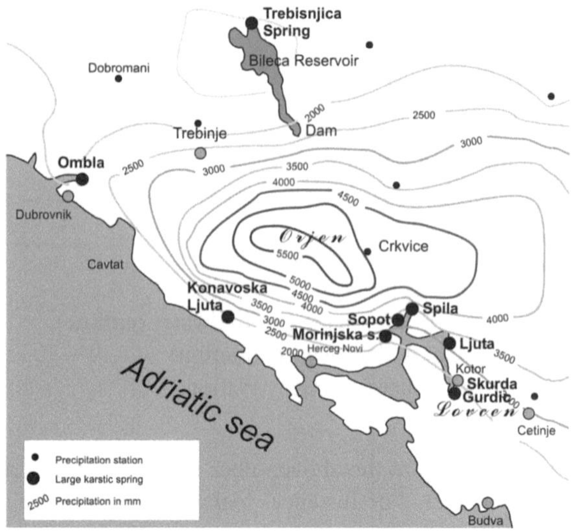

Fig. 6.3 Isohyetal map of Boka Kotorska area in Montenegro where European maximum rainfall totals are regularly recorded (*courtesy* of Petar Milanović)

As explained in Chaps. 3–5, karst waters regime is very dynamic and continual data collection is highly recommended. Effective rainfall represents the maximum volume of water available for recharge, meaning a loss of water in evapotranspiration or other processes such as interception should be deducted from total precipitation. Interception is the amount of water that does not reach the soil and mostly remains on the leaves. Some experiences show that in dense forests, a daily total of up to 3 mm of non-intensive precipitation usually does not reach their floor. In hydrology practice, there are many methods for assessment of interception as well as conversion of melted snow in effective rainfall.

I_s—*Inflow* (*surficial*). This budget component refers to the river flow of perennial streams or diffuse runoff water which enters the watershed under study. This water originates at a distance and has the potential to infiltrate in a local aquifer system, a process called *allogenic recharge* (Ford and Williams 2007; Kresic 2013). In contrast, *autogenic (autochthonous) recharge* is from water which originates exclusively inside the watershed boundaries (see also Sect. 3.4). Surficial inflow includes the infiltration of different kinds of surface waters into the ground. The surface water could percolate from sinking streams, and from the lake and reservoirs, and could also penetrate into aquifer as a sea water intrusion. The easiest way to determine inflow is to measure discharges into concentric ponors (swallow holes). Ponors often follow the direct contact between karstified and non-karst rocks. If visible and accessible ponors do not exist, the seepage of sinking streams through the riverbed can be calculated as a loss of river flow based on simultaneous measurements conducted on successive sections (Fig. 6.4).

I_g—*Inflow* (*via subsurface, groundwater inflow*). This component represents groundwater flux from neighboring or adjacent aquifer systems which recharge the studied aquifer. The connection zone could be lateral (Fig. 6.1) or vertical when ascending or gravity flow between aquifers is established.

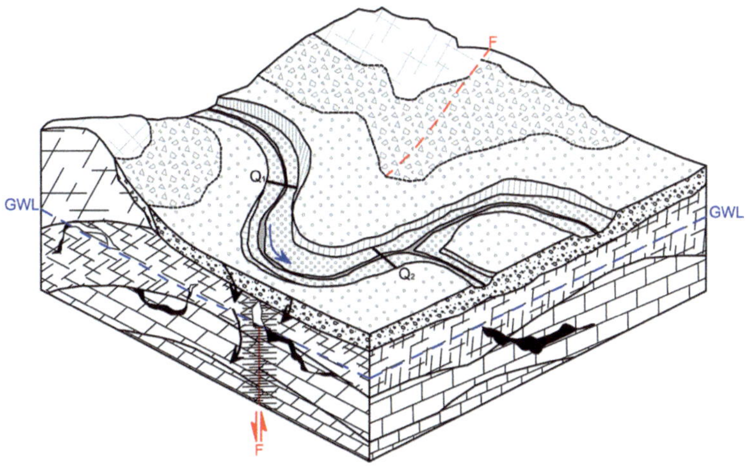

Fig. 6.4 The two sections in a river where flow has been simultaneously measured. $I_s = Q_1 - Q_2$ (subsurface water loss is equal to the difference between registered river flows if no other drainage or seepage between river sections exists)

R_f—*Runoff*. This is the water which freely flows through streams and rivers or leaves the watershed through the vein of diffuse dispersed seepage. Although river flow can be easily recorded at the control sections on exit watershed boundaries, seepage flow is much more problematic to estimate. However, surface runoff is among the most convenient of budget parameters for determination. The methods of measuring river flows could be "classical", such as in situ hydrometry by current meters (Shaw 1994), or sophisticated, as by Doppler radar for riverbed and water velocity scanning (Fig. 6.5). Installation of gauges or weirs at the streams facilitates data collection. The rating curve $Q = f(H)$ could be drawn by standard

Fig. 6.5 Classical hydrometry with current meter (*left*) and Doppler radar for depth/velocity scanning (*right*)

Fig. 6.6 Gauging station and data transmitter on wadi (temporary stream) bank near Hargeysa in northern Somalia

readings of stage heights or from data loggers (barometric "diver", see Chap. 12) versus directly measured river flows.[1] Salt dilution and injection (Cobb and Bailey 1965) can also be used for discharge estimation, but sophisticated techniques of satellite observations and use of data transmission systems (Fig. 6.6) have become more widespread in modern hydrology practice.

E_t—*Evapotranspiration.* This budget component includes evaporation (vaporization) from surface water tables (river, lake, and sea) and transpiration from vegetation cover and soil-covered surfaces (epikarst). Although the former value is relatively easily recorded by different types of pans (Fig. 6.7) and necessary adaptations,[2] there is the problem of how to estimate the transpiration component. Specific devices such as lysimeters are also used for measuring various physical parameters or the soil water balance. As the lysimeter has to be installed into unconsolidated soil or terrains which enable excavation, its application in karst is very limited.

Although many field tests provide estimates of real transpiration rates for different plants and grass (Davis and De Wiest 1967), in hydrology the use of empirical formulae to compute evapotranspiration rates still prevails.

[1] Techniques are well described in many basic hydrology textbooks.

[2] Pan-recorded values should be corrected (factor ~0.7) to obtain real E_t which characterize open water reservoirs or lakes.

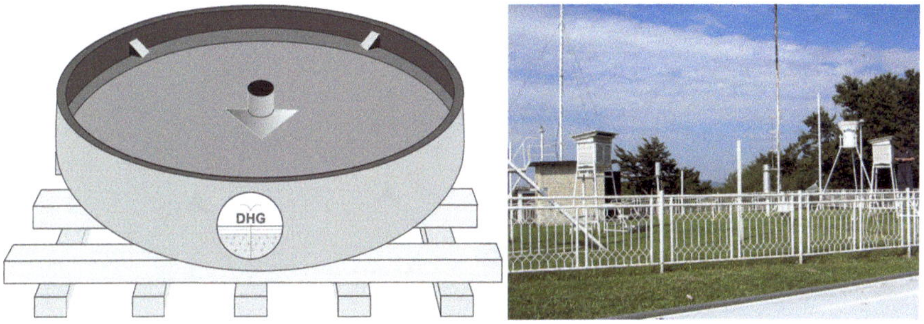

Fig. 6.7 Pan evaporimeter (*left*) and typical meteorological station (*right*)

The majority of empirical formulae incorporate the *potential evapotranspiration* as a theoretical value that may occur in the study area (Lerner et al. 1990). This parameter is based on ambient properties and does not fully consider the actual precipitation rate or other water consumers (plants, soil moisture deficit). Therefore, in arid regions during summer months, *potential evapotranspiration* is much higher than *actual evapotranspiration*. Only during the winter and wet season do these two parameters correspond.

In water budgeting practice, there are plenty of empirical solutions for evapotranspiration assessment. The well-known formulae of Thornthwaite (1948), Penman (1948) and Turc (1954) include elements such as the geographical position of the site (latitude), air temperature, humidity, winds, solar radiation, and soil moisture deficit. Several Russian authors (Kuzin, Poljakov, Semenov, explained in: Luchshewa 1976) also developed empirical equations by including the above elements. Penman (1948) combined the energy balance with the mass transfer method and derived an equation to compute the evaporation from standard climatological data of temperature, humidity, sunshine, and wind speed. During the last two decades, the use of a modified Penman–Monteith method recommended by FAO (Allen et al. 1998) became central in the calculation of evapotranspiration because standard climatic data can be easily measured or derived from commonly measured data. The Penman–Monteith form of the combination equation is (Allen et al. 1998) as follows:

$$\lambda_{\mathrm{ET}} = \frac{\Delta(R_{\mathrm{n}} - G) + \rho_{\mathrm{a}} c_{\mathrm{p}} \frac{(e_{\mathrm{s}} - e_{\mathrm{a}})}{r_{\mathrm{a}}}}{\Delta + \gamma\left(1 + \frac{r_{\mathrm{s}}}{r_{\mathrm{a}}}\right)} \tag{6.4}$$

where

R_{n}	net radiation,
G	soil heat flux,
$(e_{\mathrm{s}} - e_{\mathrm{a}})$	represents the vapor pressure deficit of the air,
ρ_{a}	mean air density at constant pressure,
c_{p}	specific heat of the air,
Δ	slope of the saturation vapor pressure–temperature relationship,
γ	psychrometric constant,

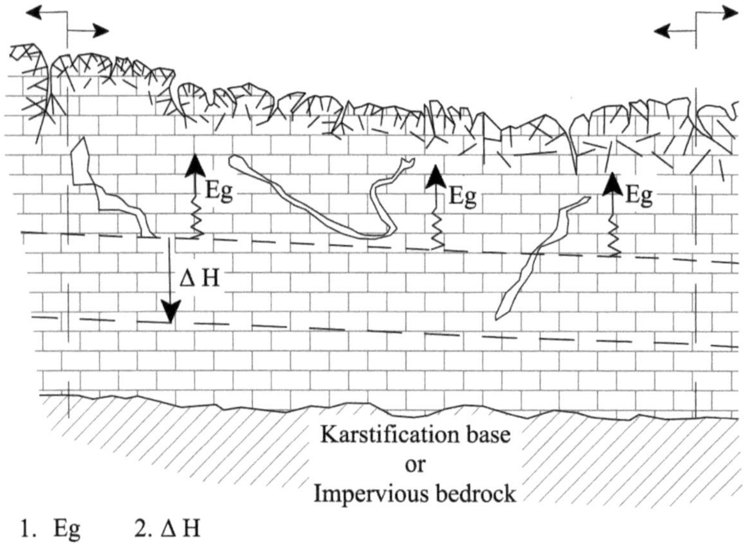

Fig. 6.8 *1* Evaporation from groundwater table (Eg) and *2* Lowering of water table/position of saturated zone (ΔH)

r_s surface resistance, and
r_a aerodynamic resistances.

The Penman–Monteith equation includes all parameters that govern energy exchange and corresponding evapotranspiration (latent heat flux). The equation can be utilized for the direct calculation of any crop evapotranspiration as the surface and aerodynamic resistances are crop specific (Allen et al. 1998).

E_g—*Evaporation from water table*. There is neither a device constructed nor an empirical equation developed to measure this budget parameter. It can be assessed only from groundwater table depletion if no discharge exists (including water extraction) and no new recharge is taking place. Such a situation is possible in the case of karstic aquifer formed in platform structures and in arid regions (Fig. 6.8). Hence, a more realistic value of E_g would be obtained in bare karst than in karst where epikarst is widely present: if there is thick soil or vegetation covering a vadose zone, then residual water may influence groundwater table fluctuations. The equation is very simple:

$$E_g = \Delta\mathrm{H} \times P_e \qquad (6.5)$$

where

ΔH depleted water table during budgeting (observation) period (in mm) (Δ is with negative sign)
P_e effective porosity (storativity) of karst aquifer.[3]

[3] The volume of interconnected voids against the total volume of rock, i.e., the interconnected fissure and void system able to absorb and to transfer water within the aquifer.

Fig. 6.9 Spring discharge observation. The three ways to record discharge: limnigraph, weir, and gauge installed at a spring

The amount of evaporated water during an observation period can be calculated by simply multiplying E_g and surface area.

Q_s—*Spring discharge* is the essential parameter of groundwater budget. It represents the ability of an aquifer to release groundwater and could be generally equalized with dynamic (replenishing) groundwater reserves. In principle, all existing springs which drain studied karst aquifer should be covered by a monitoring network, even though it is difficult to observe small and temporary springs continuously. To enable a rough estimation of total aquifer drainage (Fig. 6.9), these smaller discharge points should be measured at least in extremes, during low- and high-water periods.

Q_{sb}—*Subsurface drainage (groundwater outflow)*. This "invisible" water budget parameter is almost impossible to measure and in many cases even to estimate. In the absence of all other options, at least a very rough estimation could be made if all other budget parameters are known. If groundwater extraction from adjacent well permeable (porous) aquifer is carried out, the total discharge of tapping structures (pumping wells) and the resulting drawdown could provide an idea of the intensity and rate of subsurface flux (Fig. 6.10).

If there is subsurface drainage along the riverbeds, the concept of simultaneous measurements on successive sections similar to that explained for I_s could be applied but with an inverse effect (Fig. 6.11).

Separation of hydrographs is another commonly practiced method for assessing base flow. Along with numerous references in this field there is also software for the automatic distinction of runoff waters from waters resulting from aquifer drainage.

In the separation of aquifer drainage components, one component shows spring discharge Q_s and another shows subsurface flux Q_{sb}. The later is more problematic. One example of approximate separation is the increased base flow in a recession period under stagnant discharge of the springs and the absence of new rainy/runoff water (Fig. 6.12). These illogical peaks might be a result of delayed invisible subsurface drainage.

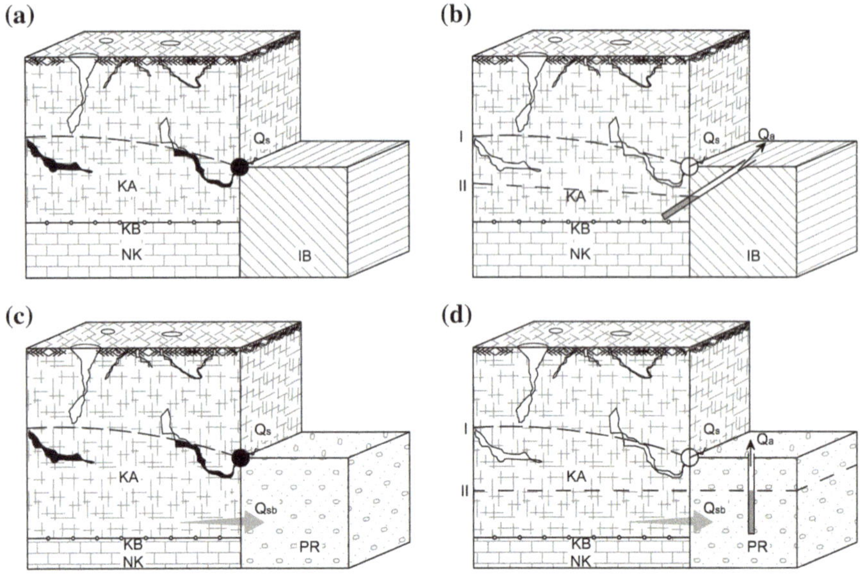

Fig. 6.10 *Subsurface flux.* Q_s spring discharge; Q_{sb} subsurface drainage; Q_a artificial drainage (extraction): *KA* karst aquifer; *KB* karstification base; *NK* non-karstified limestones; *IB* impermeable barrier (aquitard or aquifuge); *PR* permeable (porous) aquifer. *I* Dynamic water table; *II* Static water table. **a** No subsurface drainage due to presence of impermeable barrier. **b** No subsurface flux, but artificial drainage is activated, springs could dry out. **c** Active both spring and subsurface drainages. **d** Active all three drainage modes—spring, subsurface, and artificial

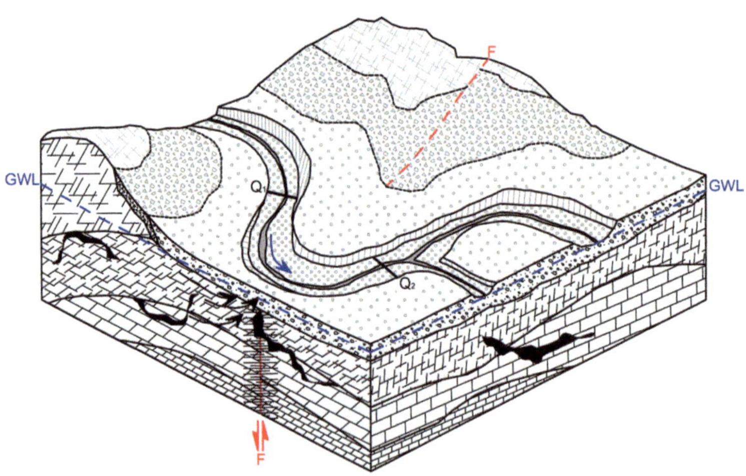

Fig. 6.11 The two sections along a river where flow has been simultaneously measured. $Q_{sb} = Q_2 - Q_1$ (Subsurface inflow is equal to the difference between registered discharges if no other drainage or seepage between river sections exists)

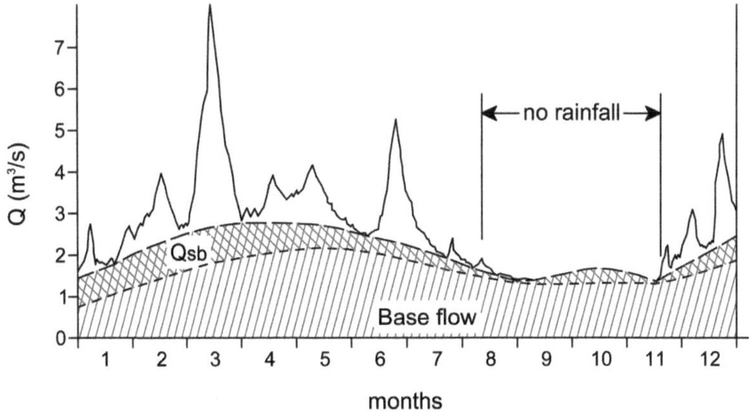

Fig. 6.12 Base flow separation for a typical karstic spring hydrograph. A small peak during recession might be caused by subsurface flux (Q_{sb})

Q_a—*Artificial withdrawal* (*extraction*). This is the disregarded parameter if tapping structures or water diversion is absent from the watershed, as is the case in remote mountains and inhabitant areas. However, in the case that water inside the watershed is widely used, but all tapped water is returning to the recipients, Q_a could be ignored as well. The rate of return flow is relatively high in urban areas: According to some experiences, it can be over 90 % (regularly over 70 %). In contrast, the return flow from irrigation water can be very small: infiltration into the ground depends on soil cover, underlying aquifer permeability, and degree of saturation (moisture), but rarely exceeds 30 %. Therefore, the urban sector generates by far the largest portion of return flows.

If proper monitoring of utilized and returned water flows is established, the Q_a parameter can be easily estimated.

I_{ef}—*Effective infiltration* is equal to the volume of water which reaches the water table and represents the replenishment potential of an aquifer. This is the sum of the three aquifer drainage components discussed above:

$$I_{ef} = Q_s + Q_{sb} + Q_a \tag{6.6}$$

Castany (1967) established the relation:

$$I_{ef} = Q_{s+sb(av)} \times T/F \times 1000 \tag{6.7}$$

where

I_{ef}	total annual infiltration in mm,
$Q_{s\,+\,sb\,(av)}$	average groundwater discharge of aquifer (in m^3/s),
T	time (year in s, 31.536×10^6),
F	aquifer's surface area (in m^2).

R—Groundwater reserves variation. Groundwater budgeting leads to water reserves assessment. Formula (6.3) could be adapted to have a solution for $\pm R$:

$$R = \left(P + I_s + I_g\right) - \left(R_f + E_t + E_g + Q_s + Q_{sb} + Q_a\right) \pm E \quad (6.8)$$

Another method to estimate R is similar to that presented in the case of E_g (6.5) but with the reverse effect: storage change is equal to groundwater table rise.

$$R = \Delta H \times P_e \quad (6.9)$$

where

ΔH augmented water table during budgeting (observation) period (in mm) (Δ is with positive sign)
P_e effective porosity (storativity) of karst aquifer.

To calculate the total volume of newly stored water within the aquifer (*V*), the storage change (*R*) should be multiplied by the aquifer autogenic surface area (*F*). In the case that the dynamic component has to be identified conversion to yield (*Q*) is very simple: the total volume of new water should be divided by the period of time (*T*) in which a new yield has to be discharged (or utilized).

$$V = R \times F (m^3) \quad (6.10)$$

$$Q = V/T \, (m^3/s) \quad (6.11)$$

Nevertheless, the obtained value of $\pm R$ implies only the variation of reserves, i.e., positive or negative change in storage throughout the budgeting period, but not the amount of total reserves. The method for their definition and assessment is discussed in the following text.

E—Error. The miscalculation or incorrect estimation of some parameters is directly reflected in the *E* component. A good method will have few errors associated with it, but due to difficulties in the accurate determination of many budget elements, an error can be significant. Moreover, it would not be easy to distinguish which element is responsible for the mistake made.

Lerner et al. (1990) classified errors as follows:

- incorrect conceptual model,
- neglected spatial and temporal variability,
- measurement error,
- calculation error.

The same authors stated that errors can be high, with the errors in all the other fluxes accumulating in the recharge estimate. For example, high river flows can often be estimated only to ±25 %. If recharge is 25 % of flow, the error in estimating it is ±100 %.

Units. Budget calculation is commonly undertaken in mm of water height or in volume of water expressed in millions of m^3 for a certain budget period.

Budgeting period. As already mentioned, this can be historical, annual, seasonal, or even monthly. The shorter the period of budgeting, the greater the necessity of attention to separate waters accumulated in previous episodes. This "delay effect" is due to a sometimes very slow infiltration process and the long travel time needed by the water to percolate to the aquifer system. Therefore, it is possible that the reaction of observed springs on slow recharge could take longer than even the entire budgeted period.

Inverse calculation of certain parameters. As explained in the case of variation in groundwater reserves (change in storage), all budget parameters can be calculated from a basic budget Eq. (6.1) if all other parameters are known. However, it is clear that many uncertainties may preclude such an approach.

Level of certainty in budget analysis. Given the regularly different levels in the quality of collected data and the many approximations required to assess groundwater reserves, the establishment of a system and criteria to grade certainty of data is recommended. The grades would depend mainly on observation of main water budget parameters, or at least observation of those that are accessible and measurable, such as spring discharge (Q_s), groundwater extraction (Q_a), precipitation (P), river flow regime (R_f), and climatic elements used for assessment of evapotranspiration (E_t). The following four categories are advised:

A Certain, based on permanent observation for at least 10 years by national water agencies or organizations certified for hydrology/hydrogeology survey.
B Relatively certain, based on permanent observation for at least 2 years and further stochastic modeling (time series).
C Uncertain, estimation based on temporary measurements and observations/or modeling.
D Completely uncertain, rough assessment based on analogy, rare visits, and measurements.

6.2 Classification of Groundwater Reserves

Aquifer geometry. The very first step in every water budget analysis should include a conceptual hydrogeological model and delineation of aquifer boundaries, both surficial and in the subsurface (see also Chap. 3). While definition of the allogenic part of the catchment (non-karstic) is not so problematic because it corresponds to the topographic watershed, dividing lines or limits of a karst aquifer are not easy to estimate even at the land surface. Along with field and cave mapping, the tracing tests are essential for definition of an aquifer's geometry, but in many cases, the results of tracings conducted in different periods of years indicate that aquifer boundaries are changeable (Fig. 6.13).

Moreover, the water table can vary considerably throughout a hydrological year, and as a result, a maximum table during the wet season and a minimum one during dry periods could orient groundwater flux in different directions. Usually, during maximal floods, the aquifer surface area becomes smaller due to an overflow toward the adjacent basin which is inactive during the dry season (Fig. 6.14). Therefore, in the case of such variable aquifer geometry, water budget parameters could be defined separately, for high- and for low-water periods. When estimating lateral and vertical underground boundaries, knowledge of geological structures and lithology and of the position of the karstification base is essential. Below the

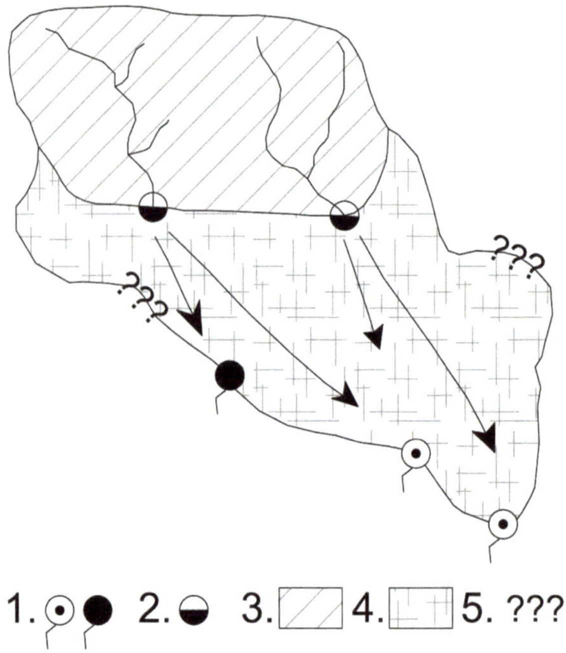

Fig. 6.13 Allogenic recharge, groundwater flow directions based on tracing tests and still questionable aquifer lateral boundaries. *1* springs, *2* ponor, *3* non-karst, *4* karst, and *5* problematic lateral aquifer boundaries

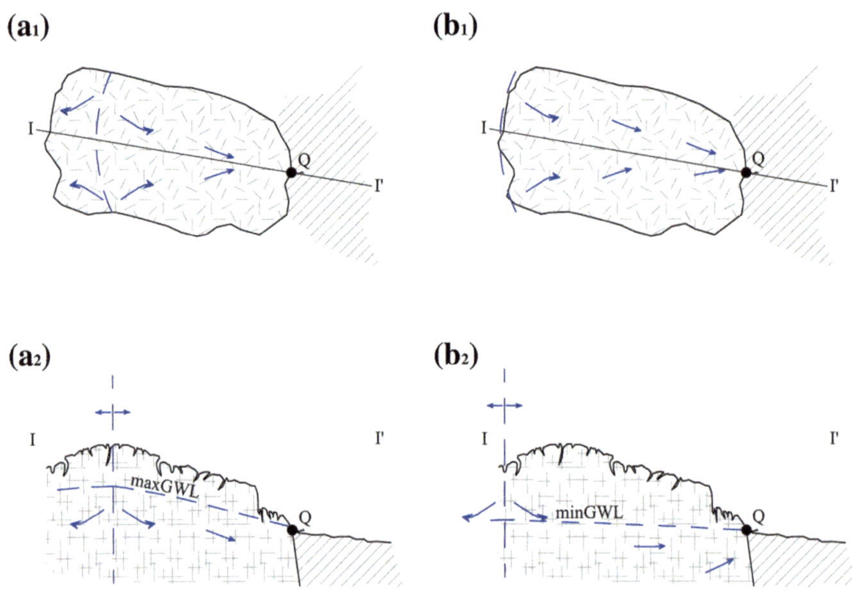

Fig. 6.14 Variable surface area of karstic aquifer. $\mathbf{a}_{1,2}$ watershed during floods and maximal groundwater level (max GWL); $\mathbf{b}_{1,2}$ watershed during recession and minimal groundwater level (min GWL)

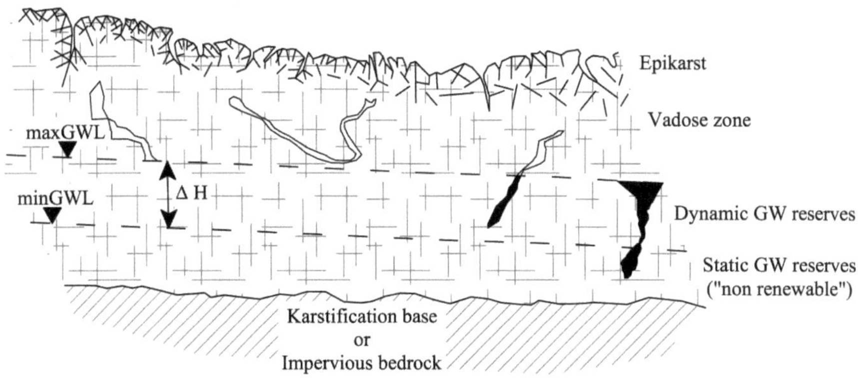

Fig. 6.15 Schematization of natural groundwater reserves in karst. *GWL* groundwater level, *ΔH* variation of water level

karstification base, the effective porosity is very small (Milanović 1981; Ford and Williams 2007) and the deeper part of the aquifer can be ignored in further analysis of groundwater reserves.

In general, there are two main types of natural groundwater reserves (Fig. 6.15):

1. Dynamic reserves
2. Static reserves

Natural means that accumulated reserves are the result of natural recharge. *Dynamic reserves* are accumulated between the maximal and minimal water table in a hydrology cycle. They could vary greatly throughout the year. Castany (1967) called these reserves *regulative*.

Static reserves are accumulated beneath a minimal groundwater table. These reserves are often called "non-renewable" or geological (Castany 1967), but this is not always the case. While it is true that the stagnant part of aquifer represents storage which would not naturally discharge out of aquifer, if drawdown resulted from forced water extraction continuing below the minimal water level, the forthcoming flood could fully cover water loss. Therefore, if available water for replenishment (rainfall, sinking surface waters) is sufficient, we would have renewable reserves even beneath the minimal water table. In such cases, even the term "static" becomes relative. And in some specific cases, such as that on Fig. 6.16, static reserves are totally absent.

Artificial reserves are the result of direct artificial recharge. Apart from artificial, there are also *supplementary reserves* which do not result from directly induced recharge. Rather, they are the result of some interventions in the watershed which change natural properties and boundaries and cause augmented recharge: for instance, building dams or canals (water table increases due to seepage from a reservoir or the canal bottom), irrigation (return flow), cone of depression.[4]

[4] Expansion of cone of depression causes creation of new free spaces, both lateral and vertical for infiltration of new water.

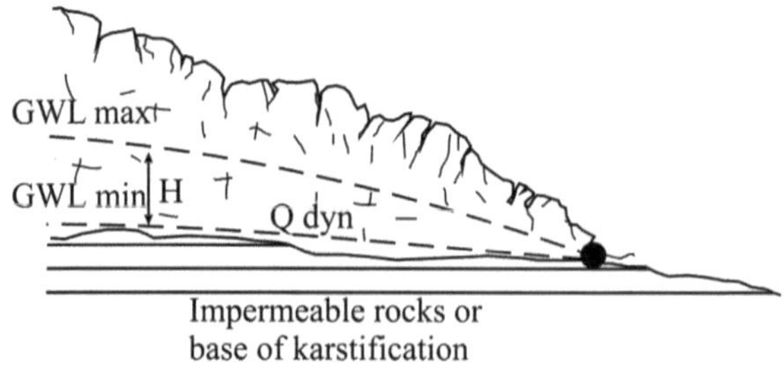

Fig. 6.16 Karst aquifer with shallow impermeable basement. Only dynamic groundwater reserves (Q_{dyn}) are available (after Stevanović 2010a, Reprinted from Groundwater hydrology of springs, Kresic and Stevanović (eds.) (2010), with permission of Elsevier)

Finally, there are *exploitable water reserves* or available water reserves. Here, we have again to distinguish the two types: (1) totally available reserves and (2) ecologically available reserves. The former include groundwater which can be extracted from aquifer without negative consequences for the aquifer itself (no enormous depletion or even drying). These reserves could be equalizing to *safe yield* the concept which is discussed in the Sect. 15.5. Ecologically available reserves are totally available minus the water needed for a water-dependent ecosystem. An approach to sustainable development of water resources is included in some important regulation acts such as the Water Framework Directive of European Union (WFD 2000/60).

6.3 Assessment of Groundwater Reserves

Many methods have been developed in hydrogeology practice to assess water reserves of karstic aquifers directly or indirectly (Goldscheider and Drew 2007). Three of them are most frequently used: (1) Spring's hydrograph (Maillet 1905; Yevjevich 1956; Drogue 1972; Milanović 1979; Mangin 1984; Kullman 1984; Bonacci 1987, 1993; Mijatović 1990; Kresic 2007; Kresic and Bonacci 2010); (2) time series analysis (Atkinson 1977; Panagopoulos and Lambrakis 2006; Kresic 2007); and (3) hydrograph separation (Rorabaugh 1964; Padilla et al. 1994; Rutledge 1998).

Other simpler methods could also be useful for rough assessment of groundwater reserves. Two of them, one for static and one for dynamic reserves, are presented here:

1. Equation for static water reserves is as follows:

$$\Sigma V = F \times H \times P_{\mathrm{e}} \quad (\mathrm{in\,m}^3) \tag{6.12}$$

where

ΣV volume of accumulated water below minimal water table (static reserves),
F aquifer surface area (in m^2),

H aquifer thickness (depth from minimal water table to karstification base or impermeable bedrock (in m),
P_e effective porosity (storativity)

2. Equation for dynamic water reserves is as follows:

Based on Darcy law, the dynamic reserves could be simulated as flow in karst aquifer passing through defined cross-sectional area:

$$Q_{\text{dyn}} = V_c \times F \times P_e \quad (\text{in m}^3/\text{s}) \tag{6.13}$$

where

Q_{dyn} dynamic groundwater reserves expressed as aquifer's flow rate,
V_c average groundwater velocity (in m/s),
F surface of cross-sectional area (in m^2),
P_e effective porosity (storativity).

6.4 Application of Groundwater Budget on Reserves Estimate

Many practical calculations of water budget and consequently of groundwater reserves of regional karst aquifers were made (Boni et al. 1984; Kullman 1984; Paloc 1992; Stevanović 1995).

In order to facilitate assessment of water reserves of regional aquifers (with larger extension) and to obtain a general picture of water reserves and availability, two tables have been created and a simplified procedure established. The recommended budgeting period is at least annual although even seasonal and monthly calculations are possible.

Table 6.1 represents the application of the formula (6.3). The first task is to enter mean monthly and annual values of the parameters. This data, which includes precipitation (P), spring discharge (Q_s), surficial inflow (I_s), or runoff (R_f or river flows generated inside watershed), can be easily collected.

There are two options for estimating the I_s parameter. If the watershed surface area (F) has autogenic and allogenic parts and total water volume is expressed as $(P) \times (F_{\text{total}})$, then I_s is equal only to river flow(s) at the entrance section(s) to the studied watershed as a whole. But, if the surface area (F) includes only the autogenic part and the total water volume is expressed as $(P) \times (F_{\text{autogenic}})$, then in addition to river flow(s) at the entrance section(s), discharges of small streams at border points between karst (autogenic) and non-karst (allogenic) should be taken into consideration and accounted. The first option is advisable.

The next task involves assessment of the budget parameters, which is complicated to determine. In order to get some preliminary examination, we approximate total "losses" (L) as the difference between the obtained input ($P + I_s$) and

Table 6.1 Groundwater budget assessment

Budget elements															
Months	I	II	III	IV	V	VI	VII	VIII	IX	X	XI	XII	Total (mm)		Total water volume in 10^6 m^3 (P × F)
P (mm)															
Months	I	II	III	IV	V	VI	VII	VIII	IX	X	XI	XII	Mean annual (m^3/s)	Minimum annual (m^3/s)	Total in 10^6 m^3 (from mean values)
I_s (m^3/s)															
ΣQ_s (m^3/s)															
Spring 1															
Spring 2															
ΣR_f (m^3/s)															
River 1															
River 2															
L															
	I	II	III	IV	V	VI	VII	VIII	IX	X	XI	XII	Total (mm)		
E_t															
Q_{sb} (L − E)															

Explanation [*see also* Fig. 6.1 *and formula* (6.1)]
F (total surface area of the watershed) = ... km^2
P precipitation (defined by isohyets or by data from representative climatic stations)
I_s surface water inflow into studied watershed
Q_s total average springflow of the springs
R_f surface runoff
L total losses: $E_t + Q_{sb} = (P + I_s) - (Q_s + R_f)$
E_t evapotranspiration (calculated by empirical Thornthwaite or other formulae/real values not potential)
Q_{sb} subsurface drainage ("invisible" groundwater drainage below surface, from deeper aquifer parts)

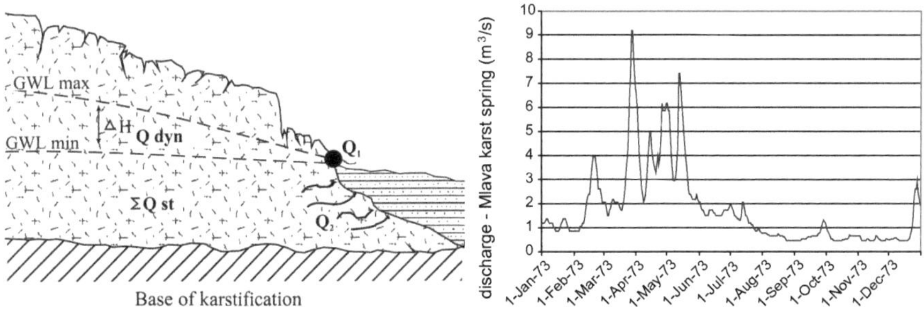

Fig. 6.17 Conceptual scheme of studied aquifer (if Q_{sb} exists) (*left, after Stevanović* 2010a, *Reprinted from Groundwater hydrology of springs, Kresic and Stevanović (eds.)* (2010), *with permission of Elsevier*) and hydrograph of main spring for characteristic year (*right*)

output ($Q_s + R_f$) values. The total losses consist of the two main elements E_t and Q_{sb}, whereas the most problematic parameters related to underground processes I_g (influx) and E_g (ex-flux) could be ignored in such rough assessments.

Evapotranspiration (E_t) could be calculated by empirical Penman–Monteith or Thornthwaite or other formulae but always with careful deduction of real values from potential ones.

Finally, the difference $L - E_t$ provides remains a parameter of subsurface drainage (Q_{sb}) as "invisible" groundwater drainage. As a large number of our engineering interventions depends on proper assessment of this parameter (preventing influx into mines or seepage from reservoirs or utilizing deeper aquifer storage), the last task is to verify the logic of the data obtained for Q_{sb} (Fig. 6.17). If a wrong number or significant error appears, an attempt to verify and calibrate other input/output budget parameters is required.

Table 6.2 is devoted to assessment of the groundwater reserves based on information abstracted from Table 6.1. The dynamic groundwater reserves (Q_{dyn}) correspond to the summary average annual discharge from all springs in the study area. The total static ("non-renewable") groundwater reserves (Q_{st}) should result from the aquifer's volume (surface area times saturated aquifer's thickness below minimal groundwater table) and effective porosity (storativity) of the deeper part of the aquifer ($F \times H_{av} \times P_e$). This deeper water storage can be calculated for unconfined karst aquifer (a), and for the confined part (b), respectively. The latter represents an extension of the karst aquifer overlain by a non-karst unit which can be permeable (then Q_{sb} is possible) or fully impermeable (no Q_{sb} could take place).

The exploitable part of static reserves ($Q_{static.expl}$) has to be defined as a sustainable flow. In Table 6.2 a "loan" of these reserves not exceeding 10 % of total static reserves for the exploitation period of 10 years ($Q_{st,\ 10\ \%/10\ years}$) is proposed. The available reserves (Q_{avail}) are equal to the dynamic reserves minus the flow required for water-dependent ecosystems. When there is an additional significant flow from downstream tributaries (outside of the aquifer's drainage area), the waters needed for the dependent ecosystem could match the average minimal discharges of springs (average monthly values). Hence, the minimal water flow would not be disturbed by exploitation. But, if there is no supplementary recharge

Table 6.2a Groundwater reserves assessment from budget elements

Groundwater Reserves		I	II	III	IV	V	VI	VII	VIII	IX	X	XI	XII	Mean annual (m^3/s)	Minim. annual (m^3/s)	Equiv. total annual water volume in 10^6 m^3
ΣQ_{dyn}																
		F (km^2)				H_{av}, karst (m)		P_e						$\Sigma Q_{stat.}$	$Q_{st,10\,\%/10\,years}$	$Q_{static.\ expl}$ (m^3/s)
ΣQ_{st}	(a)															
	(b)															
		I	II	III	IV	V	VI	VII	VIII	IX	X	XI	XII	Av. annual (m^3/s)		
$\Sigma Q_{avail\ (dyn)}$																
		$\Sigma Q_{pot.available\ (dyn\ +\ stat)}$ in m^3/s						I_{ef} (%)								
		I	II	III	IV	V	VI	VII	VIII	IX	X	XI	XII	Av. annual (m^3/s)	Min annual (m^3/s)	
$\Sigma Q_{pot\ Avail\ (avail\ dyn\ +\ stat)}$																
$\Sigma Q_{active\ use}$																
C_{use}																
Total reserves $Q_{available}$ (m^3/s)																

Table 6.2b Groundwater reserves assessment from budget elements (Fig. 6.17)

Legend	
ΣQ_{dyn} Dynamic groundwater (GW) reserves	$Q_{pot\ avail}$ Potentially available in case of aquifer regulation and overpumping for certain period of time throughout the year
$\sum Q_{dyn} \cong \sum Q_{allsprings\ average\ flow}$	$\sum Q_{potent.avail} = \sum Q_{avail} + \sum Q_{st,10\%/10years}$
Q_{stat} Static (geological GW reserves) (a) in the open (unconfined) structure below min GW table (F × H × P_e) (b) in confined part (F × H × P_e) $\Sigma Q_{stat.exploit.}$ Exploitable part of static GW reserves	$Q_{st,10\ \%/10years}$ Part of static groundwater reserves to be exploited during 10 years in an amount not to exceed 10 % of total static groundwater reserves $\Sigma I_{ef} = \frac{\sum Q_{dyn}}{\sum P}$
F Surface area in m^2; H av. aquifer thickness (saturated zone below discharge points) in P_e av. effective porosity, i.e., storage coefficient	I_{ef} Effective infiltration (groundwater replenishment potential)
$Q_{avail.dyn}$ Available dynamic reserves = dynamic reserves minus total minimal discharges of all springs as a guaranteed ecological river flow or other prescribed flow for water-dependent ecosystems: $\sum Q_{avail} \cong \sum Q_{dyn} - \sum Q_{allsprings,min}$	$Q_{active\ use}$ Total currently exploited amount of water (water supply, industry, agriculture...) C_{use} Coefficient of utilization of reserves: $\Sigma Q_{active\ use}/\Sigma Q_{pot.\ avail}$

downstream and consumers' demands are very large, then the required ecological flow has to increase to a certain level, sometimes even totally limiting groundwater exploitation. Finally, to the potentially available reserves, the exploitable part of the static waters ($Q_{static.expl}$) could be added.

Relation Q_{dyn} versus total precipitation (P) provides I_{ef}, i.e., an idea of how much water can be replenished in an average hydrological year.

The last part of the evaluation includes a comparison of the potentially available reserves ($\Sigma Q_{potent.avail.}$) and the summary flow of currently exploited water for all purposes ($\Sigma Q_{active\ use}$). This relation shows a quantitative pressure on the aquifer system, an important indicator of aquifer sustainability also included in the EU Water Framework Directive (WFD 2000/60).

6.5 Case Studies and Exercises

Case Study 1. Estimation of volume of effective rainfall, potential infiltration rate, and water utilization probability

Exercise:

Aiming to assess the exploitation potential of the Middle Miocene "Pila Spi" karstic aquifer in northern Iraq (Fig. 6.18), calculation was made of the catchment area south of Dohuk *city. The surface area (F) of the studied karstic aquifer comprises 25 km*^{2}*. How much water (*P_0*) would be potentially generated annually for infiltration, runoff, evapotranspiration, and*

Fig. 6.18 One of the wells drilled in the foothill of karstified limestones (Shaqlawa, north Iraq)

other output elements if the average annual sum of the rainfall (P) is equal to 557 mm? How much water would be potentially infiltrated in the ground (I_{ef}) if runoff (R_f) is assessed on 10 % P and total real evapotranspiration ($E_t + E_g$) on 55 % of P? Can an equivalent yield of 65 l/s be theoretically utilized from this aquifer for drinking water supply?

Result:

$$P_0 = P \times F$$

$$P_0 = 0.557 \times 25 \times 10^6 \text{ m}^3 = 13.9 \text{ x } 10^6 \text{ m}^3$$

$$I_{ef} = P_0 - R_f - \Sigma E$$

$$I_{ef} = 13.9 \times 10^6 \text{ m}^3 - 0.1\left(13.9 \times 10^6 \text{ m}^3\right) - 0.55\left(13.9 \times 10^6 \text{ m}^3\right)$$

$$I_{ef} = 4.87 \times 10^6 \text{ m}^3$$

$$Q = I_{ef}/365 \times 86,400 \text{ (s)}$$

$$Q = 154.5 \text{ l/s}$$

As an average yield of 65 l/s is equal to 42 % of the total recharge, it is very probable that exploitation of this aquifer for drinking water supply could be organized in a relatively sustainable way. In a concrete case, a linear system of 10–12 wells, pumping 10–12.5 l/s/well, has been proposed for drilling in the piedmont of Pila Spi aquifer. However, any longer-term exploitation that exceeded this limit would affect the natural compensation of the extracted groundwater.

Reference: Stevanović and Iurkiewicz (2004): Hydrogeology of northern Iraq, Vol. 2 General hydrogeology and aquifer systems.

Case Study 2. Assessment of subsurface drainage along riverbed

For the new regional water system Bogovina, planned to supply the Timok region in eastern Serbia, complex hydrogeology survey aiming to assess groundwater resources and opportunities for the engineering of regulation of aquifer was undertaken. The karst aquifer formed in the carbonate rocks of Upper Jurassic (Thitonian) and Lower Cretaceous ages is rich in groundwater and is recharged mainly from rainfall and from sinking flows which gravitate from higher altitudes and impermeable Paleozoic rocks. Despite large water reserves, natural minimal discharge of the main spring Mrljiš was not sufficient to meet water demands in a recession period and the option of temporary overpumping was evaluated. An increase in the river flow of the Crni Timok River in the Carpathian karst of eastern Serbia is supposed to be the result of subsurface drainage of local karst aquifer (Fig.6.19). The question is how the existence of this underground inflow can be verified and how it influences currently available water resources?

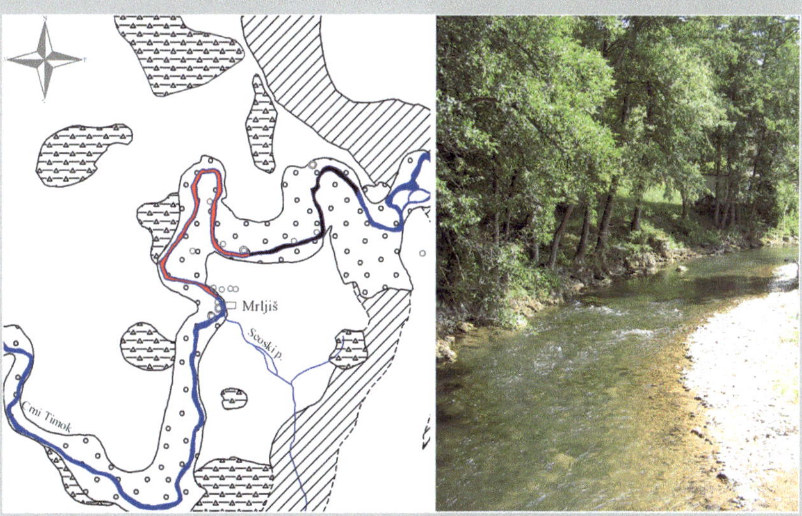

Fig. 6.19 Crni Timok River. The river sector in *blue* is subsurface inflow zones, the *red* indicates river water losses, while only the *black sector* is indifferent. The karstic terrains are marked *white*

Solution:
In order to assess the subsurface drainage component of the water budget, hydrometric prospections of the Crni Timok River as a regional erosional base were conducted.

The simultaneous measurements of river flow in the Crni Timok were carried out on several occasions from 1981–1985 at the dense network of sections along the river length of 8 km. At the upstream section of about 3 km, a total increment of flow of about 250 l/s was recorded, which was about 20 % of total river flow at that time (Fig. 6.19). The measurements were repeated on several occasions, and proof of the existence of a subsurface flow from karst aquifer into the river bed and alluvial deposits was obtained. In this project, several series of hydrometry were undertaken along with the installation of limnigraph and gauging stations.

As a result, the variable karstic water regime and the presence of sectors with subsurface inflow and losses were certified. But in general, there was a constant yield surplus. It ranged from 260 l/s in mean water periods to 140 l/s during the recessions. The linear yield coefficient varied from 17.5 to 30 l/s/km. The subsurface inflow during the high-water period was disregarded for the "safe case" reason. The budget equation for an average hydrological year was formulated as follows:

$$\text{Input} = \text{Output}$$
$$P_0 + I_s = Q_s + Q_{sb} + R_f + E_t + (\pm \mathrm{R} \pm \mathrm{E})$$
$$45.5 + 7.2 = 8.8 + 8.2 + 10.5 + 23.6 + 1.6$$

(*all values in* $10^6\ m^3$)
When minimal aquifer discharge is concerned, the separation of flow shows that more waters drain subsurficially than at the Mrljiš Spring itself. Dynamic reserves during the recession season are assumed on the basis of 0.22 m^3/s and distributed as follows:

$$\Sigma Q_{\text{dyn min.}} = Q_{\text{min Mrljiš Spring}} + Q_{\text{subsurface into C. Timok}}$$
$$0.22\ \mathrm{m^3/s} = 0.08\ \mathrm{m^3/s} + 0.14\ \mathrm{m^3/s}$$
$$\mathrm{Q_{dyn.}} = 0.22\ \mathrm{m^3/s}$$

The obtained results favoured the next survey phases which also confirmed aquifer engineering regulation opportunities.

Reference: Stevanović (2010b): Management of karstic aquifer of regional water system "Bogovina" (eastern Serbia)

Case Study 3. Rough assessment of dynamic and static groundwater reserves

Karstic aquifer is extended over some 30 km^2*. Carbonate rocks are well karstified. Based on measurements of their outcrops and sporadic pumping tests of drilled wells, an average effective porosity is estimated at 0.025. This value is*

Fig. 6.20 One of the gravity karst springs of Grza source in Carpathian karst of eastern Serbia ($Q_s = 30$–1,500 l/s). During recession periods, the discharge point lowers and the water flow is at the lower level in close proximity to the installed weir

simulated along a whole vertical section in the saturated aquifer part. According to geophysical survey and drilling, the base of karstification is at the average depth of 80 m. The drainage of the aquifer occurs mostly over a group of springs located along the contact with non-karst (Fig. 6.20). This connection zone is sharply represented by overthrusted low-permeable Permian sandstones. The distance between springs is around 120 m. They are all characterized by gravity flow and discharge points which fluctuate up to 8 m throughout a hydrological year (average is 5 m). The tracing test was conducted in a mid-water period at one ponor in the upper watershed at a distance from the springs of around 11.5 km. The tracer was registered at the spring after 8 days and 19 h.

In order to assess the potential of this aquifer, assessment of its dynamic and static groundwater reserves has been proposed. The request was that exploitable static reserves should not exceed 10 % of the total static reserves while their utilization is limited to maximum 10 years.

Exercise:

Rough assessment of dynamic water reserves should be obtained by application of the formula (6.13):

$$Q_{dyn} = V_c \times F \times P_e$$

Groundwater velocity calculated from the tracing test is 0.015 m/s. Groundwater in the drainage zone is moving through a sectional area ca. 120 m wide and 5 m high. Effective porosity is 0.025.

$$Q_{dyn} = 0.015 \times 120 \times 5 \times 0.025\ (m^3/s)$$

$$Q_{dyn} = 0.225\ m^3/s$$

The obtained result corresponds well with an average discharge of the group of springs (30–1,500 l/s). By applying the relation $\sum Q_{\mathrm{avail}} \cong \sum Q_{\mathrm{dyn}} - \sum Q_{\mathrm{allsprings,min}}$ assessment of the available dynamic reserves is possible: 0.225 m³/s – 0.03 m³/s = 0.195 m³/s. Due to the very small minimal discharge of around 10 % of dynamic reserves, an increase of the ecological flow to up to 30 % of dynamic reserves is proposed. Therefore, after applying calculation: 0.225 m³/s – 0.0675 m³/s = 0.157 m³/s, for utilization, a discharge of around 0.15 m³/s should be available.

For calculation of static water reserves, the following Eq. (6.12) can be applied:

$$\sum V = F \times H \times P_{\mathrm{e}} \quad (\text{in m}^3)$$

$$\sum V = 30 \times 10^6\ \mathrm{m}^2 \times 85\ \mathrm{m} \times 0.025$$

$$\sum V = 63.75 \times 10^6\ \mathrm{m}^3$$

Exploitable part of 10 % of static reserves ($\Sigma Q_{\mathrm{stat.exploit}}$) is 6.375×10^6 m³ or converted to average yield during 10 years 0.02 m³/s (20 l/s)

Finally, an average yield of 0.17 m³/s can be extracted from this aquifer without negative consequences on the environment, because ecological flow of 0.0675 m³/s is ensured for downstream consumers. Although minimal discharges are lower than ecological flow groundwater pumping would be necessary during critical drought periods not only for water utility but also for guaranteed downstream flow.

Case Study 4. Calculation of episodic recharge and effects on groundwater reserves

After a stormy rain episode in the watershed of Oued M'zi near Laghouat (Algeria, northern edge of Sahara), the groundwater table in Turonian limestone aquifer rose on average some 1.85 m (Fig. 6.21). The aquifer's surface area covers around 39.6 km². Based on previous pumping tests of existing wells and other experiments, an average effective porosity is assumed to be 0.018.

Fig. 6.21 Oued M'zi after heavy rains (*left*) and typical landscape of Turonian karstic aquifer near Laghouat (Algeria)

The task was to calculate the new water storage in terms of effective water height and the total additional water volume. Finally, recalculation of this additional water into an equivalent yield to cover the period of the next five presumably dry months was requested.

Solution:
By using Eq. (6.9), storage change expressed as new water height can be calculated as follows:

$$R = \Delta\mathrm{H} \times P_{\mathrm{e}}$$
$$R = 1,850\ \mathrm{mm} \times 0.018 = 33.3\ \mathrm{mm}$$

The total volume of newly stored water (V) within the aquifer from Eq. (6.10) is as follows:

$$V = R \times F\ (\mathrm{m}^3)$$
$$V = 0.0333\ \mathrm{m} \times 39.6 \times 10^6\ \mathrm{m}^2 = 1.518 \times 10^6\ \mathrm{m}^3$$

Finally, the yield (Q) which can be expected from the last recharge episode to contribute to existing dynamic reserves is calculated based on Eq. (6.11), as follows:

$$Q = V/T\ (\mathrm{m}^3/\mathrm{s})$$
$$Q = V \times T = 1.518 \times 10^6\ \mathrm{m}^3/150 \times 86,400\ \mathrm{s} = 0.117\ \mathrm{m}^3/\mathrm{s}$$

Eventually, the aquifer is recharged with a certain amount of new water, which could, in the case of no natural drainage, contribute to the existing water supply system with an average yield of around 117 l/s during a five-month period.

Reference: Stevanović and Simić (1985): The problems of groundwater extraction in Laghouat area (Algeria)

Case Study 5. Calculation of groundwater reserves of one regional karst aquifer

Beljanica is one of the largest karstic massifs and aquifer systems in the Carpathian Mountain arch of eastern Serbia. Its northern part is drained by three large springs situated at the edge of the massif and tectonic contact of Lower Cretaceous limestones and Neogene lacustrine sediments of the Žagubica basin. The Mlava Spring (locally: Vrelo Mlave) is a typical vauclusian spring (Fig. 6.22), the largest in eastern Serbia, with great potential for regional water supply. The typical average springflow is 1.85 m³/s, the absolute extreme maximum is 19 m³/s, while the recorded extreme minimum is equal to 0.215 m³/s *The second largest spring, Belosavac, has an average discharge of 0.52 m³/s. During high-water periods, it activates a temporary spring for the overflow of its aquifer. The third largest spring, Krupaja, discharges at*

the NW Beljanica edge. Its minimal and average discharges are 0.2 and 1.49 m³/s, respectively. It is very probable that groundwater is also partly discharged underground into porous integranular Neogene aquifer.
Of these three springs, Belosavac is the only one currently exploited for drinking water supply, but only to a small degree. A small amount of water from all three springs is also used for the nearby fishponds. This mean utilized yield is equal to 0.45 m³/s.
The karst aquifer of Beljanica Mountain and its foothills has a variable surface of around 261 km². The catchment area varies depending on the water table position. During high-water periods, temporary streams in the eastern and NW part of the massif are activated. At that time, maximal floods can exceed 1 m³/s, but calculated annual average runoff is around 190 l/s.
Beljanica Mt. represents one of the still well-preserved pure ecological oases in Serbia. The area also has potential for tourism development. It is unpolluted, rich in clean waters, and boasts beautiful landscapes and features such as caves, waterfalls, and springs, including a few thermal water occurrences.
In order to assess groundwater reserves and available yield for regional water supply, the budget equation was imposed, and procedures presented in Tables *6.1 and 6.2 were implemented.*

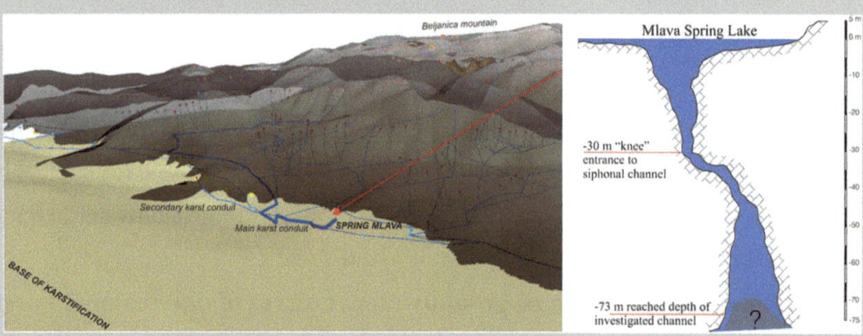

Fig. 6.22 Physical model of Beljanica Mt. and cross section of Mlava Spring based on cave diving (after Milanović S. from Stevanović et al. (2010), Reprinted with permission of the Slovenian Academy of Science—SAZU)

Exercise:

Budget elements F (km^2) = 261.85															
Months	I	II	III	IV	V	VI	VII	VIII	IX	X	XI	XII	Total (mm)		Total water volume in 10^6 m^3 (P x F)
P (mm)	42.2	42.2	49.0	66.2	92.5	103.8	77.1	55.7	63.4	53.8	57.9	54.5	758.3		198.561
Months	I	II	III	IV	V	VI	VII	VIII	IX	X	XI	XII	Mean annual (m^3/s)	Mini-mum annual (m^3/s)	Total in 10^6 m^3 (from mean values)
I_s (m^3/s)	0	0	0	0	0	0	0	0	0	0	0	0			
ΣQ_s (m^3/s)	3.73	5.00	5.70	7.08	5.59	3.98	2.95	1.77	1.47	2.84	2.56	3.87	3.883	0.475	122.45
Mlava sp.	1.75	2.26	3.17	3.67	2.54	2.02	1.35	0.93	0.84	1.00	1.21	1.65	1.870	0.215	58.972
Belo-savac	0.44	0.51	0.59	0.83	0.85	0.60	0.72	0.34	0.23	0.39	0.28	0.44	0.522	0.060	16.462
Krupaja	1.54	2.22	1.93	2.58	2.20	1.35	0.86	0.48	0.39	1.44	1.07	1.77	1.491	0.200	47.020
ΣR_f (m^3/s)													0.190	0.0	6.041
L															70.07
	I	II	III	IV	V	VI	VII	VIII	IX	X	XI	XII	Total (mm)	Total (m^3/s)	
E_t	0.6	1.0	3.8	15.2	36.6	48.8	51.1	37.8	25.1	11.3	3.7	0.9	236		61.797
Q_{sb} (L–E)													31.5	0.262	8.272

The rainfall rate and snow (P) were observed at the meteorological station at the top of the massif. Based on 30 years of observation, the annual sum is equal to 758.3 mm. The evapotranspiration (E_t) rate was calculated by the empirical Thornthwaite formula. The calculated average percentage of E_t is 31 % of precipitation, which is similar to values obtained in other karstic areas in that region. There is no surface water inflow (I_s) into the study catchment. There was no systematic monitoring of the runoff component, but based on sporadic measurements, the surface outflow was equalized to average 0.19 m^3/s which is only 3 % of the total precipitation. All three major springs are under permanent observation, Mlava for 30 years so far, and the other two for 4 years so far. Their summary average discharge is 3.88 m^3/s. The balance equation resulted in large total losses of 70×10^6 m^3 or 2.2 m^3/s. Considering that evapotranspiration consumes around 88 % of that value, the possible average subsurface discharge (Q_{sb}) including some small springs discharges has been found to be around 0.26 m^3/s.

Groundwater reserves

	I	II	III	IV	V	VI	VII	VIII	IX	X	XI	XII	Mean annual (m^3/s)	Minim. annual (m^3/s)	Equiv. total annual water volume in 10^6 m^3
ΣQ_{dyn}	3.73	5.00	5.70	7.08	5.59	3.98	2.95	1.77	1.47	2.84	2.56	3.87	3.883	0.475	122.445
	F (km^2)				H_{av}, karst (m)		Pe av.						ΣQ_{stat}. × 10^6 m^3	$Q_{st,10\,\%/20\,years}$ × 10^6 m^3	$Q_{static.expl}$ (m^3/s)
ΣQ_{st} (a)	261.85				100		0.015						392.8	1.96	0.062
(b)	30.3				100		0.02						60.6	0.3	0.01
													Av. annual (m^3/s)		0.072
ΣQ_{avail} (dyn)	3.25	4.52	5.22	6.60	5.1	3.50	2.47	1.30	0.99	2.36	2.08	3.40	3.408		
	$\Sigma Q_{pot.available\ (dyn + stat)}$ in m^3/s						I_{ef} (%)								
	3.480 m^3/s (109.7 × 10^6 m^3)						62 %						Av. annual (m^3/s)	Min annual (m^3/s)	
ΣQ_{active} Use													0.45	0.15	
C use	13 %														
Total reserves $Q_{available}$ (m^3/s): 3.48															

Total reserves $Q_{available}$ (m^3/s): 3.48

Dynamic groundwater reserves are equal to average summary discharges of major springs. The dynamic reserves of 3.88 represent 62 % of total annual precipitation. To ensure the sustainable flow of the Mlava River downstream of the springs, the summary minimal discharge of all three springs of 0.475 m^3/s was subtracted from the dynamic reserves. In this way, available dynamic reserves have been reduced by 3.40 m^3/s.

Concerning the possible supplementary extraction of static reserves during a 20-year period, this yield was calculated not to exceed 10 % of their total volume. The karstification base is at a depth of around 100 m, while an averaged storage coefficient to that depth was approximated as 0.015 within the entire watershed (a) and as 0.02 in the saturated confined part under Neogene sediments (b) Hence, the calculated yield from static reserves was around 72 l/s which, together with dynamic reserves, resulted in a total of 3.48 m^3/s of potentially available reserves. Although this average exploitation potential was 3.5 times greater than the minimal average monthly discharge of 1 m^3/s, the engineering aquifer regulation was proposed. In favor of that idea are ascending discharge, deep karstification and results of speleo diving exploration of the Mlava and Krupaja Springs. Considering that demands of the local water-dependent ecosystem are relatively low, there is a good prospect for karstic groundwater use for the wider region.

References: Stevanović et al. (2007): Management of karst aquifers in Serbia for water supply;
Milanović (2007): Hydrogeological characteristics of some deep siphonal springs in Serbia and Montenegro karst;
Stevanović et al. (2012): Climate changes and impacts on water supply.

References

Allen R, Pereira LA, Raes D, Smith M (1998) Crop evapotranspiration. FAO Irrigation and Drainage Paper 56, Rome, p 101

Atkinson TC (1977) Diffuse flow and conduit flow in limestone terrain in Mendip Hills, Somerset (Great Britain). J Hydrol 35:93–100

Bonacci O (1987) Karst hydrology with special reference to the Dinaric karst. Springer, Berlin

Bonacci O (1993) Karst springs hydrographs as indicators of karst aquifers. Hydrol Sci 38(1):51–62

Boni CF, Bono P (1984) Essai de bilan hydrogeologique dans une region karstique de l'Italie Centrale. In: Burger A, Dubertret L (eds) Hydrogeology of karstic terrains. Case histoires. International Contributions to Hydrogeology, IAH, vol 1. Verlag Heinz Heise, Hannover, pp 27–31

Boni CF, Bono P, Kovalevsky VS (1984) Evaluation of water resources. In: Burger A, Dubertret L (eds) Hydrogeology of karstic terrains. Case histories. International Contributions to Hydrogeology, IAH, vol 1. Verlag Heinz Heise, Hannover, pp 9–17

Brown M, Wigley T, Ford D (1969) Water budget studies in karst aquifers. J Hydrol 9:113–116

Castany G (1967) Traite pratique des eaux souterraines. Dunod, Paris

Cobb E, Bailey J (1965) Measurement of discharge by dye-dilution methods, Book 1, Chap 14. US Geological Survey, Surface-water techniques, p 27

Davis S, De Wiest R (1967) Hydrogeology. Wiley, New York (reprint: Krieger Pub. Co. 1991: p 463)

Drogue C (1972) Analyse statistique des hydrogrammes de decrues des sources karstiques. J Hydrol 15:49–68

Ford D, Williams P (2007) Karst hydrogeology and geomorphology. Wiley, Chichester

Goldscheider N, Drew D (eds) (2007) Methods in karst hydrogeology. International contribution to hydrogeology, IAH, vol 26, Taylor and Francis/Balkema, London

Healy RW, Winter TC, LaBaugh JW, Franke OL (2007) Water budgets: foundations for effective water-resources and environmental management: US Geological Survey Circular 1308, p 90

Kresic N (2007) Hydrogeology and groundwater modeling, 2nd edn. CRC Press/Taylor and Francis, Boca Raton, FL

Kresic N (2013) Water in karst: management, vulnerability and restoration. McGraw Hill, New York

Kresic N, Bonacci O (2010) Spring discharge hydrograph. In: Kresic N, Stevanović Z (eds) Groundwater hydrology of springs. Engineering, theory, management and sustainability. Elsevier Inc. BH, Amsterdam, pp 129–163

Kullman E (1984) Evaluation des changements des reserves en eau souterraine dans la structure hydrogeologique du complexe calcaire dolomitique des Petites Carpates (Tchecoslovaquie) en vue du bilan hydrologique. In: Burger A, Dubertret L (eds) Hydrogeology of karstic terrains. Case histoires. international contributions to hydrogeology, IAH, vol 1. Verlag Heinz Heise, Hannover, pp 46–53

Leontief W (1986) Input-output economics, 2nd edn. Oxford University Press, New York

Lerner D, Issar A, Simmers I (1990) Groundwater recharge, a guide to understanding and estimating natural recharge. International contributions to hydrogeology, vol. 8. Verlag Heinz, Heise. Hannover

Luchshewa AA (1976) Prakticheskaya gidrologija (Practical hydrology, in Russian). Gidrometeoizdat. Leningrad (St. Petersburg)
Maillet E (ed) (1905) Essais d'hydraulique souterraine et fluviale, vol 1. Herman et Cie, Paris
Mangin A (1984) Pour une meilleure connaisance des systèmes hydrologiques à partir des analyses corrélatoire et spectrale. J Hydrol 67:25–43
Mijatović B (1990) Karst. Hydrogeology of karst aquifers. Spec. ed. Geozavod, Belgrade
Milanović P (1979) Hidrogeologija karsta i metode istraživanja (Karst hydrogeology and exploration methods, in Serbian). HE system Trebišnjica, Trebinje
Milanović P (1981) Karst hydrogeology. Water Resources Publications, Littleton CO
Milanović S (2007) Hydrogeological characteristics of some deep siphonal springs in Serbia and Montenegro karst. Envi Geol 51(5):755–760
Miller RE, Blair PD (2009) Input-output analysis: foundations and extensions, 2nd edn. Cambridge University Press,
Padilla A, Pulido-Bosch A, Mangin A (1994) Relative importance of baseflow and quickflow from hydrographs of karst spring. Ground Water 32(2):267
Paloc H (1992) Caracteristiques hydrogeologiques specifiques de la region karstique des Grands Causses. In: Paloc H, Back W (eds) Hydrogeology of selected karst regions, vol 13. Heise, Hannover, pp 61–88
Panagopoulos G, Lambrakis N (2006) The contribution of time series analysis to the study of the hydrodynamic characteristics of the karst systems: application on two typical karst aquifers of Greece (Trifilia, Almyros Crete). J Hydrol 329(3–4):368–376
Penman HL (1948) Natural. evaporation from open water, bare soil and grass. Proc R Soc Lond A 193:120–145
Ranković J (1979) Teorija bilansa (Budget theory, in Serbian). University of Belgrade, Belgrade
Rorabaugh M (1964) Estimating changes in bank storage and groundwater contribution to stream flow. Int Assoc Sci Hydrol Publ 63:32–441
Rutledge RT (1998) Computer programs for describing the recession of groundwater discharge and for estimating mean groundwater recharge and discharge from stream flow records-update. USGS water-resource investigations report 98-4148. Reston, VG
Scanlon BR, Healy RW, Cook PG (2002) Choosing appropriate techniques for quantifying groundwater recharge. Hydrogeol J 10:18–39
Shaw EM (1994) Hydrology in practice, 3rd edn. Routledge, Abingdon
Stevanović Z (1984) Primena bilansno-hidrometrijskih metoda za određivanje rezervi karstnih izdanskih voda (Application of budget and hydrometry methods in definition of karstic groundwater reserves, in Serbian). Vesnik Serb Geol Surv B, vol XX, Belgrade, pp 1–13
Stevanović Z (1995) Karstne izdanske vode Srbije–korišćenje i potencijalnost za regionalno vodosnabdevanje (Karst groundwater of Serbia–use and potential for regional water supply; in Serbian). In: Stevanović Z (ed) Water mineral resources of lithosphere in Serbia. Spec. ed. of Fac. Min. Geol. University of Belgrade, pp 77–119
Stevanović Z (2010a) Utilization and regulation of springs. In: Kresic N, Stevanović Z (eds) Groundwater hydrology of springs. Engineering theory, management and sustainability. Elsevier Inc. BH, Amsterdam, pp 339–388
Stevanović Z (2010b) Regulacija karstne izdani u okviru regionalnog vodoprivrednog sistema "Bogovina" (Management of karstic aquifer of regional water system "Bogovina", Eastern Serbia). University of Belgrade—Faculty of Mining and Geology, Belgrade
Stevanović Z, Iurkiewicz A (2004) Hydrogeology of Northern Iraq general hydrogeology and aquifer systems, Spec edition TCES, vol 2. FAO, Rome, p 175
Stevanović Z, Jemcov I, Milanović S (2007) Management of karst aquifers in Serbia for water supply. Environ Geol 51/5:743–748
Stevanović Z, Milanović S, Ristić V (2010) Supportive methods for assessing effective porosity and regulating karst aquifers. Acta Carsologica 39/2:313–329
Stevanović Z, Ristić-Vakanjac V, Milanović S (eds) (2012) Climate changes and water supply. Monograph. SE Europe cooperation programme. University of Belgrade, Belgrade, p 552

Stevanović Z, Simić M (1985) Problematika eksploatacije podzemnih voda na širem području grada Laghouata (Alžir) (The problems of groundwater extraction in Laghouat area, Algeria, in Serbian), Vesnik Serb Geol Surv B, vol XXI, Belgrade, pp 35–43

Thornthwaite CW (1948) An approach toward a rational classification of climate. Geograph Rev 38:55

Turc L (1954) Le bilan d'eau des sols: relations entre les precipitations, l' evaporation et l' ecoulement. Ann Agron 5:491–596

Water Framework Directive WFD 2000/60, Official Journal of EU, L 327/1, Brussels

Yevjevich V (1956) Hidrologija, 1 deo. (Hydrology, Part 1, in Serbian). Ins Wat Res Develop Jaroslav Černi, Spec.ed. vol 4, Belgrade, Serbia

Yevjevich V (ed) (1981) Karst water research needs. Water resources publications, Littleton CO

Chapter 7
Evaluating Discharge Regimes of Karst Aquifer

Peter Malík

Karst springs are typical for abrupt changes of discharge immediately following recharge events. Monitored discharges of springs are used to determine quantitative variability over the period of time, showing their reliability as dependable water sources.

Karst aquifers also exhibit (at least) dual groundwater flow regimes, that is, fast (conduit-dominated) flow and slow (diffuse) flow. This is something that can be observed in nature as the fast change of water amount outflowing from the groundwater source, or described by rapidly responding hydrographs, recording water levels or discharges. Selection of proper investigative techniques characterising discharge regime properties of a karst aquifer is therefore important in order to identify possible theoretical background models describing this behaviour. On this basis, we can also find a particular method of hydrograph separation into flow components linked to the fast-flow regime, slow-flow regime or intermediate regimes as well. With this point in mind, several quantitative methods that might be particularly useful in hydrograph analysis of water outlets from the karst aquifer system are briefly discussed here.

7.1 Discharge Regime: Definition, Typical Karstic Manifestations

In hydrogeology and hydrology, a term "regime" refers to the changing conditions of (ground) water phenomenon, its characteristic behaviour or prevailing system of natural processes which are usually observed in regular pattern of occurrence.

P. Malík (✉)
Geological Survey of Slovak Republic, Mlynská dolina 1, 817 04, Bratislava, Slovakia
e-mail: peter.malik@geology.sk

Z. Stevanović (ed.), *Karst Aquifers – Characterization and Engineering*,
Professional Practice in Earth Sciences, DOI 10.1007/978-3-319-12850-4_7

Nature of temporal changes during time-varying aquifer characteristics, in particular groundwater level, but also physical and chemical properties of groundwater, studied in relation to the factors that affect and determine it is a frequent subject of hydrogeological studies. Regular pattern of discharge changes, a discharge regime, typical for individual springs or river basins, is used for their characterisation and distinction. The term "discharge regime" means the regular, expected discharge of flowing water within a year. For example, watersheds in alpine regions may have snowmelt-dominated or glacier-dominated discharge regime, while in great lowland basins the discharge regime can be described as rain or monsoon dominated. Karst aquifers, due to their ability of groundwater flow concentration, are clearly recognisable in their discharge regime characteristics.

Immediate and ultra-intensive discharge response to recharge impulses manifested in hydrographs as steep peaks with enormous amplitudes is typical for wide range of karstic springs (Fig. 7.1). This feature is caused by a significant shift of transmissivity values of karstic rocks towards high levels contrary to other aquifer types (intergranular or fissure permeability). On the other hand, specific yield (storativity) values of rock types prone to karstification are not different, or are even lower than in other aquifers.

High values of hydraulic diffusivity parameter (ratio of transmissivity and specific yield) are then reflected in typical spiky shape of karst hydrographs. Even spring situated in granites and its debris, with apparently shallow groundwater circulation, can better absorb and buffer occasional recharge inputs due to relatively higher storage capacity and relatively lower hydraulic conductivity (Fig. 7.2). We should note that the two springs for which hydrographs were depicted in Figs. 7.1 and 7.2 are approximately 15 km away from each other, and hydrographs show

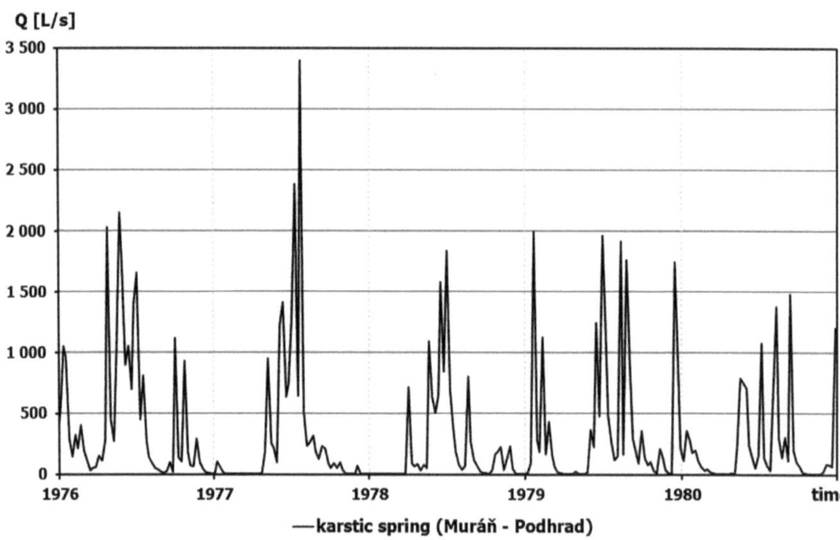

Fig. 7.1 Typical quantitative behaviour of a karstic spring (Podhrad spring in Muráň municipality, Central Slovakia)

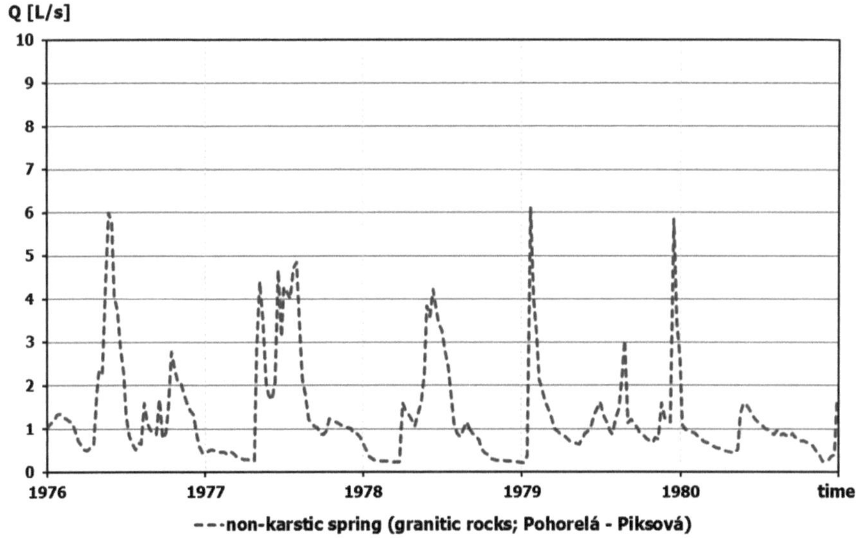

Fig. 7.2 Typical quantitative behaviour of a non-karstic spring (Piksová spring in Pohorelá municipality, Central Slovakia)

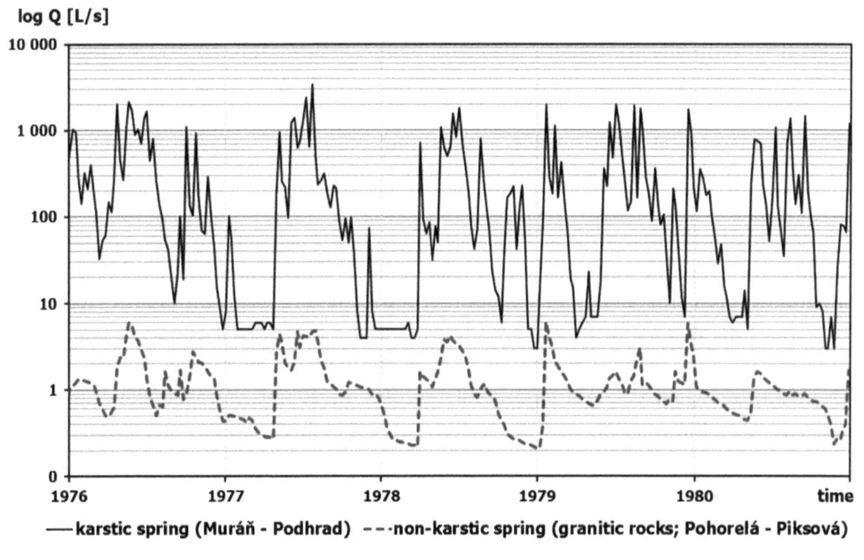

Fig. 7.3 Comparison of discharge regime patterns of a karstic (Fig. 7.1) and non-karstic (Fig. 7.2) spring on a semilogarithmic plot

response to nearly the same course of precipitation episodes. The karstic spring (Fig. 7.1) is the Podhrad spring in Muráň municipality (Central Slovakia), and the one with groundwater circulation in crystalline rocks is named Piksová near the village of Pohorelá, both with weekly measured data in the period of 1976–1981 period. To compare the responses of springs with different regimes, sometimes, it is better to use a logarithmical plot of discharge versus time as in the case of Fig. 7.3, so in one view, you can see the comparison of the two values without having to adjust the scale on the y axis. The usefulness of logarithmical plots of discharge (Q) values will be presented also later, but here in Fig. 7.3 one can see an unstable behaviour of karstic spring with strong reaction to every precipitation impulse, while weathered crystalline aquifer response only to major recharge events.

7.2 Spring Discharge Variability

Discharge of spring changes in time, due to the recharge and emptying of its natural reservoir. The amplitude and frequency of discharge changes are also dependent on aquifer geometry and physical properties, and karstic springs are typical for abrupt and significant changes of discharge. However, from the water management point of view, such instability is less welcome and temporal changes of spring's discharge should be classified before undertaking a serious water management investment. One should keep in mind that proper quantitative description of spring cannot be based on a single measurement of discharge or several sporadic data from different periods. Estimation of discharge variability parameter requires multiple measurements. Systematic measurements of discharge should be performed at least for one complete hydrological cycle where both regular major recharge events and recession are observed. Gauging frequency of groundwater sources was recommended to be at least once a week, but only daily measurements are capable of recording rapid changes of discharge in the case of karstic springs. Having a possibility of automatic electronic discharge recording nowadays, the problem of gaps in discharge time series is less actual than before. More pronounced problem today is the resolution of measurement. The shape of gauging weirs of karstic springs should enable precise reading in both low-water stages, when the discharge falls to several litres per second, and the high-water stages when several cubic metres per second flow through the same gauging object. The V-notch weirs (Thomson weir is a V-notch weir of 90° angle) are better in low-water stages, but these are not able to cope with high flow rates. For high-water stages, rectangular shapes of weirs (Poncelet weirs) of sufficient width can be used, and sometimes, the combination of both V-notch and rectangular shapes is used (Fig. 7.4a, b) for enhancing the capability of both extreme readings. As a better option, combination of several (two or three) V-notch weirs with gradually more opened angle towards outer edges (Fig. 7.4d) can be used. Hydraulic behaviour of this type of weir is said to be better theoretically described than that in Fig. 7.4a, b. Parshall flume (Fig. 7.4c) can be used if there is limited vertical space for achieving downflow gradient, but its accuracy is lower than that of previously mentioned sharp-crested weir allowing water to fall cleanly down from the weir.

According to their changes of discharge, we can distinguish perennial springs that will never get dry (where their discharges are always higher than zero) and

Fig. 7.4 Various types of weirs used for discharge gauging of karstic springs **a** combined Thomson (*V-notch*) and Poncelet (*rectangular*) weir—dry period; **b** combined Thomson (*V-notch*) and Poncelet (*rectangular*) weir—wet period; **c** Parshall flume; and **d** combined V-notch weir

ephemeral (intermittent) springs that discharge groundwater in irregular time intervals. Intermittent springs discharge only for a period of time reflecting the aquifer recharge pattern, while at other times they stay dry. An interesting case is the phenomenon of periodic springs (ebb-and-flow springs; "rhythmic" springs). These springs are regularly discharging approximately the same amount of water in short time intervals; their discharge time series are oscillating for at least a certain period of time. In the past, these were connected to existence of a siphon that fills up and empties out with certain regularity, irrespective of the recharge pattern (Kresic 2007). Mangin (1969) explained this phenomenon as a result of emptying a water reservoir through two differently situated pipes, which fits also to the model proposed by Oraseanu and Iurkiewicz (2010), and Bonacci and Bojanic (1991) had suggested a mathematical working model of such a spring function consisting of two reservoirs joined by a siphon. We can classify the spring as being periodic if its discrete emptying mechanism is working at least in some specific hydrological stages. Appearance of periodic springs is typical especially for karstic regions elsewhere in the world.

Classification of springs based on average discharge rate can be geographically dependent as springs—natural groundwater sources—are differently perceived and understood according to water scarcity or profusion in different parts of the world. Meinzer's classification (1923b) shown in Table 7.1, however, can serve as the first reference point.

Table 7.1 Classification of spring's magnitude according to average annual discharge after Meinzer (1923b)

Magnitude of spring's discharge	Average annual discharge (Q) in [L/s]
1st—first	>10,000
2nd—second	1,000–10,000
3rd—third	100–1,000
4th—fourth	10–100
5th—fifth	1–10
6th—sixth	0.1–1
7th—seventh	0.01–0.1
8th—eighth	<0.01

Without specifying other discharge parameters, classification based on average discharge is not very useful as the statistical distribution of discharge values (especially in karst) is often lognormal. The overall discharge average value may be influenced by several massive flood events, while for the rest of the period, the spring discharges only small amounts of water or may dry out. In many countries, spring classification is based on minimal discharge data (Kresic 2007), but sometimes maximum discharges are useful for karst hydrological modelling (Bonacci 2001). However, when evaluating availability of water for potential spring utilisation, it is important to estimate a measure of spring's discharge variability. Spring discharge monitoring data should be used to determine spring discharge variability over the period of study for individual spring. Classification of spring's discharge variability is used to characterise trends in low-flow periods and in combination with mean annual discharge to estimate the category of total annual discharge. Discharge variability can be also used to estimate regional hydrogeological processes and hydraulic properties of aquifers. Springs with high discharge variability can indicate high degree of transport properties of the aquifer and quick response of the system to recharge.

Presented methods of classification of spring's discharge variability are based on statistical parameters of regularly measured discharges. The simplest measure of spring's discharge variability is the ratio of maximal and minimal recorded discharge ($Q_{\max}/Q_{\min}$), what may be defined as index of variability I_v (Eq. 7.1)

$$I_v = \frac{Q_{\max}}{Q_{\min}} \tag{7.1}$$

Springs with the I_v values greater than 10 are considered highly variable, and those where index of variability is less than 2 are sometimes called constant or steady springs (Kresic 2007). Based on comparison of maximal and minimal recorded discharge, some other (classical) classifications of springs by discharge variability (Dub and Němec 1969; Netopil 1971) may be as follows. Index of variability here can be classified into several degrees of spring's reliability (Table 7.2).

Slovak Hydrometeorological Institute, responsible for observations of more than 1,300 springs since 1960s, uses the same index of variability I_v (Eq. 7.1) for characterisation of spring's discharge stability as shown in Table 7.3.

Table 7.2 Degrees of spring's discharge reliability based on index of variability I_v value (Dub and Němec 1969; Netopil 1971)

Degree of spring's reliability	I_v (Q_{max}/Q_{min})
Excellent	1.0–3.0
Very good	3.1–5.0
Good	5.1–10.0
Modest	10.1–20.0
Bad	20.1–100.0
Very bad	>100.0
Ephemeral spring	∞

Table 7.3 Degrees of spring's discharge reliability based on I_v (index of variability) value as applied in the Slovak Hydrometeorolgical Institute

Degree of spring's discharge stability	I_v (Q_{max}/Q_{min})
Stable	1.0–2.0
Unstable	2.1–10.0
Very unstable	10.1–30.0
Totally unstable	>30.0

It is clear that with longer observation period, the more extreme hydrologic phenomena can be recorded (higher floods, longer droughts) and accordingly, spring's reclassification can lead only to "less reliable" degrees.

To reduce the influence of extreme outliers on spring's classification, other statistical parameters of discharge time series can be employed. We should always keep in mind that a simple arithmetical mean (average) is the "worst characterization parameter" as in the case of high discharge amplitude, typical for karstic springs, it emphasizes the large discharges that occur only several times a year. Instead of using arithmetical mean, the use of median value is more recommended in the case of karstic springs, together with other parameters characterizing the variability of discharge changes.

Meinzer (1923b) proposed a measure of spring variability V expressed in percentage (Eq. 7.2):

$$V = \frac{Q\text{max} - Q\text{min}}{\emptyset} \times 100\,(\%) \tag{7.2}$$

where

V is the spring variability index expressed in %,
Q_{max} and Q_{min} are maximal and minimal recorded discharge and
Φ is the arithmetical mean of spring discharge values.

If $V < 25$ %, we can speak about *constant* spring discharge, for *variable* spring the $V > 100$ %.

The spring variability coefficient (SVC; Eq. 7.3) is based on comparison of the discharges with 10 and 90 % exceedence; spring coefficient of variation parameter (SCVP; Eq. 7.4) is based on standard deviation and arithmetical mean of spring discharge values.

$$\text{SVC} = \frac{Q_{10}}{Q_{90}} \tag{7.3}$$

where SVC is spring variability coefficient, Q_{10} is a discharge value which is exceeded in 10 % of the time and Q_{90} is discharge which is exceeded in 90 % of the time (see part on flow duration curves (FDCs) for definition of exceedence). Spring's classification according to SVC (Flora 2004; Springer et al. 2004) shown in Table 7.4 is based on works by Meinzer (1923a), Netopil (1971) and Alvaro and Wallace (1994).

Also Slovak technical standard STN 751520 (SÚTN 2009; Table 7.5) quantifies "spring's discharge stability" according both to Q_{max}/Q_{min} ratio (index of variability I_v) and to the *SVC* (Q_{10}/Q_{90}).

Flora (2004) and Springer et al. (2004) proposed also SCVP, calculated according to Eq. 7.4.

$$\mathrm{SCVP} = \frac{\sigma}{\emptyset} \tag{7.4}$$

where SCVP is spring coefficient of variation parameter, σ is standard deviation of spring discharge values and Φ is the arithmetical mean of spring discharge values. Variability classes according to SCVP are then classified as shown in Table 7.6.

Table 7.4 Classification of springs by discharge using *SVC* (Flora 2004; Springer et al. 2004)

Spring's classification	Spring variability coefficient (SVC)
Steady	1.0–2.5
Well balanced	2.6–5.0
Balanced	5.1–7.5
Unbalanced	7.6–10.0
Highly unsteady	>10.0
Ephemeral	∞

Table 7.5 Degrees of spring's discharge stability based on index of variability (I_v) or spring variability coefficient (*SVC*) (SÚTN 2009)

Degree of spring's discharge stability	Value of SVC or I_v
Very stable	1.0–3.0
Stable	3.1–10.0
Unstable	10.1–20.0
Very unstable	20.1–100.0
Extremely unstable	>100.0

Table 7.6 Classification of springs by discharge using spring coefficient of variation parameter (SCVP) (Flora 2004; Springer et al. 2004)

Spring's classification	Spring coefficient of variation parameter (SCVP)
Low	0–49
Moderate	50–99
High	100–199
Very high	>200

7.3 Flow Duration Curve

The FDC is a measure of the range and variability of a stream's flow or spring's discharge. The FDC represents the percentage of time during which specified flow rates/discharges are exceeded at a given location (Foster 1924, 1934; Searcy 1959). This is usually presented as a graph of flow rate (discharge) versus percentage of time at which flows are greater than, or equal to, that flow. Although the FDC does not show the chronological sequence of flows, it is useful for many studies. The FDC submits one of the most fundamental pieces of information about the discharge regime of karstic springs. While many past discharge regime classifications were using only comparison discharge minima and maxima to estimate the "stability of discharge", the FDC enables more detailed insight into the mode of spring's quantitative behaviour. To construct a reliable FDC, one needs sufficiently long set of regular observations, at least to cover the whole annual hydrological cycle (both recharge period and period of recession). The data in the discharge time series should be then simply ordered according to the size, from highest to lowest. The data are then plotted against a "percentage exceedence" scale. Each percentage exceedence increment is 100 % divided by the number of data points (dataset population, or number of measurements). If there are 365 discharge measurements, each regularly taken every day within 1 year, and if the discharge data are organised from highest to lowest, the percentage exceedence of the first datum (maximum) will be 1/365 = 0.27 %. The twelfth highest discharge from the dataset would have percentage exceedence of 12/365 = 3.29 %, while for the 279th discharge from the maximum it is 279/365 = 76.44 %. Percentage exceedence can be added as a new data column to show at what percentage exceedence each discharge occurred.

After the FDC was constructed, it can serve as a reference object for the spring discharge. If we look at the discharge value at "50 % exceedence", we will see the value that corresponds to the median value of discharge. If the value of "70 % exceedence" is perhaps 147 L/s, this does not mean that the discharge is 147 L/s for 70 % of the time, but that the discharge is equalled or exceeded for 70 % of the time. In other words, the discharge is at this value or at a higher value for 70 % of the time. If we look at the discharge at 20 % exceedence and it is 700 L/s, this is a higher flow rate, so the discharge is only at or greater than this value for a smaller proportion of the year. If we look at 100 % exceedence and it is perhaps 25 L/s which is the lowest discharge recorded, so by definition, the discharge of the spring is at this value or more for 100 % of the time.

Discharge (flow rate) is often referred to as Q, and the exceedence value as a subscript number, so Q_{95} means the discharge equalled or exceeded for 95 % of the time. Q_{50} is then equal to median value of discharge, but the average or mean flow rate Q_{mean}, and the arithmetic mean of all of the discharges in the dataset usually occurs between Q_{20} and Q_{40}, depending on how "flashy" or "steady" the spring being analysed is. Flow rates between Q_0 and Q_{10} are considered high flow rates, and Q_0 to Q_1 would be extreme flood events. Discharges from Q_{10} to Q_{70} would be of the medium range, while discharges from Q_{70} to Q_{100} are the "low flows" when

Table 7.7 Example of the discharge exceedence table for two karstic springs

Spring	Q_1	Q_5	Q_{10}	Q_{20}	Q_{30}	Q_{40}	Q_{50}	Q_{60}	Q_{70}	Q_{80}	Q_{90}	Q_{95}	Q_{99}
Brúsik	70.0	29.0	20.0	14.4	10.8	8.4	7.1	6.1	4.6	3.8	2.9	2.0	0.9
Vlčie bralo	29.1	26.2	24.0	21.6	19.7	15.6	15.2	14.7	14.7	13.7	10.6	9.7	9.3

Brúsik with pure limestones in its recharge area, Vlčie bralo dewaters dolomitic limestones. Discharge data are in L/s

waterworks will be in a need of water, and should take into account these values for securing available groundwater amounts. As we move further to the right on the FDC, water supply systems will begin to shut down due to low flow. As discharges move from Q_{95} towards Q_{100}, we would be facing the low-flow droughts. It is usual to present the discharge exceedence values in the tables where the most interesting reference points are more densely scaled at both the sides of the exceedence percentages (Q_1; Q_5; Q_{10} … Q_{90}; Q_{95}; Q_{99}), while the bulk is shown with the 10 % step. An example of the discharge exceedence table for the two karstic springs with similar values of the mean discharge is shown in Table 7.7. Available (exploitable) discharge amounts are then related—based on legislation or regional experience—as discharges with 70–90 % exceedence, i.e. Q_{70} or Q_{90}, usually Q_{80}, for securing water availability in the respective period of the year.

Another form of showing the discharge exceedence value is the so-called M-day discharge or M-day continuous discharge during low-water period where the exceedence value is given in the number of days throughout the year: 300-day exceedence then corresponds to 82.19 % exceedence (as is equal to 300/365) or 355-day exceedence corresponds to 97.26 % (=355/365). This means that 330-day discharge is statistically secured for 330 days annually—in other words that during 330 days within a year, the discharge of the spring is higher or at least equal to this value. We should bear in mind that many authors use the same way of showing the exceedence values as they were shown in percentages (e.g. Q_{90} for 90-day exceedence), and we should carefully check the author's attitude to distinguish the meaning of the values. Usually, the description of discharge exceedence exceeding the value of 100, such as Q_{300}, reveals that the "M-day discharge" format was applied.

The shape of the FDC is determined by hydraulic characteristics and geometrical shape of the aquifer and the recharge area, and the curve may be used to study these characteristics or to compare the characteristics of spring with those of another. A curve with a steep slope throughout denotes a highly variable spring whose discharge is largely from karstic conduits, whereas a curve with a flat slope reveals the presence of better groundwater storage capacity, which tends to equalise the flow. The slope of the lower end of the FDC shows the characteristics of the perennial storage in the recharge area; a flat slope at the lower end indicates a large amount of storage, and a steep slope indicates a negligible amount. Springs whose high discharges come mainly from snowmelt tend to have a flat slope at the upper end. The same is true for springs with large epikarst storage or those that are connected to surface water inputs draining swamp areas. An example of the FDC for the two karstic springs shown in Table 7.7 is shown in Fig. 7.5.

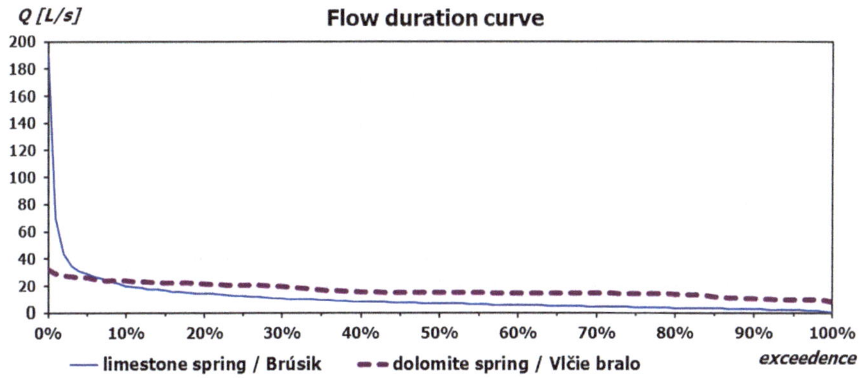

Fig. 7.5 Example of the FDC for two karstic springs—Brúsik spring dewaters pure limestones, Vlčie bralo dolomitic limestones. Discharge data are also listed in Table 7.7

It is of course better if we can construct a FDC using thousands of data points measured over many years or even decades. Here, the help of a function PERCENTILE( ) that is found in spreadsheet programs like MS Excel can be usefully exploited. Using this function, there is no need to order the discharge data according to the size (from highest to lowest). We simply reference the dataset field of all the measured discharge values («dataset») at first. We should have in mind then that we have to input the exceedence value as a residue between 1 and exceedence in a decimal format (0.7 for 30 % exceedence or 0.95 for 5 % exceedence). As an example, the PERCENTILE («dataset»; 0.8) would give the value of Q_{20} and the PERCENTILE («dataset»; 0.01) returns the value of Q_{99}.

7.4 Discharge Regime: Sub-regimes Versus Flow Components

Two simple examples of different discharge regimes shown of Fig. 7.3—the conduit-dominated karstic one and the diffuse discharge regime with much slower groundwater flow—are not only representing differences between the two groundwater circulation environments (aquifer types), but can be present within one aquifer. In traditional concept of "diffuse flow" and "quick-flow" part of discharge regime of karstic aquifers, the karst discharge regime can be composed of at least two (or more) parts recognised as sub-regimes. Complete discharge regime is then composed of superimposed sub-regimes, and majority of researchers prefer to describe these as "flow components". The two contrasting flow components typical for karst aquifers, the fast (conduit-dominated)-flow component and slow (diffuse)-flow, component are shown in Fig. 7.6. Discharge in summer and autumn period of 1979, again on Podhrad spring in Muráň, is quickly (within a day or two after precipitation) reaching the peak of several hundreds of litres per second (a tenfold discharge than the discharge before

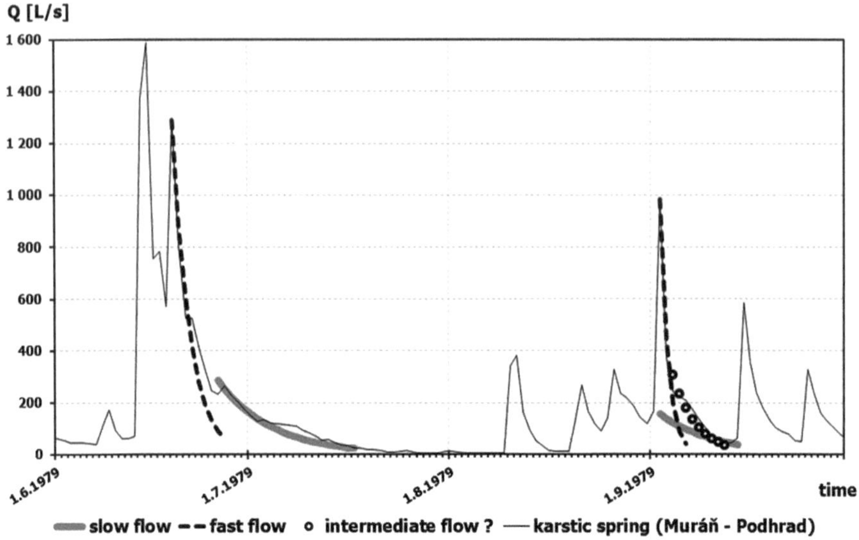

Fig. 7.6 Delineation of differently shaped sub-regimes of discharge on the recessional limb of hydrograph as different flow components—example of the Podhrad spring in Muráň municipality (Central Slovakia) during the summer period of 1979

this precipitation impulse), and then within next 4–6 days reaches the value between 100 and 200 L/s. This can be described as the quick-flow component. However, the decrease in discharge starting from the values of 100–200 L/s is not so rapid, and the discharge then reaches the previous value within the next approximately 20 days (left part of Fig. 7.6, discharge after 15 June 1979). Both fast-flow and slow-flow components here are represented by exponential function line, with different input parameters. On the right side of Fig. 7.6, however, the recessional limb of hydrograph since 2 September 1979 is different. The fast-flow component, before its complete transition to the slow-flow type of recession, is followed by another recessional type, here described as "intermediate flow". The different way of discharge declination in the period of "intermediate flow" domination, with its slope steeper than slow-flow slope and less steep than fast-flow slope, can be observed in some cases, while it is missing in other hydrologic situations. In any case, each karstic spring has a typical form of recessional hydrograph that makes it different from all other springs.

7.5 Mathematical Description of Recession and Flow Components

Boussinesq (1877) laid down the first theoretical principle of aquifer drainage and spring's discharge recession in time. His diffusion Eq. (7.5) describes flow through a porous medium:

$$\frac{\partial h}{\partial t} = \frac{K}{\varphi} \frac{\partial}{\partial x} \left(h \frac{\partial h}{\partial x} \right) \tag{7.5}$$

where

K represents hydraulic conductivity,
φ is the effective porosity (specific yield/storage coefficient) of the aquifer,
h stands for hydraulic head and
t is the time.

Using simplifying assumptions of homogeneous and isotropic intergranular unconfined aquifer of rectangular shape with concave floor, of depth H under the outlet level, where variations of h are negligible compared to aquifer depth H, neglecting capillarity effect above the water table, Boussinesq (1877) obtained an approximate analytical solution described by exponential Eq. (7.6):

$$Q_t = Q_0 \mathrm{e}^{-\alpha t} \tag{7.6}$$

where

Q_0 is the initial discharge,
Q_t is the discharge at time t and
α is the recession coefficient—an intrinsic aquifer parameter, expressed in reciprocal time units (day^{-1}) or (s^{-1}).

Maillet (1905) described similar aquifer recession curve by observations on analogous model of water-filled reservoir emptying through a porous plug—Eq. (7.6) is therefore also known as Maillet's formula. Considering the same simplifying assumptions of intrinsic aquifer properties (homogeneous, isotropic, intergranular, unconfined, rectangular shape, no capillarity), but of its shape limited by an impermeable horizontal layer at the outlet level, with curvilinear initial shape of water table (incomplete inverse beta function) where all flow velocities within the aquifer are horizontal (Dupuit–Forchheimer assumption), Boussinesq (1903, 1904) developed the analytical solution shown in Eq. (7.7):

$$Q_t = \frac{Q_0}{(1 + \alpha t)^2} \tag{7.7}$$

This quadratic formula (Eq. 7.7) of discharge recession was also in accordance with results of Dewandel et al. (2003), performing numerical simulations of shallow aquifers with impermeable floor at the outlet level. Moreover, according to Boussinesq's solution (1903, 1904), physical properties of the aquifer (hydraulic conductivity K, effective porosity φ) influence values of initial discharge Q_0 and recession coefficient α parameters as shown in Eqs. (7.8) and (7.9).

$$Q_0 = 0.862 K l \frac{h_m^2}{L} \tag{7.8}$$

$$\alpha = \frac{1.115 K h_m}{\varphi L^2} \tag{7.9}$$

where

L is the width of the aquifer and
h_m is the initial hydraulic head at the distance of L,
other parameters as mentioned above.

According to Eq. (7.9), the recession coefficient α as well as initial discharge Q_0 are dependent on the initial hydraulic head h_m (in other words, on the degree of aquifer saturation). On the other hand, apart from "Maillet's exponential formula", an approximate analytical solution (linearisation) by Boussinesq (1877) has more convenient constant value of recession coefficient α. Recession coefficient here is dependent only on aquifer properties—hydraulic conductivity K, effective porosity φ, width of the aquifer L and depth of the aquifer under the outlet level H (Eq. 7.10):

$$\alpha = \frac{\pi^2 K H}{4 \varphi L^2} \tag{7.10}$$

Compared to mathematically less convenient quadratic formula (Eq. 7.7) of the Boussinesq's solution (1903, 1904), Maillet's exponential formula (Eq. 7.6; Boussinesq 1877; Maillet 1905) is widely used by hydrologists and hydrogeologists due to its simplicity and linearisation in logarithmical plots. However, various shapes of hydrograms lead many authors to formulate other recession equations, e.g. exponential reservoir model (Hall 1968) in Eq. (7.11):

$$Q_t = \frac{Q_0}{(1 + \alpha Q_0 t)} \tag{7.11}$$

Griffiths and Clausen (1997) proposed two models, one for surface water accumulations (Eq. 7.12) and one for karstic channels (Eq. 7.13)

$$Q_t = \frac{\alpha_1}{(1 + \alpha_2 t)^3} \tag{7.12}$$

$$Q_t = \alpha_1 + \alpha_2 t \tag{7.13}$$

Kullman (1990) proposed a linear model for supposed turbulent flow in conduits, analogous to discharge recession in open surface channels (Eq. 7.14), where β is the recession coefficient for quick flow (similar to linear reservoir coefficient by Bonacci (2011), or linear decrease in discharge of discharge in Torricelli reservoir by Fiorillo (2011). Kovács (2003) proposed hyperbolic model of discharge recession of karstic springs (Eq. 7.15).

$$Q_t = \left(\frac{1}{2} + \frac{|1 - \beta t|}{2(1 - \beta t)}\right) Q_0 (1 - \beta t) \tag{7.14}$$

$$Q_t = \frac{Q_0}{(1+\alpha t)^n} \tag{7.15}$$

Still, it is not easy to find the single equation that can entirely describe the recession hydrograph—this is the reason for considering participation of various sub-regimes (flow components) in the process of discharge recession. The first studies (starting from Maillet 1905) recognised only two basic flow components (e.g. Barnes 1939; Schöeller 1948; Werner and Sundquist 1951; Forkasiewicz and Paloc 1967; Hall 1968; Drogue 1972; Kullman 1980; Milanović 1981; Padilla et al. 1994), and later studies (e.g. Kullman 1990; Bonacci 2011; Tallaksen 1995) describe presence of more than two flow components in springs' hydrographs. In order to interpret the entire recession hydrograph, the recession limb of a karst spring hydrograph can be approximated by a function that is the sum of several exponential segments of the total recession (Eq. 7.16), or taking into account other descriptions of recession, also as Eq. (7.17) where several Kullman's Eq. (7.14) for linear discharge decrease are applied.

$$Q_t = \sum_{i=1}^{n} Q_{0i} \mathrm{e}^{-\alpha_i t} \tag{7.16}$$

$$Q_t = \sum_{i=1}^{n} Q_{0i} \mathrm{e}^{-\alpha_i t} + \sum_{j=1}^{m} \left(\frac{1}{2} + \frac{|1-\beta_j \mathrm{t}|}{2(1-\beta_j t)} \right) Q_{0j}(1-\beta_j t) \tag{7.17}$$

Every ith or jth member of Eq. 7.16 or 7.17 describes one flow component, such as shown in Figs. 7.7 and 7.8.

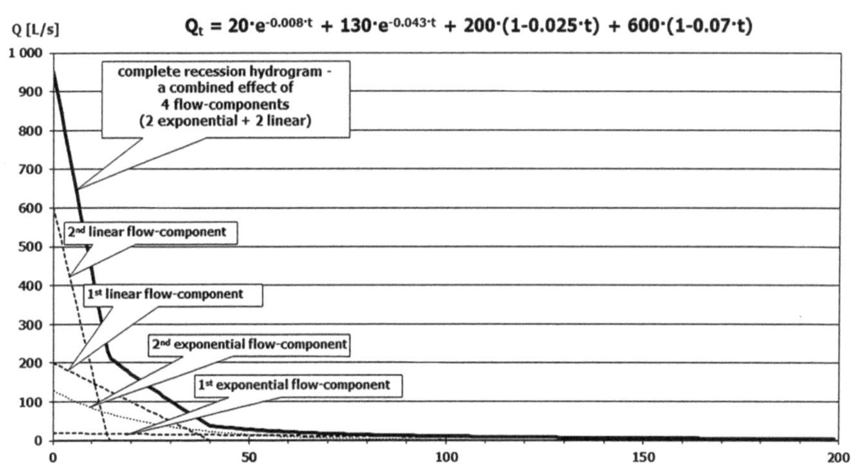

Fig. 7.7 Four flow components in ideal recession hydrograph (master recession curve)—normal plot

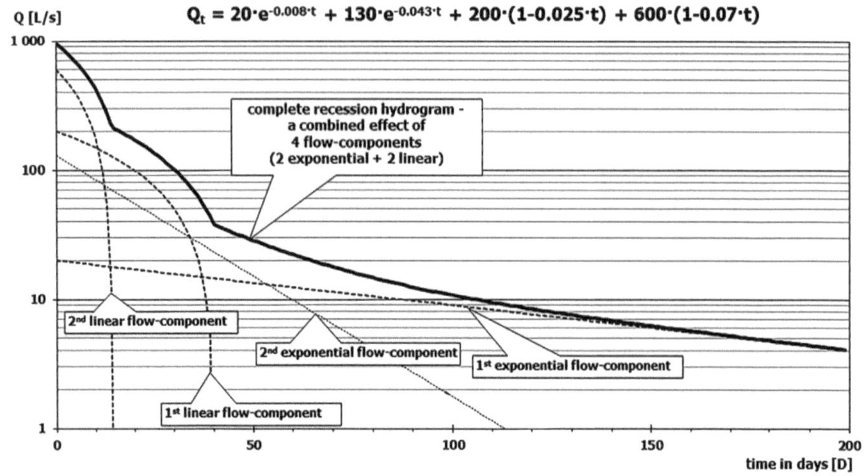

Fig. 7.8 Four flow components in ideal recession hydrograph (master recession curve)—semi-logarithmical plot

7.6 Identification of Flow Components in Recession Curves

Recession curves of karstic springs can be processed in many ways, starting from manual interpretation of the selected part of hydrograph on a paper used in matching strip method by Toebes and Strang (1964), through its digital processing (Lamb and Beven 1997; Rutlege 1998; Posavec et al. 2006; Gregor 2008) or even assembling recession discharge time series using genetic algorithms (Gregor and Malík 2012).

There are many methods for recession curve analysis, but in all of them we should select a part of the hydrograph showing the whole recessional period or its part. Evaluation of discharge threshold value, from which the recession starts (not always maximum), and the evaluation of recession period are often subjective; various authors show various criteria (Tallaksen 1995). In particular, in regions with groundwater recharge distributed within the whole hydrological cycle (e.g. moderate climate with many rainfall periods), it is not easy to distinguish the recession that is not influenced by additional recharge. The shape of recession curve is then changed—in order to avoid this problem, several methods have been developed to construct a master recession curve (MRC) from a set of shorter recessions (Tallaksen and van Lanen 2004). In this part, we should concentrate only on the analyses of the already selected recessional part of hydrograph, not taking into account whether it was assembled from several shorter recessions or selected as a single recessional event. In the hydrograph analyses, we should rely on the better visibility of linear elements by human eye and use both normal and semilogarithmical representation of the discharge time series.

The exponential flow components should be more visible in semilogarithmical graphic representation; the normal plots are more suitable for describing the linear recessional models (fast-flow components). We can also use both graph types, as shown in Fig. 7.9. The recessional part of hydrograph selected from the discharge data on Machnatá spring near Závadka and Hronom (Slovakia) is shown both in normal plot (left; Fig. 7.9a) and semilogarithmical plot (right; Fig. 7.9b). In the process of hydrograph decomposition, it is useful to start from the slow-flow (base-flow) component, which is usually of exponential nature and more visible on the semilogarithmical plot (Fig. 7.9d). This flow component is the last one that remains in the whole recessional process, and therefore, we should perform the analysis "from right to left", staring from the minimal discharge. We can derive the recession parameter of this flow component as a slope of the line that copies its shape, and starting discharge value as the discharge on the y axis that is cut by the prolongation (grey line) of this line towards the y axis showing discharge values (Fig. 7.9d). The main problem to be solved is the length of the period of basic slow-flow component domination. It should be limited by both the last and lowest discharge value on the right side, but the "left side" limitation is to be either visually estimated or computed, e.g. as the period with the best correlation coefficient of exponential regression. After the first interval was interpreted, we received the first couple of parameters: starting discharge of the first flow component Q_{01} and the recession coefficient α_1 (or β_1). In Fig. 7.9c, d, the interpreted values are 20 L/s for Q_{01} and -0.008 D^{-1} for α_1. For the next analysis step, it is better to subtract the interpreted flow component from the measured data to make the other flow components more visible (Fig. 7.9e, f). This means that from the value measured in the 48th day (here 29.83 L/s), the value $20{\cdot}e^{-0.008{\cdot}48}$ (=13.62 L/s) is subtracted giving the result of 16.20 L/s. The residual values are also shown in Fig. 7.9e, f. The interpretation then continues in the same way as in the case of the first exponential flow component (Fig. 7.9g, h), but the exponential regression (black line) and its prolongation (grey line) are described as $y = 130{\cdot}e^{-0.043x}$, and thus, $Q_{02} = 130$ L/s and $\alpha_2 = 0.043$ D^{-1}. Still, the second exponential flow component is more visible on the semilog plot (Fig. 7.9h). Figure 7.9i, j shows the recessional time series to be analysed after the next subtraction of both the first and second exponential flow components from the original values. Again, for the value measured in the 48th day (29.83 L/s), the values $20{\cdot}e^{-0.008.48}$ (=13.62 L/s) and $130{\cdot}e^{-0.043{\cdot}48}$ (=16.50 L/s) are subtracted—the result is -0.30 L/s. Similar residual values, both positive and negative, are plotted in Fig. 7.9i, j. We can see that the absolute value of these residua increases with higher discharge values showing us the uncertainties of measurement or of the chosen model. From this point, it is clear that the higher flow components are following the linear recessional model (fast-flow components). It is more convenient to interpret them on the normal plot—as in Fig. 7.9k in comparison with Fig. 7.9l. The black line of the interval analysed by linear regression and the grey line of its prolongation are described as $y = -5x + 200$ which can be in the sense of Eq. 7.14 translated into the form $y = 200(1-0.025x)$. Therefore, we interpret the parameters of the first linear flow component as $Q_{03} = 200$ L/s and $\beta_1 = 0.025$.

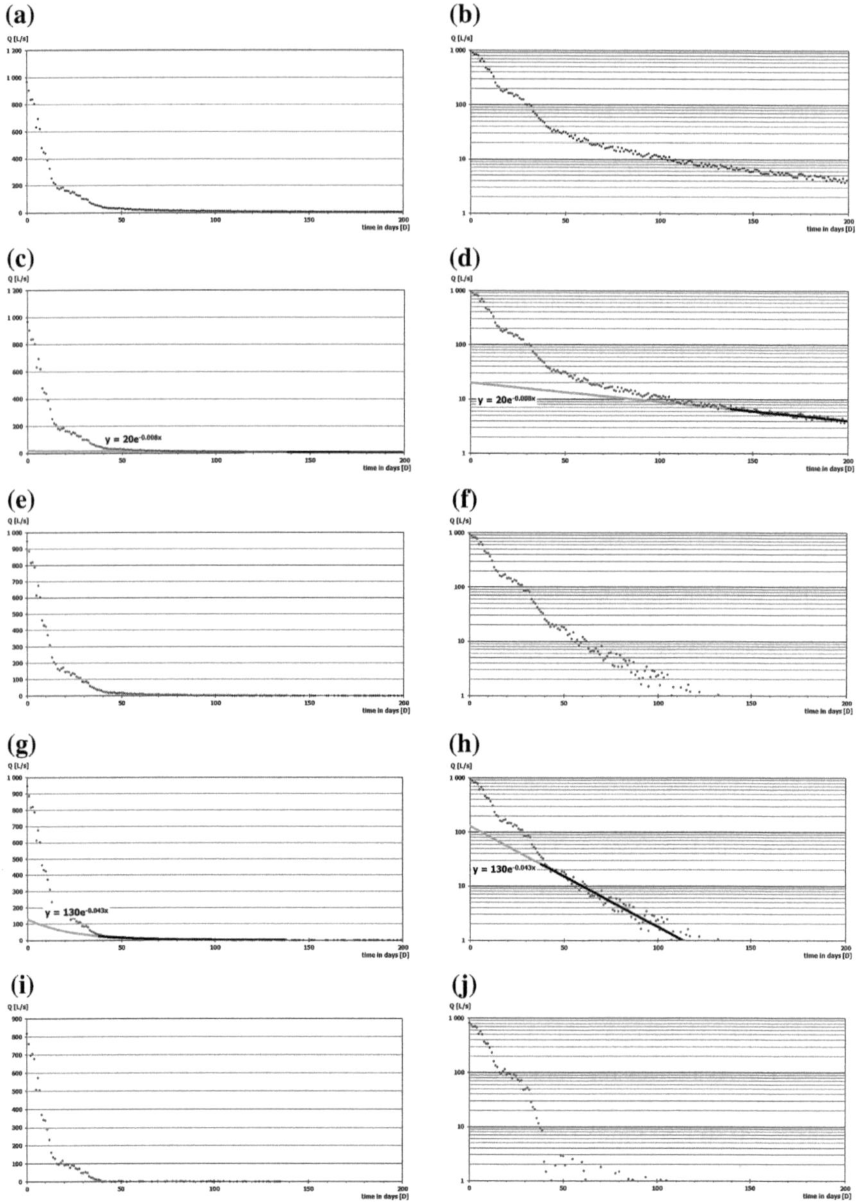
(a)
(b)
(c)
(d)
(e)
(f)
(g)
(h)
(i)
(j)
Q [L/s]
time in days [D]
y = 20e-0.008x
y = 130e-0.043x

Fig. 7.9 Gradual decomposition of recessional hydrograph into flow components. **a** original recessional hydrograph—normal plot; **b** original recessional hydrograph—semilog plot; **c** the first exponential flow component—normal plot; **d** the first exponential flow component—semilog plot subtraction of the first exponential flow component values—normal plot; **e** recession hydrograph after the subtraction of the first exponential flow component values—semilog plot; **f** recession hydrograph after the subtraction of the first exponential flow component values—semilog plot; **g** the second exponential flow component—normal plot; **h** the second exponential flow component—semilog plot; **i** recession hydrograph after the subtraction of the both first and second exponential flow component values—normal plot; **j** recession hydrograph after the subtraction of the both first and second exponential flow component values—semilog plot; **k** the first linear flow component—normal plot; **l** the first linear flow component—semilog plot; **m** recession hydrograph after the subtraction of the two exponential and one linear flow component values—normal plot; **n** recession hydrograph after the subtraction of the two exponential and one linear flow component values—semilog plot; **o** the second linear flow component—normal plot; **p** the second linear flow component—semilog plot; **q** the whole recession curve decomposed into four flow components (2 exponential + 2 linear; thin lines)—normal plot; and **r** the whole recession curve decomposed into four flow components (2 exponential + 2 linear; thin lines)—semilog plot

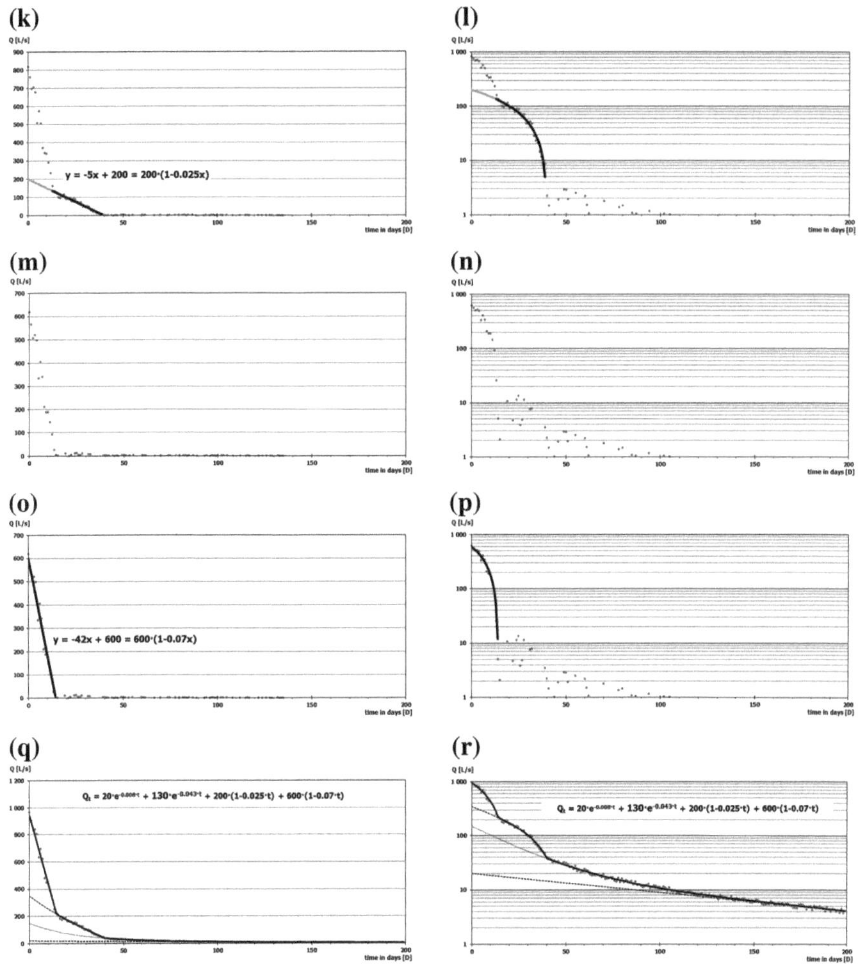

Fig. 7.9 (continued)

The values of the possible first linear flow component ought to be subtracted from the measured data together with the previous two exponential flow components, as shown in Fig. 7.9m, n. On the 7th day from the peak of 969 L/s, the measured discharge was 621 L/s. From this, 18.91 L/s belongs to the first exponential flow component (=20·$e^{-0.008 \cdot 7}$), 96.21 L/s to the second exponential flow component (=130·$e^{-0.043 \cdot 7}$), 165 L/s to the first linear flow component (=200−200·0.025·7), and only the rest of 341.17 L/s should be analysed. The result of this last partial recession analyses is shown in Fig. 7.9o, p described as $y = -42x + 600$, according to Eq. 7.14 translated into $y = 600(1-0.07x)$. The second linear flow component parameters are therefore $Q_{04} = 600$ L/s and $\beta_2 = 0.07$.

We should keep in mind that the duration of the exponential flow components lasts for the whole recessional period. To the left, we can prolong the interpreted duration towards the y axis ($t = 0$). To the right (higher time values), due to the nature of the exponential equation, we can still find the presence of all exponential flow components unless we set some artificial threshold value (e.g. 0.01 L/s or 1 L/s) from which we regard the presence of particular exponential flow component to be negligible. According to Eq. 7.14, the duration of each linear flow component is given by the parameter linear recession parameter β as $t_{DUR} = 1/\beta$ where t_{DUR} is the time of duration of the particular linear flow component. In the case of the recessional hydrograph shown in Fig. 7.9o, p, the first linear flow component lasts for 14 days (as 1/0.07 = 14.3). The second linear flow component diminishes after 40 days (1/0.025 = 40).

There are several methods how we can place the interpretation line to the graph—in the computer, we can create a linear regression line for the part selected for interpretation or create a line input manually, having a possibility to influence its position by changed input parameters.

7.7 Calculations of Flow Component Volumes

In the sense of Eq. 7.17, we can suppose that within a complete recessional process, different volumes of particular flow components are present (Figs. 7.10 and 7.11). Total change of groundwater volume in the aquifer during the discharge recession ΔV_{exp} between the discharge Q_{t1} measured in the time t_1 and later discharge Q_{t2} measured in the time $t_2 (Q_{t1} > Q_{t2})$ can be described due to the superposition principle as the sum of volume changes within individual flow components. If the discharge recession can be described merely by one flow component (exponential Eq. 7.6), the change of groundwater volume would be as follows:

$$\Delta V_{exp} = \int_{t_1}^{t_2} Q_0 e^{-\alpha t} dt = \frac{Q_{t_1} - Q_{t_2}}{\alpha} \tag{7.18}$$

Total volume of groundwater V_{exp} that was discharged during the complete recession process within one flow component with the initial discharge of Q_0 and

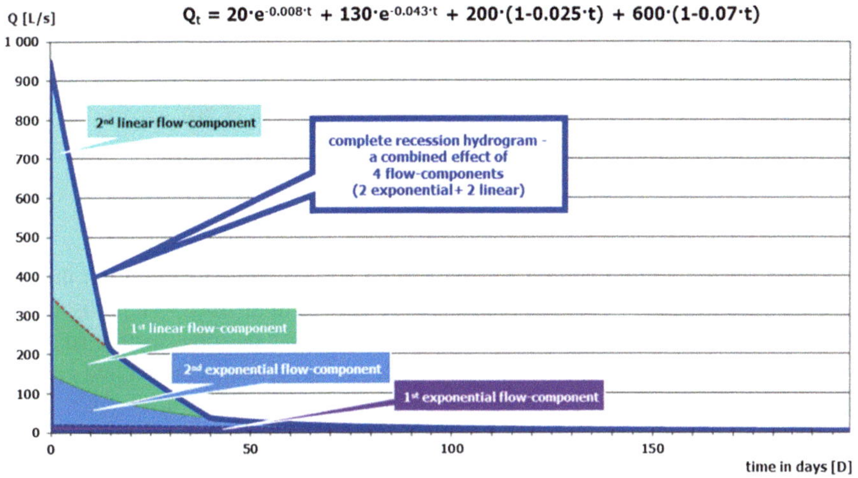

Fig. 7.10 Representation of particular volumes of flow components within a complete recessional hydrograph—normal plot

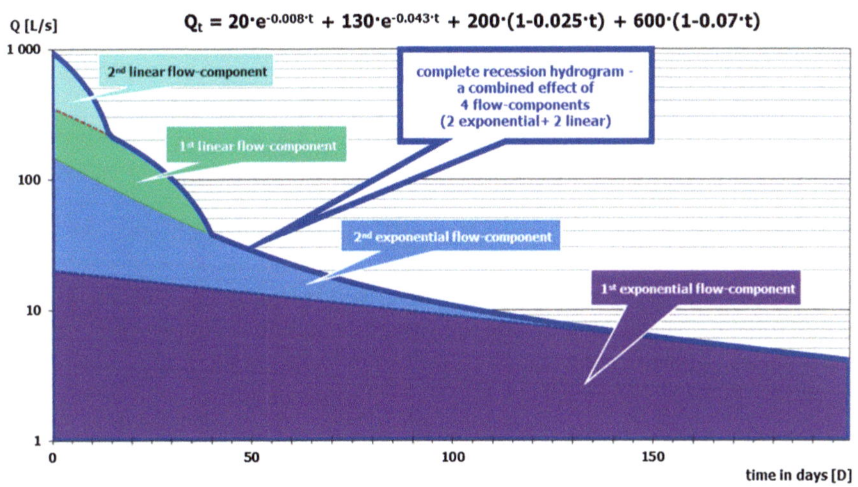

Fig. 7.11 Representation of particular volumes of flow components within a complete recessional hydrograph—semilogarithmical plot

the recession coefficient of $\boldsymbol{\alpha}$ can be then calculated as simple ratio of Q_0 and α (Eq. 7.19) as for the whole recession duration $Q_{t1} = Q_0$ and $Q_{t2} = 0$.

$$V_{\exp} = \frac{Q_0}{\alpha} \tag{7.19}$$

For more (n) flow components that can be described by exponential Eq. 7.6, the total groundwater volume change ΔV_{exp} between the discharge Q_{t1} measured in the time t_1 and discharge Q_{t2} measured in the time t_2 ($t_2 > t_1$ and $Q_{t1} > Q_{t2}$) is the summation of volume changes in all flow components (Eq. 7.20).

$$\Delta V_{exp} = \sum_{1}^{n} \int_{t_1}^{t_2} Q_0 e^{-\alpha t} dt = \frac{Q_{1t_1} - Q_{1t_2}}{\alpha_1} + \cdots + \frac{Q_{nt_1} - Q_{nt_2}}{\alpha_n} \tag{7.20}$$

While performing aforementioned calculations, one should have in mind the correct use of units: recession coefficients are usually in units (day^{-1}) and discharges in (L s^{-1}), so if we want to know the volume change in cubic metres (m^3), we have to convert discharges into (m^3 day^{-1}) or (m^3 s^{-1}), but in the second case we need also the conversion of recession coefficients from (day^{-1}) into (s^{-1}).

Change of groundwater volume in linear recession model (Eq. 7.14) ΔV_{lin} in the similar solution, where β is the linear recession coefficient, can be for one linear sub-regime described as in Eq. 7.21. The discharge Q_{t1} corresponds to the time t_1 and discharge Q_{t2} to the time t_2; $t_2 > t_1$ and $Q_{t1} > Q_{t2}$; both time t_1 and time t_2 are fulfilling the condition of being $<1/\beta$ to obtain only positive Q_{t1} and Q_{t2} values. The whole volume discharged in each individual flow component of linear recession model (where $Q_{t1} = Q_0$ and $Q_{t2} = 0$) can be calculated according to Eq. 7.22.

$$\Delta V_{lin} = \int_{t_1}^{t_2} Q_0 (1 - \beta t) dt = \frac{Q_{t_1}^2 - Q_{t_2}^2}{2 Q_0 \beta} \tag{7.21}$$

$$V_{lin} = \frac{Q_0}{2\beta} \tag{7.22}$$

For several quick-flow components (m) described by linear Eq. 7.14, the groundwater volume change ΔV_{lin} discharged (usually quick-flow components) between the moment of time t_1 and t_2 can be calculated as shown in Eq. 7.23.

$$\Delta V_{lin} = \sum_{1}^{m} \int_{t_1}^{t_2} Q_0 (1 - \beta_m t) dt = \frac{Q_{1t_1}^2 - Q_{1t_2}^2}{2 Q_{01} \beta_1} + \cdots + \frac{Q_{mt_1}^2 - Q_{mt_2}^2}{2 Q_{0m} \beta_m} \tag{7.23}$$

Also here, using Eqs. 7.22 and 7.23, we should be careful about the units, as β is usually given in (D^{-1}) and Q in L/s or m^3/s so we should express β in (s^{-1}) by dividing it by the number of seconds within a day (86,400). If both linear flow components and exponential flow components were identified in the discharge recession process, calculation of the total change of groundwater volume in the aquifer ΔV is as in Eq. 7.24.

$$\Delta V = \sum_{1}^{n} \int_{t_1}^{t_2} Q_0 e^{-\alpha t} dt + \sum_{1}^{m} \int_{t_1}^{t_2} Q_0 (1 - \beta_m t) dt \tag{7.24}$$

and then

$$\Delta V = \frac{Q_{1t_1} - Q_{1t_2}}{\alpha_1} + \cdots + \frac{Q_{nt_1} - Q_{nt_2}}{\alpha_n} + \frac{Q_{1t_1}^2 - Q_{1t_2}^2}{2Q_{01}\beta_1} + \cdots + \frac{Q_{mt_1}^2 - Q_{mt_2}^2}{2Q_{0m}\beta_m}$$

7.8 Hydrograph Separation into Flow Components

Hydrograph separation into flow components is a tool for distinguishing basic proportions of individual flow components present in total discharge. It enables quantitative referencing of flow components for further interpretations, e.g. for the purposes of estimation of the quick-flow duration or securing the exploitable amounts of karstic groundwater for longer periods. Or, at least, by the help of hydrograph separation, the diminishing points (in time) of individual flow components can be defined and proportions of individual flow components present at every stage of discharge can be quantified.

Hydrograph separation can be performed manually, by creation of stencil that is step by step laid directly to the hydrograph, while the lines of the respective flow component are gradually depicted on the same paper. The same process can be similarly performed by creation of MRC using the set of equations and input parameters that enable creation of virtual replica of MRC. The main idea behind this method is based on a simplified understanding of a hydrologic system reality: the same discharge should reflect the same water saturation (piezometric) level in the system. Although this assumption is a gross simplification, this method still may be helpful in quantitative referencing of flow components for further interpretations. In reality, temporary unequal distribution of saturation levels is usual in quantitative behaviour of karstic aquifers. Within karstified rock masses, several piezometric levels should exist at least for each saturated system (small fissures, medium fissures, karst conduits), if not for their different parts. Time dependency of these individual piezometric levels then substantially differs one from another. In spite of all, facing practical problems, the discharge data are in most cases the only values that may provide quantitative reference describing the whole system.

Principles of hydrograph separation based on master recession are demonstrated in Fig. 7.12. On the left side of Fig. 7.12, there is a typical MRC (violet line) delineated by superposition of 3 flow components—2 exponential and one linear. On the right side of Fig. 7.12, there is a part of real discharge hydrograph measured on the same spring. Individual flow components on the left side are highlighted by different patterns. Two horizontal lines starting on points Q_a and Q_b at the real discharge hydrograph from the right side of Fig. 7.12 are intersecting the MRC at corresponding two discharge levels of Q_A and Q_B. Each of these two discharges (Q_A and Q_B) on MRC is composed of different proportional representation of flow components, as visible on vertical bars drawn down from the intersections. In the case of the discharge Q_A, 3 flow components are present, while only

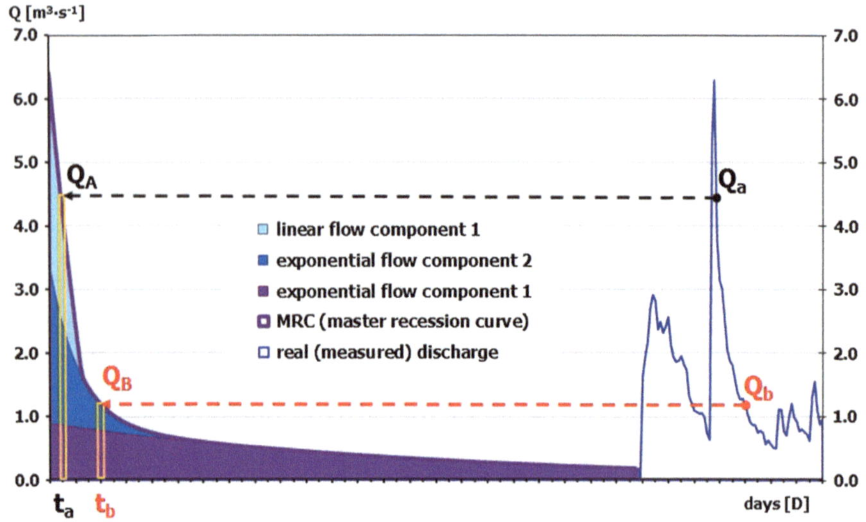

Fig. 7.12 Principles of hydrograph separation into flow components using master recession curve parameters

two flow components are found to sum up the discharge Q_B. This is to demonstrate that for every discharge on the right side of the figure, relevant value on recession curve is found (by calculation, as described later). Figure 7.12 also shows that each measured discharge value can be divided into several sub-regimes, depending on its position on the MRC. Also, every discharge value can be described by representative time t_R that had theoretically elapsed from the maximum discharge value $Q_{\max}$: discharge Q_A by the time t_a and discharge of Q_B by the time t_b as described in the following text.

Every karstic spring can be described by its own MRC, or—in other words—by unique set of parameters, individual constant values of starting discharges $Q_{01} \ldots Q_{0n}$ and $Q_{01} \ldots Q_{0m}$ and recession coefficients ($\alpha_1 \ldots \alpha_n$ and perhaps $\beta_1 \ldots \beta_m$). These parameters should be determined for each detected flow component (sub-regime). Theoretically, every equation from the aforementioned set of recession Eqs. (7.6, 7.7, 7.11–7.14 or 7.15) or also other recession equations can be applied—here, it is sufficient to use Eqs. 7.6 and 7.14.

In the process of hydrograph separation into individual flow components, each measured discharge value is understood to be composed of at least one flow component or superposition of two or more flow components. In the following text, several slow-flow components (exponential recession) and eventually also quick-flow components expressed by linear recession model are considered. According to Eq. 7.17, every measured discharge value Q_t is determined merely by ***representative time*** t_{R}, i.e. time that had theoretically elapsed from the absolute (overall) maximum discharge value $Q_{\max}$. This means that every discharge can be written just by substitution of the representative time t_{R} into Eq. 7.17. Subsequently, using the representative time t_R substituted into partial flow component equations, amounts discharged in these flow components can be calculated. For every flow component of exponential recession model, the representative time t_R can be calculated according to Eq. 7.25:

$$t_R = \frac{\ln Q_t - \ln Q_0}{-\alpha} \tag{7.25}$$

and for time t fulfilling the condition $t < 1/\beta$, the representative time t_R of (fast-flow) linear model flow components can be calculated according to Eq. 7.26

$$t_R = \frac{1}{\beta}\left(1 - \frac{Q_t}{Q_0}\right) \tag{7.26}$$

Having in mind that the karst spring recession hydrograph can be composed by summation of several exponential segments and several linear segments (Eq. 7.17), it is convenient to perform the computation of each theoretically elapsed time t_R by iteration process. Iterative method is a mathematical procedure that generates a sequence of improving approximate solutions up to some termination criteria which here, discussing the discharge of springs, can be related to the accuracy of discharge measurement. In practice, 10 iteration procedures were usually sufficient to give result within the discharge reading accuracy. Iterative solution is based on comparison of two solutions influencing the next iteration step (procedure). It is convenient to set the two starting t_R time inputs as $t_{R1} = 0$ (minimum value) and $t_{R2} = 1/\alpha_1$ (maximum value). In the next iteration step, the value of t_R achieved in the previous solution is substituted into Eq. 7.17 and the result is compared to the measured discharge Q_t. If it is higher, half of the interval between the two previous t_R values is added to the t_R in next iterative solution step. If the substituted Q_t value is lower than measured, the t_R in the next solution is lowered in one half of the interval between the two previous calculations of t_R values. In the next iteration step again, if after the substitution the Q_t value is lower than real, the proposed t_R value in the next solution is lower in one half of the interval between the two previous calculations of t_R values and vice versa. Gradual development of iterative solutions is repeatedly compared to the measured value and can be arbitrarily stopped if the difference is negligible enough, or—as stated before—is set to stop perhaps after the 10th or 20th iteration procedure. As the sequence of Eqs. 7.17 converges for given initial approximations, this iterative method is convergent.

In this way, for each measured discharge value Q_t, we can calculate the representative time t_R that enables its decomposition into flow components. Eq. 7.17 suggests that there are periods, when all flow components are present in the spring's discharge, and also periods when they gradually, one after another, diminish. In Eq. 7.17, all flow components are sufficiently described by their partial starting (maximal) discharges $Q_{0n} \ldots Q_{0m}$ and recession coefficients $\alpha_n \ldots \beta_m$. We should note that the representative time t_R that had theoretically elapsed from the maximal discharge $Q_{\max}$ is the same for each flow component. We can receive the actual partial discharge for each of the flow components (sub-regimes) by substitution of the t_R to its partial Eq. 7.6 or 7.14, but also other recession equations, e.g. Eqs. 7.7, 7.11–7.13 or 7.15. To check the calculation, the total discharge Q_t has to be the sum of these partial discharges.

Knowing the representative time t_R for each discharge value, proportional amounts of different discharging flow components can be calculated in one moment (Fig. 7.12), or for the whole evaluated period (Figs. 7.10 and 7.11). Presence of individual flow components can be expressed in discharge units (in L/s or m^3/s) for one moment of spring's discharge or as average discharges (average discharge of slow-flow component, etc.). The MRC shown in Fig. 7.12 can be described by Eq. 7.27. Discharge units here are m^3/s.

$$Q_t = 0.9e^{-0.007t} + 2.5e^{-0.09t} + 3.0(1 - 0.08t) \tag{7.27}$$

Another way of presentation of the results of hydrograph decomposition is to show the flow components in volumes discharged within the duration of evaluated periods. For example, the discharge $Q_a = Q_A$ of 4,661 L/s in Fig. 7.12 according to Eq. 7.27 consists of the slow-flow discharge of 875 L/s, "another slow-flow" component (exponential flow component 2) of 1,744 L/s (flow component with higher recession coefficient α) and the quick-flow component of 2,040 L/s. During the complete recession process in Fig. 7.12, using Eqs. 7.19 and 7.22, the volume of 11,108,571 m^3 was discharged within the slow-flow component, next 2,400,000 m^3 was discharged within the flow component with higher recession coefficient, and 1,620,000 m^3 was discharged as quick-flow component. The total volume of water discharged during the whole recession is then 15,128,571 m^3.

The advantage of the use of MRC parameter's hydrograph separation method is the clear solution for every discharge value it allows. However, described understanding of underground hydrologic system functions (assumption that the same discharge reflects the same water saturation or piezometric level in the aquifer) is a gross simplification. In reality, temporary unequal distribution of saturation levels is usual in aquifer's quantitative behaviour. Within karstic aquifers, several piezometric levels should exist at least for each saturated system (small fissures, medium fissures, karst conduits), if not for their different parts. Time dependency of these individual piezometric levels then substantially differs one from another. Király (2003) and Kovács et al. (2005) described recession coefficient as a global parameter depending on global configuration of the karst aquifers (also form and extension) and do not recommend its use to for aquifer hydraulic properties calculations. The same author underlines the role of mixing processes and dilution within the aquifer and shows that improperly used chemical or isotopic hydrograph separation methods may lead to invalid inferences regarding the groundwater flow processes. In spite of all, facing practical problems, the discharge values are in most cases the only data that represent quantitative reference describing the whole system. Simplified hydrograph separation method, based on proper recession curves analyses of the whole discharge time series in such cases, can help to distinguish and quantitatively express basic proportions of individual flow components. The aforementioned method is at least useful as quantitative reference of flow components for further interpretations. Also, by its help, the diminishing or starting point of individual flow components can be properly quantified, such as knowledge that the quick-flow component often connected with unwanted turbidities in the water source diminishes within 12.5 days after the peak maximum discharge (Eq. 7.27, Fig. 7.12), which might be useful from the water management point of view.

Case study 1. Assessment of spring discharge variability

Dolné Veterné is a spring far away from the inhabited areas of the Veľká Fatra Mts. (Slovakia). It was not possible to perform regular observations of its discharge more frequently than once a month. Within a year period, the observer provided discharge measurement results as follows (Figs. 7.13 and 7.14).

Try to apply classification criteria as degree of spring's reliability, degree of spring's discharge stability, SVC, degree of spring's discharge stability and SCVP *on the Dolné Veterné spring.*

Date	Q—discharge in L/s
07.11.2001	11.00
05.12.2001	17.20
02.01.2002	14.00
06.02.2002	27.70
06.03.2002	36.40
03.04.2002	33.20
01.05.2002	30.10
05.06.2002	25.30
03.07.2002	17.20
07.08.2002	13.20
04.09.2002	16.00
02.10.2002	15.60

Fig. 7.13 Spring Dolné Veterné in Gaderská dolina valley (Blatnica municipality) in the second decade of the twenty-first century—photograph from the database of the Slovak Hydrometeorological Institute

Exercise

Using the twelve discharge data above, we receive

$Q_{\min} = 11.00$ L/S (Eq. 7.1)
$Q_{\max} = 36.40$ L/S (Eq. 7.1)
$\Phi = 21.41$ L/S (Eq. 7.2)
$Q_{10} = 32.89$ L/S (Eq. 7.3)
$Q_{90} = 13.28$ L/S (Eq. 7.3)
$\sigma = 8.65$ L/S (Eq. 7.4)

Solution
By using Eq. (7.1), the value of variability index I_v is 3.31—according to this, degree of spring's reliability (Table 7.2) is *very good*. According to Table 7.3, degree of spring's discharge stability is *unstable*.
Equation (7.2) gives value of spring variability V as 119 %, and the spring should be supposed as *variable*.
SVC according to Eq. (7.3) equals 2.48, and its classification using this coefficient should be "*steady*" (Table 7.4), or the degree of spring's discharge stability can be described as "*very stable*" (Table 7.5).
According to Eq. (7.4), the value of SCVP is 0.40. Its variability is then classified as "*low*" (Table 7.6).

P.S.: From this spring (Dolné Veterné), weekly discharge data were taken for a period of more than 30 years since 1978; for preparation of this exercise, only few of them were selected. Based on the 1357 readings of weekly gauging results for the period 1978–2004, the complete dataset would give results as $Q_{\min} = 3.16$ *L/s;* $Q_{\max} = 82.40$ *L/s;* $\Phi = 21.46$ *L/s;* $Q_{10} = 34.20$ *L/s;* $Q_{90} = 8.93$ *L/s; and* $\sigma = 9.83$ *L/s. The spring would then reach bad reliability* (Table 7.2) *and very unstable discharge stability* (Table 7.3) *as variability index* $I_v = 26.08$*; but according to* Table 7.4 *would be well balanced and stable* (Table 7.5) *as* $SVC = 3.83$*; with low SCVP* (Table 7.6; $SCVP = 0.46$), *while* $V = 369$ % (Eq. 7.2). *This is to illustrate the importance of discharge gauging time span.*

Case study 2. Construction of FDC duration curve, calculation of exceedence

Try to calculate discharge exceedence Q_1; Q_5; Q_{10}; Q_{20}; Q_{30}; Q_{40}; Q_{50}; Q_{60}; Q_{70}; Q_{80}; Q_{90}; Q_{95}; *and* Q_{99} *using the data from the Dolné Veterné spring in the Case study 1. The dataset here is quite limited, but using the PERCENTILE () function in MS Excel, you will be able to cope with the task. Based on the exceedence data, draw a graph of the FDC.*

Exercise
In MS Excel, use the function PERCENTILE («dataset» ;«percentage») with references to «dataset» as «A2:A13» and «percentage» derived from column «C» as follows:

	A	B	C	D
1	Q (L/s)		%	Q_x
2	11.00		1	=PERCENTILE(A2:A13;1-C2)
3	17.20		5	=PERCENTILE(A2:A13;1-C3)
4	14.00		10	=PERCENTILE(A2:A13;1-C4)

	A	B	C	D
5	27.70		20	=PERCENTILE(A2:A13;1-C5)
6	36.40		30	=PERCENTILE(A2:A13;1-C6)
7	33.20		40	=PERCENTILE(A2:A13;1-C7)
8	30.10		50	=PERCENTILE(A2:A13;1-C8)
9	25.30		60	=PERCENTILE(A2:A13;1-C9)
10	17.20		70	=PERCENTILE(A2:A13;1-C10)
11	13.20		80	=PERCENTILE(A2:A13;1-C11)
12	16.00		90	=PERCENTILE(A2:A13;1-C12)
13	15.60		95	=PERCENTILE(A2:A13;1-C13)
14			99	=PERCENTILE(A2:A13;1-C14)

Solution

We can derive following exceedence values from the data given in the Case study 1 (also in the A2:A13 database above):

Q_1	Q_5	Q_{10}	Q_{20}	Q_{30}	Q_{40}	Q_{50}	Q_{60}	Q_{70}	Q_{80}	Q_{90}	Q_{95}	Q_{99}
36.05	34.64	32.89	29.62	26.98	22.06	17.20	16.48	15.72	14.32	13.28	12.21	11.24

From these, we can plot a FDC like this (in thick line, data from the table above are used).

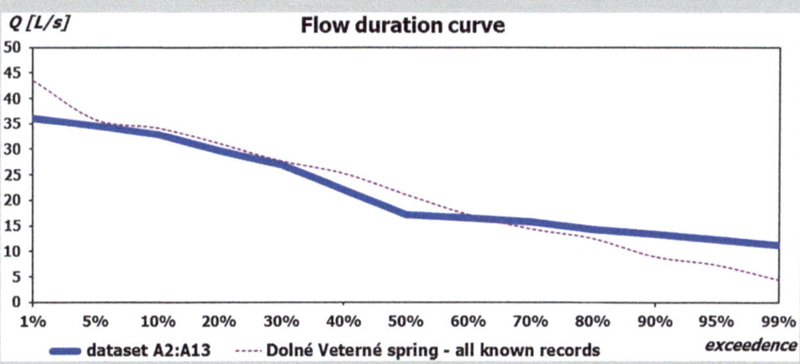

Fig. 7.14 Flow duration curve—spring Dolné Veterné in Gaderská dolina valley (Blatnica municipality); *thick line* is for the exercise dataset A2:A13; *scattered thin line* is derived from all known discharge records (1,357 records taken weekly): note that the curves differ mainly at the beginning (*maximal values*) and the end (*minimal values*) of the lines

Case study 3. Plotting of theoretical recession curve, calculation of flow component volumes and duration of flow components

For spring XY, after recession curve analyses, MRC was created with two exponential flow components (following Eq. 7.6) *and two linear flow components (following* Eq. 7.14). *Following parameters were identified for*

individual flow components: $Q_{01} = 58.7$ *L/s and* $\alpha_1 = 0.005\ D^{-1}$ *for the first exponential flow component and* $Q_{02} = 174.9$ *L/s and* $\alpha_2 = 0.08\ D^{-1}$ *for the second exponential flow component. For the two flow components following linear depletion model* (Eq. 7.14), $Q_{03} = 629$ *L/s and* $\beta_1 = 0.04\ D^{-1}$ *were for the first and* $Q_{04} = 2450$ *L/s and* $\beta_2 = 0.3\ D^{-1}$ *for the second linear flow component. Try to calculate decrease in individual flow components using* Eqs. 7.6 *and* 7.14, *as well as to find values for the whole recessional process.*

Exercise

In MS Excel, use the EXP ( ) function for the first two "slow-flow" components. Input parameters in the following table are marked as bold and shaded. Calculate values of the flow components up to 160 days after the maximum. Then, plot the data of both flow components into one graph with normal plot of discharge values on another graph with logarithmical *y* axis for discharge. Try to play with the input data in the fields $B$1, $B$2, $B$6 and $B$7 (make recession coefficients or starting discharges bigger or smaller), and see how the plotted curves change their shape.

	A	B	C	D	E	F
1	**alpha1 [1/D]**	**0.005**		**time [days]**	**exponential 1**	**exponential 2**
2	**alpha2 [1/D]**	**0.08**		0	=$B$6*EXP(-$B$1*$D2)	=$B$7*EXP(-$B$2*$D2)
3	**beta1 [1/D]**	**0.04**		1	=$B$6*EXP(-$B$1*$D3)	=$B$7*EXP(-$B$2*$D3)
4	**beta2 [1/D]**	**0.3**		2	=$B$6*EXP(-$B$1*$D4)	=$B$7*EXP(-$B$2*$D4)
5				3	=$B$6*EXP(-$B$1*$D5)	=$B$7*EXP(-$B$2*$D5)
6	**Q01 [L/s]**	**58.7**		4	=$B$6*EXP(-$B$1*$D6)	=$B$7*EXP(-$B$2*$D6)
7	**Q02 [L/s]**	**174.9**		5	=$B$6*EXP(-$B$1*$D7)	=$B$7*EXP(-$B$2*$D7)
8	**Q03 [L/s]**	**629**		6	=$B$6*EXP(-$B$1*$D8)	=$B$7*EXP(-$B$2*$D8)
9	**Q04 [L/s]**	**2450**		7	=$B$6*EXP(-$B$1*$D9)	=$B$7*EXP(-$B$2*$D9)
10				8	=$B$6*EXP(-$B$1*$D10)	=$B$7*EXP(-$B$2*$D10)
...				...	...	...
160				158	=$B$6*EXP(-$B$1*$D160)	=$B$7*EXP(-$B$2*$D160)
161				159	=$B$6*EXP(-$B$1*$D161)	=$B$7*EXP(-$B$2*$D161)
162				160	=$B$6*EXP(-$B$1*$D162)	=$B$7*EXP(-$B$2*$D162)

For the linear flow components, use another expression but the same input parameters:

	A	B	C	D	…	G	H
1	**alpha1 [1/D]**	**0.005**		**time [days]**	…	**linear 1**	**linear 2**
2	**alpha2 [1/D]**	**0.08**		0	…	=IF($B$8*(1-$B$3*$D2)<0;0;$B$8*(1-$B$3*$D2))	=IF($B$9*(1-$B$4*$D2)<0;0;$B$9*(1-$B$4*$D2))
3	**beta1 [1/D]**	**0.04**		1	…	=IF($B$8*(1-$B$3*$D3)<0;0;$B$8*(1-$B$3*$D3))	=IF($B$9*(1-$B$4*$D3)<0;0;$B$9*(1-$B$4*$D3))
4	**beta2 [1/D]**	**0.3**		2	…	…	…
5				3	…	…	…
6	**Q01 [L/s]**	**58.7**		4	…	…	…
7	**Q02 [L/s]**	**174.9**		5	…	…	…
8	**Q03 [L/s]**	**629**		6	…	…	…
9	**Q04 [L/s]**	**2450**		7	…	…	…
…				…	…	…	…
161				159	…	=IF($B$8*(1-$B$3*$D161)<0;0;$B$8*(1-$B$3*$D161))	=IF($B$9*(1-$B$4*$D161)<0;0;$B$9*(1-$B$4*$D161))
162				160	…	=IF($B$8*(1-$B$3*$D162)<0;0;$B$8*(1-$B$3*$D162))	=IF($B$9*(1-$B$4*$D162)<0;0;$B$9*(1-$B$4*$D162))

In the linear flow components, their values would quickly fall below zero—to eliminate the influence of the negative values on the final result, the simple formula = $B$8*(1−$B$3*$D2) for the cell G2 has to be blocked out for values bellow zero. In Eq. 7.14, the part of the formula $\left(\frac{1}{2}+\frac{|1-\beta t|}{2(1-\beta t)}\right)$ serves the same purpose, but in spreadsheet we can do it manually—e.g. by formulae = IF($B$8*(1-$B$3*$D2) < 0;0;$B$8*(1-$B$3*$D2)) for the cell G2 and so on.

Add the data of both linear flow components into the previous graphs (normal and semilog one). Try to change the input data in the fields $B$3, $B$4, $B$8 and $B$9 (make them bigger or smaller), and see how the plotted curves change their shape.

For creating the whole MRC, do the sum of all partial flow components: count together columns «E» + «F» + «G» + «H». Try to change the input data in the fields \$B\$1–\$B\$4 and \$B\$6–\$B\$9 (make them bigger or smaller), and see how the main plotted (MRC) and the recession curves of partial flow components change the shape with different combination of input values both in normal and semilogarithmical plot.

Exercise

Now, having the recession described by Eq. (7.28), *we should try to calculate volumes of individual flow components that are depleted within the whole recessional process.*

$$Q_t = 58.7\mathrm{e}^{-0.005t} + 174.9\mathrm{e}^{-0.08t} + 629(1 - 0.04t) + 2450(1 - 0.3t) \qquad (7.28)$$

Solution

Volume of water discharged within individual flow components is illustrated as an area of different colour, delineated by axes and curves or lines of flow component functions of Figs. 7.10 and 7.11. For exponential flow, Eq. 7.19 and for linear model flow component, Eq. 7.22 are used for calculation of the whole volume discharged within the complete recession cycle. Summation of partial volumes of individual flow components also represents the maximal water storage within the dewatered aquifer or within the recharge area. Partial volumes can be linked to storage in small or bigger fissures (slow-flow/exponential components) or conduits and karst channels (quick-flow/linear model components).

$$V_{\mathrm{exp1}} = \frac{Q_{01}}{\alpha_1} = \frac{58.7/1{,}000}{0.005/86{,}400} = 1{,}014{,}336\,\mathrm{m}^3$$

$$V_{\mathrm{exp2}} = \frac{Q_{02}}{\alpha_2} = \frac{174.9/1{,}000}{0.08/86{,}400} = 188{,}892\,\mathrm{m}^3$$

V_{exp1} is the volume discharged within the first exponential flow component, V_{exp2} within the second one. Note that number of 86,400 s within a day was used in the denominator, as the recession coefficients were in the (D^{-1}) units. Also, divided by 1,000, discharge in the numerator was recalculated from (L/s) units into (m^3/s) units. We can see here that although starting discharge of the second exponential flow component is three times bigger than that of the first one, due to small recession coefficient, the water volume discharged in first exponential flow component in more than five times prevails over the second one. From this, we can also judge on the volume of joints and fissures with different apertures.

$$V_{\mathrm{lin1}} = \frac{Q_{03}}{2\beta_1} = \frac{629.0/1{,}000}{2 \cdot 0.04/86400} = 679{,}320\,\mathrm{m}^3$$

$$V_{\mathrm{lin2}} = \frac{Q_{04}}{2\beta_2} = \frac{2{,}450.0/1000}{2 \cdot 0.3/86{,}400} = 352{,}800\,\mathrm{m}^3$$

Volume discharged in the first quick-flow (linear) component V_{lin1} is nearly two times as big as in the second case (V_{lin2}). Recalculation for units (from (L/s) to (m^3/s) and from (D^{-1}) to (s^{-1})) is applied also here. The total discharged volume is then

$$\begin{aligned} V_{\mathrm{TOTAL}} &= \frac{Q_{01}}{\alpha_1} + \frac{Q_{02}}{\alpha_2} + \frac{Q_{03}}{2\beta_1} + \frac{Q_{04}}{2\beta_2} \\ &= \frac{58.7/1000}{0.005/86400} + \frac{174.9/1000}{0.08/86400} + \frac{629.0/1{,}000}{2 \cdot 0.04/86{,}400} \\ &\quad + \frac{2{,}450.0/1{,}000}{2 \cdot 0.3/86{,}400} \\ &= 2{,}235{,}348\,m^3 \end{aligned}$$

Exercise

Try to calculate volume of water discharged within the first exponential flow component during the period while its partial discharge decreased from 40 to 30 L/s. For the second linear flow component, try to calculate volume of water discharged within this particular sub-regime while it had fallen down from 2,000 to 1,000 L/s. Apply on the same spring described by the same recessional Eq. (7.28).

Solution

The change of groundwater volume discharged within single recessional flow component of exponential nature (Eq. 7.6) between two given discharge values is described by Eq. 7.18. In our case, $Q_{t1} = 40.0$ L/s and $Q_{t2} = 30.0$ L/s. Then, using Eq. 7.18,

$$\Delta V_{\mathrm{exp}} = \frac{Q_{t_1} - Q_{t_2}}{\alpha} = \frac{40/1{,}000 - 30/1{,}000}{0.005/86{,}400} = 172{,}800\ \mathrm{m}^3$$

Again, note that L/s had to be converted into (m^3/s) and (D^{-1}) into (s^{-1}) to receive the result in (m^3).

Equation 7.21 is used to calculate the change of groundwater volume in linear recession model; in our task, the discharge in time t_1 was $Q_{t1} = 2{,}000.0$ L/s and Q_{t2} in time t_2 was 1,000.0 L/s. The partial starting discharge for the second linear flow component (see it in Eq. 7.28) $Q_{04} = 2{,}450.0$ L/s. Also here, we have to take care about the units. Then,

$$\Delta V_{\mathrm{lin}} = \frac{Q_{t_1}^2 - Q_{t_2}^2}{2Q_0\beta} = \frac{(2{,}000/1{,}000)^2 - (1{,}000/1{,}000)^2}{2 \cdot (2{,}450/1{,}000) \cdot (0.3/86{,}400)} = 58{,}776\ \mathrm{m}^3$$

We can see that although discharges within the second linear flow component are enormous in comparison with those we have in the first exponential flow component, the total volume discharged in the slow-flow component while it drops from 40 to 30 L/s is nearly 3 times as big as discharged within quick flow.

Exercise

Try to calculate duration of individual flow components, both exponential and linear, having the same recession described by Eq. (7.28).

Solution

Duration of each linear model flow component can be described by Eq. 7.26; we should just assume that it is the moment when Q_t reaches zero. For the first linear flow component, then,

$$t_{\mathrm{DUR-linear1}} = \frac{1}{\beta_1}\left(1 - \frac{Q_t}{Q_{03}}\right) = \frac{1}{0.04}\left(1 - \frac{0}{Q_{03}}\right) = \frac{1}{0.01} = 25 \text{ days}$$

Note that there is no recalculation from (D^{-1}) unit of the recession coefficient into (s^{-1}), and therefore, the result is expressed in days. For the second linear flow component, then, the time of its duration $t_{\mathrm{DUR-linear2}}$ is calculated as

$$t_{\mathrm{DUR-linear2}} = \frac{1}{\beta_2} = \frac{1}{0.3} = 3.3 \text{ days}$$

The exponential formula used for the slow-flow description (Eq. 7.6) causes that each slow-flow discharge component should be steadily present in each moment of the recession. Sometimes, it is useful to use some conventional threshold to delineate the duration interval of an exponential flow component. Let us define this threshold here to be 1.0 L/s. Then, we can use Eq. 7.25 and set Q_t to be 1.0 L/s. For the two exponential flow components from Eq. 7.28, their duration ($t_{\mathrm{DUR-exponential1}}$ and $t_{\mathrm{DUR-}l\ \mathrm{exponential2}}$) then can be delineated as

$$\begin{aligned} t_{\mathrm{DUR-exponential1}} &= \frac{\ln Q_t - \ln Q_{01}}{-\alpha_1} = \frac{\ln(1.0) - \ln(58.7)}{-0.005} \\ &= 814.5 \text{ days} \\ t_{\mathrm{DUR-}e\mathrm{exponential2}} &= \frac{\ln Q_t - \ln Q_{02}}{-\alpha_2} = \frac{\ln(1.0) - \ln(174.9)}{-0.08} \\ &= 64.6 \text{ days} \end{aligned}$$

Theoretically, the base flow (as the first exponential flow component is the most persistent flow component in the hydrograph) of this spring should last for 814.5 days until its discharge falls below 1.0 L/s. The second exponential flow component diminishes much more sooner; let us suppose within 65 days when it participates on the total discharge by less than 1.0 L/s.

Case study 4. Hydrograph separation into flow components—calculation of the representative time t_R, theoretically elapsed from the maximum discharge value Q_{max} for discharge value of Q_t

Hydrograph separation based on parameters of MRC described in Eq. 7.29 *is shown on the example of karstic spring Vítek in Chtelnica (Brezovské Karpaty Mts., Slovakia). As it is evident from* Eq. 7.29 *and* Fig. 7.15, *karstic groundwater discharged in the Vítek spring has a strong portion of the 1st exponential (slow-flow) component constantly present in the spring's discharge, while the quick-flow component (1st linear model component) appears only at high-flow periods when the total discharge exceeds ~30 L/s. The 2nd exponential flow component that is probably linked to groundwater circulation in opened fissures is regularly present in the total discharge, but only in small proportion. Its volume steeply rises in discharges of more than ~30 L/s. Still, nearly 85 % of the total volume of water is discharged in the 1st exponential flow component (sub-regime) that points to groundwater circulation in strongly fissured aquifer in dolomites.*

$$Q_t = 25.68e^{-0.0030t} + 7.66e^{-0.01t} + 9.52(1 - 0.04t) \tag{7.29}$$

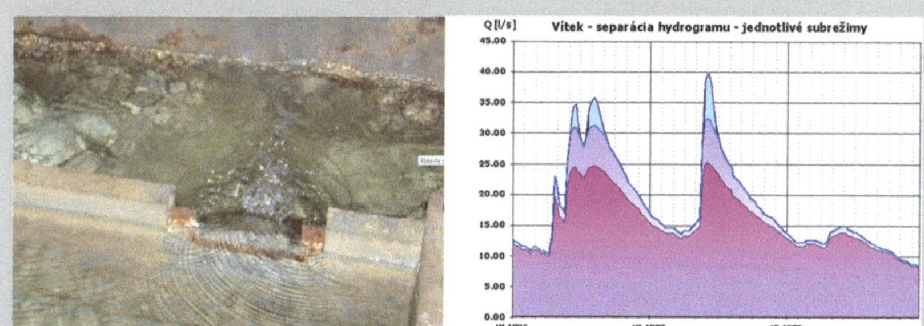

Fig. 7.15 Spring Vítek in Chtelnica (Brezovské Karpaty Mts., Slovakia), captured as a drinking water source (*left*, photograph from the database of the Slovak Hydrometeorological Institute). Its hydrograph separation into flow components using Eq. 7.29 (*right*)

Try to find out the how much of each flow component will be present in the discharge of 37.0 L/s and what relative portion of the second exponential flow component you will find in the discharge of 25.0 L/s in the spring Vítek.

Exercise

In MS Excel, use functions as in Case study 3, but change the input values in the «B» column to 0.003 ($B$1); 0.01 ($B$2); 0.04 ($B$3); and 0 ($B$4) for the recession coefficients as the second linear flow component is not described by Eq. 7.28. For starting discharges, change «B» column values in the following way: \$B\$6 to 25.68; \$B\$7 to 7.66; \$B\$8 to 9.52; and \$B\$9 to zero (only one linear fast-flow component is present). Then, make the summation of all 3 existing flow components columns perhaps in the column «I», so that «I» = «E» + «F» + «G» in every row. Then, compare the values in the column «I» with the thresholds of 37.0 and 25.0 L/s.

In the first case, you will see that for time of 11 days the total discharge is 37.04 L/s, while for time of 12 days after the maximum, the total discharge is 36.52 L/s. This means that the representative time t_R for the discharge of 37.00 L/s ($t_{37.0}$) is somewhere between 11 and 12 days. Then, try to change the value in the column «D» unless the total discharge (column «I») in the same row is within an acceptable limit (the second decimal place). Here, $t_{37.0} = 11.08$ D. You were working manually instead of the described iteration process. You will see then that for $t_R = 11.08$ days, the first exponential component (column «E») would be 24.84 L/s (as it is $25.68{\cdot}e^{-0.003\,\cdot\,11.8}$), the second exponential component (column «F») would be 6.86 L/s (as it is $7.66{\cdot}e^{-0.01\,\cdot\,11.8}$) and the relative discharge of the only linear flow component would be 5.30 L/s (as it is equal to 9.52–9.52·0.04·11.08).

In the second case, we will find that for time of 66 days, the total discharge is 25.03 L/s, and for 67 days after the maximum, the total discharge would be 24.92 L/s. Again, the representative time t_R for the discharge of 25.00 L/s ($t_{25.0}$) should be somewhere between 66 and 67 days. After several trials to change the value in the column «D» unless the total discharge (column «I») in the same row would be within an acceptable limit (the second decimal place), we can estimate $t_{25.0}$ as 66.25 days. Then, that for $t_R = 66.25$ days, the first exponential component (column «E») would be 21.05 L/s ($=25.68{\cdot}e^{-0.003\,\cdot\,66.25}$), the second exponential component (column «F») would be 3.95 L/s ($=7.66{\cdot}e^{-0.01\,\cdot\,66.25}$) and the linear flow component would not exist at this moment.

The relative portion of the second exponential flow component would be then

$$3.95/25.00 = 16\%$$

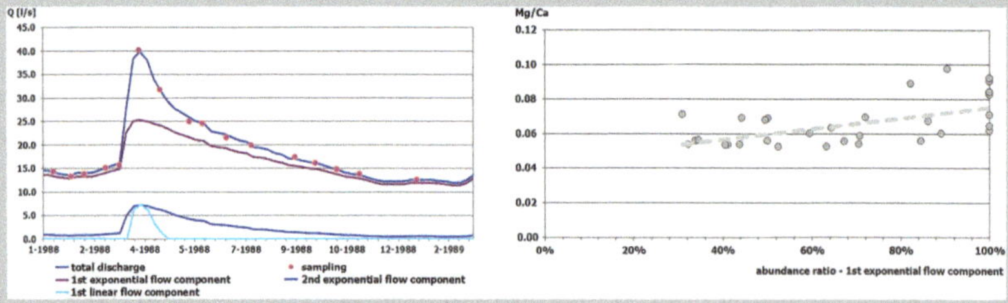

Fig. 7.16 Groundwater sampling of spring Vítek at various water stages with individual flow components differently represented (*left*). Correlation of Mg/Ca content ratio with relative abundance of the 1st exponential flow component (*right*)

In majority of cases, groundwater in a single sample can be considered as being a mixture of several water types of slightly different origin. This is even more evident in the case of karstic springs. Knowing that at every moment each measured discharge of a karstic spring is composed of superposition of two or more individual flow components, while only one flow component is usually present only during the driest period, we can try to link the results of water quality analyses with quantitative parameters. Proportional amounts of individual flow component discharges during the moment of sampling can be linked to the content of various (chemical) components present in the water sample. To obtain the end members of the theoretical mixture, if having sufficient number of samples taken at various water stages, we can perhaps try to link the relative representation of flow component in the total discharge with the analysed parameter (Fig. 7.16*). To obtain the end member of the theoretical mixture corresponding to the "pure" (100 %) representation of the certain flow component, we only draw the forecast line based on statistical regression of flow component relative representation and parameter content up to the 100 % value. This is shown in Fig.* 7.16*, where the dissolved magnesium* (Mg^{2+})*/calcium* (Ca^{2+}) *ratio is correlated with the relative representation of the 1st exponential flow component to obtain estimation of the end member of 0.075 Mg/Ca ratio in this flow component, if existing as pure solution.*

Reference:

Malík, P., Michalko, J., 2010. Oxygen Isotopes in Different Recession Subregimes of Karst Springs in the Brezovské Karpaty Mts. (Slovakia). Acta Carsologica 39, 2, 271–287

Different shapes of the spring's recession are attributed to drainage from different components of the groundwater system, reflecting karstification degree. For example, flatter parts of the recession curve may represent slow groundwater drainage of pores and micro fractures, which is characteristic of the 1st exponential flow component. This is described with an exponential equation having a smaller exponent. Portions of the hydrograph with a steeper slope are characteristic of enhanced karstification degree, with groundwater circulating in increasingly widened joints, bedding planes, fissures, and conduits, reflected by increasingly larger exponents of the exponential equations. Enhanced karstification degree and circulation in conduits is described by one or several linear equations. Several properties of the aquifer can be evaluated by recession curve analysis: the type of rock disruption or karstification degree, and the anticipated character of attenuation (self-purification) processes (Kullman 2000*; Malík* 2007*; Malík and Vojtková* 2012*). The karstification degree of a recharge area derived from recession curve analysis can be an important feature determined. In the* Table 7.8 *karstification degree is classified using a 10th-degree scale, 1st* *degree for the lowest karstification, 10th degree for best developed karstification. Differences in character of individual depletion hydrographs and resulting karstification degrees are also listed here. Karstification degree was primarily described by Kullman (*1990*) in 10 categories, later applied (Kullman* 2000*) for assessment of groundwater vulnerability in 10-degree ranking, later supplemented by Malík (*2007*) and Malík and Vojtková (*2012*) to cover all results of hydrograph analyses of springs and to refine more precisely defined parameters of depletion equations with the intention not to disturb previously defined classification.*

References:

Kullman, E., 2000. Nové metodické prístupy k riešeniu ochrany a ochranných pásiem zdrojov podzemných vôd v horninových prostrediach s krasovo-puklinovou priepustnosťou [New methods in groundwater protection and delineation of protection zones in fissure-karst rock environment; in Slovak]. Podzemná voda 6, 2,31–41.
Malík, P., 2007. Assessment of regional karstification degree and groundwater sensitivity to pollution using hydrograph analysis in the Velka Fatra Mts., Slovakia. Water Resources and Environmental Problems in Karst. Environ Geol 2007, 51. 707–711.
Malík, P., Vojtková, S., 2012. Use of recession-curve analysis for estimation of karstification degree and its application in assessing overflow/underflow conditions in closely spaced karstic springs. Environ Earth Sci 2012, 65, 2245–2257.

Table 7.8 Karstification degree in recharge areas of springs according to recession curves parameters (after Kullman 2000; Malík 2007; Malík and Vojtková 2012; supplemented and modified)

Karstification degree	Groundwater flow component type	Characteristic recession curve equation	Characteristics of recession curve parameters	Characteristics of karstification degree
0.5	Single exponential flow component, lower values of α_1	$Q_t = Q_{o1} \bullet e^{-\alpha 1 \bullet t}$	$\alpha_1 < 0.001$	Usually tectonic faults and shear zones filled with crushed material, with high buffering capability in relation to discharge. Mainly deeper groundwater circulation
1.0			$\alpha_1 = 0.001$ to 0.0025	
2.0	Single exponential flow component, higher values of α_1		$\alpha_1 = 0.0025$ to 0.007	Tectonic faults filled with crushed material with higher permeability and lower buffering capability in relation to discharge. In some cases deeper groundwater circulation
2.3			$\alpha_1 > 0.007$	
2.5	Combination of two or more merely exponential flow components characterised by different recession coefficients, lower values of α_1 and α_2	$Q_t = Q_{o1} \bullet e^{-\alpha 1 \bullet t} + Q_{o2} \bullet e^{-\alpha 2 \bullet t}$	$\alpha_1 < 0.0024$ and $\alpha_2 < 0.033$	Aquifer with dense, prevailingly regular fissure network, with majority of microfissures and small fissures. Increase in values α_1 and α_2 accompanies increasing permeability and also higher heterogeneity of fissured rock environment
2.7			$\alpha_1 < 0.0024$ or $\alpha_2 < 0.033$	
3.0			$\alpha_1 = 0.0024$ to 0.0045; $\alpha_2 = 0.033$ to 0.067	
3.5	Combination of two or more merely exponential flow components' flow characterised by different recession coefficients, higher values of α_1 and α_2		$\alpha_1 = 0.0024$ to 0.0043 and $\alpha_2 = 0.060$ to 0.16	Aquifer with irregularly developed fissure network, with majority of open macrofissures, also with possible presence of karst conduits of limited extent. In extreme cases, even short-term turbulent flow might occur in this type of rock environment
3.7			$\alpha_1 > 0.0043$ and $\alpha_2 < 0.060$	
4.0			$\alpha_1 = 0.0041$ to 0.018 and $\alpha_2 = 0.055$ to 0.16	

(continued)

Table 7.8 (continued)

Karstification degree	Groundwater flow component type	Characteristic recession curve equation	Characteristics of recession curve parameters	Characteristics of karstification degree
4.3	Discharge hydrograph is composed of linear model flow component and unexponential flow component. Substantial role in groundwater discharge plays the exponential flow component	$Q_t = Q_{01} \bullet e^{-\alpha 1 \bullet t} + Q_{04} \bullet (1 - \beta_1 \bullet t)$	$\alpha_1 > 0.018$ or $\alpha_2 > 0.16$	Aquifer with anticipated existence of crushed water-bearing zone (e.g. fault zone) or by dense network of open small fissures in combination with simple, partly (or occasionally) phreatic conduit system of considerable extent (e.g. with open karstified fault in the vadose zone)
4.7			$\alpha_1 > 0.018$ and $\alpha_2 > 0.16$	
5.0			β and α—low values	
5.5	Complex discharge regime, a combination of one linear model flow component and two exponential flow components. Discharging of linear model flow component is of short-term influence in comparison with overall groundwater discharge	$Q_t = Q_{01} \bullet e^{-\alpha 1 \bullet t} + Q_{02} \bullet e^{-\alpha 2 \bullet t} + Q_{04} \bullet (1 - \beta_1 \bullet t)$	$\alpha_1 > 0$ and $\alpha_2 > 0$ and $\beta_1 > 0$	Extensive disruption and disintegration of rock environment, with majority of open, medium-sized, both not karstified and karstified fissures in the phreatic zone of the fissure karst aquifer (according to α_1) and with smaller influence of connected conduits (groundwater of large karstic channels, according to $ß_1$)
6.0	Very complex discharge regime, a combination of two linear model flow components and two exponential flow components. Discharges of linear model flow components are of short-term influence in comparison with overall groundwater discharge	$Q_t = Q_{01} \bullet e^{-\alpha 1 \bullet t} + Q_{02} \bullet e^{-\alpha 2 \bullet t} + Q_{04} \bullet (1 - \beta_1 \bullet t) + Q_{05} \bullet (1 - \beta_2 \bullet t)$	β_2, β_1 and α_1, α_2—high values	Extensive disruption and disintegration of rock environment, with majority of open, medium-sized, both not karstified and karstified fissures in the phreatic zone of the fissure karst aquifer (according to α_1) and with smaller influence of connected conduits (groundwater of large karstic channels, according to $ß_1$ and $ß_2$)

(continued)

Table 7.8 (continued)

Karstification degree	Groundwater flow component type	Characteristic recession curve equation	Characteristics of recession curve parameters	Characteristics of karstification degree
7.0	Discharge regime is a combination of one exponential flow component and two to three flow components with linear flow model. Substantial role in groundwater discharge plays the exponential flow component	$Q_t = Q_{01} \bullet e^{-\alpha 1 \bullet t} + Q_{04} \bullet (1\text{-}\beta_1 \bullet t) + Q_{05} \bullet (1\text{-}\beta_2 \bullet t) + Q_{06} \bullet (1\text{-}\beta_3 \bullet t)$	β_3, β_2, β_1 and α—high values, $\beta_1 > \beta_2$	Developed karstification of the aquifer, formed by large open tectonic faults and karstic channels, as well as by significant portion of karstified and non-karstified fissures and microfissures in rock blocks, generating phreatic zone with karstic groundwater table
8.0	Discharge regime is a combination of unexponential flow component with two to three linear model flow components. Substantial role in groundwater discharge is played by linear model flow components. Exponential flow component is less significant		β_3, β_2, β_1 and α—high values	Highly developed karstification of the aquifer, formed by large open conduits (karst channels). Presence of open active small fissures and microfissures is reduced. Circulation of substantial part of groundwater is mainly via preferred pathways of channel systems. Phreatic zone is missing or its role is insignificant
8.5	Flow regime that is represented merely by linear flow model, as the only one flow component present, which represents linear circulation in channel systems (conduits) without hydraulic connection to groundwater in adjacent rock blocks. Groundwater circulation is mostly connected to vadose zone	$Q_t = Q_{04} \bullet (1 - \beta_1 \bullet t)$	α_1; $\alpha_2 = 0$ and $\beta_1 > 0$	Karstic aquifer with well-developed conduit system, without any significant connection to open phreatic fissure systems in adjacent rock blocks. Extensively developed conduit system can ensure permanent groundwater replenishment

(continued)

Table 7.8 (continued)

Karstification degree	Groundwater flow component type	Characteristic recession curve equation	Characteristics of recession curve parameters	Characteristics of karstification degree
9.0	Flow regime, represented merely by two linear flow components. These represent linear circulation in channel systems (conduits) without hydraulic connection to groundwater in adjacent rock blocks. Groundwater circulation is mostly connected to vadose zone	$Q_t = Q_{04} \bullet (1 - \beta_1 \bullet t) + Q_{05} \bullet (1 - \beta_2 \bullet t)$	β_1, β_2—low values	Karstic aquifer with well-developed conduit systems of karstic groundwater pathways, without any significant connection to open phreatic fissure systems in adjacent rock blocks. Replenishment of groundwater resources can still be ensured by extensively developed conduit system
10.0	Complex discharge regime, consisting of three different linear flow components. Probability of a very complex groundwater circulation by occasional flows in the vadose zone. Documented only in perennial flows	$Q_t = Q_{04} \bullet (1 - \beta_1 \bullet t) + Q_{05} \bullet (1 - \beta_2 \bullet t) + Q_{06} \bullet (1 - \beta_3 \bullet t)$	β_1, β_2 and β_3—high values	Karstic aquifer with extensive conduit system, without any significant connection to open phreatic fissure systems in adjacent rock blocks. In its spatial extent, the conduit system is bound only to the vadose zone, without enabling permanent groundwater replenishment

References

Alfaro C, Wallace M (1994) Origin and classification of springs and historical review with current applications. Environ Geol 24:112–124

Barnes BS (1939) The structure of discharge recession curves. Trans Am Geophys Union 20:721–725

Bonacci O (2001) Analysis of the maximum discharge of karst springs. Hydrogeol J 2001(9):328–338

Bonacci O (2011) Karst springs hydrographs as indicators of karst aquifers. Hydrol Sci (Journal des Sciences Hydrologiques) 38(1–2):51–62

Bonacci O, Bojanic D (1991) Rythmic karst spring. Hydrol Sci (Journal des Sciences Hydroloqiques) 36(1–2):35–47

Boussinesq J (1877) Essai sur la théorie des eaux courantes do mouvement non permanent des eaux souterraines. Acad Sci Inst Fr 23:252–260

Boussinesq J (1903) Sur un mode simple d'écoulement des nappes d'eau d'infiltration à lit horizontal, avec rebord vertical tout autour lorsqu'une partie de ce rebord est enlevée depuis la surface jusqu'au fond. C R Acad Sci 137:5–11

Boussinesq J (1904) Recherches théoriques sur l'écoulement des nappes d'eau infiltrées dans le sol et sur le débit des sources. J Math Pure Appl 10(5):5–78

Dewandel B, Lachassagne P, Bakalowicz M, Weng Ph, Al-Malki A (2003) Evaluation of aquifer thickness by analysing recession hydrographs. Application to the Evaluation Oman ophiolite hard-rock aquifer. J Hydrol 274:248–269

Drogue C (1972) Analyse statistique des hydrogrammes de décrues des sources karstiques. J Hydrol 15:49–68

Dub O, Němec J (1969) Hydrologie. Česká matice technická, LXXIV (1969), 353, Technický průvodce 34, SNTL – Nakladatelství technické literatury, Praha, p 378

Fiorillo F (2011) Tank-reservoir drainage as a simulation of the recession limb of karst spring hydrographs. Hydrogeol J 2011(19):1009–1019

Flora SP (2004) Hydrogeological characterization and discharge variability of springs in the Middle Verde River watershed, Central Arizona. MSc thesis, Northern Arizona University, p 237

Forkasiewicz J, Paloc H (1967) Le régime de tarissement de la Foux de la Vis. Etude préliminaire. AIHS Coll. Hydrol.des roches fissurées, Dubrovnik (Yugoslavia) 1: 213–228

Foster HA (1924) Theoretical frequency curves and their application to engineering problems. Am Soc Civil Eng Trans 87:142–303

Foster HA (1934) Duration curves. Am Soc Civil Eng Trans 99:1213–1267

Gregor M (2008) Vývoj programov na analýzu časových radov výdatností prameňov a prietokov vodných tokov. (Software development for time-series analysis of springs yields and river discharges; in Slovak). Podzemná voda 14/2:191–200

Gregor M, Malík P (2012) Construction of master recession curve using genetic algorithms. J Hydrol Hydromechanics 60(1):3–15

Griffiths GA, Clausen B (1997) Streamflow recession in basins with multiple water storages. J Hydrol 190:60–74

Hall FR (1968) Base-flow recessions—a review. Water Resour Res 4(5):973–983

Király L (2003) Karstification and groundwater flow/Speleogenesis and evolution of karst aquifers. In: Gabrovšek F (ed) Evolution of karst: from prekarst to cessation. Zalozba ZRC, Postojna-Ljubljana, pp 155–190

Kovács A (2003) Geometry and hydraulic parameters of karst aquifers—a hydrodynamic modelling approach. PhD thesis, La Faculté des sciences de l'Université de Neuchâtel, Suisse, p 131

Kovács A, Perrochet P, Király L, Jeannin PY (2005) A quantitative method for the characterisation of karst aquifers based on spring hydrograph analysis. J Hydrol 303:152–164

Kullman E (1980) L'evaluation du regime des eaux souterraines dans les roches carbonatiques du Mésozoique des Carpates Occidentales par les courbes de tarissement des sources.

Geologický ústav Dionýza Štúra, Bratislava, Západné Karpaty, sér. Hydrogeológia a inžinierska geológia 3:7–60
Kullman E (1990) Krasovo-puklinové vody (Karst-fissure waters; in Slovak). Geologický ústav Dionýza Štúra, Bratislava, p 184
Kullman E (2000) Nové metodické prístupy k riešeniu ochrany a ochranných pásiem zdrojov podzemných vôd v horninových prostrediach s krasovo-puklinovou priepustnosťou (New methods in groundwater protection and delineation of protection zones in fissure-karst rock environment; in Slovak). Podzemná voda 6/2:31–41
Kresic N (2007) Hydrogeology and groundwater modeling, 2nd edn. CRC Press/Taylor and Francis, Boca Raton
Lamb R, Beven K (1997) Using interactive recession curve analysis to specify a general catchment storage model. Hydrol Earth Syst Sci 1(1):101–113
Maillet E (ed) (1905) Essais d'hydraulique soutarraine et fluviale vol 1. Herman et Cie, Paris, p 218
Malík P (2007) Assessment of regional karstification degree and groundwater sensitivity to pollution using hydrograph analysis in the Velka Fatra Mts., Slovakia. Water Resources and environmental problems in karst. Environ Geol 51:707–711
Malík P, Michalko J (2010) Oxygen isotopes in different recession subregimes of karst springs in the Brezovské Karpaty Mts. (Slovakia). Acta Carsologica 39(2):271–287
Malík P, Vojtková S (2012) Use of recession-curve analysis for estimation of karstification degree and its application in assessing overflow/underflow conditions in closely spaced karstic springs. Environ Earth Sci 65:2245–2257
Mangin A (1969) Etude hydraulique du mecanisme d'intermittence de Fontestorbes (Belesta, Ariege). Annales de Speleologie 24(2):253–298
Meinzer OE (1923a) Outline of ground-water hydrology. USGS Water-Supply Paper, p 494
Meinzer OE (1923b) The occurrence of ground water in United States with a discussion of principles. USGS Water-Supply Paper, 489, Washington DC, p 321
Milanovic PT (1981) Karst hydrogeology. Water Resources Publications, Littleton
Netopil R (1971) The classification of water springs on the basis of the variability of yields. Studia Geographica 22:145–150
Oraseanu I, Iurkiewicz A (2010) Calugari ebb and flow spring. In: Oraseanu I, Iurkiewicz A (eds) Karst hydrogeology of Romania. Belvedere Publishing, Oradea, pp 262–274
Padilla A, Pulido Bosch A, Mangin A (1994) Relative importance of baseflow and quickflow from hydrographs of karst spring. Ground Water 32:267–277
Posavec K, Bačani A, Nakić Z (2006) A visual basic spreadsheet macro for recession curve analysis. Ground Water 44(5):764–767
Rutledge RT (1998) Computer programs for describing the recession of groundwater discharge and for estimating mean groundwater recharge and discharge from stream flow records-update. USGS water-resource investigations report 98-4148. Reston, VG
Searcy JK (1959) Flow-duration curves. manual of hydrology, part 2. Low-flow techniques. Geological survey water-supply paper 1542-A, Methods and practices of the geological survey, United States Government Printing Office, Washington, p 33
Schöeller H (1948) Le régime hydrogéologique des calcaires éocénes du Synclinal du Dyr el Kef (Tunisie). Bull Soc Géol Fr 5(18):167–180
SÚTN (2009) Slovak technical standard STN 751520 Hydrológia, Hydrologické údaje podzemných vôd, Kvantifikácia výdatnosti prameňov. (Hydrology, hydrological data on groundwater, quantification of spring's discharge; in Slovak), Slovenský ústav technickej normalizácie (SÚTN) Bratislava, p 14
Springer AE, Stevens LE, Anderson DE, Parnell RA, Kreamer DK, Flora SP (2004) A comprehensive springs classification system: integrating geomorphic, hydrogeochemical and ecological criteria. In: Aridland springs in North America: ecology and conservation, pp 49–75
Tallaksen LM (1995) A review of baseflow recession analysis. J Hydrol 165:349–370

Tallaksen LM, van Lanen HAJ (eds) (2004) Hydrological drought, processes and estimation methods for streamflow and groundwater. Developments in Water Science, vol 48. Amsterdam, Elsevier Science B.V., p 579

Toebes C, Strang DD (1964) On recession curves 1: recession equations. J Hydrology (New Zealand) 3/2:2–15

Werner PW, Sundquist KJ (1951) On the groundwater recession curve for large watersheds. IAHS Publ 33:202–212

Chapter 8
Vulnerability to Contamination of Karst Aquifers

Ana I. Marín and Bartolomé Andreo

The karst aquifers are especially vulnerable to pollution due to their hydrologic behavior derived from karstification. The vulnerability mapping is one of the most applied tools to protect them. There are a wide range of methodologies for vulnerability mapping that have been developed for karst aquifer, to consider the specific characteristics of karst into the vulnerability assessment, such as EPIK, PI, COP, Slovene Approach, and PaPRIKa, among others.

The vulnerability map can help to the water stakeholder for making decision and to promote a land-use management compatible with the water protection. So the maps should have reliable accuracy. Many works highlight that the maps of groundwater contamination vulnerability obtained by the different methods differ significantly, although they were all obtained by methods developed for karst aquifers, even if they are obtained from the same sources of information and applied by the same person. So, the validation is an essential element of any contamination vulnerability assessment. The current challenge of researchers is to obtain versatile and easy to apply methods to test and validate vulnerability maps.

A.I. Marín (✉)
Centre of Hydrogeology, Department of Geology,
University of Málaga (CEHIUMA), 29071, Málaga, Spain
e-mail: aimarin@uma.es

B. Andreo
European Topic Centre for Spatial Information and Analysis (ETC/SIA),
University of Málaga, 29071, Málaga, Spain
e-mail: andreo@uma.es

Z. Stevanović (ed.), *Karst Aquifers – Characterization and Engineering*,
Professional Practice in Earth Sciences, DOI 10.1007/978-3-319-12850-4_8

8.1 Introduction

The concept of the contamination vulnerability of an aquifer has been defined by many authors (Margat 1968; Foster 1987; Zaporozec 1994, among others). It is possible to differentiate between intrinsic and specific vulnerability. The intrinsic vulnerability of an aquifer concerns its sensitivity to contamination, taking into account its geological, hydrologic, and hydrogeological characteristics, independent of the nature and scenario of the contamination (Daly et al. 2002; Zwahlen 2004). For specific vulnerability assessment, the pollutant characteristics should be considered in addition to the variables considered for the intrinsic vulnerability. Although the nature of pollutant influence to contamination pattern throughout the aquifer, generally, the vulnerability mapping is based on intrinsic concept to simplify the vulnerability schemes. Thus, the intrinsic vulnerability mapping is a more useful tool for land-use planning.

The European COST Action 620 (Vulnerability and Risk mapping for the Protection of Carbonate (Karst) Aquifer) was a relevant milestone into the development of sustainable long-term protection strategies for karst groundwater bodies bringing together experts from across Europe (Zwahlen 2004).

The European COST Action 620 proposed conceptual approach, named the origin-flow-target (Fig. 8.1), to define two general approaches for groundwater protection: resource protection and source protection. Concerning the resource protection, the element to be protected is the groundwater stored in the aquifer (the water table). The flow of the contaminating agent from the origin is considered to be practically vertical, thorough the soil and the unsaturated zone. However, in the case of source protection, the target to be protected is the discharge point of the aquifer (well or spring). Conceptually, the contaminant is assumed to be transported from the origin (the surface) to the water table and from there to the spring or well. This displacement has firstly a predominantly vertical component (origin—water table surface) and secondly horizontal one (water table surface—source), the latter being in the saturated zone of the aquifer.

Karst groundwater is particularly sensitive to contamination, due to aquifer structure and hydrologic behavior which determines rapid recharge of the infiltrating water and its fast distribution over large distances, high flow velocities and short residence time, and consequently, the self-cleaning capacity of the karst groundwater is very low (Ford and Williams 1989; Dörfliger and Zwahlen 1998). In consequence, karst aquifers require specific methodologies for vulnerability mapping which take into account their properties (Zwahlen 2004). Multiple methods, specifically for karst, have been developed on the basis of the guidelines of the European COST Action 620, including PI (Goldscheider et al. 2000), the COP method (Vías et al. 2006), the Slovene Approach (Ravbar and Goldscheider 2007), and PaPRIKa (Kavouri et al. 2011). Nowadays, there is no international consensus regarding which method should be used to assess the vulnerability of carbonate aquifers. The application of different methods on the same test site, using the same database, often leads to significant differences in mapping results

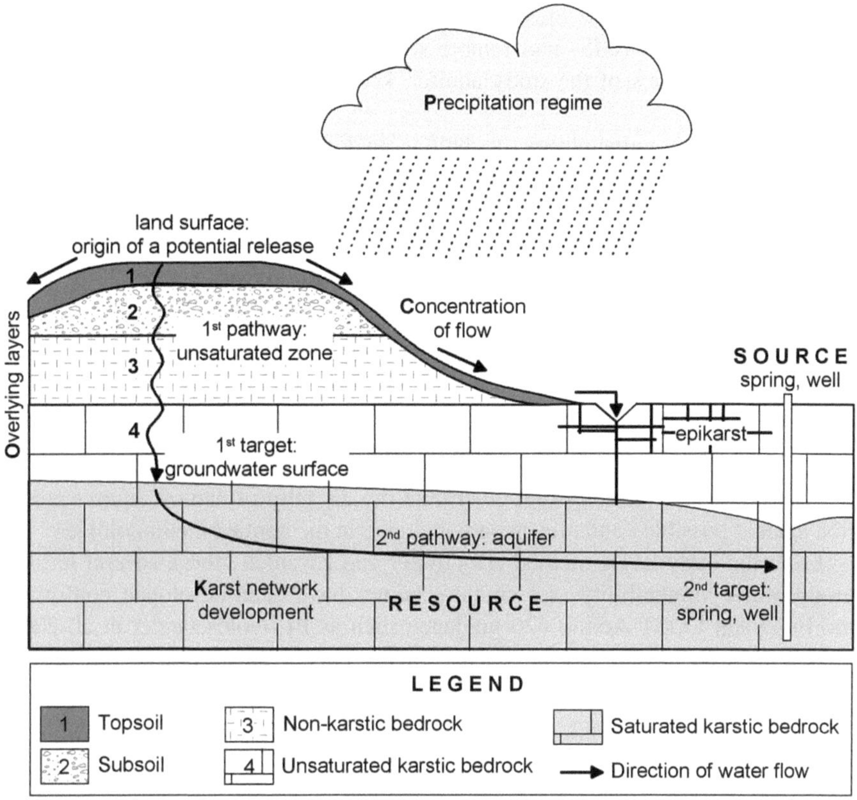

Fig. 8.1 The origin–pathway–target model for groundwater resource and source protection after European COST Action 620 (Goldscheider et al. 2000; Daly et al. 2002; Zwahlen 2004)

(Vías et al. 2005; Neukum and Hötzl 2007; Ravbar and Goldscheider 2009; Marín et al. 2012), affecting the uncertainty about the applicability of the vulnerability maps on land-use decision making.

8.2 Vulnerability Mapping

The objective of contamination vulnerability mapping is to identify the most vulnerable zones of catchment areas and to provide criteria for protecting the groundwater. Vulnerability to pollution is not a characteristic that can be directly measured in the field, so indirect methodologies for assessing are required. The widely used methodologies are based on multiparameter systems. Each parameter represents the variables involved in groundwater vulnerability that are discretized using scored intervals according to the relative degree of sensitivity to contamination.

Vulnerability maps are most often created with the assistance of geographic information systems (GIS) and remote sensing. The GIS permits to match data on the characteristics of the study aquifer, keeping the geographic framework as reference.

Previously to vulnerability mapping, hydrogeological characterization of the studied aquifer is required. The considered variables depend on the method selected for vulnerability mapping. However, there are a set of variables generally included into the assessment. These variables represent physical characteristics of the study area and can be grouped into the following:

- external factors (climatic, which influences the recharge),
- topography, including altitude, slope, and additionally vegetation (influence to runoff capacity),
- related to the unsaturated zone: rock permeability, karstification, fracturing, thickness, and soil development (texture and thickness).

In general, all the variables that determine the capacity of natural aquifer protection against possible contaminants are included in the applied methodologies.

From the early EPIK method (Doerfliger and Zwahlen 1998), several intrinsic groundwater vulnerability assessment schemes have been developed conform to the European COST Action 620 guidance such as PI (Goldscheider et al. 2000), VULK method (Jeannin et al. 2001), COP (Vías et al. 2006), or the Slovene Approach (Ravbar and Goldscheider 2007), among others.

8.3 EPIK Method

The EPIK method was the first one developed specifically for karst environment by Doerfliger and Zwahlen (1998). The method name is an acronym for epikarst (E), protective cover (P), infiltration conditions (I), and karst network development (K). The method assesses the protection factor (F) as inverse characteristic of vulnerability to pollution. So, lower values of F correspond to higher vulnerability. This method considers the karst network development, so it is applicable for source vulnerability mapping.

Briefly, the method is applied by division of the four parameters into rating intervals. Weighting factors are used to balance their importance of each parameter. The protection factor (F) is the result of a weight sum of the rating values. F is divided into vulnerability classes, from "low" to "very high" by conversion from protection factor (Fig. 8.2).

The EPIK method, widely applied (Gogu and Dassargues 2000; Vías et al. 2005; Neukum and Hötzl 2007; Barrocu et al. 2007; Ravbar and Goldscheider 2009), has been integrated in Swiss legislation for karst groundwater protection.

The Fig. 8.3 shows an example of contamination vulnerability maps obtained by EPIK method in the carbonate massif of Sierra de Mijas (S Spain), formed by Triassic marbles highly fractured but poorly karstified. From the hydrogeological

Epikarst		Characterisation of karst morphological features
Highly developed	E_1	Shafts, sinkholes or dolines (from all kinds of genesis), karrenfields, cuesta, outcrops with high fracturing (along roads and railways, quarries)
Moderately developed	E_2	Intermediate zones in the alignment of dolines, dry valleys. Outcrops with medium fracturing.
Small or absent	E_3	No karst morphological phenomena. Low fracture density.

Protective cover		Characterisation of Protective cover	
		A. Soil lying directly on limestone or on some high permeability coarse detritus layers, e.g. rock debris, lateral glacial tills	B. Soil lying on low permeability geological layers, e.g. lake silt, clays
Absent	P_1	0–20 cm of soil	0–20 cm of soil on layers that have a thickness of less than 1 m
	P_2	20–100 cm of soil	20–100 cm of soil on layers that have a thickness of less than 1 m
	P_3	100–200 cm of soil	<100 cm of soil or >100 cm of soil and >100 cm of layers of low permeability
Present	P_4	>200 cm	>100 cm of soil and thick detritus layers of very low hydraulic conductivity (point-information needs to be checked) or >8 m of clay and clayey silt

Infiltration conditions		Characterisation Infiltration conditions
Concentrated	I_1	Perennial or temporarily losing streams – perennial or temporarily stream feeding a swallow hole or a sinkhole (doline) – water catchment areas of these above mentioned streams, including artificial drainage system.
	I_2	Water catchment areas of streams in I1 (without artificial drainage system) with a slope greater than 10% for cultivated areas and 25% for meadows and pastures.
	I_3	Water catchment areas of the I1 stream (without artificial drainage system) whose slope is less than: 10% for cultivated areas and 25% for meadows and pastures. Low relief areas collecting runoff water and slopes feeding those low areas (slope higher than: 10% for cultivated sectors and 25% for meadows and pastures).
Diffuse	I_4	The rest of the catchment.

Karst network		Characterisation of Karst network development
Well developed karst network	K_1	Presence of a well developed karst network (network with decimetre to meter sized channels that are rarely plugged and are well connected).
Poorly developed karst network	K_2	Presence of a poorly developed karst network (small conduits network, or poorly connected or filled network, or network with decimetre or smaller sized openings).
Mixed or fissured aquifer	K_3	Presence of a spring emerging through porous terrain. Non-karst, only fissured aquifer.

$$F_p = 3\,E_I + 1\,P_j + 3\,I_K + 2\,K_l$$

Epikarst			Protective cover				Infiltration conditions				Karst network		
E_1 1	E_2 3	E_3 4	P_1 1	P_2 2	P_3 3	P_4 4	I_1 1	I_2 2	I_3 3	I_4 4	K_1 1	K_2 2	K_3 3

Vulnerability classes	Protection factor F	Protection areas Si
Very high	F lower or equal to 19	S1
High	F between 20 and 25	S2
Moderate	F higher than 25	S3
Low	Presence of P4	The rest of the water catchment

Fig. 8.2 Diagram of EPIK method (Doerfliger et al. 1999)

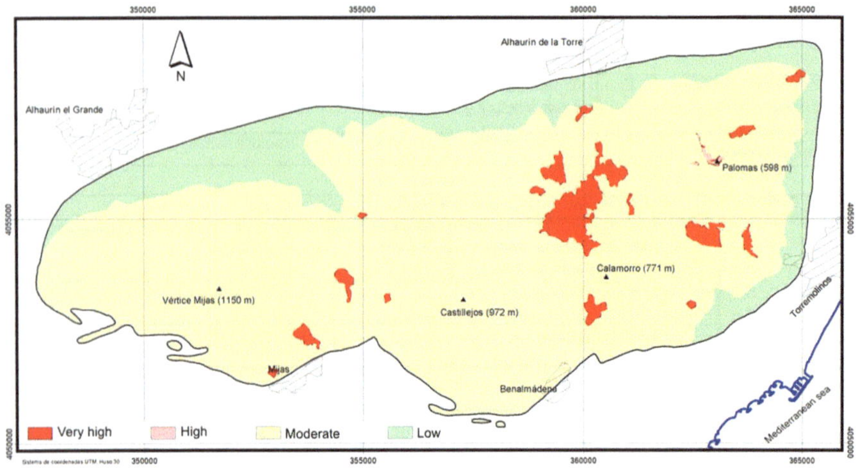

Fig. 8.3 Contamination vulnerability map of Sierra de Mijas (S Spain) carbonate aquifer obtained by EPIK method (Vías et al. 2005)

standpoints, the aquifer has a diffuse flow behavior (Andreo et al. 1997). The aquifer is recharged by diffuse infiltration of rainwater. In this case, the EPIK method permits to determine moderate vulnerability in the Triassic marbles identifying with higher vulnerability upper zones where an important degree of fracturing occurs, such as in quarries, and consequently, the infiltration conditions are supposed favorable (Vías et al. 2005).

8.4 PI Method

The PI method was developed in the frame of Action COST 620 but before the European approach was proposed (Zwahlen 2004). Although the method follows the guidance and conceptual model of the Action, the terminology is different (Goldscheider et al. 2000).

The PI method considers two factors (Fig. 8.4): the protective cover (P factor) and the infiltration conditions (I factor). They correspond basically to the O and C factors of the European approach, respectively. The P factor describes the protective function of the unsaturated zone—the soil, the subsoil, the nonkarst rock, and the unsaturated zone of the karstic bedrock. The I factor takes into account the infiltration conditions and how the protective functions of unsaturated zone is bypassed as a result of lateral surface and subsurface flow in the catchment area of streams sinking in swallow holes. The PI method does not consider the karst network development (K factor of European approach), and thus, it is only applicable for resource vulnerability mapping.

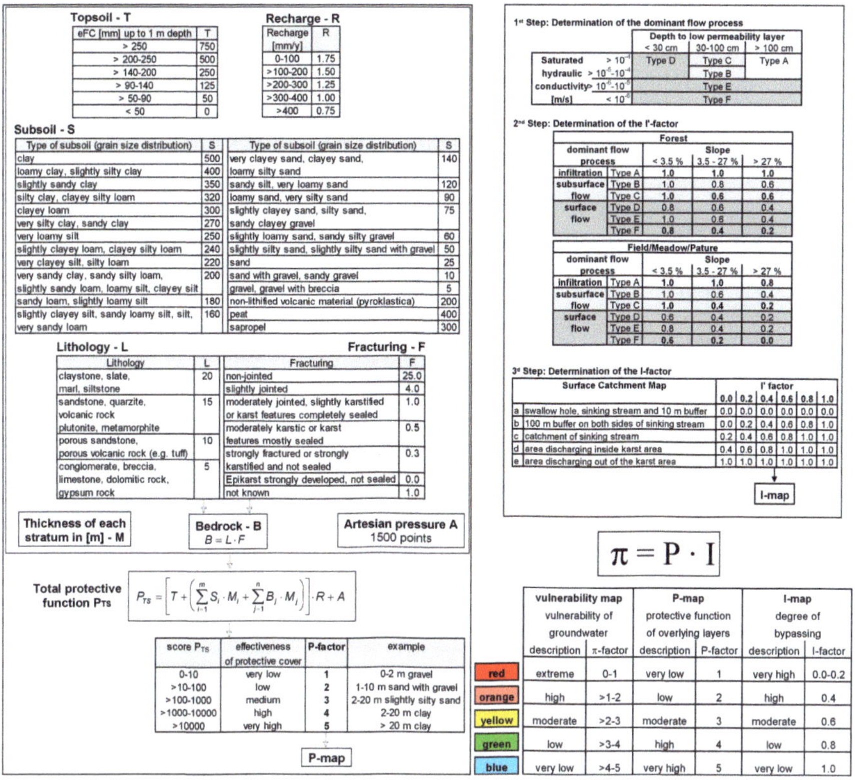

Fig. 8.4 Diagram of PI method (Goldscheider et al. 2000)

Based on PI method, a simplified method (Nguyet and Goldscheider 2006) was developed for assessing the vulnerability map in areas where the hydrogeological knowledge and geoinformation are limited.

PI method has been tested in different sites (Goldscheider 2002; Andreo et al. 2006; Neukum and Hötzl 2007; Ravbar and Goldscheider 2009).

Box 8.1

An example of application of PI method is shown in Fig. 8.5; it is the contamination vulnerability map for catchment area of the Podstenjšek springs in Slovenia (Ravbar 2007). The aquifer drained by this spring is formed by Paleocene limestones and Cenomanian limestones and breccias moderately karstified. Spring discharge shows fast and strong rise as response to recharge events. The recharge is both concentrate and diffuse. The contamination vulnerability map obtained by PI method shows a general high vulnerability of the aquifer. The extreme vulnerability class embraces areas where thickness of the unsaturated zone is very shallow and the limestones are uncovered, as well as the intermittent lake of Šembijsko Jezero.

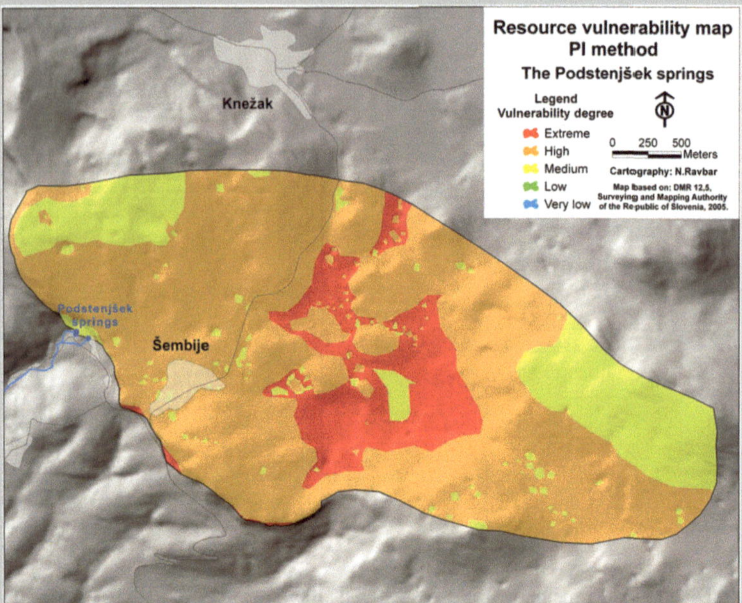

Fig. 8.5 Contamination vulnerability map for catchment area of the Podstenjšek springs (Slovenia) obtained by PI method (Ravbar 2007)

8.5 COP Method

The COP method (Vías et al. 2006) was developed in the framework of European COST Action 620. This method has been applied, among others, by Vías et al. (2006, 2010), Yildirim and Topkaya (2007), Polemio et al. (2009), Plan et al. (2009), and Marín et al. (2012). The COP method considers the characteristics of layers overlying the water table (O factor), the parameters which control the flow concentration (C factor), and the precipitation (P factor). The O factor reflects the protective capacity of the overlying layers provided by soils (texture and thickness) and the lithology of the unsaturated zone (fracturing, the thickness of each layer, and the confining conditions). Karst geomorphology, slopes, and vegetation cover are taken into account in the C factor, which discriminates between areas where the infiltration is concentrated via swallow hole and where the overlying

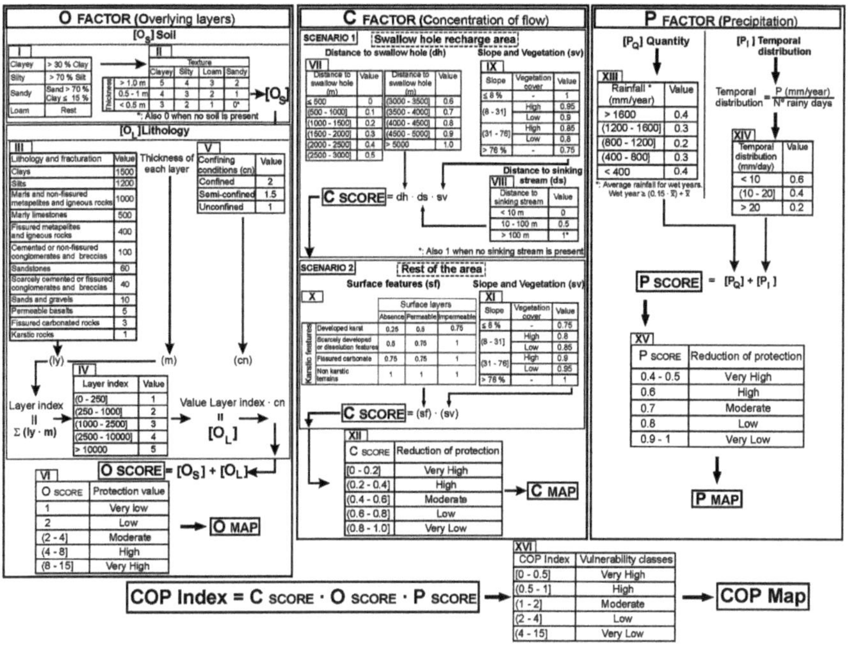

Fig. 8.6 Diagram of the COP+K method (modified of Vías et al. 2006 and Andreo et al. 2009)

layer might be bypassed, and the rest of the area. Karst features may be present in each of these infiltration scenarios. In the first case, the existence of swallow holes is usually associated with drainage of the karst landforms (poljes, dolines, etc.). In the rest of the area, karst features and the presence or absence of surface layers are considered because these parameters control the relation between runoff and infiltration processes. The P factor considers both the spatial and temporal variability of precipitation, which plays a role in the transfer of contaminants, especially in large aquifers. The COP vulnerability index value is obtained by multiplying the C, O, and P scores (Fig. 8.6).

Box 8.2

The Fig. 8.7 shows an example of vulnerability maps obtained by COP method in the Alta Cadena aquifer (S Spain). This aquifer is constituted by Jurassic dolostones and limestones, with a basement of Triassic clays and evaporites. The limestones are moderate-highly karstified with small dolines, karrenfield and swallow holes only active during high rainfall events, but with flow velocities exceeding 200 m/h (Mudarra et al. 2014). Recharge to the aquifer takes place entirely by rainfall infiltration, both in the concentrated form (via swallow holes) and diffuse form. Regarding the vulnerability map resultant from COP method, the aquifer is mainly highly vulnerable rising to very high in the karst landform zones (karrenfields and swallow holes) and at flat uncovered carbonate outcrops. In the aquifer sectors protected by impermeable marls overlie the aquifer, the vulnerability is lower.

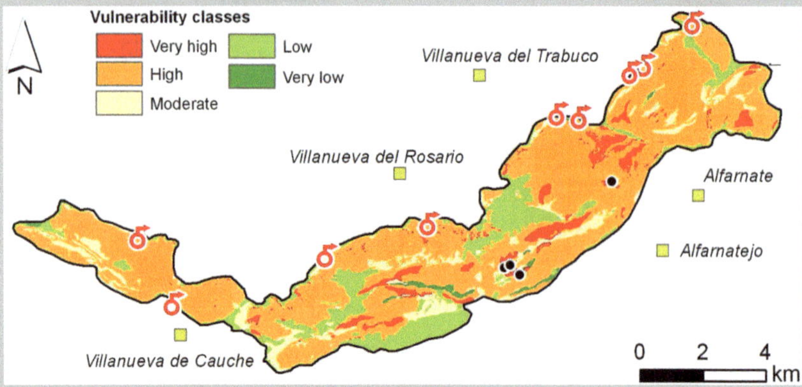

Fig. 8.7 Vulnerability map of Alta Cadena aquifer obtained by COP method (Marín et al. 2012)

The COP method was designed to assess the resource vulnerability, but adding the K factor related to saturated zone results in source vulnerability assessment method. The COP+K method (Andreo et al. 2009) is an extension of the COP method that requires, firstly, the resource vulnerability map (using COP) and, additionally, to assess the flow path thorough the saturated zone (K factor).

Box 8.3

The example of application of COP+K is shown in Fig. 8.8 is about Sierra de Líbar (Andreo et al. 2009) aquifer which is made up by highly karstified Jurassic limestones. There are abundant karst features including karren, sinkholes, and poljes with swallow holes in which sinking streams infiltrate. The springs have typical karst behaviors of quick and strong responses to recharge events. The source vulnerability map obtained by COP+K method represents as highly vulnerable the more karstified areas and areas close to the springs.

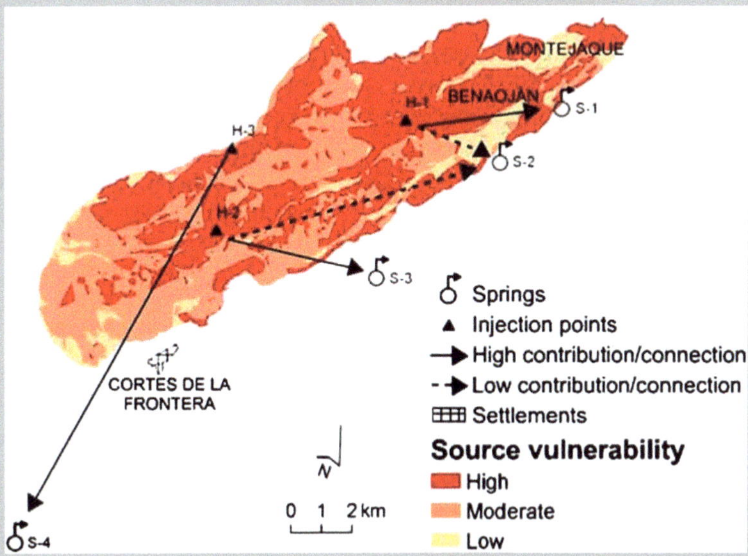

Fig. 8.8 Source vulnerability map of Sierra de Libar aquifer (S Spain) obtained by COP+K method (Andreo et al. 2009)

8.6 Validation

The vulnerability maps have become routine procedure to support the land-use planning as measure to protect the groundwater quality. However, about vulnerability maps is necessary to know what really means, for groundwater protection, the vulnerabity grade shown on the map. Using the vulnerability guidance, the vulnerability mapping by one of the existing methods is relatively easy to implement after knowledge on GIS and on hydrogeology.

The weaknesses of groundwater contamination vulnerability mapping, implicit in the choice of method and the subjectivity of its application, are obvious. Nevertheless, it remains a tool with great potential for groundwater quality protection. It is relatively simple to apply, if supported by appropriate hydrogeological studies and baseline maps, and its implementation within land-use planning policies is intuitive, since the outcome of this technique is a map that shares a common territorial basis with the working environment.

One key question is to know if the vulnerability map obtained under the assessment process is coherent with the hydrological knowledge of the study area, as well as to know the meaning of low, moderate or low vulnerability in terms of groundwater protection. Several authors (Vías et al. 2005; Neukum and Hötzl 2007; Ravbar and Goldscheider 2009; Marín et al. 2012) highlight that the maps of groundwater contamination vulnerability obtained by the different methods differ significantly, although they were all obtained by methods developed for karst aquifers, using the same sources of information and applying by the same person. So, the validation is an essential element of any contamination vulnerability assessment.

The current challenge is facing researchers to obtain versatile and straightforward methods to test and validate vulnerability maps. For a comprehensive validation of the maps, a wide range of methods and techniques can be applied for characterize fast and slow flows within the system, the responses in both high and low water condition, the overall response and the response to short-term signal, etc.

There is a range of variables that could be used to validate the vulnerability map (Zwahlen 2004) such as analysis of the hydrochemical response at springs, tracers (natural or artificial), or hydrodynamic modeling. Neither of them is concluding, each of these instruments informs about certain processes within the aquifer. The key is combining several of them to support the resulting map.

Analysis of hydrochemical responses, especially the natural tracers coming from the aquifer surface (TOC, NO_3^-, natural fluorescence and bacteria), constitute a useful tool to know transit times and as proxy of vulnerability of the aquifer. Natural tracers as chemical and isotopic composition of the water are transported within the aqueous solution, and they permit to deduce flow conditions, speeds, and transit times (Batiot et al. 2003; Baker et al. 2004; Mudarra et al. 2011). The evolution of the concentration of these groundwater components during recharge periods is a descriptor of the system overall response to water entry distributed over the aquifer surface (Fig. 8.9). Analyses of the evolution of TOC, NO_3^- and natural fluorescence, together with the hydrographs and chemographs of other components, permit to characterize the general response of the aquifer and the vulnerability to contamination (Marín et al. 2012).

The main limitation in the application of natural tracers as tool to validate the groundwater vulnerability mapping is that the source point (the origin) in aquifer can be not accurately located in most cases. Therefore, it is not possible to estimate flow velocities, although the arrival of water infiltration at the spring can be identified and estimated average transit times.

Dye tracers are widely applied to validate the vulnerability maps (Perrin et al. 2004; Andreo et al. 2006; Ravbar and Goldscheider 2007; Goldscheider 2010; Marín et al. 2012). This tool permits to know the flow path as well as the connections and the contribution between surface and discharge points and thus enhance the knowledge for delineating the boundaries of the catchment area. The tracer tests allow to estimate the transit time and to deduce the concentration decline of a potential contaminant (artificial tracer) from the injection point (origin) to a sampling point (target), following the origin—pathway—target model. The intrinsic vulnerability is usually validated by conservative tracers such as uranine or several other fluorescent dyes or salt tracers, such as chlorides, whereas for validating specific vulnerability, reactive tracers should be used. To validate a source vulnerability map, the tracer breakthrough must be recorded at the spring or well. For resource vulnerability, the tracer breakthrough should be recorded at the basis of the unsaturated zone (water table), which is, however, often not possible.

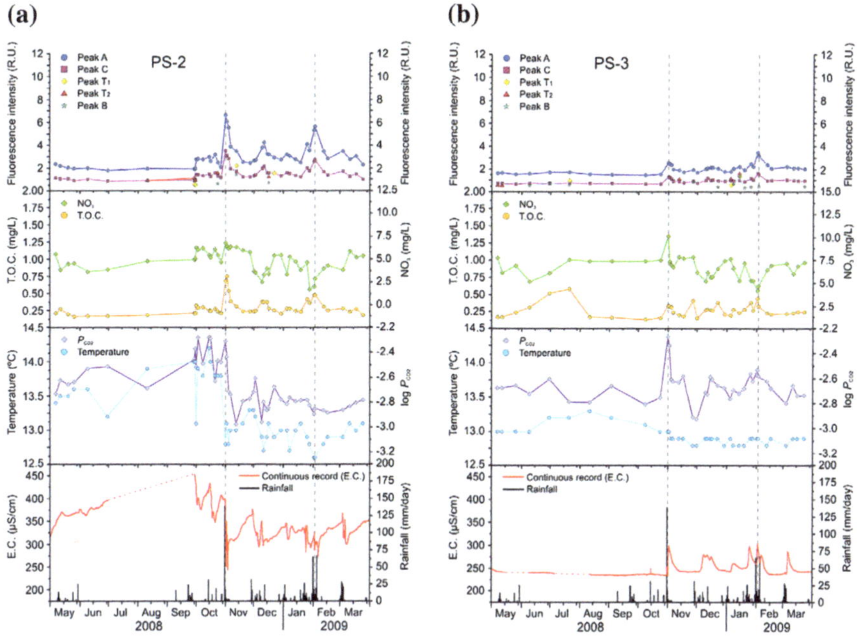

Fig. 8.9 Temporal evolution of electrical conductivity, PCO_2, temperature, and natural tracers (NO_3^- content and organic matter parameters, including TOC and natural fluorescence peaks) of the karst spring water, with respect to precipitation events (Alta Cadena aquifer, S Spain; after Mudarra et al. 2011)

The validation of vulnerability map by tracer tests is based on the analyses of the breakthrough curves which permit to estimate the following:

- Travel time of a contaminant from the origin to the target as the mean transit time, the time of the first arrival and the time of the maximum concentration.
- Relative quantity of the contaminant that can reach the target as the recovery rate which shows the degree of "attenuation".
- Duration of a contamination event when tracer concentration exceeds a given limit.

Tracer tests are a powerful tool to validate vulnerability maps, although this technique is not exempted from problems and limitations (Goldscheider et al. 2001). First at all, the tracer test only allows to check the vulnerability from the injection points and not in the large areas where the vulnerability has been assessed. The results obtained during the tracer tests depend on the hydrologic conditions and on the injection points due to the anisotropy and heterogeneity of karst aquifer. So, the results should be carefully analyzed in the frame of the hydrogeological context, and they can be not extrapolated to the whole of the catchment area due to the high heterogeneity of karst environment.

The natural and artificial tracers are useful techniques to validate the vulnerability maps. Despite the potential and the utility of both types of tracers

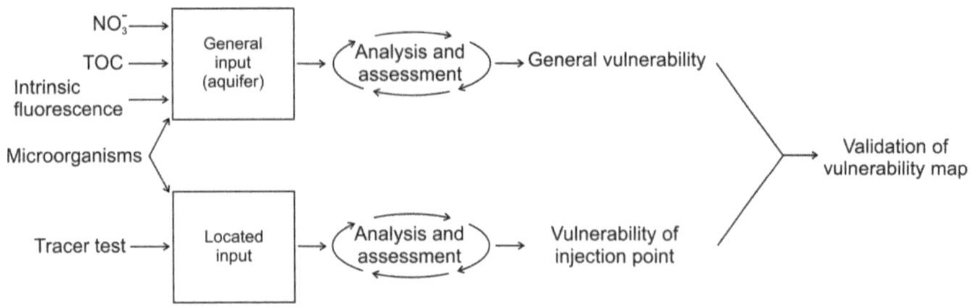

Fig. 8.10 Natural and artificial tracers as techniques for validation of the vulnerability maps

(Fig. 8.10), they are not commonly used in conjunction with validate the maps (Marín et al. 2012; Mudarra et al. 2014). However, the natural and artificial tracers complement each other enhancing the knowledge of karst aquifer (infiltration processes, recharge mechanisms, and vulnerability).

Acknowledgments This work is a contribution to the projects CGL2012-32590 of DGICYT and IGCP 598 of UNESCO and to Research Group RNM-308 funded by the Regional Government of Andalusia (Spain).

References

Andreo B, Carrasco F, Sanz de Galdeano C (1997) Types of carbonate aquifers according to the fracturation and the karstification in a southern Spanish area. Environ Geol 30(3/4):163–173

Andreo B, Goldscheider N, Vadillo I, Vias JM, Neukum C, Sinreich M et al (2006) Karst groundwater protection: first application of a Pan-European approach to vulnerability, hazard and risk mapping in the Sierra de Libar (Southern Spain). Sci Total Environ 357(1–3):54–73

Andreo B, Ravbar N, Vías JM (2009) Source vulnerability mapping in carbonate (karst) aquifers by extension of the COP method: application to pilot sites. Hydrogeol J 17(3):749–758

Baker A, Ward D, Lieten SH, Periera R, Simpson EC, Slater M (2004) Measurement of protein-like fluorescence in river and waste water using a handheld spectrophotometer. Water Res 38(12):2934–2938

Barrocu G, Muzzu M, Uras G (2007) Hydrogeology and vulnerability map (EPIK method) of the "Supramonte" karstic system, north-central Sardinia. Environ Geol 51(5):701–706

Batiot C, Liñán C, Andreo B, Emblanch C, Carrasco F, Blavoux B (2003) Use of total organic carbon (TOC) as tracer of diffuse infiltration in a dolomitic karstic system: the Nerja Cave (Andalusia, southern Spain). Geophys Res Lett 30(22)

Daly D, Dassargues A, Drew D, Dunne S, Goldscheider N, Neale S, Popescu I, Zwahlen F (2002) Main concepts of the "European approach" to karst-groundwater-vulnerability assessment and mapping. Hydrogeol J 10(2):340–345

Doerfliger N, Zwahlen F (1998) Practical guide, groundwater vulnerability mapping in karstic regions (EPIK). Swiss Agency for the Environment, Forests and Landscape (SAEFL), Bern, p 56

Doerfliger N, Jeannin PY, Zwahlen F (1999) Water vulnerability assessment in karst environments: a new method of defining protection areas using a multi-attribute approach and GIS tools (EPIK method). Environ Geol 39(2):165–176

Ford DC, Williams PW (1989) Karst geomorphology and hydrology. Chapman and Hall, London

Foster S (1987) Fundamental concepts in aquifer vulnerability, pollution risk and protection strategy. In: Van Duijvenbooden W and Van Waegeningh HG (eds) Vulnerability of soil and groundwater to pollutants, TNO committee on hydrogeological research, proceedings and information. The Hague, 38:69–86

Gogu RD, Dassargues A (2000) Sensitivity analysis for the EPIK method of vulnerability assessment in a small karstic aquifer, southern Belgium. Hydrogeol J 8(3):337–345

Goldscheider N (2002) Hydrogeology and vulnerability of karst systems—examples from the Northern Alps and Swabian Alb. PhD thesis, University of Karlsruhe, Faculty for Bio- and Geoscience, Karlsruhe, p 236

Goldscheider N (2010) Delineation of spring protection zones. In: Kresic N and Stevanović Z (eds) Groundwater hydrology of springs. Engineering, theory, management and sustainability. Elsevier Inc. BH, Amsterdam, pp 305–338

Goldscheider N, Klute M, Sturm S, Hötzl H (2000) The PI method: a GIS based approach to mapping groundwater vulnerability with special consideration of karst aquifers. Z Angew Geol 463:157–166

Goldscheider N, Hötzl H, Fries W, Jordan P (2001) Validation of a vulnerability map (EPIK) with tracer tests. In: Mudry J, Zwahlen F (eds) 7th conference on limestone hydrology and fissured media, Besançon, 20–22 Sept 2001, pp 167–170

Jeannin PY, Cornaton F, Zwahlen F, Perrochet P (2001) VULK: a tool for intrinsic vulnerability assessment and validaton. In: Mudry J, Zwahlen F (eds) 7th conference on limestone hydrology and fissured media, Besançon, 20–22 Sept 2001, pp 185–190

Kavouri K, Plagnes V, Tremoulet J, Dörfliger N, Fayçal R, Marchet P (2011) PaPRIKa: a method for estimating karst resource and source vulnerability—application to the Ouysse karst system (southwest France). Hydrogeol J 19(2):339–353

Margat J (1968) Vulnérabilité des nappes d'eau souterraine à la pollution: Bases de la cartographie: Orléans, France, Bureau de Recherche Géologique et Minière, Document 68 SGL 198 HYD

Marín AI, Dörfliger N, Andreo B (2012) Comparative application of two methods (COP and PaPRIKa) for groundwater vulnerability mapping in Mediterranean karst aquifers (France and Spain). Environ Earth Sci 65(8):2407–2421

Mudarra M, Andreo B, Marin AI, Vadillo I, Barberá JA (2014) Combined use of natural and artificial tracers to determine the hydrogeological functioning of a karst aquifer: the Villanueva del Rosario system (Andalusia, southern Spain). Hydrogeol J 22(5):1027. doi:10.1007/s10040-014-1117-1

Mudarra M, Andreo B, Baker A (2011) Characterization of dissolved organic matter in karst spring waters using intrinsic fluorescence: relationship with infiltration processes. Sci Total Environ 409(18):3448–3462

Neukum C, Hötzl H (2007) Standardization of vulnerability maps. Environ Geol 51(5):689–694

Nguyet VTM, Goldscheider N (2006) A simplified methodology for mapping groundwater vulnerability and contamination risk, and its first application in a tropical karst area, Vietnam. Hydrogeol J 14:1666–1675

Perrin J, Pochon A, Jeannin P, Zwahlen F (2004) Vulnerability assessment in karstic areas: validation by field experiments. Environ Geol 46:237–245

Plan L, Decker K, Faber R, Wagreich M, Grasemann B (2009) Karst morphology and groundwater vulnerability of high alpine karst plateaus. Environ Geol 58(2):285–297

Polemio M, Casarano D, Limoni PP (2009) Karstic aquifer vulnerability assessment methods and results at a test site (Apulia, southern Italy). Nat Hazards Earth Syst Sci, 9(4):1461–1470

Ravbar N (2007) Vulnerability and risk mapping for the protection of karst waters in Slovenia: application to the catchment of the Podstenjšek springs (in English). PhD thesis, University of Nova Gorica, Slovenia

Ravbar N, Goldscheider N (2007) Proposed methodology of vulnerability and contamination risk mapping for the protection of karst aquifers in Slovenia. Acta Carsologica 36(3):461–475

Ravbar N, Goldscheider N (2009) Comparative application of four methods of groundwater vulnerability mapping in a Slovene karst catchment. Hydrogeol J 17:725–733

Vías JM, Andreo B, Perles MJ, Carrasco F (2005) A comparative study of four schemes for groundwater vulnerability mapping in a diffuse flow carbonate aquifer under Mediterranean climatic conditions. Environ Geol, 47:586–595

Vías J, Andreo B, Perles M, Carrasco F, Vadillo I, Jiménez P (2006) Proposed method for groundwater vulnerability mapping in carbonate (karstic) aquifers: the COP method. Hydrogeol J 14(6):912–925

Vías JM, Andreo B, Ravbar N, Hötzl H (2010) Mapping the vulnerability of groundwater to the contamination of four carbonate aquifers in Europe. J Environ Manage 91(7):1500–1510

Yildirim M, Topkaya B (2007) Groundwater protection: a comparative study of four vulnerability mapping methods. CLEAN—Soil, Air, Water Pollution 35(6):594–600

Zaporozec A (1994) Concept of groundwater vulnerability. In: Vrba J, Zaporozec A (eds) Guidebook on mapping groundwater vulnerability. International contributions to hydrogeology, vol 16. Verlag Heinz Heise, Hannover, pp 3–8

Zwahlen F (ed) (2004) Vulnerability and risk mapping for the protection of carbonate (karst) aquifers. Final report of COST Action 620. European Commission, Directorate-General XII Science, Research and Development, Brussels

Chapter 9
Physical Modeling of Karst Environment

Saša Milanović

9.1 Introduction

It is very difficult to define hydrogeological parameters in an anisotropic aquifer. Even after many different methods are employed, the result in the case of karst aquifer cannot be exact and sure. Still, in the past few years, a whole new branch of karst investigations has become widely used in the earth sciences: 3D modeling of karst aquifers and karst conduits.

Physical modeling of karst environment generally combines state-of-the-art solids and parameter modeling with advanced survey technologies to produce decision support tools that are vastly superior to standard groundwater flow models. Physical model refers to a model which is constructed in a 3D computer environment using real physical parameters from the field (from the surface and karst interior). In general, this approach uses innovative and advanced technologies in order to produce predictions of spatial position of karstic conduits that are based on real world data rather than idealized assumptions as happens in global parameter optimization, stochastic models, finite element numerical models, or other types of hydrogeological mathematical modeling.

Physical models visually convey the real morphology and hydrogeology data of a conduit system in as much detail as was recorded by the geological, hydrogeological, speleological, cave diving, and other necessary surveys. On the other hand in karst research, two-dimensional maps of karst aquifer and cave maps offer limited ability to visualize and interpret the data they contain. Because conduits

S. Milanović (✉)
Centre for Karst Hydrogeology, Department of Hydrogeology,
Faculty of Mining and Geology, University of Belgrade, Belgrade, Serbia
e-mail: sasamilanovic@rgf.bg.ac.rs

Z. Stevanović (ed.), *Karst Aquifers – Characterization and Engineering*,
Professional Practice in Earth Sciences, DOI 10.1007/978-3-319-12850-4_9

are three-dimensional, understanding conduit morphology and deciphering correlations with hydrogeologic and geologic variables is made more difficult by the simplicity of two-dimensional maps (Kincaid 2006).

The main topic of this chapter is 3D reconstruction and modeling of karst physical interior as a new approach in karst hydrogeology. This is often the only way to visualize the geometry of cave networks. Shown together with surface topography and the main elements of the geological setting, this type of model can provide a very clear understanding and visualization of the cave system under investigation (Jeannin et al. 2007). Powerful tools such as ArcGIS and other software for three-dimensional modeling enable fast quantitative analysis and reconstruction of karstic conduits from the surface (sinkholes, pits, dolines) to the discharge zone (karstic springs).

Furthermore, an approximate calculation of the size and volume (storage) of open active channels could also be easily computed. This analysis can include both saturated and unsaturated parts. A variety of software packages for karst conduit modeling makes it possible today to outline measured data in three dimensions and display data trends and conduit morphology from various angles and inclinations, as well as to calculate the volume and finally obtain some hydrogeological data such as effective porosity etc.

9.2 Background

Creating 3D (physical) models of karst conduits as well as 3D models of karst aquifer has been the interest of several researchers for many years (Ulfeldt 1975; Schaecher 1986; Fish 1996; Gogu 2001; Ohms and Reece 2002; Springer 2004; Strassberg 2005; Kincaid 2006; Butscher and Huggenberger 2007; Jeannin et al. 2007; Kovács 2003; Filipponi and Jeannin 2008; Filipponi 2009; Borghi et al. 2010, 2012). All of these authors and many others have generally practised a method of forming a 3D model of karst interior based on large scale parameters that are collected first by the ground survey and further by special speleological or cave diving survey. The modeling technique described in this chapter is based on a multiparameter approach with 3D shape of conduits as output which is connected to the database with all collected spatial data. Generally, the field of physical or 3D modeling of karst interior is for now at the "first stage of evolution," a forming of 3D model depends mainly on the development of software and underground karst survey techniques.

Today laser scanners and computer 3D visualization make it possible to represent a cave in 3D with a very high degree of precision. 3D laser scanners allow for a detailed and precise survey of the walls and floor. By scanning the cave from many viewpoints, it is possible to locate cave walls with millimeter precision. Photographs of the walls (or floor) can then be projected onto those virtual surfaces, producing an almost exact reproduction of the cave, and allowing any virtual analysis without damaging the real cave (Jeannin et al. 2007).

9.3 Overview and Methodology

This chapter proposes a methodology that aims at simulating the geometry of karst conduits which is defined by different data collected during field investigations as well as from remote data analyzing.

Analyzing the geometry of the main karst conduits in the saturated zone and the connection with conduits in the aeration zone, "insight" into karst channels has been gained and has enabled the creation of a 3D model of karst channels. Analysis of various parameters obtained by quantitative and qualitative monitoring of groundwater characteristics and their analysis through the established physical model provides data on the relationship between the water recharge and discharge zone. Such a model can be further used for an analysis of speleogenesis and hydrogeology of the area.

Generally, there are three zones of vertical research in karst aquifer spatial modeling. The first is the recharge zone and all underground forms that are related to this area such as dolines, sinkholes, caves, pits. The third zone is the phreatic zone with main forms such as karst springs and their submerged karst channels that often lead to the central areas of karst aquifers. The most problematic is the zone between these two sections, but with the support of different softwares for 3D modeling and the knowledge of karstification development and intensity, this part could be reconstructed and the pathways (conduits) connections from inlet to outlet could be estimated.

In general, the proposed methodology consists of five main steps:

- Forming of conceptual model,
- Defining the level of input data and forming of model layers,
- Establishing of physical model and its relation base,
- Creating of the three-dimensional karst conduit,
- Analyzing of physical model—through case studies.

The basic problem in the determination of the methodology was how to perform a quality analysis or how to state the problem whose goal was to develop a model which would be an analysis of the geometry of karst conduits, integrating hydrogeological laws, and geological characteristics. So far, only a few techniques have been developed to structure a karst conduit networks model in a 3D environment. Some of them are based on modeling the physics and chemistry of the speleogenesis processes, as well as simulating the geometry of karstic conduits on a regional scale constrained by the regional geology, regional hydrology and hydrogeological knowledge, and by a conceptual knowledge of the genetic factors (Borghi et al. 2012). Filipponi (2009) in his doctoral thesis, "Spatial analysis of karst conduit networks and determination of parameters controlling the speleogenesis along preferential lithostratigraphic horizons," puts forth a good approach to deal with the formation of a 3D model of a karst system.

The modeling problem was approached from four parallel directions:

- Theoretical approach (Conceptual model), which initially played a major role and provided guidelines for field activities,
- Detailed and highly complex field investigations,
- Development of a basic input for a 3D model, then an empirical approach and later also a mathematical approach aimed at producing the final form of the model, and
- Establishment of correlation to monitoring data.

Based on the above problem statement, it was safe to assume that interactive work and integrated use of known 2D and partly defined 3D parameters would produce an output of a three-dimensional nature. By way of solution, evidence was needed that the method to be applied for the construction of a 3D geometrical (physical) model and a parametric model of karst aquifer, aided by an incomplete data series, is feasible.

A 3D geological model was developed for the purpose of generating a network of potential karst conduits running from the sinkhole zones and dolines to the discharge zones, or from the determined potential infiltration zone to the accurately defined karst discharge zone.

The model presented in this chapter was developed using ArcGIS software and its 3D Analyst, Spatial Analyst, and Network Analyst extensions, as well as COMPASS software, and finally through some special programs made especially for reconstruction of conduit distribution. All spatial data, such as geological maps and cross sections, caves, sinkholes, springs, and karst channels, were converted into digital form, and each spatial unit was defined by its x-, y-, and z-coordinates. The compilation of all elements in a 3D environment produced a real, spatially oriented network of potential karst conduit pathways.

Finally, the methodology is illustrated by the case study of Beljanica karst aquifer (area of around 300 km^2). This case study generally shows the procedure and results of 3D modeling methodology of karst conduit, and further analysis for the purpose of watershed delineation, water pathways determinations, and definition of karst reservoir storativity as static and dynamic reserves.

Step 1. Forming of conceptual model

Developing a conceptual model for the physical modeling of some complex karstic aquifer is reflected through the hierarchy, technologically neutral organizing of geological and hydrogeological digital recording, and their interpretation.

Simplification of a karst hydrogeological system is performed through creating a conceptual model, where the basic knowledge about groundwater pathways such as the fault, channels, and caverns is merged with all available data. As karst is a very complex hydrogeological system, the principle behind the formation of a conceptual model is to simplify it as much as possible, while still retaining all the complex elements of the model that are necessary to simulate all the essential features of the system. Development of a conceptual model is of great importance since it directs and defines the future approach to forming the physical 3D model

(Milanović 2010). Schematic representation of the conceptual model of a karst aquifer in order to create a physical model is shown in Fig. 9.1.

Step 2. Defining the level of input data and forming of model layers

Data entry is the starting point for the establishment of a karst aquifer model and includes the conversion of available data into a digital format (graphic and alphanumeric). Such a model-designed system includes four main elements:

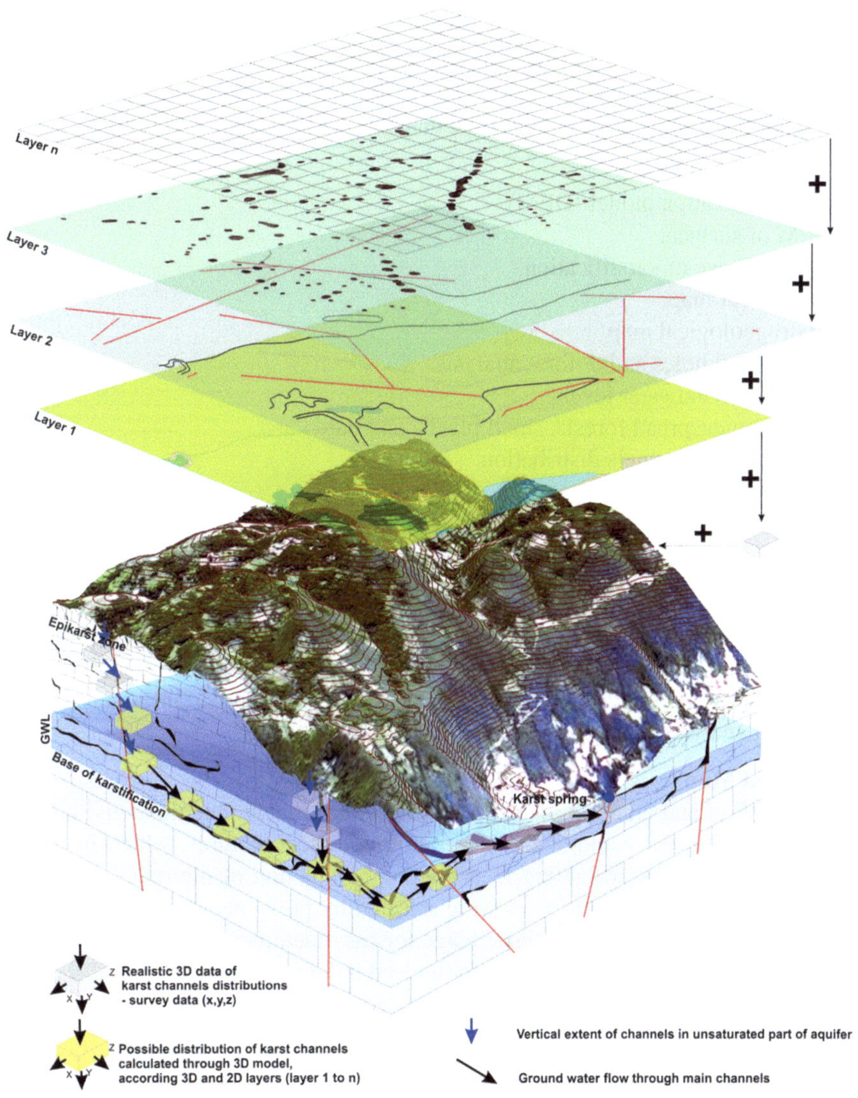

Fig. 9.1 Conceptual model of karst aquifer and its 3D physical model forming

1. Preparation and data entry,
2. Logical grouping and rational connection between relational of data connections,
3. Data analysis, and
4. Visualization and interpretation.

Data preparation includes the following:

- ensuring of appropriate graphic layers—topographic, hydrographic, geological, hydrogeological, speleological, geomorphological etc.,
- geocoding of available layers or "merging" in the appropriate map projection.

The following text provides an overview of the most basic layers necessary for the creation of 3D models of distribution of karst channels and generally the functioning of the karst aquifers. Layers got from combinations of obtained cabinet research (collected raster and vector entities) and detailed field studies are as follows:

- Topographic layer,
- Digital elevation model (DEM),
- DEM of surface,
- DEM of base of karstification,
- Geological map,
- Hydrogeological map,
- Map of sinkholes and dolines analysis,
- Map of groundwater levels,
- Map of widespread forests, small plants, and bare karst,
- Map of cave channels distribution,
- Tectonic map,
- Map of base of karstification,
- Map of the hypsometric location of the sinkholes, caves, and karst springs.

Topographic Layer
The topographic layer is one of the basic foundations for the development of 3D models as it is a vector layer of izoline with an equidistance usually of 10 m (depending on the investigation area size), which is very important in any kind of 3D spatial analysis. The main output of this layer is DEM.

Geological Map
The formation and analysis of geological maps accurately determine and allocate zones with different geological and geomorphological features, which in further analysis have a large weighting factor. Upon completion of digitizing, the entire area and all the specific attributes were selected by polygons which automatically generate and classify the investigated area for easy identification of zones with different geological features important for understanding karst geology.

Hydrogeological Map
The hydrogeological map that was created for the purpose of forming a 3D physical model of the terrain is actually a digital map with a related geodatabase. The most important information which should be located on the map (defined by x, y, z

coordinate position) and explained in detail in the geodatabase is issued with all types of hydrogeological characteristics, the thickness distribution of karst, sinkholes (the swallowing capacity, types, etc.), springs (types of spring, yield), direction and velocity of groundwater flow (tracing tests), boreholes (wells logo with all data obtained), and so on. After its formation, such a conducted hydrogeological map makes it possible to predefine the map usable for spatial analysis and 3D.

Map Analysis of Dolines and Sinkholes
One of the essential bases for the definition of karst aquifer is certainly the distribution of dolines and sinkholes in the karst area. The exact location of dolines and sinkholes is a good indication of the actual and possible existence of karst channels and caverns in the karst massif. As well, it indicates the privileged directions in the development of karst processes (channels) which can be observed through a series of sinkholes, while a large group of sinkholes in a small area suggests a well-developed karstification but not some other exact factors, in which case the spatial parameters of distributions of karst channels cannot be easily determined.

Map of Groundwater Levels (GWL)
Probably, one of the most complicated layers which has a very significant impact on the determination of the distribution of groundwater is the map of groundwater levels. For the purposes of the GWL map, a substantial fund of point data and good coverage of the investigation area are required. Formation of the base map of the underground water level is related to a number of complex problems. Probably, the biggest problem is the interpolation of point entities if it is known that neighboring points can be in completely different formations according to the level of karstification and that the GWL differs by tens of meters vertically, while the model points are almost slightly distant.

Map of Widespread Forests, Small Plants, and Bare Karst
It is necessary to analyze the type of vegetation cover, which represents the major parameters that affect the speed of infiltration of precipitation in the aquifer, of the different types of soil that affect the infiltration of water in karst aquifer.

Map of Cave Channels Distribution
One of the most complex tasks in the development of physical models is clearly defined distribution of karst channels especially in the deeper parts of the aquifer and its relationship with the circulation of groundwater in karst. The link between the geometry of the network conduits and the geological and hydrogeological conditions is certainly best defined by the characteristics of the karst underground objects.

In order to analyze such a complex system as karst aquifer, it is necessary to create precision layers which through 3D model of karst channels will provide the possible development of channels which cannot be observed by direct explorations.

Before making the map of cave distribution with all the necessary elements, it is necessary to model each objects in a 3D environment accurately.

Tectonic Map
Tectonics are represented in the model by spatial linear entities that are classified through length and spatial orientation (providing azimuths and angle of decline), determined by weight, and statistically analyzed. This process forms the layer of 3D tectonic and gets the full integration of the model in particular as a factor for potential distribution of karst channels in a 3D direction.

Map of Base of Karstification
Karstification is a vertical element in the model forming and is used for separation of those parts of the aquifer in which circulation varies quickly from the recharge zone to the discharge zone expiration of the parts where the circulation is very slow. That zone represents the transition between the high karstified rocks and nonkarstified rocks and is usually the main zone of groundwater circulation especially when it is siphonal circulation or a zone where the largest karst conduits (deep phreatic zone) form. Such a map must be developed as the base of karstification is precisely what it represents: the basement of the hydrogeological physical model of karst aquifers.

The problem of determining the base of karstification is often very complex. Often, only exploration through drilling in the background of drainage zones or very expensive cave diving surveys can get more specific information about the depth of karstification. Unfortunately, drilling and diving surveys often provide information only within limited and narrow areas of research, yet it is generally the only data we have to put into the model to analyze.

Map of the Hypsometric Location of the Sinkholes, Caves, and Karst Springs
In the process of forming a 3D physical and hydrogeological karst aquifer model, it is necessary to establish the quality hypsometrical model established of hypsometric position recharge areas, underground conduits, (which are the connection between the recharge and discharge zone) and the hypsometric position of karst springs as a discharge area.

Step 3. Establishing of physical model and its relation base

Model schema with all layers can be generally shown through the relationship of spatial entities into a single model as is shown in Fig. 9.2.

Components of the model shown in Fig. 9.2 include all of the classes and entities which, through a series of computational network and geostatistical analyses, form the 3D output of karst aquifer which is actually the result of this physical model. The result of the output data is the connection of surface spatial-oriented data with a defined position of karst channels in karst systems that are a logical connection between the implemented database and the spatial 2D and 3D entities.

Network spatial data 2D and 3D entities actually represent their field of numerical clusters that are 2D data transformed into a 3D object through a simulation model. 3D objects belong to the "multipatch feature class" and contain three-dimensional information.

Each network field that is integrated into the model and contains all the data that appear in this geographical unit and its numerical description is prepared for any calculation of the known extensions and thus finally forms a numerical

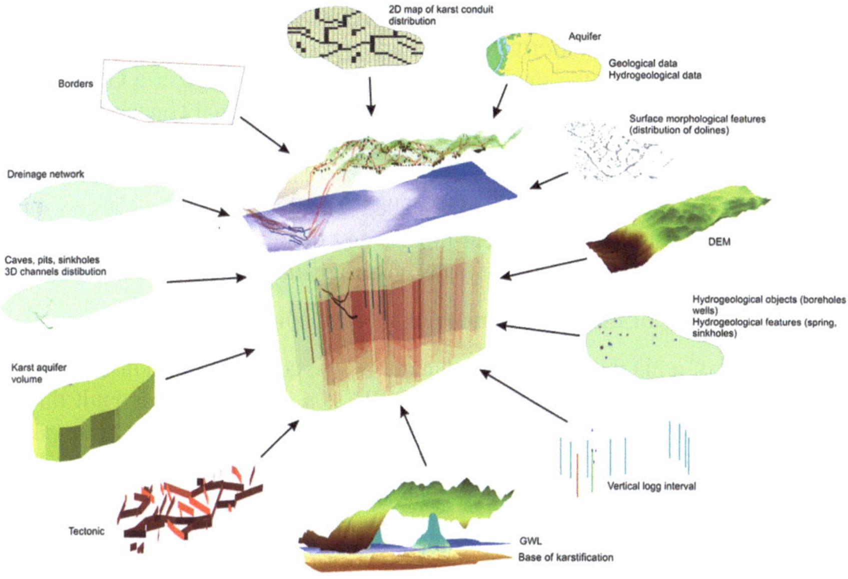

Fig. 9.2 Scheme of creation of 3D karst model

network as a two-dimensional surface. Such a system is used to map the 2D distribution of karst channels with all the integrated data transformed into a 3D view of the same image data (Fig. 9.3).

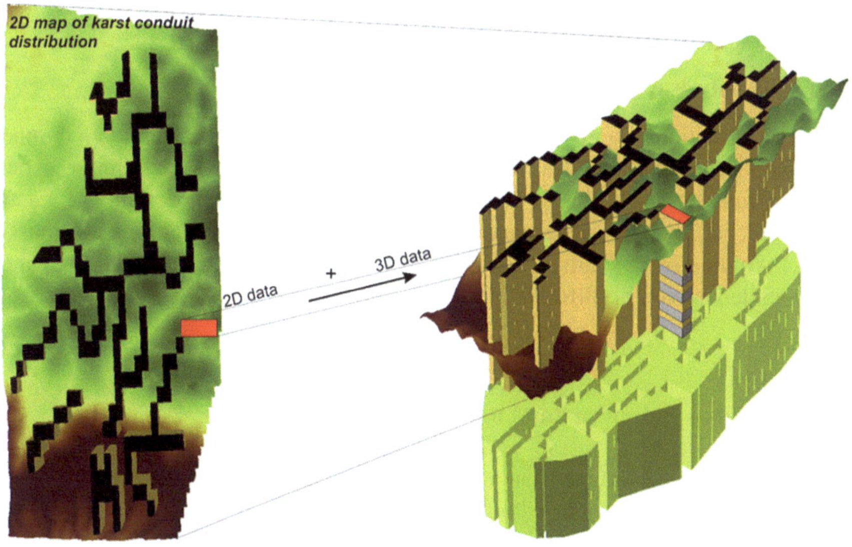

Fig. 9.3 Transformation of spatial layers from 2D to 3D model

The database is one of the most important components of the physical model as part of the Geographical Information System and includes the connection among all available data within the model. This is the main advantage of a Geographical Information System which is provided with complex relational connection maps to the database. The application of this concept is oriented layers, where each layer contains the appropriate graphic layout (map) associated with the database.

Relation graphs and a database can be called "small-world" as they largely reflect what is in the "real world" (physical model) and therefore through spatial display can describe and present fully the physical characteristics of a karst aquifer.

Step 4. Creating of the three-dimensional karst conduit

The forming of a 3D model of karst channels is the conversion of known nodes of groundwater and conduit directions from a 2D to a 3D model. It is built through the known factors that could be used to create a 3D mesh. One of the basic parameters of the formation of a 3D model is faults and other ruptural features and their inter-sections, together with dolines, sinkholes, and speleological features. This parameter converts 2D to 3D points and determines the z-coordinates and possible correction of the x-, y-coordinates to the new position.

Given the earlier statement about the set of problems, it can be concluded that the interactive work and interface of well-known 2D and 3D partially defined parameters get the output with the three-dimensional character. Once more, it must be reiterated that the resulting surface potential distribution of karst channels is the basis for the definition of most major directions, and the orientation of a channel of karst aquifers follows the deterministic low that can be established with a small percentage of mistakes.

Figure 9.4 presents the basic steps of physical modeling of karst conduits formed from an initial 2D model map of distribution of karst channels to the final 3D model of karst channel network.

Step 5. Analysis of physical model

After the formation of a 3D network of karst conduits it is possible to do a number of different analyses aimed at defining the functioning of a karst aquifer better. These include the following: providing the area of karst massif, defining of watershed of recharge zone, sub-groundwater catchments, volume of the karst massif (block), volume of conduits, dissolution porosity, main direction of groundwater flow, and delineation of different local watershed (Box.9.1).

Box 9.1
Case study—Beljanica karst aquifer
The Beljanica karst massif is located in the eastern part of Serbia. The Beljanica mountain range is an anticline (Kučaj-Beljanica structure) composed of Jurassic and Cretaceous carbonate rocks generally inclined from the central to peripheral parts of the massif (Stevanović 1991, 1994). The carbonate rock complex of the Kučaj and Beljanica massif is the result of the Alpine orogenic

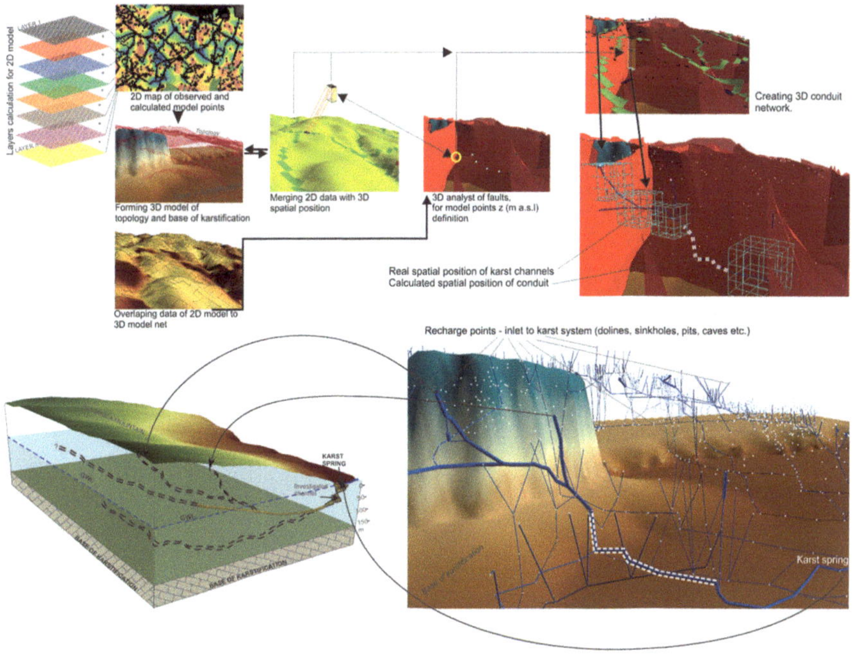

Fig. 9.4 Schematic procedure of 3D physical model forming

phase, with the most intensive tectonic movements during the Tertiary. Tectonic events resulted in a complex system of faults and fractures as privileged subterranean water paths. Moreover, climatic conditions, particularly the successions of wet and warm periods, significantly contributed to karstification. Karst of Beljanca Mountain is characterized by a large number of surface and underground morphological forms. Surface forms are well developed and widely distributed. The main forms are dolines, caves, pits, sinkholes, and karstic springs, and their classification and exploration were of great importance in the process of creating a Beljanica karst aquifer physical model. The main hydrogeological characteristic of the aquifer is its deep siphonal circulation. The cave divers explored channels of Krupaja Spring to a depth of 120 m and Mlava Spring, whose channels in fact go much deeper (studies have not been completed), to a depth of 73 m (Milanović 2007).

This case study includes an analysis of the created 3D ArcGIS physical model of karst interior. The input data of the model are represented by the 69 caves, 15 sinkholes, 1,682 dolines, 7 major karst springs and around 70 minor springs explored as well as all geological, morphological, and hydrogeological data connected with the area of investigation (Fig. 9.5a). As 2D output from the model, more than 6,000 data points (registered in database) were calculated. There are 60 different data for each point (coordinates, groundwater level, type of channel, dimensions of channels, next point of conduit, connection with conduit, orientation of channel, hydrogeological function etc.). Eventually, the model is determined by more than 360,000 data (Milanović 2010). 2D model mesh is shown on Fig. 9.5b.

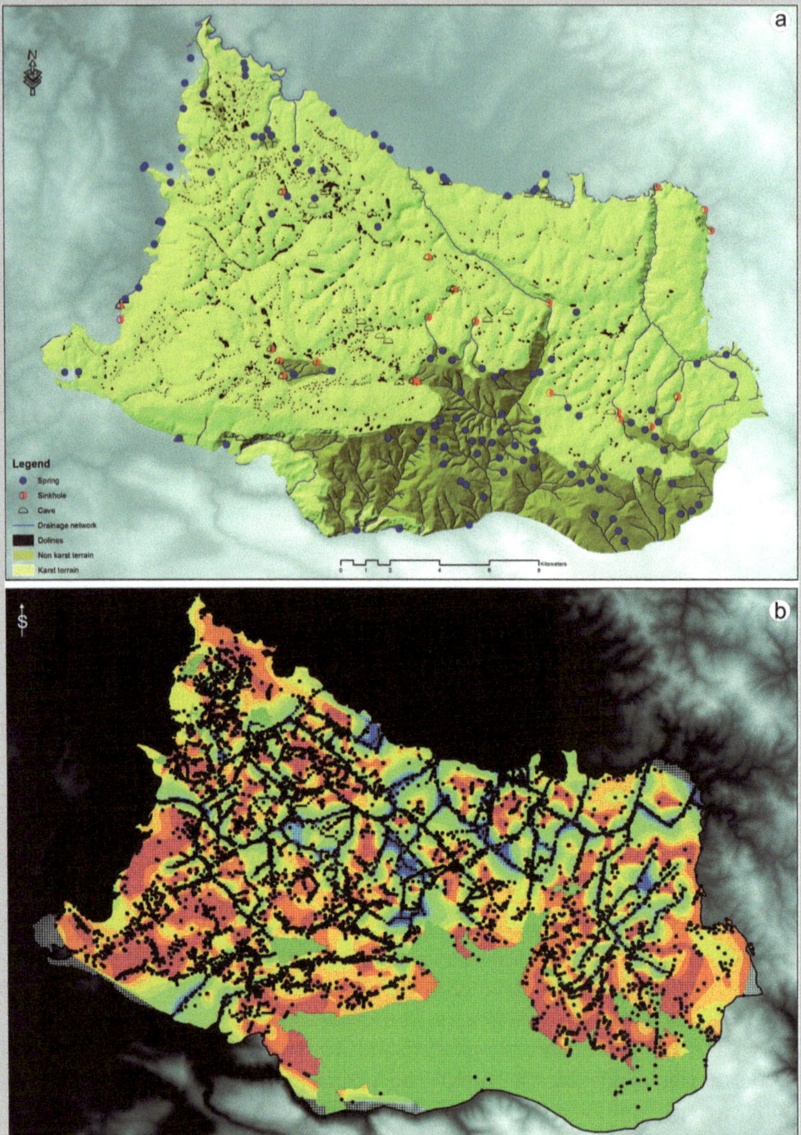

Fig. 9.5 **a** Map of Beljanica karst massif including position of hydrogeological, geological, geomorphological, and speleological features, **b** 2D model mesh is created in accordance with Beljanica physical model data

3D model of karst channels distribution on Beljanica aquifer

For the 3D model of the karst underground, the ArcGIS program was used along with some additional associated special software (Milanović 2010). To set the model in a "real" spatial environment, some detailed spatial and surface layers were created. Linked layers and an associated base form a basis for further analysis of karst aquifers through different model components. By defining the main directions and orientation of karst conduits (Fig. 9.5b) (followed by possible small percentage error), layers of potential 3D distribution of karst channels were created. The model based on deterministic laws included input parameters as the main channels with the other channels as a function of water transfer from the top surface into the deep ground and toward output parameters approximated with all zones and points of discharges (Milanović et al. 2013).

The total length of the karst channels network which is calculated using this model and presented in a 3D environment is 647.29 km. Detail of the model is shown on Fig. 9.6.

Discussion and Conclusion

On the basis of the modeling and simultaneous field studies, it was noted that the circulation of groundwater depends on the hydrogeological situation and

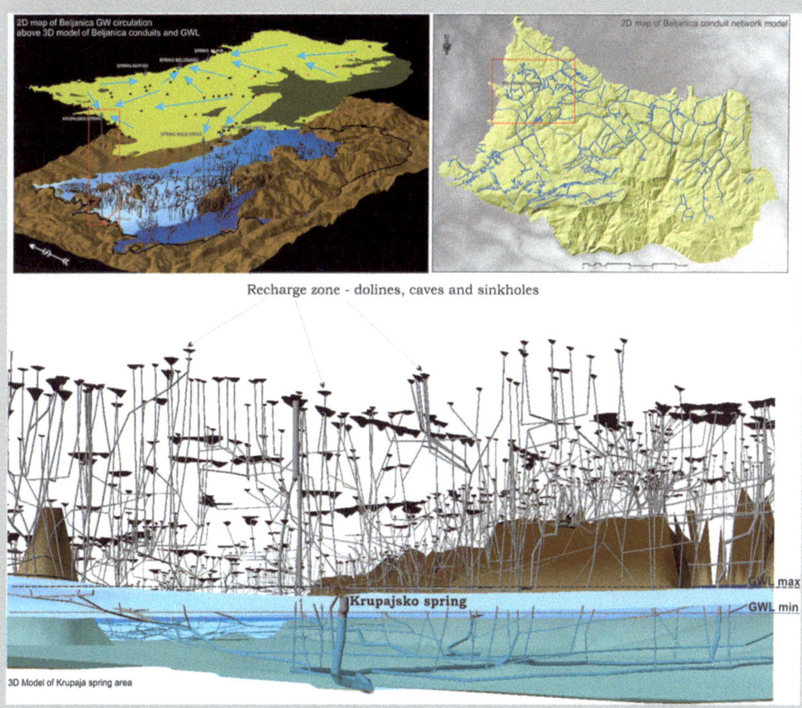

Fig. 9.6 2D and 3D model of conduit system of Beljanica karst aquifer with Krupaja Spring physical model detail

piezometric level within the Beljanica aquifer. Groundwater flows in different directions throughout the year. Therefore, redistribution of water within the aquifer results in no constant watersheds. The total length of the karst channels network, calculated using the physical model and presented in a 3D environment, is 647 km. The groundwater exploitable reserves of Beljanica karst aquifer are estimated to be over 4 m^3/s. The waters are low mineralized, unpolluted and have great potential for water supply (Milanović et al. 2013).

Despite such inconsistencies, the created complex 3D model enables an approximation of the hydrogeological watersheds. As watersheds have been proclaimed "realistic," they represent a scenario for the periods of medium waters (neither extreme maximums nor minimums). In this case, the largest is the Mlava Spring catchment which comprises some 124 km^2. The Belosavac Spring watershed covers 27 km^2, while 85 km^2 belongs to the Krupajsko Spring. Malo Vrelo Spring is the smallest catchment in Beljanica covering some 7 km^2, while Veliko Vrelo Spring comprises some 24 km^2 (Fig. 9.7). The complete network of karst channels is shown on Fig. 9.7 (Milanović 2010).

A spatial-oriented network, i.e., potential karst channels providing a base for the assessment of storativity, is a crucial component for groundwater resources analysis. The underground karst development along with the

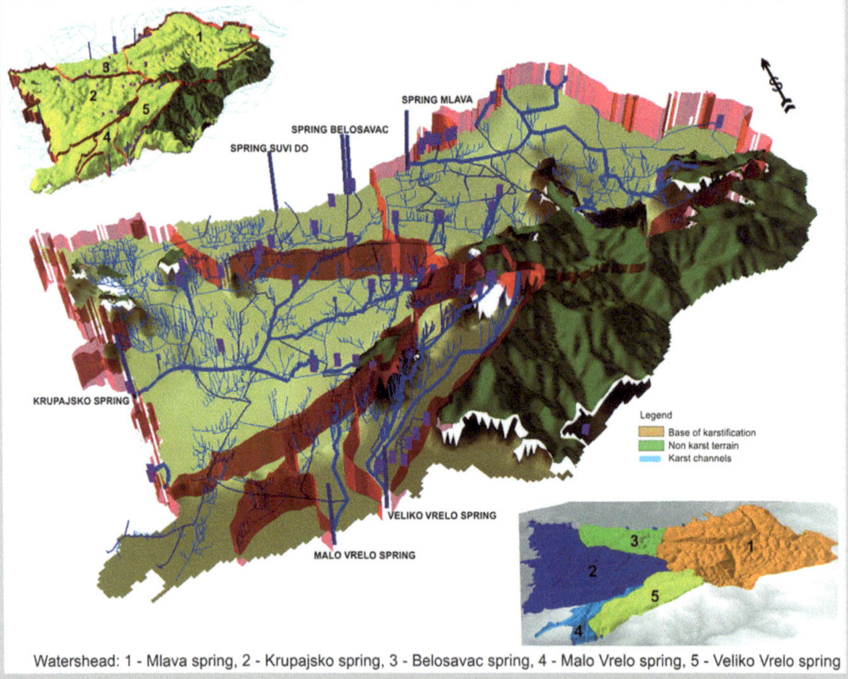

Fig. 9.7 Delineation of watershed of Beljanica karst aquifer through analysis of karst channels distribution (3D analyses)

saturation status were important elements from the 3D model to be correlated with data collected from simultaneous groundwater quantity (discharge) observations (Milanović et al. 2010, 2013).
The total volume of karst aquifer from the surface to the impermeable bottom (karstification base) is 482 km^3. The thickness of the saturation zone (from maximal groundwater table to karstification base) is regularly over 200 m. The created model resulted in an average storativity value of 0.0125 for the massif as a whole. Static and dynamic water storage (reserves) for Beljanica karst aquifer calculated through physical model is shown on Table 9.1 (Milanović 2010).

Table 9.1 Results of 3D analysis of Beljanica karst aquifer model (static and dynamic water storage)

Spring name	Watershed area F (km^2)	Static storage ΔV (10^6 m^3)	Average dynamic storage ΔV (10^6 m^3)	Q_{static} (m^3/s) if discharge all within one year	$Q_{dynamic}$ (m^3/s)
Spring Mlava	124	256.255	61.48	8.125	1.95
Krupajsko spring	85	187.05	52.03	5.9	1.64
Springs: Belosavac, Živkova rupa, Suvi Do	27	31.42	19.20	0.99	0.60
Veliko and Malo Vrelo (springs)	31	7.7	22.18	0.255	0.70

Finally, we should emphasize that the complex field research and its results, along with continuous monitoring and the correlation of those results with the created 3D model of karst conduit network, enable the reconstruction of complex karst aquifer, including forecast of its spatial, temporal, quantitative and qualitative characteristics. The obtained results favor wide application of such an approach as a base for sustainable water management and appropriate utilization of aquifers with large groundwater reserves.

References

Borghi A, Renard P, Jenni S (2010) How to model realistic 3D karst reservoirs using a pseudo-genetic methodology—example of two case studies. In: Andreo B, Carrasco F, Duran J, LaMoreaux JW (eds) Advances in research in karst media. Springer, Berlin, pp 251–256

Borghi A, Renard P, Jenni S (2012) A pseudo-genetic stochastic model to generate karstic networks. J Hydrol 414–415:516–529

Butscher C, Huggenberger P (2007) Implications for karst hydrology from 3D geological modeling using the aquifer base gradient approach. J Hydrol 342:184–198

Fish L (1996) Compass. http://members.iex.net/lfish/compass.html

Filipponi M, Jeannin PY (2008) Possibilities and limits to predict the 3D geometry of karst systems within the inception horizon hypothesis, geophysical research abstracts, vol 10. EGU General Assembly 2008, EGU2008-A-02825

Filipponi M (2009) Spatial analysis of karst conduit networks and determination of parameters controlling the speleogenesis along preferential lithostratigraphic horizons. Thèse no 4376, École Polytechnique Fédérale de Lausanne, Suisse

Gogu RC, Carabin G, Hallet V, Peters V, Dassargues A (2001) GIS-based hydrogeological databases and groundwater modelling. Hydrogeol J 9:555–569

Jeannin PY, Groves C, Häuselmann P (2007) Speleological investigations. In: Goldscheider N, Drew D (eds) Methods in karst hydrogeology, vol 26, International contribution to hydrogeology, IAH, Taylor & Francis/Balkema, London, pp 25–44

Kincaid TR (2006) A method for producing 3-d geometric and parameter models of saturated cave systems with a discussion of applications. In: Sasowskiy I, Wicks C (eds) Groundwater flow and contaminant transport in carbonate aquifers. Balkema, Rotterdam, pp 169–190

Kovács A (2003) Geometry and hydraulic parameters of karst aquifers—a hydrodynamic modelling approach. Ph.D. thesis, La Faculté des sciences de l'Université de Neuchâtel, Suisse, p 131

Milanović S (2007) Hydrogeological characteristics of some deep siphonal springs in Serbia and Montenegro karst. Environ Geol 51(5):755–759

Milanović S (2010) Creation of physical model of karstic aquifer on example of Beljanica mt (Eastern Serbia). Ph.D. thesis, University of Belgrade, Faculty of Mining and Geology, Belgrade, Serbia

Milanović S, Stevanović Z, Vasić Lj (2010) Development of karst system model as a result of Beljanica aquifer monitoring. Vodoprivreda, Belgrade, 42 (2010)/ 246–248, 209–222

Milanović S, Stevanović Z, Vasić Lj, Ristić-Vakanjac V (2013) 3D Modeling and monitoring of karst system as a base for its evaluation and utilization—a case study from Eastern Serbia. Environ Earth Sci 71(2):525–532

Ohms R, Reece M (2002) Using GIS to manage two large cave systems, wind and jewel caves, South Dakota. J Cave Karst Stud 64(1):4–8

Stevanović Z (1991) Hidrogeologija karsta Karpato-Balkanida istočne Srbije i mogućnosti vodosnabdevanja (Hydrogeology of Carpathian-Balkan karst of eastern Serbia and water supply opportunities; in Serbian). Faculty of Mining and Geology (spec. ed.), Belgrade, p 245

Stevanović Z (1994) Karst ground waters of Carpatho-Balkanides in Eastern Serbia. In: Stevanović Z, Filipović B (eds) Ground waters in carbonate rocks of the Carpathian—Balkan mountain range. CBGA, Allston Hold., Jersey (spec.ed.), pp 203–237

Schaecher GR (1986) 3D cartography for the rest of us. Compass and tape 4(1):20–23

Strassberg G (2005) A geographic data model for groundwater systems. Ph.D. thesis, The University of Texas at Austin, USA

Springer G (2004) A pipe-based, first approach to modeling closed conduit flow in caves. J Hydrol 289:178–189

Ulfeldt S (1975) Computer drawn stereo three-dimensional cave maps. Proceedings of the national cave management symposium. Albuqerque, New Mexico

Chapter 10
Mathematical Modeling of Karst Aquifers

Alex Mikszewski and Neven Kresic

10.1 Introduction

Groundwater models vary in complexity from simple analytical models to numerical models representing full three-dimensional heterogeneity and anisotropy of porous media. Models enable hydrogeologists to better understand groundwater systems and test hypothesis and assumptions. With the proliferation of high-speed computing technology, numerical modeling has become a useful and reliable (when appropriately applied) decision-support tool. The basis for almost all of these models is the equivalent porous media (EPM) formulation using Darcy's Law. In karst, however, the physical complexity of the subsurface environment makes the equivalent porous media approach highly questionable. The reality is that Darcy's Law, an equation based on a simple homogeneous sand column experiment (Darcy 1856), does not apply to flow in conduits (cavities), sinks, and springs. A visual representation of the discrepancy between Darcian flow and karst systems is presented in Fig. 10.1.

This challenge is well understood by karst hydrogeologists, many of whom believe that numerical modeling of karst flow is not feasible. This problem is compounded by the fact that there is a general distrust of any groundwater model by some professionals and many regulators, even a model representing simple unconsolidated porous media. Often, regulators believe groundwater model results to be entirely non-unique, meaning that modelers tweak parameters to fulfill a preordained purpose while still satisfying a calibration standard. There is some truth

A. Mikszewski (✉)
Amec Foster Wheeler, Kennesaw, NC, USA
e-mail: alex.mikszewski@amecfw.com

N. Kresic
Amec Foster Wheeler, Kennesaw, VA, USA
e-mail: neven.kresic@amecfw.com

Z. Stevanović (ed.), *Karst Aquifers – Characterization and Engineering*,
Professional Practice in Earth Sciences, DOI 10.1007/978-3-319-12850-4_10

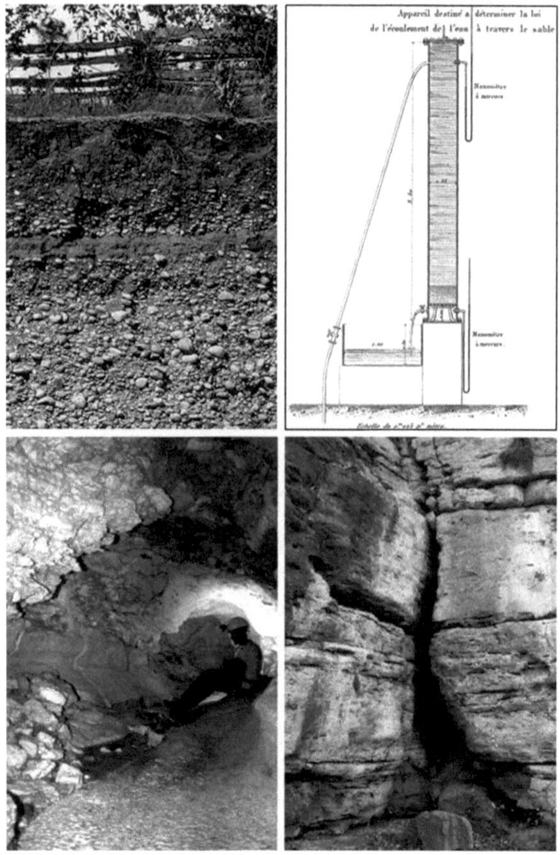

Fig. 10.1 *Top left* Example of unconsolidated sediments where Darcy's Law does apply—glacial outwash sands and gravels (*courtesy* of USGS 2011). *Top right* Sketch of the original apparatus used by Darcy (1856). *Bottom* A large portion of groundwater flow in karst takes place in solution cavities and dissolutionally enlarged fractures of many different shapes and sizes. (from Kresic 2013, copyright McGraw Hill, printed with permission)

to this, as a hydrogeologist has tremendous influence over model results, and can sway things one way or another by modifying variables that are difficult to trace. For this and other reasons, some professionals hired by parties with high-stake agendas, such as winning lawsuits, may bend the limits of professional ethics and develop groundwater models that cannot be comprehended even by experts. Because of the complexity and randomness of sinks and conduits, karst models are especially susceptible to this "expert" manipulation.

As presented by Kresic and Mikszewski (2013), all of this uncertainty leads to the questions "Why do we need to do groundwater modeling in the first place?" The answer is hydrogeologists need groundwater models for two primary reasons: One is to predict future conditions, and the other is to understand how current conditions came to be. Nearly all projects in hydrogeology require future projections, and the hydrogeologist is relied upon for expert opinion in this matter. Most commonly, the hydrogeologist must evaluate the efficacy of selected interventions in water supply development or hazardous waste remediation (particularly groundwater remediation). Examples include the installation and operation of new water supply wells or extraction wells associated with pump and treat remediation, the injection of chemical reagents for groundwater remediation (in situ chemical oxidation), or the stimulation of contaminant biodegradation. Groundwater models are the best available means of simulating these interventions and most simply, determining if they will work. Modeling can also fill in the gaps between data that have been or will be collected at discrete time intervals, thus helping the hydrogeologist better understand and simulate transient processes.

Despite the uncertainty inherent to karst hydrogeology advocated by the skeptics, we believe numerical modeling has a role in karst projects. For example, a karst site may have several mapped conduits with varying importance in the transport of water and contaminants. A numerical model representing both the rock matrix and these conduits will enable a hydrogeologist to better understand the transfer of water and solutes within the conduits, between the conduits, and between the conduits and the rock matrix. This has many practical applications including the design of groundwater supply or containment of contaminants. In contrast, an EPM model of the same system using cells of unreasonably high hydraulic conductivity and low effective porosity (the common practice) would not be able to simulate these interactions.

This chapter presents an overview of numerical modeling techniques for EPM and conduit flow formulations, including a case study illustrating the value of conduit presentation.

10.2 Summary of Numerical Modeling Techniques

10.2.1 Equivalent Porous Media Formulation

Numerical models describe the entire flow field of interest at the same time, providing solutions for as many spatial data points as specified by the user. The porous media volume ("model domain") is subdivided into many small volumes (referred to as cells or elements; see Fig. 10.2), and the basic differential

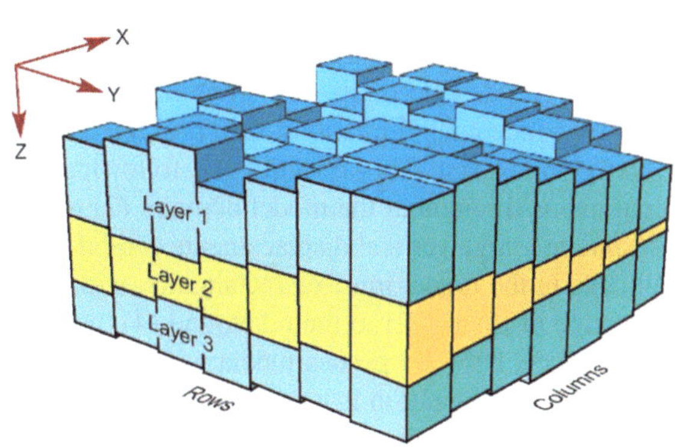

Fig. 10.2 Three-dimensional model domain in MODFLOW divided into layers and cells (from Kresic and Mikszewski 2013)

groundwater flow (and fate and transport) equation is defined for each cell. This system of differential equations, based on Darcy's Law, is replaced (approximated) by the systems of algebraic equations so that the flow field is represented by x equations with x unknowns where x is the number of cells. The system of algebraic equations is solved numerically, through an iterative process, leading to the "numerical model" designation. Finite difference and finite element methods are two common numerical methods of solving the groundwater flow equations. Which method will be used depends largely on the type of problem and the knowledge of the modeler. Finite elements more easily describe irregular model boundaries and internal boundaries such as faults. They are also more appropriate in handling point sources and sinks and large variations in water table, such as those found in karst. As a result, many believe that finite element techniques are more applicable to karst than finite difference models. However, finite difference models are easier to program, require less data, are friendlier for data input, and generally exhibit better mass conservation than finite element alternatives.

Unlike time series, statistically based input–output models, numerical models can answer the following questions:

- At what locations and how many wells are needed to provide a desired flow rate?
- What is the impact of current or planned abstraction on groundwater and surface water levels?
- How long would it take a contaminant to reach the water table or potential receptors, and at what concentration?
- How would a remedial intervention affect contaminant concentrations in the source area and in the downgradient plume?

Once these questions are addressed by the model(s), many new ones may pop up, which is exactly the purpose of a site-specific groundwater model to answer all kinds of possible questions related to groundwater flow and fate and transport of contaminants.

"A modular three-dimensional finite difference groundwater flow model" (MODFLOW) developed at the USGS, with the latest version documented in Harbaugh (2005), is considered by many to be the most reliable, verified and utilized groundwater flow computer program today. There are several integrated, user-friendly pre- and post-processing graphical software packages (GUIs) for MODFLOW that greatly facilitate data input and visualization of modeling output (results).

However, despite of its numerous advantages, classic MODFLOW has some serious conceptual limitations that prohibit accurate representation of complex geology and karst. This includes the following: (1) all model layers have to be continuous throughout the model domain, (2) model instability arises when trying to simulate large vertical displacements caused by faults or artificial structures, (3) all cells in the model must be rectangular, with rows and columns extending from one edge of the model to the other, and (4) model instability arises when simulating contacts between porous media with highly contrasting hydraulic conductivities. This last limitation is a common problem when the EPM approach is applied to conduits. The authors have seen modelers use hydraulic conductivities varying by over six orders of magnitude in adjacent cells, without assessing whether the converged solution is reasonable.

To overcome this and other limitations of MODFLOW as applied to karst settings, the USGS has developed a new package called the Conduit Flow Process (CFP) for MODFLOW-2005 (Shoemaker et al. 2008). The CFP package is described in more detail in the next section.

10.2.2 Modeling for Karst, the Conduit Flow Process (CFP)

The Conduit Flow Process has the ability to simulate turbulent groundwater flow conditions by the following: (1) coupling the traditional groundwater flow equation with formulations for a discrete network of cylindrical pipes (Mode 1 or CFPM1), (2) inserting a high-conductivity flow layer that can switch between laminar and turbulent flow (CFPM2), or (3) simultaneously coupling a discrete pipe network while inserting a high-conductivity flow layer that can switch between laminar and turbulent flow (CFPM3). As explained by the USGS, CFPM1 may represent dissolution or biological burrowing features in carbonate aquifers, voids in fractured rock, and/or lava tubes in basaltic aquifers and can be fully or partially saturated under laminar or turbulent flow conditions. Preferential flow layers (CFPM2) may represent the following: (1) a porous media where turbulent flow is suspected to occur under the observed hydraulic gradients; (2) a single secondary porosity subsurface feature, such as a well-defined laterally extensive underground cave; or (3) a horizontal preferential flow layer consisting of many interconnected voids.

As stated by the USGS, the CFP was designed to be flexible enough for use in locations with both limited and abundant field data. In some geologic environments, such as Mammoth Cave, Kentucky, detailed information is available (or could be derived) on the location, diameter, tortuosity, and roughness of the subsurface caverns. CFPM1 was designed with these locations in mind. In other locations, such as the Biscayne aquifer of southern Florida, void connections and distributions are so complicated within preferential flow layers that a complete characterization is not possible. CFPM2 was designed with these locations in mind; specifically, laminar and turbulent flow through complicated void connections is represented with a limited number of effective or bulk layer parameters.

One of the options in the CFP is that, in cases with abundant field data on the void architecture and hydraulic behavior, complex two- or three-dimensional networks of conduit flow pipes and nodes can be designed to represent interconnected or dead-end voids in the subsurface. Flow calculations assume pipe nodes are located in the center of MODFLOW cells. An exception is in the vertical direction, for which, there are two options. First, pipe nodes can be assigned elevations above a datum, and therefore, are not restricted to center elevations of MODFLOW cells. Second, pipe nodes can be assigned a distance above or below the center of the MODFLOW cell. With this second option, if the distance is set to zero, pipe nodes are assumed to exist at the vertical center of the MODFLOW cell.

Pipes can connect diagonally between two finite difference cells within the same layer or within and adjacent model layers (Shoemaker et al. 2008).

Data requirements for simulations with CFPM1 are more complex than using CFPM2; they include conduit pipe locations, length, pipe diameter, water temperature, tortuosity, internal roughness, critical Reynolds's numbers, and exchange conductances between pipes and the matrix. The parameters for CFPM2 simulations include water temperature, mean void diameter, and critical Reynolds's number. Simulations with CFPM3 include the combination of CFPM1 and CFPM2 input parameters. Boundary and initial conditions (for transient simulations) are the same as for classic porous media MODFLOW models. Concentrated aquifer recharge, such as via sinks, can be assigned directly to pipe nodes.

Figures 10.3, 10.4 and 10.5, from Mikszewski and Kresic (2012), illustrate several conceptual models developed with the MODFLOW CFP module using Groundwater Vistas GUI (Rumbaugh and Rumbaugh 2011). A comparison of EPM and coupled continuum pipe-flow (CCPF)-model-calculated hydraulic head contours is shown in Fig. 10.3. The flow direction is from the right to the left as prescribed with two constant head boundaries. The hydraulic conductivity assigned to all finite difference cells is the same in both cases. The CCPF example includes a conduit (thick line) for which the flow equation is solved separately and then coupled with the Darcian flow solution for the finite difference cells. The influence of networks of conduits on the hydraulic head contours in the surrounding aquifer matrix is shown in Fig. 10.4. The example in Fig. 10.4-right shows the influence of a focused recharge applied via a conduit node directly to conduit #2 which is more conductive than conduit #1. Because of this difference, the hydraulic head in conduit #1 increases, and the conduit backs up and discharges to the surrounding porous media matrix as indicated by the shape of the contours. The flow mass balance results saved by MODFLOW-2005 can be processed externally

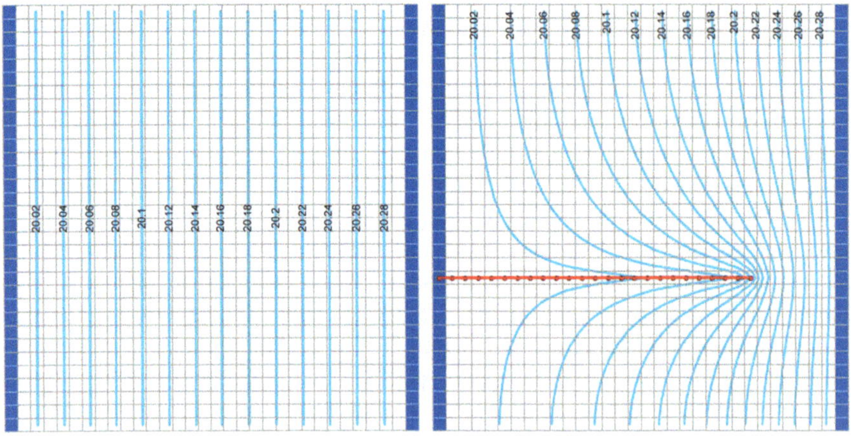

Fig. 10.3 Comparison of EPM (*left*) and coupled continuum pipe-flow (CCPF) model-calculated hydraulic head contours. (From Mikszewski and Kresic 2012)

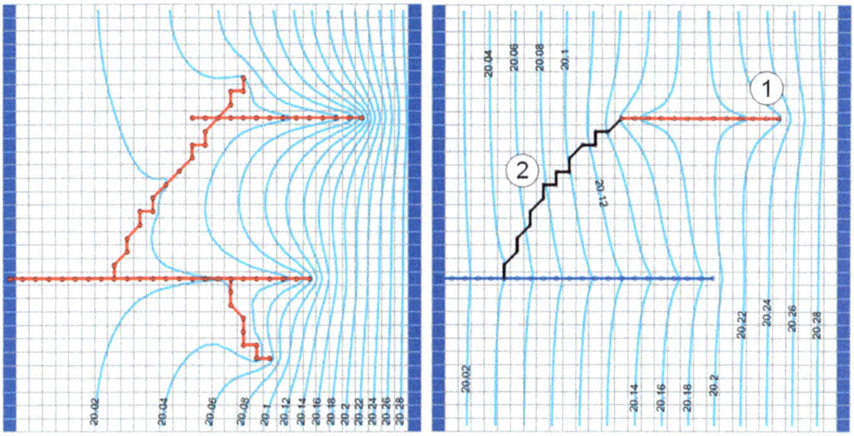

Fig. 10.4 Influence of networks of conduits on the hydraulic head contours in the surrounding aquifer matrix in coupled continuum pipe-flow (CCPF) models. (From Mikszewski and Kresic 2012)

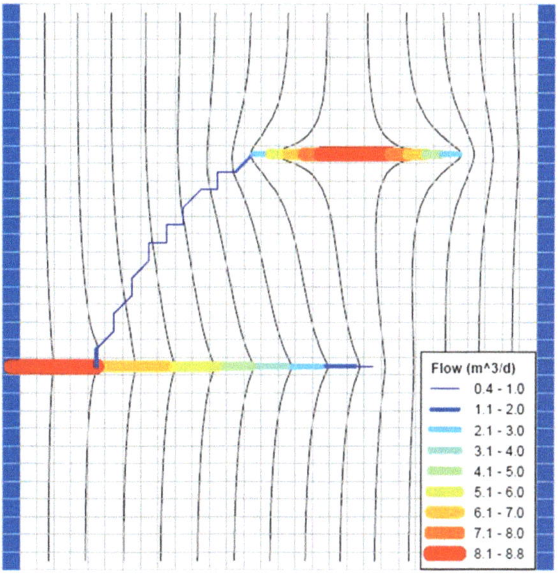

Fig. 10.5 Flow mass balance results for the conduits in Fig. 10.4-right processed externally to the GUI. (From Mikszewski and Kresic 2012)

to the GUI and visualized for either the matrix portion of the model or for the conduits as illustrated in Fig. 10.5.

The MODFLOW-2005 CFP module is still a work in progress, and some useful add-ons available for traditional porous media simulations are not yet available. This includes fate and transport modeling and particle tracking. For example, particle tracks and related groundwater flow velocities, which can be optionally marked by the Groundwater Vistas GUI using tickmarks, stop short of conduits and completely ignore them. The user must process the model results externally in order to analyze and visually present these features. In any case, encouraging continuing efforts of improving the CFP by both USGS and non-USGS researchers will hopefully increase its utility and facilitate its adoption by the modeling community.

10.3 Case Study in Karst Modeling

10.3.1 Conceptual Site Model

The case study site is located in a karst region of the United States. A topographic terrain map with exaggerated vertical scale is presented in Fig. 10.6. The geology is characterized by faulted and gently folded carbonate rocks of Paleozoic age.

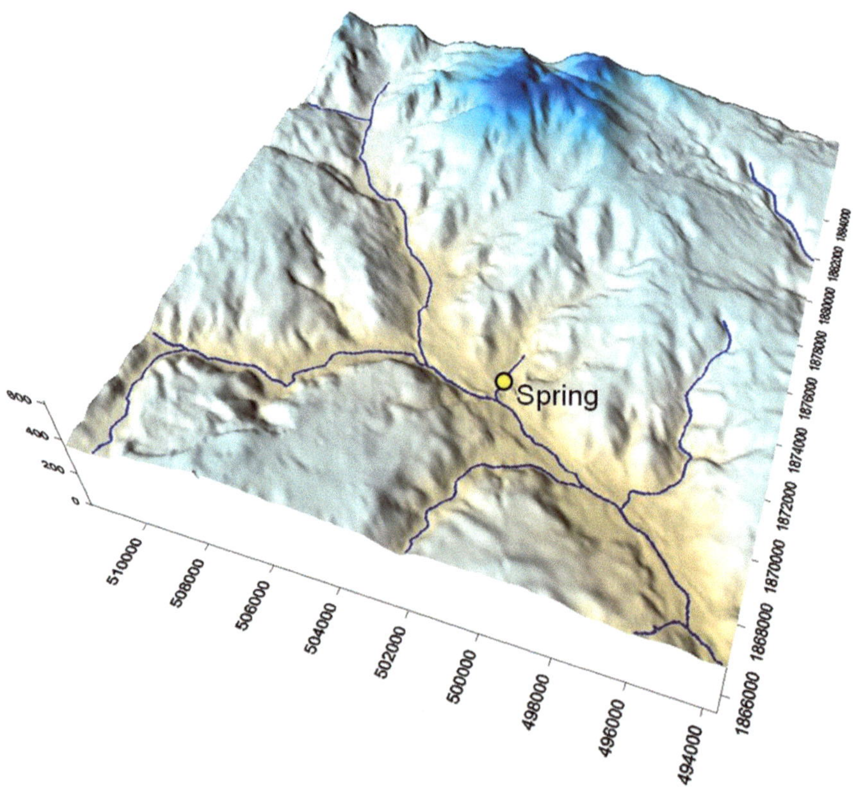

Fig. 10.6 Topographic terrain map of the numeric model area with exaggerated vertical scale

There are three primary geologic units, shown on Figs. 10.7 and 10.8 and listed from bottom to top as:

- Unit A: Low-permeable, non-carbonate metamorphic rocks
- Unit B: Karstified permeable limestone, permanently saturated
- Unit C: Highly karstified and permeable limestone, variably saturated
- Unit D: Low-permeable, non-carbonate sediments

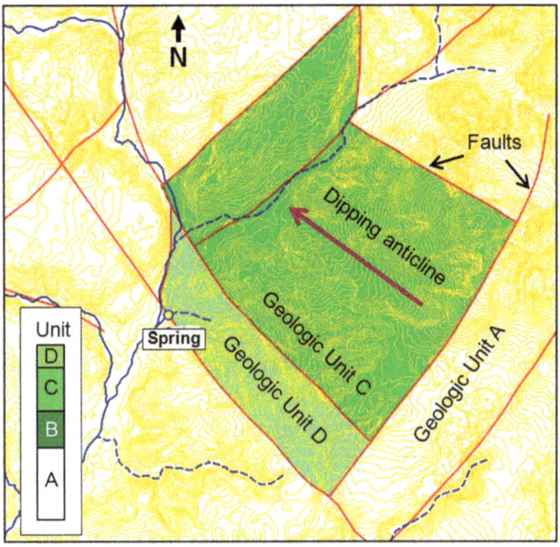

Fig. 10.7 Simplified geologic map of the model area. Explanation in text

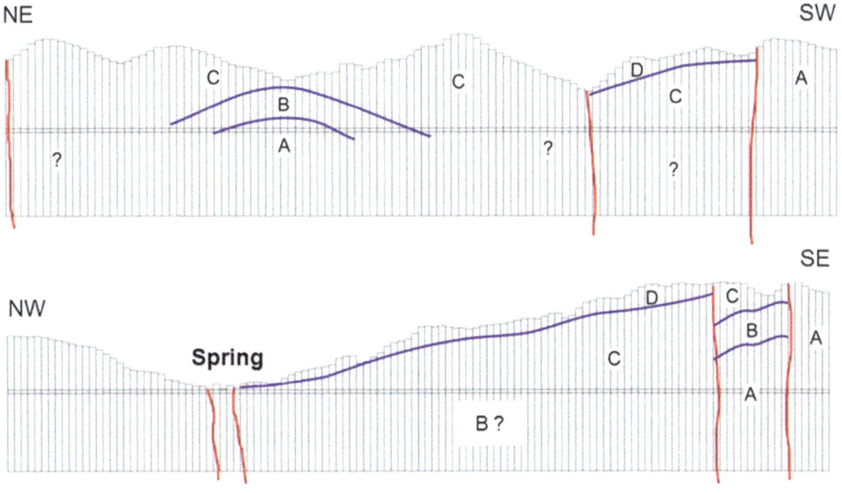

Fig. 10.8 Representative cross sections of the model area

The site receives significant rainfall, with an estimated 508 mm/year of groundwater recharge. Water supply comes from a large permanent spring that drains this watershed without contribution from permanent streams.

10.3.2 EPM Model Formulation

10.3.2.1 Model Geometry

The model domain is presented on Fig. 10.9 and encompasses the major features that govern groundwater flow and discharge to the spring. While not mandatory, in this case, it is advantageous to rotate the model grid such that the rows and columns are parallel to major structural features, such as boundary faults and the axis of the dipping anticline. In some cases, it is also helpful to align the grid with the primary directions of hydraulic conductivity tensor.

The layering of this model is complicated because of vertical displacements associated with faulting (see Fig. 10.8). As stated previously, all model cells within a layer must be interconnected in classical MODFLOW. Therefore, the elevations and geometry (slope) of model layers cannot always correspond to actual stratigraphy at the site. To overcome this limitation, the hydraulic conductivity of the distinct porous media is used together with the model layering to represent the discontinuity, as shown in Fig. 10.10. This allows the model to more accurately simulate the groundwater flow pattern, or hydrostratigraphy, associated with the discontinuity.

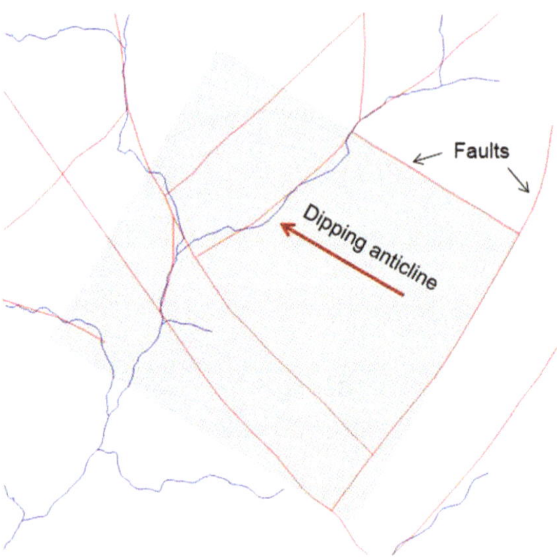

Fig. 10.9 Model domain showing rotated grid

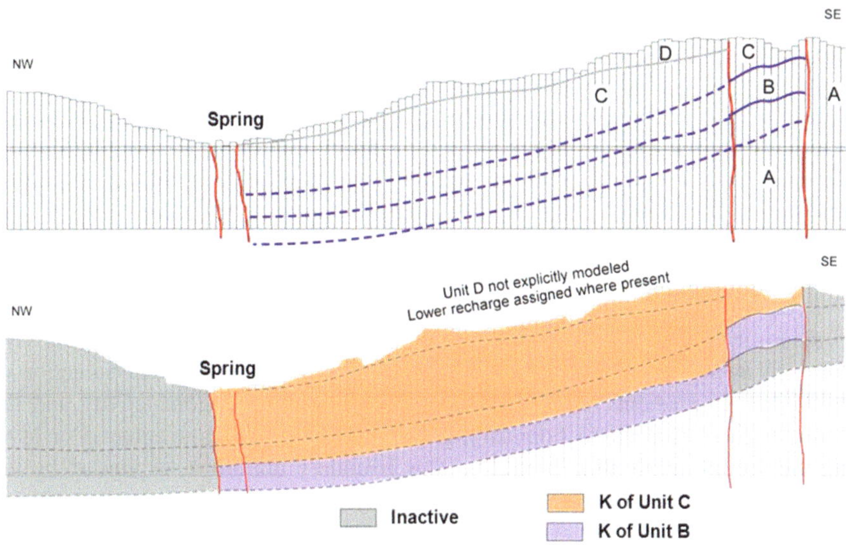

Fig. 10.10 Principle of continuous model layers which do not have to correspond to actual geologic layers. Complicated tectonics including offset of geologic layers at faults is simulated with a combination of representative hydraulic conductivity and the extent of model layers

As a rule of thumb, a uniform grid (i.e., equal cell size across the model) is used to initialize the model. This facilitates transfer of model input and output parameters between different platforms such as Geographic Information Systems (GIS) or gridding programs such as Surfer.

10.3.2.2 Model Boundary Conditions

It has become standard practice in hydrogeology and groundwater modeling to describe the inflow and outflow of water from the model domain with three general boundary conditions: (1) known flux, (2) head-dependent flux, and (3) known head, where "head" refers to the hydraulic head. These conditions are assigned to both external and internal boundaries, or all locations and surfaces where water is entering or leaving the model. One example of an external boundary, sometimes overlooked as such, is the water table of an unconfined aquifer which receives recharge coming from the vadose (unsaturated) zone. This estimated or measured vertical flux of water into the model is applied as recharge rate over certain surface area. It is expressed in model-consistent unit of length (e.g., feet or meters) per unit of time (e.g., days), which, when multiplied by the area, gives the flux of water as volume per time. A large spring draining an aquifer is another example of an external boundary with a known flux. An example of an internal boundary with a known flux, where water is also leaving the model, is water well with the pumping rate expressed in model-consistent units (e.g., cubic feet per day or cubic meters per day).

It is obvious that water can enter or leave the model in a variety of natural and artificial ways, depending upon hydrogeologic, hydrologic, climatic, and anthropogenic conditions specific to the system of interest. In many cases, these water fluxes cannot be measured directly and have to be estimated or calculated externally to the model. The simplest boundary condition is one that can be assigned to a contact between an aquifer and a low-permeability or impermeable porous medium, such as an "aquiclude." Assuming that there is no, or very limited, groundwater flow across this contact, it is called a zero-flux or no-flow boundary.

Recording hydraulic heads at external or internal boundaries, and using them to determine water fluxes indirectly, rather than assigning them directly, is very common in modeling practice. The hydraulic heads provide for determination of the hydraulic gradients which, together with the hydraulic conductivity and the cross-sectional area of the boundary, give the groundwater flow entering or leaving the model across that boundary. This boundary condition, expressed by the hydraulic heads on either side of the boundary and the hydraulic conductance of the boundary, is called head-dependent flux. One example of the head-dependent flux boundary would be a drain. A drain removes water from a model based on a conductance term representing the permeability of the drain and the difference in hydraulic head between the aquifer cell and drain.

Boundary conditions for the model are presented on Fig. 10.11. No-flow boundaries are used along the contacts with the low-permeability Unit A. The stream boundary is represented by drain cells due to its intermittent nature. An important distinction between the drain package and the similar river package is that drain cells become inactive when the water table drops below the drain elevation. In this manner, the drain functions intermittently, just like the stream it simulates. One limitation of the drain package is the inability to simulate losing segments of the stream, which would require use of MODFLOW stream package.

10.4 EPM Model Results

The flow solution for the EPM model is presented on Fig. 10.12. Groundwater flows toward the intermittent stream and the spring, as expected. The purple cells indicate dry cells in the top layer at the low-permeable core of the anticline. Blue cells indicate flooded areas where the calculated potentiometric surface is above the ground surface elevation. While for some models this may correspond to wetland areas, at this site there is no observed standing water. The flooding in the model is therefore erroneous and there is an alternate explanation.

A review of hydrogeologic data in the vicinity of the spring suggests that there may be karst conduits transmitting high quantities of water in the subsurface. This

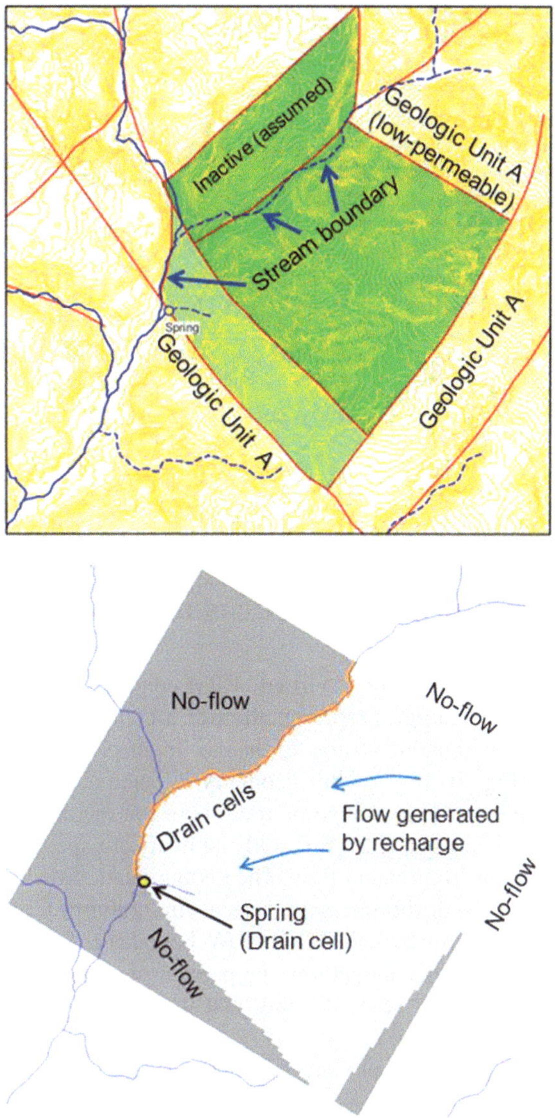

Fig. 10.11 Boundary conditions of the model

supports the inclusion of a conduit approximately along the axis of the flooded cells. A traditional (EPM) approach is to incorporate a "virtual" conduit by assigning extremely high hydraulic conductivity to model cells representing the conduit. In this way, a virtual conduit may successfully reduce flooding and convey more water to the spring for discharge. However, as previously discussed, there are serious limitations of this approach under transient conditions when simulating high recharge directly into the conduit via a sink.

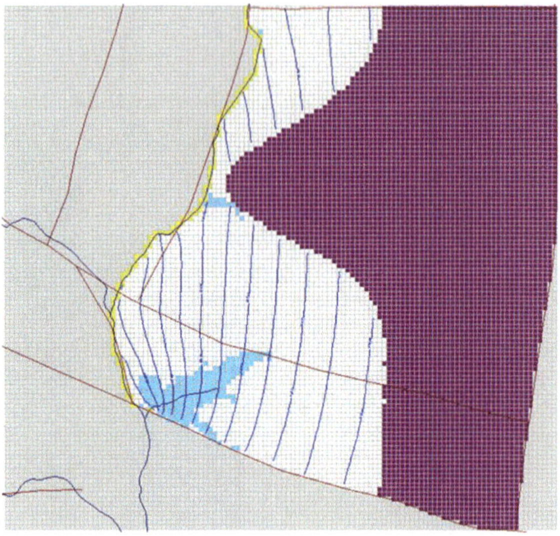

Fig. 10.12 Flow solution for the EPM model. Groundwater flows toward the intermittent stream and the spring. The *purple cells* indicate dry cells in the top layer at the low-permeable core of the anticline. *Blue cells* indicate flooded areas where the calculated potentiometric surface is above the ground surface elevation

10.5 Integrating the Conduit Flow Process

Rather than using a "virtual" EPM conduit, the CFP was used to include a physically meaningful representation of a conduit capable of turbulent flow and more realistic conduit-matrix exchange. Input parameters for the conduit are presented on Fig. 10.13. Conduit geometry is entered, such as diameter and bottom elevation, and the exchange of water between the conduit and matrix is determined by a conductance term. Reynolds numbers are specified to dictate the transition from laminar to turbulent flow. The ending head at the conduit is fixed to represent the elevation of the spring. A current limitation of CFP is the inability of conduits to connect to external MODFLOW boundaries in the outside matrix, such as drain cells. For this simulation, the percent of aerial recharge directly entering the conduit is set to zero, meaning that there is no sink connected to the conduit from ground surface.

Modeling results are depicted on Fig. 10.14, and the resulting potentiometric surface clearly shows the hydraulic influence of the CFP conduit. While this surface may also be approximated by an EPM virtual conduit, there is a fundamental difference in the flow process that becomes apparent when implementing transient recharge through a sink. An alternate simulation was performed with focused recharge directly into the conduit system. This focused recharge occurs through a smaller, feeder conduit connected to a karst depression receiving runoff. The recharge area, feeder conduit, and resulting potentiometric surface are shown on Fig. 10.15. The hydraulic head contours demonstrate a unique feature of karst that cannot be replicated by any EPM approach: The conduits discharge to the matrix due to higher hydraulic heads in the conduits themselves. In other words, the conduit backs up due to excessive pressure caused by the intensity of recharge.

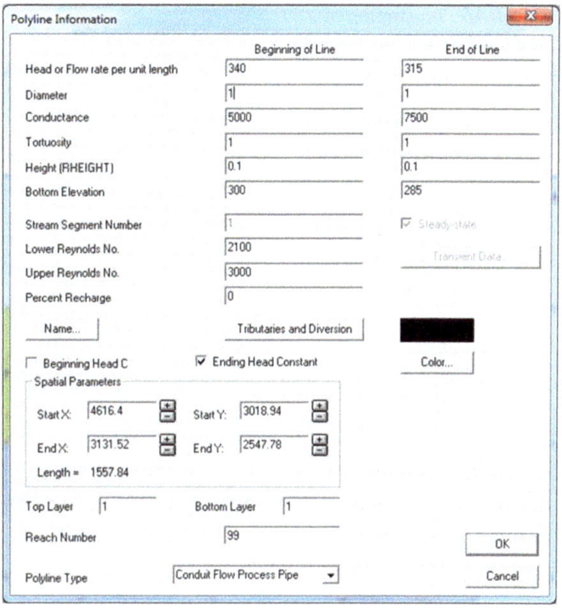

Fig. 10.13 Input parameters for the Conduit Flow Process (CFP)

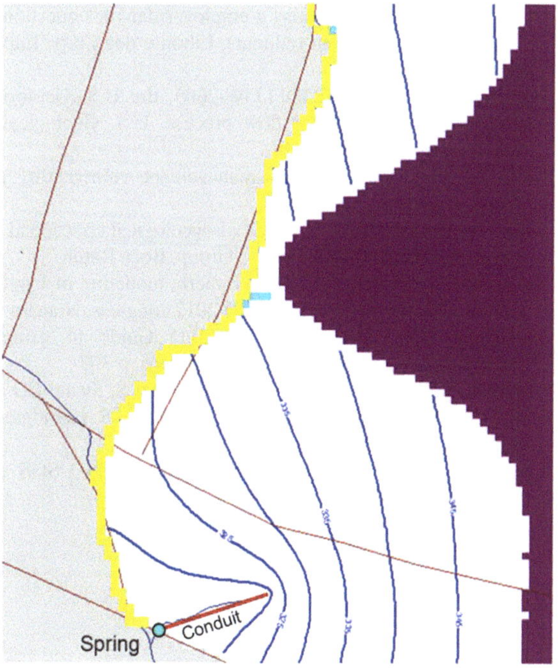

Fig. 10.14 Model output for the Conduit Flow Process (CFP)

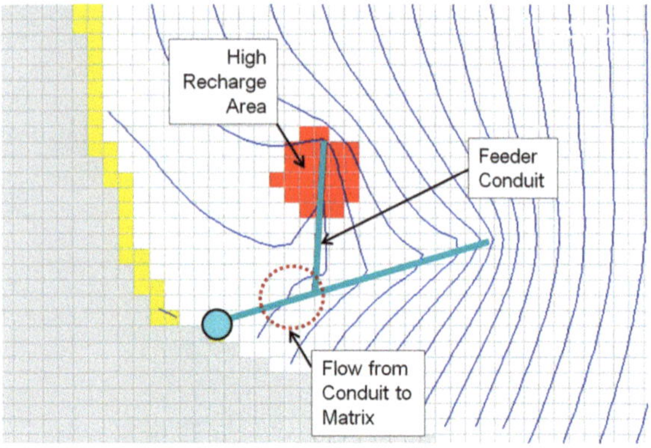

Fig. 10.15 Illustration of a simulated flow from conduits to surrounding matrix due to high recharge rate into the conduit and the resulting high hydraulic head in the conduits

There are numerous situations where exchange of water from the conduit to the matrix is of critical importance for engineering design. This includes plume containment/capture for a pump and treat system, or sizing a dewatering system as part of a construction project.

References

Darcy H (1856) Les fontaines publiques de la ville de Dijon; Exposition et application des principes a suivre et des formulas a employer dans les questions de distribution d'eau; Appendice, Note D. Victor Dalmont (editeur), Libraire des Corps Imperiaux des Ports et Chausses et des Mines, Paris

Harbaugh AW (2005) MODFLOW-2005, the U.S. Geological Survey modular ground-water model—the ground-water flow process. U.S. Geological Survey Techniques and Methods 6-A16 (variously paginated)

Kresic N (2013) Water in karst: management, vulnerability, and restoration. McGraw Hill, New York

Kresic N, Mikszewski A (2013) Hydrogeological conceptual site models: data analysis and visualization. CRC/Taylor & Francis Group, Boca Raton

Mikszewski A, Kresic N (2012) Numeric modeling of karst aquifers: comparison of EPM and CFP models. In: Proceedings IAH 2012 congress, Niagara Falls, Canada

Rumbaugh JO, Rumbaugh DB (2011) Guide to using groundwater vistas, Version 6. Environmental Simulations, Inc., Reinholds, p 307

Shoemaker WB, Kuniansky EL, Birk S, Bauer S, Swain ED (2008) Documentation of a Conduit Flow Process (CFP) for MODFLOW-2005. U.S. Geological Survey Techniques and Methods, Book 6, Chap. A24

USGS (United States Geological Survey) (2011) USGS photographic library. Available at: http://libraryphoto.cr.usgs.gov

Chapter 11
Tapping of Karst Groundwater

Zoran Stevanović

Tapping groundwater is never an easy task largely because we are dealing with an "invisible resource." The situation is even more complicated, however, when we deal with a non-homogenous and anisotropic aquifer such as karst: It is very often necessary to reach, catch, and capture groundwater which circulates through voids or through porous blocks isolated from the rest of the rock mass.

We may distinguish the two main kinds of tapping groundwater in karst:

1. Tapping karstic groundwater flow at discharge points—springs.
2. Tapping karstic groundwater flow within the aquifer catchment—artificial structures such as wells, galleries, or other similar structures.

Groundwater is tapped not only to exploit it for drinking, industrial, heat extraction, or irrigation purposes, but also for dewatering in the case of mine pits, urban areas, or cultivated land which should be protected from a high groundwater table. And in both cases, we do *extraction of groundwater* either by utilizing only the natural flow without any engaged energy (by gravity), as is the case of most spring intake structures and artesian wells, or by pumping out groundwater from drilled wells using some energy source. When it comes to extraction of groundwater, this process should be differentiated from *engineering regulation of aquifer*, which also implies groundwater extraction but whose primary target is aquifer control and management of its flow. This issue is discussed in Chap. 14 and Sect. 15.5 of this book.

Z. Stevanović (✉)
Centre for Karst Hydrogeology, Department of Hydrogeology,
Faculty of Mining and Geology, University of Belgrade, Belgrade, Serbia
e-mail: zstev_2000@yahoo.co.uk

Z. Stevanović (ed.), *Karst Aquifers – Characterization and Engineering*,
Professional Practice in Earth Sciences, DOI 10.1007/978-3-319-12850-4_11

11.1 Tapping Karstic Groundwater Flow at Discharge Points—Springs

A spring is in fact an "eye" of groundwater, the place where this invisible and precious resource reaches the land surface and creates continuous flow. In ancient China, Babylon, Persia, and Egypt, there were many remnants of intake structures around large springs. Famous Greek philosophers and mathematicians such as Pythagoras and Archimedes developed principles for the functioning of hydraulic structures. The Roman times are a golden age of the utilization and long-distance delivery of springwater (Bono and Boni 1996). Rounded arches appeared first in Mesopotamian brick architecture but were further developed and systematically applied by Romans enabling the construction of long aqueducts for water transportation. For instance, for the historical city of Rome, 11 long aqueducts delivered more than 13 m^3/s of springwater from distances ranging from 16 to 91 km (Lombardi and Corazza 2008).

Tapping springwater is not always an engineering job. It is clear that many intake structures worldwide were constructed by local inhabitants that were semiskilled or just experienced enough to perform this task. But it is also true that in many places, we witness seepage around intakes due to inappropriate tapping or we can see submerged structures during high-water periods or extreme floods. There are also many such structures where we may see only pipes which take out and divert all springwater, leaving no flow for local populations or a dependent ecosystem. In addition, many springs, although essential for supplying water to many consumers, have no fence or established protective zones. To avoid such deviations, it is thus necessary to take many steps in design, construction work, and utilization to ensure technically proper and environmentally friendly tapping of springwater (Stevanović 2010).

The evaluation of spring potential should start from *spring classification* (Krešić 2010, and also Sect. 3.5 and Table 3.3 of this book) and *analysis of springflow regime*. These are the two most crucial points in the decision to tap or not to tap the spring. In addition, the following aspects should also be considered (from Stevanović 2010, modified):

1. Proximity of consumers or point of use,
2. Topography of spring site, pipeline route, and usage area,
3. Geology and hydrogeology of the drainage zone and entire catchment,
4. Groundwater quality (in detail, especially if purpose is potable water tapping),
5. Water demands,
6. Accessibility to the spring site (machines, trucks, workers, electricity),
7. Existing and potential pollutants in spring, and in entire catchment area,
8. Modes of spring and catchment protection from pollution,
9. Runoff component (estimated maximal surficial flood around the spring),
10. Environmental and aesthetic requirements,
11. Cost of the project (design, construction) and financing sources,
12. Cost estimates for operation and maintenance.

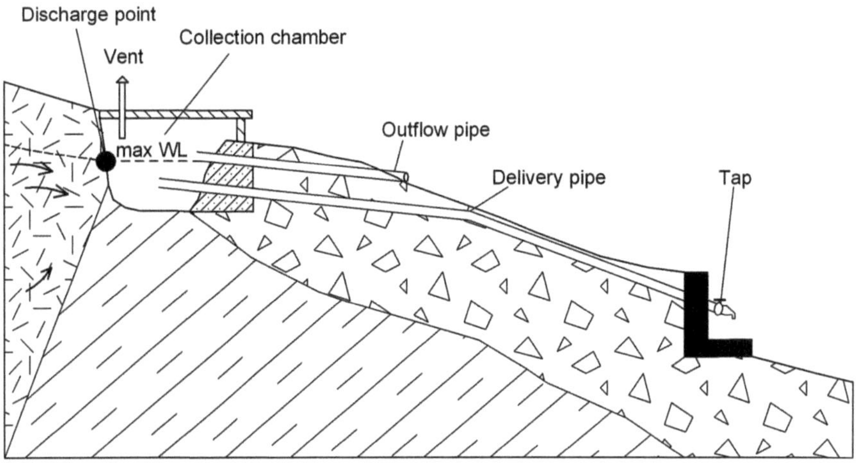

Fig. 11.1 Simple spring box at the primary contact type of spring issuing from karst aquifer (Reprinted from Groundwater Hydrology of Springs, Krešić and Stevanović (eds) (2010), with permission from Elsevier)

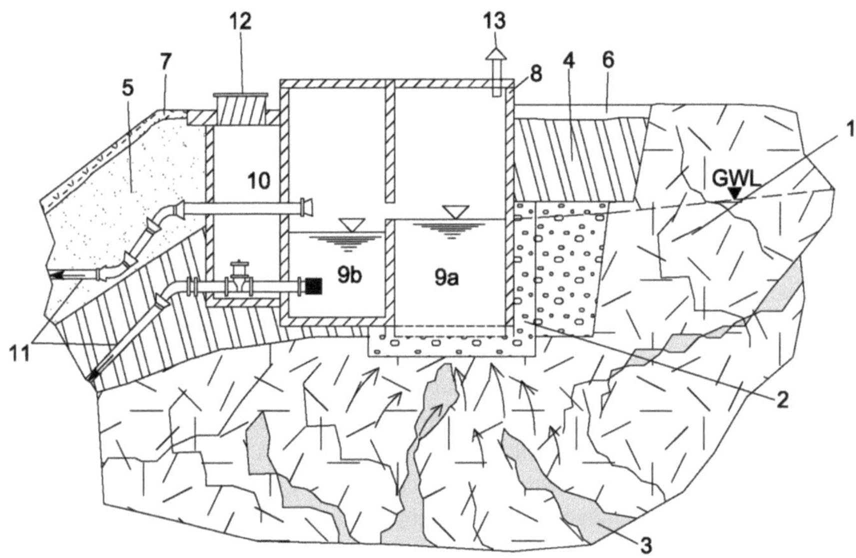

Fig. 11.2 Tapping structure for ascending spring with concentric flow in a karstic aquifer. *Legend 1* karst aquifer; *2* backfill; *3* caverns; *4* clay seal; *5* silty sand cover; *6* top soil; *7* grassed cover; *8* reinforced concrete for two chambers and access room; *9a* sediment box; *9b* reservoir of clean water; *10* maintenance room with valves; *11* overflow and delivery pipes; *12* entrance; *13* vent (Reprinted from Groundwater Hydrology of Springs, Krešić and Stevanović (eds) (2010), with permission from Elsevier)

The elements of a simple spring capture are shown in Fig. 11.1. They are collection chamber (spring box), pipes (delivery and evacuation/outflow), and tap (at usage points). Additionally, the following elements could also be included: a

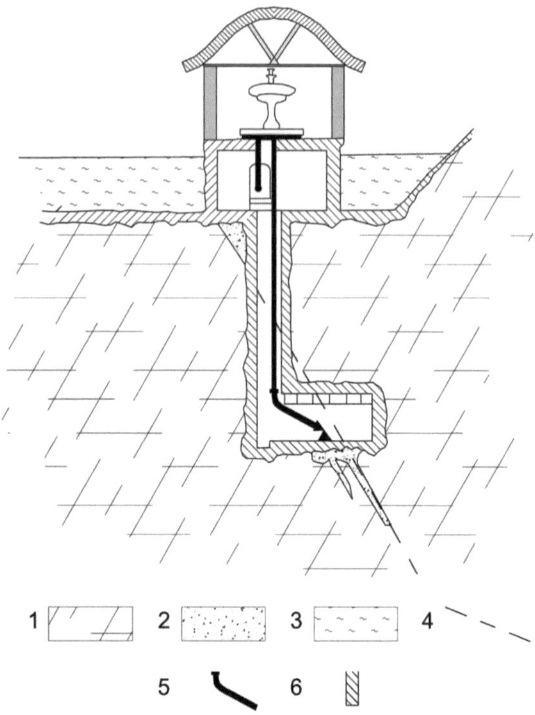

Fig. 11.3 Tapping structure for fracture spring with ascending gas-lift flow. *Legend 1* karstic fissured rocks; *2* fault; *3* clay plug; *4* supposed fault; *5* small pipe inserted directly into the fault preventing escape of CO_2 losses; *6* concrete (modified from Filipović and Dimitrijević 1991, printed with permission)

small storage box (reservoir), maintenance room (with chlorinator and monitoring equipment), pumps (if gravity use and distribution are not possible), cutoff and retention walls (to collect and channel groundwater, but also to protect from debris and landslides), ventilation, fence, etc.

The type of spring and the discharge rates directly influence the design and size of the tapping structure. For instance, it is much easier to tap a gravity spring with a concentric flow than an ascending spring with numerous diffuse discharge points, or a spring with a gas-lift system. Certainly, due to the necessary separation of ground and surface (sea, lake, river) water, it is most difficult to tap an impounded spring. And due to large fluctuations in the groundwater table and variable recharge/discharge function, tapping an estavelle would be no less complicated.

Figures 11.2 and 11.3 show typical tapping structures of an ascending spring with concentric flow and a fracture type spring with gas-lift flow, respectively.

The first step before preparing the final design is to clean the spring site by removing debris, loose fragments, blocks, and soil cover in order to make visible and accessible the spring orifice(s) and provide evidence of the origin of flow. If water is issuing from one single opening, the task to capture it would be much easier (Fig. 11.4). Otherwise, cutoff walls (collection trenches) should be expanded to

Fig. 11.4 Detail from Jadro Spring (Split, Croatia): concrete barrier placed on main spring orifice (old Roman intake reconstructed in 1885/1886)

capture the complete flow, or some of the smaller rock outlets from which groundwater is issuing should be plugged with clay or cement. The concept is to centralize flow for its easier capture.

Where separated outlets are not very close to each other, perforated drainage pipes or cutoff walls have to be installed among them to collect and divert springwater. Deeper digging or cutting of soil (rocks) is often required, but should always be done very carefully in order not to damage the underlying impermeable layer which directs the flow. Otherwise, water could "escape" (Stevanović 2010).

Clay plugging and sealing around or over the spring is the cheapest way to avoid infiltration of contaminants. The mixture of clay and a little water should be compacted. The use of a rolling machine for the top layer is recommended. Keeping the clay moist most of the time will prevent cracking. Additional humus with grass and flowers above the compacted layer will make this a pleasant place.

The first protection zone around the spring should be properly fenced and access to the spring house allowed only to water utility employees. The fence posts should be 2–3 m apart. The posts' height of 3 m should be sufficient (Fig. 11.5).

Fig.11.5 Spring intake: top soil covers basic karstic rocks; properly posted fence prevents uncontrolled access to spring; current meter installed in channel measures spring overflow

Box 11.1: General design and spring box volume

The volume of the engineered reservoir at the spring very much depends on the amount of water that needs to be controlled. For small discharge springs, a very simple structure such as a spring box may be sufficient, while in some other cases, big and complicated structures such as reinforced concrete basins or even dams are required (Krešić 2009). But whatever the size, the main task of the design is to try to control as much water as possible. Only then are the engineers in a position to manage the water, use the amount needed, and enable surplus water to flow freely out downstream (Stevanović 2010).

How big should the collector chamber be and is it necessary to have a sediment box?

1. Case example: A gravity karstic spring with concentric flow discharges between 0.45 and 6.5 l/s. The groundwater quality is stable, and the turbidity value is regularly less than 0.8 NTU. Water should be used locally and access to tap should be free.

   - For such a case, a simple structure consisting of a single small collector chamber (spring box) with dimensions of 3 × 2 × 2 m would be sufficient. A sediment box is not needed. A chamber volume of 12 m^3 is large enough to be completely filled with water in 8 h during minimal flows or in 30 min during maximal floods. To prevent flooding and avoid having to make a special overflow opening at the chamber, the diameter of the delivery pipe has to be large enough to allow a flow of 6.5 l/s. Therefore, a pipe with an open end (external tap posted on the chamber wall) has to have a diameter of 2″ or around 50 mm to enable a flow of 6.5 l/s in the case where pressure is at least 0.7 Ba. The chamber should be founded in the bedrock and constructed from reinforced concrete with a watertight additive, which is the best option to prevent leakage from the spring box (Fig. 11.6).

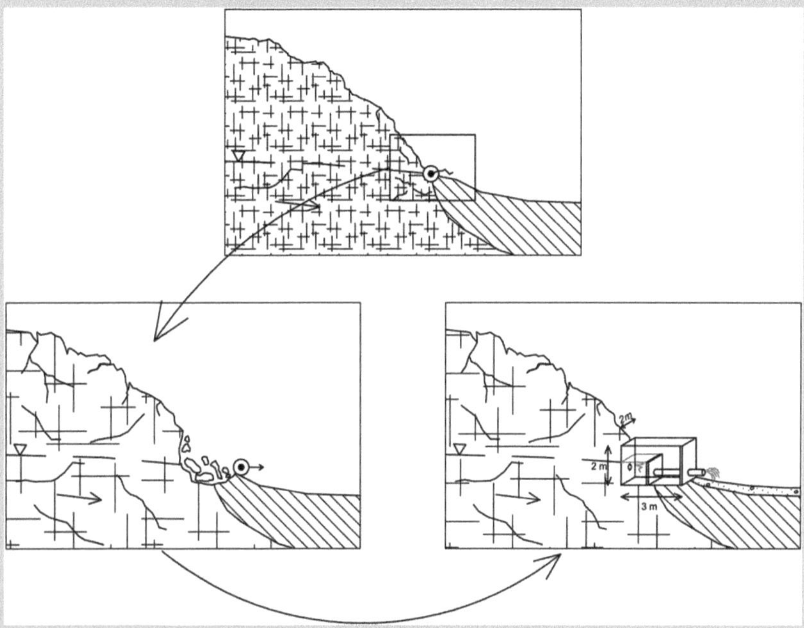

Fig. 11.6 Small gravity spring captured by simple collector chamber with one tap. Prior to construction (I), debris is removed, and after construction (II), thin soil is used to cover bedrock

2. Case example: A large gravity spring drains karstic carbonate aquifer and has one main orifice but also a few small outlets. The minimal spring discharge is 50 l/s, while the maximal is 360 l/s (on average, 80 % of the water flows through the main orifice). The maximal flows regularly result in increased turbidity and bacterial pollution. The spring should be used for centralized water supply of the nearby village located downstream.

   - For such cases, a large tapping structure (intake house) is needed (Fig. 11.7). It should consist of two boxes. The primary box in which groundwater from the main orifice enters through several holes on the

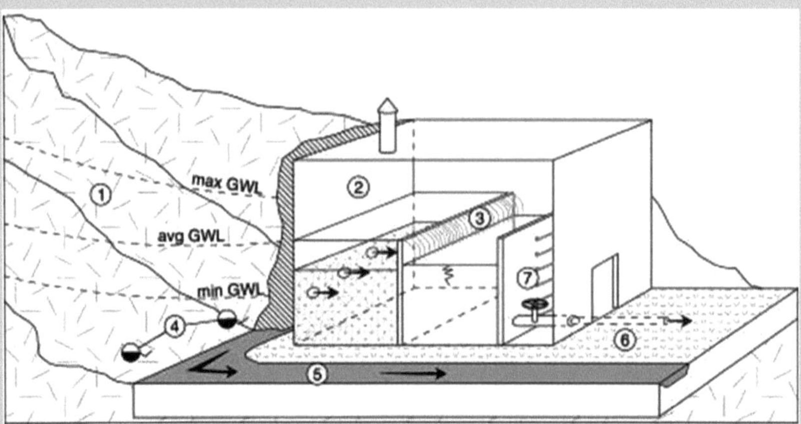

Fig. 11.7 Intake house for large gravity spring. *Legend 1* karstic aquifer; *2* filter room; *3* reservoir of clean water; *4* small seepage springs; *5* evacuation canal; *6* delivery pipe; *7* valve

back wall is filled with sandy gravel filter. This filter amortizes improper water quality in periods of floods. From this primary box, water flows over the wall crest to the secondary box representing the clean water storage. The delivery pipe and bottom outlet are equipped with a valve which controls the flow to the main reservoir where the final water treatment occurs before the water reaches the end-users. Seepage from smaller spring outlets is drained into a circular canal surrounding the main intake building and evacuates these waters along with untapped water from the main orifice. This canal provides water to the downstream natural riverbed and ensures water for dependent biocenosis. The size of the two boxes should be adapted to receive the minimal flow deducted for water needed for guaranteed downstream flow, while all other drain springwater should be regularly evacuated. The volume of the reservoir of clean water could be approximately 2,500 m^3 (e.g., 20 × 18 × 7 (height) m).

Box 11.2: Construction work

Table 11.1 could be used as a checklist for construction work at the spring site. Considering specific local conditions, it is clear that not all proposed work would be implemented everywhere, and some sites will need some other specific work.

Table 11.1 Construction works at the spring site—general specification

No.	Activity	Objective	Work description and outputs	Volume of work planned (units)
1	Conducting precise geodetical survey and mapping[a]	Create detailed map of the site (preferably in GIS)	…map scale 1:…	… (km^2 or m^2)
2	Ensuring or constructing access road	Transport people and materials and enable further access to the site	(Un)paved road…	…(km' or m')
3	Cleaning the site[b]	Make direct access to discharge point and free and visible spring flow	Excavator, shovel, use…	…(m^3), working hours…
4	Providing electricity to the site	Facilitate construction work and future operation and maintenance	Generator, cables, transmission line…	(m', fuel, kwh…
5	Diverting water from the site	Prevent floods at construction site	Temporary trenches, ditches, culvert…	…(m') …(m^3)
6	Constructing spring box and chambers[c]	Complete main part of the structure	Saw, wooden panels, reinforced concrete, mortared rocks…	…(m^3), working hours…
7	Backfilling the space between spring box and rocks or spring box itself	Clean groundwater initially	Gravel, sand…	…(m^3)
8	Covering the object and sealing and site	Prevent pollution as precautionary measure	Top roof, covers (concrete, metal), seal (clay, plastic)…	…(m^2, m^3)
9	Evacuating runoff water	Avoid infiltration of rainfall or runoff water	Permanent drainage ditch…	…(m')
10	Fencing the narrow spring site	Limit access to the site	Pillars, wire, camera…	…(m'), working hours…
11	Installing all facilities inside the tapping structure	Enable start of operation	Pipes, valves, water treatment equipment (chlorinator, etc.)	… working hours…
12	Getting back or transporting all excavated material to the landfill	Recultivate landscape to its state before construction or to an improved state	Excavator, shovel, truck use…	…(m^3)
13	Providing additional ambient improvement	Make site more pleasant than before	Soil cover, planted trees, greenery…	…m^2 …pcs. working hours…

(continued)

Table 11.1 (continued)

No.	Activity	Objective	Work description and outputs	Volume of work planned (units)
14	Managing water utilization	Create detailed map of the site (preferably in GIS)	…	Working hours…
15	Monitoring water quantity and quality	Transport people and materials and enable further access to the site	…	Working hours…

[a]If not implemented in the earlier stage of the project
[b]Moving vegetation, excavating unconsolidated rocks, etc
[c]*Bearing in mind* maximal water control, adequate storage, free overflow, water losses prevention, adequate space for infrastructure facilities, ventilation of the space, easy access into rooms and chambers for maintenance and cleaning, easy emptying of reservoir, architectural compatibility with environment, possible expansion of the structure, etc.

11.2 Tapping Karstic Groundwater Flow Within the Aquifer Catchment—Drilled Wells

Vertical tube wells have so far yielded the best results in tapping deeper karstic groundwater flow because the drilling technology is known everywhere. But before a well with a successful yield is achieved, there are several steps to be undertaken:

1. *Feasibility*—An analysis of water demands and alternative technical solutions. Although drilling the well in karst is not cheap and is often technically demanded, a cost-benefit analysis should support the final decision.
2. *Drilling site(s)*—Geology and a hydrogeology survey should identify the most appropriate site for drilling. In addition, many other methods can also be used to ensure the most convenient location for drilling, such as remote sensing, geological prospection, geophysical survey (resistivity, tomography, VLF, georadar, gravimetry, microseisms, etc.), water points inventory, speleology, simultaneous hydrologic measurements, and tracing tests (Stevanović and Dragišić 1998; Milanović 2000; Goldsheider and Drew 2007; Krešić 2009). These and many other suitable methods are discussed in Chap. 4 of this book. However, hydrogeological factors do not always define the drilling site, and in many cases, the land property, road access, the presence of water, and such have an active role in the choice of an optimal drilling site.
3. *Well design*—Before any drilling, an appropriate project document and/or well design should be prepared. Due to the specificity of drilling works and karst environment, deviations from the initial design concerning the chosen drilling technology, final depth, casing, or position of screen (filter) are always possible. It is thus advisable to prepare as flexible a contract document

as possible for drilling which considers the final cost as a result of the actual state (achieved depth, working hours, length of casing, and similar concerns). However, the main issues that should be determined in the well design document and in the contract are the drilling technology, estimated final well depth, drilling diameters (starting and final), and casing/grouting works.

Well depth. The fully penetrated wells are those finishing in the bottom of the aquifer layer or, in the case of karst, in the karstification base. Driscoll (1986) considers these wells more appropriate than shallower wells for two reasons: (1) More of the aquifer thickness can be utilized, resulting in higher specific capacity, and (2) More drawdown can be made available permitting greater well yield.

Drilling diameters. Most wells do not have the same diameter from top to bottom. Very deep holes in particular require far fewer diameters before the well end. If several diameters are planned, they are chosen in reverse manner looking from the bottom to the top. The bottom part should accommodate the pump and supported cables and enable its proper operation and maintenance. In addition, the diameter of the intake section, i.e., the well's screen, should be large enough to assure good hydraulic efficiency. Driscoll (1986) suggests the diameter of the casing be double the largest diameter of the pump. But modern technology provides efficient pumps even with small diameters, meaning the problem of the space is not as delicate as it was in the past. Drilling diameters are standardized in accordance with the size of drilling bits and hammers.

Casing/Grouting. In order to protect groundwater from surficial pollution, an introductory column ("upper casing") should be inserted into the hole, and the annular space between the casing and wall cemented (grout) or sealed with clay (e.g., bentonite powder mixed with water). Similarly, the bottom drilled intervals could also be isolated by grouting. This can be due to unstable walls or sequential discharge of groundwater of improper quality. In this case, in order to seal annular space properly between the casing and the wall and to prevent any inflow, a high-pressure pump should be used for cementing. Every sealed interval requires a reduced diameter for drilling of the next deeper segment sometimes causing a very complex well design with a very large start diameter and conversely a very small end diameter.

Driscoll (1986) stated that "good design aims to assure an optimum combination of performance, long service life, and reasonable cost." These objectives should be considered together.

For the rationalization of the drilling process and reducing the cost of the tapping structure, it is advisable first to conduct a small diameter drilling (test hole, Fig. 11.8) and in the case of its successful result, to expand the diameters of the same hole or to drill another hole of larger diameter nearby. Hereafter, the term "*borehole*" will be commonly used for a smaller diameter hole, while the term "*well*" implies the larger diameter hole which enables groundwater pumping. The borehole is often used for installation of perforated pipes, and in this case, we speak of an *observation borehole* or *piezometer*. If the walls of a drilled hole are stable, the installation of pipes is not necessary. In the case of a well, the *open-hole* system results in much higher well productivity, simply because there is no resistance from a screen when water enters the well.

Fig. 11.8 A small drilling rig used for drilling test holes in upper Cretaceous limestones at Ourkiss dam site (North Algeria)

Many classical books and manuals describing the drilling of water wells were written in the past. Probably, the most famous and most cited is Driscoll's *Groundwater and Wells* (1986). It is also true that few universities worldwide include in *Hydrogeology* courses topics such as drilling technologies, construction of wells, and specific technical issues related to the installation of casing, gravel packing, organization of pumping tests, and so on. Such issues are usually devoted to employment practice or to engineers and technicians of other specializations (e.g., mining, mechanics, drilling courses). We do not support such an approach, because most hydrogeology surveys aiming to utilize or protect groundwater involve the construction and exploitation of wells. However important, it is impossible to cover all drilling topics in our practical guidebook and the following pages provide an overview and summary of the subject.

11.2.1 Drilling Technology

Many water wells, especially those in unconsolidated rocks, are constructed by digging or by simple devices and machinery. But given that karstic media belong to hard rock groups (even chalk and limes of vuggy porosity), the single method to drill the proper hole is by hydraulics and drilling machines—rigs mounted on various kinds of vehicles. The following drilling techniques are commonly practised:

1. Rotary drilling with direct circulation of the fluids.
2. Rotary drilling with reverse circulation of the fluids.
3. Down-the-hole hammer drilling using air, foam, or other lubricants.
4. Combined rotary–hammer drilling.
5. Cable-tool percussion drilling.
6. Other drilling techniques rarely applied in karst such as auger drilling, driven wells, or blasting.

For all these techniques, a pump which drives fluids or air and supports the removal of cuttings is necessary. In addition, the sedimentation tank or excavated pond should be filled by drilling fluids.

1. In direct rotary drilling, the drilling fluid is pumped down the drill rods and through the attached bit on their end (Fig. 11.9a). The fluid circulates back by moving up through annular space between the drill rods and the walls of the hole. This is also the way for cuttings to be taken out the hole. Direct rotary is the most convenient method for core drilling (core inserted into drill rod or taken out by wire line system).
2. The reverse circulation method (Fig. 11.9b) is more efficient than any other when it comes to cutting removal. Water or air circulates down between the casing and walls or between two casing walls ("dual wall"), and after picking up the cuttings from the bottom, it circulates upward through the inner casing pipes. However, this method is applied mostly in alluvial sediments and unconsolidated rocks and is rarely used for drilling karstified rocks especially when they are very hard, such as dolomitic formations.
The role of fluid is to (1) cool and lubricate the bit, (2) stabilize the borehole wall and prevent collapsing, (3) seal the wall to prevent fluid loss and inflow of drilled formation fluids, and (4) remove cuttings. The fluid tends to infiltrate into permeable zones and chemically interact with formation fluids, and for this reason,

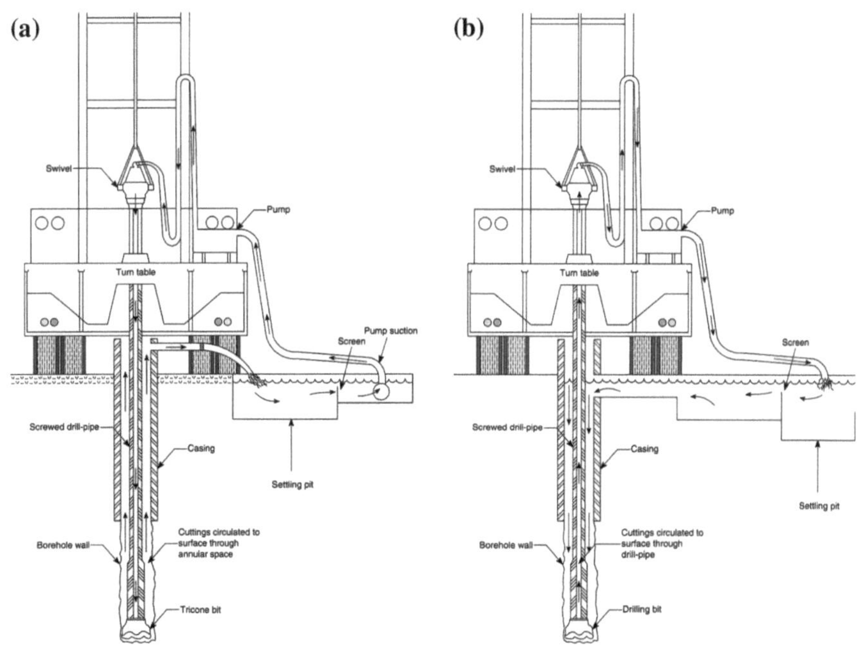

Fig. 11.9 Rotary drilling hydraulic methods. **a** Direct rotary circulation system (modified from the National Water Well Association of Australia 1984). **b** Reverse rotary circulation system

it should be removed after borehole (well) completion. The fluids can be mud, muddy water, water, air, and foam. The viscosity of fluid is a very important factor, and many wells yield much less water than their potential capacity simply because mud remnants have filled voids and porous space and wells are improperly developed after drilling completion. When heavy mud made from bentonite or even more dense fluids such as barite (barium sulfate) is used, the development process must be extended until all mud cake is removed from the hole. Both bentonite and barite are extremely surface active and form clay/organic complexes, and their mixture with original groundwater is always problematic (Aller et al. 1989). It is therefore always better to drill by water in case that stability of the walls has ensured.

3. Drilling with a hammer (Fig. 11.10) and using only air for cooling and removal of particles is a method often recommended in karst because of (1) its drilling efficiency, (2) proper identification of groundwater table position, and (3) the absence of any mud or liquid that may disrupt normal groundwater circulation. The last factor also facilitates the well development process. One of the factors for choosing down-the-hole hammer and air may be the absence of water if drilling is taking place inside the catchment and in remote mountainous areas. Furthermore, loss of fluid circulation which often follows drilling in karst is an additional factor which supports replacing rotary by hammer drilling. As well, the water level in different penetrated zones can be determined easily, which is not the case with

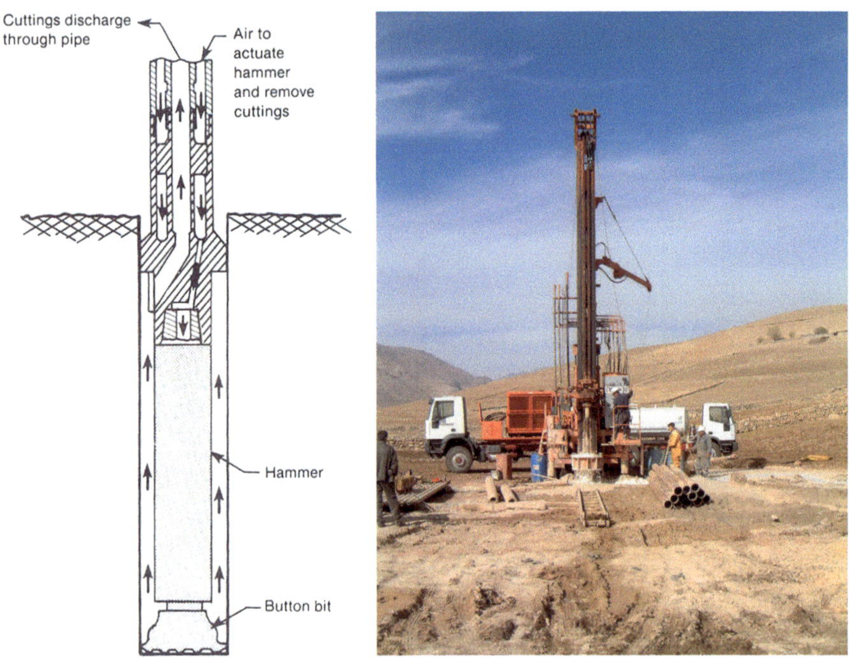

Fig. 11.10 Scheme of the down-the-hole hammer drilling and service drilling rig operating in northern Iraq

the rotary method where long development and geophysical logging have to be applied to distinct zones of inflows and major permeability in the well section.
4. Combined method—hammer drilling with small rotation and use of compressed air or foam produces the best results by far in the drilling of limestones. Foam effectively extracts the cuttings and cools the bits, but should be totally removed before it starts to react with local groundwater.

The experience from the Middle East and drilling in limestones from the Cretaceous and Tertiary ages quite commonly confirmed drilling penetration rates of 100 m/day or even faster (Stevanović and Iurkiewicz 2004).

11.2.2 Well Equipment (Casing, Screening, Gravel Packing, Protecting Well Cap)

Casing material can be very different and depends on the stability of the walls, groundwater chemistry, bacterial content, and other factors that may impact the hydraulic efficiency and aging of the well. Usually, the casing and the screen are of the same material. Although in ancient times, many materials such as wood or ceramics were used for protecting the walls of the wells, in the nineteenth and the first half of the twentieth-century, metal constructions became dominant: Heavy iron pipes were installed in many deep wells. More recently, other types of light metal (aluminum) and plastic are used. Casing, however, must be strong enough to resist the forces exerted by the surrounding geological materials. When casing power is estimated, tensile, compressive, and collapse strength components should be evaluated separately. For instance, the resistance of casing to collapse is defined by wall thickness and outside casing diameters: Casing collapse potency is proportional to the cube of wall thickness, and any increase in the thickness of the walls provides substantial collapse strength.

Aller et al. (1989) divided the three categories of casing materials used for monitoring wells:

1. Metals: carbon steel, low-carbon steel, galvanized steel, stainless steel.
2. Thermoplastic materials: polyvinyl chloride (PVC) and similar.
3. Polymer materials: fluoropolymer, polytetrafluoroethylene (PTFE), and similar.

Corrosion is a frequent problem in well exploitation. Galvanization and electroplating of carbon and low carbon improve the corrosion resistance. Plastic materials and stainless steel are resistant to corrosion, but their cost cannot be compared as stainless steel is much more expensive than plastic.

A casing is most susceptible to collapse during installation but is also vulnerable before the placement of a gravel pack or the annual seal materials around the pipes. Purdin (1980) noted a few steps to minimize the possibility of collapse:

1. drilling a straight, clean borehole,
2. uniformly distributing gravel pack material,

3. avoiding the use of high-temperature cement for thermoplastic casing installation,
4. adding sand or bentonite to cement to lower the heat of hydration, and
5. controlling negative pressure inside the well during development.

The screens or filters are hydrogeologically the most important part of the well's construction. The screen materials should be the same as those of casing pipes, and the connection between segments, commonly 3 or 6 m long, is made by welding via the application of heat or by threaded joints (Fig. 11.11). Driscoll (1986) emphasizes the desirable features of a well screen. Some of them are as follows:

1. Openings in the form of slots and uninterrupted around the circumference of the screen;
2. Close spacing of slot openings to provide maximum percent of open area;
3. V-shape slot openings that widen inwardly;
4. Adaptability to different conditions by the use of various materials;
5. Maximum open area consistent with adequate strength;
6. Full series of accessories and end fittings to facilitate screen installation and well completion operations.

Very common is the *bridge type screen* (Fig. 11.12). The slot opening is by definition vertical (when horizontal, this is a shutter-type screen) with two parallel openings longitudinally aligned to the well axis (Aller et al. 1989).

A well *screen with a bonded-on gravel pack* is more frequently applied in intergranular aquifers than in karst. Usually, quartz gravel of high purity is graded in accordance with the grain size of the aquifer layer and assembled around the perforated tube (Fig. 11.13).

The *continuous slot well screen* is one of the most efficient types of screen made from spirally wires (often called after their constructors or manufacturers "Gavrilko" or "Johnson"). The wires, usually triangular in cross-section, are continuously and spirally mounted around the circular array or longitudinal rods. The slot openings are produced by spacing the successive turns of the outer

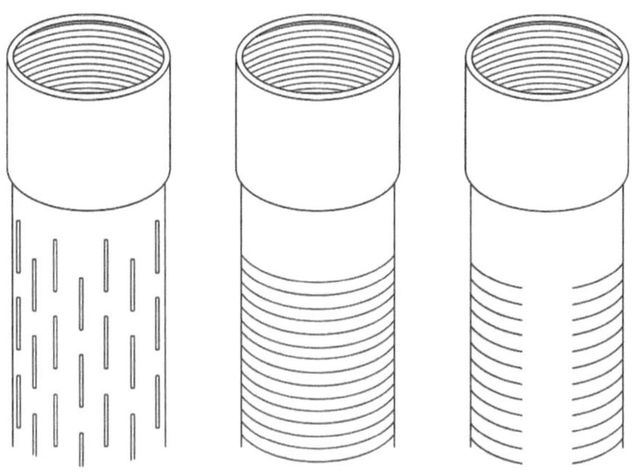

Fig. 11.11 Several types of screens with threaded joint (from *left to right* slotted screen, continuous slot wire-wound screen, and screen with plain surfaces)

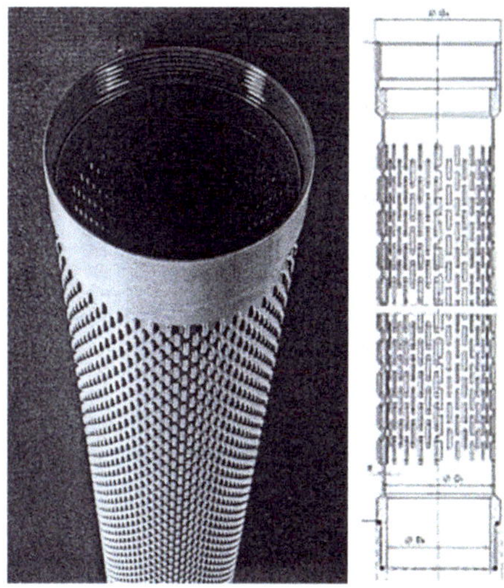

Fig. 11.12 The bridge slot screen

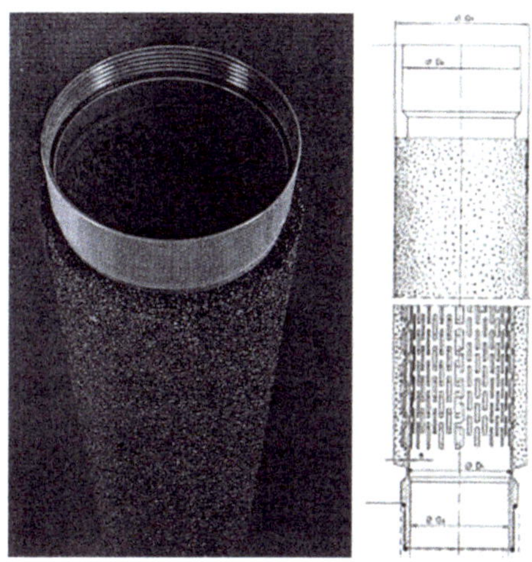

Fig. 11.13 Well screen with bonded-on gravel pack

wire. The process is thus flexible, and change in the size of V-shaped openings is possible during manufacturing. A V-shape prevents wedging of the openings: Any fine grain easily comes inside the screen and can be extracted during well development.

The *continuous slot well screen* provides more intake area per square unit than any other type (Driscoll 1986). The free flows and large open area of intake are of great importance in karstic aquifers (Fig. 11.14). Although small particles and cuttings from carbonate rocks could be present particularly in tectonic zones, the problem of screen and gravel pack clogging is very rare in karst aquifer, in contrast to intergranular aquifers. Likewise, the problem of fast aging of the screens

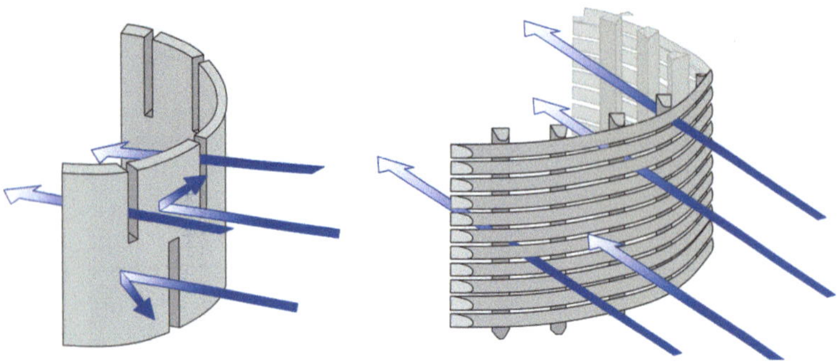

Fig. 11.14 The slotted screen (*left*) and continuous slot wire-wound screen (*right*). The difference in opening percentage reflects on streamline passage and free flow efficiency

of the wells in karst even if the screens are placed in deep confined parts and are under problematic chemical conditions is not so common.

However, in the case that the stability of the walls is secured, open-hole wells are by far the best solution in terms of efficiency and cost, (Box 11.3).

Box 11.3: Open-hole well advantage

Figure 11.15 shows two wells drilled in the same lithology section and in a confined karstic aquifer. The uppermost part of the section, which consists of clayed sands, marly sands, and gravels (2), has been drilled in diameter of 450 mm and then isolated by preventive casing tubes (323 mm) and grouting. Further drilling through the main carbonate aquifer (1) was conducted applying the narrower diameter of 250 mm, as far as the bottom impervious layer (3). Two options are possible: The left (a) shows a fully open-hole intake, and groundwater can thus freely flow into the well. On the right sketch (b), the tube with two screen intervals and bottom sedimentation end cap is installed in the intake and bottom layer, while the gravel pack is also placed in the annular space between the walls and screen. In order to reduce the total cost and to allow installation of a more powerful pump in the upper part, the well tube with screens has been installed as a so-called blind column without having access to the surface: The upper part is removed by recoupling after gravel packing. The malfunction of such an approach could complicate intervention for screen maintenance or replacement, and for this reason, many wells contain inner smaller diameter tubes extending up to the surface.

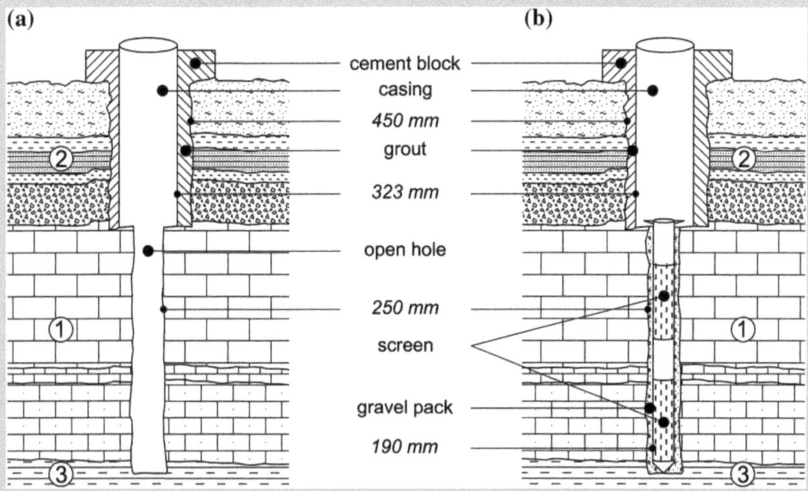

Fig. 11.15 **a** Open-hole drilled well and **b** well with installed casing/screen in intake area. *Legend 1* karstic aquifer; *2* surficial soil, clay, sand, gravel; *3* impervious bottom marls

An artificial gravel pack or *filter pack*[1] introduced from the surface by pipes or simply manually placed in the area surrounding the screen, i.e., annular space between tubes and walls, aims both to prevent large formation materials from entering into the well and also to work in conjunction with the well screen to filter out very fine materials. In contrast, in naturally developed wells, the formation material is allowed to collapse around the tube and screen and to form a *natural gravel pack*. By creating a succession of cylindrical graded zones around the screen (development removes particles smaller than screen openings), stabilization of formation material is expected as is sediment-free water at the end of the well development process.

11.2.3 Well Development

The duration and intensity of the well (borehole) development depend on many factors. In many cases, it is difficult even to predict the time needed to cleanse the well from mud or from original materials of the drilled aquifer (Campbell and Lehr 1973). In particular, in karst where cavities and joints can be filled with thick secondary silty, clayey, and sandy materials transported by groundwater, it is sometimes weeks and even months before totally clean water starts to flow out.

[1] Not always gravel, but also sand is used in drilling practice, and this is why the filter pack is the more proper term. However, gravel pack is more commonly used.

Driscoll (1986) highlighted three beneficial results of well development:

1. Correction of damages or clogging of aquifer layer as a side effect of drilling;
2. Increase in porosity and permeability of aquifer layer;
3. Stabilization of unconsolidated formation around the screen, so that the well will yield free of sand.

The conventional development of the well includes washing and air lifting. Initial washing (backwashing) can be done by the rig's pump and circulation of clean waters. The purpose is to replace fully muddy waters used for drilling but also to enable natural gravel packing. In karst, natural gravel packing is possible only if some unconsolidated material other than carbonates is present in the drilled section of the formation.

Compressed air is a very efficient method of well development. If only a small diameter exploratory borehole is drilled for monitoring purposes, air lifting is then the sole method to test and estimate aquifer productivity.[2] Use of a compressor, installation of a pressure gauge and security valve, as well as mounting air-lifting pipes or a hose into the well (flexible enough to be able to reach the bottom of the well but also movable upward) are considered air-lifting test.

A compressor providing pressure up to 10 Ba is usually sufficient to develop a well 80–90 m deep. Gradual, step-by-step development increasing pressure and raising and lowering the air line in the well should enable complete removal of drilled particles and fine secondary deposits from the formation which entered the well. The full yield of an aquifer can only be effectuated if all cavities and fissures freely supply water to the well. Total cleaning is also prerequisite for installation of pumps and further testing, although many pumps are sensitive and require purely clean water.

More advanced development techniques or the rehabilitation of a well may include surging, high-velocity jetting, chemical treatment, or even blasting. Use of chemicals such as sodium hypochlorite may support dilution of clogging materials around screens. Clogging may result from fine clays brought along with waters from caverns. Different acids (muriatic, sulfamic) are also used in drilling practice in karstic formations because they dissolve the limestones or evaporites, but also eliminate any incrustation around the screens. Long-term development is then necessary to ensure safe water quality.

11.2.4 Well Testing (Pumping)

Pumping the well is the best method to do the following:

1. complete well development;
2. estimate well quality and efficiency;
3. test aquifer productivity;
4. assess aquifer permeability; and
5. create a base for well completion (equipment).

---

[2] Testing a borehole enables the decision whether to drill the exploitation well or not and what well design to impose.

In principle, there are *short-term control pumping tests* which aim to estimate the quality of the constructed well and its elements (screen hydraulic loss) and *long-term pumping tests* which simulate real exploitation and provide a base to define optimal well capacity and type and the depth of the pump to be installed for permanent groundwater extraction. In freely flowing wells (artesian), additional pumping with a capacity higher than natural flow may provide a base for more intensive well exploitation (if and when necessary).

Short *step-drawdown tests* can be undertaken for *steady flow* or for *non-steady flow* conditions. Steady flow considers constant hydraulic conditions: Drawdown is stable for chosen and fixed pump capacity; in non-steady flow, the water table is depleting throughout the test for one constantly pumped yield. The "steps" mean that pump capacities are enlarging from smallest to highest (usually two to four different yields) with no break during pumping. The effects on groundwater table fluctuation should be observed continually during a pumping test.

Calculation of hydraulic parameters and well losses (resistance of well intake) is based on observation of drawdown or recovery values and in the case of non-steady-state flow use of Theis, Jacob, Hantush, Coper, or other formulae (Castany 1967, 1982).

Although it looks simple, pumping is an expensive test (due to consumable energy and manpower) and requires proper organization. It is necessary to ensure the presence of qualified mechanics and spare parts in order to reduce to a minimum the extent of the possible risk of damages and operation breaks. In the case of unconfined karst, aquifer-pumped waters should be evacuated as far as possible from the pumped well to prevent any return flow and infiltration in the tested aquifer (at minimum out of the radius of the well). The pump capacities should be determined based on assessed aquifer potential. For that purpose, indications collected from drilling such as registered cavities (intervals without cores), mud dilution, direct groundwater inflows, as well as geological and geophysical logging and air lifting (well development) should be sufficient to obtain this first picture of the aquifer potential. Furthermore, after starting the pumping operation with the chosen smallest capacity, other planned capacities could be corrected in accordance with water table fluctuations. For instance, if the water table is more or less stable for a certain yield, more intensive pumping can be allowed during the following test stages. And in contrast, significant drawdown requires a reduction in further pump capacities. It is recommended that drawdown exceeds not more than 1/3 of the total aquifer thickness: If drawdown reaches 2/3 or more, there is substantial threat of depletion of the total aquifer layer, and when a critical yield is reached, there is also the threat of pump damage (Fig. 11.16). However, when operated with caution, it is practical to carry out higher pumping capacities during testing than during later exploitation. The construction of curve drawdown (recovery) versus time ($d = f(\log t)$ is essential for the calculation of parameters in non-steady-state flow conditions (Fig. 11.17).

The groundwater table could be observed automatically (by installed "driver") or, if it is not possible to install proper instrumentation into the well, then by manually recording needs measurements frequency which ensures collecting as much as possible (see Box 11.5), especially at the beginning of pumping and during recovery, as well.

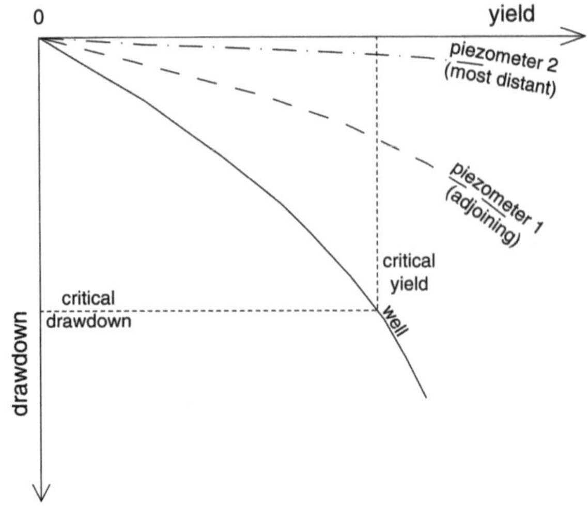

Fig. 11.16 Critical yield/drawdown in pumping well and water table in farthest and nearest piezometers (after Castany 1967, modified)

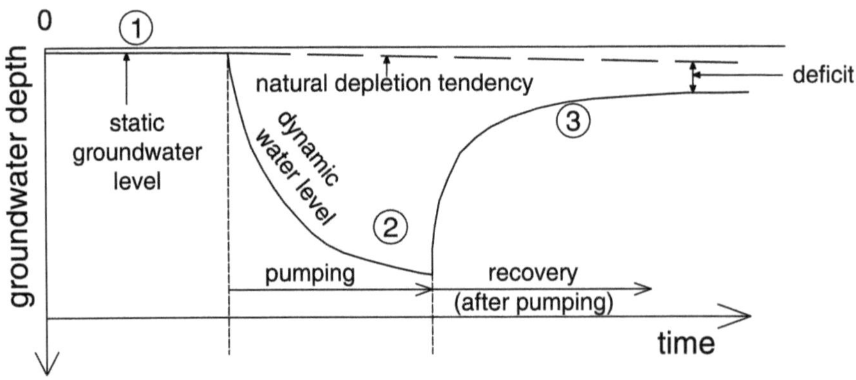

Fig. 11.17 The groundwater table curve before (*1*), during (*2*), and after (*3*) pumping. A small difference between the natural static water table and final recovery water table resulting from pumping is evident

11.2.5 Optimizing Yield, Install Pump, and Protect Well

The pumping test should provide sufficient information on the optimal well yield and accordingly on the suitable pump and its installation depth. The main interconnected parameter for the well yield is drawdown. Figure 11.18 shows the schemes of confined and unconfined karstic aquifers and drawdown resulting from the well pumping.

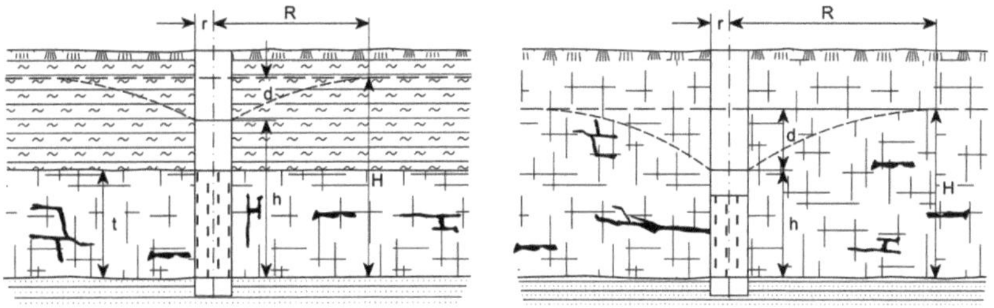

Fig. 11.18 Scheme of pumping confined (*left*) and unconfined (*right*) karstic aquifers. *Legend H* piezometric level or thickness of saturated zone; *h* piezometric level or thickness of saturated in intake area during pumping, *t* thickness of confined aquifer, *d* drawdown, *R* radius of well (cone of depression), *r* semi-diameter of well

Box 11.4: Optimal pumping rate

For determining hydraulic parameters for steady-state flow, the "classical" Dupuit–Forchheimer equation or assumption (1863, 1886) can be used, while the radius of the well could be approximately determined by the use of Castany's method (1967). If the hydraulic conductivity (K) and radius of the well (R) are known from previous survey in the area and from pumping tests, while the desired drawdown (d) is fixed to a certain value, then the optimal pumping rate (Q) can be very roughly estimated from the inverse solution of the Dupuit–Forchheimer equation:
For confined aquifer:

$$Q = (Ktd)/(2\pi \ln R/r) \tag{11.1}$$

For unconfined aquifer:

$$Q = K(2H - d)d/(4\pi \ln R/r) \tag{11.2}$$

Symbols are the same as in the legend for Fig. 11.18.
As an example, for the desired fixed drawdown not to exceed 10 m (1/5 of the total aquifer thickness) for the 50-m-thick confined karstic aquifer for which hydraulic conductivity is roughly approximated on 5×10^{-4} m/s and the radius of the well on 100 m (well's semi-diameter is 0.2 m), the calculated optimal pumping rate should be 6.4 l/s or less. The pump should be installed to the depth of around 20 m (definitely deeper than 10 m, but not too deep because of the need to efficiently pump water out). Given that hydraulic parameters are very problematic in karst, as discussed in Chap. 3 of this book, all results obtained in this way should be carefully compared with those obtained from the test and further verified through well exploitation.

Pumping capacity can also be dictated by the position of the end-user. In the case that the water is not pumping to a nearby reservoir tank but instead much further and directly to the consumer which is located on a higher ground, then the calculation of the head must consider this additional value.

Pump capacity lowers proportionally with water depth. Pump manufacturers always produce several types which fit with various local conditions. A pump able to pump water from 300 m or even deeper can be found at the market, which in the case of deeply developed karstification and a water table at a great depth is a very important fact for hydrogeologists. It is also better to install a *variable displacement pump* which delivers water in variable quantities depending on head than a *constant displacement pump* with a fixed yield. The external pump diameter is also an important issue, and the casing diameter must accommodate the pump. But manufacturers have significantly improved technology to produce less and less in dimension, but still be powerful pumps. Figure 11.19 shows the diagram: Yield (Q) $= f$ (Total dynamic head H) for seven pumps with power ranging between 10 and 50 HP. By working at the same performance of 3,000 rotations/min, we may see a great difference in their efficiency. Even the one same pump as that of 50 HP can produce more than twice the largest yield when installed at a depth two times shallower (70 m^3/h from depth of 90 m instead 30 m^3/h when at the depth of 180 m).

Fig. 11.19 The diagram $Q = f(H)$ for seven variable displacement pumps of different power

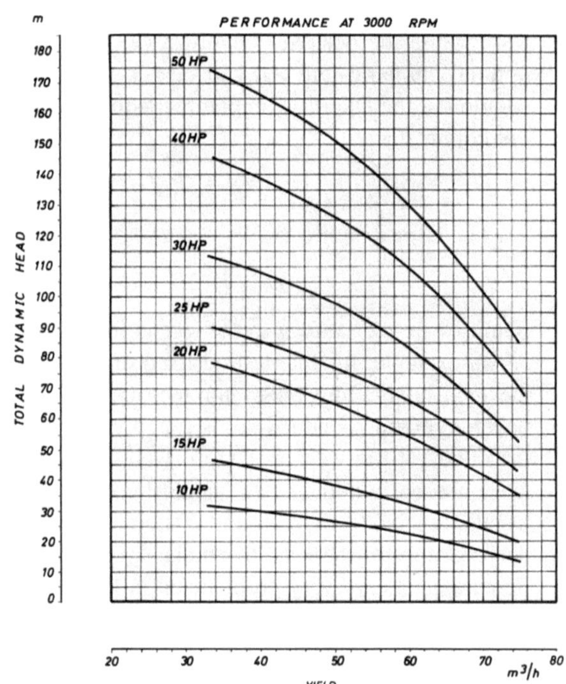

Driscoll (1986) distinguishes three groups of variable displacement pumps: (1) centrifugal pumps, (2) jet pumps, and (3) air-lift pumps. In his classic book *Groundwater and Well,* the principle of centrifugal force which causes water to inflow into the pump and be driven up is well explained. Although not the same, in groundwater, practice term "turbine pump" is equalized with all the types of centrifugal pumps. They could be single or multistage depending on the number of vans used in an impeller.

In contrast to *turbine pumps* (Fig. 11.20) where the motor and most of the elements are not submerged in water (which somehow limits installation depth), the *submersible pumps* are as their name implies fully plunged in water. The motor of submersible pumps is usually directly below the intake of the pump. The number of vanes of a submersible pump can be very large and increase with a required depth (Fig. 11.21).

Because pumps are constructed in such a way that they do not have their own power, another local energy source is obligatory. Pumps may either be connected to an electricity net or operate connected to a generator.

After the proper installation of the tubes, screens, gravel packs, and the testing of the well, a cap and lock should be installed on each completed well. This common above-ground protection measure should prevent any damage, uninhibited use, or contamination.

Fig. 11.20 Installation (*left*) and operation of turbine vertical pump in well tapping Qamchuga limestone Fm. in northern Iraq

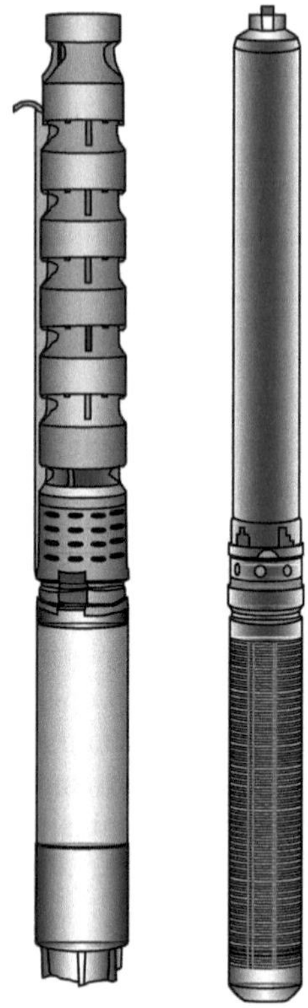

Fig. 11.21 Submersible pumps

When the saturated zone is under pressure (caused by overlain impervious rocks), the flow is commonly artesian. Therefore, if the usage point is lower than the piezometric level of the aquifer, it is enough just to install the pipes to get free flow. Otherwise, pumping is necessary, which increases the operation costs.

In conclusion, searching for water in a catchment of karstic or fissured aquifer is always delicate. Anisotropic and non-homogenous media dictate water circulation along privileged pathways, in this case, joints, fissures, faults, caverns, or even bigger features such as cave channels. Therefore, to obtain a maximal flow, the wells, tubes, or pumps must reach those veins in hard rocks. Krešić stated (2013) that the best-kept secret is that a little bit of luck is always needed when installing a well in karst for any use. One exception may be in rocks with vuggy porosity where diffuse flows and not only concentrated flow along privileged paths are present.

Finally, depending on aquifer effective porosity, recharge, and storage as well as on the performances of drilled wells, there are small capacity wells which yield less than 1 l/s, sometimes drilled for pure humanitarian purposes, but also functioning as highly productive wells. The most productive one, capable of yielding more than 2,000 l/s, was drilled in the Edward aquifer, in Texas near San Antonio (USA).

Box 11.5: Instructions for drilling and pumping the well—Case example

Method of statement for drilling and testing the well Bari Bahar 2 (Northern Iraq)

The following general conditions should be accepted by the contractor (hereafter driller and pump operator):

- The driller and pump operators must check the site conditions before installation of drilling rig and testing the well.
- The driller and pump operators are responsible to ensure prompt and timely writing of the daily reports for all operations executed.
- The driller and pump operator shall submit their reports of the completion of well drilling and pumping and prepare the invoices exclusively in accordance with the work done and certified by the supervisor. The payment will be executed only if the work is fully completed according to this method of statement and specific technical instructions provided in the field by the appointed supervisor. The certificate of supervisor will include bill of quantities for all kinds of work done.
- The investor is not responsible for any "compensation to workers in case of accident or injury."

1. *Technical conditions for drilling of one exploratory well*

The well Bari Bahar 2 will be drilled up to a total depth of 153 m (Fig. 11.22) by the direct circulation method of rotary drilling. The location of the well is determined in the vicinity (3–4 m) of piezometer IB-2 on the left wadi bank as shown on the sketch map. The minimal site correction is possible only in the case that a drilling rig cannot approach the exact site due to very narrow distance between the embankment and riverbed. In this case, the closest site will be selected and approved by the supervisor.
The driller is responsible to adapt the rig and drilling technique to the following conditions:

- No drilling fluid except clear water is to be used.
- Drilling starts in moderately hard marls, while deeper sections involve drilling in limestones, marly limestones and marls.
- The groundwater table is at a depth between 16 and 18 m from the ground surface.

The diameter of the well will be ø 12″ (301 mm) from the surface as far as the bottom of 153 m. Samples will be collected at every meter and properly labeled and saved until checked by the supervisor.

The well casing, delivered in advance to the drilling site, is plastic, 3-m-long PVC pipes ø 6 5/8″ (168 mm) resistant to the pressure of 15 Ba. Only the first two intervals (or last mounted when installing) are PVC pipes ø 8″ (203 mm).

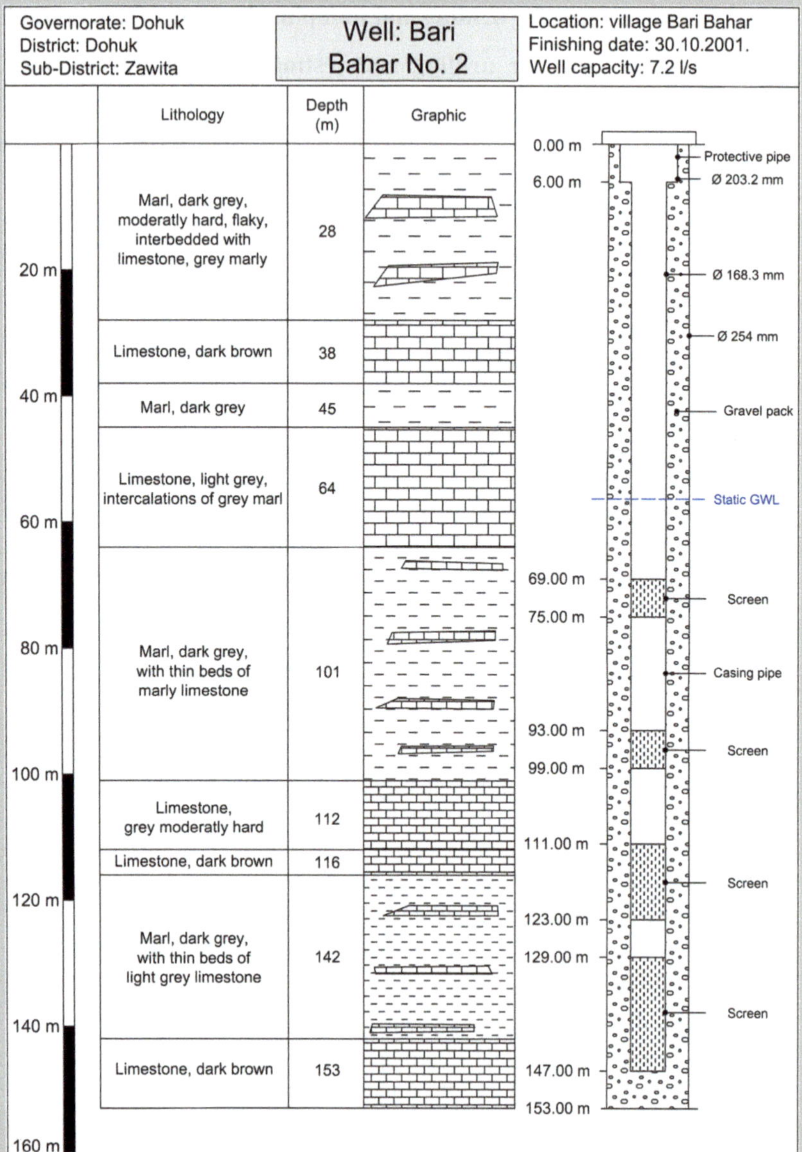

Fig. 11.22 Geological log and construction of the well Bari Bahar 2

The pipes will be jointly coupled by male and female threads. The total length of the screen (filter) is 42 m and consists of the slotted intervals (or holes) on the PVC pipes which guarantee a porosity of about 12–15 %.

The upper part of the casing (over the filter) is 69.5 m long, of which 0.5 m will stay over the ground surface. The filter intervals are distributed as shown in Fig. 11.22. The segments are as follows: 69–75, 93–99, 111–123, and 129–147 m. The plain pipes are located between the segments (Fig. 11.22).

The sedimentary bottom casing PVC pipe is 6 m long, i.e., from 147 to 153 m (not shown on Fig. 11.22).

During the casing installation, the two centralizers at 50 and 100 m depth will ensure proper vertical positioning.

The filter pack is sorted sandy gravel of 1–3 mm in diameter, installed by tremie pipe from the bottom of the well to the top. The water circulation with filter pack will ensure proper segregation at the annular space. The filter pack is installed within the whole drilled interval with the exception of the first 6 m, which will be the protective clay tampon.

Clay (bentonite) slices will be put into place dry by gravity method from the ground surface. Then, the material will be tamped by tamping rod. The bentonite will be placed to the depth of 6 m from the ground surface around the protective casing pipe.

The rubber packer and end cup will be mounted so as to prevent possible inflow of the surface water into the well to the maximal extent possible.

The well development includes 2–3 h of initial washing by clear water and then air lifting lasting a minimum of 24 h. The air-lift pipes of 2 inches (51 mm) will be mounted to the well bottom. Air-lift development will be gradually conducted with initial small pressures (1–2 Ba) to ensure additional segregation and a small inflow velocity. The packer installed on air-lift pipes will ensure washing of each interval 3 m long.

The pumped water from the well will be directly evacuated to the river at least 60 m downstream from the well Bari Bahar 2.

In the case of well failure due to cutting casing pipes inside the well or accordance of collapse, the driller shall remedy the situation on his own account.

If the poor capacity of the well is the result of the failure of the drilling and pipe installation, the driller will rectify the entire situation at his expense or a new well will be drilled.

2. *Technical conditions for pumping and recovery test*

The Bari Bahar 2 will be tested by step method with three pump capacities for non-steady-state flow conditions.

During the pumping test, the following water points will be constantly observed: piezometers IB1–IB2, wadi gauge and the well Bari Bahar 2 itself. The fixed flow rate counter will be mounted at the evacuation pipes on the well mouth. The volumetric method by the barrel of minimum 100 l will be used at least once per day for checking the discharge rate.

All pumped water will be evacuated by 4″ or 5″ plastic tubes outside of the estimated radius of the well (minimum 60 m downstream of the well). Complete measures of discharge and the GW table during and after the test will be executed following the instructions provided and demonstrated by the supervisor.

The pump operators' staff consists of workers and at least one qualified mechanical technician responsible for the pump, generator, and other technical matters. Adequate equipment and tools on location will ensure quick and efficient interventions, if required.

The test duration is 30 continuous days. The first two pump capacities will last for 24 h, while the third will last for 28 days, i.e., until the end of pumping.

The pump and cables will be suitable for installation into 168 mm diameter of the casing pipes of Bari Bahar 2. The pump and valves will enable optimization of pumping rates which rank between 3 and 12 l/s. The electricity supply for the source (generator) will be at least 15 kW in order to ensure the pump will start.

The pump operator will ensure there is an adequate amount of the fuel on site. Necessary changes of the oil, filters, or similar will be done in such way not to disturb the operation or pollute the drilling site.

The bottom of the pump will be mounted to the depth of 50 m.

The pumping rates are as follows:

$$Q1, 2, 3 = 3 - 6 - 12 l/s,$$

but the third capacity could be accordingly modified (possibly increased) by the order of the supervisor.

A break of a maximum of 3 h can be tolerated and only for the third pumping capacity. Otherwise, the test is fully repeated. During the break, all measurements will be done at all water points in the exploitation field: four piezometers, river, and well.

The pump operator staff will cover by measurements all the above-mentioned water points. The water table recording will be done by electrical and mechanical sounders. The measurement intervals for each pumping capacity and recovery are in the following order:

- static level during the 3 days before the test,
- first 3 min: every 15 s,
- until 10 min: every 1 min,
- until 30 min: every 5 min,
- until 1 h: every 10 min,
- until 2 h: every 15 min,
- until 4 h: every 30 min,
- until 24 h: every 1 h.

For the third pumping capacity, the continual observation of all water points includes the following:

- until the end of the third day: every 2 h
- until the end of the seventh day: every 4 h
- until the end of the fourteenth day: every 6 h
- until the end of the test, i.e., twenty-eight day: every 8 h (three times per day).

In the case of significant changes of hydrology and/or meteorological conditions (river flow, intense rainfall, or snow melting), the supervisor will provide new instructions for the measurement frequency. In the case that the supervisor provides an automatic recorder ("diver"), the pump operator staff is responsible for its installation to the requested depth.

The recovery test includes the same frequency as the pumping test during the first 24 h. The water table should be followed until the return to the initial one (static). A deficit of about 5 % concerning the initial level is tolerable. In the case of general water decrease or increase, the consultant will decide to extend the observation but for no longer than 7 days. In this case, measurement intervals are as follows:

- until the end of the third day: every 2 h
- until the end of the seventh day: every 4 h

The pump operator is responsible for entering all data into the pumping test diary, such as water level, pumping capacities, water physical changes (temperature, turbidity), breaks and reasons, water sampling, and similar items.

After the test completion, the well mouth will be properly closed and locked.

All other obligations possibly not described hereto will be managed and professionally fulfilled by the three parties, i.e., the supervisor, the driller, and the pump operator, in order to obtain the best results from the operation.

Box 11.6: Drilling diary example

General information:	**Drilling Well Diary**
	Well / Borehole code:
District:	
Village:	
Location: (address, owner or descriptive explanation)	
X,Y,Z :	
Drilling rig (type):	
Drilling company:	
Aim:	
Project (Tender):	
Driller:	
Geologist:	
Supervisor:	

	Drilling Report
Time & operation	
Date:	
Working hours:	
Kind of operation (technical remarks):	
Drilling Method:	
Drilling diameter (mm or "):	

Depth (m) :						
Drilling fluid :						
Fluid losses (%) :						
Reservation						
Protective casing interval (m)						
Protective casing diameter (mm or "):						
Cementation Interval (m):						
Sampling & coring						
Samples (number of using samples and intervals):	No.	Interval	No.	Interval	No.	Interval
Cores (coring interval (m) / % of cores):	Coring interval (m)			% of cores		
Lithological profile and stratigraphy (intervals and determination):	Intervals			Determination		
Technical construction						
Well Casing (interval / diameter (mm or "), type):	Type		Interval (m)		Diameter (mm / ")	
Well Screen (interval / diameter (mm or "), type):	Type		Interval (m)		Diameter (mm / ")	
Gravel pack (interval, diameter sort):	Interval (m)			Diameter (mm)		
GW information and development of well						
Water level (first registration depth /m):						
Mud Dilution Indications	Interval (m):			Estimated dilution (%)		

Static water level (m): (Before development)			
Development method (washing, air-lifting, jetting, pumping):			
Duration (h):			
Indications (drawdown (m)/ time / capacity):	Time (h / min)	Capacity (l/s)	Drawdown (m)
Recovery (water level / time)	Time (H / min)	Water table (m)	
Water level after test (m / after …h):			
Technical remarks Detailed explanations of all technical operation, problems, stoppages, troubles and other			
Date of completion:			
Signatures of responsible personnel:	Driller	Geologist	Supervisor

Tapping karstic groundwater flow within the aquifer catchment—Galleries, shafts Slightly inclined or sub-horizontal galleries (to enable gravity flow) are also frequently applied tapping structures, especially in the case of barrier (contact) springs. They are often constructed below natural discharge points by opening the barrier of impervious rocks (Fig. 11.23). Drilling or excavating the hole as classical mining work to reach a confined or trapped part of the aquifer and enable free gravity discharge is the main task of such structures. They are sometimes called *bottom outlets* due to their function which is similar to cleanout drainage pipes from the spring boxes. By piping and installing the valve on the gallery, opening the full control over the flow can be achieved. This is important if water utilization

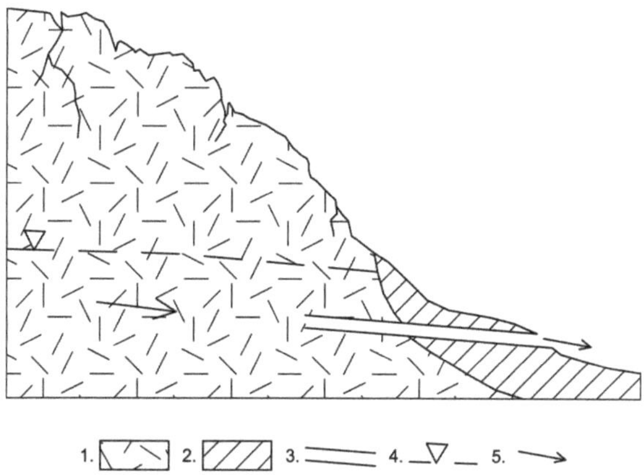

Fig. 11.23 Sub-horizontal gallery. *Legend 1* karstic aquifer; *2* impervious rocks; *3* gallery; *4* groundwater table; *5* groundwater flow direction

is seasonal and water can be stored until it is needed. For temporary closing and blocking the flow, a barrage door could also be constructed, similar to those used for preventing floods or for reasons of safety inside many mining galleries (Stevanović 2010). All this should be done with many precautionary measures and good geological background knowledge to avoid possible inappropriate consequences of such interventions (Milanović 2000).

Some of the intake represents a combination of above-described tapping structures. The most common is the case of a well or battery of wells placed around the spring in order to enlarge capacity. Although the issue of permitted aquifer control and sustainable aquifer development is discussed in Sect. 15.5, it is worth noting here that uncontrolled drilling of wells in many places in the world has resulted in aquifer depletion along with drying out of natural springs.

The case of Ras el Ain spring in Syria near the Turkish border is one of the most and best described (Burdon and Safadi 1963; Hole and Smith 2004; Stevanović 2010). This spring, one of the largest in the world with a natural yield in the range of 34.5–107.8 m^3/s, no longer flows during low-water periods. Groundwater over-extraction from numerous drilled irrigation wells in the catchment covered by cotton fields on the Turkish side of the border caused serious depletion of the groundwater table. Today, this is one of the most fertile and intensively cultivated regions in the Near East, but it is questionable if such intense development is justifiable from an environmental and transboundary (equitable share of the resource) point of view.

Another example of wells drilled in karstic "veins" which transfer groundwater toward a very large spring but without significant deterioration is from the classical Carso area. Perhaps the most famous and the largest spring on the northern Italian coast is the submarine spring Timavo in the Trieste Gulf. In the Slovenian part of Carso near the border city of Sežana, several highly productive wells were drilled in the 1980s just in main channels diverting water to Timavo. For example, the pumping well VB-4 discharged 0.05 m^3/s for a drawdown of only 0.45 m (Stevanović 2010).

References

Aller L, Bennett WT, Hackett G, Petty JR, Lehr HJ, Sedoris H, Nielsen MD, Denne EJ (1989) Handbook of suggested practices for the design and installation of Ground-water monitoring wells. EPA 600/4-89/034. Publication National Water Welfare Association, Dublin

Bono P, Boni C (1996) Water supply of Rome in antiquity and today. Environ Geol 27(2):126–134s

Burdon D, Safadi C (1963) Ras-El-Ain: the great karst spring of Mesopotamia. An hydrogeological study. Hydrol J, 1:58–95

Campbell MD, Lehr J (1973) Water well technology. McGraw-Hill, NewYork

Castany G (1967) Traité pratique des eaux souterraines, 2nd edn. Dunod, Paris

Castany G (1982) Principes et méthodes de l'hydrogéologie. Dunod Université, Paris

Driscoll FG (1986) Groundwater and wells. Johnson Filtration systems Inc., St. Paul

Dupuit J (1863) Estudes thèoriques et pratiques sur le mouvement des eaux dans les canaux dècouverts et à travers les terrains permèables, 2nd edn. Dunod, Paris

Filipović B, Dimitrijević N (1991) Mineral waters (in Serbian). University Belgrade, Faculty Mining Geology, Belgrade, p 274

Forchheimer P (1886) Über die ergiebigkeit von brunnen-anlagen und sickerschlitzen. Z Architekt Ing Ver Hannover 32:539–563

Goldsheider N, Drew D (ed) (2007) Methods in Karst hydrogeology. International contribution to hydrogeology, IAH, vol 26. Taylor & Francis/Balkema, London

Hole F, Smith R (2004) Arid land agriculture in northeastern Syria—will this be a tragedy of the commons? In: Gutman G et al. (eds) Land change science. Kluwer Academic Publication, The Netherlands, pp 209–222

Krešić N (2009) Groundwater resources: sustainability, management and restoration. McGraw-Hill, New York

Krešić N (2010) Types and classification of springs. In: Kresic N, Stevanović Z (eds) Groundwater hydrology of springs: engineering theory management and sustainability. Elsevier BH, Amsterdam, pp 31–85

Krešić N (2013) Water in Karst management vulnerability and restoration. McGraw Hill, New York

Lombardi L, Corazza A (2008) L'acqua e la città in epoca antica. In: La Geologia di Roma, dal centro storico alla periferia, Part I, Memoire Serv. Geol. d'Italia, vol LXXX. S.E.L.C.A, Firenze, pp 189–219

Milanović P (2000) Geological engineering in Karst. Dams, reservoirs grouting, groundwater protection, water tapping, tunneling. Zebra Publ Ltd., Belgrade, p 347

National Water Well association of Australia (1984) Drillers training and reference manual. NWWA, St. Yves, South Wales, p 267

Purdin W (1980) Using nonmetalic casing for geothermal wells. Water Well J 34(4):90–91

Stevanović Z, Iurkiewicz A (2004) Hydrogeology of Northern Iraq general hydrogeology and aquifer systems, Spec edn TCES, vol 2. FAO, Rome, p 175

Stevanović Z (2010) Utilization and regulation of springs. In: Kresic N, Stevanović Z (eds) Groundwater hydrology of springs: engineering theory, management and sustainability. Elsevier, BH, Amsterdam, pp 339–388

Stevanović Z, Dragišić V (1998) An example of identifying Karst groundwater flow. Environ Geol 35(4):241–244

Chapter 12
Monitoring of Karst Groundwater

Saša Milanović and Ljiljana Vasić

12.1 Introduction

Both quantitative and qualitative characteristics are included in the monitoring of groundwater which is of great importance for its abstraction and its long-term and safe use, particularly for the purposes of water supply. The quantity and quality of water is often not constant, and in order to obtain relevant information about the functioning of an aquifer, the implementation of a series of actions and measures to track the quantity and quality is necessary. This is especially important when dealing with karst groundwater. Consideration of the capacity and chemical composition of groundwater, as well as the establishment of a system of monitoring, is the first step toward sustainable use of groundwater in karst (Milanović et al. 2010a; Stevanović et al. 2010).

In order to ensure high-quality groundwater monitoring, a monitoring network that will define the number of observation points, observation frequency, and the parameters that need to be monitored have to be established. This kind of monitoring provides data that give us a basic outline of the groundwater and the ability to define recommendations and measures for its use or for the improvement of its condition. Reliable monitoring of groundwater quality and quantity in any terrain is difficult, but in karst terrains, it is especially hard.

S. Milanović (✉) · L. Vasić
Centre for Karst Hydrogeology, Department of Hydrogeology, Faculty of Mining and Geology, University of Belgrade, Belgrade, Serbia
e-mail: sasamilanovic@rgf.bg.ac.rs

Z. Stevanović (ed.), *Karst Aquifers – Characterization and Engineering*,
Professional Practice in Earth Sciences, DOI 10.1007/978-3-319-12850-4_12

The main points of groundwater monitoring in karst terrains can be grouped into six major categories depending on different tasks. These categories are as follows:

1. Location of monitoring
    (a) Sinkhole
    (b) Cave
    (c) Spring
    (d) Well
2. Type of monitoring
    (a) Quantity
    (b) Quality
3. Equipment used for monitoring
    (a) Equipment for water flow measurement
        - Mechanical devices
        - Semiautomatic
        - Digital devices
    (b) Equipment for water quality analysis
    (c) Equipment for pumping tests
    (d) Equipment for groundwater level measurement
        - Mechanical devices
        - Digital devices
4. Intervals of monitoring
    (a) Continual
    (b) Daily
    (c) Weekly
    (d) Monthly
5. Monitoring data base
    (a) Spatial display
    (b) Time series
    (c) Data interpretation
    (d) Reporting
6. Monitoring network range
    (a) Local
    (b) Regional
    (c) National
    (d) Transboundary

12.2 Location of Monitoring

Each karst aquifer is unique in its individual characteristics, but some structural components are widely found, although they vary in relative significance in different systems (Ford and Williams 2007). Still, data for karst aquifer characterization can be obtained by monitoring main hydrogeological occurrences and objects such as sinkholes, cave streams, springs, and wells (Fig. 12.1).

Sinkholes (ponors) are one of the best places for monitoring "input" to the karst system. Generally, ponors can be classified according to their swallowing capacity and by aspects of their surface appearance, for example, ponors that are permanently and temporarily active; ponors with a big entrance (shaft or cave); or ponors with a system of narrow joints along the stream. Depending on those characteristics, monitoring equipment can be located at the entrance of ponors or in the surface stream close to the sinking area.

Caves (underground flow/streams) are the only place where groundwater in the karst system can be monitored directly. According to Ford and Williams (2007), caves are solution channels that are greater than 5–15 mm in diameter or width, but the most common definition is that a cave comprises underground channels, chambers, or other different caverns large enough for human activity (survey). The placement of monitoring equipment in a cave depends on hydrological and hydrogeological characteristics. Generally, a monitoring location is connected on the one side for observation of groundwater level (GWL) fluctuation ("pool water", Toomey 2009) or on the other side for observation of underground flow in different levels of karst aquifer.

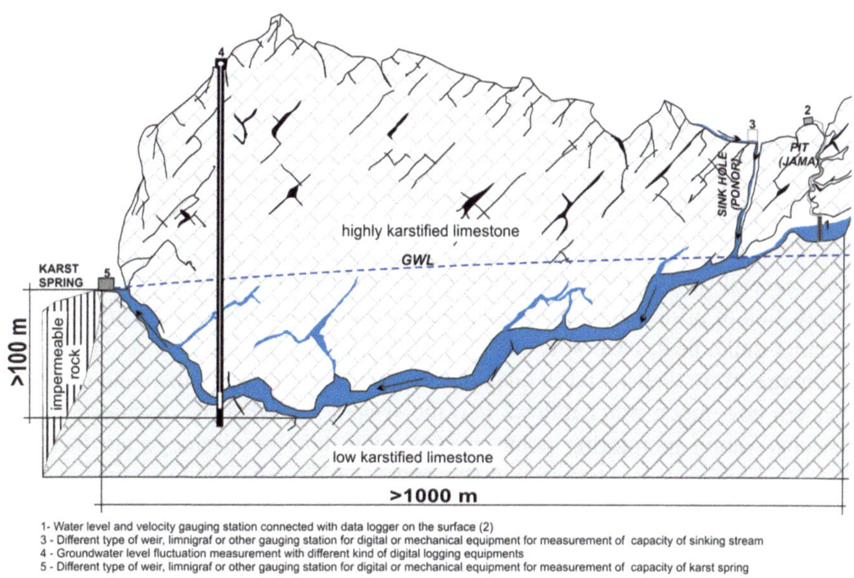

Fig. 12.1 Different locations for groundwater monitoring in karst aquifer

Springs are probably the most important hydrogeological occurrence in karst. Monitoring of spring waters is a basic part of hydrogeological survey of karst aquifer. The majority of springs are located along the perimeter of the erosion base, river valleys, and the sea coast (Milanović 2000). In general, the capacity and hydrogeological character of karst springs as well as the drainage area determine the way to install monitoring equipment. Finally, depending on those characteristics, monitoring equipment can be located just at the entrance of springs or downstream from the spring, but generally, it is placed in the area close to the spring channel or discharge zone (see Fig. 6.9 in Chap. 6).

Monitoring *wells* in karst aquifers may be specifically constructed for the purpose or may be preexisting. The cost of this kind of monitoring is very high and mostly depends on whether existing wells can be used or new wells must be drilled for this purpose (Toomey 2009). There are, generally, two different types of wells in karst: those with casing pipe or those with just an open hole (i.e., a well without any kind of construction).

Some data necessary to be collected prior to the monitoring are as follows:

- name or ID number
- location
- coordinates (x, y, z)
- geology settings of the narrow area
- type of monitoring site
- depth of sinkhole or cave
- depth of spring (if is siphonal character)
- depth of well
- type of well construction, diameter
- short description of monitoring site

12.3 Type of Monitoring

12.3.1 Quantity

The general objective of monitoring of karst groundwater quantity is the acquisition of information about quantity of groundwater flows, including recharge (ponors–sinkholes), cave streams, discharge over springs, as well as GWLs and discharges from wells. And in quantitative monitoring in karst, the main questions concern how much water inflows into the system and what the discharge rate on the observation monitoring points is. Common units of discharge are liters per second or cubic meters per second (l/s, m^3/s), which, concerning springs, ponors, and cave streams, is on the one hand generally calculated as the product of the stream's cross-sectional area and mean velocity at the point of measurement (Groves 2007) and, on the other hand, by measuring discharge from wells using different measurement devices.

12.3.2 Quality

Monitoring the groundwater quality provides data and information which can be used to characterize the water quality of the karst aquifer, as well as to describe variables of the water chemistry and its spatial variability, temporal trend, and any changes during circulation from the inlet to the outlet of the karst system. Basic parameters for monitoring quality of karst waters include temperature, pH, dissolved oxygen, specific conductivity, hardness, calcium, magnesium, potassium, sodium, bicarbonate, sulfates, chlorides, nitrates, and phosphates. In addition, microbiological parameters, such as coliform bacteria, fecal coliform bacteria, and *E. coli,* are very important for detecting the connection between surface waters and its velocity through the system. Also depending on land use in the watershed of interest, it is important to monitor for a variety of organic and inorganic pollutants as well as contamination from some waste disposal systems. Collecting samples for groundwater chemistry is fairly straight forward and inexpensive. Today, most of the chemical analyses can be done in situ, without specialized laboratory equipment.

12.4 Equipment Used for Monitoring

Accurate discharge estimation is crucial for sustainable karst aquifer management. There are generally four types of discharge measurement, the first two of which are traditional ways of estimating the discharge in hydrological practice. The first of these is to measure the water stage and to convert the recorded water stage values into discharge by using the single-valued rating curve (Brilly et al. 2008). The second is to measure discharge using a weir with a stick gauge, usually used for small streams (discharges). The third is a semiautomatic to automatic gauging station which is a combination of limnigraph, loggers, weirs, and water stage gauge. Finally, the fourth is a digital type of measuring based on the Doppler effect.

12.4.1 Mechanical Devices

The most used and most traditional way of estimating the discharge in hydrological monitoring is to measure the discharge using the current meter and water stage and to convert the data into discharge by using a single-valued rating curve (Fig. 12.2). The measurement must be taken at convenient times such as low, medium, and high water periods. The current meter measures the flow velocity on the basis of the conversion table which relates the number of current meter rotor revolutions per time unit.

Portable weirs can be made of steel plates and are used for measuring the flow of small springs or sinking streams. Generally, they can give accurate

Measurement of water level / interval daily h (cm)

	Jan	Feb	March	Apr	May	June	Jule
1	85	33	100	49	49	55	40
2	96	30	85	48	48	56	40
3	83	30	75	48	46	56	40
4	70	30	68	45	46	53	40
5	60	30	58	44	46	50	40
6	60	30	50	50	44	46	40
7	60	31	49	60	43	51	40
25	70	30	48	71	44	50	40
26	80	30	46	70	44	49	39
27	74	30	44	68	44	47	38
28	68	30	44	65	41	45	38
29	60	30	42	59	40	42	38
30	56	30	40	55	40	42	38
31	48	30	38	55	38	38	38

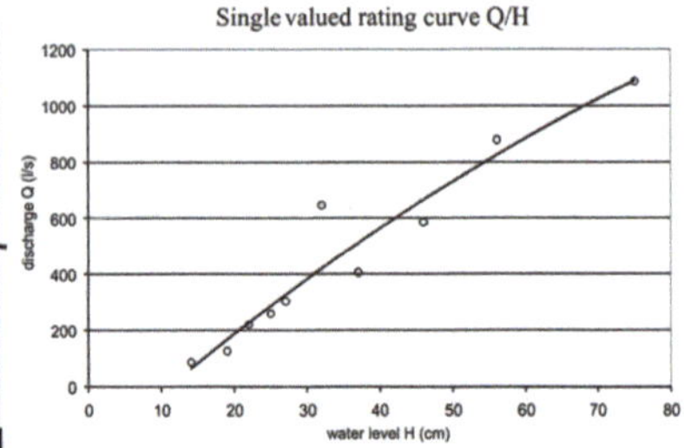

Dicharge - measured and calculated Q m3/s

	Jan	Feb	March	Apr	May	June	Jule
1	1.210	0.441	1.362	0.718	0.718	0.812	0.566
2	1.324	0.385	1.210	0.701	0.701	0.828	0.566
3	1.187	0.385	1.091	0.701	0.668	0.828	0.566
4	1.027	0.385	1.000	0.652	0.668	0.781	0.566
5	0.887	0.385	0.858	0.635	0.668	0.734	0.566
6	0.887	0.385	0.734	0.734	0.635	0.668	0.566
7	0.887	0.404	0.718	0.887	0.618	0.750	0.566
25	1.027	0.385	0.701	1.040	0.635	0.734	0.566
26	1.152	0.385	0.668	1.027	0.635	0.718	0.549
27	1.079	0.385	0.635	1.000	0.635	0.685	0.531
28	1.000	0.385	0.635	0.959	0.584	0.652	0.531
29	0.887	0.385	0.601	0.873	0.566	0.601	0.531
30	0.828	0.385	0.566	0.812	0.566	0.601	0.531
31	0.701	0.385	0.531	0.812	0.531	0.531	0.531

Fig. 12.2 Measurement and calculation of spring discharge

Fig. 12.3 Installation of triangular weir (*left*). Two typical weirs (*right*)

measurements, if they are installed properly. The weir should be perpendicular to the stream, vertically, and with the horizontal position of the crest (Fig. 12.3). A free fall is required over the weir crest. Also, it is important that leaking be prevented around the weir sides, possibly by grouting or using a polythene sheet.

A stick gauge marked in centimeters and used to measure the head, h (cm), (difference between the upstream water level and the crest of the weir) must be set vertically upstream of the weir. There are different sizes and shapes of weir, each having a standard formula for calculation of discharge. Some of them are shown in Fig. 12.3. Again, in this case, as in the first case outlined, measurement must be taken by the monitoring observer at specified intervals.

12.4.2 Semiautomatic–Automatic

Discharge measurement of springs and sinking streams can be taken through different combinations of limnigraph or data logger for water level measurement with different types of weirs, together with measurement of discharge by a current meter for establishing *Q*/*H* curve. Those are the most useful types of discharge monitoring in karst areas, especially at remote and inaccessible monitoring points, where to have a surveyor daily is impossible. Still, this equipment for monitoring is not cheap and not easy to install.

12.4.3 Digital Devices

Modern measuring equipment such as the one based on the Doppler effect has enabled us to measure water velocity continuously. There is a wide range of instruments that enable the measurement of water depth, velocity, and temperature at the same time (Fig. 12.4). Usually, it integrates an automatic data logger with the possibility of averaging at the chosen or default time interval. It measures flow velocity in the main direction of the water course by detecting the frequency change of sent and received ultrasonic signals due to the reflection of instream moving particles and air bubbles in the water volume (Brilly et al. 2008).

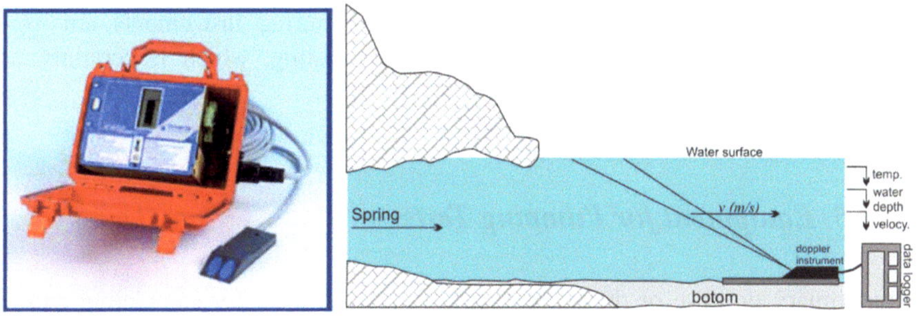

Fig. 12.4 *Left* one of the instruments based on Doppler effect. *Right* schematic figure of equipment installation in location of spring discharge

Fig. 12.5 *Left* monitoring of groundwater quality (field work). *Right* groundwater monitoring equipment

12.4.4 Equipment for Water Quality Analysis

As mentioned before, establishment of a monitoring network, determination of monitoring frequency as well as definition of the parameters that need to be monitored are all essential for good groundwater monitoring, either in the field or in the laboratory. Also very important for groundwater monitoring and precision of data is accurate and quality equipment. The equipment used to determine the physical and chemical parameters is generally divided into field and laboratory equipment.

Since the value of certain physical and chemical parameters changes during transport of the sample to the laboratory and its disposal to a protected place, their measurement is best carried out in situ (Fig. 12.5), when conditions allow. In recent years, practical and portable devices that can specify parameters directly in the field and with sufficient precision have been produced. They can monitor parameters that should always be monitored in the field such as temperature, pH, specific conductance, redox potential, alkalinity, and dissolved oxygen. In addition, with newer portable equipment, nitrates, nitrites, carbonates, ammonium ion, iron, manganese, chlorine, calcium, and many others can also be measured in the field (Fig. 12.5). Of course, this kind of equipment has certain disadvantages, such as a relatively high price and somewhat less precision.

In the laboratory, however, more precise measuring instruments are used, so that the laboratories conduct more complex testing, which is certainly not accessible in the field.

12.4.5 Equipment for Pumping Tests

The primary purpose of the pump test is to provide data for calculating aquifer properties, such as discharge capacity, transmissivity, and storage coefficient. Pump tests are also used to assess the impacts of boundaries on well production.

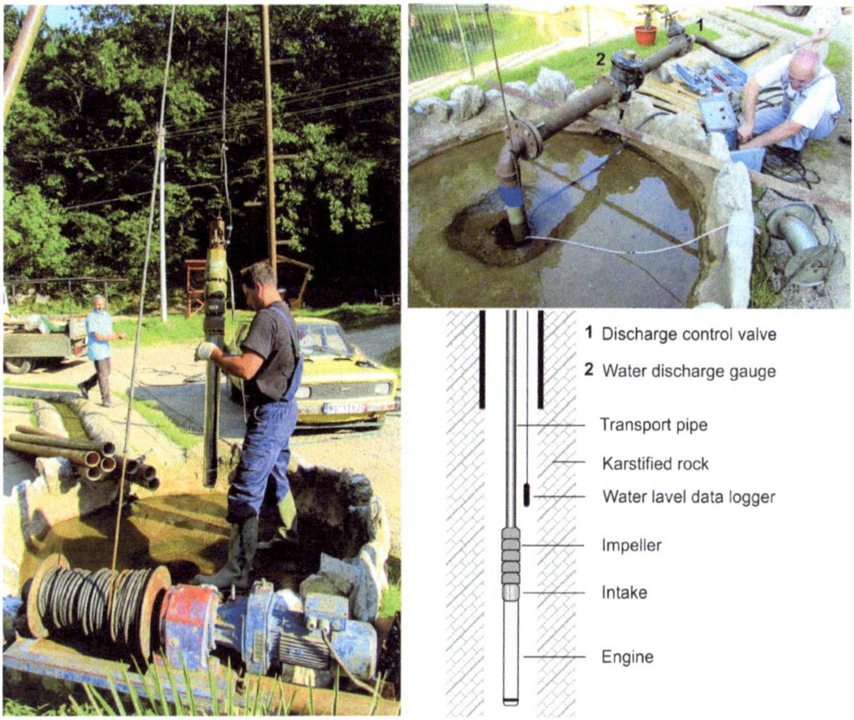

Fig. 12.6 Equipment for pumping test

The results of aquifer tests provide fundamental input data that can be used in calculations that predict drawdown in an area. Equipment for the pumping test is comprised of pump, transport pipe, and equipment for GWL measurement and for pumping capacity measurement (Fig. 12.6).

12.4.6 Equipment for Groundwater Level Measurement

The type of equipment for measurement of GWL in observation wells depends not only on the objectives of the monitoring network for groundwater and temporal variations, but also on aspects such as site accessibility, available personnel, and sampling interval. Measurement of GWLs in wells can be accomplished either mechanically or with digital recording instruments (data logger). Depending on those conditions, it is necessary to choose equipment that can fully meet the assigned task. Generally, there are two types of devices used for GWL measurement: mechanical devices and digital devices.

A mechanical GWL meter is generally capable of measurement within 1 cm, and information about the GWL is achieved by the use of sound or light during the contact of water and probe (Fig. 12.7). Usually, GWL meters are made for a depth from 50 up to 300 m and more. The limits to depths of measurement are essentially associated with the length of cable, the design of electrical circuitry, and the weight of the equipment.

At present, many different types of digital GWL recorders (dip meter) are in use. Generally, they record data concerning water pressure and temperature. These instruments are portable, easily installed, and both capable of recording under a wide variety of climatic conditions and capable of operating unattended for varying periods of time. Also, they have the facility to measure high ranges in groundwater fluctuation at different times, such as from 0.5 s, 1 min, 1 h, and 1 day etc. Usually, the ranges of water level fluctuation which can be observed with these instruments are from 50 to 100 m and sometimes even higher than 100 m (Fig. 12.8).

Fig. 12.7 Using groundwater dip meter for groundwater monitoring in Berbera (Somalia), see Box 12.1

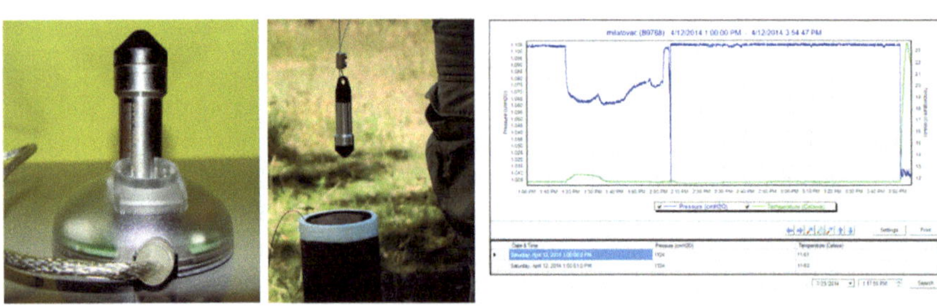

Fig. 12.8 One type of groundwater data logger, with graph of groundwater level and temperature fluctuation

12.4.7 Intervals of Monitoring

The frequency of monitoring depends on the hydrogeological environment and aquifer regime, so in areas with rapid circulation, such as karst, intensity measurements will be more frequent than in areas with slow circulation, such as intergranular aquifers. Depending on the frequency of measurement, monitoring can be divided into four groups: continuous, daily, weekly, and monthly. The choice of measurement interval mainly depends on four conditions: accessibility of the location, the equipment used for the monitoring, the task, and the desired results.

Monitoring on a daily level is normally sufficient for most springs, sinking streams, ponors, and streams. For measurement in small or flashy streams as well as in special cases, the stage has to be recorded more frequently in order to obtain a sufficiently accurate hydrograph (continual); still, there is the limitation of the available battery capacity and data memory if we use digital equipment (data loggers).

According to the interval of measurements we can divide monitoring as follows:

(a) Continual (hourly level)—monitoring of spring discharge or sinking stream capacity. Not common; used just in special cases such as before, during, and after heavy rains.
(b) Daily—flow measurement mostly provided in karst areas. Generally provides sufficient data for further analyses.
(c) Weekly—flow measurement and water quality analysis. Mostly provided for springs with stable regime. Not usual for chemical analyses, except in special cases.
(d) Monthly—very rare flow measurement on a monthly basis, but for quality analysis, it is usual at intervals in hydrogeological survey.

12.5 Monitoring Database

The last phase in the groundwater monitoring is the creation of a database and its connection to GIS. It is essential that all the data collected during monitoring certain areas be put on a specific database for future research and analysis. The basic elements that constitute this database are as follows: time series of the collected data, whether the characteristics are quantitative or qualitative, and the spatial position or location of the monitoring points. The basic elements that should be contained in any database related to monitoring of groundwater in karst are given in Table 12.1. Schematic representation of the formation of a GIS for monitoring of groundwater in karst is shown in Fig. 12.9.

Table 12.1 Main elements for database related to monitoring of groundwater

Quality	Well	Spring
OBJECTID	OBJECTID	OBJECTID
Shape	Shape	Shape
District	District	District
SOURCE_COD	SOURCE_COD	SOURCE_COD
SOURCE_TYP	SOURCE_TYP	SOURCE_TYP
LATITUDE	LATITUDE	LATITUDE
LONGITUDE	LONGITUDE	LONGITUDE
ELEVATION	ELEVATION	ELEVATION
EC__μS_cm_	BH_Depth__	NEAREST_SE
PH	SWL__m_	DISCHARGER min.
TEMP__0C_	DWL _m	DISCHARGER max.
Chloride__	YIELD__m3_	Water_Use_
Sulfate	Yield__l_s	FUNCTIONIN
Nitrate	PUMPLEVEL	Water_Use_
Nitrite	FUNCTIONIN	
Ammonium	LIFTINGTEC	
Fluoride	Water_Use_	
Hydrogen		
Carbonate		
Aluminum		
Arsenic		
Calcium		
Iron		
Potassium		
Magnesium		
Sodium		
Phosphorus		
Silicon		
Zinc		
Classification_ Number		

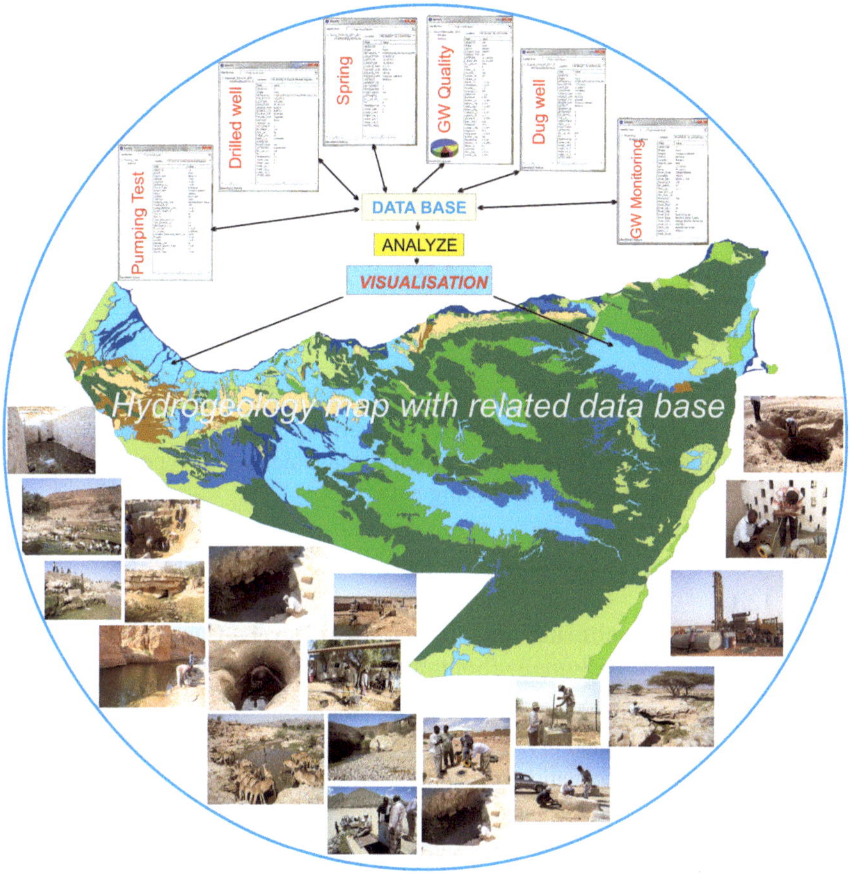

Fig. 12.9 Schematic representation of the formation of a GIS for monitoring of karst groundwater in SWALIM project (Stevanović et al. 2012b), see Box 12.1

12.6 Monitoring Network Range

One of the important characteristics of groundwater monitoring in karst is the size of the monitoring network area. In relation to the size, monitoring can be classified as local, regional, national, and transboundary.

Local monitoring refers to monitoring certain occurrences or objects, generally at distances no greater than a few kilometers. It mainly relates to the local monitoring of a spring or sinkholes or the monitoring of one or two exploitation wells. This type of monitoring is the most common monitoring of groundwater in karst, since it usually observes only one karst spring whose data can indicate a lot about the functioning of karst aquifers.

Regional monitoring refers to those areas which cover usually the entire karst aquifer whose surface extends over several hundred square kilometers (Box 12.2). Often regional monitoring of such huge space involves simultaneous monitoring of several karst springs, sinkholes, and often numerous wells. Such monitoring is generally carried out only in a large regional survey with a high degree of detail. This monitoring is one of the most demanding and expensive works.

For the "typical" karst countries in the Mediterranean basin, the national monitoring of karst aquifer areas should be related to karst springs whose minimal yield exceeds 50 l/s, or which are very important for bottled water or water supply. National monitoring is the monitoring that is done through the country or, in the case of groundwater in karst, is performed on entire karst terrains (Box 12.3).

Transboundary monitoring and information exchange should be regulated by the agreement of two or more countries concerning shared observation of karst groundwater under their protection and joint management. This type of monitoring is usually part of the national monitoring network.

Box 12.1
Groundwater monitoring on the regional level—Case example of project SWALIM in Somalia (Somaliland and Puntland)

This case example gives an outline of the monitoring surveys carried out in Somaliland and Puntland in order to collect all the data necessary to make conclusions and recommendations concerning groundwater potential at a regional scale. After desktop studies and field reconnaissance, one of the most important parts of the project was monitoring of groundwater as well as examination of the local groundwater conditions at the sites and analysis, evaluation, and interpretation of the available data from the previous SWIMS database.
Finally, to enable proper operation and sustainable water utilization of the utilities in Somaliland and in Puntland, managers need to know whether GWL is falling, steady, or rising, and, even more important, what is the quality of groundwater used for water supply. This information is only possible by measuring depth to GWL and water quality over time. Collection of data from monitoring network is the only way to get reliable information (Stevanović et al. 2012b).
The purpose of these investigations (important part was monitoring) is to prepare a preliminary assessment of the comprehensive groundwater resources of Somaliland and Puntland. Some of the main specific objectives of groundwater monitoring were as follows:

- Identify and survey as many as possible of Somaliland and Puntland's hydrogeological conditions, water points, and aquifers as potential monitoring places

- Identify and make an inventory of all drilled wells, dug wells, and springs as a base layer for further database and monitoring network
- Make a database of all investigated points
- Search and identify aquifers with good water potential and quality
- Assess groundwater aquifers and their potential for meeting domestic, livestock, agricultural, and commercial water demands for both Somaliland and Puntland
- Form a viable base for the identification of major priority areas within which detailed investigations will be conducted

Developing a hydrogeological monitoring database and related map involves a lot of initial field work, analyzing the results and interpretation data from a hydrogeological point of view. The monitoring was conducted through the two following steps:

- Monitoring of groundwater quality (springs, wells)
- GWL monitoring (wells and investigation boreholes)

Monitoring of groundwater quality

Concentrations of chemical components in groundwater vary widely within the Somaliland and Puntland areas, depending on the location and type of hydrogeological objects (spring, dug well, and drilled well).

Generally, groundwater quality in the study area is not high and usually does not meet all WHO guidelines for acceptable drinking water. Based on the chemical analyses of 511 monitoring points, hardness, calcium, magnesium, sodium, potassium, and other components are usually above the WHO standard (Stevanović et al. 2012b).

Generally, in Somaliland and Puntland, very few groundwater sources will conform to international standards. The salt content of the water commonly exceeds 1 g/l, which under normal circumstances is the upper limit for human consumption, but acceptance of water with relatively high ion concentrations is a necessity, as there is usually no alternative.

Monitoring data about the chemical analysis of investigation points (boreholes, shallow wells, and springs) are on the one hand analyzed in terms of the spatial position of districts and related quality and on the other hand in terms of the relationship between the aquifer and the groundwater quality. The main range of chemical parameters analyzed in during monitoring are presented in Table 12.2. The location of monitoring points is shown in Fig. 12.10.

As conclusion of groundwater quality monitoring, we can conclude that water quality acceptability is subject to the habits of the people. Consumption of very hard water with high EC is customary and tolerated in northern Somalia. The quality of the groundwater varies by region and depends mainly on the underlying rock formations and their thickness. For instance, the EC for the Bari region varies between approximately

Table 12.2 The concentrations of analyzed components (Stevanović et al. 2012b)

	Min.	Max.	Average
EC (μS/cm)	271	9480	3539.07
pH	6	10	7.68
Temp (^{0}C)	19	38	26.40
Chloride (mg/l)	7.6	4330	600.83
Sulfate (mg/l)	24	8930	1653.39
Nitrate (mg/l)	0.9	168	31.38
Nitrite (mg/l)	0.02	0.48	0.12
Ammonium (mg/l)	0.07	58	15.01
Fluoride (mg/l)	0.2	8.4	1.82
Hydrogen carbonate (mg/l)	112	567	272.41
Carbonate (mg/l)	5	10	7.86
Aluminum (mg/l)	0.09	0.19	0.13
Arsenic (mg/l)	0.005	0.026	0.01
Lead (mg/l)	0.005	0.009	0.01
Boron (mg/l)	0.07	4.5	0.92
Cadmium (mg/l)	0.001	0.002	0.00
Calcium (mg/l)	21.6	825	305.03
Chromium (mg/l)	0.005	0.025	0.01
Iron (mg/l)	0.02	0.12	0.07
Potassium (mg/l)	1.7	111	21.54
Copper (mg/l)	0.005	0.012	0.01
Magnesium (mg/l)	8.41	2330	180.18
Manganese (mg/l)	0.006	4.2	0.89
Sodium (mg/l)	9.6	2530	353.28
Phosphorus total (mg/l)	0.05	0.11	0.07

1,139–4,000 μS/cm with pH ranges of 7.4–7.8. In the Sanaag, Nugal, Sool, and Mudug regions, the EC varies between approximately 2,100–5,500 μS/cm with a pH in the range of 7.6–8.0. In the Awdal, Galbeed, and Togdheer regions, the EC ranges from 417 to approximately 4,000 μS/cm with a pH in the range of 7.0–7.6. Water quality in most parts of northern Somalia is below the international standard (WHO guideline).

Groundwater level monitoring

The monitoring of GWL was planned with the aim to verify the behavior of general aquifers under natural variations, in terms of the relationship between rainfall and the recharge of the aquifers. The monitoring activity was supposed also to improve the degree of responsibility of the relevant Somaliland and Puntland Water Authorities.

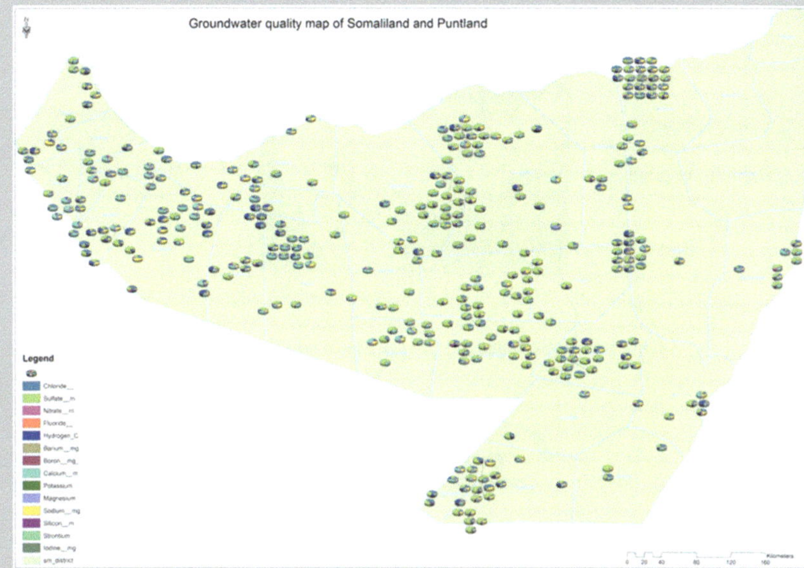

Fig. 12.10 Location map of all groundwater quality sampling points in Somaliland and Puntland provinces, Somalia (Stevanović et al. 2012b)

The main target is the possibility of future depletion of aquifers. Due to the high risk of groundwater overexploitation in major towns of Somaliland and Puntland where the demand for water is growing as the population increases, the towns which were prioritized for the monitoring of network installation included Hargeisa, Borama, Berbera, and Burco in Somaliland; and Garowe, Bosaso, Galkayo, and Qardho in Puntland (Stevanović et al. 2012b).

Divers were installed in the following order: Hargeisa, Boramo, Berbera, Galkayo, Garowe, Qardho, Bosaso, and Burco. A baro diver was installed at the FAO offices in Hargeisa and Garowe to measure the atmospheric pressure which is used to compensate data with the regular divers. Selection of installation boreholes was done in consultation with the water utility agencies in the respective towns. For all the divers, the data collection interval was set to 6 h (4 readings a day).

The schematic overview of monitoring points is presented in Fig. 12.11.

Generally speaking, on the basis of monitoring data, there are daily oscillations of water level in some boreholes in Somaliland and Puntland, an indication that it is slightly affected directly by water pumping in the nearest wells. There is a gradual decline in water level in some boreholes up to a magnitude of 1.2 m in the 38 days of data recording.

The future completed Groundwater Monitoring Network of Somaliland and Puntland must provide the information necessary to assess groundwater

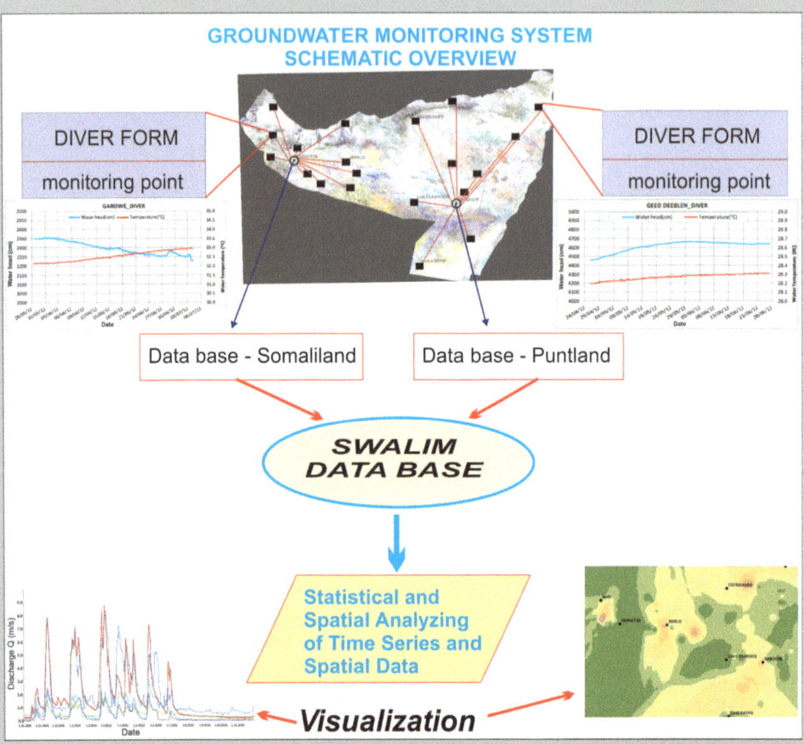

Fig. 12.11 Map of monitoring points and groundwater monitoring schema of Somaliland and Puntland

quantitative status, chemical status, and significant, long-term trends in natural conditions and trends in groundwater resulting from human activity.

For a whole territory of Somaliland and Puntland according to current assessment and collected and evaluated data on hydrogeology and hydrochemistry, it would be necessary to establish about 100–130 monitoring water points in Somaliland and similar number in Puntland. The total projected number of about 250 monitoring points is still providing lesser average density than 1 object per each 1,000 km^2 (Stevanović et al. 2012b):

- About 30–40 % of monitoring points should be monitored by divers, while for the other objects, simple mechanical water level meter can be used (man monitoring).
- In case of sparsely populated area for every 500–1,000 km^2, at least one monitoring water point should be included in GW Quality Monitoring Network. In case of urban settlements or presence of more vulnerable aquifers and potential pollutants, more observation points would be required.
- Recommended monitoring interval for groundwater level recording should be on the daily basis, while for water quality tests at least on a monthly basis.

Box 12.2

Case Example—Groundwater monitoring of karst springs on Beljanica Mt. (Eastern Serbia)

The test area of Beljanica Mt. is a part of the Carpathian–Balkan mountain range in the central part of eastern Serbia. This is one of the several mountain massifs in the region that is built mostly from carbonate rocks. The study area comprises the Beljanica mountain and its foothills with a surface of around 370 km^2. The altitudes vary between 1,339 m a.s.l. at the highest peak and 300 m a.s.l. in the surrounding plains. The major basin in the northern edge of the mountain is Žagubica, one of the typical intermountain depressions of eastern Serbia filled with lacustrian Neogene and Quaternary sediments (Stevanović 1991; Milanović 2010; Stevanović et al. 2012a).

Monitoring of the quantity and physical and chemical characteristics of groundwater in situ and sampling for chemical as well as bacteriological analyses in the area of Beljanica have been carried out since August 2009.

The data obtained by simultaneous monitoring of quantitative and qualitative characteristics of karstic springs that drain the Beljanica karstic aquifer significantly contributed to defining the separate hydrogeological catchment areas of five karstic springs. They also had a significant role in the monitoring of fluctuations in the GWLs and defining the thickness of the saturation zone, especially in the period of low and extreme high water (Fig. 12.12).

Fig. 12.12 Krupajsko spring during min. discharge, average max. discharge, and discharge during extreme high water period

Monitoring data of the regime for five springs that drain Beljanica aquifer, Vrelo Mlava Spring, Belosavac Spring, Krupajsko Spring, and Malo Vrelo and Veliko Vrelo Spring, for the period (2009–2012), made possible the comparative analysis of regime, precipitation, and water quality. Physical and chemical characteristics of the springs were observed in situ monthly, using the different field instruments for pH, temperature, dissolved oxygen, conductivity, and turbidity (Fig. 12.13). Some previous data have also been taken into consideration, particularly to compare characteristics of actual waters with those that are a few decades older (Stevanović 1991, 1994; Stevanović et al. 2012a).

Results of monitoring show that the total mineralization is at intervals from 160.9 to 374.7 mg/l, which classifies them in the group of water with low mineral content. Depending on the season and spring discharge, water temperature varies from a minimum of 6.8 °C to a maximum of 11.9 °C. It was concluded that with higher discharge of the spring, the water temperature decreases, and conversely, smaller discharge results in higher temperatures. pH value of the siphonal springs (Mlava, Belosavac, Suvi Do, and Krupajsko springs) ranged from 6.8 to 7.9; the water, therefore, could be classified as neutral to slightly basic. Increased pH values were observed in the gravity springs (Veliko and Malo Vrelo Springs), where the maximum recorded

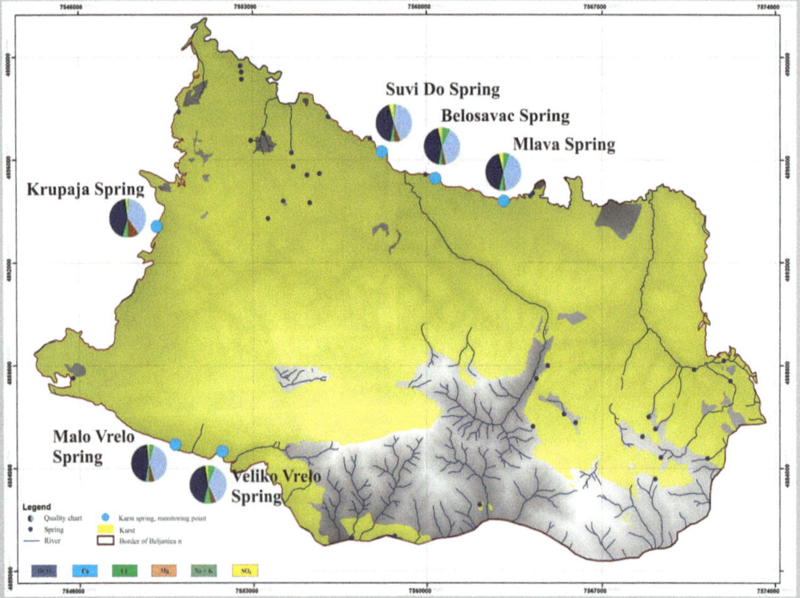

Fig. 12.13 Map of Beljanica test area with main chemical characteristics of groundwater of main karstic springs

value was 8.4. All springs show a decrease in pH values with increasing capacity of the spring, except the Belosavac Spring which showed an inverse trend. In terms of hardness, the waters belong to the soft to hard water (according to Alekin), with a value from 6.2 to 19.4 °dH. The greatest turbidity of 5.1 NTU was observed at the Mlava Spring. Genetically, these waters are of the hydrocarbonate class, calcium group. HCO_3^- ranges from 146.4 to 390.4 mg/l (Milanović et al. 2013).

During continuous monitoring in the years 2009–2011 (daily monitoring of regime and monthly physical and chemical characteristics), it was concluded that the mean yield of springs which drain the northern part of Beljanica averages $Q_{av} = 6.113$ m^3/s, while the average yield for the southern rim of Beljanica or for Malo and Veliko Vrelo springs is $Q_{av} = 1.248$ m^3/s, which leads to a total sum of mean yield of the whole mass of Beljanica of $Q_{av.tot} = 7.361$ m^3/s. Comparative diagrams (Fig. 12.14) show that change in pH and temperature depends on the yield of springs: with increasing yield decreases the pH, while in periods of recession the pH value increases up to 8.5. This increase in pH values in the extremely dry period indicates a considerable influence of recharge from the central part of Beljanica mountain and from Paleozoic rocks that generate waters downstream directly infiltrated into karstified rocks (epikarst and open karst). With increasing precipitation and significant infiltration of water into aquifer, pH declines (Milanović et al. 2010b, 2013).

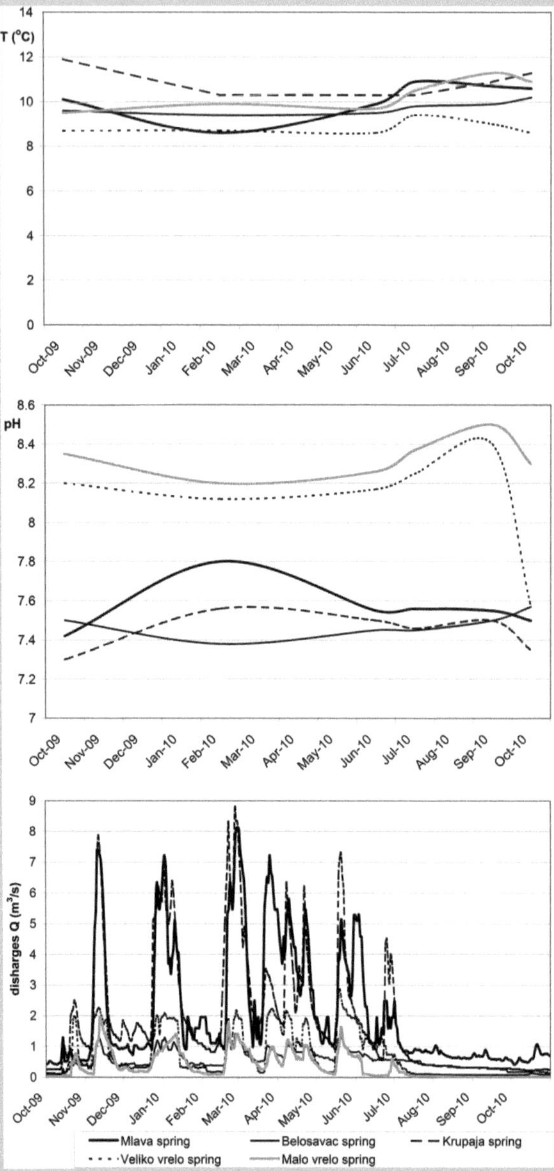

Fig. 12.14 Comparative diagrams: discharge—water temperature Q/T and discharge—pH value Q/pH for analyzed springs of Beljanica mountain (Milanović et al. 2013)

Box 12.3
Groundwater Monitoring Network—northern Iraq case example

In the year 2000, the Food and Agriculture Organization of the United Nations (FAO/UN) together with local authorities in the three northern governorates of Iraq, Duhok, Erbil, and Sulaimani designed a comprehensive research project targeting the assessment of groundwater resources and reserves. Geological and geophysical investigations, hydrogeological mapping, groundwater monitoring (wells and springs), remote sensing analysis, and groundwater quality assessments were conducted in the region from 2000 to 2003 (Stevanović and Iurkiewicz 2004). The monitoring network, meant to observe groundwater fluctuations and to prevent the consequences of possible overexploitation, has been designed and operational for more than two years (December 2000–February 2003) and at the last stage included 180 deep wells and 150 springs monitored on a daily or weekly basis (Stevanović and Iurkiewicz 2009).

Of 180 wells, 54 were located in karstic and fissured aquifers and were observed more frequently than those in other aquifer systems, especially during the wet season. The arid climate in northern Iraq results in a total absence of rainfall in most of the region during the period of May–October, which is the period of groundwater stagnancy or more often permanent depletion of the water table. Figure 12.15 shows a typical correlative diagram of a rainfall/groundwater table of one of the observed piezometers in

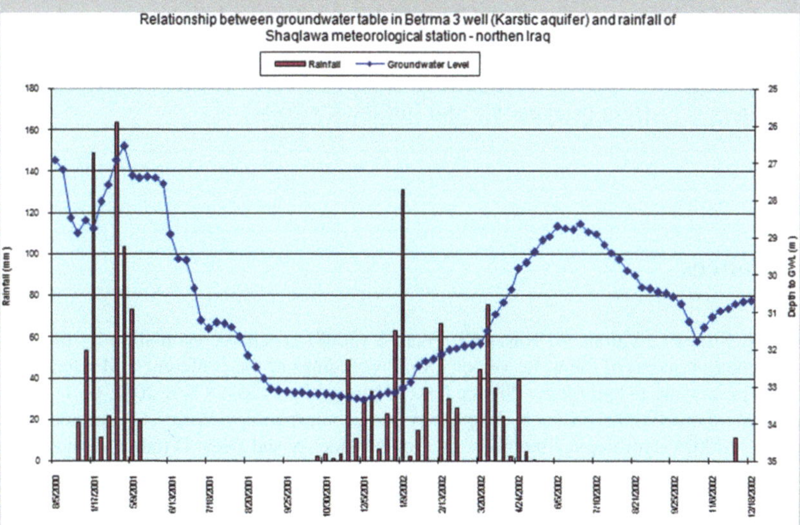

Fig. 12.15 Correlative diagram of rainfall and water table in the Betrma well situated in the Bekhme aquifer of Safin Mt. (northern Iraq)

karst aquifer. The water table's annual amplitude is 7 m, and groundwater depletes relatively slowly in the recession period, but also rises slowly during the rainy season. This is a typical situation for aquifer which is characterized by good storage capacity.

Table 12.3 Annual groundwater table variation of karstic aquifers Bekhme and Pila Spi (northern Iraq)

Group/ variation	Annual amplitude value (m)	Bekhme aquifer/% (out of 15 wells)	Pila Spi aquifer/% (out of 24 wells)
1/Insignificant	1–3	13.3	37.5
2/Moderate	3–6	20	16.7
3/Considerable	6–20	46.7	25
4/Very large	>20	20	20.8

Table 12.3 presents calculated annual amplitudes of the groundwater tables of two main karstic aquifer systems at Bekhme and Pila Spi, respectively. The recorded values considered mostly static water tables out of depression cones resulting from groundwater extraction.

Therefore, these two aquifers are characterized by generally moderate to considerable variation of resources accumulated throughout the hydrological year. Pila Spi aquifer, which is younger and formed in Tertiary rocks, is more stable. The absolute maximum amplitude of 48 m was recorded in the piezometer relatively close to extraction wells tapping karstic groundwater for Shaqlawa (Safin Mt.). Conversely, there are a few highly productive wells that have a stable regime with absolute amplitude less than 2 m. It may be assumed that the normal yearly variation of static water level for these aquifers is 5–10 m (Stevanović and Iurkiewicz 2004).

References

Brilly M, Stravs L, Vidmar A, Rusjan S, Petan S (2008) Discharge estimation by the continuous measurements of the water velocity. In: Proceedings of the conference: Measurement and data processing in hydrology, Plitvice Lakes National Park, 26–28 Nov 2008, pp 1–14

Ford D, Williams P (2007) Karst hydrogeology and geomorphology. Willey, Chichester

Groves C (2007) Hydrological methods. In: Goldscheider N and Drew D (eds.) Methods in karst hydrogeology, IAH Book Series. Taylor & Francis, London

Milanović P (2000) Geological engineering in karst. Monograph, Zebra Publ. Ltd., Belgrade

Milanović S (2010) Creation of physical model of karstic aquifer on example of Beljanica Mt. (eastern Serbia). PhD thesis, University of Belgrade—Fac Min Geol, Belgrade, Serbia

Milanović S, Stevanović Z, Jemcov I (2010a) Water losses risk assessment: an example from Carpathian karst. Environ Earth Sci 60(4):817–827

Milanović S, Stevanović Z, Vasić Lj (2010b) Development of karst system model as a result of Beljanica aquifer monitoring. Vodoprivreda, Belgrade 42(246–248):209–222

Milanović S, Stevanović Z, Vasić Lj, Ristić-Vakanjac V (2013) 3D Modeling and monitoring of karst system as a base for its evaluation and utilization—a case study from eastern Serbia. Environ Earth Sci Springer 71(2): 525–532

Stevanović Z (1991) Hidrogeologija karsta Karpato-Balkanida istočne Srbije i mogućnosti vodosnabdevanja (Hydrogeology of Carpathian-Balkan karst of eastern Serbia and water supply opportunities; in Serbian). Spec. ed. Fac Min and Geol, Belgrade p 245

Stevanović Z (1994) Karst ground waters of Carpatho-Balkanides in Eastern Serbia. In: Stevanović Z, Filipović B (eds) Ground waters in carbonate rocks of the Carpathian—Balkan mountain range, Spec. ed. of CBGA, Allston Hold, Jersey, pp 203–237

Stevanović Z, Iurkiewicz A (2004) Hydrogeology of northern Iraq. Regional hydrogeology and aquifer systems, Spec. ed FAO (Spec. Emerg. Prog. Serv.), Rome, vol. 2, p 175

Stevanović Z, Iurkiewicz A (2009) Groundwater management in northern Iraq. Hydrogeol J 17(2):367–378

Stevanović Z, Milanović S, Ristić V (2010) Supportive methods for assessing effective porosity and regulating karst aquifers. Acta Carsologica 39(2):313–329

Stevanović Z, Ristić-Vakanjac V, Milanović S (eds) (2012a) CCWaterS—climate changes and impact on water supply. University of Belgrade—Faculty of Mining and Geology, Belgrade

Stevanović Z, Balint Z, Gadain H, Trivić B, Marobhe I, Milanović S et al (2012b) Hydrogeological survey and assessment of selected areas in Somaliland and Puntland. Technical report no. W-20, FAO-SWALIM (GCP/SOM/049/EC) Project (http://www.faoswalim.org/water_reports), Nairobi

Toomey RS (2009) Geological monitoring of caves and associated landscapes. In: Young R, Norby L (eds) Geological monitoring: boulder. Geological Society of America, Colorado, pp 27–46

Chapter 13
Catalog of Engineering Works in Karst and Their Effects

Petar Milanović

13.1 Introduction

Any engineering activity in karst, particularly construction of large structures (dams, reservoirs, tunnels, and man-made caverns) is sensitive and risky task. The risk of constructing in karst cannot be eliminated completely even the best engineering practices are followed. Because of the specific nature of karst, it is not easy to select dam site, reservoir space, tunnel route or tailings in karstified rock that would be acceptable without risk. Reservoir in karst may fail to fill despite an extensive investigations and remedial works. Successful solutions require serious and complex investigations including long time monitoring and (in many cases) remedial works during operation. Remedial works require a lot of patience and perseverance and adequate investment. As a result of appropriate treatment in a number of cases the problem was successfully overcome. However, in some cases hydrogeological and geotechnical properties are too complicated and problem could not be overcome by applying the available technology. In extreme cases seepage problem could not be overcome and consequently the dams had to be abandoned. On the base of long time experience it is obvious that in a karstified rock mass the final design of large structures in karst can only be definitely defined during the execution phase. These structures need modifications and adaptations on the basis of the geological findings in the underground.

Of course, building a dam or reservoir in karst does not automatically mean there will be leakage or failure. In number of cases if dam sites and reservoir areas are properly selected in the rock mass with suitable geological conditions, the leakage problem does not exist because is solved during construction, or is negligible and without influence on dam stability.

P. Milanović (✉)
National Committee of IAH for Serbia, Belgrade, Serbia
e-mail: petar.mi@eunet.rs

Z. Stevanović (ed.), *Karst Aquifers – Characterization and Engineering*,
Professional Practice in Earth Sciences, DOI 10.1007/978-3-319-12850-4_13

At the text below is presented the list of dams and reservoirs situated in karstified rocks (carbonates and evaporates). Of course, is not possible to catalogue all water leakage examples, because, for instance, in Guangxi Province China, only, 644 reservoirs in karst suffer due to leakage (Yuan 1991). Some of prominent case studies are briefly explained and grouped according to the some generalized seepage problems and success of remedial works: not necessary additional remedial works; successful corrective (remedial) works; unsuccessful or partially successful remedial works; re-designed anti-seepage structures; dams and reservoirs with acceptable leakage; abandoned dams; abandoned or temporary frozen projects; underground dams in karst; and dams and reservoirs in evaporites.

Explanation of some specific terms is presented at the end of textual part. The terms also include short definition of some of the works form engineering practice such us grouting curtain, impervious blanket, cutoff wall, shotcrete and similar applied to prevent or to mitigate water leakage from reservoirs built in karstic terrains. The most of these interventions aiming to seal or plug underground conduits, but their success depend on the following (Milanović 2000):

- position of karst channel or concentrated zone flow;
- channel dimensions (cross-sectional area);
- underground flow, characteristics in the channel (permanent or intermittent flow under pressure or with free surface, direction and velocity);
- possible presence of clay and silty cave deposits and their thickness.

13.2 Building Dams and Reservoirs in Karst

Box 13.1: List of dams in different karst regions

A number of case histories show problems of old dams situated in karstifid rock because of inadequate investigations and weak anti-seepage measures or due to inadequate anti-seepage technology in that time. Because of that many dams and reservoirs long time suffer due to unacceptable leakage in spite of frequent remedial measures: **McMillan Dam** (USA, 1893); **Miller Dam** (USA, 1900); **Wolf Creek Dam** (USA constructed in 1938–1952); **Center Hull Dam** (USA, constructed 1938–1948); **Biver Dam** (USA, 1966); **Clearwater Dam** (USA, 1948); **Walter F. George Dam** (USA, 1963); **Mississinewa Dam** (USA, 1967); **Camarassa Reservoir** (Spain, 1927–1931); **Kentacky Dam** (USA); **Salanfe Reservoir** (Switzerland, 1952); **Maria Cristina and Sichar dams** (Spain); **Ira Price Dam** (USA); **Jinlong Reservoir** (China); **Anchor Dam** (USA); **Civitella Liciana** (Italy); **Cheurfa Dam, Ouizert Dam, Foum El Gherza Dam, Tichy Haf Dam** (Algeria); **Fremont Reservoir** (USA)**; North Dike** (USA, Florida); and **Cuber Dam** (Spain).

Well known case studies related to the dams and reservoirs situated in the karstified rock are: **Mornos Reservoir** (Greece, 135 high rock-fill dam); **Perdikas Resevoir** (Greece, 40 m high earth-filled dam); **Nolin River Dam** (Kentucky, USA, 40 m high, rock-fill); **Green River Dam** (Kentucky, USA,

43 m high, rock-fill dam); **Patoka Dam** (Indiana, USA, 46 m high, embankment dam); **May, Apa, Cevisli** and **Onac Reservoirs** (Turkey); **Lone Pine Reservoir** (USA, 31 m high dam); **Krupac Reservoir** (Montenegro); **Pueblo Vijeo Dam** (130 m high, Guatemala); **East Fork Dam** (Kentacky, USA); **La Bolera Reservoir** (Spain); **Kopili Dam** (India); **Charmine Dam** (France); **Ataturk Dam** (Turkey, 179 m high fill dam); **Dongmenhe Dam** (China); **Jinlong Reservoir** (Guangxi, China); **Shibazi Reservoir** (Hunan, China); **Vudu Reservoir** (Sichuan, China); **Beni Haroun Dam** (Algeria); **Cheboksarsky Reservoir** (Russia); **Ričica Reservoir** (Croatia); **Armagan Dam** (Turkey); **Niukouyu Reservoir** (Beijing, China); **Guangting** (Hebei, China, 45 m high); **Hoa Binh Dam** (Vietnam, earth-filled, 128 m high); **Adam Back Upper Reservoir** (Canada—Niagara Escarpment, built on top 10–40 m of glacial deposits overlaying dolomite and limestone); **Qaraoun** and **Maifedoum**, 4 km long and 250 m deep curtain (Lebanon); **Megenin** and **Wadi Gattara** (Libya); **King Talal** (Jordan,100 m high, karstified dolomites and caverns in sandstones); **Mujib Dam** (Jordan, 67 m high); **Wala Dam** Jordan); **Kasseb Dam** (Tunisia, arch dam, 70 m high); **Pengshui** (gravity dam of 120 m on the Wujiang River, China); **Okukubi Dam** (40.9 m high gravity dam at Okinawa, Japan); **Kozjak Dam** (FYR Macedonia, concrete arch dam, 110 m high, situated in karstified marbles, dolomitic marbles and schists); **Scrivener Dam** (Canberra, Australia, 33 m high); **Yumaguzinsk Reservoir** (Russia, karstified limestone and dolomite, problems with karstification and ecology). According the Kutepov et al. (2004) the following dams and reservoirs in Russia are also situated at karstified rock: **Volhovskaya** (1926, first dam constructed in karstified rock in Russia, 10 m high); **Atbashinskaya** (rock filled, 80 m high); **Volzskaya** (concrete gravity, 40 m high); **Zubcovskaya** (earth-filled, 15 m high); **Kahovskaya** (earth-filled, 16 m high); **Miatlinskaya** (rock/earth filled, 90 m high); **Pavlovskaya** (earth-filled, 56 m high); **Plyiavinskaya** (earth-filled, 40 m high); **Rizhkaya** (earth-filled, 18 m high); **Charvakskaya** (rock/earth filled, 168 m high); and **Shaorskaya** (earth-filled, 13 m high).

Very high leakage from reservoirs, after first filling or during operation was registered in following cases: **Hales Bar**, USA, leakage of 50 m^3/s; **Vrtac**, Montenegro, 25 m^3/s; **Taoqupo**, China, 27.8 m^3/s; **Keban**, Turkey, 26 m^3/s; **Zaixiangkou**, China, 20 m^3/s; **Maotiao 4th cascade**, China, 20 m^3/s; **Višegrad**, Bosnia and Herzegovina, 14.5 m^3/s; **Iliki**, Greece, 13 m^3/s; **Camarassa**, Spain, 11.2 m^3/s; **Lar**, Iran, 10.8 m^3/s; **Salakovac**, Bosnia and Herzegovina, >10 m^3/s; **Ataturk**, Turkey, >10 m^3/s; **Mavrovo**, FYR Macedonia, 9.5 m^3/s; **Marun**, Iran, 9.5 m^3/s; **Great Falls**, USA, 9.5–12.7 m^3/s; **Canelles**, Spain, 8–10 m^3/s; **Slano**, Montenegro, 8 m^3/s; **Dokan**, Iraq, 6 m^3/s; **Buško Blato**, Bosnia and Herzegovina, 5 m^3/s; **Montejaque**, Spain, 4 m^3/s; **Fodda**, Morocco, 3–5 m^3/s; **Contreas**, Spain, 3–4 m^3/s; **Gorica**, Bosnia and Herzegovina, 3–4 m^3/s; **Kowsar (Tange Duk)**, Iran ~3 m^3/s; **Hutovo**, Bosnia and Herzegovina, 3 m^3/s; **Karun III**, Iran, 2–3 m^3/s; **Samanalawewa**, Sri Lanka, 2.8 m^3/s (max. –7 m^3/s); **Špilje Dam**, FYR Macedonia, 2 m^3/s; **Vrtac**, Montenegro, 2 m^3/s; **Shuicaozi**, China, 1.8 m^3/s.

In some, very risky and hazardous cases, investigation takes 30 or more years before final decision and selection of proper dam type and proper ant-seepage works: **Ilarion Project—Elati** (Greece, seepage area within the reservoir); **Geheyan Project** (China); **Kavar Dam site** (Iran); **Rilja Dam site** and **Reservoir** (Herzegovina); and **Kavsak Dam** (Turkey).

Large caverns were detected below dam foundation in the case of: **Keban Dam** (Turkey), huge caverns below dam body and at left dam foundation; **Ourkis Dam** (Algeria), number of large empty and interconnected caverns at depth 45–52 m below foundation surface; **Grabovica Dam** (Bosnia and Herzegovina), long and wide cavern at the dam foundation filled with clayey-sandy material, locally deep about 40 m; **Seymareh Dam** (Iran), caverns at the both side of the dam banks close to the foundation surface; **Karun IV** (Iran), cavern beneath the left dam foundation surface; **Salman Farsi Dam** (Iran), large cavern system at the right bank; **Sklope Dam** (Croatia), huge caverns at both banks close to the foundation surface; large cavern below the **Lar Dam** foundation (Iran); **Wudongde** dam site (China), large cavern (vol. 369,000 m^3) filled with plastic clay and blocks at the right bank; and in the case of number other dam sites and along grout curtain alignments.

At some dam sites and reservoirs the leakage is consequence of hydrothermal water (hypogene karst features) or coupled influence of hydrothermal and meteoric waters: **Salman Farsi dam site** (Iran), **Francisco Morozan dam site** (Honduras); **Višegrad dam site** (Bosnia and Herzegovina); **Hamam Grouz** (Algeria); **Salanfe** (Switzerland), **Samanalawewa** (Sri Lanka) and **Chicik Dam** (Uzbekistan).

Box 13.2: Dams with not necessary additional corrective works

If dam site and reservoir areas are selected on the basis of very detail geological and hydrogeological analysis the risk is low, however never can be absolutely eliminated. Of course proper adaptations and modifications during grout curtain execution are crucial. At many examples leakage during the first filling and later during reservoir operation is acceptably or negligible low. Some examples are: **Castillon, Genissiat, Greoux, Sainte Croix, Ceirac, La Rouviere, Quinson and Conqueyrc** (France); **Ekbatan** (Iran); **Panix** and **Punta Del Gall** (Switzerland); **Rama** and **Grabovica** (Bosnia and Herzegovina); **Štikada** (Croatia); **Qued Fodda** (Algeria); **Altapina** (Turkey); **Nebaana** (Tunisia); **Hsinankiang** and **Chinwuo** (China); **Santa Guistina** (Italy); **Bin al Ouidance** (Morocco); **Beniner** (Spain); **La Angostura** (Mexico); **Pueblo Vijeo** (Guatemala); **Doosti**, border between Iran and Turkmenistan; **Storglomvatn** (Norway); and **Zhai Xiang Kou** (Kueizhou Plateau, China).

The most prominent examples of successful watertightness works during construction are:

Oymapinar Dam (Turkey) is double curvature arch dam, 185 m high, situated in karstified Permian limestone. The grout curtain was laterally and under river bed tight to upstream contact between limestone and impervious schist formations making the successful bath-tab structure. The important karstic open joints were plugged by concrete. The large Dumanli spring (Q $\approx$ 20–100 m^3/s), close upstream from the dam, is submerged with 120 m without any influence in reservoir watertightness (Altug 1999). Based on the calculation of Günay et al. (1985), the storage rate at the Oymapinar Reservoir banks ranges between 35 and 40 %.

Grančarevo Dam (Herzegovina). Double curvature concrete dam, high 123 m, create the Bileća Reservoir, $V = 1.27 \times 10^9$ m^3. Dam is located in the Jurassic limestone. The entire reservoir is in karstified Mesozoic limestone and dolomite. Watertightness of the dam site and reservoir is based on the positive (impervious) hydrogeological role of the large anticline structure, with core composed of grusified dolomite, between reservoir and lower erosion base level. Leakage as consequence of filtration at the dam site is only 50–100 l/s. However, the deep excavation for power house foundation has endangered stability of one part of the left slope (Fig. 13.1). This part of rock mass was stabilized by 200 pre-stressed anchors of 40–60 m in length (Stojić 1966).

Fig. 13.1 Grančarevo Dam, Herzegovina. Stabilization of *left slope* by pre-stressed anchors. *Photos* P. Milanović

Peruća Dam (Croatia). Rock-fill dam is high 65 m from limestone foundation was constructed 1958. Grout curtain at dam site is deep 200 m and long 1,600 m. Total surface of curtain is 242,122 m^2. According long time monitoring it is estimated that constant leakage from the reservoir at full storage level is about 1 m^3/s (1.5 % of the average river flow).

Geheyan Dam (China) is concrete gravity arch dam, high 151 m from foundation. Reservoir water head is 122 m, and reservoir volume 3.12 billion m^3. More than 600 caves and caverns were detected on the surface and in the underground, with total volume of 60,000 m^3. During investigation of the dam site 79 caverns (volume 22,000 m^3) were detected. Based on these data the grout curtain alignment was proposed. Sixty caverns along the curtain route (volume 6,000 m^3) have been filled with concrete before grouting. Curtain length is 1,400 m within an area of 180,000 m^2 (Ruichun and Fuzhang 2004). The serious problem of groundwater intrusion during excavation of the diversion tunnel (1.67 m^3/s) was successfully done. During operation time the leakage was not detected.

Khao Laem Dam (Thailand) is 90 m high rock-fill structure. Karstic features were developed down to 50 m below the river bed. To prevent filtration through the number of karst channels (0.2–10 m wide) the concrete diaphragm (cut-off) wall was constructed down to a depth of 15–55 m. A five level gallery system for cut-off wall was applied. The diaphragm wall surface is about 15,900 m^2. Depth of conventional grout curtain was between 100 and 150 m. A total anti-seepage structure is 77,000 m^2 at the area of dam foundation and about 360,000 m^2 in the right abutment (Bergado et al. 1984).

Poliphyton Dam (Greece) is 105 m high rock-fill structure. Poliphyton Reservoir is partially in contact with karstified marble and limestone. The large Neraidha Spring (10 m^3/s) was flooded by 40 m of water head. In spite of intensive karstification no negative influence of the spring submergence was observed.

Karacaoren II Dam (Turkey) is arch gravity dam, 49 m high from the foundation. To prevent seepage through the karstified Mesozoic limestone three different protective measures, at surface, have been applied (approximately 1 m thick concrete lining close to the dam; 7 cm, layers of reinforced shotcrete over the limestone, 40 × 650 m; and clayey blanket at slope). The sealing treatment is successful and justified the investment in protection (Okay and Soidam-Bas 1999).

Normandy Dam (Tennessee, USA) the 34 m high earth-fill dam is an example of successful treatment. Dam foundation is situated at the horizontal limestone including a weak zone, 80 m wide. After detail investigations the cut-off wall (overlapping wells) has been constructed and reinforced with upstream and downstream grout curtain.

Berke Dam (Turkey) is arch concrete structure, 201 m high situated in locally karstified Mesozoic formations. Karst channels (aperture 20–200 cm) are developed along the discontinuity planes. In heavily karstified zone a triple-row array with the cross pattern holes was used. The uppermost grouting gallery is at elevation of 346 m and deepest grout curtain contour reach elevation of 50 m below zero. Deepest grouting holes were 225 m. It is one of most complicated and largest grout curtains in the world (Altug and Saticoglu 2001).

Wujiangdu Dam (China, Guizhou). The dam structure, situated in karstified limestone, is 165 m high. The many karst cavities, particularly in the left bank, down to the depth of about 250 m beneath the river bed, are distributed. Cavities were filled with soft clay (water content—56 %). To increase the resistance of cave filling an unconventional grouting mechanism was applied. The soft clay in the caverns was subject to four actions: hydraulic fracturing, extrusion, consolidation, and chemical hardening. The maximum applied pressure was 60 kg/cm^2 (Zuomei and Pinshow 1986).

Box 13.3: Case studies with successful corrective works

Great Falls (Tennessee, USA). Leakage through the karstified left reservoir bank increased rapidly from 0.47 m^3/s in 1926 to more than 6.6 m^3/s in 1939 and to 12.7 m^3/s in 1945. Velocity of underground leakage flows ranges between 60 and 180 cm/s. Combined asphalt and cement grouting has been applied. By construction grout curtain, 1,750 m long and an average depth of 15.25 m the water losses from the reservoir were reduced to only 2 %.

Guanting Dam (China, Hebei) is a 45 m high dam. Immediately after its construction in 1955 leakage and collapses occurred. Three karst channels were discovered and the treatment results provided at the Guanting Dam site were successful (Lu 2012).

Ekbatan Dam (Iran). Concrete buttress dam, 53 m high, for irrigation. Large caverns at grout curtain route are successful plugged (Soletanche, France).

Marun Dam (Iran). The rock-fill Marun Dam (165 m high) is situated at karstified Asmari limestone. Seepage occurs through two mutually interconnected karst systems: along the interbedding joints and along a thin layer of intensively vuggy limestone of the Middle Asmari. In a few days leakage of only 40 l/min increased to 1.5 m^3/s and finally to 10.0 m^3/s. The main conduit was speleologically investigated and mapped. After extension of grout curtain up to the section with shaley interbeds and plugging of the main karst channel by concrete plug the leakage decreased to the negligible amount.

Francisco Morazan Dam (Honduras), formerly El Cajon, arch Dam (236 m high) is situated at Cretaceous limestone which was karstified before being covered by lava. Presence of thermal water was detected during excavation of underground power plant. In spite of bath tab antifiltration structure during the reservoir filling (1993) a seepage water of 1.65 m^3/s was observed. Voluminous cavities of up to 5,000 m^3, about 170–180 m below the gallery level, were grouted. To reach deep cavities, the maximum borehole length of 250 m was executed. Total leakage was reduced to the negligible amount of 0.1 m^3/s (Guifarro et al. 1996).

Canelles Dam (Spain) is a 151 m high arch dam, with a crest length of 203 m. Dam is situated in karstified Cretaceous limestone. During the first filling the leakage of approximately 6 m^3/s shortly increase up to 8.0 m^3/s. Due to reservoir level of about 100 m (means very high pressure) the underground flows were very fast (over 16 m/s). To stop these flows a 300 mm borehole was drilled and channel was filled with 2,130 pieces of 240 mm granite cobbles to form prefilter plug. Before final grouting the gravel was injected over the plug. After speleological investigations at the left bank the karst channel (surface 17 m^2) was plugged with an 11 m long concrete plug. To fill the caves, 1,380 tons of dry materials were used. Finally the leakage was reduced to a negligible quantity of 68 l/s (Weyermann 1977).

Dokan Dam (Iraq) is a concrete arch dam, 116 m high. Dam site consists of the karstified dolomite overlaying by thinly bedded limestone. After reservoir impounding of two thirds of its capacity (1960) a leakage of 6.0 m^3/s occurred bypassing the end of curtain. After grout curtain extension (336 m) to a depth of 150–160 m leakage from reservoir was eliminated.

Hutovo Reservoir (Bosnia and Herzegovina) is situated at the end section of extremely karstified Popovo Polje. In natural conditions leakage through the 75 registered ponors at the reservoir bottom area ranged between 7.0 and 10.0 m^3/s. To achieve required watertightness several sealing methods were applied. Exposed limestone reservoir banks were protected by shotcrete. The alluvial bottom was compacted and ponor zones were filled and blanketed with plastic foil. After reservoir filling (1975) new ponors and wide (5–30 cm) fissures has as consequence leakage of 3.0 m^3/s (Fig. 13.2). During three cycles of reservoir filling and empting more than 120 ponors and 1,300 m of fissures were sealed with clay-cement mixture. Finally, the leakage rate was reduced to acceptable 1.0 m^3/s (Milanović 2004).

Fig. 13.2 Hutovo reservoir, Herzegovina. Sinkhole and cracks created at the reservoir bottom and beneath protective plastic foil after first reservoir filling. *Photos* P. Milanović

Douglas Dam (USA) is concrete gravity, 62 m high, built in 1940–1943. To prevent leakage the large cavity has been plugged by placing 4,965 tons of concrete.

Župica Dam (Bosnia and Herzegovina) is only 24.75 m high (depth of water is 23 m). However, due to very deep karstification, the average depth of the grout curtain is 185 m. It is 8 times more than dam height. The length of the curtain is 620 m, and surface is 127,777 m^2.

The ***plugging of single karst channels with underground flows*** to stop seepage was effectively applied in number of cases. The examples are: Honeycomb Cave in the **Guntersville Reservoir,** USA; Krupac ponor (4 m^3/s) in **Krupac Reservoir,** Montenegro; karst channel (0.8 m^3/s) in the **Charmine Reservoir,** France, to reduce loss from reservoir from 800 l/s to 20 l/s; number of karst channels at foundation of the **Khao Laem Dam**, Thailand; in the case of **Wujiangdu Dam,** China, (early 1980s, concrete gravity, 165 m high) specific grouting mechanism has been used to plug cavities felled with soft clay (pressure up to 60 kg/cm^2); **Salman Farsi** dam site, Iran, to plug six large caverns 3,125 m^3 (7,500 tons) of concrete was used; **Khoabin Dam,** Vietnam (17,600 m^3 of caverns have been cleaned for construction of concrete wall, 10.5 km long); **Buško Blato Reservoir**,

Bosnia, three large plugs along the one channel (former underground river). In the case of power plant **Čapljina** (BiH) plugging of the karst channel with flow (300 l/s) at an elevation of 32 m below zero level, was applied using a 146-mm diameter boreholes.

Box 13.4: Dams with unsuccessful or partially successful remedial works

Lar Dam (Iran) is earth-fill structure, 105 m high above foundation. The geological conditions are very complex: lava, tuff, lake deposits over intensively karstified Jurassic and Cretaceous limestone. In natural conditions groundwater level was at depth of approximately 200 m. Immediatelly after impounding started (1980) the seepage was registered at two springs, as well as collapses at reservoir bottom (Fig. 13.3). For maximal realized reservoir level during past 30 years (24 m lower than designed) measured discharge was 9.85 m^3/s. During the past 30 years massive remedial works were performed including intensive grouting and filling of the large cavern (volume—90,000 m^3) however without success. The reservoir was never completely filled up, and quantity of leakage does not change.

Fig. 13.3 Lar Dam and reservoir, Iran. One of the sinkholes at the reservoir bottom. *Photos* P. Milanović

Salakovac Dam (Bosnia and Herzegovina). Concrete gravity dam at Neretva River is 70 m high, from foundation, constructed in the mid 80s. Subhorizontal thick bedded limestone is intensely tectonized and karstified. Discharge of the downstream springs at the left river bank in natural conditions was $\approx 5\ m^3/s$. Immediately after reservoir impounding the massive leakage through the left bank has been registered. Discharge of spring zone increased to 15 m^3/s (with tendency of increasing). Extensive grouting to intersect and plug karst channels with leakage flows and shotcrete blanket over the left reservoir bank were not successful.

Keban Dam (Turkey) is made of two parts: rock-fill and concrete gravity section. The dam height is 211 m and reservoir volume is $30.6 \times 10^9\ m^3$ (Božović et al. 1981). The dam is situated in a heavily karstified marble and limestone. During the dam construction about 30 caverns were treated including Crab Cavity with volume of 105,000 m^3. During the reservoir filling (1976), at elevation of 10 m below maximal reservoir level huge karst shaft was opened near the dam body and losses from reservoir sharply increased to 26 m^3/s. This water discharges in the Keban Creek, 2.5 km south of the dam site (Fig. 13.4). The karst shaft was connected with huge Petek Cavern. After the Petek Cave was filled with about 605,000 m^3 of limestone blocks, gravel, sand and clay the flow rate in the Keban Creek decreased to 8–9 m^3/s.

Fig. 13.4 Keban Dam, Turkey. Leakage through the cavern at the left bank. *Upper* Dam body and location of shaft excavated for cavern filling. *Lower* One of the springs (leakage water) at the Keban creek. *Photos* P. Milanović

Samanalawewa Reservoir (Sri Lanka). Samanalawewa is 107 high rock-fill dam with a central clay core. Dam site and reservoir are composed of Gneisses with limestone layers. Limestone is karstified by solution and hydrothermal action. Large cavities were detected. A 100 m deep and 1,800 m long (hanging) grout curtain has been constructed. During first reservoir impounding seepage from 75 l/s increased to the more than 7 m^3/s (1992). The erosion process washed away from caverns about 25,000 m^3

of cave deposits. After short time seepage decrease and stabilized at 1.8–2.2 m^3/s. As addition measure, about 250,000 m^3 of clay have been used to provide blanket over the seepage zones at the reservoir bottom, however, without success. In 2006, after heavy monsoonal rains leakage increased up to approx. 5 m^3/s and suddenly stabilized at 2.2–2.5 m^3/s. However, new collapses are registered in the reservoir area.

Višegrad Dam (Bosnia and Herzegovina). The gravity concrete dam is 50 m high (70 from foundation) is situated at Drina river bed. Foundation formations are limestone and dolomite rocks karstified by solution and hydrothermal action. During the first filling of the reservoir (1986) seepage of 1.4 m^3/s, below the dam body, indicate intensive washing of clay from caverns. In spite of intensive re-grouting the leakage water gradually increased to 6.5 m^3/s (1996), 9.4 m^3/s (2003) and 13.92 m^3/s (2008). Active karst channels are at depth more than 130 m below the dam. Deepest karst features are detected 283 m below the river bed. During investigations in 2009 and 2010 the large karst opening (ponor—diameter 3.5 m) with massive water inflow was discovered by divers at the reservoir bottom, approximately 150 m upstream from the dam (Milanović and Vasić 2014). Presently the new remediation program is under implementation (Fig. 13.5).

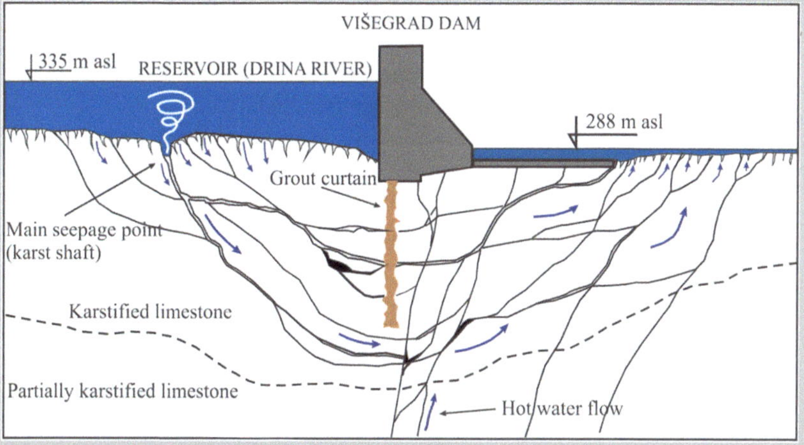

Fig. 13.5 Višegrad Dam, Bosnia and Herzegovina. Seepage and progressive erosion beneath the dam (final leakage flow Q ~ 14 m^3/s). Milanović S., modified

Salanfe Reservoir (Switzerland) is created by 52 m high gravity dam (1952). Dam and reservoir area consists of quartzite, dolomites and limestone. Because of high seepage (>1 m^3/s) through the left reservoir bank, during many years of operation, only 50 % of its volume has been used.

Together with seepage from reservoir appear new thermal sprigs in valley, 9 km from reservoir (20 l/s, 30 °C). Filling of reservoir and intensive seepage were accompanied with considerable induced seismicity. Remedial works (1992–1994) included construction of additional grouting gallery (600 m long) and execution of the one row grout curtain (Bianchetti et al. 1992, Dawans et al. 1993).

Slano Reservoir (Montenegro) is situated in the extremely karstified and temporary flooded Nikšić karst polje. In natural conditions, during maximum flood level the registered losses were 34 m^3/s. Reservoir depths −12 m. By construction of a few grout curtains, with total length of 7.011 m and average depth of 65 m, the losses have been reduced to an acceptable amount of 3.5 m^3/s. Grouting works takes 11 years (1961–1971). During the reservoir operation from 1971 till 2001 the grout curtain was heavily damaged and new ponors were opened between reservoir bank and grout curtain. Leakage increased up to 7 m^3/s, mostly though the uppermost part of curtain (Vlahović 2005). To prevent groundwater flow through the cavern at depth 90–100 m plugging take period of 4 years and 5 months however without success (injected—4,124 m^3 of gravel and 4,500 tons of cement).

Perdikas Reservoir (Greece), 40 m high earth filled dam. The groundwater level was 70 m below the reservoir bottom, in the karstified Upper Cretaceous limestone. During reservoir filling number of subsidence (ponors) occurred inside the reservoir. The additional loam blanket and suspended silt streams at leakage area provided to be ineffective. As consequence only lower part of the reservoir can be filled.

May Reservoir (Turkey). The rock-fill May Dam is 28 m high (flood control and irrigation). The reservoir bottom is covered with alluvium (15–20 m thick) deposited over karstified limestone, conglomerate and marl. As consequence of reservoir filling number of subsidence (ponors) were created at the reservoir bottom as well as along the right bank close to the dam. Reservoir is never completely filled with water.

Beni Haroun in Algeria is 107 m high RCC dam, founded on karstified limestone. At a head of water of about 90 m leakage close to 1 m^3/s occurs. In spite of extensive investigations and analysis position of concentrated underground flows was not exactly detected. Leakage water, most probably, flows through diffuse karst porosity. Extensive grouting was only partially successful.

Kalecik Dam (Turkey) is 77 m high. The right bank of dam foundation consists partially of karstified Paleocene limestone. The grout curtain on the right abutment, 200 m long and 60 m deep, was not enough successful to prevent leakage towards the downstream springs. In spite of three phases of intensive remedial works, including new deep grouting, the seepage from reservoir is still not eliminated (Turkmen 2003).

Box 13.5: Re-design of anti-seepage structures

Salman Farsi (Iran). The concrete arch-gravity dam is high 125 m above the foundation. The dam foundation area is composed of heavily karstified Asmari limestone. Due to huge cavern at the grout curtain route (cavern length 130 m, height more than 70 m and width 15–25 m) including number of secondary caverns the original curtain route has been re-designed. The new upstream bypass route was selected as technically and economically most feasible solution (Fig. 13.6). For plugging of 6 largest karst systems at grout curtain route 3,125 m^3 of self-compacting concrete were used (Dolder et al. 2009 Salm an Farsi Dam project: final report phase 3. Poyry Energy Ltd, Zürich).

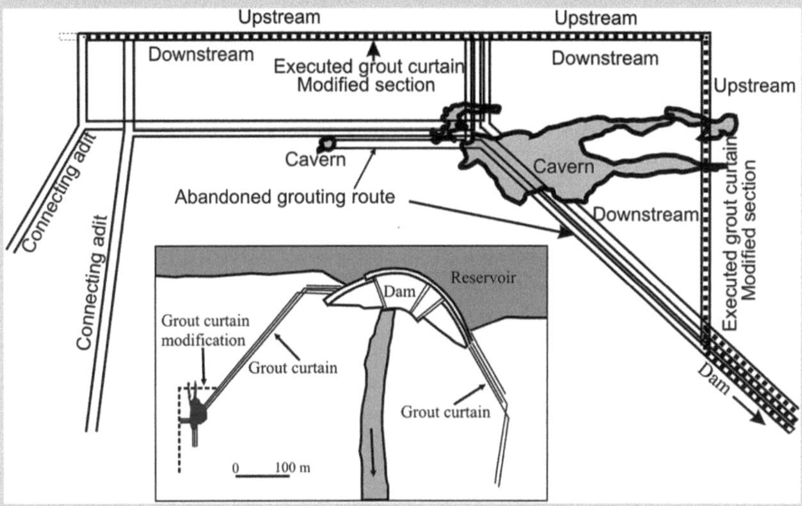

Fig. 13.6 Salman Farsi, Iran. Modification of grout curtain alignment due to presence of large caverns along the grout curtain

Freeman Dam (Kentucky, USA). Originally it was clay embankment dam, 15.25 m high, constructed in the mid 1960s. Significant seepage through the large solution features below the dam was reported 1997. The entire dam was removed and a positive cutoff wall, rock treatment, and filters were included in the re-designed structure.

Sklope Dam (Croatia) is rock-fill structure, 74 m high. The dam foundation consists of stratified and extremely karstified Cretaceous limestone. During construction of grout curtain large caverns were discovered at both abutments. The volume of cavern inside the left bank was about 25,000 m^3.

After detail speleological investigations the grout curtain route has been re-designed. The new route made a detour around the left bank cavern from downstream side and around cavern in the right bank from the upstream side. Leakage from reservoir (at full reservoir) is estimated of about 1 m^3/s.

Seymareh Dam (Iran) is concrete arch dam 135 m high (from foundation 185 m high) situated at the karstified limestone. During grout curtain construction, due to findings in underground, the curtain was improved (redesigned, direction and length). Problem appears, also, during excavation of the Power Plant foundation pit, situated inside the large spring zone. Excavation of the Power Plant access tunnel was in trouble due to enormously high groundwater intrusion (from 1.2 m^3/s in dry period up to 2.0 m^3/s in wet period of year). Because of that the downstream section of Head Race Tunnel and the Power Plant foundation pit were re-designed, and particular anti-filtration treatment has been applied.

Kavar Dam (Iran) is 60 m high rock-fill irrigation dam. Dam is located in marl (bottom and left bank), and limestone (right bank) formations. After many years of investigations project was considerably re-designed, particularly technology of reservoir watertightness (watertight surface blanket at the right bank close to the dam site).

Clarence Cannon Dam (Missouri, USA), 46 m high (concrete/earth fill) is located at karstified Carboniferous limestone. After a few large cavities were additionally discovered the modification of originally consolidation and grout curtain design has been done.

Chichik Dam (Uzbekistan), 168 m earth fill structure is located in the karstified Lower Carbon limestone. Large cavities are detected at depth below 100 from the river bottom. Karstification process occurred at various stages of karst evolution process. Deepest parts are karstified by thermal water. During the first filling of reservoir the grout curtain structure was modified, i.e. adapted to existing hydrogeological situation by extension and locally deepen (Kagan and Krivonogova 1999).

Darwin Dam (Tasmania), 21 m high, gravel fill embankment is situated at gravels, sandstone and silts overlying karstified limestone. After detail investigations and project re-design (dam axis is shifted) the Darwin Dam has been constructed (Guidici 1999).

Akkopru Dam (Turkey) is embankment structure, 110.5 m high. The small part of reservoir (2 km upstream from the dam site) is in direct contact with karstified limestone (Fig. 13.7). A few ten of large diameter karst shafts were discovered at an area of 250,000 m^2. Groundwater level is very deep, 100–116 m below the reservoir bottom (close to the sea level). To prevent leakage a thick reinforced concrete slab is constructed at the reservoir bottom and left bank (design by DSI). To prevent leakage below the concrete slab a cut of wall (overlapped concrete piles, 800 m long and 40 m deep) has been constructed.

Fig. 13.7 Akkopru reservoir, Turkey. Construction of heavy reinforced concrete slab to prevent leakage through the number of large vertical karst shafts. *Photos* P. Milanović

Box 13.6: Dams and reservoirs with (acceptable?) seepage

Kowsar (Tang-e-Duk) Dam (Iran). Arch concrete dam, high 125 m above the river level. Asmari limestone at the dam site is selectively karstification prone rock mass. Reservoir operates with constant leakage, of approximately 3 m^3/s. The seepage water flows through the karstified zone at the right bank and discharges at the large spring (right bank) close downstream from the dam site.

Karun I Dam (Iran) is a double-arch concrete dam, 200 m high. Asmari limestone at the dam site is well karstified. Leakage through the most critical clay-filled fault zone in the right bank is prevented by a concrete cut-off wall, 100 m high and 30 m wide. However, discharge of Big Spring at right abutment close to the powerhouse increased from 3 to 15 m^3/s in natural conditions to 10–16 m^3/s after reservoir filling and later during reservoir operation.

Buško Blato Reservoir (Bosnia and Herzegovina). The Buško Blato Reservoir is locate in the extremely karstified Livanjsko polje (Cretaceous limestone). Reservoir surface is 57 km^2, volume 800 hm^3 and average depth

14 m. To prevent leakage through the number of ponors and estavelles, with swallowing capacity of 35 m^3/s, two embankment dams (length 3,000 + 600 m) and 6,206 m long grout curtain was constructed along the perimeter of reservoir. Large karst channels are plugged by appluing vertical shafts from surface. The loss of water from reservoir (seepage + evaporation) ranges between 2 and 6 m^3/s. Average annual inflow into the reservoir is 11 m^3/s (Nikolić et al. 1976).

Camarassa Dam (Spain). Dam is gravity-arch structure, 92 m high constracted between 1927 and 1931. Foundation consists of karstified Jurassic dolomites and Cretaceous limestone. During first reservoir filling seepage increased from 3 to 11.26 m^3/s. Grout curtain (long 1,400 m), locally deep between 112 and 394 m down to the impervious formation. After a few steps of additional regrouting the leakage was reduced to 2.6 m^3/s.

Špilje Dam (FYR Macedonia) is an earth-filled dam, high 101 m. During the filling of the reservoir (1969/1970) the seepage of about 2 m^3/s has been registered immediately downstream of the dam. Discharge of the largest single spring was 350 l/s. The deepest underground flows are estimated at depth of 250 m below the dam body. According to long time monitoring these deep flows do not produce progressive erosion. Possibility of hypogene karstification can not be excluded.

Ataturk Dam (Turkey) is the fill dam, 179 m high, with crest length of 1,800 m. Reservoir volume is 49 billion m^3. Dam foundation is composed of karstified limestone and dolomitic limestone, extending to a depth of about 600 m. Size of the karst conduites ranges from 0.1 to 1.5 m including a few caverns of 1,000 m^3. Grout curtain is 5.5 km long and to 300 m deep. During reservoir filling (1990) started leakage through the dam foundation and discharge from grouting galleries. Seepage was 2.5 m^3/s for low reservoir level and about 14 m^3/s for reservoir level 15 m lower than operational level. Seepage losses are torelable because an average flow of Euphrates River is 850 m^3/s, in average (Riemer et al. 1997).

Karun III Dam (Iran) is 200 m high concrete arch dam. The dam site is composed of slightly karstified Asmari limestone formation. From reservoir the seepage occurs along the near vertical bedding planes through the right reservoir bank to the Karun River downstream from the dam at distance of about 2 km (Dehzir valley area). This seepage, between 0.5 and 2.1 m^3/s, is declared as acceptable. In the same time discharge of the downstream Abolghasem Spring increased from 0.3 m^3/s to, roughly, 1.0 m^3/s. The water collected in the drainage galleries reach 1.2 m^3/s.

Alakir Dam (Turkey) is constructed at karstified limestone. Because the role of dam is flood control protective measures were not necessary.

Mornos Dam (Greece) is 135 m high rock-fill structure. The groundwater level is about 200 m below storage level. The seepage protection measures (rockfill structure along the reservoir bank and asphalt blanket at lower part of reservoir flank) are not fully successful. Some leakage (≈0.5 m^3/s) still exists (Pantzarzis et al. 1993).

15 Khordad Dam (Iran). At the right reservoir bank the karstified limestone is in direct contact with pervious alluvial deposites (natural collapses were detected at that area). Filtration occurs through the wide area beneath a lateral dike founded on alluvium. Role of dike is to prevent overflow from reservoir. Part of reservoir is in touch with evaporites (gypsum) provoking an additional water quality problem.

Gorica Dam (Herzegovina) is 35.5 m high concrete gravity dam built at karstified limestone. Intentionally grout curtain was constructed partially (below the dam only). In the right bank the grout curtain was not constructed. The reason was to make possible by-pass filtration through the right bank and to keep downstream constant flow. From stability point of view the dam was not endangered because of groundwater filtration. Due to process of progressive washing of clayey deposits from caverns and joints the seepage increased gradually: from 1.5 m^3/s in 1965, and 3.0 m^3/s in 1983, up to 4.0 m^3/s in 2003.

Piva (Mratinje) Dam (Montenegro) is the concrete arch dam 220 m high. Reservoir volume is 790 million m^3. The dam site is heavily karstified including cavern 100 × 100 m filled with partially lithify clay. Depth of the three row grout curtain below the river bed is 250 m. Depending on reservoir level leakage vary from 1.0 to 2.0 m^3/s. This leakage is declared as acceptable.

Karun IV Dam (Iran) is the double curvatore arch dam 230 m high situated at the karstified Asmari limestone. Caverns were detected at different levels within the left bank. Some of them are detected at elevation lover than the river bed. After reservoir filling the seepage flow 0.6 m^3/s occur through the left bank.

Box 13.7: Abandoned dams or with limited use

Montejaque Dam (Andalusia, Spain). The concrete arch dam (72 m high), was constructed at the Gaduares River in 1920s. The dam site is located in karstified limestone in the front of the entrance (cave opening) of 8 km long Hundindero-Gato underground karst system. Due to extremely karstified limestone at the dam site and upstream banks the reservoir was never filled in spite 1942 tons of cement used for grouting. Due to an inacceptable leakage of about 4 m^3/s, the Montejaque Dam was abandoned in the 1950s (Fig. 13.8).

Fig. 13.8 Montejaque Dam, Spain. Abandoned dam due to huge leakage through the heavily karstified dam site and reservoir banks. *Photo* P. Milanović

Hales Bar Dam (Tennessee, US), 34 m high and 706 m long. The dam was built in period 1905–1913. Seepage started during first impounding. Remedial works takes till 1944 using cement and hot asphalt however without required success. Dam was abandoned (1968) and replaced by building Nickajack Dam 9.7 km downstream.

Wolf Creek Dam (Kentucky, USA) consists of combined earth and concrete gravity structure, 1,748 m long and 79 m high (Flagg 1979). The dam construction takes from 1941 till 1951. After 17 years of operation several ponors (subsidences) were formed in the reservoir and seepage water discharges downstream area of the dam. Collapses are developed at the toe of the earth embankment. Over 300 piezometers have been installed to monitor the seepage since late 1960s. Depth of karstification is estimated 80–100 m beneath the dam body. The remediation program (deep positive cut-off wall) was still in progress in the year 2011.

Liverovići Dam (Montenegro) is situated at Triassic dolomites and limestone. The concrete arch dam, high 45.5 m, was constructed at 1955/1957. The grout curtain was constructed along the both sides of river (reservoir) banks at length of 714 m. A few detected ponors were grouted also, however, not all. Immediately after impounding the new large ponor has been opened at the right bank of reservoir, and leakage between 0.5 and 2.0 m^3/s has been detected. Underground leakage flows are very deep. Additional remedial works were never done.

Apa Reservoir (Turkey) is used for irrigation. The reservoir leaks water into the limestone which covers the upstream portion of reservoir. About 20–25 % of the total annual inflow into the reservoir leaks into the karstified limestone.

Vrtac Reservoir (Montenegro). Height of the earth-filled dam is 16.5 m. Watertightness of Vrtac Reservoir is based on the concept to isolate main ponors and estavelles by cylindrical dams; to construct non-return valves at estavelles; and to cover karstified limestone banks with concrete blankets. After the first filling more than a hundred new ponors were formed and new leakage at a rate of about 2.0 m^3/s occurred (Fig. 13.9). Additional solution was grout curtain. Along 400 m of grout curtain (depth 30–130 m) 32 caverns were discovered. Finally Vrtac Reservoir is abandoned as reservoir and has been redesigned for temporary retention space, only.

Fig. 13.9 Vrtac reservoir, Montenegro. New estavelles developed during first reservoir filling. *Upper left* one way valve constructed at estavelle opening. *Photos* P. Milanović

Abandoned dams, or with limited use are also: **Lone Pine** (USA) 31 m high; **Cevizli** (Turkey) 21 m high; and **Onac** dams also in Turkey.

Box 13.8: Reservoir failures after many years of operation

Mavrovo Reservoir (FYR Macedonia) is partly situated in karstified marblised limestone. At time of first filling (1960), two large and a few of small collapsing ponors provoking seepage of about 9–12 m^3/s. Seepage zone was successfully protected with impervious blanket. After 25 years of operation several new subsidence (swallow holes) were formed during 10 days only (Fig. 13.10). Remedial solution was to isolate porous reservoir bank (end section of the reservoir) by construction of shallow dike.

Fig. 13.10 Mavrovo reservoir, FYR Macedonia. Collapses created after many years of reservoir operation. *Photos* P. Milanović

Hammam Grouz (Algeria) is concrete gravity dam, high (from foundation) 49.5 m. Foundation is limestone karstified by influence of hydrothermal water. To achieve necessary watertightness of the reservoir bottom an overlying protective impervious clay blanket (thick 1.8 m) and reinforced shotcrete blanket (thick 0.17 m) at limestone banks were constructed. At 2003, after 17 years of operation, the seepage from reservoir has been detected. Seepage gradually increased and reservoir dried up completely in 2007 (Fig. 13.11). Leakage place was treated by structure which consists of channel plugging, reinforced slab at bottom over the plug and compacted blanket.

Fig. 13.11 Hammam Grouz reservoir, Algeria. Collaps at reservoir bank occurred after 17 years of operation. *Photo* P. Milanović

Box 13.9: Abandoned sites or temporary frozen projects after complex geological investigations

In few cases, after many years of detail geological investigations, some dam projects situated at karstified rocks were abandoned. At those cases it has been estimated that necessary sealing works are too expensive and expected results too questionable, and consequently the entire project too risky. Construction of dams and reservoirs are abandoned or temporary frozen in the cases of: **Steno-Calaritico Project** (Greece); **Magnum dam site** (Oklahoma, USA); **Taka Lake** (Peloponnesus, Greece); **Cernica Reservoir** (Herzegovina); **Behestabad Dam and Tunnel, Sazbon Dam, Paalam Site, Kuhrang 3 Dam** (Iran); and **Subansiri Dam Site** (India). **Bogovina Dam** project (Serbia) is temporary frozen due to possible seepage flows from the reservoir area in three different directions (Stevanović 2010).

Due to intensive karstification of the dam site and left reservoir abutment the **Havasan** dam site, in Iran, was abandoned and dam site shifted a few kilometers upstream.

The proposed **Upper Magnum** dam site (Oklahoma, USA), for a 33 m high earth-filled dam, has been investigated from 1937. Foundation rocks, consists of 60 m of gypsum, with thin interbeds of dolomite and shale. Finally, in 1999 the project was abandoned (Johnson 2003).

Lower Gordon Dam (Tasmania) would have flooded a large karst area containing caves of great archeological importance. The project was abandoned for environmental and legal reasons (Kiernan 1988).

Box 13.10: Dam and reservoir problems due to the very soluble and karstified evaporites

Worldwide more than 60 dams and reservoirs need rehabilitation because of problems related to the solubility of evaporites. A number of cases are reported in USA, China, Russia, Iran, Iraq, Argentina, Guatemala, Switzerland, Peru, Venezuela and some other countries.

According Cooper and Calow (1998) and Johnson (2004) several dams in the USA were seriously endangered or were abandoned because of solution problem in gypsum: **Upper Magnum Dam** (Oklahoma, abandoned before construction); **McMillan, Avalon**, and **Rio Hondo** dams in New Mexico; **San Fernando, Dry Canyon, Buena Vista, Olive Hills, Rattlesnake** and **Castaic** dams in California; **Stanford Dam** (Texas); **Quail Creek Dike** (Utah, catastrophic failure 1989); **Red Rock Dam** (Iowa); **Fontanelle Dam** (Oklahoma); and **Moses Saunders Tower Dam** (New York).

Dams with problems or concern because of gypsum dissolution are also: **El Isiro Dam** (Venezuela); **Poechos Dam** (Peru); **San Loran, Estremera** and **Alloz Dams** (Spain); **Hessigheim Dam**, (Germany); **Casa de Piedra Dam** (Argentina); **Puebla de Pava** (head race tunnel, Guatemala; **Tbiliskaya Dam** (Georgia); **Erevanskaya Dam** (Armenia); **Caspe Dam** (Spain); **Lower Kafirniganskiy Dam** (Tajikistan); **Chardarinskaya Dam** on Syr-Daria River (Central Asia); **Baipazinskaya Dam** (Tajikistan); **Yangmazhain, Langdai, Baiyanjiao, Mahuangtian Reservoirs** (China); **Khersan III Dam** and **Tang-E-Shemiran Reservoir** and **Marun Reservoir** (Iran); **Hatra Dam** (Wadi Tharthar—Iraq); **Jarreh Dam** (Iran); and few others elsewhere in the world.

Box 13.11: Catastrophic failures

A few prominent case studies are as follow:

The worst American civil engineering failure of 20th century was failure of the **St. Francisco Dam** (California, USA) killing 450 people along the St. Francisco Canyon and St. Clara valley. Foundation rock of the concrete gravity dam (60 m high) was conglomerate with gypsum intercalation. At March 12/13 1928 the St. Francisco dam collapsed due to paleo-landslide at left abutment. According Cooper and Calow (1998) strong uplift was partly attributed to gypsum dissolution.

Catastrophic failure of the **San Juan** earth Dam (Spain) occurred during the first filling of reservoir (2001). Due to intensive dissolution of gypsum the part of dam collapsed provoking a huge flood at downstream area.

The earthquake $M = 6.6$ (February, 1971) caused failure of the **Lower San Fernando Dam** (California). After impounding the water seepage occurred through the gypsum intercalated between shale, siltstones and sandstones. Remedial grouting has been applied. To prevent possible consequences about 80,000 people were temporary evacuated.

The most complicated and the large scale problem is related to the **Mosul Dam** in Iraq (110 m high embankment dam, 3,600 m long). Dam foundation consists of well-bedded clayey and marly rocks, gypsum, anhydrite and limestone. Immediately after impounding started (1986) the leakage of 1.4 m^3/s was reported (Guzina et al. 1991). Massive dissolution intensity ranging from 42 to 80 tons/day was measured. To replace the volume of dissolved evaporates by grout mix injection continue more than 20 years but without success. In period of reservoir operation some sinkholes were opened and (after filling) reopened due to solution process and reservoir fluctuation. Finally as only solution the deep (down to 200 m) cutoff wall is recommended.

Bratsk Dam (Russia) is earth-filled dam high 125 m. Bratsk Reservoir is situated in the heavily karstified area with prevailing gypsum-anhydrite rocks. During reservoir filling (1963–1966) about 200 collapses/km^2 has been developed in the reservoir area.

Kamskaya Dam (Russia) was constructed in 1954. Dam height is 21 m, and length 2.5 km. The dam site and reservoir consist of argillites, sandstone, gypsum, limestone, dolomites and anhydrites. After filling of the reservoir seepage started through the foundation, and collapses in the vicinity of reservoir occurred. By using oxalaluminosilicat solution as additive successful results were reported by Maximovich (2006).

Huoshipo Reservoir (China) is constructed on karstified limestone including 48 layers of gypsum (Wuzhou 1988). The height of this earth-filled dam is 23 m. The leakage of 237 l/s occurred in the initial stage of the reservoir impounding. A number of sinkholes have been opened at the reservoir bottom. Because the reservoir capacity is only 4.7 million m^3 reported the leakage was declared as very high. By sealing works (grout curtain, plastic membranes, clay cover) seepage was reduced to $\approx$80 l/s, however, the solution process is not eliminated.

McMillan Dam (USA) has been constructed in 1893. The leakage problem was observed after 12 years of operation. The reservoir dried up through cavities developed in gypsum in the left abutment.

Horsetooth Reservoir and Carter Lake Reservoir (Colorado, USA, 1940) are situated in formation which consists of interbedded siltstone, shale, limestone and karstified gypsum. In period 1980–1990 sinkholes are formed in reservoir and considerable seepage occurred.

Gotvand Dam (Iran) is 178 m high rock-fill structure. Dam is located in Bakhtiary conglomerate. About three kilometers upstream from dam the reservoir is in touch with gypsum/salt diapiric structure. Length of this part of reservoir bank is approximately 2 km (Fig. 13.12). Due to very fast dissolution of salt two problems appears as crucial: stability of reservoir bank and water pollution. Because this water has to be used for irrigation the large chloride content in water is problematic.

Anchor Dam (Wyoming, USA) is 60 m high arch dam situated at karstified limestone/gypsum formation (1960). As consequence "only a small amount of water has been held in reservoir" (Johnson 2004).

Fig. 13.12 Gotvand reservoir, Iran. Left reservoir bank (salt). *Small photo*—development of sinkhole in salt. *Photos* P. Milanović

Kavsak Dam Site (Turkey) is located in the Permian karstified limestone. Cavities are developed along the fault zones (tectonic breccia) or along the subvertical joints. Additional problem is the presence of gypsum (in some boreholes 30 m thick). More than 30 years of intensive investigations and analysis has been done before dam construction (88 m high) started.

Salt is present in the foundation rock of **Nurekskaya** and **Rogunskaya** dams in Tajikistan. At foundation of the world highest **Rogunskaya Dam** (335 m, earth filled) in Tajikistan, the thick salt layer was treated by a combination of grout and salt curtains.

13.3 Underground Dams

Hundreds of underground dams have been built on the subterranean streams in south China. According Yuan (1990) these dams may be classified: for storing water at surface; for storing water in underground and to divert water in different directions for irrigation or power production. Some of these dams are: **Yidong** (Guangxi); **Yuzhai** (Guizhou); **Longwangdong** (Sichuan); and **Jijiao** (Guangxi).

Wulichong Underground Dam (China) is one the of largest underground structures. The dam is constructed inside the extremely karstified Triassic limestone at the confluence of two underground rivers. The underground dam consists of the grout curtain, long 1,333 m with maximal depth of 260 m and surface of 262,000 m^2. To block the cavern with underground flow the concrete plug was constructed. The plug is 33.46 m high, 13.9 m wide, and 2–10 m thick. The reservoir was successfully impounded and this project operates with full capacity (Kang and Zhang 2002).

Linlangdong underground dam (China, Yunnan, 1955/60). Artificial storage in karstic channel system was realized by constructing the 15 m high dam. The annual average flow of Linlangdong ground river is 23.8 m^3/s.

Siziguan underground dam (Hubei province, 2004) constructed on the underground Zongjian River. Dam is high 192 m and wide 16 m. The total volume of reservoir with a major part on the surface gorge is 141 million m^3 (personal communication Yuan Daoxian).

Among the number of underground reservoirs in China the most prominent are: **Yuhong** (Hunan Province), concrete plug 10.5 m high, 7 m wide and 3 m thick; **Beilou** (Guangxi Province), 24 m high underground dam; and **Shuanglung underground dam** (Chekiang Province). At some cases by plugging underground channels are made reservoirs partially at underground and partially at surface: **Fengfa** (Guizhou Province) and **Fu Liulang** (Guangxi Province, Fig. 13.13).

Fig. 13.13 Fu Liulang reservoir, Guangxi Province, China. Surface reservoir established by construction of underground dam. *Upper left* simplified cross-section with position of underground dam. *Sketch and photos* P. Milanović

Sunagava and **Fukuzato underground dams** at Miyakojima Island (Japan) are successful underground structures (cut-off walls, 50 and 27 m deep) completed in 2001. About 20 Million m^3 of water is stored in limestone aquifer (underground reservoirs) and used for irrigation.

Port-Miou (France, Marseille) is submarine spring with deepest point of 147 m bsl at distance 2,200 m from the spring outlet. Spring discharge varies between 2.6 m^3/s and 45 m^3/s. The underground structure was constructed in the karst channel about 500 m from the subsea outlet, at depth 10–15 m bsl. The main purpose of underground dam is to reduce concentration of chlorides in the spring water (Potie et al. 2005).

Ombla (Croatia). The proposed underground dam is designed to raise the water within karstified limestone behind the large Ombla Spring to create an underground reservoir (Fig. 13.14). The Ombla spring discharge ranges between 3 and 150 m^3/s at elevation of sea level. The main channel of Ombla Spring has a form of deep siphon with depth of 130–150 m below sea level. Underground dam (grout curtain) will stretch from 280 m below sea level to 130 m above sea level. Length of upper contour of grout curtain will be approximately 1,500 m.

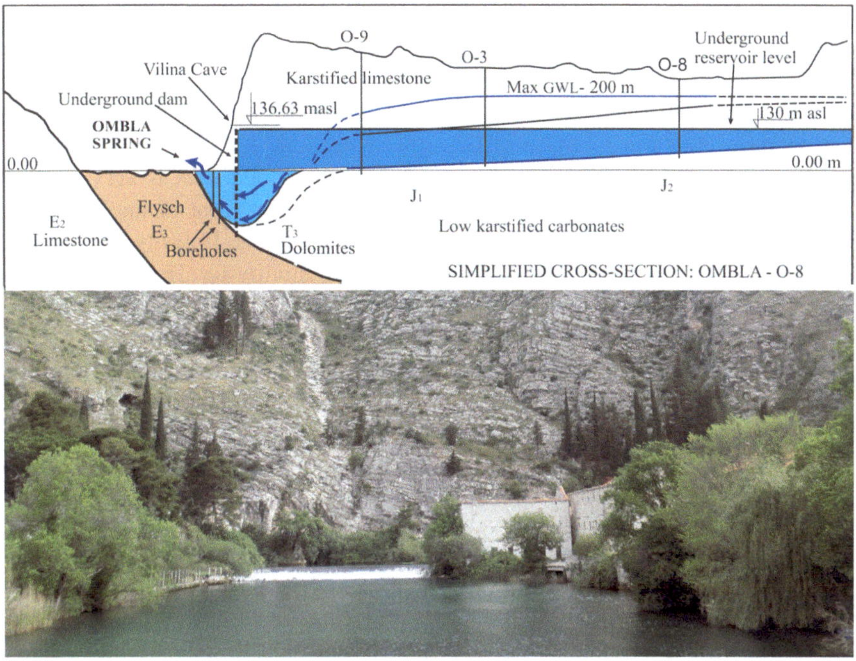

Fig. 13.14 Ombla underground dam, Croatia (general idea). Simplified cross-section along the underground reservoir (*upper*), and Ombla Spring (*lower*). *Sketch and photo* P. Milanović

13.4 Tunneling in Karst

From the ancient time (3,000 years ago) galleries have been used as successful tapping structures. Thousands of galleries (khanats) in Iran, Iraq and other Middle East countries were excavated to improve water supply and irrigation conditions. Some of them are excavated in karstified limestone. The **Siloam Tunnel** in Jerusalem (533 m long) is one of the oldest engineering structures built for the karst spring tapping (Ghion Spring, mentioned in Bible), dated about 700 BC, (Frumkin and Shirmon 2006). The first undersea tunnel was the **Severn Railway Tunnel**, cut in the 1860s in thick to massive bedded limestone beneath the Severn River estuary of Britain (Ford and Williams 2007).

Tunnel in karst appear to be the most vulnerable structure, does not matter is it traffic tunnel, headrace tunnel (high-pressure tunnel), water transmission tunnel or dewatering tunnel. During underground excavation in karst the crucial problems result from the presence of caverns and large inflow of underground water. The presence of karst channels and caverns at the tunnel route is unpredictable. There are many examples where the excavation progress was delayed because of a cavern at the tunnel route or due to massive groundwater burst into the tunnel.

Particular problems may be encountered during the excavation of tunnels below the water table. Defects during the tunnel operation are very common in karst, also. During all life (operational time) tunnels in karst must be subject of permanent and severe maintenance and frequently are subject of complicated remediation works.

Usage of a tunnel-boring machine (TBM) in karstified rock is very sensitive and requires special care. Every cavern or fault zone in front of the TBM head means excavation delay of a few days up to a few months. If cavity is filled with plastic clay efficiency of TBM is very low and possibility for TBM head sinking is very high. The incident in 1982 in Guatemala ended with the loss of the entire TBM.

Some of tunnels with karst related problems are: **Giona** (Greece); **Steinbühl** (Germany,); **Vue-des-Alpes, Lopper, Lötschberg, Engelberg** (Switzerland); **Učka, Vrata** (Croatia); **Montelungo** (Italy); and **Kastelec** (Slovenia). According Han (2010) some case-studies with water gushing and cavernosity during tunneling in Chinese karst are: **Yesanguan Tunnel** (3.69 km), **Baziling Tunnel** (3.55 km), **Zujiayan Tunnel** (1.32 km), **Biandanya Tunnel** (3.36 km), **Jiahuoyan Tunnel** (5.23 km), **Wuzchiba Tunnel** (6.69 km), **Bashuisi Tunnel** (1.38 km), **Yanwan Tunnel** (2.27 km), **Hanpeling Tunnel** (1.80 km) and **Qiyueshan Tunnel** (4.09 km). The **Long Tan Tunnel** (8.69 km) is situated at area of the Tanchunguan underground river. More than 770 m of tunnel is located in the huge cavern filled with clay and gravel deposits. The excavation of **Wanjiazhai Yellow River diversion tunnel** (China, four tunnels of total length 88.7 km) has been affected by cavities filled with soft and sticky clay. The headrace tunnel **Plat—PP Dubrovnik** (Bosnia and Herzegovina—Croatia, 16.5 km) crossing twelve karstified zones. In spite of the tunnel is situated above the maximal groundwater level, after heavy rain (more than 100 mm/12 h), the tunnel lining is frequently demolished due to strong concentrated water pressure from outside. Remedial works, including grouting, are necessary every 5–7 years.

The large inflow of underground water considerably retarded excavation of a number of tunnels: headrace tunnel **Zakučac** (Croatia, 6.85 km, 2 m^3/s), transmission tunnel **Dabar—Fatnica** length 3.24 km between two temporary flooded karst poljes (Bosnia and Herzegovina); and traffic tunnel **Sozina,** length 6.17 km (Montenegro) was flooded by an sudden groundwater intrusion at a rate of 6.5 m^3/s. Huge water inflow and floods made enormous problem in the case of following tunnels in karst: in Switzerland—**Twann** (up to 4,000 l/s), **Sauges** (up to 1,000 l/s), **Flims** (800 l/s, Jeannin et al. 2007); **Karawanken tunnel** between Slovenia and Austria; and **Gothard base tunnel** (Switzerland).

The problems related to the collapses at surface occurred above the **Čapljina** headrace tunnel and **Bileća—Fatnica** conveyance tunnel (both in Bosnia and Herzegovina). According Marinos (2005), above the **Dodoni tunnel** (Greece, 3.3 km) two major collapses occurred and produce sinkholes at the surface with outcropping chimneys almost 100 m of height.

Caverns, empty or filled, may have dimensions and features that represent a problem for which technical solutions may be proved to be questionable.

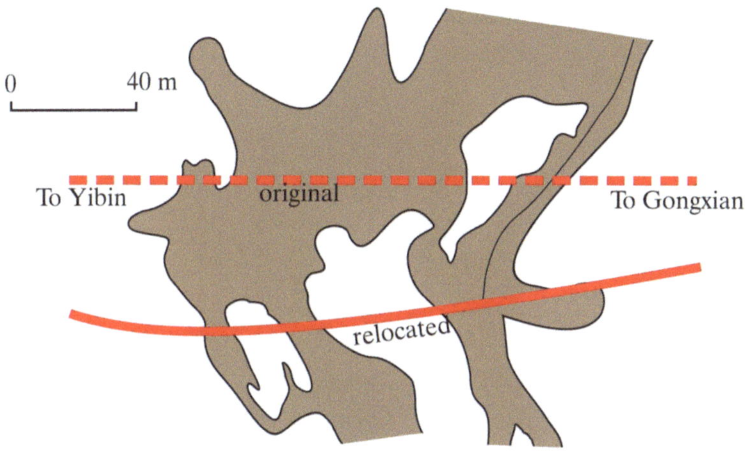

Fig. 13.15 China. Relocation of railroad tunnel because of Jiaodinshan Grand Cavern at the road Yibin—Gongxian, (Guoliang 1994)

In certain cases, the only viable solution would be to change the tunnel route locally in order to by-pass a cavern: **Nanning—Kunming Railway Tunnel** and **Jiaodingshan Railroad Tunnel** (China), **B. Bašta Headrace Tunnel** (Serbia), **Čapljina Headrace Tunnel** (8 km, Bosnia and Herzegovina; Milanović 1997), and **Ybing—Gonxian Railroad Tunnel** (China; Guoliang 1994) (Fig. 13.15).

Transmission tunnel **Fatnica—Bileća** (Bosnia and Herzegovina), length 15.6 km, (Fig. 13.16) located between the temporarily flooded Fatnica karst polje and one of the largest spring zones in Dinaric karst (Trebišnjica Springs) is one of the technically most complicated tunnels excavated in karstified rocks (Milanović 1997). Tunnel is situated in the largest underground flow zone with capacity of more than 300 m^3/s in the rainy period. This water provoked floods at the mid section of already excavated tunnel. The tunnel's inlet, situated at the sinking zone with capacity of 100 m^3/s, is more than 200 days per year submerged by flood water. A few ten of caverns, mostly filled with clay, considerably delayed excavation. Passing through the eight caverns at one section, of 70 m only, require complicated and time consuming manipulation. After heavy rain (more than 100 mm/24 h) the groundwater rises above the upstream section of the tunnel. During excavation period this section was flooded 120 days average/annually.

Nowsoud Conveyance Tunnel (Iran), 50 km long, (two Sects. 24.2 + 25.6 km) for irrigation of Khouzestan area. Karst features are developed along joint intersection with beddings (Ilam limestone). Water table in natural conditions is 100 m above the tunnel route. As a consequence of drainage role of the tunnel (inflow of about 200 l/s) some springs in vicinity of tunnel dried up.

Kouhrang III Tunnel (Iran), 23 km long, is one of technically most complicated tunnels excavated in the karstified rock. Purpose of tunnel is water transfer from Karun River to the Zayandeh River at elevation of 2,200 m. More than 70 %

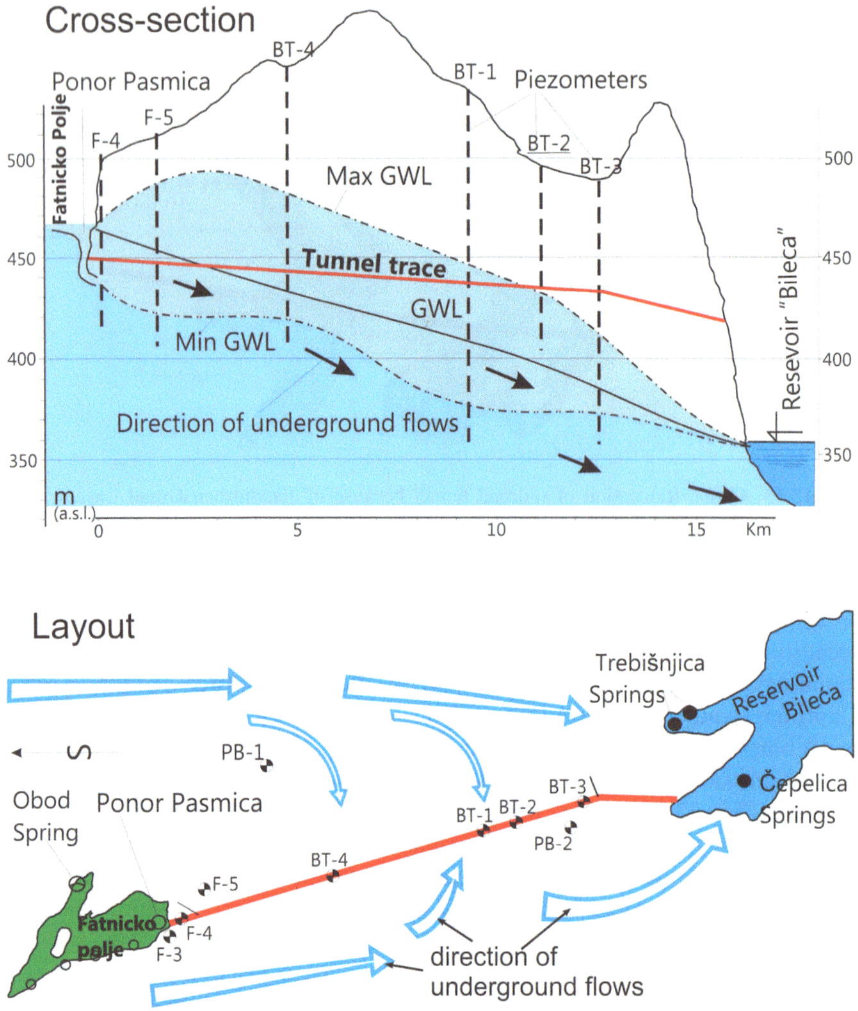

Fig. 13.16 Tunnel Fatničko Polje—reservoir Bileća

of tunnel route is situated at the intensively karstified limestone and dolomite. Two connection adits were flooded a few times by groundwater inflow of more than 1,000 l/s. The huge groundwater inflow and floods happened many times during excavation of the main tunnel section. For instance only at chainage 5 + 605 (overburden—1,100 m, water pressure 26 bars) a large water inflow up to 1,167 l/s washed from caverns into the tunnel more than 1,000 m^3 of boulders, gravel, sand and silty material plus about 500 tons of suspended material for 24 h, only. This area was stabilized by a 7 m thick concrete plug and massive grouting in front of TBM head with the total injected grout of 446 tons of dry cement (Fig. 13.17).

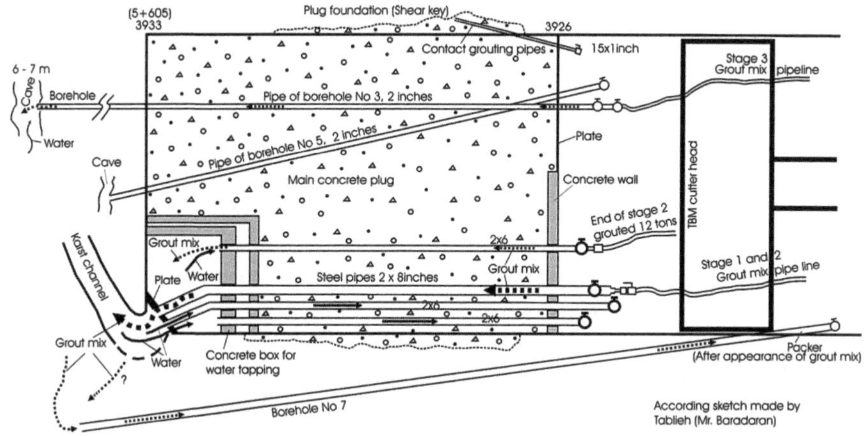

Fig. 13.17 Kuhrang III Tunnel, Iran. Plugging technology of the massive water burst in front of TBM cutter head, at chainage 5 + 605. According Baradaran, modified

Similar situation happened at a few sections of tunnel. Due to enormous problems with karstified rock excavation time was more than doubled.

Commonly applied protective measures against water inflow are: lining and grouting; construction of massive concrete plugs; cone grouting ahead or excavation of drainage galleries. To resolve the flood problem of the **Sozina** traffic tunnel the drainage tunnel (1.75 km long) was excavated below the main tunnel.

The limited and time consuming investigations of the karstified rock in front of TBM head threat a high risk of missing important caverns with water before they discharge into the tunnel. Enough precise investigation method for detection of cavern in front of TBM head does not exists, so far. Still, the horizontal probe holes in front of the tunnel face are fairly well reliable.

13.5 Lessons Learned and Recommended Approach

The most prominent hydrogeological property of karstified rocks is *hydrogeological singularity*—the underground flows occur through the channels hardly predictable by investigations from surface. In karst the elemental hydrogeological volume (EHV) does not exist. It means application of the common hydrogeological rules and laws is limited. Because each karst region is unique, the nature of problem is different, and similar situation are seldom, if ever, repeated. Due to this fact simple classification of engineering problems and remedial methods in karst require lot of flexibility. A good geological maps; type of karstification (by epigenic or hypogenic karstification); depth of karstification (base of karstification); groundwater regime; relation between surface and groundwater and monitoring (precipitation, piezometery, spring discharge) are key starting point for competent hydrogeological analyze in karst.

In general, the narrow canyons are less risky places for dam sites and reservoirs comparing by wide valleys and karst poljes. Tunnels below the base of karstification are less risky than tunnels situated above the base of karstification. Hydraulically, saturated base flows are system under pressure (piston effect). Kinetic energy of concentrated underground flows is high. Seismicity triggered as consequence of karstification and its hydrogeological properties is one of important karst specificity, also (Lu and Duan 1997; Milanović 2004).

Deep groundwater levels beneath the dam sites and reservoirs, estavelles at reservoir bottoms, presence of evaporates, hypogene karstification, undiscovered caverns and karst channels at dam sites and along the tunnel routes, floods of underground structures and erosion by turbulent flows may cause extreme problems, even failures.

To prepare a comprehensive set of geological and hydrogeological data some unconventional and specific investigation methods have to be applied: tracer tests (dye, smoke and radioactive tracers); speleology (conventional and diving); radioactive logging; geothermal logging; TV logging; borehole radar; ground penetrating radar (LOZA Terravision); echo-sounding; and different geophysical methods adapted for application in karst (geoelectrical methods, particularly mise à la masse, geoelectrical sounding and different seismic methods including geo-bomb method).

The nature of karst presents a great variety of risks associated with any kind of construction at the surface or in underground. The risk in karst cannot be totally eliminated by increasing the investigation programs and time of investigation. However, it can be minimized to an acceptable level by applying different geotechnical remediation methods in underground or at surface.

To deal with construction of the large structures in karst successfully, understanding of geology and karst evolution process, innovations, engineering practice and execution feasibility are basic prerequisites. This basic objectives leads to a number of design and execution alternatives. Solution and alternatives should be developed following the findings of a thoughtful geological and hydrogeological diagnosis. Grout curtain (cut-off) is the most common underground remedial measure. However, in the case of high developed karst grouting alone is not adequate technology. Other measures may be applied. The common approaches for treatment of the large karst features (caverns, channels and wide tectonized-karstified zones) are as follows:

- plugging of individual caverns and channels by reinforced concrete plugs or by self compacting concrete;
- construction of positive cutoff walls (diaphragm walls—deep trenches or overlapped concrete piles filed with concrete or clay); and
- bat-tab grouting structures.

In the case of extremely large caverns at curtain or tunnel route the best solution is to by-pass the cavern by re-routing. The conical grouting ahead is only solution if the tunnel route is situated within entirely saturated karstified rock mass.

Horizontal and vertical extensions of the grout curtain are common in karstified rocks. The possible appearance of karstic features might call for special and time consuming treatment from adits which have to be excavated into a karstified area prior to start drilling and grouting. Because of that the vertical spacing between grouting galleries should not exceed 30 m. Extreme non-homogenity of the karstified rocks leads to great variability of grout mix consumption. Consumption higher than 100 kg/m, in average, has been registered at more than 80 % of curtains in karst. A stable grout mix (water/cement ratio 0.6–1.0) and a high grouting pressure (max. 40–50 bars) are required. Fast water flows, under high pressure, require special grouting technology (including preventers), materials (sand, mortar, foam, bitumen, large aggregate fraction) and chemical additives.

To reduce losses from reservoirs and river beds the following structures and technologies at the surface commonly being applied: surface compacting in alluvial overburden; different kind of watertight blankets (clay layers, different kind of geosintetics); sandwich type of blanket (clay and geomembranes); shotcrete (over bare rock); dental treatment; isolation of large ponor zones with dikes or cylindrical dams; closing estavelles with concrete plugs equipped with nonreturn valves and construction of grouting carpets. To prevent air-hammer or water-hammer effect (sudden and strong uplift) installation of drainage system beneath the blankets or aeration pipes connected with karst conduits are required.

Final design of a grout curtain can only be definitely defined during the implementation phase. The curtain in karst needs modifications and adaptations on the basis of the geological findings in underground. Modifications during construction of watertight structures or during tunnel excavation in karst are the rule and not the exception.

The final proof of the curtain's or tunnel's satisfactory water tightness will be after first reservoir impoundment or even later. Some re-grouting and plugging during the large structure operation is quite common in karst and each project needs to be prepared for such a work.

13.6 Explanation of Some Specific Terms

Air-hummer effect — Explosion of the captured air in the karst caverns and channels as consequence of fast water level rising.

Bat-tab structure — Grout curtain constructed of vertical, inclined, horizontal and sub-horizontal segments creating tab structure beneath dam foundation. Entire structure is, usually, connected with impervious geological formation.

Compaction blanket — Upper layer of alluvial deposits artificially compacted and (locally) blanketed with geosyntetic.

Cut-off wall (Diaphragm wall) — A watertight vertical underground wall constructed as cut-off trench or line of overlapped piles filled with clay puddle or concrete (reinforced or not).

Grout curtain (Cut-off)	A row of holes filled with grout mix at a high pressure until the fissures in the surrounding rock are all filled making an impervious barrier for water filtration.
Impervious blanket	Artificial impervious layer over very porous surface (consists of clay, impervious membrane or combination of different impervious layers).
Induced subsidence	Collapse of the ground surface due to human activities, mostly as consequence of reservoir operation and intensive pumping of groundwater.
Positive cut-off	Grout curtain tight into the impervious geological formation. Opposite hanged grout curtain—without contact with impervious formation.
Self Compacting Concrete	is concrete that does not require vibration for placing and compaction.
Shotcrete	Sprayed concrete enables construction of thin impervious lining over the porous karstified rock. To prevent micro-cracks, steel reinforcement meshes or different fiber types are being used.
Triggered seismicity	Seismic phenomena associated with impounding of reservoir, or by fast rising of groundwater level provoking explosions of trapped air in caverns and channels.
TBM	Tunnel Boring Machine.
Water-hammer effect	Strong, sudden and concentrated water pressure to the tunnel lining or to the watertight blanket (sudden up-lift) because the underground flow along the channel is cut.

References

Altug S (1999) Oymapinar arch dam, Turkey: foundation treatment in karstic limestone and reservoir curtains. International Commission on Large Dams (ICOLD), Antalya, pp 193–212

Altug S, Saticioglu Z (2001) Berke arch dam, Turkey: hydrogeology, karstification and treatment of limestone foundation. In: Gunay G, Johnson K, Ford D, Johnson IA (eds) Proceedings of the 6th international symposium and field seminar, Marmaris, Turkey. Technical documents in hydrology, vol I(49). UNESCO, Paris, pp 315–323

Bergado TD, Areepitak C, Prinzl F (1984) Foundation problems on karstic limestone formation in western Thailand: a case of Khao Laem Dam. In: Back BF (ed) Proceedings of the first multidisciplinary conference on Sinkholes, Orlando, pp 397–401

Bianchetti G, Roth P, Vuataz FD, Vergain J (1992) Deep groundwater circulation in the Alps: relations between water infiltration, induced seismicity and thermal springs. The case of Val d'Illiez, Wallis, Switzerland. Eclogae Geol Helv 85(2):291–305

Božović A, Budanur H, Nonveiller E, Pavlin B (1981) The Keban dam foundation on Karstified Limestone—a case study. Bull Int Assoc Eng Geol 24:45–48
Cooper AH, Calow RC (1998) Avoiding gypsum geohazards: guidance for planning and construction. British geological survey. Technical report WC/98/5 overseas geological series
Dawans P, Gandais M, Schneider TR, Waldmeyer JP (1993) Sealing of the Salanfe reservoir (Switzerland)—grout curtain. In: Widmann (ed) Grouting in rock and concrete, Balkema, pp 259–268
Flagg ChG (1979) Geological causes of dam incidents. Bull Int Assoc Eng Geol 20:196–201
Ford D, Williams P (2007) Karst hydrogeology and geomorphology. Wiley, Chichester
Frumkin A, Shirmon A (2006) Tunnel engineering in the Iron Age: geoarcheology of the Sioam Tunnel, Jerusalem. J Archeol Sci 33:227–237
Guidici S (1999) Darwin Dam design and behaviour of an embankment on karstic foundations. Int Comm Large Dams (ICOLD), Antalya, pp 619–698
Guifarro R, Flores J, Kreuzer H (1996) Francisco Morozan Dam, Honduras: the successful extension of a grout curtain in karstic limestone. Reprint from Int J Hydropower and Dams 3(5):1–6
Günay G, Arikan A, Bayari S, Ekmekci M (1985) Quantative determination of bank storage in reservoirs constructed in karst areas: case study of Oymopinar Dam. IAHS 161:179
Guoliang C (1994) Prediction of covered karst halls. 7th International IAEG congress, Balkema, Rotterdam, pp 1841–1845
Guzina B, Sarić M, Petrović N (1991) Seepage and dissolution at foundations of a dam during the first impounding of reservoir. Congress des Grandes Barrages, Q66, R78, Vienna, pp 1459–1475
Han X (2010) Prediction and engineering treatment of water gushing and caves for tunneling in karst. Guangxi Normal University Press, Guangxi
Jeannin PY, Hauselmann P, Widberger A (2007) Modellierung des einflusses des flimserstein-tunnel auf die karstquelle des Lag Tiert (Flims, GR). Bull Angew Geol 12(2):39–48
Johnson KS (2003) Gypsum karst and abandonment of the Upper Magnum damsite in southwestern Oklahoma. In: Johnson KS, Neal JT (eds) Evaporite karst and engineering/environmental problems in the United States, Oklahoma geological survey circular, p 109
Johnson KS (2004) Problems of dam construction in areas of gypsum in karst. In: Proceedings of the international symposium, Karstology—XXI century: theoretical and practical significance. Perm, pp 236–240
Kagan A, Krivonogova N (1999) Impact of engineering geological conditions on decision making during the design and construction of hydraulic structures in areas of karst development. In: Turfan M (ed) Dam foundations, problems and solutions. Int Comm Large Dams (ICOLD). Antalya, pp 673–680
Kang YR, Zhang BR (2002) Karst and engineering handling to the karst in Wulichong reservoir, Yunan Province. Carsologica Sinica, vol 21(2)
Kiernan K (1988) Human impacts and management responses in the karsts of Tasmania. Resources management in limestone landscape. Department of geography and oceanography, University College, The Australian Defense Force Academy, Cambera (Special publication no 2), pp 69–92
Kutepov VM, Parabuchev IA, Kalin YA (2004) Karst and its influence on territory development. In: Proceedings of the international symposium, Karstology—XXI century: theoretical and practical significance, Perm, pp 192–198
Lu Y, Duan G (1997) Artificially induced hydrogeological effects and their impact of karst of North and South China. In: Fei J, Krothe NC (eds) Proceedings of the 30th international geological congress, vol 22. Hydrogeology, VSP Utrecht, Tokio, pp 113–120
Lu Y (2012) Karst in China—a world of improbable peaks and wonderful caves. Ministry of Land and Resources and China Geological Survey, Beijing

Maximovich NG (2006) Safety of dams on soluble rock (The Kama hydroelectric power station as an example). In: The Russian Federal Agency for science and innovations. Federal state scientific institution "Institute of Natural Science", Perm (Russian language)
Marinos P (2005) Experiences in tunneling through karstic rocks. In: Stevanović Z, Milanović P (eds) Water resources and environmental problems in karst. Proceedings of international conference KARST 2005, University of Belgrade, Institute of Hydrogeology, Belgrade, pp 617–644
Milanović P (1997) Tunneling in karst: common engineering-geology problems. In: Marinos P, Koukis G, Tsiambos G, Stournaras G (eds) Engineering geology and the environment. Balkema, Roterdam
Milanović P (2000) Geological engineering in Karst—Dams, reservoirs, grouting, groundwater protection, water tapping, tunneling. Zebra Publishing Ltd, Belgrade
Milanović P (2004) Water resources engineering in karst. CRC Press, Boca Raton
Milanović S, Vasić LJ (2014) 3D modeling of karst conduit; case example Višegrad Dam. In: Kukurić N, Stevanović Z, Krešić N (eds) Proceedings: international conference and field seminar, Karst without boundaries, DIKTAS Project, Trebinje, pp 301–306
Nikolić R, Raljević B, Franić M, Zidar M (1976) Possibility for solving of underground areas gaps and cavities in the area of storage "Buško Jezero", In: Working papers of the 1st Yugoslav symposium for soil consolidation, JUSIK, Zagreb, pp 91–96
Okay G, Soidam B-A (1999) Experience on two karstic sites. Int Comm Large Dams (ICOLD), pp 709–722
Pantzartzis P, Emmanuilidis G, Krapp L, Milanović P (1993) Karst phenomena and dam construction in Greece. In: Gunay G, Johnson I, Back W (eds) Hydrogeological processes in Karst Terrains, IAHS 207:65–74
Potié L, Ricour J, Tardieu B (2005) Port-Miou and Bestouan freshwater submarine springs (Cassis—France) investigations and works (1964–1978). In: Stevanović Z, Milanović P (eds) Water resources and environmental problems in Karst. Proceedings of the International conference KARST 2005, University of Belgrade, Institute of Hydrogeology, Belgrade, pp 249–257
Riemer W, Gaward M, Sourbrier G, Turfan M (1997) The seepage at the Ataturk fill dam. In: Q73, R38, Comm Int Des Grandes Barrages (ICOLD), Florence, pp 613–633
Ruichun X, Fuzhang Z (2004) Karst geology and engineering treatment in the Geheyan project on the Qingjiang River, China. Eng Geol, pp 155–164
Stevanović Z (2010) Regulacija karstne izdani u okviru regionalnog vodoprivrednog sistema "Bogovina" (Management of karstic aquifer of regional water system "Bogovina", Eastern Serbia; in Serbian). University of Belgrade, Faculty of Mining and Geology, Belgrade
Stojić P (1966) Bearing capacity of abutment and improvement of stability of left slope of Grančarevo Dam. In: Proceedings of the VII congress of the Yugoslav national committee for large dams, Sarajevo, pp 177–184
Turkmen S (2003) Treatment of the seepage problems at the Kalacik Dam (Turkey). Eng Geol 68(3–4):156–169
Vlahović M (2005) Hydrogeological properties for construction of reservoirs in karst environment. Case study Nikšićko Polje. MS thesis, University of Belgrade—Faculty of Mining and Geology, Belgrade
Weyermann WJ (1977) The karstic rock mass of Canelles Dam. In: Rock conditions improved through pressure grouting, RODIO in collaboration with The Institute for Engineering Research, Zurich, pp 16–23
Wuzhou H (1988) A study on the formation of Triassic "Gypsum—dissolved—strata" in Guizhou Province and the seepage prevention for reservoirs. In: Proceedings of the IAH 21st Congress, Geological Publishing House, Beijing, pp 1117–1126

Yuan D (1990) The construction of underground dams on subterranean streams in South China Karst. Institute of Karst Geology, Guilin, pp 62–72
Yuan D (1991) Karst of China. Geological Publishing House, Beijing
Zoumei Z, Pinshow H (1986) Grouting of the karstic caves with clay fillings. In: Proceedings of the conference on grouting in geotechnical engineering. Published by the American Society of Civil Engineers, New Orleans, pp 92–104

Part III
Regulating and Protecting Karst Aquifer—Case Studies

Chapter 14
Managing Karst Aquifers—Conceptualizations, Solutions, Impacts

Zoran Stevanović

14.1 Introduction

The management of an invisible resource such as groundwater is not an easy task and managing karst aquifer is even more delicate and problematic. As often repeated in this book, this is due primarily to this aquifer's heterogeneity and unstable regime However, this task attracts many researchers worldwide, sometimes simply challenging them, but in many cases pressing them to find appropriate solutions because karstic aquifer might be the single available water source in a locale.

There are three major groups of problems when dealing with karstic aquifers:

1. Watering—Effective utilization of karstic water resources for drinking water supply and other purposes;
2. Dewatering—Mitigation of karst inflow in mining, urban areas, dams or other construction works;
3. Protecting—Prevention of pollution and remediation of karst water quality.

The contributions in this Guidebook are similarly grouped into the three chapters which deal with specific problems. Numerous experiences obtained from practical engineering works in karst are also included. But before presenting them, this introductory chapter provides an overview recommended as a common engineering approach to the definition of problems, the required conceptualization of and research into the aquifers, possible solutions and alternatives, and finally optimization, implementation and assessment of any environmental impact this engineering work may have.

Z. Stevanović (✉)
Centre for Karst Hydrogeology, Department of Hydrogeology,
Faculty of Mining and Geology, University of Belgrade, Belgrade, Serbia
e-mail: zstev_2000@yahoo.co.uk

Z. Stevanović (ed.), *Karst Aquifers – Characterization and Engineering*,
Professional Practice in Earth Sciences, DOI 10.1007/978-3-319-12850-4_14

14.2 Problem Definition and Research Procedure

Dealing with a problem usually includes the following three steps:

1. Problem detection and formulation;
2. Hypothesis for solving the problem (programme);
3. Work on the problem (research).

In engineering hydrogeological practice the problem is commonly defined by the investor[1] but when purely scientific, the problem can be defined by researchers themselves. For instance, most speleological surveys were undertaken by enthusiastic speleological groups, both professional and amateur, but collected information and data, if publically available, could be used in further practical engineering projects. Or specific thematic problems can be formulated by research groups and proposed for financing to the national or international institutions and funds.

First of all, the related literature should be consulted in order to formulate the problem and assess the opportunity for resolution through the investigation process. Local climate, hydrology, geological setting, and main groundwater occurrences should be evaluated in order to get a *conceptual idea* which through further survey will have to be transformed into a hydrogeological *conceptual model*.

Checking if that or a similar problem has already been elaborated and perhaps "solved" in other areas and what experiences have been collected is very important for proper formulation and orientation of the research.

Then, a pragmatic assessment of the capability to do the work and solve the problem should include evaluation of the following (Stevanović 1990):

- capability of human resources (professional capacity of the staff);
- available equipment and methods;
- ensured finance to cover expenses;
- sufficient time for work completion.

Once the problem is formulated the *research programme* or other technical documents such as the *research project proposal* (combination of surveys and designs) can be prepared and submitted for approval for implementation. In some countries there are laws and regulations imposed to standardize the procedures and they strictly define the content of the projects, but in many countries these are "free style" documents delegating responsibility to the authors to provide appropriate and justifiable proposals. When international institutions open calls for funding some research they apparently define the required format of the project document which further facilitates their review and assessment of suitability.

[1] Various stakeholders such as water utility, agro industry, water bottling company, mining company, or similar investors.

In line with the above, scientific-research work in geological science usually includes the following phases (Stevanović 1990):

1. Selecting and formulating the problem;
2. Collecting and evaluating literature;
3. Programming research;
4. Conducting the survey;
5. Elaborating obtained results;
6. Reporting;
7. Applying results and verifying solution in the practice.

It is clear that all these phases are interrelated and repeated during survey. As such, reading and checking referent literature is a continual process and follows a whole research process. For instance comparison of achieved and previous results, which again requires studying of collected literature, is necessary. The last stage, verification of the results and data collected through monitoring, provides a base for possible further surveys, corrective measures or new ideas, which may reopen a new research process.

14.3 Kinds of Hydrogeological Surveys

Several authors have discussed the issue of hydrogeological surveys and have provided related classifications (Klimentov 1967; Milanović 1979; Filipović 1980; Fetter 2001; Moore 2002; Goldsheider and Drew 2007).

In addition to "classical" classification of *laboratory* and *field surveys* which is common for all technical sciences, the kinds of surveys listed below, and at the same time phases in hydrogeological surveys can also be distinguished:

1. Reconnaissance survey;
2. Basic survey;
3. Detail survey;
4. Monitoring.

1. *The reconnaissance survey* has to be short in duration, efficient, and low-cost but nevertheless must enable a realistic picture of the general hydrogeological setting, i.e. the *conceptual idea,* which should be used in the formulation of further survey phases. No reconnaissance phase is required if sufficient data exists, or just an improvement or remedy of the situation is required: for instance, to drill a new well in an already formed well field, or to slightly modify established sanitary protection zones.

   The reconnaissance survey should include a collection of information on previous surveys, their initial evaluation; collection and analysis of the geological and hydrogeological maps, climate and hydrology data; remote sensing; and a field visit. The field prospection should cover most important sites (springs, well-fields, representative outcrops of the present geological formation, karst features, and similar).

2. *The basic survey* is more complex, consumes more time and funds, and should prepare an adequate base for final solutions and orient the detail survey. The result of this stage should be a hydrogeological map in medium scale (i.e. 1:25,000–1:100,000).[2] This phase, along with classical hydrogeological mapping (water points inventory), should include appropriately selected hydrogeological survey methods (described in Chap. 4). Among them are commonly a geophysical survey, exploratory drilling, permeability and water quality tests, speleology, tracing tests, etc.
   The main outcome of this basic survey should be a *conceptual model*, with estimated basic elements of groundwater distribution, movement (recharge-discharge), and available water resources. The hydrodynamical modeling could be carried out in all survey phases, but it is quite logical that it be linked with this exact stage because of the existence of an adequate set of information and a created conceptual model. Although modeling is often applied in very first stage of the survey it is clear that the certainty and validity of the model rank in importance alongside the quality of collected information. Therefore, as much as *input* information is available, model tends to be legitimate.
3. *The detail survey* should focus on technical solutions and possible alternatives. If, for instance, a new well field should be open for water supply or for dewatering of a mine field, in this stage drilling of exploratory—exploitation wells, long-term pumping tests, hydrodynamical analyses and forecast modeling take place. An example of proposed content of hydrogeological surveys aiming to regulate the regime of a karstic aquifer is presented at the end of Sect. 15.5
4. *Monitoring* is a key research phase for collection of information which has to verify if the applied solution is technically correct, successful for the end-user and at the same time environmentally sound. As stated earlier, monitoring is also important of new data collection and its possible use for additional interventions and improvements of applied solutions (Fig. 14.1).

It can thus be concluded that hydrogeological survey is by definition a complex task. Surveys should be gradual, their results should provide knowledge which grow from elementary to sophisticated. The attempt to overcome some steps in surveys, to ignore procedures, and to jump straight to final answers regularly results in unsuccessful designs and implemented inappropriate solutions.

[2] The scale very much depends on the size of the study area and its accessibility, but also on the complexity of the geology and hydrogeology settings. In large arid areas such as deserts or basins with confined aquifers even smaller scale maps 1:250,000–1:500,000 may provide sufficient basic information.

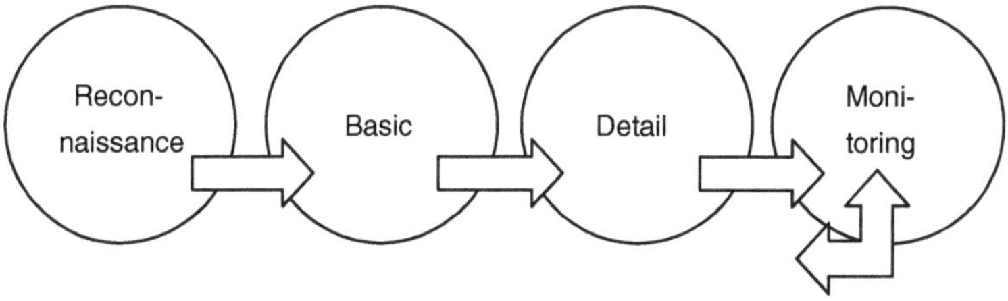

Fig. 14.1 Cycle of hydrogeological survey

14.4 Conceptual Model and Solutions

The conceptualization is an essential step in hydrogeological works. It should emerge from *basic surveys* and collected relevant and sufficient information should preferably be based on an assessment done by qualified and experienced experts. The qualifications are important because in many places, and especially in the undeveloped world, many people proclaim themselves as expert for all water and sanitation issues, including hydrogeology, and this is a reason why many projects dealing with our "invisible resources" fail.

In Chap. 4 the conceptualization of a karst system by the KARSYS method is briefly explained. This method, created by Jeannin et al. (2013), was initially developed within the framework of the SWISSKARST project but its application on terrains outside of Switzerland has also started recently. Kresic and Mikszewski (2013) wrote the book "Hydrogeological conceptual site models—data analysis and visualization" in which they define the hydrogeological conceptual site model (CSM) as a "description of various natural and anthropogenic factors that govern and contribute to the movement of groundwater in the subsurface". They also consider the main purpose of CSM to be a product composed of text, pictures and possibly animations that can be used for decision-making at any stage of project implementation. In this context Kresic and Mikszewski highlighted the importance of data visualization as an essential step in presenting a project to stakeholders of varying technical backgrounds in order to obtain their support.

The creation of a conceptual model requires a number of steps and elements and needs recognition and study. As discussed in Chap. 3 several karst aquifer properties have to be evaluated[3]:

- Geometry
- Permeability and storativity
- Recharge
- Groundwater flow
- Drainage

[3] Only elements for an aquifer quantitative model are considered in this conceptualization. Normally, in hydrogeology studies groundwater quality and vulnerability from pollution should also be evaluated.

- Hydrodynamic conditions
- Relationship ground—surface waters
- Relationship of adjacent aquifers
- Regime of groundwater.

Almost all of these elements are discussed and evaluated in this book, especially in Chaps. 3, 5–7. By collecting information on all the above properties we in fact obtain an *Identity Card* of the aquifer system—a necessary base for its further manipulation and maintenance.

The conceptual model should be able to explain the system and to provide responses to central questions such as: where the waters originate; where the water flow comes from; how the water flow can be directed and limited; where it is appropriate to tap water; what the best kind of intake is; how the water source can be protected and pollution prevented; and many others.

The following few examples show different conceptual models that have been launched and solutions in groundwater utilization and maintenance (Box 14.1).

Box 14.1

Case study—Utilizing semi-confined karstic aquifer

Karstic aquifer is under complex hydrodynamic conditions. Its limited outcrop in a hilly area is a single recharge zone, while the confined part is covered by younger lacustrine deposits and totally impervious flysch sediments. Groundwater is tapped by fully penetrated deep wells, whose upper section is cemented to prevent any leakage from overlain lacustrine sediments. The confined structure offers an (sub)artesian pressure in the aquifer and facilitates water tapping (Fig. 14.2).

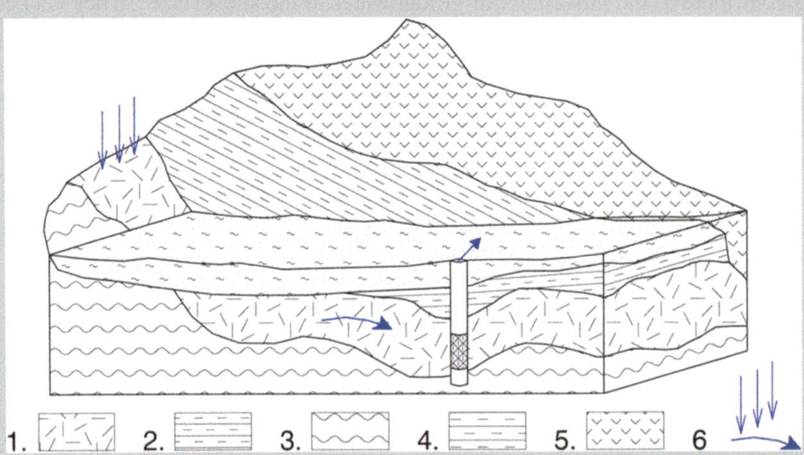

Fig. 14.2 Semi-confined karst aquifer in complex geological setting. An artesian well is drilled in an oblique bottom structure. Legend: *1* karst aquifer, *2* impervious flysch sediments, *3* folded schists, *4* basin's lacustrine sediments, *5* eruptive rocks, *6* recharge and groundwater movement zones

Case study—utilizing unconfined karstic aquifer

The karstic aquifers may consist of a well-fissured matrix, well-developed karstified channels as privileged groundwater paths, or both. Figure 14.3 shows the two cases: (a) a fully penetrated well with casing and screen placed in the main and large cavernous channel, and (b) an *open hole* well which is drilled in karstic-fissured aquifer without large cavities.

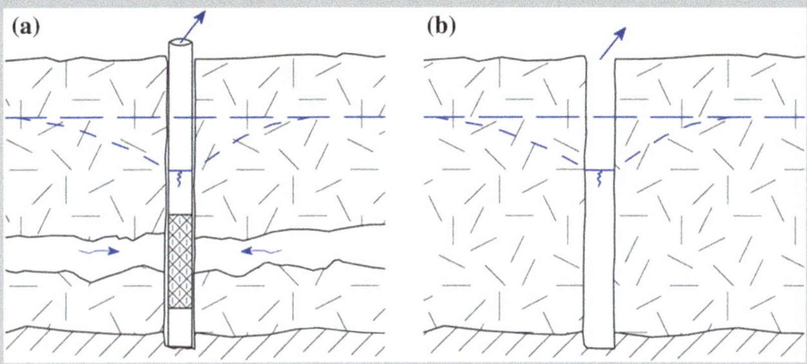

Fig. 14.3 Drilled fully-penetrated well in dominantly **a** cavernous and **b** fissured aquifer

Case study—utilizing littoral karstic aquifer

The sites where karstic aquifer rich in groundwater is outcropped while an impervious flysch barrier prevents sea water intrusion can be an ideal place for drilling the well and utilization of fresh water resources (Fig. 14.4). Searching for such places is a primary task of hydrogeological survey in coastal areas. Similar topics are discussed in Sect. 16.4.

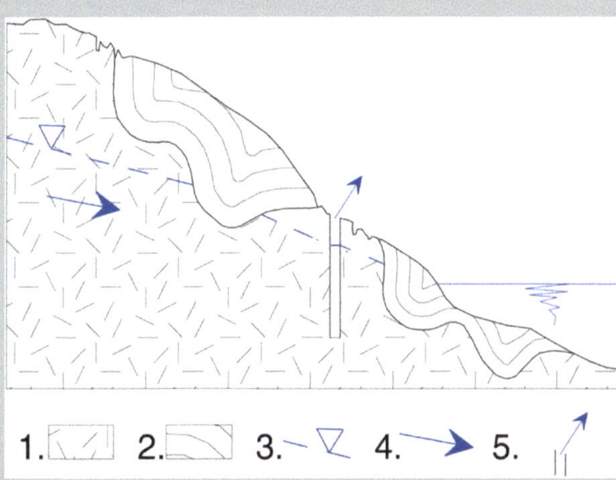

Fig. 14.4 Littoral karstic aquifer covered by impervious flysch rocks. Legend: *1* karstic aquifer, *2* flysch sediments, *3* groundwater table, *4* groundwater flow direction, *5* pumping well

Case study—deplete groundwater table and prevent inflow from karstic aquifer into coal pit mine

Sand and coal lacustrine sediments are surrounded by karstified rocks which also lie in the basin bottom. The common complex aquifer (groundwater body) is formed in these two formations. To make possible normal coal exploitation in an open mine pit it is obligatory to deplete the water table below the coal layers. The example on Fig. 14.5 shows a model of such terrain and four extraction wells tapping both formations. The intensive pumping of the wells results in an interfered drawdown dropped below the bottom coal layer. The controlled and continual pumping thus enables drying of coal layers and their normal withdrawal. Some similar examples are discussed in Sect. 16.2.

Case study—prevent leakage from reservoir built in karstic environment

Building a dam in karst always needs insurance of reservoir watertightness or at least its maximally diminished porosity. Leakage from reservoirs and preventive/remedial measures are discussed in Chap. 13 and Sect. 16.1. Figure 14.6 shows an example of a dam and reservoir built in karst and a grout curtain positioned all along the reservoir bank and dam foundation. The dense grouting material should prevent water losses or minimize them to an acceptable level.

Although the first group of presented models and solutions aims to utilize karstic water resources the two last case studies are fighting against the negative

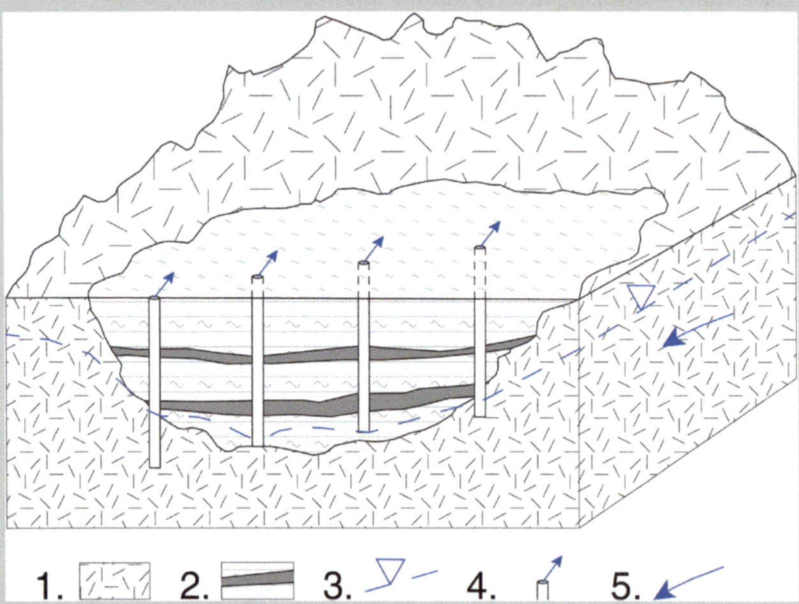

Fig. 14.5 Coal basin filled with lacustrine sediments surrounded by karstic massif and bedrock. Four extraction wells support coal mining. Legend: *1* karstic aquifer, *2* sand and coal sediments, *3* groundwater table, *4* pumping well, *5* groundwater flow direction

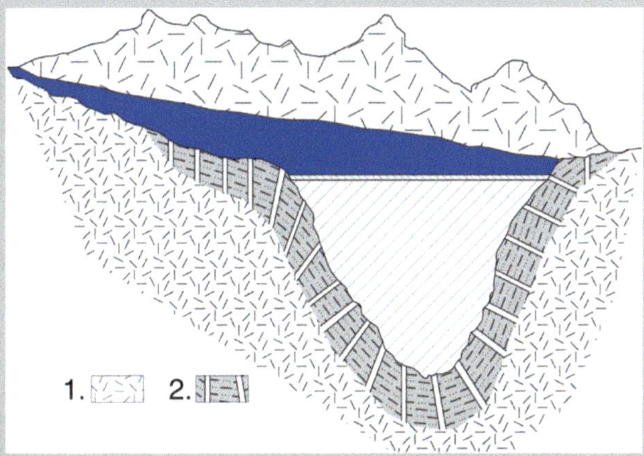

Fig. 14.6 Dam and reservoir constructed in karstic environment. The grout curtain follows complete dam shape and reservoir banks in its vicinity. *Legend 1* karst aquifer, *2* grouting boreholes

impact of karst waters. Many other structures such as tunnels, bridges or houses constructed in a karstic environment may also face difficulties: Induced collapses in urban areas or along construction sites for railways and freeways, massive debris, big subterranean floods are just some incidents which require anti-karst measures.

14.5 Environmental Implications of the Engineering Works in Karst

In contrast to the first half of the twentieth century when characteristically "A man changed nature according to his needs", today it is more apt to say "A man adapts his technical solutions to the environmental requirements". This great change results from the concept of *Sustainable development* introduced widely into social and political life in the 1990s with the Bruntland Commission report "Our common future" (1987) and followed by the Rio Submit in 1992 (Stevanović 2011).

Almost every technical solution should today conform to environmental conditions. Assessing the impact of serious interventions has become an obligatory part of technical documentation and projects, and of hydrogeological works and solutions. The environmental impact assessment study (EIAS) was introduced into engineering practice in New Zealand and Australia in the 1970s. In the USA engineering practice is regulated by the national environmental policy act (NEPA), while the European Union (1985) introduced the "Directive on the assessment of the effects of certain public and private projects on the environment".

The UK Environmental Agency (2003) defines the term "Environmental Impact Assessment" as a "technique and a process by which information about the environmental effects of a project is collected by the developer, both as new data and from other sources. This information is taken into account by the appropriate regulatory body in deciding whether the development should go ahead." For instance, the British legislation requires an EIA to be carried out for all water management projects including irrigation projects if the amounts of abstracted water exceed 20 m^3 in any 24 h.

EIAS is not obligatory for some small projects, but the application and justification have to be submitted for most, and then the responsible regulatory body determines the need for EIAS. The content of EIAS (Box 14.2) varies from country to country and is usually regulated by national legislation. It should, for instance, include and not be limited to the following topics:

- Project's investor (stakeholder),
- Description of project location,
- Description of project proposal (objectives, beneficiaries, solutions proposed, cost…),
- Alternatives to proposed solutions (expected end of project situation with and without project implemented),
- Environmental conditions—actual status,
- Environmental impacts (positive, negative) (Box. 14.3),
- Possible accidental situations resulting from the project and their direct and indirect impacts,
- Environment protection,
- Mitigation measures,
- Monitoring programme of the effects of proposed solutions on environment.

It is common that planned environmental protection, mitigation measures and monitoring of environmental impacts are part of the environmental management plan (EMP). This plan should also include detailed cost estimates, in-kind compensation for lost environmental resources, a proposal as to how to enhance environmental resources, description of a planned public consultation process as well as responsibilities and authorities for implementation of mitigation measures and monitoring.

Box 14.2

Case study—The content of EIAS

The Asian Development Bank (2003) (http://www.gdepp.cn/ewebeditor/uploadfile/20140505170216272.pdf) prescribes the following outline for the projects requesting financial support of the bank:

A. *Introduction*

The purpose of the report, extent of the EIA study and brief description of any special techniques or methods used.

B. *Description of the Project*

The type of and need for project, location, size or magnitude of operation and proposed schedule for implementation.

C. *Description of the Environment*

The physical and ecological resources, human and economic development and quality of life values in the area affected by the project. Where available, environmental standards will be used as the baseline for comparative purposes.

D. *Alternatives*

For each alternative, a summary of the probable adverse impacts and its relation to the project, and other alternatives will be discussed determine whether the project minimizes the environmental impact over all other alternatives and is within acceptable environmental impact limits. In most cases, environmental impacts "with" and "without" project alternatives should be examined.

E. *Anticipated Environmental Impacts and Mitigation Measures*

Environmental impacts, both direct and indirect, on different environmental resources or values due to project location, as related to design, during construction and regular operation will be discussed and mitigation, offsetting or enhancement measures will be recommended.

F. *Economic Assessment*

(a) Costs and benefits of environmental impacts; (b) Costs, benefits and cost effectiveness of mitigation measures; and (c) For environmental impacts that have not been expressed in monetary values, a discussion of such impacts, if possible, in quantitative terms (e.g. weight or volume estimates of pollutants). This information should be integrated into the overall economic analysis of the project.

G. *Environmental Management Plan*

The EMP will describe the impacts to be mitigated, and activities to implement the mitigation measures, including how, when, and where they will be implemented. The environmental monitoring plan will describe the impacts to be monitored, and when and where monitoring activities will be carried out, and who will carry them out.

H. *Public Consultation and Disclosure*

This section will describe the process undertaken to involve the public in project design and recommended measures for continuing public participation; summarize major comments received from beneficiaries, local officials, community leaders, NGOs, and others, and describe how these comments were addressed; list milestones in public involvement such as dates, attendance, and topics of public meetings; list recipients of this document and other project related documents; describe compliance with relevant regulatory requirements for public participation; and summarize other related materials or activities, such as press releases and notifications. This section will provide of summary of information disclosed to date and procedures for future disclosure.

I. *Conclusions*

This section will describe the gains which justify implementation of the project; explain how significant adverse environmental impacts will be mitigated or offset and compensated for; explain/justify use of any irreplaceable resources and; describe follow-up surveillance and monitoring.

Box 14.3

Case study—Environmental impacts of Bakerman dam and Khazer—Gomel irrigation projects (northern Iraq)

An assessment of the environmental impacts that may result from the construction and operation of the proposed process for the whole irrigation project has been made in reference to the baseline of the existing environmental situation. Emphasis is given to the likely significant environmental impacts, both adverse and beneficial. Temporary, short-term impacts and permanent effects are considered both for the *construction* and *operational phases*. Table 14.1 gives an overview of all impacts connected with dam construction and operating as well as the irrigation project based on water from this dam and possibly supplementary drilled wells.

Table 14.1 Summary of impacts assessed—Bakerman dam and Khazer—Gomel irrigation projects (northern Iraq)

Impact on		Construction phase	Operating phase
Microclimate		–	–
Water regime		–	++
Sedimentation transport		–	–
Neighbour countries		0	0
Water quality	Reservoir	–	–
	Downstream	–	–+
Groundwater		–	–+
Water temperature, downstream		0	–
Air quality		–	–+
Noise		–	0
Soil quality		–	–+
Fishes	Lake fishes	0	++
	Downstream river fishes	–	–+
Birds		–	–+
Mammals		–	–+

(continued)

Table 14.1 (continued)

Impact on	Construction phase	Operating phase
Ecosystem	−	−+
Landscape	−	−+
Archaeological remains	N/A	N/A
Cultural heritages	N/A	N/A
Agricultural—irrigation lands	−	+++
Fisheries	−	++
Transportation network	−	−
Waste disposal	−	−
Climate/meteo parameters	−	−+
Public health	−	+
Source of livelihood	−+	−++
Goods and services	−	−++
Population	−	−++
Tourism	−	+++
Quality of life	−	−++

Source Stevanović Z, Maran A, 2009, Environmental impact assessment study—Khazer—Gomel planning report (Iraq); IK Cons. Eng. & ITSC Ltd. UK, Belgrade—London, unpublished

Legend

0 no impact

− minor negative impact

−− moderate negative impact

−−− significant negative impact

+ minor positive impact

++ moderate positive impact

+++ significant positive impact

N/A non applicable

Minor and moderate negative impacts are acceptable, in principle; significant negative impacts cannot be mitigated and will certainly disturb the environment. When marks − and + are used together this indicates both negative and positive impacts on the issue

14.6 Environmentally Safe Groundwater Extraction and Indicators

What is the main concern in hydrogeology works and engineering regulation of the karst aquifer? First and foremost it is the *ecological flow* as the amount of water (yield) which should be in both a quantitative and qualitative manner secured for the existing eco system and which is essential for its normal functioning. Although the issue of *safe yield* is widely elaborated in Sect. 15.5 an overview of possible options of *exploitable reserves* (see Chap. 6 for explanation of the term) is presented in Table 14.2. They can be limited by ecological flows defined as biological minimum or by special water act (in quantity and/or in time).

Table 14.2 Formulae for definition of exploitable reserves of groundwater (Stevanović 2011)

Title	Formula	Ecological acceptability
1. Q_{exp} as only static (geological) reserves	$Q_{exp} = n \times Q_{st} / T$	no
2. Q_{exp} as total dynamic (renewable) and part of static reserves	$Q_{exp} = Q_{dyn} + n \times Q_{st} / T$	Very problematic
3. Q_{exp} as part of dynamic reserves but limited by biological minimum (minimal discharge or riverflow)	$Q_{exp} = Q_{dyn} - \mathrm{BM}$	Usually problematic
4. Q_{exp} as total dynamic and part of static reserves but limited by water act (in quantity and/or in time)	$Q_{exp} = Q_{dyn} + n \times Q_{st} / T - \mathrm{WL}$	Might be problematic, but Very common
5. Q_{exp} as part of dynamic reserves but limited by water act (in quantity and/or in time)	$Q_{exp} = Q_{dyn} - \mathrm{WL}$	Ecologically acceptable
6. No Q_{exp}	$Q_{exp} < \mathrm{WL}$	Drastic case, ecologically perfect but no water extraction permitted

where
Q_{exp} exploitable groundwater (GW) reserves
Q_{st} static (geological) GW reserves
Q_{dyn} dynamic GW reserves
T limited period of water extraction
N correction factor (percentage of total)
BM biological minimum (as minimal registered or calculated flow in springs or rivers)
WL water limitations defined by issued permit (can be periodical, or in quantity of extracted water)

Box 14.4

Case study—Environmental impact indicators

Karst geodiversity in many places in the world provides also habitat to different eco-systems as well as many threatened species. For instance, in the Alpine karst, including its branches Dinarides, Apennines, Hellenides and Carpathians there are numerous protected areas, and many were established within the karst. For instance, in all five Serbian national parks carbonates are present or even dominant rocks. The Water Framework Directive of European Union (2000) has established basic principles for securing water provision to dependent eco-systems.

Indicators are powerful tools for making important dimensions of the environment and society visible and enabling their management (Dahl 2012). There are many references and projects related to environmental indicators which cover different components of aquatic systems (including springs, streams, rivers, lakes, wetlands, coastal lagoons and estuaries). Some of the more recent, such as Vrba and Lipponen (2007) or UNECE (2007), pointed to a group of indicators helping to evaluate pressures on water quantity and on water quality.

In the GENESIS project, Preda et al. (2012) classify the following indicator packages:

- indicators of hydrogeomorphological units including groundwater: environmental tracers, water balance components, GW level and pressure, GW vulnerability, GW quality, river flow;
- indicators of physico-chemical components or even physico-chemical parameters as indicators: temperature, electrical conductivity, chlorophyll, concentration of different chemical compounds, dissolved oxygen, NO_3, NO_2, NH_4, PO_4, metals;
- indicators of biological compartments/trophodynamic modules: species richness of phytoplankton, macroinvertebrates, fish, diversity indices, indicator species, multimetric indices.

The International Sava River Basin Commission (2011) which is also responsible for water management of the Inner Dinarides proposed the *List of monitoring parameters* adjusted to the WFD requirements. Core parameters are: oxygen content, pH value, conductivity, nitrate, ammonium, plus parameters which put GW bodies at risk of failing to achieve good chemical status.

Similarly, the diagnostic analysis prepared under the project DIKTAS resulted in an initial list comprising 23 different parameters for assessing pressures on groundwater quantity and quality and resulting pressures on dependent ecosystems in selected aquifers of transboundary concern. Their knowledge and observation should support sustainable water use and the protection of nature and ecosystems (Stevanović 2014).

14.7 Conflicts from Karst Water Utilization

The problem of engineering works in karst aquifer and its artificial regulation can lead to the fear of potential over-extraction and accordingly to different types of conflicts, such as Users versus Users, Purpose versus Purpose, and Water versus Other Social Priorities. Such conflict can be individual and very local, but it can also be at a regional, national or even an international level (Stevanović 2011). Local conflict typically arises when drilling causes a lowering of the water table, which in turn can reduce the spring's yield or even cause nearby springs to dry out. Thus, the effects of the pumping of a well on a spring or other wells located within the cone of depression must be taken into consideration if both kinds of groundwater sources are to be used. Regional conflict may often result from different interests of water users (Burke and Moench 2000) in the same river basin.

Inappropriate compensation for water abstracted and delivered to privileged consumers out of the catchment where the water originates, for instance to the large towns, can also result in frustration of the local population. The issues of transboundary concern are discussed in Sect. 17.6.

To minimize any negative effects of the proposed interventions and to optimize the undertaken survey of technical solutions and environmental impact studies, very careful explanation of the tasks and of the benefits to the local water population and consumers should always be provided.

References

Asian Development Bank (2003) Content and format environmental impact assessment (EIA). In: Environmental assessment guidelines, Appendix 2. http://www.gdepp.cn/ewebeditor/uploadfile/20140505170216272.pdf. Visited 10 July 2014

Brundtland Commission (formally the World Commission on Environment and Development, WCED) (1987) Our common future, report. Oxford University Press. Published as annex to General Assembly document A/42/427, Development and international co-operation: environment, 2 Aug 1987

Burke JJ, Moench HM (2000) Groundwater and society: resources, tensions opportunities. Special edition of DESA and ISET, UN public, ST/ESA/265, New York p 170

Dahl AL (2012) Achievements and gaps in indicators for sustainability. Ecol Ind 17:14–19

European Union (1985) Council directive 85/337/EEC on the assessment of the effects of certain public and private projects on the environment. Official Journal No. L 175, 27 June 1985

European Union (2000) Water framework directive WFD 2000/60. Official Journal of EU, L 327/1, Brussels

Environmental Agency (UK) (2003) Environmental impact assessment in relation to water resources authorisations. Guidance on the requirements and procedures, Bristol. www.environment-agency.gov.uk

Fetter CW (2001) Applied hydrogeology, 4th edn. Prentice Hall, Upper Saddle River

Filipović B (1980) Metode hidrogeoloških istraživanja (Methods of hydrogeological research, in Serbian). Naučna knjiga, Beograd

Goldsheider N, Drew D (eds) (2007) Methods in karst hydrogeology. International contribution to hydrogeology, IAH, vol 26. Taylor & Francis/Balkema, London

International Sava River Basin Commission (2011) Sava River basin management plan. Background paper no. 2: Groundwater bodies in the Sava River Basin, v 2.0, Zagreb, p 37. www.savacommission.org

Jeannin PY, Eichenberger U, Sinreich M, Vouillamoz J, Malard A, Weber E (2013) KARSYS: a pragmatic approach to karst hydrogeological system conceptualisation. Assessment of groundwater reserves and resources in Switzerland. Environ Earth Sci 69:999–1013

Klimentov PP (1967) Metodika gidrogeologicheskih isledovaniy (Methods of hydrogeological research, in Russian). Vishaya skola, Moscow

Kresic N, Mikszewski A (2013) Hydrogeological conceptual site model: data analysis and visualization. CRC Press, Boca Raton

Milanović P (1979) Hidrogeologija karsta i metode istraživanja. HET, Trebinje

Moore JE (2002) Field hydrogeology: a guide for site investigations and report preparation. CRC Press LLC, Boca Raton

Preda E, Kløve B, Kværner J et al. (2012) New indicators for assessing groundwater dependent eco systems vulnerability. Delivereable 4.3. GENESIS FP 7 project: groundwater and dependent eco systems, p 84. www.thegenesisproject.eu

Stevanović Z (1990) Uvod u naučno-istraživački rad u oblasti hidrogeologije—sa osnovama opšte naučne metodologije (Introduction into scientific-research work in hydrogeology—with basics of general scientific methodology; in Serbian). University of Belgrade—Faculty of Mining & Geology, Belgrade, p 112

Stevanović Z (2011) Menadžment podzemnih vodnih resursa (Management of groundwater resources; in Serbian). University of Belgrade—Faculty of Mining & Geology, Belgrade, p 340

Stevanović Z (2014) Environmental impact indicators in systematic monitoring of karst aquifer—Dinaric karst case example. In: Kukurić N, Stevanović Z, Krešic N (eds) Proceedings of the DIKTAS conference: "Karst without boundaries", Trebinje, 11–15 June 2014, pp 80–85

United Nations Economic Commission for Europe (UNECE) (2007) Environmental indicators and indicators-based assessment reports: Eastern Europe, Caucasus and Central Asia. United Nations Publication. ECE / CEP 140, New York; Geneva p 93

Vrba J, Lipponen A (2007) Groundwater resources sustainability indicators, IHP—VI series on groundwater no. 14. UNESCO, Paris

Chapter 15
Karst Groundwater Availability and Sustainable Development

Francesco Fiorillo, Vesna Ristić Vakanjac, Igor Jemcov, Saša Milanović and Zoran Stevanović

15.1 Hydraulic Behavior of Karst Aquifers

Francesco Fiorillo

Centre for karst Hydrogeology, Department of Hydrogeology, Faculty of Mining and Geology, University of Belgrade, Belgrade, Serbia

15.1.1 Introduction

The main hydraulic characteristic of karst aquifers is their heterogeneity, being formed by a complex conduit network, which develops in a fractured carbonate rocks with lower permeability (White 1988; Kiraly 2002; Ford and Williams 2007). The karstification processes lead to a hierarchical conduit network into the

F. Fiorillo (✉)
Department of Science and Technology, University of Sannio, Benevento, Italy
e-mail: francesco.fiorillo@unisannio.it

V. Ristić Vakanjac · S. Milanović · Z. Stevanović
Centre for Karst Hydrogeology, Department of Hydrogeology, Faculty of Mining and Geology, University of Belgrade, Belgrade, Serbia
e-mail: vesna_ristic2002@yahoo.com

S. Milanović
e-mail: sasa.milanovic@rgf.bg.ac.rs

Z. Stevanović
e-mail: zstev_2000@yahoo.co.uk

I. Jemcov
Department of Hydrogeology, Faculty of Mining and Geology, University of Belgrade, Belgrade, Serbia
e-mail: igor.jemcov@rgf.bg.ac.rs

Z. Stevanović (ed.), *Karst Aquifers – Characterization and Engineering*,
Professional Practice in Earth Sciences, DOI 10.1007/978-3-319-12850-4_15

aquifer (Kaufman 2003; Kresic 2010) and cause the drainage to be converged to specific points: the karst springs. The karst conditions act as a filter that regulates the groundwater paths to the springs depending on the interconnections present in the karst network (Fiorillo and Doglioni 2010).

As a consequence of the karstification process, the groundwater flow presents a characteristic duality: *slow* and *fast/quick* flow, function of the hydraulic conductibility of the medium. This duality extends also to the infiltration processes, which can occur both in concentrate and diffuse types. The first is connected to the sinkhole/shafts and swallow holes, which allows rapid drainage of the runoff and causes a fast transit into and through the vadose zone and subsequently to the saturated zone. Generally, shafts and swallow holes drain after intense storms or during the wet season or snowmelt; they can also drain the runoff coming from the non-karst neighboring area (*allogenic recharge*). Diffuse infiltration occurs via the soil mantle or fractures of the outcropping carbonate rocks, and it is the main origin of percolation in the vadose zone; the percolation processes can take large time before water reaches the water table, according to the thickness and conductibility characteristics of the vadose zone.

In general, the internal structure of a karst aquifer is only partially known (often, it is totally unknown), and the main hydraulic aspects are deducted from spring and well hydrograph analyses.

Systematic records of spring discharges allow definition of the regime of springs, and shape of the hydrograph provides a useful tool to evaluate the aquifer karst condition (Bonacci 1993a; White 2002). A well-developed karstification into the aquifer favors spring hydrographs characterized by pronounced peaks, which follow the main storms, and indicate both a good connection of the conduit networks with the spring and presence of active shafts. In these systems, flow primarily occurs rapidly in large fissures through irregular conduits and is known as quick flow, fast flow, or conduit systems (Atkinson 1977a; Gunn 1986; Bonacci 1993a). On the other hand, a smoothed shape of the spring hydrographs, without sharp peaks and characterized by one or few flood peaks during the hydrological year, indicates a coarse development of karst conduits or their poor connection. In these systems, flow primarily occurs in fractures and minor fissures as laminar regime and are known as diffuse-type karst systems; they are characterized by slow flow (Atkinson 1977a; Bonacci 1993a) and generally retain water for a much longer length of time than conduit karst aquifer.

A smoother, less structured hydrograph may indicate that an aquifer is primarily recharged by diffuse infiltration (White 2002), but an important role is due to mantling deposits that limit runoff and concentrated infiltration. Also the epikarst, a zone of increased weathering near the land surface, determines the distribution of recharge to a karst aquifer in both space and time (Bauer et al. 2005). Generally, karst systems contain quick flow and slow flow components that are both reflected on the hydrographs produced by the springs (Box 15.1.1). These components vary thought the hydrological year, depending mainly on recharge conditions.

In the following sections, a general overview of hydraulic behavior will be given; further hydraulic behaviors, under droughts and earthquakes, will be given for a karst area of southern Apennine (Picentini Mountains, Italy).

Box 15.1.1

Figure 15.1 shows spring hydrographs provided by different flow types, during wet and dry hydrological year. Prolonged periods of poor rainfall can reduce the response of the spring discharge, with springs characterized by a predominant quick flow component showing a diminished frequency and/or peak value of impulses, and springs characterized by a predominant slow flow component showing a reduced annual maximum value (Fig. 15.1).

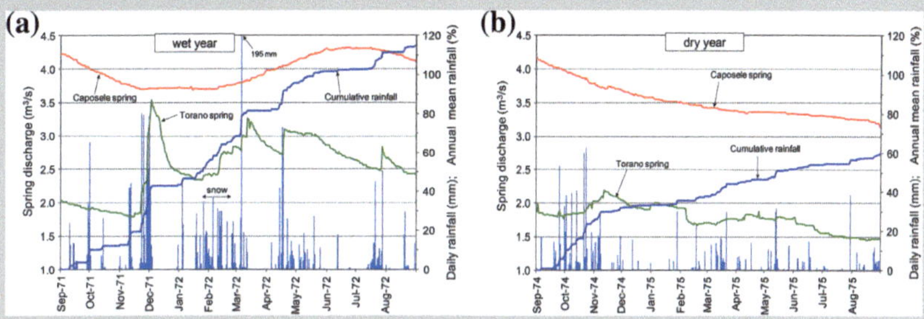

Fig. 15.1 Hydrographs and rainfall corresponding two Campania karst springs during a wet (**a**) and dry (**b**) hydrological year (modified from Fiorillo 2009). Torano spring hydrograph (Mt. Matese, Northern Campania), which is characterized by quick flow impulses overlapping the base flow. Caposele spring hydrograph (Mt. Picentini), which is characterized by slow flow

15.1.2 Hydraulic Behavior Under Different Hydrological Conditions

During heavy rain, concentrated infiltration favors a rapid inject water into an aquifer, and shafts can be temporarily filled by water, causing an increase in the hydraulic head in the conduit network. Then, a rapid pressure pulse is forced through the phreatic conduits giving a hydrographic peak at the spring (Ford and Williams 2007).

Later, after the recharge period, karst conduit network drains the saturated aquifer zone, and the water table decreases; from the hydraulic point of view, this process constitutes the *slow flow* or *base flow.*

Ford and Williams (2007) provided a discussion on the applicability of the Darcy's law to karst and highlighted how the large part of the flow passes through the conduit system; thus, according to Mangin (1975a), they assumed that the range of condition under which Darcy's law is valid is very restricted. The restriction has to refer to different parts of the aquifer (spatial meaning), but also to the different recharge conditions of the aquifer, which vary during the hydrological year (temporal meaning). A *Darcian* flow prevalently occurs during the latest part of the recession, when the mean water flow velocity in the aquifer is the lowest respect to all previous stages. The *Darcian* hydraulic behavior has to be connected to the drainage of the water from the minor fissures, even if the conduit network provides the final water transfer up to spring.

To highlight the main points of the aquifer hydrodynamics, and to associate the different parts of a hydrograph with specific hydraulic conditions, several sketches have been arranged in Fig. 15.2. In this example, non-allogenic recharge

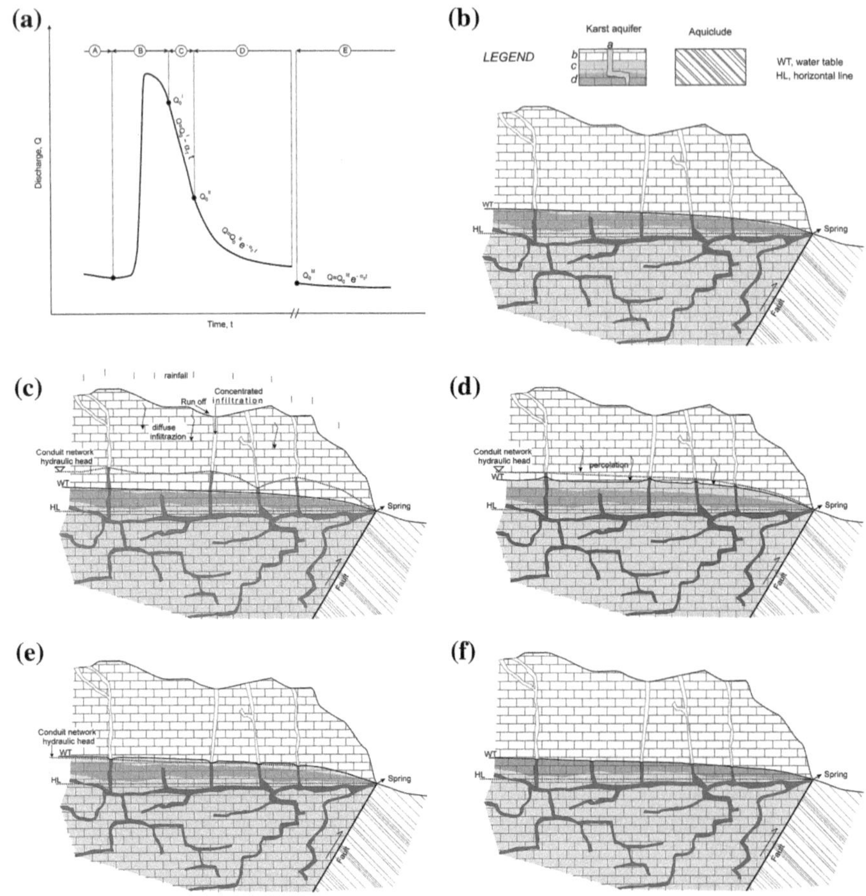

Fig. 15.2 Spring hydrograph triggered by a storm pulse, where the different parts are associated with sketches **b–f** (modified from Fiorillo 2014), showing the hydrodynamics in a karst aquifer. Recharge occurs by diffuse percolation through vadose zone, and in a concentrated manner in points (sinkholes/swallow holes and shafts); a basal spring drains the water table by a conduit network. The hydrodynamics is shown by different sketches where the slope of the water table is exaggerated. **b** After a period of no recharge (*a* conduits and shafts; *b* vadose zone; *c* and *d* saturated zone with different effective porosity). **c** During a period of strong recharge, concentrated infiltration causes the temporarily filling of shafts; later the aquifer water table rises as a consequence also of the diffuse infiltration. **d** After a period of strong recharge, water level in the shafts falls, and the water table could still rise. **e** Lowering of the water table during no-recharge condition; **f** lowering of the water table after a long period of no recharge

has been considered, and a single outlet (spring) drains the aquifer. In each sketch, the hydraulic head into the conduit network is marked by the water level in the shafts; this level can be different from the saturated aquifer zone, in function of the hydrologic condition of the aquifer.

The phase A shows the hydrological condition after a non-recharge period, characterized by a decreasing trend in the discharge, and the contribution from vadose zone can be neglected; in this phase, the saturated zone is drained by the conduit network, which has a hydraulic head similar or little lower than the saturated zone.

Phase B occurs after an intense rainfall event, characterized by several rainy hours or a few days of rainfall. Thanks to the presence of shafts and sinkholes, surface runoff infiltrates and rapidly reaches the saturated zone of the aquifer. The amount of rainfall which causes the concentrate infiltration depends on the rainfall intensity and its distribution over time, but it is strongly controlled by the hydraulic characteristics of the vadose zone, such as its thickness, the presence of shafts, and the morphological characteristics of the ground surface (presence of endorheic areas, slope angle distribution, etc.). During phase B, shafts are temporarily filled with water, and an increase in hydraulic head in the karst conduit network occurs, up to the spring (Fig. 15.2). As the concentrated recharge can vary according to the catchment features, it provides a different water level for each shaft, which can cause a temporary locally different water flow direction into the saturated aquifer zone. The rapid response of a spring to an intense rainfall event, with a typical peak in the spring hydrograph, can be associated with the rise and lowering of the water level inside conduits above the saturated zone as for example observed by Drogue (1980), Bonacci and Živaljevic (1993), Bonacci (1995), Halihan et al. (1998), terminating into the phreatic zone. However, the very high hydraulic conductivity of the conduit network can limit the rise (and lowering) in the water level in conduits.

After the beginning of the increase in the flow, a decrease in the hardness of the spring water is observed, which may mark a large part of peak hydrograph zone; Ashton (1966) provided a simple procedure to evaluate the volume of storage in fully submerged conduits feeding a spring using discharge and the hardness of the spring water. These observations indicate that the first part of recession (the steepest part) could be still characterized by "freshwater," which come from the drainage of shafts and conduits, thanks to strict connection between shafts and spring by conduit network.

The exchange of water between conduits and matrix occurs during floods (Martin and Dean 2001; Bailly-Comte et al. 2010), and it is caused by the higher conduit head in the conduit network than the matrix (Atkinson 1977a; Drogue 1980; Bailly-Comte et al. 2010; Fiorillo 2011a). During phase B, shafts and conduit network feed the water table, and diffuse infiltration as well; under such hydraulic conditions, spring discharge is connected to concentrate recharge processes and depends on the hydraulic head in the conduit network. Some karst springs (for example, weakly karstified systems) do not show this hydraulic behavior, and their hydrograph shape may appear completely smooth; in this case, spring discharge increases or decreases in function of long wet or dry periods, respectively (as Caposele spring, Fig. 15.1).

In well-karstified systems, when recharge reduces or ends, a fall of discharge can be observed at spring. This process can be interpreted as a rapid drainage of conduit network which is generally forced by temporarily filled shafts (phase C on Fig. 15.2). The phase C has a linear equation (which represents the nonlinear part in the semi-logarithmic plot), but it may also have another analytical form (Malík and Vojtková 2012). Mangin (1975a) explained this initial part of the recession as the *influenced stage*, which contributes to the unsaturated zone, which decreases over the time.

During phase C, the hydraulic head in the conduit network decreases, but it is still higher than the water table, which could increase due to percolation from diffuse recharge.

When the water level in the shafts decreases and reaches the water table level, the hydraulic head in the conduit network reaches that of the matrix/minor fractures, causing the drainage of the saturated zone by the conduit network, and phase D begins. Many karst springs show an abrupt change of the hydrograph slope from phase C to phase D; the gently sloped part of phase D highlights the higher water volume stored in the saturated zone, which causes a lower decrease in the discharge through time. During this stage, the emptying processes cause the lowering of the water table and involve water from minor conduits and the matrix, where viscosity and friction forces control the hydrograph recession shape. The concave shape of this part of the hydrograph highlights, at least in part, the energy lost by the water flow. Figure 15.2 shows the recession during phase D by a single exponential term, but the first part could be still influenced by the arrival of the diffuse infiltration, and only later the exponential form could appear. Besides, following the theoretical models (Rorabaugh 1964; Brutsaert 1994; Kovacs et al. 2005), this first part of the phase D could be characterized by the sum of several exponentials.

The phase E refers to the hydrological condition after a long period of no recharge, during which the water table reaches the minimum height, following a different recession coefficient, α_3 (or α_n). If the exponential form is expected, the changing of the recession coefficient, from α_2 (phase D) to α_3 (phase E), may depend on several factors and suggests whether anisotropy of the aquifer or its non-constant area during the drainage (Fiorillo 2011a).

The effective porosity, which can be approximate to storativity for unconfined aquifers (Stevanović et al. 2010a), is an important factor controlling the recession coefficient, as it can vary in a wide range due to heterogeneity of karst media (Fiorillo 2014). The change in the effective porosity inside the epiphreatic zone could be connected to the different ways of development in the caves, conduits, and voids, according to the main theories on karst development (Ford and Ewers 1978). Numerical models have shown how the phenomenon of mixing corrosion causes the development of conduits and porosity just below the water table of an unconfined aquifer (Gabrovšek and Dreybrodt 2010; Dreybrodt et al. 2010). So just below the water table level, there would be a strong variation in void distribution. This means that during the recession, the effective porosity can change along the water table level, conditioning the shape of the hydrograph, independently from other hydraulic parameters and catchment area (Fiorillo 2011a).

To investigate the aquifer under phase E, it could be necessary to analyze the spring discharge recession after a very dry hydrological year (Fiorillo et al. 2012). Many highly elevated springs (non-basal springs) dry up during the recession and remain dry for some time of the year. These springs follow a recession law which is different from the exponential form, as the discharge decreases faster than exponential decay. The simplest explanation of this behavior is the progressive reduction of the water table area during the aquifer emptying (Fiorillo 2011a).

Commonly, basal karst spring tends to have a progressive decrease in the recession coefficient during the recession (as phase E on Fig. 15.2), whereas other spring types may present a different hydrograph shape. Analyzing the spring recession in the semilogarithmic plot, a deviation from the straight line produced by a simple exponential decay of discharge through time provides information on the actual emptying rate of the aquifer compared to a simple exponential decline. If the recession coefficient, α, decreases under drought conditions, springs guarantee water during a long dry period and can be considered *drought resistant* (Fiorillo et al. 2012); this hydraulic behavior could be connected to the increase in the effective porosity in depth. If the recession coefficient increases under drought conditions, a different hydraulic behavior occurs, as the aquifer is drained more quickly than expected, and the springs can be considered as *drought vulnerable* (Fiorillo et al. 2012); this hydraulic behavior could be connected to a decrease in the water table area, or to the fact that the aquifer is also drained by other springs with lower ground elevation.

15.1.3 Geological and Hydrological Features of Terminio and Cervialto Karst Aquifers

The Picentini Mountains, similar to other reliefs of southern Apennine chain (Figs. 15.3 and 15.4), are made up of series of limestone and limestone–dolomite (Late Triassic–Miocene) characterized by a thickness ranging between 2,500 and 3,000 m. Along the northern and eastern sectors, these massifs are tectonically overlapped on the terrigenous and impermeable deposits, constituting argillaceous complexes (Paleocene) and flysch sequences (Miocene). The northern sector of Picentini Mountains is characterized by high mountain peaks where elevation reaches 1,809 and 1,806 m a.s.l. in correspondence with Cervialto and Terminio massifs, respectively. Pyroclastic deposits of Somma–Vesuvius volcano activity cover the Picentini Mountains, with thickness up to several meters along gentle slopes of the Mt. Terminio zone, and of few decimeters along steep slopes and entire area of Mount Cervialto. These deposits play an important role in the infiltration of water into the karst substratum.

Typical karst features of these massifs are wide endorheic areas, which have an important role in recharge processes. In particular, the Terminio massif is characterized by several endorheic areas (Fig. 15.4), where the largest is the Piana

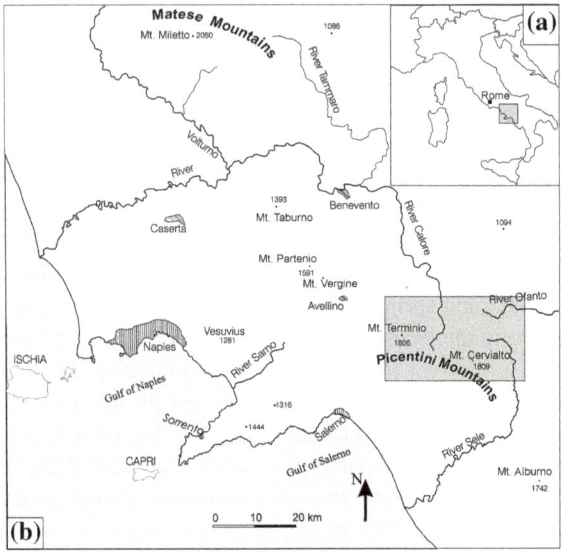

Fig. 15.3 **a** Italian peninsula; **b** map of the western Campania region. *Shaded area* is detailed in Fig. 15.4

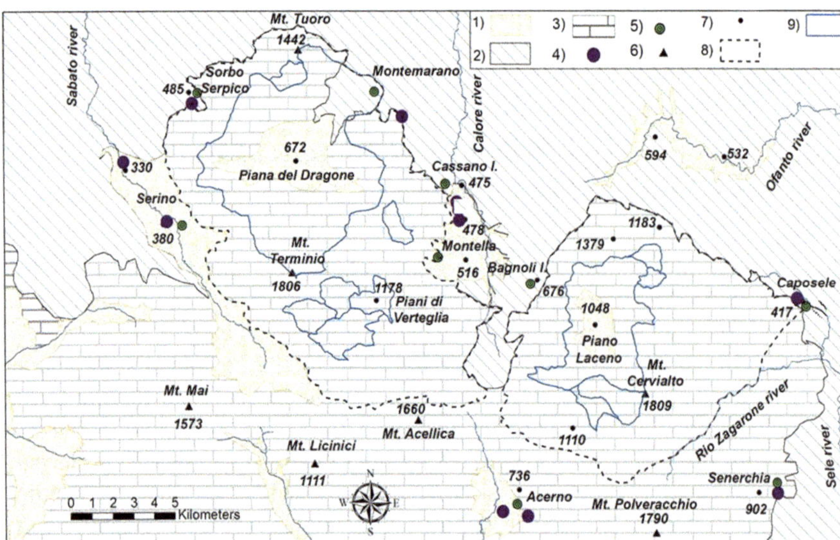

Fig. 15.4 Hydrogeological sketch of northeastern sector of Picentini Mountains (modified from Fiorillo et al. 2014); *legend* (*1*) slope breccias and debris, pyroclastic, alluvial, and lacustrine deposits (Quaternary); (*2*) argillaceous complex and flysch sequences (Paleogene–Miocene); (*3*) calcareous–dolomite series (Jurassic–Miocene); (*4*) main karst spring; (*5*) village; (*6*) mountain peak; (*7*) elevation (m a.s.l.); (*8*) spring catchments; (*9*) endorheic area

Table 15.1 Main hydrological parameters of springs and catchments (annual mean values)

Spring group and discharge, Q_s (m^3/s)		Catchment, main features		Mean	Max	Min	*F* ($m^3 \times 10^6$/ year)	*R* (%)
Serino	2.25	Terminio	*F*, (mm/year)	1,887	2,547	1,463	334.3	50.1
Cassano I.	2.65							
Sorbo Serp.	0.43		*T*, (°C)	10.3	15.5	4.4		
Others	0.15		*AET* (mm)	587	703	413		
Total output	**5.48**							
Caposele	4.00	Cervialto	*F*, (mm/year)	2,109	2,620	1,529	238.8	54.3
Others	0.17		*T*, (°C)	8.5	14.0	4.4		
Total output	**4.17**		*AET*, (mm)	529	688	410		

F afflux on the massif, *T* temperature, *AET* actual evapotranspiration, *R* recharge ($R = Q_s/F$). The mean, maximum, and minimum values express the spatial variation of parameters

del Dragone (55.1 km^2); several sinkholes drain this endorheic area, and hydraulic works were carried out to limit the flooding during the wet period, connecting a drainage system to the *Bocca del Dragone* sinkhole. Tracer tests testified the connection between this sinkhole and Cassano springs, indicating that this area belongs to the recharge area of these springs. The largest endorheic area of the Cervialto massif is the Piano Laceno (20.5 km^2); here, a permanent lake exists, which is limited by the activation of several sinkholes during the wet period. As the Caposele spring can be considered the only spring draining the Cervialto Massif, this endorheic area constitutes entirely part of the recharge area of this spring. Due to a local cave system (*Grotta del Caliendo*) in the west side of the Piano Laceno, only a limited water escape exists from this endorheic area and it is drained in the Calore valley.

These karst massifs feed many karst springs with discharge up to thousands liters/second and constitute a fundamental water resource in Southern Italy (Table 15.1).

The Serino group is located in the valley of the Sabato River, along the north-western boundary of the Picentini massif and is formed by the Acquaro–Pelosi springs (377–380 m a.s.l.) and the Urciuoli spring (330 m a.s.l.). These springs are fed by the Terminio massif (Civita 1969), with an overall mean annual discharge of 2.25 m^3/s. Roman aqueducts (first century AD) were supplied by these springs, and the Urciuoli spring was retapped between 1885 and 1888 by the Serino aqueduct, which is comprised of a gravity channel followed by a system of pressured conduits that is used to supply water to the Naples area. Additionally, the Aquaro and Pelosi springs were also retapped in 1934 by the Serino aqueduct.

The Cassano group is located in the Calore river basin along the northern boundary of the Picentini Mountains and is formed by the Bagno della Regina, Peschiera, Pollentina, and Prete springs (473–476 m a.s.l.). Also these springs are primarily fed by the Terminio massif (Civita 1969), with an overall mean annual discharge of 2.65 m^3/s. In 1965, these springs were tapped to supply the Puglia region with water, and a gravity tunnel was joined to the Pugliese aqueduct.

The Caposele group is formed by the Sanità spring (417 m a.s.l.), which is located at the head of the Sele river basin along the northeastern boundary of the Picentini Mountains. This spring, which is primarily fed by the Cervialto mountain (Celico and Civita 1976), has a mean annual discharge of 3.96 m^3/s. The spring was tapped in 1920 by the Pugliese aqueduct, which passes through the Sele-Ofanto divide via a tunnel and supplies the Puglia region with water.

In the entire Italian peninsula, a typical Mediterranean climate locally exists, characterized by dry and warm summers, and wet periods occurring during the fall, winter, and spring. Monthly rainfall reaches its highest annual peak during November–December, while the minimum occurs during July–August. In the highly elevated zones (above 1,000 m a.s.l.), snow can cumulate for several weeks/months during winter, providing a time shift of the infiltration process.

Figure 15.5 shows the distribution of the effective rainfall for a high ground-elevated station and expresses the amount of rainfall which is free to charge the soil moisture, to percolate and charge the groundwater, or runoffs. As can be seen, recharge processes begin generally in October, as the earlier rainfall of August, September, and part of October is completely adsorbed by the evapotranspiration processes or retained as soil moisture. Runoff also occurs after the middle October and ends in May–June.

The regime of Caposele spring appears almost opposite with respect to rainfall distribution, as the minimum occurs in November–December and the maximum in June. Cassano and Serino spring hydrographs also have floods several months later monthly rainfall peak.

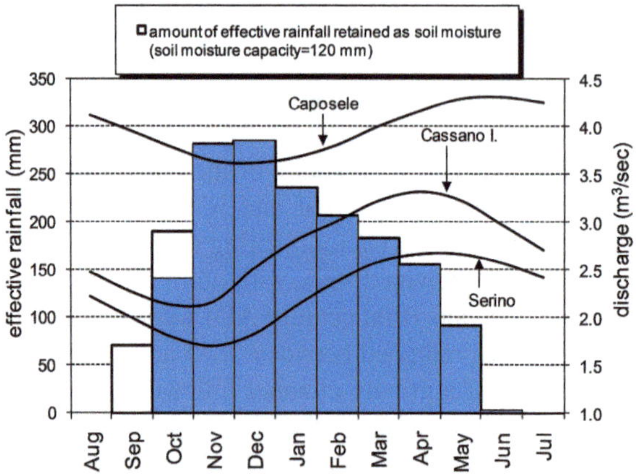

Fig. 15.5 Mean monthly effective rainfall (Montevergine stations 1,270 m a.s.l.), and mean monthly spring discharges of Caposele, Cassano, and Serino groups. The effective rainfall has been computed as the difference between monthly mean rainfall and the potential evapotranspiration computed by the method of Thornthwaite and Mather (1957)

Based on rainfall and temperature distributions with ground elevation, the annual mean afflux and the actual evapotranspiration have been evaluated in GIS environment (Fiorillo 2011b), and results are shown in Table 15.1.

15.1.4 Hydraulic Behavior During Droughts and Earthquakes

The lack of precipitation over a large area and for extensive periods of time is known as *meteorological drought*. This water deficit propagates through the hydrological cycle and gives rise to different types of droughts. If groundwater recharge and stream flow will be reduced, a *hydrological drought* may develop, and it is temporally shifted with respect to *meteorological droughts* (Tallaksen and Van Lanen 2004). A long meteorological drought can induce the lowering of piezometric levels and spring discharges, causing a *groundwater drought*, which is a specific aspect of the *hydrological drought*.

As shown in Fig. 15.2, spring hydrograph reflects the hydrological conditions of karst aquifers. During a dry hydrological year (Fig. 15.1b), as consequence of limited recharge processes, spring discharges have values lower than the mean; practically, the water volume provided by recharge processes is almost always less than that discharged at spring. When this condition occurs, a continuously decreasing discharge can be observed during the entire hydrological year, which highlights *groundwater droughts* (Fiorillo 2009).

During a *groundwater drought*, the spring discharge falls to minimum values and highlights the behavior of the spring under such extreme hydrological condition (Box 15.1.2).

Box 15.1.2

Figure 15.6 shows the groundwater drought of 1949 was induced by the poor annual rainfall of 1948–1949. The annual rainfall is shown in Fig. 15.6b, where the values are standardized (mean = 0; standard deviation = 1) from long time series; the annual rainfall of 1948–1949 was 2.65 standard deviations below the mean and was one of the intense meteorological droughts in southern Italy (Fiorillo and Guadagno 2010, 2012). During the following year 1949–1950, the annual rainfall was almost as the mean value, but Caposele spring still had a discharge one standard deviation below the mean, indicating that it maintains the memory effect of the antecedent drought year. In the same year (1949/1950), Serino spring had discharge value near the mean and did not show a memory effect. This different hydraulic behavior highlights the influence of hydrological conditions of the antecedent year on large springs, as Caposele, responding to long accumulation of rainfall.

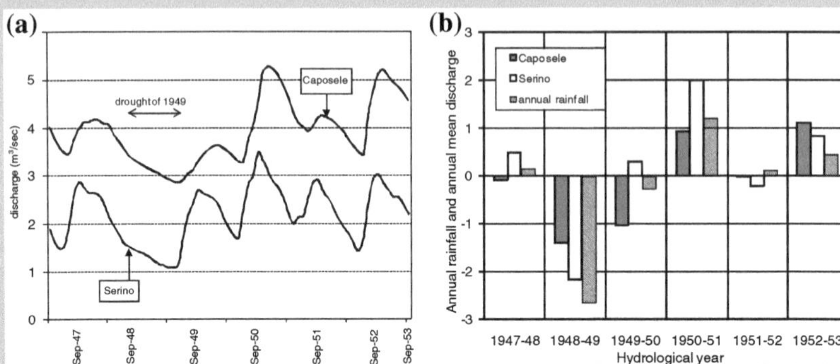

Fig. 15.6 **a** Caposele and Serino spring discharge during the period September 1947–September 1953; the groundwater drought of 1949 is marked by the continuous decreasing discharge. **b** Standardized annual rainfall and annual mean discharge (period 1920–2009); the groundwater drought of 1949 was induced by the meteorological drought of 1948–1949, and continued during 1949–1950 in Caposele spring

For water management purposes, it is useful to forecast the spring discharge, especially during dry hydrological year, which can induce later a groundwater drought. At the beginning of the hydrological year, it would be very difficult to forecast spring discharge several months later, but the shift between rainfall and aquifer response (Fig. 15.5) allows groundwater droughts to be forecast by monitoring rainfall. Because the combined effect of increasing evapotranspiration and decreasing rainfall toward the summer season results in the effective rainfall after April being negligible. Therefore, in Mediterranean climates, the rainfall that occurs up to March–April of each hydrological year recharges karst aquifers, and subsequent rainfall, up to September–October, generally does not recharge the aquifers (Fiorillo 2009; Box 15.1.3).

Box 15.1.3

Figure 15.7 shows the cumulative effective rainfall for the some hydrological years: The lower paths are associated with poor rainfall, which result in a poor aquifer response, characterized by groundwater droughts. Due to memory effect of Caposele spring, two distinct thresholds have been fixed; if the annual rainfall of the antecedent years is below the mean, to produce a spring flood, the system needs higher amount of rainfall (2° threshold). Based on historical series, cumulative rainfall recorded at a specific time can be associated with a probability value; also the amount of rainfall to reach specific threshold could be evaluated statistically (Fiorillo 2009). A groundwater drought could forecast since January–February, providing a very useful tool for water management.

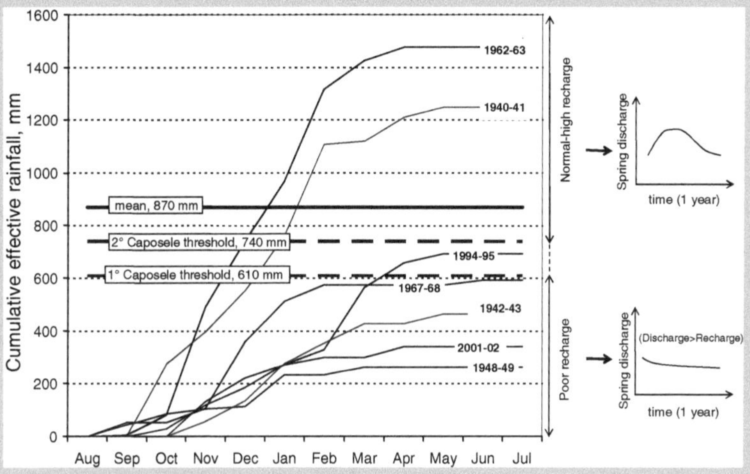

Fig. 15.7 Trend of cumulative effective rainfall for some wet and dry years (data of Serino rain gauge). The path below the threshold results in groundwater droughts (modified from Fiorillo 2009)

As karst aquifers can involve large portion of continental crust, outcropping over large areas or extending below other non-karstic rocks, they can reflect the effects induced by earthquakes. Earthquakes can cause variation of spring discharges in function of seismic magnitude and epicenter distance from aquifers (Box 15.1.4), but can also induce variation of chemical and physical proprieties of the groundwater. Most of these variations occur after earthquake occurrence and reflect the consequences of seismic trouble; preearthquake chemical and physical characteristics variations of the groundwater, including spring discharge, are a controversial topic.

Box 15.1.4

On November 23, 1980, a wide area of southern Italy was hit by disastrous earthquake (about 3,000 deaths, 280,000 homeless), with the epicenter zone close to Caposele spring and seismic magnitude Ms = 6.9. During this earthquake, several springs showed anomalous behavior, characterized by a rapid increase in discharge (Celico 1981; Cotecchia and Salvemini 1981).

Figure 15.8a shows spring hydrographs during the 1980–1985 periods. From a minimum reached 2 days after the earthquake (Nov 24, 1980–Nov 25, 1980; 4,210 l/s), since Nov 26, 1980, the Caposele spring discharge rapidly increased up to a maximum value of 7,320 l/s on Jan 29, 1981, which was the highest of the historical series. Besides, this spring never had a maximum in January as it occurs generally during May–August period (Fig. 15.5). The earthquake modified the hydraulic and hydrologic behavior of Caposele spring, causing an extraordinary discharged water volume during 1980–1981, and a decrease in spring discharge during the following 2 years (Celico and Mattia 2002). For other springs, the discharge increased several days before the earthquake and continued for several weeks later (Fig. 15.8a). The maximum was reached on Dec 10, 1980, for Bagno della Regina spring (3,060 l/s), between January 22 and February 25, 1981, for Pollentina spring (1,650 l/s), between Dec 15, 1980, and Dec 28, 1980, for Peschiera spring (510 l/s), and between Dec 14, 1981, and Jan 9, 1981, for Prete spring (380 l/s). Also in these cases, maxima were reached before the historical period of March–April for these springs.

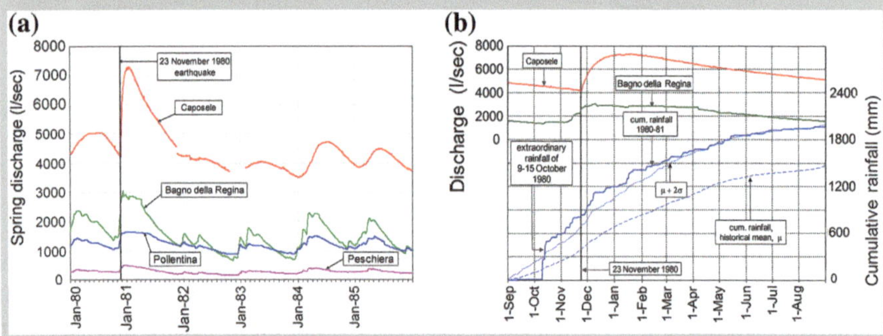

Fig. 15.8 **a** Daily discharge of several springs during 1980–1985 period. Caposele spring shows an increase in discharge after the earthquake; other springs increased discharge before earthquake. **b** Daily spring discharge and daily cumulative rainfall since Sep 1, 1980; the historical mean of cumulative rainfall, μ, and the historical mean of cumulative rainfall plus 2 standard deviations, $\mu + 2\sigma$, are also shown (period 1963–2003, n. 40 years)

Figure 15.8b shows the rainfall of 1980–1981 at Serino rain gauge; it also shows the historical mean, and historical mean + 2 st. dev. Big storms occurred between Oct 9, 1980, and Oct 15, 1980, and provided a strong recharge of karst aquifers; the following rainfall up to Nov 23, 1980, was also intense and maintained the cumulative rainfall well above the historical mean. These data show how the discharge increase in several springs occurred before the earthquake was induced by the extraordinary rainfall fell in October 1980, and could not be used as possible "anomalous" hydrological signal connected with the following earthquakes. The earthquake of Nov 23, 1980, caused the further increase in the Cassano spring discharge during the following days. The cause of spring discharge increase is not easy to find; Muir-Wood and King (1993) provided a model to explain the hydraulic behavior based on the style of faulting during earthquake. However, no previous "anomalous" hydrological signal can be deducted from spring discharge before the 1980 earthquake, and the possibility to prevent any earthquake by spring discharge measurement could provide misleading results.

15.2 Forecasting Long-Term Spring Discharge

Vesna Ristić Vakanjac

Centre for Karst Hydrogeology, Department of Hydrogeology, Faculty of Mining and Geology, University of Belgrade, Belgrade, Serbia

15.2.1 Introduction

The significance of karst water resources lies in the fact that the water quality is generally high and does not require major spending to achieve compliance with drinking water standards (Stevanović et al. 2011). As a result, karst water resources have become or are becoming increasingly important sources of drinking water supply (Nikolić et al. 2012). The type of porosity, ground morphology, inaccessibility, and absence of human settlements, as well as the places in which karst springs tend to occur, largely determine the karst groundwater monitoring regime (Stevanović et al. 2014).

Karst spring discharge is a highly complex process of streamflow formation. This complexity is the result of a very elaborate geological setting through which infiltrated rainfall passes on its way to the point of discharge via a karst spring or other type of karst aquifer drainage. In this regard, it is assumed that in high-porosity karst formations, there are underground voids, which are either large (cave conduits, caverns) or small (fractures and fissures) (Ristić 2007). The flow through this porous environment depends on the size, shape, nature, filling, and interconnection of such voids. "Watershed responses" to rainfall runoff patterns related to karst aquifers are diverse (Ristić 2007; Ristić et al. 2012a). The transformation of rainfall into karst spring discharge can be very rapid (Fig. 15.9), exhibiting steep hydrographs of short duration. After a rainfall event, some of these springs even dry out. Conversely, the process can be rather slow and long lasting, due to the storage potential and retardation capacity of the underground voids (Fig. 15.10). In such cases, the slope of the hydrograph's recession limb tends to be very gentle (Stevanović et al. 2010b). There are many variants and combinations between these two extremes.

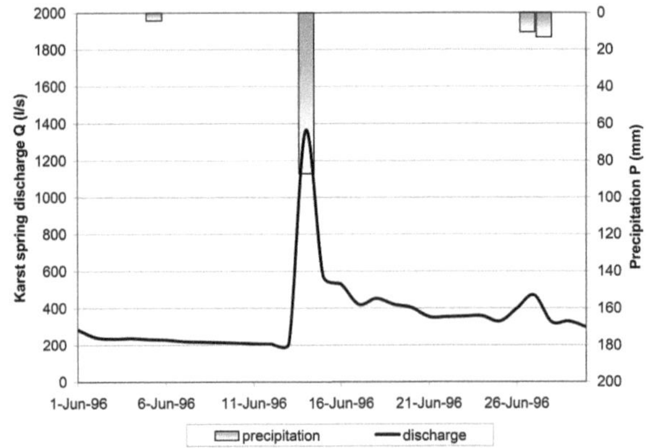

Fig. 15.9 Banja spring near Valjevo: rapid precipitation propagation and rapid rise and recession in karst spring discharge hydrograph

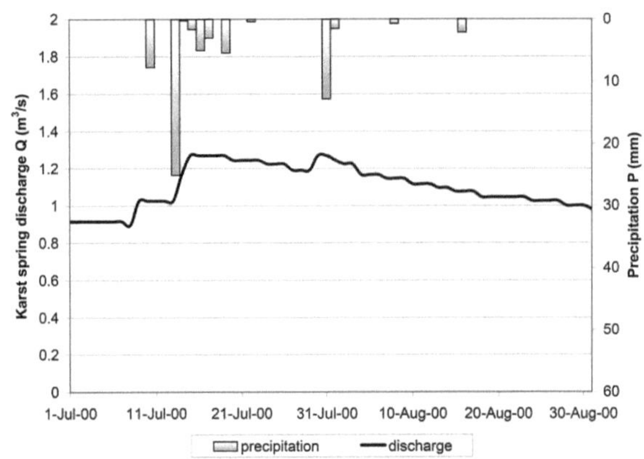

Fig. 15.10 Vapa spring near Sjenica: slow and protracted recession of karst spring discharge hydrograph

15.2.2 Autocorrelation and Cross-Correlation Analyses

Autocorrelation and cross-correlation analyses can be very useful when either the rising or recession stage of a karst spring discharge hydrograph is assessed.

Autocorrelation represents the effect of a random variable X (e.g., karst spring discharge rate, water level of the spring pool, and groundwater level) on itself for a time lag of 1, 2, 3, 4, …, n. The time lag can be expressed in days, weeks, months, etc., while the strength of the correlation between the data sets established in this way is expressed via the correlation coefficient obtained from (Kresic 2010):

$$r_k = \frac{\frac{1}{n-k}\sum_{i=1}^{n-k}(x_i - x_{\text{av}})\cdot(x_{i+k} - x_{\text{av}})}{\frac{1}{n}\sum_{i=1}^{n}(x_i - x_{\text{av}})^2} \tag{15.1}$$

where

n is the number of recorded data points,
x_{av} is the average value of the samples, and
x_i is the value of the random variable at time $t = i + k$.

The correlation coefficients for different time lags as a function of the time lags constitute an autocorrelogram.

When a karst spring discharge hydrograph is assessed, it is evident that the discharge rate recorded today is definitely affected to some extent by the water discharged yesterday, the day before yesterday, 3 days ago, etc., in the same way as the discharge rate observed today will affect the amount of water discharged tomorrow, the day after, etc. The degree of interdependence is largely affected by the discharge regime, the rate of hydrograph rise or recession, and the ratio of absolute maximum to minimum discharges ($Q_{abs,max}$:$Q_{abs,min}$). The closer this ratio is to unity, the degree of interdependence will be higher and the autocorrelation function will exhibit a gentler slope. In the opposite case, if the ratio is greater than 100 or more, the autocorrelograms are steep. When autocorrelograms are extremely steep, the analyzed time series are considered to be non-autocorrelated and are referred to as independent (there is no persistence, and the time series have no memory) (Kresic 2010).

The term "memory" refers to the time period up to the point where the values of the autocorrelation function become insignificant (or, in other words, when they are no longer statistically significant). In general, the statistical significance threshold depends on the length (extent) of the analyzed time series, where 0.2 is usually taken as the reference value (Mangin 1984a). Alternatively, this can be the time when the autocorrelation function intersects the abscissa, or the time when autocorrelation coefficients become negative (Jukic 2005). A weak memory is generally indicative of highly karstified formations with a low storage capacity. Conversely, when the memory is strong (long lasting), the karst formations feature large reserves (Kresic and Stevanović 2010). Given that karst spring discharge largely depends on precipitation, the shape and slope of the autocorrelogram also depend on the type of precipitation (rain, snow), and the frequency, amount, and intensity of precipitation, as well as (in the case of snow cover) temperature (Eisenlohr et al. 1997) (Box 15.2.1).

Box 15.2.1

The Mlava spring is a karst spring that features the longest discharge time series in Serbia (also described and discussed under case study 5 of Chap. 6, and Box 9.1 of Chap. 9). Discharge monitoring at this spring began in 1966 and continues to the present day. The spring drains the northern periphery of Mt. Beljanica and is one of the most powerful springs in Serbia (Q_{max} = 19 m^3/s (March 24, 1986), Q_{av} = 1.85 m^3/s, and Q_{min} = 0.215 m^3/s (July 28, 1988–August 2, 1988) (Ristić Vakanjac et al. 2012b). Two calendar years were selected for the autocorrelation analysis: a typical dry year (1983, Fig. 15.11) and a typical wet year (1972, Fig. 15.12). The analysis is summarized by the autocorrelograms shown in Fig. 15.13. In both cases, the autocorrelation coefficients declined with increasing time lag. All the dry years exhibited a gentler, relatively even autocorrelogram slope. During the selected dry year, the autocorrelation coefficient began to slightly differ from 0.2 after as many as 65 days. As such, the memory was up to 65 days and until then, the karst discharge process could not be deemed independent. The autocorrelogram typical of wet years looked entirely different.

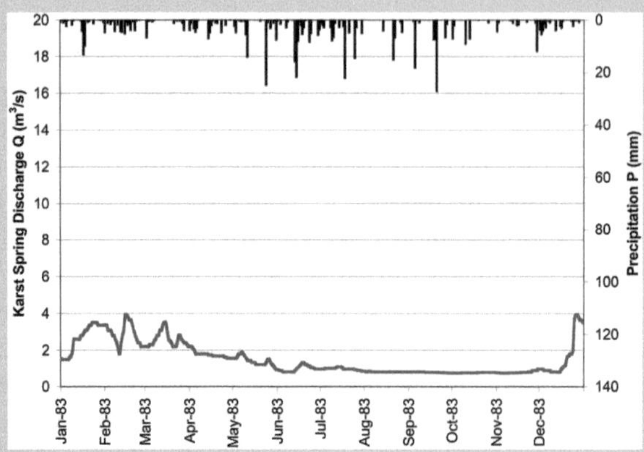

Fig. 15.11 Mlava spring: typical dry year (1983)

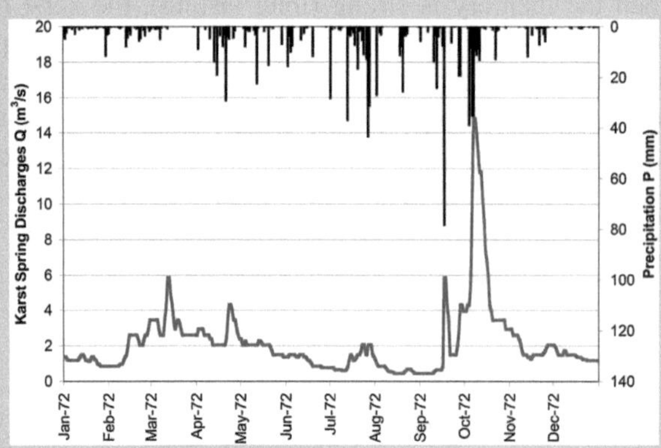

Fig. 15.12 Mlava spring: typical wet year (1972)

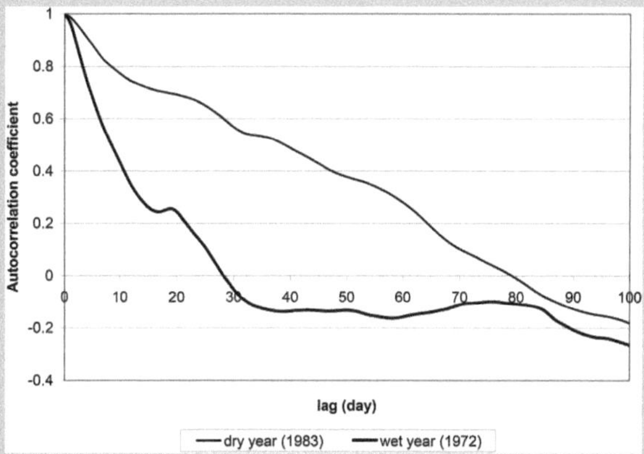

Fig. 15.13 Autocorrelograms of the Mlava spring in typical dry and wet years

The slope was much more pronounced, and the memory was up to 20 days, at which time the autocorrelation coefficients exceeded the statistical significance threshold. This was attributable to considerable aquifer recharge on account of precipitation, which was relatively continuous during the wet year and indicative of groundwater flow through fractures or relatively large karst conduits.

The result was rapid propagation of precipitation, demonstrated by a rapidly rising discharge hydrograph. Therefore, the karst spring discharge regime is affected to a large extent by the pluviographic regime of the given area. A raindrop that falls on a karst formation needs some time to reach the aquifer and then, as part of the aquifer, to arrive at the karst spring along privileged pathways of groundwater flow. Keeping this in mind, cross-correlation analysis is used to assess time-dependent random variables (specifically discharge and precipitation in the present case) (Box 15.2.2). In cross-correlation analysis, the correlation between the time-dependent random variable (daily discharge rate, groundwater level, etc.) and the independent random variable (daily precipitation total) can be quantified by calculating the cross-correlation coefficient at different time steps (Kresic 2010). The dependency of the cross-correlation coefficients for different time lags, as a function of the time lags, is referred to as a cross-correlogram (Box 15.2.2). The cross-correlation coefficient for any time lag k is obtained from the following equation (Prohaska 2006; Kresic 2010):

$$r_k = \frac{\mathrm{COV}(x_i, y_{i+k})}{\sqrt{\mathrm{VAR}(x_i) \cdot \mathrm{VAR}(y_i)}} \tag{15.2}$$

where

COV is the covariance between the two time series, is the independent variable (specifically the daily precipitation total in the present case),

y_i is the dependent variable (average daily discharge time series), and $\mathrm{VAR}(x_i)$ and $\mathrm{VAR}(y_i)$ are the variances of the two time series.

The covariance is calculated using the following equation:

$$\mathrm{COV}(x_i, y_{i+k}) = \frac{1}{n-k} \sum_{i=1}^{n-k} (x_i - \bar{x}) \cdot (y_{i+k} - \bar{y}_{i+k}) \tag{15.3}$$

The variances of the time series are obtained from

$$\mathrm{VAR}(x_i) = \frac{1}{n-k} \sum_{i=1}^{n-k} (x_i - \bar{x})^2 \tag{15.4}$$

$$\mathrm{VAR}(y_i) = \frac{1}{n-k} \sum_{i=1}^{n-k} (y_{i+k} - \bar{y}_{i+k})^2 \tag{15.5}$$

Box 15.2.2

The data used to assess the effect of precipitation on the discharge regime of the Mlava Spring came from a climatic station on Mt. Crni Vrh, located in the immediate vicinity of this spring's catchment (Box 15.2.5 and Fig. 15.19). Because of its altitude (1,037 m above sea level), this station best reflected the pluviographic regime of the northern part of the Mt. Beljanica karst, which is drained by the Mlava Spring. The long-term average precipitation total was 783.6 mm, and the range of annual totals was from 508 mm (1990) to 1,096 mm (1974).

Here, two typical years, one dry and one wet, were selected to assess the effect of precipitation. In both cases, the cross-correlation coefficients showed that after 33 days, precipitation no longer affected the discharge (Fig. 15.14). A distinct peak was associated with time lag two in both years, indicating that precipitation generally needed 2 days to propagate through the Mlava Spring catchment.

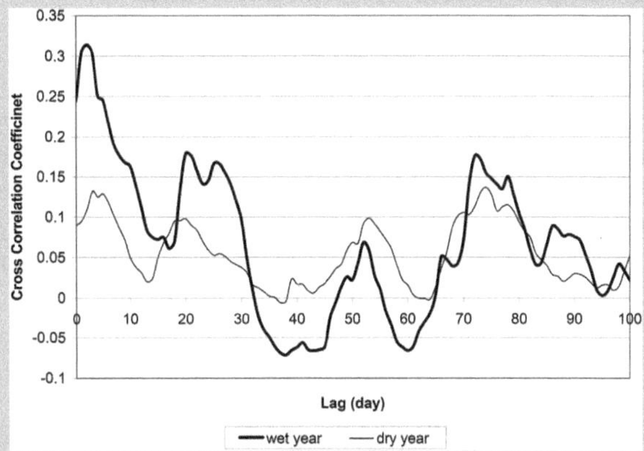

Fig. 15.14 Cross-correlogram of the Mlava spring for typical wet and dry years

Less pronounced peaks, after 20 and 72 days, were present in both cross-correlograms. They were most likely a result of the snow cover in the winter months. The peaks represented the time needed for the melted snow to reach the karst spring.

To define the shape of any karst aquifer discharge hydrograph, the most reliable approach is to apply hydrometric methods to monitor discharge and precipitation regimes in the watershed or in its immediate vicinity (Ristić 2007). However, this is generally not the case in practice. There are only a few karst springs in Serbia where discharge is systematically observed using hydrometric methods. Karst discharge regimes are generally monitored sporadically and measurement campaigns undertaken as needed by local users (water utilities), with no long-term objectives or standard data collection, processing, and dissemination. In most cases, available time series are of relatively short duration (often less than 1 year), and this has a considerable impact on the assessment of groundwater storage and budget, and consequently on the users of the aquifer (public water supply, balneotherapy, irrigation, small hydroelectric power plants, etc.).

To minimize potential errors when assessing groundwater availability, as well as to calculate the parameters of the water balance equation, it is necessary to extend available short-term time series as much as possible. Regression models, or specifically multiple regression models (linear and nonlinear), are often used for this purpose.

15.2.3 The Multiple Linear Regression Model

The method is most often used to simulate or forecast random variables. Multiple linear regression is applicable where one phenomenon is a function of two more independent phenomena. A correlation is established between a dependent variable Y, and independent variables X_1, X_2, …, X_k, based on which the dependent variable is simulated and a forecast for a certain time period produced. The correlation is described by a regression model of the form (Prohaska 2006):

$$Y_i = \beta_o + \beta_1 \cdot x_{1,i} + \beta_2 \cdot x_{2,i} + \cdots + \beta_n \cdot x_{n,i} + e_i \tag{15.6}$$

where

Y_i dependent variable of the ith order,
x_i independent variable of the ith order,
β_i unknown coefficients of multiple regression, and
e_i random error.

The least squares method is used to calculate the unknown coefficients of multiple regression, and Eq. (15.6) acquires the form:

$$\tilde{y} = a + b_1 \cdot x_1 + b_2 \cdot x_2 + \cdots + b_n \cdot x_n \tag{15.7}$$

where

$\tilde{y}$ is the analytical value of the dependent variable,
while a, b_1, b_2, …, b_n are calculated numerical values of the multiple regression coefficients.

Autoregressive (AR) models are often used to simulate (calculate) karst spring discharge, where the dependent variable is Q_t—predicted discharge at time t, and the independent variables $Q_{t-1}, Q_{t-2}, \ldots, Q_{t-k}$ are discharges 1, 2, …, k days before (Box 15.2.3). In the specific case, Eq. (15.7) acquires the form:

$$Q_t = a + b_1 \cdot Q_{t-1} + b_2 \cdot Q_{t-2} + \cdots + b_k \cdot Q_{t-k} \tag{15.8}$$

where a, b_1, b_2, …, b_k are model parameters.

Cross-regressive (CR) models can be used for these purposes applying the same principle, where the dependent variable Q_t is also the predicted discharge at time t, and the independent variables $P_{t-1}, P_{t-2}, \ldots, P_{t-n}$ are precipitation levels 1, 2, …, n days before (Box 15.2.3) Eq. (15.7) now becomes

$$Q_t = a + b_1 \cdot P_{t-1} + b_2 \cdot P_{t-2} + \cdots + b_n \cdot P_{t-n} \tag{15.9}$$

where a, b_1, b_2, …, b_n are also model parameters.

Finally, the most often used models are hybrids of AR and CR (= ARCR) (Box 15.2.3), such that Eq. (15.7) becomes

$$Q_t = a + b_1 \cdot Q_{t-1} + \cdots + b_k \cdot Q_{t-k} + c_1 \cdot P_{t-1} + c_2 \cdot P_{t-2} + \cdots + c_n \cdot P_{t-n} \tag{15.10}$$

where a, b_1, b_2, …, b_k, c_1, c_2, …, c_n are model parameters as well.

Box 15.2.3

All three models—the autoregressive (AR) model, the cross-regressive (CR) model, and the hybrid of the two (ARCR)—were used to simulate the discharge of the Mlava Spring over the entire period of monitoring (1966–2010). The correlation coefficients obtained with the AR and CR models showed that those produced by the AR model were much higher than in the case of the CR model (Fig. 15.15). The AR model coefficients were from 0.980 (for order 1) to 0.982 (for order 10) and resulted from the previously mentioned long memory (Box 15.2.1 and Fig. 15.13). Conversely, the CR model coefficients were low, from 0.08 to slightly above 0.2 for order 10 (Fig. 15.15). Given that the correlation coefficients were statistically insignificant, this model could not be used to simulate karst spring discharge. Such low values were a result of the following circumstances: (1) The simulation was based on gross precipitation, not effective precipitation, and (2) snowfall was taken at the time of occurrence, excluding the time needed for snowmelt to take place.

The hybrid ARCR model generally produced the best results. The correlation coefficients were in the interval from 0.981 (order 1) to 0.983 (order 5). Figure 15.16 shows parallel hydrographs of observed and simulated discharges of the Mlava Spring using the ARCR model, for order 5.

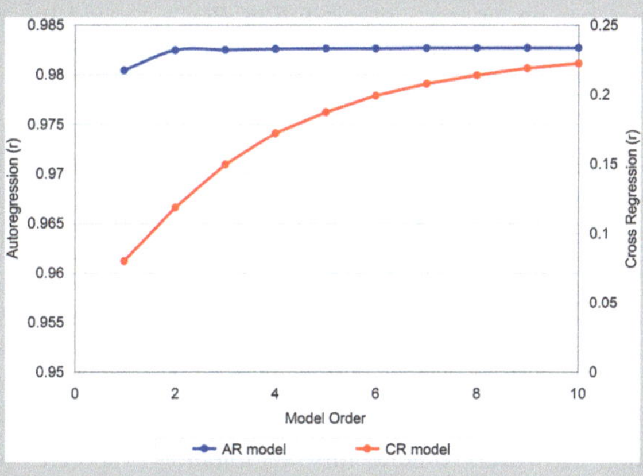

Fig. 15.15 Parallel representation of simple autoregressive and cross-regressive models

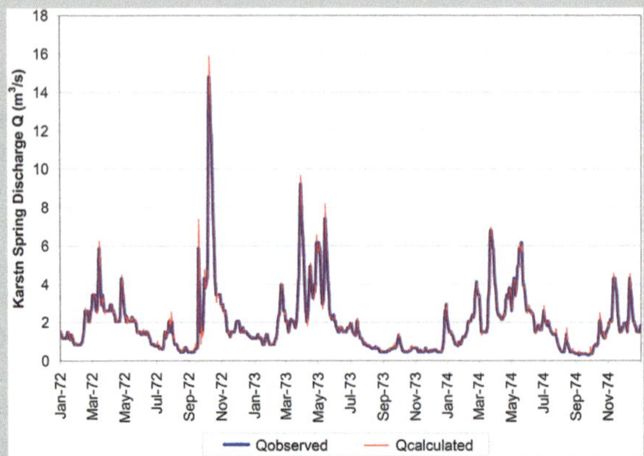

Fig. 15.16 Parallel hydrographs of observed and computed (using ARCR model) discharges of the Mlava spring

Apart from regressive models, **transform functions** can be used to simulate karst spring discharges. In the specific case, to transform precipitation into a hydrograph of daily karst spring discharges, it was assumed that the karst aquifer was an open system, variable over space and time (Ristić 2007). The input quantities of the system were weather parameters, precipitation $P(x, y, t)$, and the output was the spring discharge, distributed over time, to which the following relation applies:

$$Q(t) = H(P(x, y, t)) \tag{15.11}$$

where, H is the functional operator that defines how the system transforms inputs into an output, and y represents the spatial coordinates of the weather station at time t.

The functional operator comprises a series of functions that depend on the number of transforms that can describe the considered process. The entire precipitation-to-discharge transformation process is generally divided into three subprocesses, so three functions need to be established to describe the system.

The first subprocess is related to the transformation of gross precipitation into effective precipitation, namely the overall precipitation that reaches the watershed is gross precipitation. However, some of this precipitation evaporates (evapotranspiration), some is infiltrated into the ground to replenish soil moisture, and the remainder reaches the water table and recharges the aquifer. The portion of precipitation that reaches the karst spring via the aquifer is net or effective precipitation.

The other two subprocesses involve the transformation of effective precipitation into a karst spring discharge hydrograph, comprised of two functions (Ristić 2007):

- Distribution function, by which the volume of effective precipitation is distributed over time ("elementary hydrograph"), and
- Propagation function, by which the elementary hydrograph is transformed into the karst spring output hydrograph.

The distribution and propagation functions are often replaced with a single transform function.

In the specific case, gross precipitation was transformed into effective precipitation by means of the runoff coefficient φ_j, on a monthly basis:

$$\varphi_j = \frac{h_j}{P_j} = \frac{P_{\mathrm{ef}_j}}{P_j} \tag{15.12}$$

or

$$P_{\mathrm{ef}_j} = \varphi_j \cdot P_j \tag{15.13}$$

where

P_j is the gross precipitation,
P_{ef_j} is the effective precipitation,
j is the given month (1, 2, 3, …, n), and
h_j is the runoff layer derived from

$$h_j = \frac{Q_j \cdot T}{F}\ (\mathrm{mm}) \tag{15.14}$$

where

Q_j is the karst spring discharge,
F is the surface area of the karst spring catchment, and
T is the time in seconds for the given month.

The distribution transform function that transforms net precipitation to a hydrograph of daily karst aquifer discharges is defined by the relation (Box 15.2.4):

$$Q(k) = F \cdot \sum P_{\mathrm{ef}_j} \cdot \mathrm{TF}_{(k-s+1)} \tag{15.15}$$

where

$Q(k)$ is the ordinate of the discharge hydrograph on the kth day,
P_{ef_j} is the effective precipitation in the jth month,
$\mathrm{TF}_{(k-s+1)}$ is the transform function at time $k - s + 1$ (1/day), and is the surface area of the catchment (km^2).

The following equation is used to calculate the transform function at the vth point in time:

$$\mathrm{TF}_{(v)} = \frac{1}{\tau \cdot (n-1)!} \cdot \left(\frac{v}{\tau}\right)^{n-1} \cdot \mathrm{e}^{-\frac{v}{\tau}} \tag{15.16}$$

where τ and n are the parameters: τ has a time dimension (day), while n is dimensionless.

The model parameters τ and n are obtained through calibration, and the quality of the calibrations is controlled via the coefficients of correlation between real (observed) and simulated spring discharges (Box 15.2.4). The parameters are calibrated until the strongest correlation between the two time series is obtained, or, in other words, until the highest correlation coefficient is found.

Box 15.2.4

The Mlava Spring was also selected to apply transform functions to simulate discharge from karst. Taking the entire period of discharge monitoring, first the model parameters τ and n were calibrated by means of the correlation coefficient, until the strongest correlation between the observed and simulated values was obtained. Figure 15.17 shows the coefficients of correlation between calculated and real spring discharges as a function of different values of the parameters τ and n. It is apparent from the graphic that the combination of $\tau = 13$ and $n = 1$ yielded the highest correlation coefficient ($r = 0.72$). Figure 15.18 is a parallel representation of the observed and simulated discharge hydrographs, based on the combination that provided the strongest correlation ($\tau = 13$ and $n = 1$).

Relatively significant deviations of up to 20 % on a monthly basis occurred in the winter and spring months and were a logical consequence of the model itself, namely the transform function model did not account for the time during which precipitation was trapped in the snow cover in winter, or the subsequent snowmelt in the spring months when such precipitation took part in the formation of runoff. The model only transformed precipitation (either rain or snow) at the time it was registered, and this obviously resulted in higher positive deviations ($Q_{\text{real}} - Q_{\text{simulated}}$) in spring and negative deviations in winter.

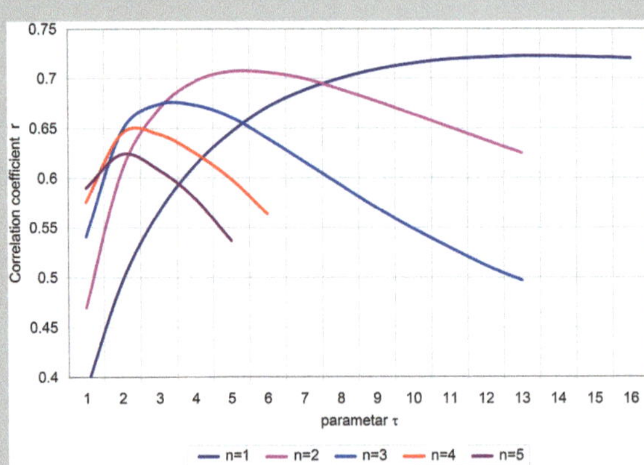

Fig. 15.17 Mlava spring: correlation coefficients as a function of varying τ and n (1966–2010)

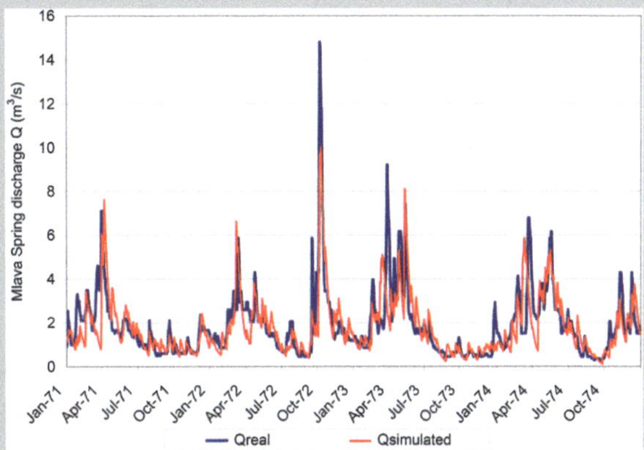

Fig. 15.18 Parallel representation of real and simulated (using transform function) discharges of the Mlava spring (1972–1974) for $t = 13$ and $n = 1$

15.2.4 Model for Filling the Data Gap and Assess Catchment Size and Dynamic Storage

For the purposes of assessing partially (insufficiently) gauged karst springs, at the Department of Hydrogeology at the Faculty of Mining and Geology of the University of Belgrade has been developed a model that apart from filling the gaps and extending recorded discharge time series computes the real evapotranspiration, catchment size, and the dynamic storage volume of the karst aquifer (Ristić 2007; Ristić Vakanjac et al. 2012b, 2013a). The model is comprised of four levels:

- Level 1: Filling the gaps of average monthly discharge time series using one of the previously mentioned models;
- Level 2: Water budgeting of the karst aquifer;
- Level 3: Identification of transform function model parameters; and
- Level 4: Simulation of long-term daily discharges

Level 1 comprises the establishment of linear correlation functions between standardized variables of the average monthly discharge time series and the hydrometeorological parameters relevant to the formation of karst spring discharge. The parameters of the established correlations are then spatially analyzed, and in regions where they are homogeneous, they are used inversely to find corresponding long-term discharge time series, where information about the "analog" watershed and meteorological data on the considered watershed are available.

In the specific case, the multiple nonlinear correlation (MNC) model was used to establish a correlation between the standardized variables of the following hydrometeorological parameters:

$$U\left(Q_{ij}\right) = a_{01} \cdot U_1\left(Q_{ij}^a\right) + a_{02} \cdot U_2\left(P_{ij}\right) + a_{03} \cdot U_3\left(T_{ij}\right) + a_{04} \cdot U_4\left(V_{ij}\right) + a_{05} \cdot U_5\left(N_{ij}\right) \tag{15.17}$$

where

$U_k\left(X_{ij}\right)$ standardized value of the corresponding variable X_{ij};
Q_{ij} average monthly discharge of the karst aquifer;
Q_{ij}^a average monthly discharge of the "analog" river;
P_{ij} average monthly precipitation totals in the considered karst watershed;
T_{ij} average monthly air temperatures in the considered karst watershed;
V_{ij} average monthly humidity in the considered karst watershed;
N_{ij} average monthly vapor pressure in the considered karst watershed;
a_{0l} unknown regression coefficient (parameters)
$j = 1, 2, \ldots, 12$ serial number of the month
$i = 1, 2, \ldots, N$ serial number of the year (in the sequence)
$l = 1, 2, m, \ldots M$ number of independent variables–regression

The procedure for defining model parameters is described in detail in the literature (Prohaska et al. 1977, 1979, 1995). The basis for calculating the model parameters are the mutual correlation coefficients between the standardized variables of all simple combinations of dependent and independent quantities Eq. (15.17) and the corresponding standardized variable as a function of its real value $U_l\left(X_{ij}\right)$. The values of the standardized correlation coefficients r_{lm} are used to obtain the coefficients of linear regression $a_{01}, a_{02}, \ldots, a_{0M}$ and the corresponding weights $\delta_{01}, \delta_{02}, \ldots, \delta_{0M}$.

At Level 2, the basic water balance equation for a karst aquifer is used, with a monthly time step:

$$P_j = h_j + E_j + \left(V_j - V_{j-1}\right) = h_j + E_j \pm \Delta_j \tag{15.18}$$

where

P_j monthly precipitation total in the karst watershed in the *j*th month,
h_j average monthly karst spring discharge layer in the *j*th month,
E_j monthly sum of actual (real) evapotranspiration from the karst watershed in the *j*th month,
V_j water volume in the considered karst aquifer in the *j*th month, and
Δ_j variation in the amount of water stored in the karst formation in the *j*th month.

Given data availability, the presented water balance Eq. (15.18) has two unknown quantities, E_j and Δ_j. To assess the first approximation of the real monthly sums of evapotranspiration from karst, analytical values of daily sums of potential evapotranspiration (PET) are used, applying a modified Thornthwaite method (Thornthwaite 1948; Ristić Vakanjac et al. 2013a).

To determine the daily sums of actual, "real" evapotranspiration (RET), the water balance Eq. (15.18) is solved in stages, assuming that (Ristić Vakanjac et al. 2013a):

- The initial volume of water stored in the karst formation is equal to the volume at the end of the predefined analytical period, $V_0 \cong V_K$, where 0 and K denote the beginning and end of the analytical period;
- The distribution of daily sums of real evapotranspiration is nonlinear, such that on rainy days $\mathrm{RET}_i = \mathrm{PET}_i$, and in the subsequent days, the real daily sum of evapotranspiration declines according to the rule:

$$\mathrm{RET}_{i+\tau} = \Theta^{2\tau} \cdot \mathrm{PET}_i \tag{15.19}$$

where

$\mathrm{RET}_{i+\tau}$	real daily sum of evapotranspiration, and
$\tau = 1, 2, 3, \ldots, m$	time lag in days;

- The catchment size will be determined for different values of Θ ($\Theta = 0, 0.1, 0.2, \ldots, 0.9$, to which the rule $V_0 \cong V_K$ applies, and
- by constructing the function $\Theta = f(F)$, the real catchment size (apex of the nonlinear function $\Theta = f(F)$) will be defined by the intersection of the function $\Theta = f(F)$ with the axis of symmetry of the angle formed by the tangents of function $\Theta = f(F)$ (Box 15.2.5 and Fig. 15.23).

This yields the real catchment size, and the defined values of parameter Θ are used to calculate the daily values of real evapotranspiration. Now that real evapotranspiration is a computed quantity, Eq. (15.18) can be applied to calculate the dynamic storage volume variation during a given period (where the gaps in the karst spring discharge time series have been filled) (Box 15.2.5 and Fig. 15.24).

Box 15.2.5

Veliko Vrelo (Big Spring), along with Malo Vrelo (Small Spring), drains the southern edges of the Mt. Beljanica karst (Fig. 15.19). It is situated near the top of a gorge. The water emerges at the bottom of a long scree slope between large limestone blocks at the top of a short and shallow doline, filled with large limestone rocks (Fig. 15.20). The spring shifts upstream and downstream, depending on the water table. The point of discharge can be as much as 50 m higher during wet years, compared to dry years (Milanović et al. 2014). At times of maximum discharges, the spring yields more than 10 m^3/s of water. When discharges are minimal, some 50 l/s of water emerges from the scree and blocks (Ristić 2007). In 1978 and 1979, Geozavod from Belgrade conducted hydrogeological and hydrological investigations in the areas of Mt. Kučaj and Mt. Beljanica, which included discharge monitoring of Veliko Vrelo once in 5 days (Stevanović 1992). The National Hydrometeorological

Service of Serbia (RHMZ Serbia) initiated river-stage gauging and spring discharge monitoring on January 1, 1995, which were discontinued after 2003. During that period, the average discharge was 0.584 m^3/s. Average monthly discharges of Veliko Vrelo ranged from 0.048 m^3/s (July 2001) to 2.054 m^3/s (April 1999). The absolute maximum was recorded on June 11, 2002, and amounted to 6.1 m^3/s (Ristić Vakanjac et al. 2013b).

Because of the short period of monitoring of this spring, historical data from the nearest watershed, of similar physical and geographical characteristics, were used to assess long-term water availability. Specifically, data on the Resava River from the gauging station at Manastir Manasija (Manasija Monastery, catchment size 388 km^2) were used. The coefficients indicative of the strength of the correlation between daily discharges of Veliko Vrelo and the Resava River at Manastir Manasija were in the interval from 0.482 (wet years) to as much as 0.888 (dry years) (Fig. 15.21).

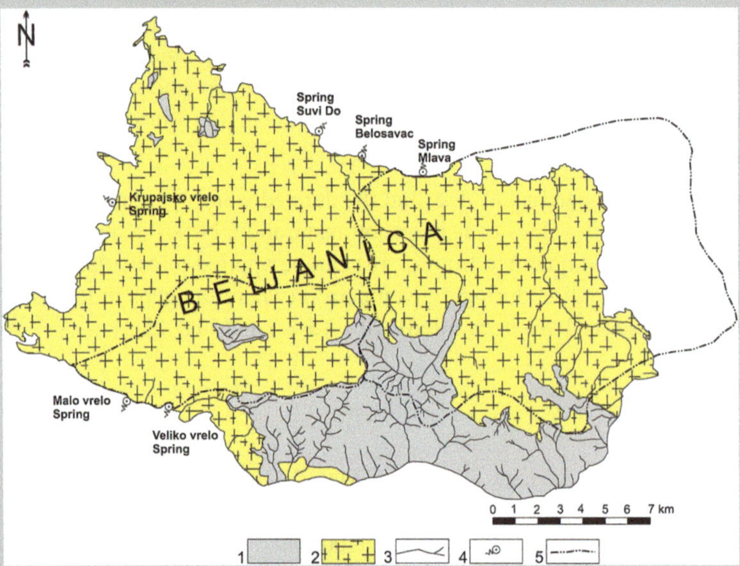

Fig. 15.19 Hydrogeological schematic map of the Beljanica karst massif (*1*) Non-karst area, (*2*) karst area, (*3*) watercourse, (*4*) karst spring, (*5*) hydrogeological watershed

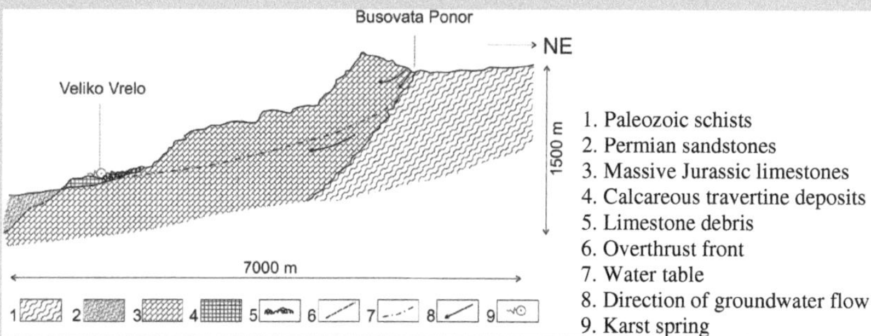

Fig. 15.20 Hydrogeological section through the Veliko Vrelo catchment (after Stevanović 1987)

In the immediate vicinity of the spring's catchment, there are two rain gauge stations (at Despotovac and Strmosten) and two climatic stations (at Žagubica and on Mt. Crni Vrh), installed by RHMZ Serbia. The correlation coefficients between precipitation totals and annual average spring discharges ranged from 0.523 (at Žagubica) to 0.851 (at Strmosten). On a monthly basis, they were between 0.444 (Žagubica) and 0.563 (Strmosten). The cross-correlation analysis (Jemcov et al. 1998; Prohaska et al. 2006) was used to define the effect of daily precipitation totals on the daily discharges of Veliko Vrelo. Hundred-day cross-correlograms were produced with a time step of $\tau = 1$ day, at all four stations, for the analytical period of nine years (1995–2003) (Fig. 15.22).

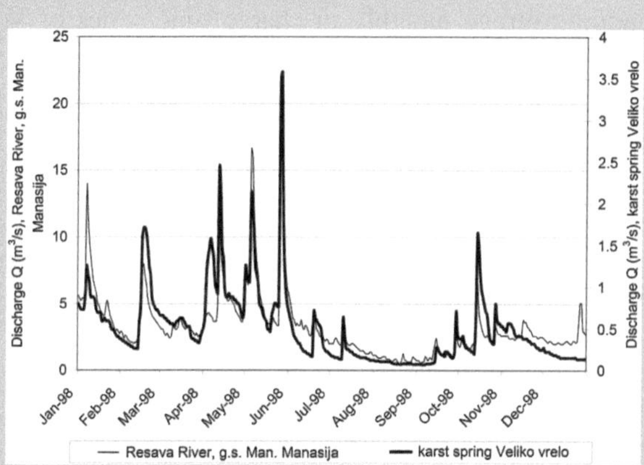

Fig. 15.21 Hydrographs of the Resava River at Manastir Manasija and Veliko Vrelo (Ristić Vakanjac et al. 2013b)

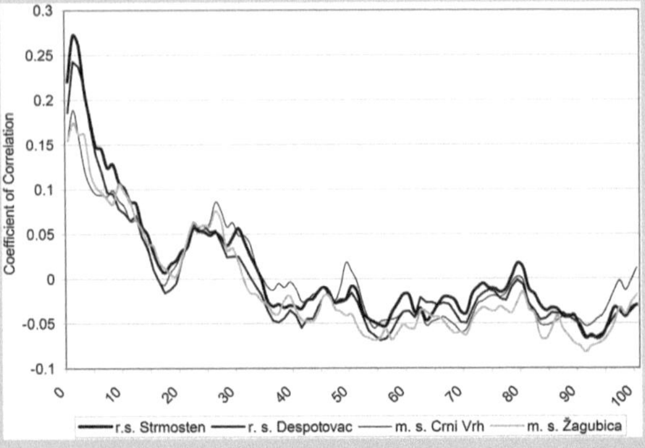

Fig. 15.22 Cross-correlogram (Stevanović et al. 2010a; Ristić Vakanjac et al. 2013b). Printed with kind permission of Acta Carsologica, 2010

The cross-correlogram shows that at all four stations the catchment responded to daily precipitation totals with a time lag of one day and that the highest correlation coefficient was 0.27 (precipitation at Strmosten, $\tau = 1$ day). All the studied stations exhibited this peak. There was another peak after 26 days. It mostly likely reflected the time needed for the water from snowmelt to reach the spring. This peak was the most pronounced on Mt. Crni Vrh because this station was at the highest altitude of the four stations and therefore situated in the part of the catchment that had the deepest snow cover.

In view of all the above, the following data were used at Level 1 of the model (extending average monthly discharge time series of Veliko Vrelo, Eq. 15.17):

- Dependent variable: time series of average monthly runoff layer of Veliko Vrelo (observation period),
- Independent variables:
  - Time series of average monthly runoff layer of the "analog" river (the Resava at Manastir Manasija),
  - Monthly precipitation totals at Strmosten,
  - Average monthly air temperatures on Mt. Crni Vrh,
  - Humidity on Mt. Crni Vrh, and
  - Vapor pressure on Mt. Crni Vrh.

Based on the established correlations, the time series of average monthly discharges of Veliko Vrelo was defined for the analytical period from 1960 to 2000, where the long-term average discharge was 0.581 m^3/s.

Then, at Level 2, it was possible to calculate daily sums of real evapotranspiration, catchment size, and dynamic storage volume of the karst aquifer.

Simulation of real (actual) daily sums of evapotranspiration: Monthly sums of potential evapotranspiration were calculated applying the Thornthwaite method and using the data recorded by the weather station on Mt. Crni Vrh (Ristić Vakanjac et al. 2013b). Then, taking the values of parameter $\Theta = 0, 0.1, 0.2, \ldots, 0.8$ and 0.9, the water balance equation was established for the catchment of Veliko Vrelo, by calibrating the potential catchment size such that the condition $V_0 \cong V_K$ was fulfilled. Using the obtained ranges of values of Θ and F, which fulfilled the set criterion, the functional dependencies $\Theta = f(F)$ were formed. The graphical interpretation of a resulting function is shown in Fig. 15.23 (Ristić Vakanjac et al. 2013b). Analysis of the functions $\Theta = f(F)$ revealed that the correlation between the parameter Θ and the catchment size was nonlinear over the entire range of Θ. At low values of Θ, variations in this parameter did not have a substantial effect on the catchment size or, in other words, the differential change in catchment size δF was insignificant such that the catchment size remained virtually unchanged. Conversely, when the values of Θ were close to unity (generally greater than 0.8), or when the variations in this parameter were small, there was a significant increase in the catchment size. The border point was also the point of inflection (apex) of the nonlinear function $\Theta = f(F)$. This approach was followed to define the catchment size (Veliko Vrelo, 38.1 km^2—Fig. 15.23), as well as the real evapotranspiration.

Determination of the dynamic storage volume of Veliko Vrelo: Using the generated input time series of monthly precipitation totals, actual (real) evapotranspiration, and runoff layers, Eq. (15.18) was applied to compute the volumes of water stored in the karst aquifer on a monthly basis. The results are graphically represented in Fig. 15.24, showing that the overall dynamic volume of water in the Veliko Vrelo catchment is approximately 26×10^6 m^3.

The output of Level 2 of the model was a tabular representation of the water budget components, characteristic of the Veliko Vrelo catchment. Table 15.2 shows the following water budget components: catchment size F (km^2), long-term average discharge rate Q (m^3/s), discharged volume of water W (10^6 m^3), long-term average runoff modulus q ($l/s/km^2$), runoff layer h (mm), average annual precipitation P (mm), average annual evapotranspiration E (mm), and long-term average runoff coefficient φ. All these parameters are related to an analytical period of 40 years (1960–2000).

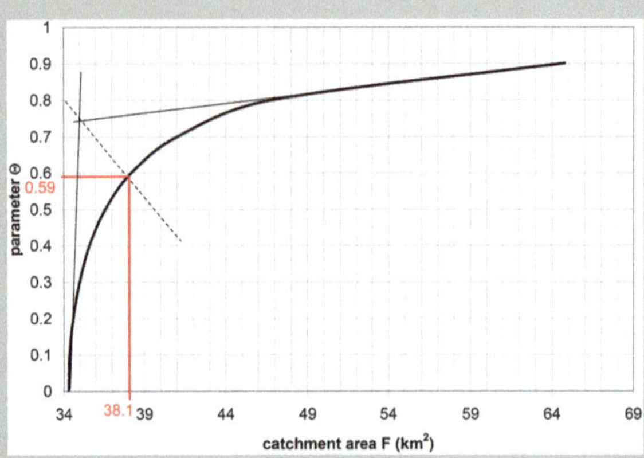

Fig. 15.23 Function $\Theta = f(F)$ of Veliko Vrelo catchment (Stevanović et al. 2010a; Ristić Vakanjac et al. 2013b). Printed with kind permission of Acta Carsologica, 2010

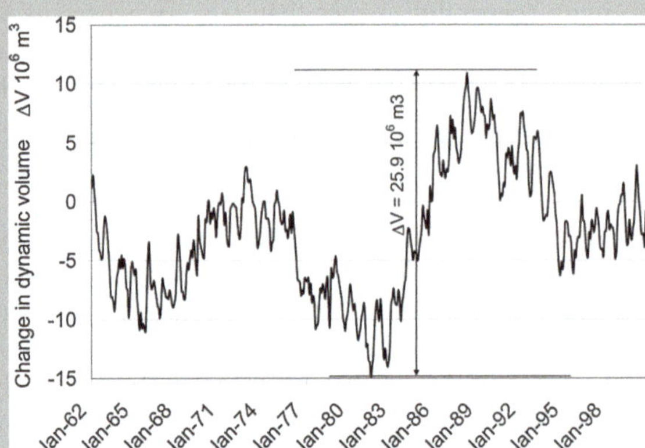

Fig. 15.24 Variation in Veliko Vrelo aquifer storage volume during the analytical period (Ristić Vakanjac et al. 2013b)

Table 15.2 Summary of the main water budget components of the studied catchment (Ristić Vakanjac et al. 2013b)

F (km^2)	P (mm)	E (mm)	h (mm)	Q_{av} (m^3/s)	q (l/s/km^2)	W (10^6 m^3)	φ
38.1	760.3	279.4	480.9	0.581	15.249	18.322	0.633

In order to simulate the daily discharges of Veliko Vrelo using daily discharge data for the period 1995–2003 (observation period), it was necessary to first determine the parameters τ and n. During the observation period (model parameters τ and n calibrated), following the procedure described above [Level 3, Eqs. (15.11–15.16) and Box 15.2.4], the highest correlation coefficient between the observed and simulated discharges of Veliko Vrelo ($r = 0.686$) was obtained for the combination $\tau = 10$ and $n = 1$ (see Fig. 3.50). Finally, at Level 4, the computed parameters τ and n were used to calculate the daily discharges of Veliko Vrelo over the entire study period, 1960–2000 (Fig. 15.25).

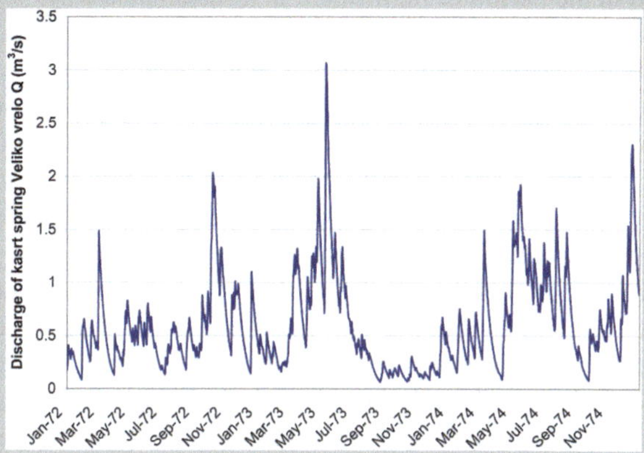

Fig. 15.25 Simulated discharge hydrograph of Veliko Vrelo during a period without monitoring

15.3 Model for Assessing Extraction Effects in an Aquifer System

Igor Jemcov

Department of Hydrogeology, Faculty of Mining and Geology, University of Belgrade, Belgrade, Serbia

15.3.1 Introduction

Rational use of karst water resources depends considerably on the existence of a successful approach to estimating karst groundwater potential for exploitation. Furthermore, the conception of assessment of groundwater potential for exploitation has to involve the application of engineering regulatory measures which affect natural outflow regime.

The most common method used nowadays is based on water extraction from static reserves (Bonacci 1987; Stevanović 1994), while the most effective one is the exploitation using wells (Pulido-Bosch 1999). Unlike previous methods, more complex technical solutions of tapping structures involve the construction of underground reservoirs. Due to their complexity, there are not many examples of underground dams. Of course, there are lot of variations in between, ranging from conceptually more or less sophisticated solutions, all of them being based on implementing active karst groundwater management system. Active groundwater management considers the application of different regulatory measures in order to activate storage. Important characteristics of the active management can be supported with the fact that it provides the required flow, greater than natural, that will be compensated in the period of high water, in a sustainable way.

Generally, a regulatory system may follow one of the two basic concepts. The first concept of the active groundwater management relies on an extensive utilization of natural storage so-called static reserves, and the "borrowing" of karst groundwater during periods of recession. The second one is based on the increase in the dynamic groundwater storage by the construction of underground dam and reservoir in order to provide artificial water storage. The selection between the two depends on the nature of each karst aquifer system. The first one is widely applied in practice due to favorable conditions such as the existence of subsurface outflow below main outlet zone (spring)—Box 15.3.1. On the other hand, the construction of an underground dam and reservoir to provide an artificial water storage and control groundwater discharge regime presents much more complex systems of tapping structure. Some studies have demonstrated that an artificial underground storage system may be a realistic way for active groundwater management (Milanović 1988). One of the important prerequisites for artificial underground reservoir construction is a hydrogeological structure with outlet zone confined with impervious formations. Several concepts of underground dams can be established—Box 15.3.2.

Box 15.3.1

Case example: St. Petka source

Example of a hydrogeological exploration for estimation of suitability of the aquifer system applying the concept of partial extraction of groundwater from the static reserves mainly based on the extensive exploration:

The St. Petka karst aquifer system is developed mainly in Jurassic, and partially Cretaceous limestone of the "Ravanica zone"—the western part of the Carpatho-Balkanic arch in Serbia. The karstic source is located at the contact with less permeable Neogene and alluvial sediments (Fig. 15.26)

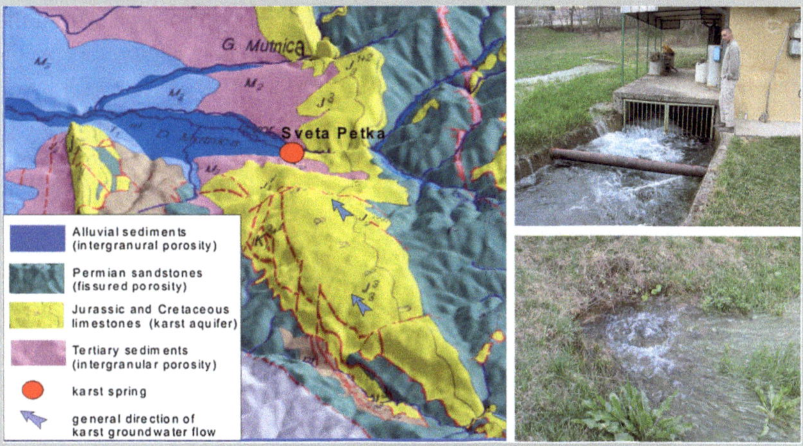

Fig. 15.26 Hydrogeological map of St. Petka karst aquifer system (*left*); St. Petka spring during flood (*right up*); subsurface drainage: during the high-water season, part of the subsurface outflow appeared downstream, as a consequence of the high water pressure (*right down*)

The assessment of groundwater potentiality is based on estimation of karst development deeply below the present main outlet. Using different methods, e.g., geophysics, it is possible to interpret depths of the karstified rock beneath less permeable beds. Gravimetry—map of the Bouger anomalies and horizontal gradients of Bouger anomalies—gravity acceleration; seismic refraction—two layers have been identified: clay 3–7 m thick and marlstone min. 30 m thick; and resistivity sounding −AB/2−400 m, with interpreted cross section (Fig. 15.27).

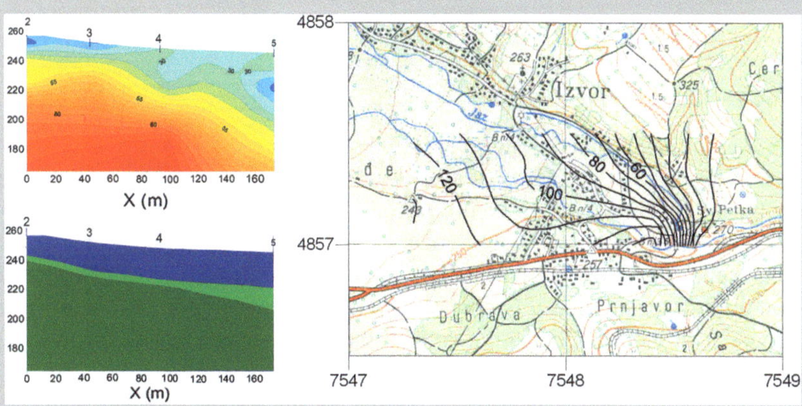

Fig. 15.27 Resistivity profile of the St. Petka source (*left up*), interpreted cross section according to resistivity sounding (*left down*); contours of the karstified limestone beneath less permeable layers

Additionally, successive hydrometry and thermometry downstream from the source, as well as borehole logging present indispensible exploratory methods for subsurface outflow detection (Fig. 15.28).

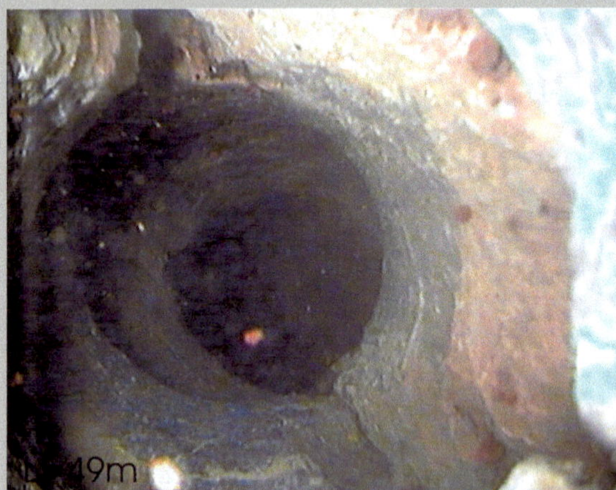

Fig. 15.28 Presence of the cavernous system, detected deeply below alluvial and Neogene deposits, about 25 m thick. Photograph was taken at the depth of 49 m in Jurassic limestone of the "Ravanica zone"

Application of the presented hydrogeological exploration and construction of proper tapping structure—the well—allows the use of larger groundwater capacity than natural to meet the demands of potential users, particularly during the period of recession.

Box 15.3.2

Case example: Perućac source

The Perućac karst aquifer system is developed mainly in Triassic limestone and dolomite, surrounded by impervious Paleozoic rocks. The spring outlet is situated at the contact of sandstone and overlying karstified limestone. Sandstone barrier plays a particularly important role since it provides a good condition for building an underground dam. Another important feature is a pronounced vertical fluctuation zone at the outlet, depending on hydrological conditions (Fig. 15.29).

Fig. 15.29 Perućac spring, during low (*left*)- and high (*right*)-water season. *Note* The elevated discharge zone is clearly visible during the high-water season

The outlet zone was explored in detail, by means geophysical survey, analysis of several boreholes situated in the zone route of the underground dam, exploration gallery above the spring, etc. The results show that the Perućac karst spring has favorable conditions for the construction of an artificial underground storage, i.e., an underground reservoir. Generally, three possible concepts for the underground reservoir can be applied (Fig. 15.30).

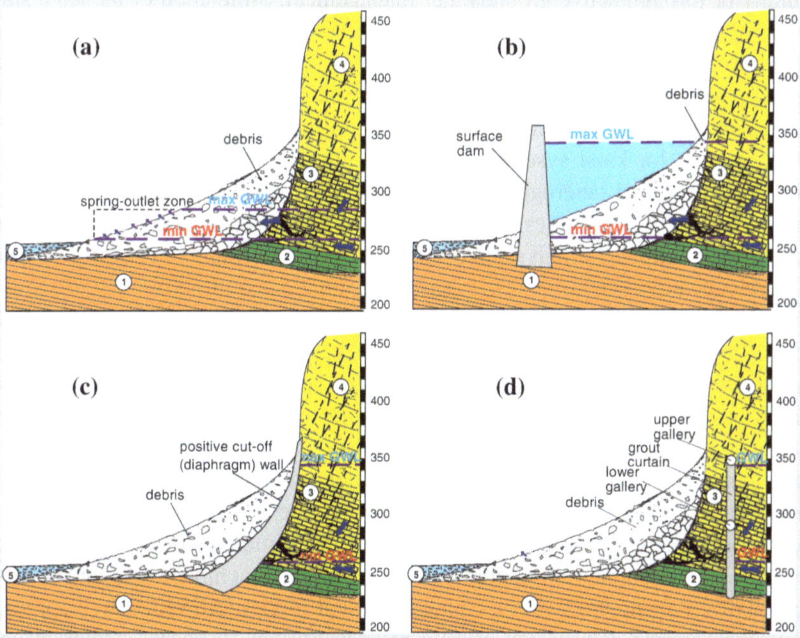

Fig. 15.30 Schematic cross section of different tapping structure for Perućac spring natural condition (**a**); surface dam in front of the main outlet zone (**b**); cutoff wall structure (**c**); plugging channel and grout curtain deep in rock masses, connected to the impermeable rocks (Jemcov et al. 2011, copyright Springer)

Along with many successfully applied concepts of active karst groundwater management, there were also less successful concepts. Some of negative effects are inevitable due to the nature of the karst system, but in some cases, they were based on inappropriately applied concept. The impacts of exploitation can be direct or indirect (Pulido-Bosch 1999; Milanović 2004), with possible irreversible effects. Therefore, the selection of a proper concept and consequently proper tapping structure requires thorough and scrupulous research. This is the main drawback of such endeavors. Even the most meticulous explorations may not eliminate the risk as there are lots of uncertainties related to the specificities of the nature of karst hydrogeological system. Due to the above, developing the model for assessment of the extraction effects of future exploitation may facilitate future research and give further important directions. This fact is particularly significant at early stages of hydrogeological exploration in order to implement adequate tapping structure.

15.3.2 Current Concepts for Assessing Extraction Effects

A necessity to simulate karst spring hydrographs derives from a need to analyze hydraulic behavior of karst aquifer, as well as to predict flow rate fluctuation whether naturally or through active groundwater management. Simulation of the karst spring hydrographs can be performed applying different approaches: physical, distributed model (Mohrlok et al. 1997; Cornaton and Perrochet 2002; Hugman et al. 2013) and lumped-parameter (back or gray box) or reservoir (conceptual) model (Fleury et al. 2007; Jemcov and Petrič 2009). The first group of simulated models provides reasonable results for karst system with known internal geometry and physical properties. Due to spatial heterogeneity of karst aquifer, mathematical modeling can be used with a lot of difficulties. The second group presents a more robust approach, treating karst aquifer through systemic approach considering the relation of recharge and discharge components as input and output functions of the system.

Due to high complexity and heterogeneity of karst aquifer, the examples of karst aquifer modeling under the conditions of active water management are relatively rear. A characteristic example based on three reservoir models (saturated, slow infiltration, and rapid infiltration zone reservoir) was applied on Lez karst aquifer (Fleury et al. 2009). For the same aquifer system, the semi-distributed lumped modeling approach was applied simulating the hydrogeological response of a heterogonous karst aquifer made up of different compartments (Ladouche et al. 2014). An interesting approach presents a simple model that simulates different exploitation scenarios of a fissured karstic aquifer located in the Pinchinade graben (Debieche et al. 2002). Assembled extreme value statistical model (AEVSM) was applied to obtain the extreme distribution of Niangziguan Springs discharge depletion under the effects of extreme climate variability and intense groundwater development by eliminating the trend and periodicity to acquire the residuals (Fan et al. 2013).

Presented approaches in exploitation conditions for modeling of karst aquifer hydrodynamic behavior in most cases treated already developed regulatory systems and consider the effects of active groundwater management *post-festum*. Therefore, it is impossible or unlikely to influence the structure of regulatory system. Moreover, most of these approaches consider an actively managed system, based on the principle of groundwater abstraction during low-water period (from static reserves). For the above reasons, the estimation of the groundwater potential for exploitation, either based on the extensive utilization of natural storage or based on artificially increased dynamic groundwater storage, is of utmost importance at the early stages of the regulatory system development. By simulating different scenarios of groundwater exploitation, along with knowledge of hydrogeological behavior, a realistic basis for future optimal control of karst outflow regime can be created. This implies the analyses of storage changes in karst water reservoirs under natural conditions, and calculation of the future potential exploitation.

15.3.3 Applied Model for Assessing Extraction Effects in an Aquifer System at Hydrogeological Exploration Early Stages

The quantitative identification of karst aquifers and the karst groundwater budget are important issues which are the basis for the estimation of kart groundwater exploitation potential. The relation between precipitation and effective infiltration is analyzed in further detail by Jemcov and Petrič (2009).

The study of aquifer behavior is based on the premise that karst aquifers are hierarchically organized and highly heterogeneous systems involving spatial variability of hydrogeological parameters (Bakalowicz 2005). Another important characteristic of karst aquifers is duality of groundwater flow: on the one hand the conduits of great transmissivity and on the other voluminous media with poor permeability in block matrixes or small fissures (White 1969; Atkinson 1977b). The study of the relation between these two interconnected flows provides valuable information about the behavior of karst aquifers. Therefore, the time series analyses used in the study of hydrodynamic properties and functioning of the whole karst aquifer system provide an indispensable method for karst aquifer characterization and along with "conventional" field exploratory methods further complement the knowledge of the karst aquifer systems (Box 15.3.3).

Box 15.3.3

Case example 3: Comparative study of time series analyses of the St. Petka and Perućac karst aquifers

Groundwater budget and studying of the relation between flow components (e.g., base and fast flow) provide valuable information about the hydrodynamic behavior of karst aquifers. The univariate and bivariate time series analyses demonstrate the difference in obtained parameters between those two selected karst aquifers. The differences in the analyzed karst aquifers behavior are clearly visible, particularly on the autocorrelograms and cross-correlograms (Figs. 15.31 and 15.32), considering total flow component, as well as base-flow and fast-flow components. For the St. Petka (same spring as in Box 15.3.1) karst aquifer, the obtained results indicate a significant storage and a highly structured system (Table 15.3).

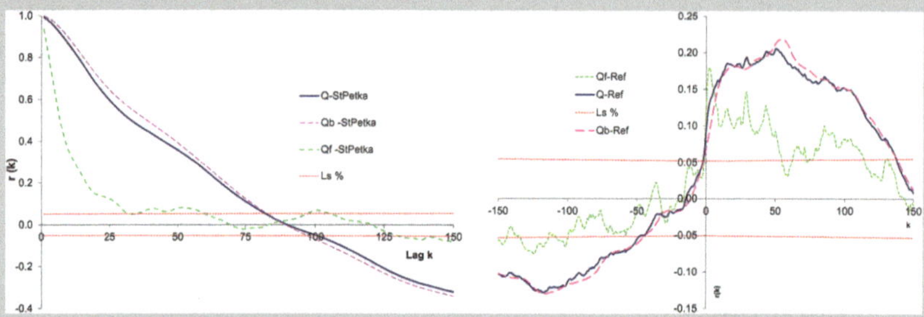

Fig. 15.31 Autocorrelograms (*left*) and cross-correlograms (*right*) for the St. Petka karst aquifer *legend* *Q* discharge, Q_b base flow, Q_f fast flow, *Ref* effective infiltration, L_s% level of significance (Jemcov 2014, modified)

Table 15.3 Calculated parameters of univariate and bivariate analyses

	St. Petka	Perućac
N (days)	4	5
BFI	0.90	0.83
AC conf. limit Q (days)	81	60
AC conf. limit Q_b (days)	83	67
AC conf. limit Q_f (days)	32	12
T_{reg} (days)	147	59.5
CC max $P - Q$ (days)	0.08	0.14
Time lag $P - Q$ (days)	4	1
CC conf. limit $P - Q$ (days)	22	136
CC max $R_{\mathrm{ef}} - Q$ (days)	0.2	0.36
Time lag $R_{\mathrm{ef}} - Q$ (days)	50	2
CC conf. limit R_{ef}–Q (days)	29	77

N Number of days over which a minimum flow is determined, *BFI* base-flow index, *AC*/*CC* conf. limit number of days in which the auto/cross-correlation functions exceed the confidence limit, (Q, Q_f, Q_b—for the total flow, fast flow, and base flow); T_{reg} regulation time, *CC* max maximum value of the cross-correlation function; time lag number of days in which the cross-correlation function exceeds the maximum value

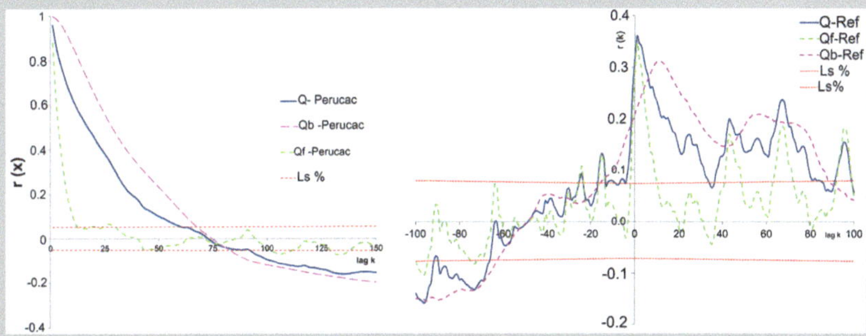

Fig. 15.32 Autocorrelograms (*left*) and cross-correlograms (*right*) for the Perućac karst aquifer (Jemcov and Petrič 2009, modified)

The differences between the results of the applied analyses above are due to their different structures and degrees of karstification. The comparison of the results enabled us to draw inferences about these characteristics and the capability of the applied analyses on the potentiality for exploitation.

Prior to defining the "exploitation" regime, an analysis of the change of water storage in karst aquifer must be completed under natural conditions. This includes estimation of the groundwater budget and storage in the karst aquifer under natural conditions. Based on the inflow–outflow relation, values of the changes in the storage (ΔV_i) may be obtained by adding continuous cumulative values to the initial storage level (Jemcov 2007a). Estimation of the initial storage (V_0) in the karst aquifer represents an important part of this analysis. It largely depends on the characteristics of a karst aquifer, and indicates the best method to increase the exploitable water resource.

Possible changes in the catchment area (or in a part of it) and the presence of the annex-to-drain system by Mangin (1975b) may also affect the storage level in a karst aquifer. Any or all of the above may result in variations of the recession coefficient and the values of discharge rate. For this reason, groundwater storage should be assessed according to the differences in values at the beginning and at the end of the period of recession, rather than according to the differences at the beginning and end of the analysis period (Jemcov et al. 2011).

The starting point is the adoption of exploitation capacity *Qe* which, under certain conditions, can potentially cause changes in storage level in the karst aquifer. Therefore, two different scenarios of storage changes can be described by simulating the exploitation potential, when it comes to the principle of "water-borrowing" from the storage in the times of recession (Jemcov and Petrič 2009):

Fictive discharge (Q_p') equal to the natural discharge (Q_p) occurs only if $\Delta Qe_i' = 0$ and $\Delta Ve_i' = 0$:

$$\left.\begin{array}{l} Qp_i' > Qe_i \\ Vp_i' = Ve_i' \end{array}\right\} \Rightarrow \left.\begin{array}{l} \Delta Qe_i' = 0 \\ \Delta Ve_i' = 0 \end{array}\right\} \begin{array}{l} Qp \\ Vp_i \end{array} \qquad (15.20)$$
$$i = 1, 2 \ldots n - \text{days}$$

where

Qp discharge of karst spring under natural conditions,
Vp_i storage level under natural (unchanged) conditions of discharge regime,
Qe_i simulated discharge regime under exploitation condition,
Qp_i' fictive (changed) discharge which will appear at karst spring as the consequence of changes in the storage level in case the exploitation is instantly cancelled,
Qe_i' fictive discharge of karst spring,
Ve_i' storage level in KHS under the exploitation regime, which corresponds to fictive discharge—Qp_i'; ↓ period of "water-borrowing" from the storage, ↑ period of cumulated water in storage.

$$\left.\begin{array}{l} Qp_i' < Qe_i \\ Vp_i' < Ve_i' \end{array}\right\} \Rightarrow \begin{array}{c} Qe_i' \\ Ve_i' \downarrow \end{array} \qquad (15.21)$$

Then starting the process of "water-borrowing" from the storage in karst aquifer, provoking cumulative lowering of the storage level and resulting in successive emptying of water reserves.

As opposed to the above, the newly infiltrated water increases the storage level (15.22), maintaining the given operating conditions Qe_i', until it reaches the condition referred to in the Eq. (15.20):

$$\left.\begin{array}{l} Qp_i' > Qe_i \\ Vp_i' < Ve_i' \end{array}\right\} \Rightarrow \begin{array}{c} Qe_i' \\ Ve_i' \uparrow \Rightarrow Vp_i' \uparrow \end{array} \qquad (15.22)$$

The state of storage in karst water reservoir under the exploitation condition based on the principle of "water-borrowing" from the storage can be described by the relation (15.23):

$$Ve_i' = Vp_i - \left[\underbrace{(Qe_i' - Qp_i')}_{\text{current\_water\_deficit}} + \underbrace{(Vp_{i-1} - Ve_{i-1}')}_{\text{previously\_formed\_water\_deficit}} \right] \qquad (15.23)$$

where $Qp_i' \downarrow = f(Ve_i')$ and $Qp_i' \uparrow = Qp_i$

The main obstacle to the application of the proposed equation is the fact that it is difficult to determine the fictive spring discharge—Qp_i'. The changed conditions, due to the exploitation of the storage level, will lead to the lowering of the discharge rate. This can be resolved by introducing the (fictive) coefficient of recession (α') which is in inverse proportion to state of storage (Milanović 1981):

$$\alpha' = \frac{Qp_i'}{Ve_i'} \tag{15.24}$$

The concept of water-borrowing is based on the presumption that the karst aquifer would retain the same recession conditions as in its natural state. The values of fictive recession coefficient represent the mean value of each recession episode. Specified fictive coefficient of recession should not be identified with real coefficients of recession, given the fact that the initial storage level in the karst aquifer affects this parameter. In addition, the value of fictive recession coefficient depends on the capacity of exploitation, because of the fact that each exploitation capacity is characterized by a potentially different condition of recession.

As previously discussed, fictive values of the spring discharge—Qp_i' during refilling karst groundwater storage for the quantity of depleted water retain the same trend as under natural conditions (15.23). The real value of water rise during resupply is likely higher. This is the consequence of potentially increased inflow of water in the karst aquifer, as well as of the induced kinetic energy due to the lower storage level than naturally.

Box 15.3.4

Case example: Concept of the active groundwater management based on extensive utilization of the natural storage—static reserves for the St. Petka karst aquifer

Detailed groundwater budget calculation for the St. Petka karst aquifer shows that a share of the effective infiltration, as percentage of total precipitation, varies in range from 32 to 38 % on annual basis.

The estimated water resources stored in the karst aquifer under natural conditions were used as a baseline point for further analyses, which included assessment of the variations in storage under artificial conditions (exploitation potential, limits, and optimal values). Simulation of exploitation potential conditions for the St. Petka aquifer was conducted for three continual capacities of exploitation—200, 250, and 300 l/s. The exploitation capacity of 300 l/s was not feasible because of rapid and almost irrecoverable storage deficit. Continuous exploitation at that level is not a realistic scenario. According to the coefficient of seasonal variation of exploitation (1.4), the optimal capacity was estimated at the values of 160–280 l/s (Figs. 15.33 and 15.34).

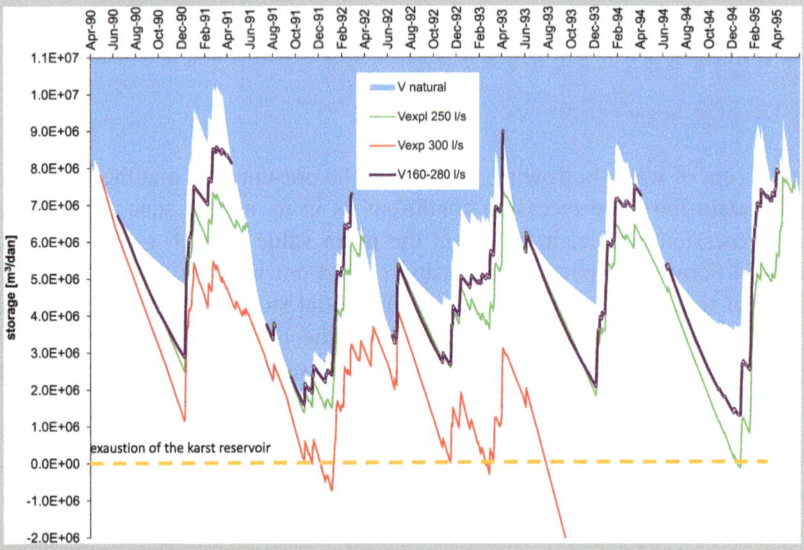

Fig. 15.33 Estimated storage level in natural (*Vnatural*) and exploitation (*Vexpl*) conditions in the karst water reservoir of the St. Petka source (Jemcov 2014, copyright Springer)

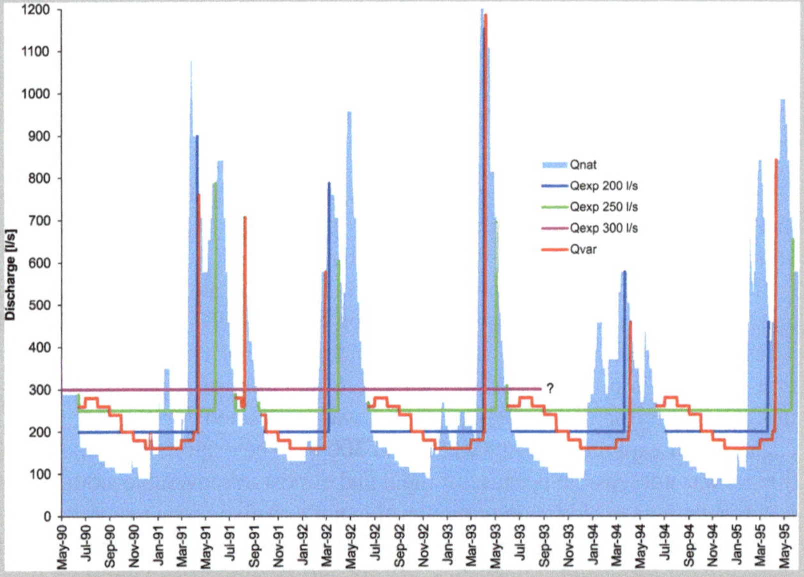

Fig. 15.34 Hydrograph of the St. Petka spring under natural and simulated exploitation conditions. *Legend Qnat* natural discharge rate of karst spring, *Qexp* simulated exploitation potential conditions with continual capacities of exploitation, 200, 250, and 300 l/s, *Qvar* simulated exploitation potential conditions with variable capacity (Jemcov 2014, copyright Springer)

Another concept is based on increasing the dynamic groundwater storage under controlled release conditions. This concept provides a possibility to construct an underground dam and a multipurpose (*Qi*) artificial underground water reservoir for both permanent (e.g., water supply and ecological criteria) and temporary users (e.g., hydropower plant):

$$Q\exp = \sum_{1}^{i} Qi \tag{15.25}$$

As for temporary user, power plant, the use of artificial storage is possible until achieving the certain level that provides sufficient difference of potential energy for electricity production. The difference between the maximum and minimum levels for power plant use is usually calculated at the level of 5–10 m, and therefore, exploitation capacity for the temporary user reaches zero, while newly formed capacity includes only permanent users of the system. According to this, the previous Eq. (15.23) becomes slightly different:

$$Ve'_i = Vp_i - \left[\underbrace{\left(\sum_{1}^{i} Qe'_i - Qp'_i \right)}_{\text{current water deficit}} + \underbrace{\left(Vp_{i-1} - Ve'_{i-1} \right)}_{\text{previously formed water deficit}} \right] \tag{15.26}$$

where $\sum_{1}^{i} Qe'_i = Ve^i_{i-1} \cdot \alpha' iQp'_i = Qp_i$ and $i = 1, 2 \ldots n$ – days

The contour of the future artificial underground water reservoirs requires extensive research, as shown on the typical example of the Ombla (Dubrovnik, Croatia) power plant (Milanović 2004). In the initial stage of the research in order to estimate maximum water level in the underground reservoir, the fluctuation zone at the outlet may be used.

The water level and storage in the karst reservoir cumulatively reduce due to different users. The key issue in the calculation might be calculating the fictive discharge in the karst spring Qp_i' because the increase in the storage level would lead to a significant change in the discharge. As in the previous case, this can be resolved by introducing a fictive coefficient of recession (α'). The representative value of the fictive recession coefficient of the future underground reservoir should be the maximum value under natural conditions (the first few days of the recession) (Box 15.3.5).

Box 15.3.5

Case example: Concept based on the dynamic groundwater storage increase by the construction of underground dam and reservoir for the Perućac karst aquifer

Detailed groundwater budget calculation for the Perućac karst aquifer shows that a share of the effective infiltration, as percentage of total precipitation, is about 60 % annually.

Groundwater dam and reservoir are complex structures, and in most cases, increase in the dynamic groundwater storage would facilitate the needs of permanent and temporary users (water supply, hydropower plant, etc.). A power plant would be partially operational in decrements of 5 m, from the maximum water level. The minimum operational value used for the power plant was estimated at 6.15×10^6 m^3. This part of the artificial water reservoir discharges relatively fast due to highly karstified zone, as the main prerequisite for successful power plant operation. The rest of the exploitation capacities of $Qexpl$ = 450 l/s (which includes flow under ecological criteria) could be used permanently. The analysis shows the possibility of power plant continuous operation varying from 18 to 56 days. The time needed to refill the karst reservoir to the maximum level is significantly higher, and varies from 35 to 164 days (Figs. 15.35 and 15.36).

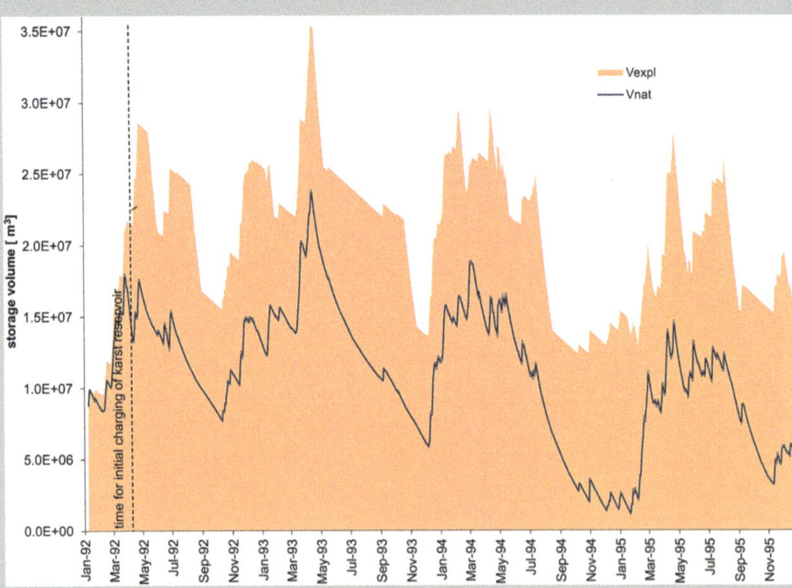

Fig. 15.35 Estimated storage level, in natural (*Vnat*) and exploitation (*Vexpl*) conditions, in the karst water reservoir of the Perućac karst spring. (Jemcov 2014, copyright Springer)

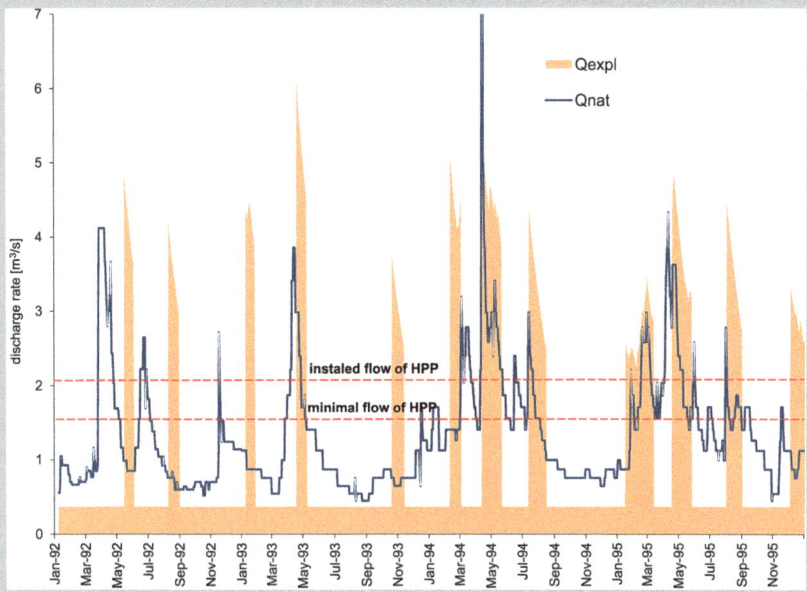

Fig. 15.36 Hydrograph of the Perućac spring under natural (*Qnat*) and simulated exploitation (*Qexpl*) conditions. *HPP* Hydropower plant (Jemcov 2014, copyright Springer)

Quantitative and budget analyses of karst aquifers are complementary to the traditional techniques of the hydrogeological exploration. These methods for the estimation of groundwater potentiality for exploitation provide valuable information for future calculation of the optimal exploitation rates at the early stages of hydrogeological exploration. The models presented do not examine the type of aquifer regulation, but rather define quantitative conditions of possible rational exploitation with the purpose of satisfying the needs of potential users for a long term. Moreover, they give important directions both for further research and groundwater management and should considerably contribute to feasibility studies.

15.4 Speleology and Cave Diving as a Base for Tapping Structure Design

Saša Milanović

Centre for Karst Hydrogeology, Department of Hydrogeology, Faculty Mining and Geology, University of Belgrade, Belgrade, Serbia

15.4.1 Introduction

The very complex hydrogeological characteristics of karst, particularly the position of caverns and karst conduits, and including the position and directions of the main zones of groundwater distribution, are not yet thoroughly enough explained. Even after detailed and complex research into the geological, hydrogeological, and geomorphologic characteristics of karst, some characteristics of karst hydrogeology and groundwater flows in karst remain unclear.

In hydrogeological and engineering–geological surveys, the only research method that allows direct observation, investigation, and exact geological mapping of karst channels and caverns is speleology and cave diving. For example, detailed use of deep pits or sinkholes can be made for monitoring fluctuations in the groundwater level, for observing objects in tracer tests, and for observation of water head, discharge, chemistry, isotopes, and all geological settings such as lithology and fracture patterns. Data collected during the exploration significantly contribute to the reconstruction of the evolution of karst processes, which is very important in assessing the depth of karstification and of water pathways leaking, and potential loss from reservoirs.

The research methods practiced set karst hydrogeology apart from other terrains: In karst hydrogeology, part of the groundwater flowpaths of karst conduits can be directly explored and mapped by speleologists or cave divers. In combination with other techniques such as tracing tests and monitoring of karst systems, this can be useful to get a more complete overview of the locations of underground flows and to determine sites for the tapping structure. For example, direct mapping of caves or karst conduits can identify locations of underground stream confluences that can only be approximated through tracing, and which can allow for sampling that represents all the underground water catchments (Jeannin et al. 2007).

Speleology and cave diving is the only exploration method which enables direct observation, investigation, and exact geological mapping of karstic channels or conduits filled with water. For instance, in the past 50 years, more than 100 water caves and springs longer than 1,200 m were discovered by cave diving, as were more than 120 water caves, springs, and karst channels deeper than 100 m (Toulomudjian 2005; GUE https://www.globalunderwaterexplorers.org). Tapping some of those springs are good examples of a modern approach to tapping the large karst springs as a method of direct abstraction of water from karst springs in the inland channels. Main approaches in the strategy of karst spring water tapping directly from spring outlets include using highly productive wells located behind but not far from the spring outlet, and using common well technology for groundwater extraction directly from the karst channel and tapping galleries (Milanović 2000a). In these cases, cave diving gives the best results as input data for further hydrogeological research. Some good examples are as follows: spring Lez (France), Krupac spring (Serbia) (Box 15.4.1), Sekerpinari spring (Turky), BeleVode spring (Serbia) (Box 15.4.2), the underground dam Ombla in Croatia, Opačica spring (Montenegro), etc. Some very special technologies were used in the case of coastal spring tapping to prevent the influence of seawater. Cave diving explorations have been applied successfully in researching deep siphonal channels in coastal areas due to salt water intrusion in springs like Gurdić and Škurda, Orahovačka Ljuta, Sopot (Box 15.4.3), Spila Risanska (Montenegro), and Port Miou (France) (Box 15.4.4) to a depth sometimes more than −140 m. Data collected during these explorations suggest that further research is necessary, with the goal to collect reliable data about salt water intrusion into the karst aquifer and new data about the possibilities of tapping the groundwater from karstic channels in the deep part of karst massif. The role of speleology and cave diving in the tapping of the large karst spring is unavoidable. The karstic springs are the most interesting phenomenon from a hydrogeological point of view, and their investigation needs particular attention.

On the other hand, speleological research is usually applied as part of complex hydrogeological investigations. A large number of cave forms, especially channels and caverns, are explored in detail in the course of hydrogeological studies. In the construction of dams and reservoirs in karst, speleology is extensively used in the stages of research and development. Successful detection and recording characteristics of potential leak paths through the dam cross section or through the reservoir area is a key task of speleology. The successful sealing of these pathways (karst channels and caverns) usually depends on the ability of project implementation on so complex an object as a dam. Accordingly, speleological data concerning local adaptation of the grout curtain are usually necessary as is adaptation of some technical solutions that provide water permeability of the grout curtain and reservoir area. Some good examples of speleological investigations due to dam and reservoir constructions were those performed for the construction of Hydrosystem Trebišnjica (Herzegovina), Salman Farsi dam (Iran) (Box 15.4.5), Bogovina reservoir and dam (Serbia) (Box 15.4.6), the reservoir of Nikšić polje, dam Beilow, Yuhongand Wulichong (China), Maroon dam (Iran), Keban dam (Turkey), reservoir Buško Blato (Bosnia), Sklope dam (Croatia), and in the case of a number of dams and reservoirs that are not listed here. Also, cave explorations are inevitable during the development of roads and tunnels in karst. For example, while building highways, cuttings, tunnels, bridge foundations, viaducts, etc., in karst regions of Croatia during the last 15 years, over 9,000 caves and pits were discovered, as well as over 767 caverns (speleological objects without natural entrance) which were discovered and thoroughly explored (Garašić 2005).

Still, the possible density of karst conduits is in the order of several hundred km per km^2. Thus, it is clear that the accessible conduits represent only a very small part of all available karst conduits in some karst massif.

15.4.2 Overview of Speleology and Cave Diving Explorations

The first phase of speleological and cave diving research includes a speleological survey to gather information in order to examine the conditions and size of the caves as well as review the existing layers of the finished cave exploration. The second phase of the research involves detailed speleological or cave diving mapping (recording and measurement of karst channels) and all other aspects of the geological and hydrogeological mapping applied, depending on the type and scope of tasks. The third phase is the office research work which involves the development of topographic maps and profiles with all the necessary data (geological and hydrogeological). The fourth and final phase uses geostatistical methods combined with computer programs, and data of certain speleological objects such as lithostratigraphical indicators and speleogenesis type, their relation toward topographic position, morphometric parameters in relation to overbeds above caves, as well as hydrogeological parameters in relation to tectonic and lithostratigraphical factors.

During the research, it is necessary to create an internal database that contains basic information about topographic measurements, as well as geological and hydrogeological data. Some main data collected during cave, cavern, or karst channel survey are as follows:

- name—ID of measuring point,
- length, azimuth, and angle of descent (standard speleological mapping),
- dimensions of the channel (left, right, up, below),
- description of the point area (features),
- geological characteristics (type of rock, rupture, stratification, collapse, etc.),
- morphological features (terraces, accumulation of deposits, etc.),
- hydrogeological characteristics (siphon depth, GWL—ground water level, Q—groundwater flow capacity, the direction of groundwater flow, physicochemical characteristics of GW), and
- photograph and video record of characteristic points.

Generally speaking, in the study of the distribution of karst channels, the most important task is the determination of their real position in the coordinate system, or an accurate determination of (x, y, z coordinates) of all measurement points. The degree of accuracy is very important for particular purposes in hydrogeological–engineering surveying, for instance, well or gallery placement, grout curtain design, installation of pump system, and channel tampon construction. One method to improve the accuracy of channel topography (standard methods are tape or a laser distance meter, the slope with a clinometer, and an azimuth with a compass) is to use a laser distance meter, or theodolites or finally, in recent years, underground positioning system (UGPS). Depending on the effort dedicated to these measurements, though, the precision ranges are usually between 1 and 0.01 % of the distance to the cave entrance (depending on the method applied).

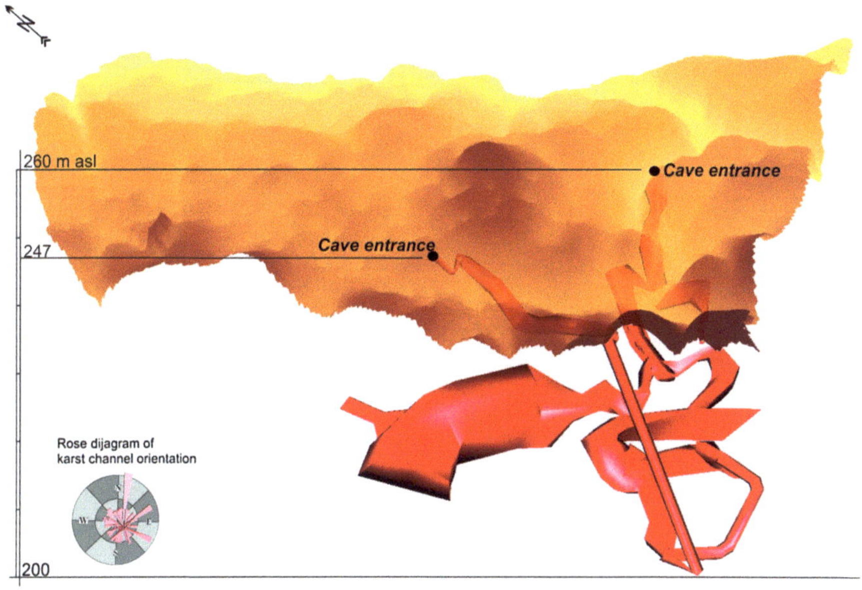

Fig. 15.37 3D model of cave system and spatial position of karst conduits

Collection of topology data during speleological or cave diving investigations is the first step in hydrogeological analysis. Through several specialized computer programs for cave maps produced in 2D and 3D view such as COMPASS, Toporobot, Visual Topo, Win Karst, Survex, and Walls, it is possible to get a good base for further complex investigations of an underground karst system. These softwares can plot plan views and cross sections of the caves as well as 3D models of karst conduits. Also some of the programs are designed to export files in formats that can be directly imported into geographic information systems (GIS) software that offers powerful data management, analysis, and visualization capabilities (Fig. 15.37). Once established, a 3D model of karst conduits and their spatial position gives a good starting position for karst groundwater monitoring survey.

Finally, when it comes to the tapping structure designed for the purpose of water supply, dam, or tunneling, forming the karst channel network monitoring is a very important step in the decision making.

A monitoring network is formed depending on the characteristics of karst aquifers. It depends on the number of features that can be used in monitoring purposes. The general area where observation points are placed is divided into three zones: recharge (sinkholes) zones, karst channels as a transport zone of groundwater (karst conduits), and discharge zones (springs). A schematic presentation of a monitoring network of a karst system as a base for a tapping structure design is shown in Fig. 15.38.

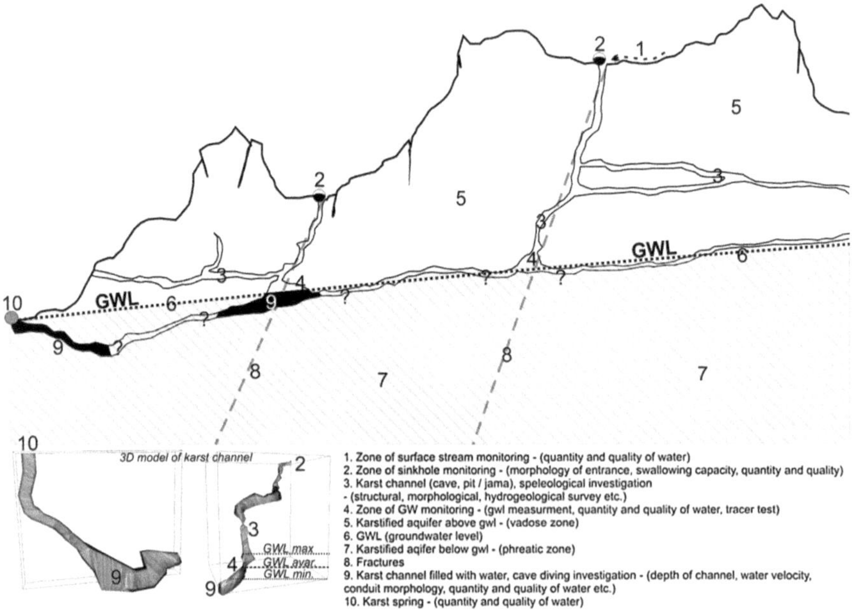

Fig. 15.38 Schematic presentation of observation points above and within the karst system

Box 15.4.1

Case study—Cave diving survey for Krupac spring tapping

The key contribution during investigative work for tapping Krupac Spring near the city of Niš (Serbia) was provided by cave diving investigation of deep siphonal channels. Cave divers investigated and outlined the main spring karstic channel to a distance of 170 m and depth of −86 m (in relation to spring overflow) (Fig. 15.39). Based on this survey, the location of the temporary tapping structure was determined, e.g., installation of exploitation equipment for testing directly in the karstic channel (Fig. 15.39).

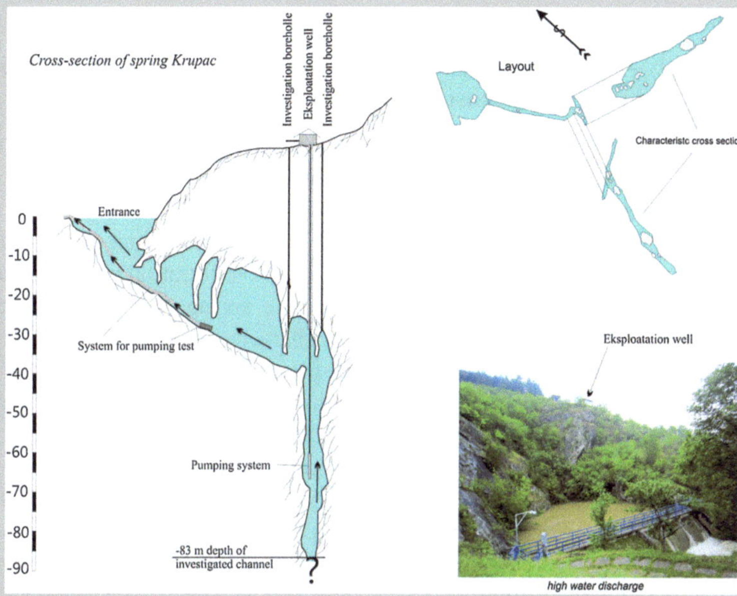

Fig. 15.39 Krupac spring cross section—positions of pumps for well testing, investigative borehole, and exploitation wells (sketches after Jevtic et al. 2005a)

Placing two pumps of 2 × 200 l/s capacity at a depth of 30 m has made pumping tests possible over a period of several years (periods of minimum discharge) and enabled the acquisition of data about the available quantity of water. This allowed us to pump static reserves during the drought period or, in other words, to enlarge minimum capacity many times and at the same time define the true reserves of the Krupac aquifer. Underwater activities (cave diving) also confirmed the technical feasibility of aquifer regulation and additional pumping of groundwater on account of the static reserves accumulated beneath the existing outflow zone (overflow threshold) (Jevtić et al. 2005a).

The very precise topography of the submerged karstic channel, combined correctly with surface geodesy, and the construction of quality investigative wells have made it possible to define the location of exploitation wells precisely. The well entered the main vertical channel and enabled placement of the pump at a depth of −65 m (Fig. 15.39). By placing the pump at this depth and by overpumping in periods of drought, a significant enlargement of minimum spring capacity has been made possible, and further implementation of such intake improved water supply of the city of Niš in periods of low water levels.

Box 15.4.2

Case study—Speleology and cave diving in Bele Vode cave-spring (Miroč karst massif) as a base for groundwater tapping

The karst massif of Miroč Mountain is situated in the northern part of the Carpathian-Balkanides of Eastern Serbia and extends to the Danube River and the Romanian border.

The area of the karst massif, composed mainly of Titonic limestones, exceeds 140 km^2. Carbonate rocks exhibit favorable filtration properties and significant groundwater reserves. Many karstic springs, deep caves, deep underwater caves, and ponors are the main characteristics of the Miroč karst area (Fig. 15.40). However, with the construction of the Djerdap reservoir and drowning of western and northern Miroč board, all significant springs draining the karst aquifer were submerged (Stevanović et al. 1996).

Step-by-step approach and investigation procedures were carried out in stages using several methods: 1. preliminary analysis of the water budget of the catchment area; 2. ruptural and geomorphological analysis; 3. geophysical investigations; 4. speleology and cave diving investigations; and 5. investigation by borehole drilling.

Results of speleology and cave diving investigations

Speleology and cave diving investigations resulted in the recording of an unexpectedly large number of speleology occurrences. So far, more than 30 caves, pits, and sinkholes in the Miroč karst oasis have been investigated, while the cave diving research was carried out in the Bele Vode cave as well

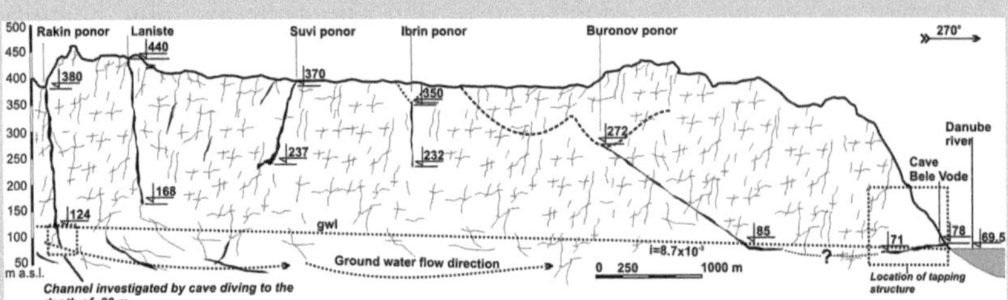

Fig. 15.40 Cross section over significant speleology occurrences with main groundwater flow direction and location of proposed tapping structure (Zlokolica et al. 1996, modified by Milanović 2005)

as in the Rakin ponor (sinkhole).

The Bele Vode cave represents the most northern object of speleology so far explored on this terrain. In the morphological sense, it is a simple cave consisting of one channel developed by a fissure system generally oriented toward the Danube gorge. The Bele Vode cave is a horizontal, temporarily active spring but for the most of the year submerged by water of the large Danube Reservoir. The total length of channels investigated so far amounts to 285 m, of which 131 m represent siphon channels (Fig. 15.41).

Fig. 15.41 Cave diving survey in Bele Vode cave

The dynamic reserve calculation obtained showed that the volume of the Bele Vode watershed is BeleVode—$V_d = 112.1 \times 10^6$ m^3. On the basis of the monitoring and estimated multiannual water budget (1946–2000), the following hydrological parameters of the Bele Vode watershed were obtained:

- Average springflow, Qsr = 0.474 m^3/s
- Average specific yield $q = 10.705$ l/s/km^2

The theoretic value of the minimal monthly average springflow of the Bele Vode amounts to

- $Q_{95\,\%} = 40$ l/s

Concept of groundwater tapping according to speleology and cave diving investigations
Results of cave diving exploration reveal that in the Bele Vode cave, starting from 128 m and ending on 253 m from the point of the cave entrance (the channels researched so far by cave diving), there are channels filled with water (Fig. 15.42), tending to flow from the Miroč karst aquifer toward the Danube, specifically to the Djerdap reservoir. Given the characteristics of the channels and general directions of the groundwater flow, potentially the best channel zones for the groundwater tapping are the extreme channels (Dragišić et al. 2004).
Both the main groundwater circulation zones and significant data concerning the further designs have been confirmed by test drilling. In the period of low water level, besides dynamic reserves of karst groundwater, part of the static reserves would be partly exploited by the well.

There are two main strategies for underground water tapping (Fig. 15.42) presented as Var. 1 and Var. 2.

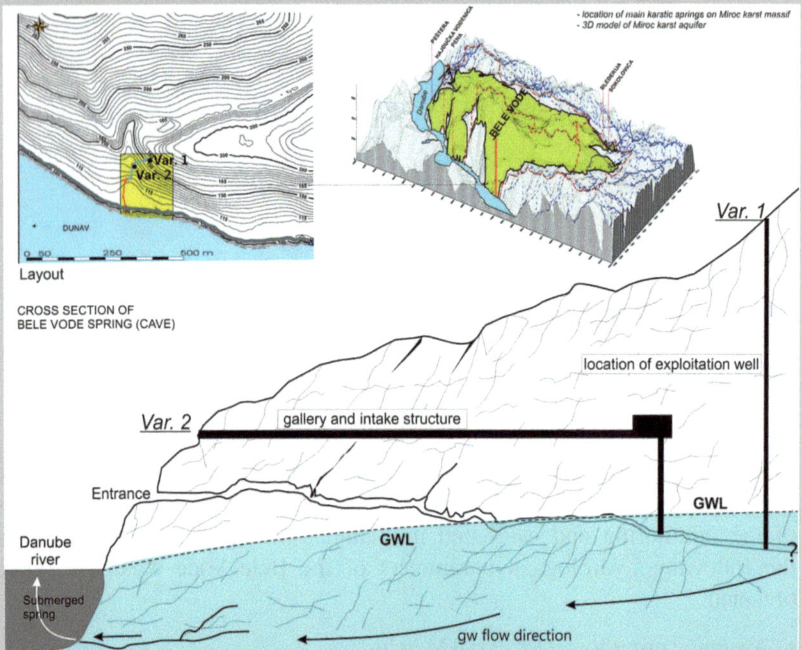

Fig. 15.42 The area of detailed cave diving, speleological and hydrogeological investigations, with position of projected tapping structure (two variants)

Var. 1. Drill and use a high productive well located behind discharge site, but not far from the spring outlet,

Var. 2. Excavate and use tapping gallery with vertical well on its end.

Both Var. 1 and Var. 2. should provide better water quality; the tapping structure is well protected; concerning the environment and the landscape, the spring zone is not disturbed, and because of use of deeper reservoir, the yield of the wells in a dry period will be higher than the corresponding natural spring discharge. The project is still under feasibility and waiting for approval for implementation.

Box 15.4.3

Case study—Problem of karst spring tapping in the coastal zone—Boka Kotorska Bay

Cave diving explorations have been applied successfully in researching deep siphonal channels, in springs like Gurdić, Škurda, Orahovačka Ljuta, Sopot (Fig. 15.43), and Spila Risanska in Boka Kotorska Bay, to a depth sometimes deeper than −130 m. Data collected during these explorations suggest that further research is necessary, with the goal to collect reliable data about salt water intrusion in the Boka Kotorska Bay background and new data about the possibilities of tapping the groundwater in karstic channels, in the deep background of these springs (Milanović 2007).

Detailed investigation much deeper into the karstic background of springs, continual sampling in the depth and distance of investigated channels, and monitoring of the Cl concentration until an acceptable quantity is found—finding zones with no sea intrusion detected—are mandatory before we can think about methods of groundwater tapping.

Note: The mechanism of discharge of this coastal spring is also discussed in Sect. 3.5 (Fig. 3.42).

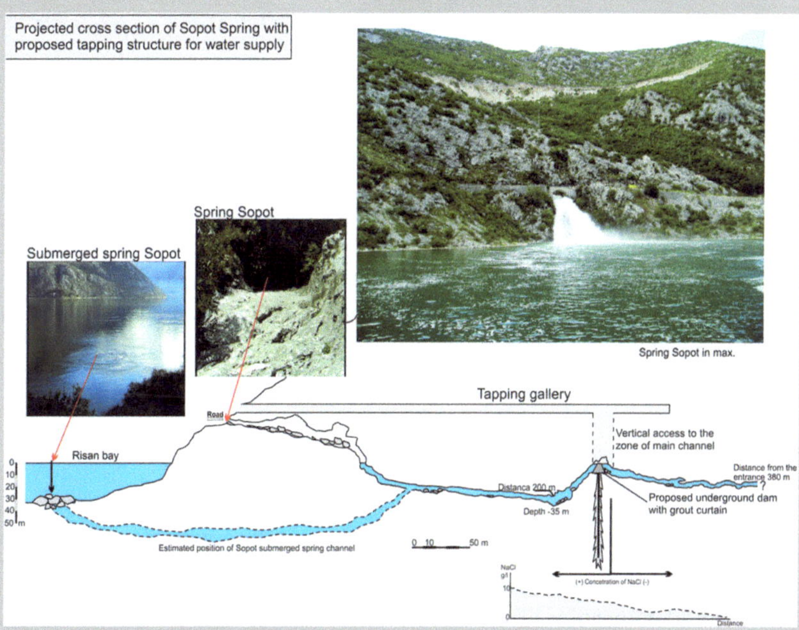

Fig. 15.43 Karst Spring Sopot with discharge in max. and min. water, as well as an option for possible tapping structure (Boka Kotorska bay, Montenegro; scheme after Milanović S, modified)

Box 15.4.4

Case study—Port Miou tapping freshwater from submarine spring

The existence of submarine freshwater springs in the Calanques range of hills on the coastline between Marseilles and Cassis has been known for many years. "Calanque" is the local name for a partially drowned coastal karst valley. One of the largest springs, Port Miou is in fact an underground drowned karst river large enough to be explored by divers (Potié et al. 2005). The roof of the entirely submerged Port Miou gallery lies between 10 and 20 m below sea level (bsl) to a point 800 m from the entrance. It then goes deeper, and about 2,200 m from the entrance, it suddenly drops into a deep shaft that has been explored to a depth of 147 m bsl. A gallery cross section varies in size and shape (Fig. 15.44).

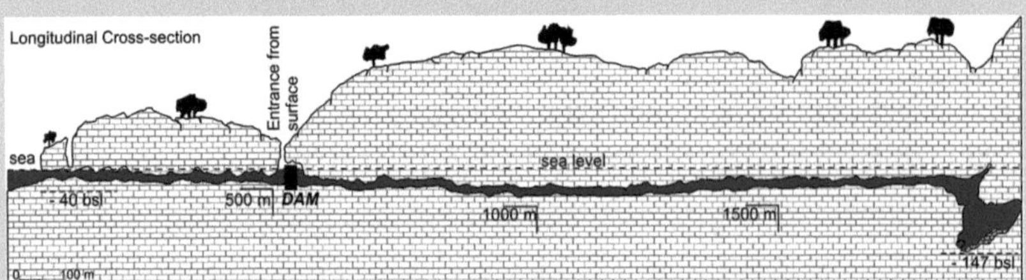

Fig. 15.44 Port-Miou longitudinal cross section of (from Crps M. Douchet, Potié, 2005)

The knowledge built up by all the scientific investigation campaigns (hydrogeology, geophysics, flow measurements, etc.) and direct observations by divers made it possible to envisage means of limiting saltwater penetration and making use of the freshwater. The first objective was to assess the freshwater resources and develop a means of improving separation of freshwater and seawater.
Note: The design of underground dam, intake gallery, and their effects on mixture of salty and freshwaters is further discussed in Sect. 16.4.

Box 15.4.5

Case study—Speleological investigation in grouting galleries of Salman Farsi dam

The exact identification of the three-dimensional development of each cavern is the key issue and prerequisite for taking final decisions how to plug or to circumvent the karst systems in the most economical and most efficient way, and how to incorporate this plugging structure into the grout curtain.

A speleological investigation of the karst also helped to identify karstified pathways and zones of potential water losses in the Salman Farsi Dam in Iran (Fig. 15.45). During the excavation of the grouting gallery in highly

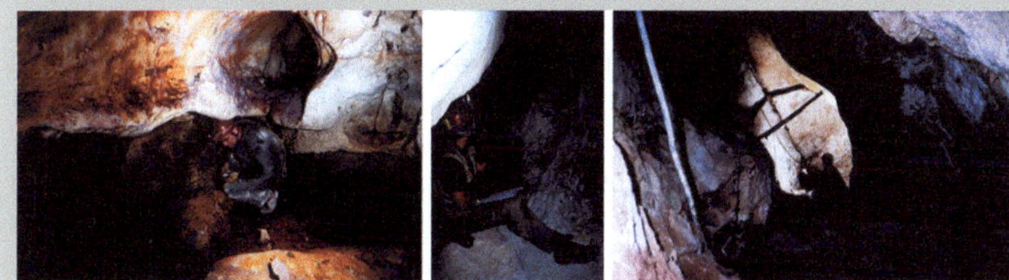

Fig. 15.45 Photographs from speleological investigations on Salman Farsi Dam site

karstified Asmari limestones, large caverns on both abutments were identified and further speleological exploration was undertaken (Fazeli 2007).
The original speleological investigation program of caverns, chimneys, and channels included the following surveys:

- Mapping according to a common coordinate system.
- A graphical presentation of results which should include the following characteristics of each investigated cave channel: layout plan, longitudinal sections, and selected cross sections on the scale 1:100.
- All important geological features (dip of faults, joints, and beddings, characteristics of cave deposits, and infilling of faults and joints) have to be mapped and graphically presented.
- The air temperature is to be measured on different levels of each cave.
- If a cavern or karst channel is filled with clay or sand deposits, the samples of these materials have to be taken for analysis.
- If caverns are filled with water, its temperature should be measured and samples for chemical analysis taken.
- Any appearance of an air current in a karst channel has to be noticed and, if possible, followed (smoke tracing).
- The main surveying points along the karst channels have to be clearly marked with resistant paint.

Results of Speleological Investigations
Speleological investigations were performed by applying conventional speleological measurements, using a geological compass and standard geodetical tape. A speleological measurement net is connected with a previously defined measurement point in the grouting galleries or in the investigation audits. The absolute error of the measurement was $\pm 0.5^{\circ}$ for the horizontal (azimuth) angle measurements, $\pm 1^{\circ}$ for the vertical (dipping) angle measurements, and ± 50 mm for the length measurements. Estimation of the error during the measurements was done by establishing the open traverse.

The quantification of the error was possible for each station in the polygon. The average vector of relative error for the positioning of the station in 3D Cartesian space was less than 4 %.

During the acquisition of the data, the base grew to a huge number of more than 2,000 measured values.

The largest and most complicated cavernous system in the right dam site is "Golshan's Cave" (Fig. 15.46). This karst system is accessible from two galleries: GG 802R and GG 769R. The main speleological units are the following: "Golshan's Hall," "Fazeli's Channel," "Parhizkary's Channel," Small Cave, and smaller units related to GG 802R. The general direction of the cavern extension is S–N. The main reason for the cavern development is the wide fault zone. Discontinuities, a few meters wide and subvertical, create the western and eastern cavern walls. This zone (main fault zone) is clearly visible at the surface, 200 m above the cavern (D. Vučković, S. Milanović, 2001, Salman Farsi (Ghir) Dam Project, Speleological Investigations, Mission Report, unpublished).

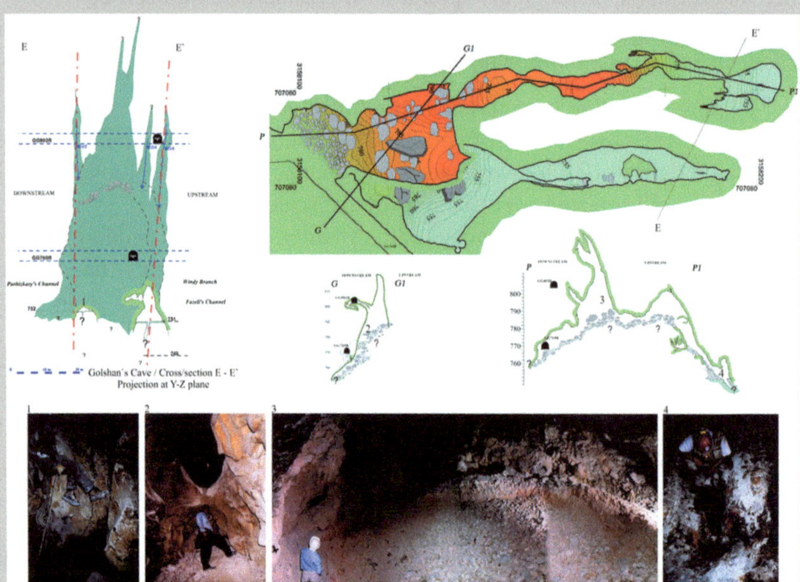

Fig. 15.46 Golshan's cavern, layout, and cross sections. Projection of Golshan's Cave at *Y–Z* plane. Photographs of Golshan's cavern investigation

The largest discovered cavern on the left side of the grout curtain route is "Saidi's Cave." This cavern is situated very close to the upstream hill side and is most probably connected to the reservoir rim. In the cave section closer to the hillside, the air current was observed during the previous reconnaissance visits to "Saidi's Cave" (Fig. 15.47). All speleologically investigated parts of the cavern are lower than the gallery elevation. Some of the inaccessible channels were investigated after hard and long excavations.

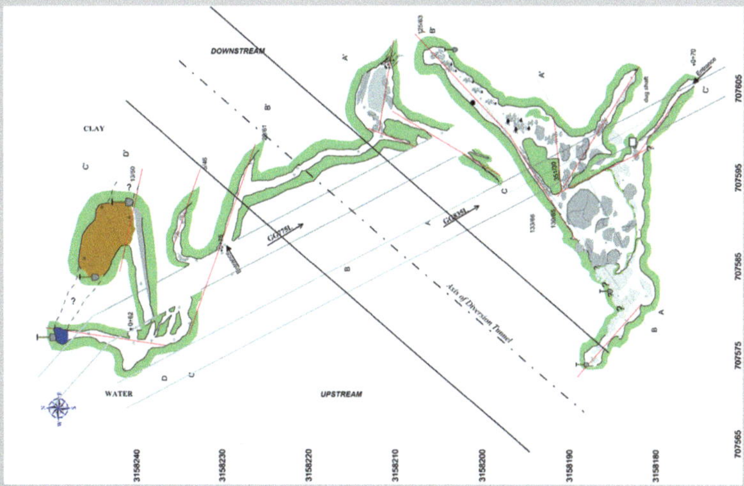

Fig. 15.47 Saidi's cavern and caverns at GG 775 L

The inlet channel from the gallery invert (0 + 70) is developed along the interbedding joint (13/52). The other connection gallery and cave is a dug shaft. The J-1 system has the primary role in the speleogenesis of this cavern. The main discontinuity (125/63) is almost perpendicular to the grout curtain route. The longest axis of the cavern coincides with the main J-1 discontinuity.

Generally, it is one large cavern space filled with a pile of limestone blocks. The higher part of the cave is characterized by corrosion features on the roof and wall side. In the lower part of the cavern, the space between blocks is filled with clay. In the main room, the size of the limestone blocks is a few cubic meters, in some cases more than 10 m^3. The exact position of the solid rock mass (bedrock below the pile of blocks) is unknown.

In the natural conditions, only a limited part of the channels and caverns was accessible for speleological investigations. Many of them were filled and plugged with clayey deposits.

Structural Analysis of Cave Development

The explored cave systems at Ghir Dam site were developed in Asmari limestone rocks (Tertiary age), under particular hydraulic and structural conditions. Therefore, cave systems are composed of many segments (channels, large halls, shafts, etc.), each of which is contained entirely by distinct discontinuities, such as bedding planes, joints and shear fractures, faults, or along interception of those structures. These structural elements are initial pathways for the groundwater filtration (Milanović 2005).

Faults, joints, and bedding planes are planar features that serve as the principal structural guides for groundwater flow in almost all karstified rocks. In hydrogeology, it is customary to categorize all of them together as fractures forming fracture aquifers. Later in the text, characteristics of those discontinuities will be briefly explained.

- The hydrogeological and speleogenetic role of faults varies with their type, size, and the diagenetic record since they were formed. At the Ghir dam site area, the fault displacements are not so distinct. Also, the difference between faults and joints at the site is not distinct. They extend from a few hundred meters vertically and around 0.5 to 1 km laterally.
- Joints are simple pull-apart breaks in previously consolidated rocks. Historically, joints were classified first by their crosscutting relationships, and more recently by their relationships to diagenesis and microtectonic. In this karst area, two main systems of more or less straight joints dominated.
- Bedding planes in sedimentary rocks are produced by some change in sedimentation or an interruption of it. Major changes are represented by big differences in grain size or homogeneity and, more often, by the introduction of clay by a storm or flood that leaves a paper-thin or thicker parting between the successive layers of karstic rock.

One of the important parts of analysis is based on the data (strike/dip) collected from caves and along the grouting galleries. The aim of collecting data in this zone is to provide analysis of cave speleogenesis. A few sets of structural geology data, processed by computer, are presented in the form of contour and rose diagrams.

On the right bank, three strongly developed discontinuity sets play a crucial role in the karstification process (J1, J2, and interbedding discontinuities). By analysis of data frequency in the area of "Golshan's Cave," it is obvious the structural system can be divided into two parts. In the upper part (elevation 860–790 m), a system of separated discontinuities is coupled in one unique system, and at the lower elevation (755–789 m) means in "Golshan's Cave" (D. Vučković, S. Milanović, 2001, Salman Farsi (Ghir) Dam Project, Speleological Investigations, Mission Report, unpublished).

Shown in the form of contour diagrams are the following:

1. Cumulative contour diagram for the right bank (Fig. 15.48). Generally, in the right bank, two maximums exist 284/81 (J2) and 110/78 (J1). The rose diagram for the same set of joints is presented on Fig. 15.48.
2. On the contour diagram for the elevation 755–789 m, the maximum 281/62 indicates the J2 system, and two submaximums, southern 6/47 and western 113/69, indicate interbedding discontinuities and the J1 system. The rose diagram is plotted (Fig. 15.49) using the same measurement data. The main discontinuity directions are clear.
3. A contour diagram was also made for the elevation 790–860 m where one stretched maximum (274/81 and 294/81) comprising J2 poles dominates. The concentrations of these poles indicate a system of individualized discontinuities. Also, two stretched submaxima (86/72 and 112/74) indicate the presence of J1 in the right bank.

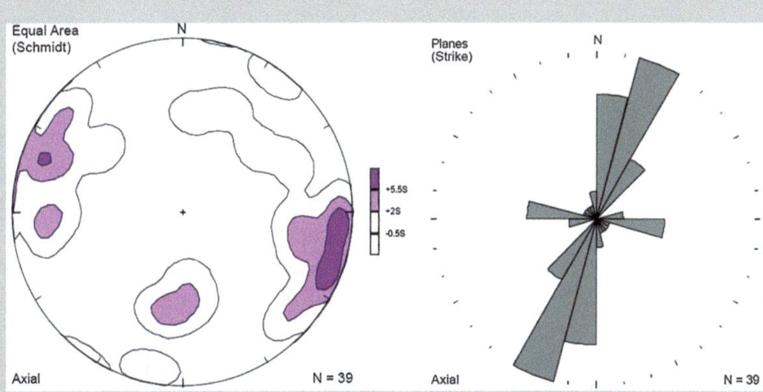

Fig. 15.48 Cumulative contour diagram and rose diagram for right bank and rose diagram of the right bank

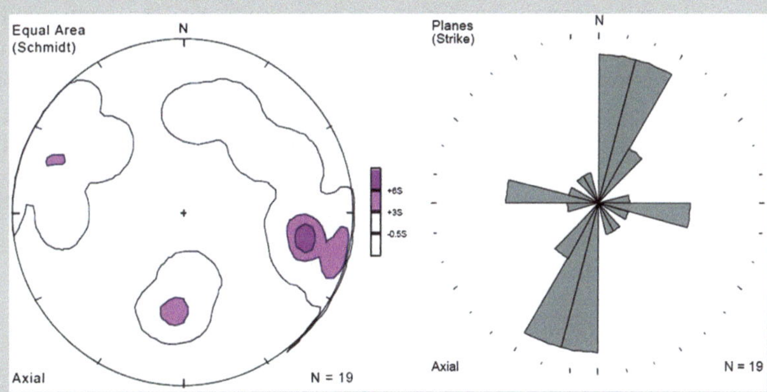

Fig. 15.49 Contour diagram and rose diagram for elevation 755–789 m (right bank)

Finally, comparing the right bank statistical analysis with that of the left bank presents a much simpler and much clearer structural picture of two distinct systems presented in the form of two maximums: 10/50 (interbedding discontinuities) and 130/65 (J1 system).

Taking into consideration analysis of all available data collected during the speleological investigations, the main directions of the local karst aquifer evolution process can be determined; that is, possible locations of larger karst caverns can be determined. Two main determining elements (factors) are the following:

- space position of main discontinuity systems, and
- position of erosion base level, i.e., the mutual influence of fluvial and karst erosion processes.

Results of investigations are graphically presented, according to a common coordinate system, in the form of layouts and selected cross sections. All important geological features, observed in caverns, are registered and graphically presented. The largest system of caverns in the right dam site ("Golshan's Cave") was developed along the J-1 discontinuity. The largest cavern in the left dam site ("Saidi's Cave") was developed predominantly along the J-2 discontinuity. The deepest karst channel is registered at the very end of "Fazeli's Channel," at an elevation of 747 m. A few subvertical (but not accessible) chimneys indicate development of karst features below the elevation 740 m (D. Vučković, S. Milanović, 2001, Salman Farsi (Ghir) Dam Project, Speleological Investigations, Mission Report, unpublished).

By comparison of findings from inside of the rock mass (in galleries and caverns) with data from the surface on the right side, it can be concluded that the huge system named "Golshan's Cave" is a center of discontinuity and karstification concentration. Consequently, the possibility of the development of karst channels was guided in two directions. The main conduit system is developed along the J1 discontinuities from North toward the South. Progressive karstfication along that system toward the South was hindered by Pabdeh and Middle Asmari formations. Because of that, the lateral branches from the main system toward the canyon bottom were developed.

Note: The solution to avoid main karstification zone is presented in Box. 16.1.1 *of* Sect. 16.1.

Box 15.4.6

Case study—Speleological investigation of Bogovina cave under conditions of "Bogovina" dam and reservoir construction

The multipurpose dam and reservoir "Bogovina" in the Crni Timok Valley (Eastern Serbia) is one of several projects that have been initiated to improve water availability in Serbia. Bogovina cave is one of the largest in Serbia and belongs to the group of caves with temporary hydrogeological function. After complex speleological and hydrogeological surveys, the potential of significant water losses from the reservoir was identified. The analyses show that once the reservoir is filled up, groundwater flow currently oriented toward the future reservoir would saturate the upper part of the karstified rocks, reactivate currently unsaturated pathways, and form a reverse discharge outside of the reservoir area (Fig. 15.50). In response to these findings, mainly to the speleological investigations, the dam design and technical details have been adapted accordingly: The dam height has been reduced by 9 m, and it is proposed that grouting and consolidation work be conducted both at the foundation of the dam and extensively on the embankments of reservoir (Milanović et al. 2010).

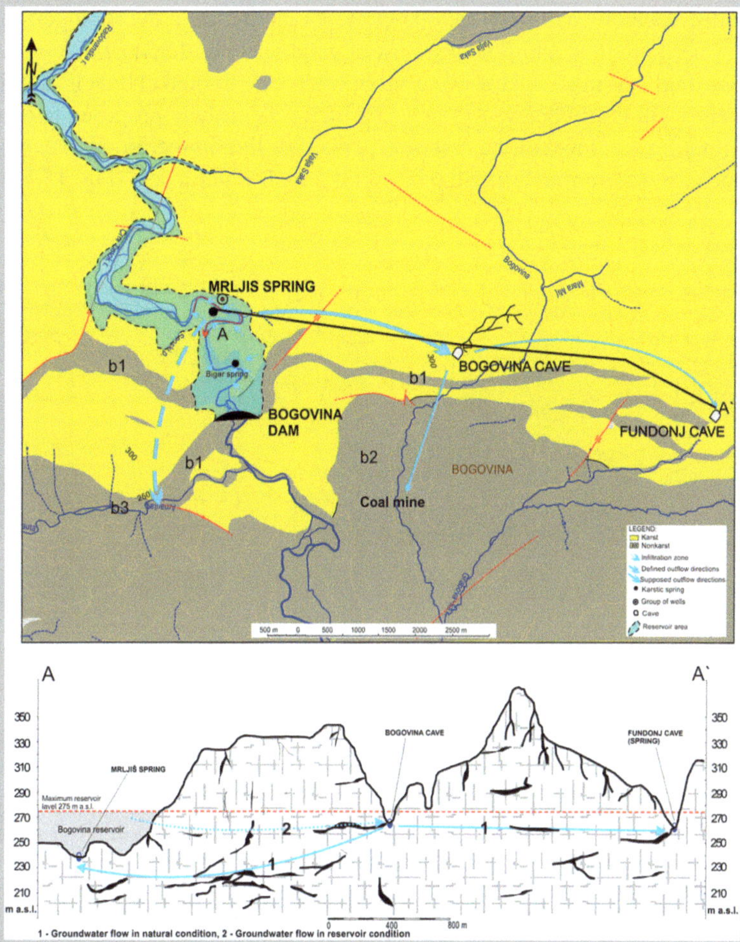

Fig. 15.50 Schematic hydrogeological map (non-karst: *b1* Aptian orbitolinae sandstones; *b2* Oligocene deposits of Bogovina basin; *b3* Upper Cretaceous volcanic rocks). Cross section along karstic spring Mrljis, Bogovina, and Fundonj cave, (*1*) groundwater flow in natural condition, (*2*) groundwater flow in reservoir condition (Milanović et al. 2010)

It was concluded that although the proposed remedial measures cannot guarantee reservoir tightness, they can reduce the risk of large-scale leaking.

According hydrogeological characteristics, Bogovina cave has generally two levels of channels: (1) lower, active channels important for the aim of the Bogovina reservoir and dam construction and (2) higher, “dry” channels (Stevanović and Dragišić 2002). At its lowest points, the level of the lower channels is under water throughout the year, and the level of the higher channels is active only in periods of highest water.

Detailed speleological explorations of the Bogovina cave during a period of low water identified an upper network of galleries (Fig. 15.51), most of them higher than 280 m a.s.l. However, out of 6 km of explored channels, approximately 1 km is still below 275 m a.s.l which was initially proposed to be the maximum water level of the reservoir. This part should therefore be exposed to intensive saturation during the reservoir's high water levels (Fig. 15.52).

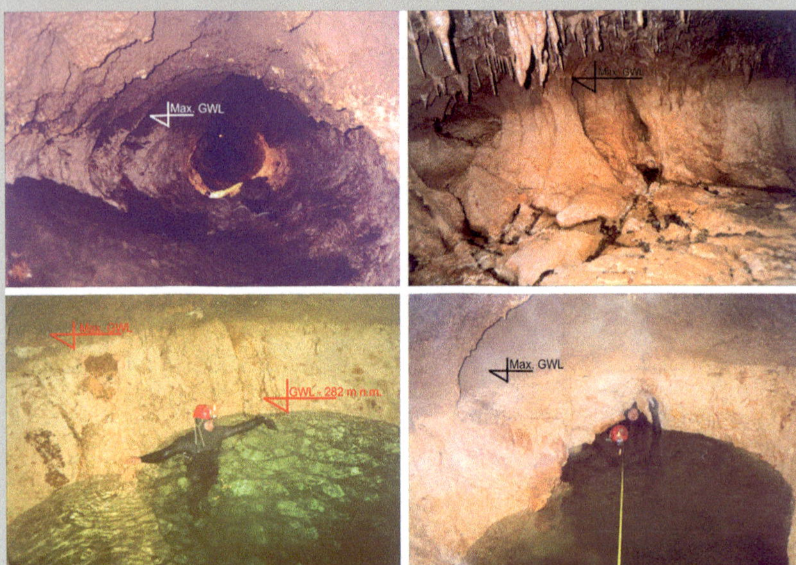

Fig. 15.51 Photographs from speleological investigations of Bogovina cave. Photographs show groundwater level during investigations and zone of maximal water level

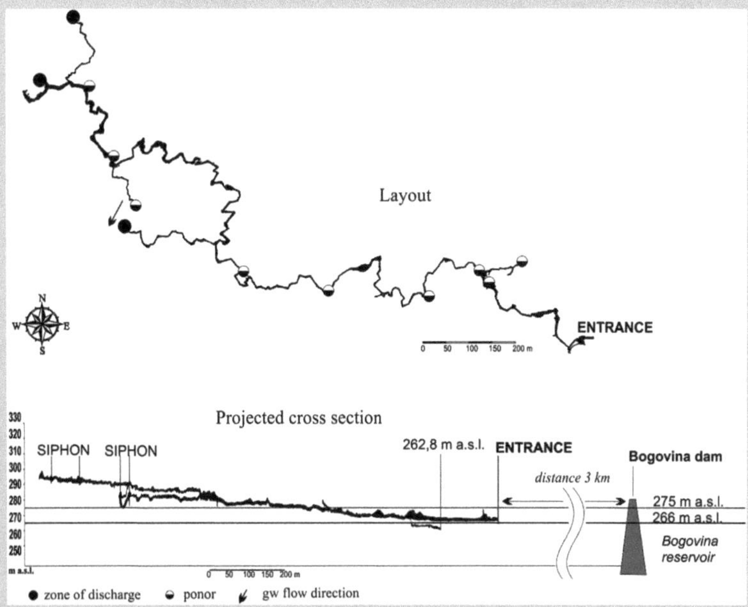

Fig. 15.52 Layout and projected cross section with proposed two water levels in reservoir

Speleological research of Bogovina cave, which was realized in the (wider) scope of complex hydrogeological research for the purpose of "Bogovina" Dam and reservoir construction, found the following:

- Reservoir construction with the initially projected max. water level of 275 m a.s.l. is very problematic according karst channel position and potential great water losses.
- During high hydrostatic pressures (ca 4 Bars), the water infiltration will be intensified through the reservoir bottom, particularly in the Mrljiš Spring area, some 2 km upstream from dam site.
- After reservoir impounding up to the elevation of cca 266 m a.s.l., infiltrated waters will flow underground toward the Bogovina cave, and further into three directions (Figs. 15.50 and 15.52).
- Although lower maximal water level in the reservoir is suggested to be at 266 m a.s.l., it will still submerge a part of the Bogovina cave channels and have a negative influence on bio- and geodiversity.

The above-given examples of research are only a part of the experiences that point to significant development of karstic channels in karst areas. Speleodiving and speleology research in defining the position and functioning of karstic aquifer have contributed significantly to solving water supply problems, and to the construction of hydrotechnical and other structures in karst.

15.5 Engineering Regulation of Karstic Springflow to Improve Water Sources in Critical Dry Periods

Zoran Stevanović

Centre for Karst Hydrogeology, Department of Hydrogeology, Faculty of Mining and Geology, University of Belgrade, Belgrade, Serbia

15.5.1 Introduction

In spite of being one of the main water sources for drinking water supply worldwide, karst aquifers are at the same time one of the most problematic resources. Due to the unstable flow and great variation of the discharge of the karstic springs, during recession (lean) periods, the local population may suffer from water shortage and it is often a problem to ensure sufficient water for various consumers (Paloc and Mijatović 1984). Another problem is that recession periods which in continental climates coincide with summer and early autumn months are also the time when water demands significantly increase, and this divergence requires an adequate technical response. In arid and semiarid karstic areas of the Near and Middle East and northern Africa, there are similar and even worse situations. During periods of minimal flow followed by increased demand, periods which could continue for months in these parts of the world, the risk of conflicts among water users significantly increases (Stevanović et al. 2005; Stevanović 2011).

It is thus the main challenge for many waterworks to ensure water supply and to avoid restrictions or total interruptions in water provision during these critical periods. The minimal discharges of the springs and the accordingly reduced minimal river flows during recession periods result in water deficit not only for water consumers but also for dependent ecosystems.

If aquifer is well karstified and has adequate storage in its deeper parts, it is often possible, just as it is in the case of open water reservoirs, to regulate and manage minimal flow by various engineering interventions. Such an option provides opportunities to satisfy water demands not only of direct consumers but of ecosystems as well by ensuring ecological flow downstream. In this section, possible engineering solutions to regulate groundwater flow in karst and physical, ecological, and economical implications of such interventions are discussed. A few successfully implemented projects of aquifer discharge control and lessons learned in management of groundwater reserves are also presented in this section.

15.5.2 Solutions to Regulating Karstic Aquifers

Although the total dynamic reserves in karst often surpass by far the exploitation capacities, most of the tapping structures are constructed simply to tap the natural discharge of the springs and thus depend solely on the natural flow regime.

Engineering regulation of karst aquifers *implies water intake or other structures that aim to create physical conditions for controlling and exploiting the* *water reserves in a controlled manner by changing the natural regime.* This is, therefore, an artificial intervention and application which the standard tapping structures do not permit. Hence, karst aquifers attract hydrogeologists as well as other engineers searching for optimal solutions to controlling karstic groundwater.

The two main groups of engineering regulation for controlling karstic groundwater can be distinguished as follows:

1. Regulation of discharge zone,
2. Regulations, i.e., interventions in wider catchment area.

Both of them, however, require available water resources. Otherwise, applied intervention may result in only a temporary or short-term effect and negative environmental impact. This interaction of *solution–impact–cost* could also be expressed as (Stevanović 2010a):

1. ***Solution***: Regulation is **physically possible**;
2. ***Impact***: Regulation is **environmentally sound** and friendly;
3. ***Cost***: Regulation is **economically feasible** and sustainable.

15.5.2.1 Physical Preconditions

There are several limitations which may impede engineering regulation projects and artificial control of an aquifer's discharge. The most common are insufficient data on aquifer systems, problematic access to water storage, and limited water resources.

Shortage of knowledge of aquifer distribution and properties could be one of the major obstacles. Knowledge of geological and hydrogeological properties is essential: Geology, i.e., lithology and tectonic patterns, is the main factor in the creation of a physical karstic environment. Aquifer geometry (position of the groundwater table and thickness of the saturated zone), permeability, storativity, and water availability (reserves) directly depend on geological conditions. However, other factors such as topography, climate, hydrography, vegetation cover, rocks, and water chemistry also influence the amount and quality of accumulated water in an aquifer. This data along with a groundwater regime derived from systematic monitoring of groundwater quantity and quality are preconditions for a successful regulation project. In contrast, without prior extensive research and monitoring, any direct attempt at their control is bound to fail.

Inaccessible aquifer cannot be properly regulated. It is due either to very thick impermeable cover (stored water is often of poor quality and with limited recharge) or because of complicated topography. Many karst aquifers are extended over high and inaccessible mountains, and many springs discharging from them are located just below very steep cliffs. In such circumstances, even simple tapping of springs can be difficult with no space for drilling or possibility to intervene on the natural flow regime. If springs are submerged by sea or lake waters, their utilization can be additionally complicated or even impossible (see Sect. 16.4).

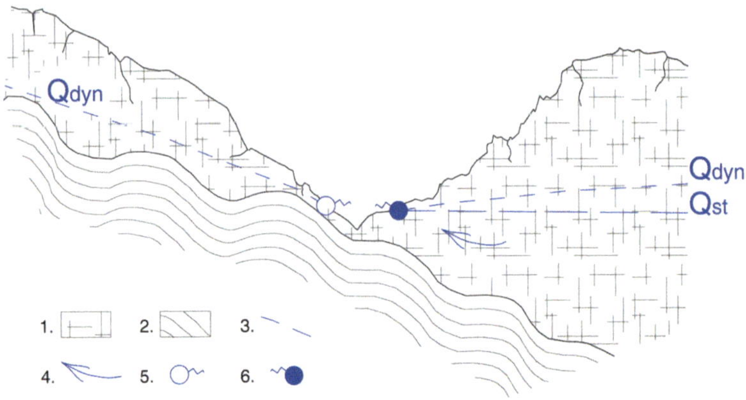

Fig. 15.53 Cross section of a composite karstic aquifer drained by the depressional springs. On the left side of the gorge, there is no storage and only dynamic groundwater reserves exist, and the spring has a temporary character. On the right gorge side drained by a perennial ascending spring, both dynamic and static reserves are present. The opportunity for managing groundwater flow is thus on the right gorge side. *Legend* (*1*) karst aquifer; (*2*) impervious rocks; (*3*) groundwater table; (*4*) groundwater flow direction; (*5*) temporary spring; (*6*) perennial spring

Limited water resources result from limited recharge or from absence of groundwater storage (Fig. 15.53). Groundwater availability and recharge component (effective infiltration) should be assessed on a long-term or multiannual basis but not on a momentary basis. In the case when there is no storage in an aquifer, there would be no option to regulate its regime. Such a situation can be easily imagined as we compare a riverflow with and without a constructed upstream reservoir. Only if a reservoir (storage in the case of aquifer) exists is manipulation by regime possible.

15.5.2.2 Environmental Requirements

Environmental requirements should be strictly respected in any kind of engineering works with groundwater resources (see Chap. 14). The only exception can be humanitarian water needs: In certain places on our planet, we cannot discuss "sustainable exploitation" but must simply follow demands of local populations and their animals. However, even then, we are obliged to try to impose an optimal artificial intervention within the aquifer or on its surface, with the aim of counterbalancing the regime, or rather equalizing the quantitative and qualitative groundwater properties (Stevanović 2010a).

What is a "sustainable" and environmentally sound engineering regulation? Many decision makers in the water sector are trying to impose the totally restrictive concept that exploitation of groundwater must be completely covered by recharge throughout the entire period of water extraction. This means no depletion in aquifer is allowed even for a short period of time. The other more "flexible" concept allows only temporary depletion and fast compensation of extracted waters. The issue of allowed water exploitation rate is closely linked with the "safe yield" concept which was introduced for the first time by Meinzer (1920). His concept of safe yield considered an abstracted rate of water without generating "undesirable effects." No further attempt to quantify the amount of extracted waters versus the depleted water table or to define a timely component was made at that time. The recent and we may say modern approach was introduced by Custodio (1992), Margat (1992), Burke and Moench (2000), and several other hydrogeologists. It can be summarized by the following citations: "…Keeping aquifer full is equally bad or even worse than one utilized and always empty…" (Custodio 1992) and "The planned mining of an aquifer is a strategic water resource management option where the full physical, social and economic implications are understood and accounted for over time. A declining water table does not necessarily indicate overabstraction of the groundwater resource. Overabstraction should not be defined in terms of an annual balance of recharge and abstraction. Rather, it needs to be evaluated over many years, as the limit between non-renewable stock and the stock that is replenished by contemporary recharge from surface percolation is usually unknown" (Burke and Moench 2000).

The aim of engineering regulation of an aquifer is to enable tapping of the necessary water volume during certain periods (usually periods of increased water demands) by being able to count on sufficient water replenishment (during the following wet seasons). It is true, however, that this aim is not achievable everywhere, and these two possible scenarios exist (Stevanović 2010a):

1. Replenishment potential is sufficient to cover a "loan" over a short period of time (within the same or the next hydrologic cycle) (Box. 15.5.1).
2. The deficit cannot be compensated, and a water table decline is envisaged.

Box 15.5.1

Figure 15.54 shows a typical hydrograph of a karst spring with sufficient replenishment potential. Results of observations made during several consecutive years revealed an average discharge from karstic aquifer of 0.8 m^3/s (as total natural springflows). The minimal recorded discharge is 0.18 m^3/s. Such a situation is repeated in the hydrological year presented on Fig. 15.54. Although total water requirements (potable water, small industry, and ecological flow) are 0.3 m^3/s, the deficit is 0.12 m^3/s but could be compensated by the high replenishment potential of this aquifer. Of course, this is in the case that other prerequisites are met and that it is physically possible to tap additional water (overpumping). In the described case, the deficit occurs for about 60 days when pumping is absolutely necessary, but this period would be further extended to cover any additional deficit resulting from pumping. The first recharge will thus compensate additional drawdown and thereafter start to refill the "natural" aquifer's storage.

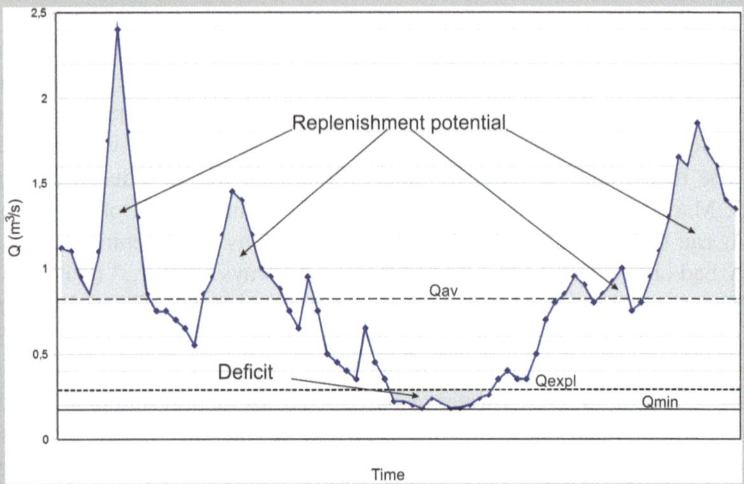

Fig. 15.54 Typical annual hydrograph of a karstic spring with potential exploitation capacity (Q_{exp}) which is bigger than natural minimal discharge and still much lower than dynamic groundwater reserves accounted as annual average discharge (Q_{av})

Figure 15.55 shows a hydrograph of Mlava Spring in the Carpathian karst of eastern Serbia (spring features discussed also in Chaps. 6, 9, and Sect. 15.2) for an average hydrological year. Tapping this spring and ensuring stable water delivery to various consumers in an amount of around 0.85 m^3/s would also necessitate planned pumping for about 2 months.

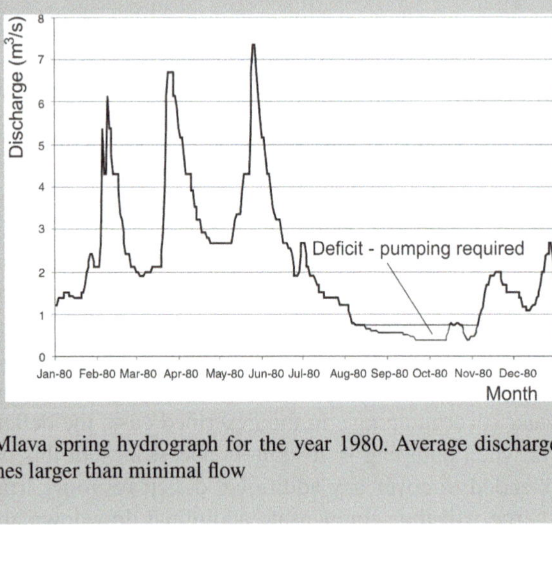

Fig. 15.55 Mlava spring hydrograph for the year 1980. Average discharge is 1.85 m^3/s, about five times larger than minimal flow

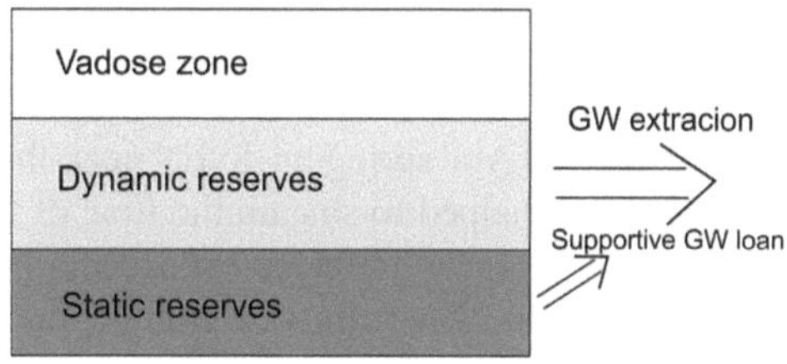

Fig. 15.56 Scheme of groundwater reserves and temporary loan of static reserves during critical drought periods

Therefore, where replenishment potential exists to overcome the problem of water shortage during critical low-water periods, the study of deeper aquifer storage, i.e., static groundwater reserves and opportunity for their temporal "loan," is necessary (Fig. 15.56).

The temporal loan of stored water when water demands are higher than natural discharge does not lead to overextraction if sufficient replenishment potential exists.

$$\text{If} \quad Q_{\text{expl}} > Q_{\text{dyn critic}}, \text{ but}$$
$$Q_{\text{expl}} = Q_{\text{dyn critic}} + Q_{\text{st "loan"}}, \text{ then}$$
$$Q_{\text{st "loan"}} = Q_{\text{expl}} - Q_{\text{dyn critic}}$$

where

Q_{expl} is exploitation request, water demands;
$Q_{dyn\ critic}$ dynamic groundwater reserves in critical dry periods; and
$Q_{st\ \text{“loan”}}$ static groundwater reserves for periodic loan (equal to the difference between demands and current natural flow).

Counting on water replenishment during the subsequent wet season, overpumping, and groundwater extraction during a limited time period is possible in the regions with moderate climate and rainfalls well distributed throughout the year (Stevanović et al. 2013). Different from arid areas where such an approach often leads to aquifer overexploitation and depletion of groundwater resources, the climatic and hydrogeological preconditions in SE Europe and the Mediterranean basin are favorable for such kinds of interventions.

Although the above scenario is not too problematic, the second can be compared to a reckless and unsustainable credit-financed spending spree. Furthermore, water management can be a very delicate task, given the fact that a large portion of the world's groundwater is accumulated in areas where an arid or semiarid climate with low rainfalls and huge evaporation prevails. Many countries have been using far more water than they have or can recover from the rate of replenishment. The result is a large water-based ecological debt that will be transferred to future generations. Case examples from Saudi Arabia, Kuwait, Yemen, and parts of India (Rajasthan) are some of the relatively recent and well-known ones (Box. 15.5.2); others are ongoing and still not addressed (Stevanović 2013). Therefore, water management in arid climates should be more oriented toward water saving and conservation in the rocks.

Box. 15.5.2

The Ras el Ain spring in Syria near the Turkish border was a very large spring that helped to sustain the flow of the Euphrates River via its tributary Khabour. Some references even declare it to have been the largest spring in the world. According to Burdon and Safadi (1963), its discharge was between 34.5 and 107.8 m^3/s. They stated that groundwater issued from thirteen springs that drained one basin of over 8,000 km^2 consisting of Eocene limestones, Miocene evaporates, and limestones as well as basaltic rocks. Unfortunately, this spring no longer flows. Overexploitation of the water resources by numerous drilled wells in the area caused serious depletion of the groundwater table. Hole and Smith (2004) discussed the environmental and landscape changes in northeastern Syria over the last 100 years, and its transformation from an open rangeland to an intensely cultivated landscape. As in most arid or semiarid zones, successful agriculture requires either supplemental or full irrigation. But very ambitious plans to develop the water resources for summer cropping, and decisions by individual farmers to install wells led to a fundamental alteration of the natural drainage in favor of groundwater extraction, storage reservoirs, and irrigation canals. In addition, the settlement of refugees and a rapid population growth have contributed to the increase in water demands in this area bordering Turkey and Syria. This is today one of the most fertile and intensively cultivated regions in the Near East, but it is questionable whether such intense aquifer regulation is justifiable from an environmental point of view.

In conclusion, a clear distinction is always required between those areas with climatic and hydrogeological conditions which favor regulation measures ("good" rainfall and snow and "good" infiltration capacity and storage) and those where adequate replenishment is impossible or questionable.

15.5.2.3 Financial Qualifications

It is a long route to an idea of engineering regulation of the source and to full implementation of such a project, and each step along the way requires justification. Along with described hydrogeological and environmental assessments, the designers should also estimate the cost of their intervention, provide alternatives, prepare a feasibility study, and finally make an optimization of technical solutions. The long-term effects of planned interventions should be taken into consideration. To forecast these effects, the analyses of spring hydrographs and stochastic models could provide valuable indirect information on the structure of karst hydrogeological systems and their behavior (Mangin 1984b; Bonacci 1993b; Jemcov 2007b;

Kresic 2009). The two previously presented contributions (Sects. 15.2 and 15.3) also deal with this problem. Optimizing the tapping and regulating structures and reducing the cost to the minimum for the same final effects is an engineering art also expected from such forecasting.

In some countries, respective water management bodies determine the environmental limitations which can also be incorporated into such models. In practice, exploitation permits include various limitations to protect both the ecosystem and downstream consumers. Often, the percentage of historical streamflow values (minimal or average) has to be determined and is required to be kept in the riverbed downstream of intake (see Box 15.5.5). Problems could arise when there is no surface water around, and then, regulatory measures usually fix the groundwater table.

Despite the fact that in many cases economic justification is obtained and the total dynamic reserves surpass by far the exploitation capacities, many constructed structures simply tap the natural discharge of the springs (and provide gravity water delivery; Paloc and Mijatović 1984) and thus depend solely on natural flow regime. It is the engineers involved here who should make a bigger effort to attract the decision makers and investors and justify the sustainability of regulation projects which, however, must be guaranteed through the many steps taken before, during and after the implementation of these projects.

The example presented in Box 15.5.2 also has a sensitive financial component. The recently prepared mitigation project by the Mediterranean Agronomic Institute from Bari (Italy) aims at increasing efficiency in agriculture and introducing economical irrigation techniques. In total, 1,390 wells were mapped throughout the project area, over half of them lacking any kind of usage restrictions. The authors stated that "insufficient water resources and the poor management (which) resulted from the use of inappropriate and inefficient techniques have caused severe economic losses and harmful social and environmental effects, over (the) last few years" (http://www.delsyr.ec.europa.eu/en/eu_and_syria_new/projects/24.htm).

Stevanović (2010a) noticed that "implementation of large groundwater control projects such as the construction of underground reservoirs can be a very expensive and delicate task. Although numerous analyses show that benefits often surpass negative consequences, the fear of uncertain (and non-visible) results and large investment has put many such projects on hold."

Some other smaller projects such as drilling the wells near the springs and the temporary pumping of a required amount of water are less costly, and their implementation can easily be approved. However, where extraction rates are concerned when providing this "tool" to the end users, *monitoring* and *limits* are always required. Otherwise, unlimited pumping not only results in a negative impact on the environment but also causes economic losses. The negative consequences might not be seen immediately, but if significant depletion of the water table starts, this can be taken as a clear sign that the water use privileges of the current generation will not be enjoyed by future ones.

15.5.3 Indicators of Prosperous Sites for Engineering Regulation

From previous discussion, it is clear that an attempt to control aquifer artificially requires the existence of *static groundwater reserves* (geological, or "non-renewable")[1] or a *deeper karstification base*. Preferably, the discharge should be over the *ascending springs*, which often indicates *thicker aquifer storage*. If the discharge is along the relative barrier (i.e., the barrier from permeable rocks), then regulation structures beside the springs can focus on the capture of the deeper underground flow or on tapping the secondary (contact) aquifer.

Along with ascending type springs, the *subsurface flow* is one of the indicators of possible considerable storage within the aquifer. Figure 6.10 in Chap. 6 shows schemes with or without subsurface drainage which strongly depends on the permeability of contact rocks.

How can this invisible flow be identified? In practice, the occurrence of underground water discharge may be identified on the basis of simultaneous fluviometric measurements of streamflow and simultaneous thermometric and conductivimetric measurements. Positive results can also be obtained by using certain geophysical methods ("spontaneous potential," "mise a la masse," see Box 15.5.3). A reliable verification of this form of discharge, however, may be obtained only from drilling and testing of boreholes placed in adjacent aquifers.

Box 15.5.3

Simultaneous hydrometric methods have been broadly applied for investigation purposes in eastern Serbia. Regardless of certain limitations of this method (instrumentation, accuracy of measurements, representativeness of the profile), favorable results on the presence of subterranean flows which was subsequently verified by exploratory works have been obtained at several sites.

The riverflow anomalies and flow increments were recorded at several streams in the Carpathian karst of eastern Serbia (Stevanović 1995). One example is the river Svrljiški Timok near Knjaževac (Tresibaba Mt.). The "input" flow in the karst system was only 132 l/s. About 4 km downstream, the flow increased 3.5 times, and at the lowest downstream profile, the total increase in flow was sixfold (at this section, no springs have been recorded). The mean subsurface inflow was 80 l/s/km. Similar results have been obtained at Toponička reka, Mirosava, and Crni Timok River (see Case study 2 in Chap. 6).

[1] See Chap. 6 for detailed explanation.

Thermometric method is based on different temperatures of groundwater (flowing into surface courses) and of surface waters in the periods of extreme air temperatures (winter season, summer season). Only in this way might temperatures of surface waters be changed and identified as considerably different from stable groundwater temperature.

Measurement of electrical conductivity of water is based on various ground and surface waters conductivities. It may be used in addition to the thermometric method.

Kullman and experts from GUDS, Bratislava (1984), were the first to have successfully used these two methods in the zone of Galmus na Hornade and Chocsky vrhy in Slovakia. The simultaneous thermometric and conductivimetric measurements have also been carried out on the Hron River. The subterranean flow from karst aquifer, dolomites, and limestones of Middle and Upper Triassic reaches the river flow by concentric movement toward a silt barrier. After the anomalies had been recorded, the borehole PNTK-11 was drilled and maximum water-yielding capacity was 100 l/s (Kullman 1990).

The electrical spontaneous potential (SP) method may also provide useful data. This geophysical method is based on the measurements of natural electrical field, and positive anomalies indicate the upward filtration of groundwater (most often into elevated aquifers with intergranular porosity). The presence of negative fields of electrofiltration proves the opposite processes—the centers of anomalies are in the area of most intensive water percolation.

In the valleys of Crni Timok and Radovanska Reka in the Serbian Carpathians, alluvial deposits 2–4 m thick overlie the karst aquifer. The SP measurements were made along the cross sections of the valleys (100–300 m wide), up to the karst edge, along about 2 km both upstream and downstream from the main zone of discharge of aquifer (Stevanović and Dragišić 1998). The profiles are at their respective distances of 50 m, while the measuring points on the profile are at 10 m.

Along the valley, a great number of fields with anomalous values were recorded. They proved the presence of descending and ascending streams (a complex hydraulic mechanism of recharge and discharge of karst aquifer). The greatest positive anomalies occur in a broader zone of the largest Mrljiš Spring ranging from 20 to 30 mV, which is in sharp contrast to the adjoining relatively "calm" fields (Fig. 15.57). The investigations of electrical resistivity in these localities proved deeper karstification and significantly marked break points of resistivity curve indicating the positions of water level in aquifers. In the centers of anomalies of the SP, several test boreholes were located and drilled. Most of them gave positive results in terms of the presence of cavernous systems and channels with active circulation of aquifer streams. Especially favorable are the profiles of boreholes HG-19 and HG-7 in the centers of maximal values of SP (Stevanović 2010b).

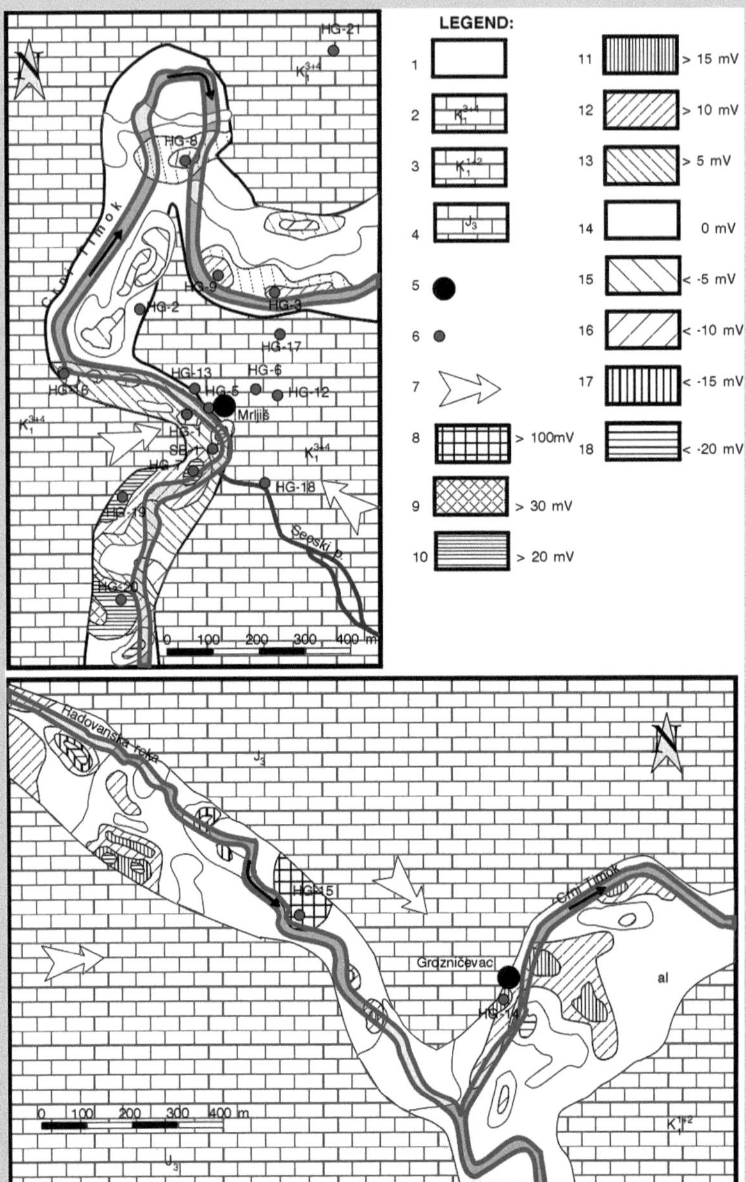

Fig. 15.57 The map of Crni Timok and Radovanska reka valleys—results of measurements of electrical spontaneous (self) potential (SP). *Legend* (*1*) alluvial deposits; (*2–4*) Jurassic and Cretaceous limestones; (*5*) karst spring; (*6*) borehole; (*7*) direction of groundwater flow; (*8–18*) values of spontaneous electrical potential in millivolts (Stevanović and Dragišić 1998, copyright Springer)

In borehole HG-19 up to the depth of 60 m, a series of cavernous intervals were recorded (Stevanović 2010b). The total effective cavernous porosity is as high as 17 %. Especially marked is a cavernous system 49–58 m which is assumed to represent the main influent channel to the discharge zone of the main spring Mrljiš. In borehole HG-7, several cavernous intervals were also recorded, with one at a depth of 67.5–69.5 m. By applying the "mise a la masse" method, the aquifer stream in the direction of the karst spring was verified. The results obtained represented the basis for determining the drilling site and construction of production wells from which a direct aquifer overpumping during the recession period was made possible.
Some other methods described in Chap. 4 and especially remote sensing discussed more in Sect. 16.3 may also be useful for determining points or zones with subsurface flows.

15.5.4 Regulation of Discharge Zone

Paloc and Mijatović (1984) distinguish three kinds of karstic groundwater tapping:

1. Tapping groundwater by gravity intakes,
2. Tapping by decrease in water table, and
3. Tapping by increase in water table.

Although the first kind is in fact simple flow capturing without compulsory changing water regime, the two others imply artificial intervention and direct impact on groundwater regime. Additionally, the two groups of consumers, *exploitive* and *sustainable,* can be distinguished.

We may also introduce the following categorization of groundwater engineering interventions to regulate the flow:

1. The spring overpumping,
2. Drilling the wells or other supplementary intakes,
3. Constructing subsurface (underground) dam,
4. Artificial recharge.

Tapping groundwater by decreasing water table involves the former two options, while the two latter options consider increase in water table within the aquifer and further utilization of this additionally stored water.

- *The spring overpumping* considers the earlier discussed water "loan" from deeper aquifer parts and the use of that "loan." In the case of deep vauclusian springs or even potholes without outflow but in zone with active groundwater flow, it is possible to install pumps into siphon channels and pump the required amount of water. It is the simplest way of aquifer regulation but not the cheapest one because heavy centrifugal pumps (rarely submersible) may require a lot of generated energy for their operation.

Several successfully implemented projects worldwide promote such a method as very efficient at least in the first exploration and exploitation phase (Avias 1984; Stevanović et al. 1994; Stevanović et al. 2007). In many cases, conducted pumping tests of the siphon confirm the feasibility of this kind of aquifer regulation and prospect for further design and technically improved solutions (Box 15.5.4).

Box 15.5.4

One of the most successful projects in engineering regulation of karst aquifers was completed in southern France for the water supply of the city of Montpellier (Avias 1984). The survey conducted in the 1960s included speleodiving, geophysical, geoelectrical, and geomagnetic prospection, and pumping tests of deep siphon discovered in the karst interior. Based on the survey results, a shaft 80 m deep and 5 m wide including a large pumping room was designed and constructed. The minimal natural spring discharge is only 0.4 m^3/s below the actual pumping capacity of 2.2 m^3/s. What is more important from an ecological point of view, a horizontal gallery connects the natural spring site and the shaft at its depth of 23 m and delivers water to the spring orifice thereby ensuring minimal flow of the Lez River (Fig. 15.58). This excellent example of finding a compromise between water utilization and ecology motivates many engineers worldwide to search for similar solutions in karst environment.

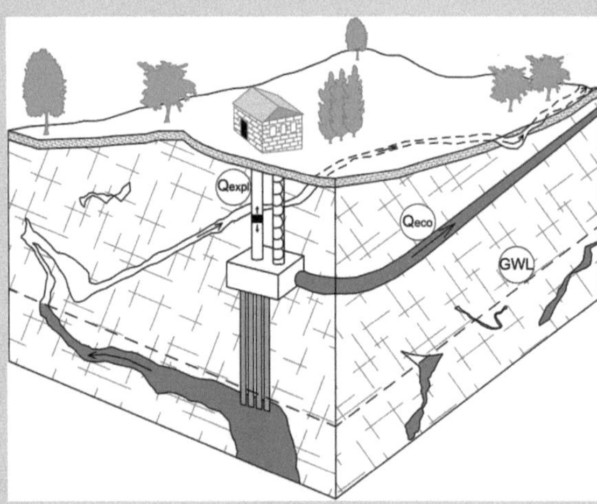

Fig. 15.58 Lez source and intake scheme (from Montpellier waterworks leaflet, modified by Stevanović)

Similar projects implemented as temporary solutions or conducted tests are described in Stevanović and Dragišić (1995), Milanović (2000b) and Jevtić et al. (2005b; presented also in Box 15.4.1).

- *Drilling the wells or other supplementary regulation intakes* is also frequently applied in karst. Large-diameter vertical tube wells have so far yielded the best results because the drilling technology is known everywhere. Aiming to control the groundwater flow, the well can be placed down- or upstream of the discharge points. The local topography and hydrogeology are the main factors in such a decision. This additional intake can be in close proximity to or far away from the spring, depending on the decision to disturb or not to disturb the natural spring's regime. As described in the case of Lez, planned drying of a spring can also be accounted in a regulation project.

In the case of covered and inaccessible karstic aquifer, instead of a vertical well, the inclined well or horizontal galleries can also be drilled or constructed (by excavators). If such intake taps the lowest part of the aquifer, it functions as *bottom outlet* of the aquifer, similar to such structures in open reservoirs.

Figure 15.59 shows all three above-mentioned options, while in Box 15.5.5 four case studies of aquifer regulation by the wells and galleries constructed under very different hydrogeological settings are described.

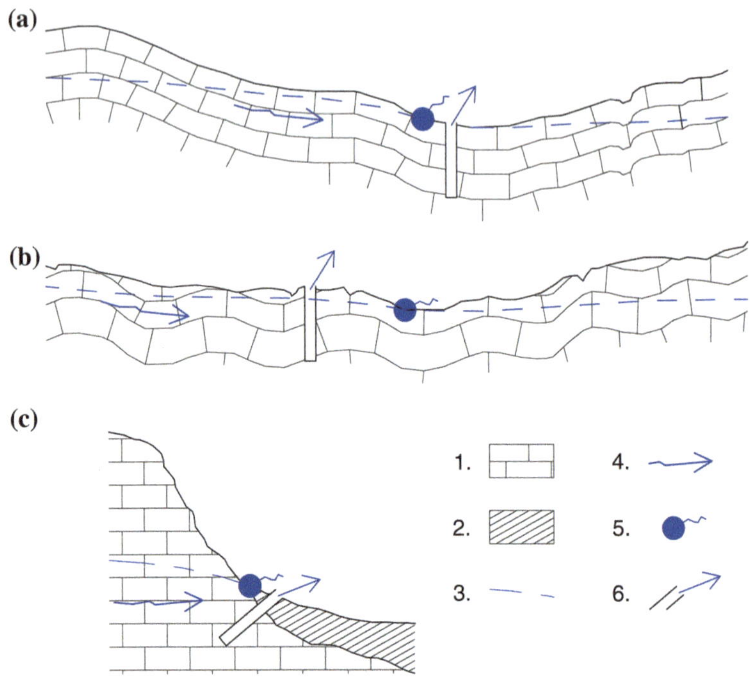

Fig. 15.59 Controlling the groundwater flow in drainage area. **a** Vertical well downstream (beneath) the spring; **b** vertical well upstream (above) the spring; **c** inclined well drilled through impervious cover. *Legend* (*1*) karst aquifer; (*2*) impervious rocks; (*3*) groundwater table; (*4*) groundwater flow direction; (*5*) spring; (*6*) discharge from well

Box 15.5.5

Case study—Jergaly source, Slovakia

In the 1970s, Kullman and experts of the Slovakian Geological Survey undertook extensive hydrogeological research in the basin of Jergaly Spring, the main source for the water supply of the town Banska Bystrica (Kullman 1984a; Kullman and Hanzel 1994). The Jergaly spring actually represents a wider discharge zone that drains karstified Jurassic rocks of Velka Fatra Mts (Fig. 15.60). Despite the deeper karst water circulation, the natural spring discharge is variable, from 123 l/s to 1,315 l/s, while retardation of precipitation is assumed on 7–14 days. In the first stage of the survey, the well HJ-1 was drilled to a depth of 44 m in the vicinity of the main spring. An ascending flow was documented and successfully tapped. Accumulated groundwater reserves were calculated according to the results of pumping tests and of long-term systematic measurements of the spring over the course of 13 years (1962–1974).

The pumping test in the low-water season in 1971 started from the HJ-1 well when the spring discharge was *Q*—178 l/s. After 26 days, a depression of 4.5 m was reached and stabilized with a discharge of 236 l/s. The volume of additional pumped accumulated groundwater reserves was about 261,000 m^3. The diagram of the pumping test at the depression of 4.5 m is shown in Fig. 15.61. After this successful pumping test, three new wells were drilled in the spring area to a depth of 50 m.

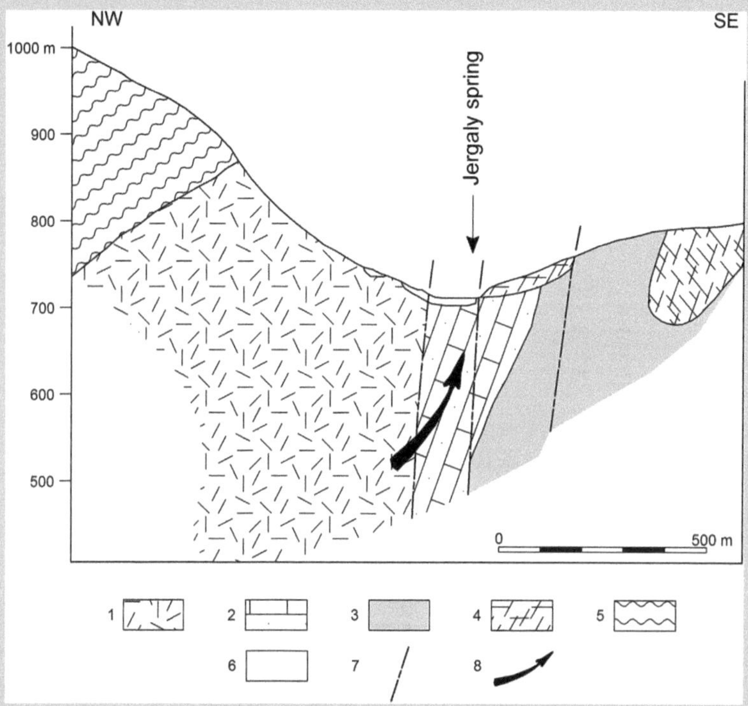

Fig. 15.60 Geological cross section of Jergaly source (modified from Kullman 1984b). *Legend* (*1*) Mid-Triassic limestones and dolomites; (*2*) Liassic limestones; (*3*) lower Triassic schists; (*4*) Mid-Triassic massive dolomites and limestones; (*5*) upper Triassic clayed schists, marls, marly limestones, (*6*) Quaternary deposits; (*7*) fault; (*8*) groundwater flow direction

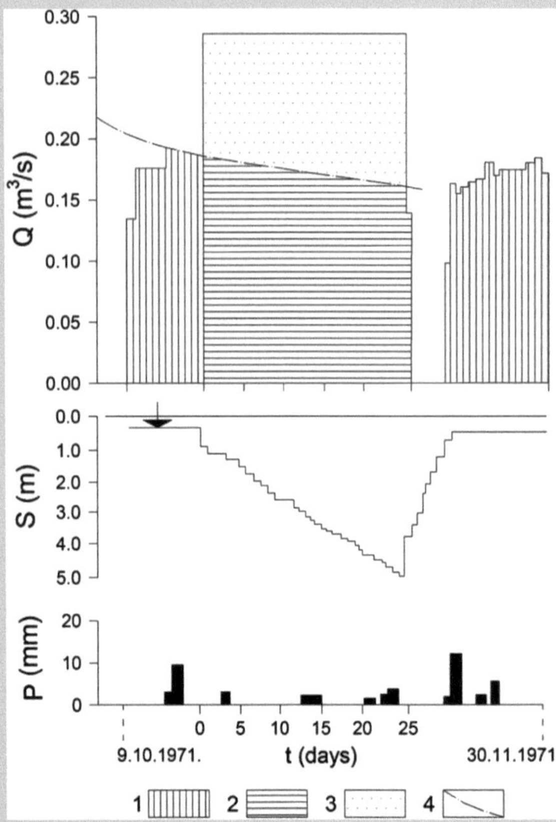

Fig. 15.61 Diagram of pumping test of HJ 1 well near "Jergaly" spring from Oct 19, 1971, to Nov 13, 1971. (Kullman and Hanzel 1994, printed with permission). *Legend* (*1*) natural spring discharge, (*2*) pumped amount corresponding to natural discharge during the test, (*3*) pumped amount from accumulated groundwater reserves of deeper hydrogeologic structure, (*4*) groundwater depletion curve; *Q* discharge and pumping rate, *S* drawdown during the pumping; *P* precipitations during the pumping

The pumping test from the group of wells lasted 54 days. In the initial phase, the pumped amount was 536 l/s, while in the final phase, it decreased to 257 l/s. The same method as in the HJ-1 test was applied and obtained, a maximal depression of 19 m found to correspond with 1,134,200 m^3 of the exploitable accumulated groundwater.

On the basis of these results, three stages of exploitation have been suggested and implemented as most suitable and economical. At a high spring discharge (above 250 l/s), only natural discharge water is exploited. At a lower discharge (below 250 l/s), the water is exploited to the depression at 4.5 m by the siphon system (overflowing), while at the depression below 4.5 m, submersible pumps are activated.

Kullman and Hanzel (1994) concluded that in the West Carpathians the best option is the combined exploitation of the natural discharge of a spring and the utilization of deeper groundwater reserves from the adjacent hydrogeologic structure in the periods of low natural discharge of the spring. According to personal communication with Dr Peter Malik from the Slovakian Geological Survey, the pumping of the Jergaly source nowadays is irregular first because of the high operation cost for pumping, and secondly due to decreased water consumption resulting mainly from imposed higher water taxes.

Case study—Bogovina source, Serbia. The hydrogeological surveys undertaken from 1987 to 1997 have made possible the construction of several successful systems for the artificial control of karst aquifer in Serbia (Stevanović et al. 2007). The largest system for aquifer control was constructed as part of the regional water system "Bogovina" that supplies several towns of the Timok region in eastern Serbia (Stevanović 2009, 2010a, 2010b).

The hydrogeological setting of Bogovina area is partly explained through several examples presented in this book (Box 3.5 in Chap. 3; Case study 2 in Chap. 6; Boxes 15.4.6 and 15.5.3 of this Chapter). The greatest part of the catchment area of ca 100 km^2, which consists of karstified Lower Cretaceous limestones, drains mainly through the permanent vauclusian spring Mrljiš (yielding 80–1,000 l/s). In order to estimate the physical conditions for aquifer control, a pumping test lasting several days was conducted at Mrljiš Spring. Prior to pumping, natural springflow had been 172 l/s, while during the test 325 l/s were constantly pumped out (Stevanović et al. 1994). The maximal drawdown at the discharging siphon was 2 m.

After successful drilling and testing, the final design of the Mrljiš source included four exploitation wells (Fig. 15.62). Several performed pumping tests verified their large capacities ranging from 50 to 110 l/s. A total amount of 240–320 l/s was pumped out during the long-term simultaneous testing in a low-flow water period, producing a small drawdown of less than 2 m in the Mrljiš zone. Therefore, the optimal extraction rate of the source, as compared to the natural minimal springflow of Mrljiš, had increased almost fourfold.

Fig. 15.62 One of the drilled big-diameter wells at Mrljiš source

However, the Ministry of Agriculture, Forestry and Water of Serbia, responsible overall for the water sector, provided the Exploitation permit which limited extraction depending on the riverflow regime of the nearby Crni Timok River. The ecological limitation is related to the riverflow, and the fixed maximal extraction of 350 l/s is set for riverflows over 850 l/s (measurements are taken at the downstream control section), while when the riverflows fall below 95 % the guaranteed flow of 250 l/s extraction should be completely stopped (an extremely rare case). This permit allows extraction from five drilled wells around Mrljiš Spring, while one well IE-3 (Q ~ 50 l/s) located downstream should be used for pumping groundwater to the Crni Timok stream during its critical minimal flows (Fig. 15.63). This system has functioned well for the more than 10 years that it has been in operation. During 2005 and 2006, the two analyzed years, the exploited water in a total annual amount of $\sum Q = 6 \times 10^6$ m^3 caused a flow reduction for Crni Timok of only 7.5 % (Stevanović et al. 2007, 2010b).

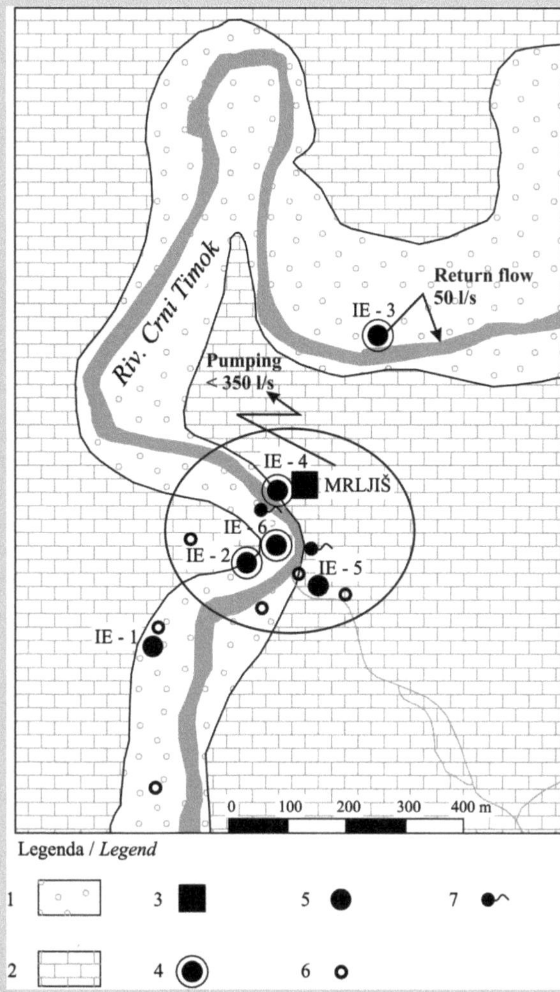

Fig. 15.63 Mrljiš source and corrective well IE-3. *Legend* (*1*) alluvium, (*2*) karst aquifer, (*3*) Mrljiš Spring, (*4*) extraction well, (*5*) reserve well, (*6*) borehole, (*7*) small spring (after Stevanović 2013)

Case study—Reževići gallery, Montenegro. The water supply of Petrovac on the Sea is an example of successful aquifer regulation by horizontal gallery in the Adriatic coastal area, where seasonal consumption is conspicuously uneven (coincides with yield minimum). The gallery, which is 374 m in length and built under the primary point of the Reževica Rijeka Spring overflow at 67 m a.s.l., has terminated in karstified limestones at 73 m a.s.l. By passing through Triassic flysch sediments, the gallery reached the primary limestone's aquifer and the discharge point was lowered by about 16 m

(Radulović 2000). This intake has enabled an increase in source capacity by ca 50 l/s and, more importantly, stabilization of water distribution from this source (Fig. 15.64).

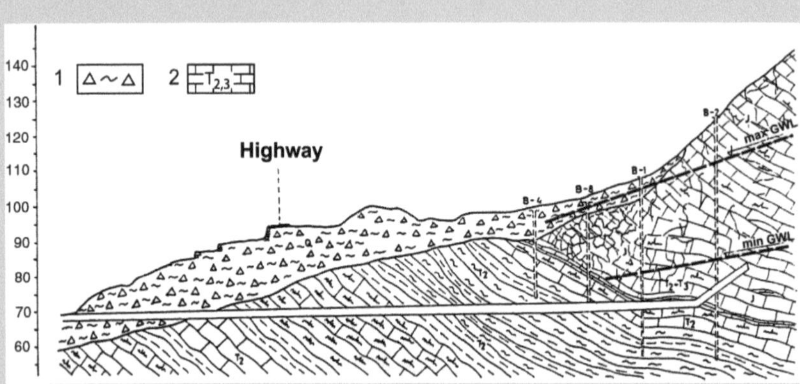

Fig. 15.64 Geological cross section and Reževići gallery. *Legend* (*1*) recent unconsolidated deposits, (*2*) flysch and limestones of Middle and Upper Triassic age (after Burić, from Radulović 2000; Printed with permission)

Case study—Swarawa source, Iraq. This example presents regulation of a confined karstic aquifer. Many such aquifers are discovered in the world by drilling deep wells, and this example is typical of those where by drilling, the released karst reservoir resulted in a very high pressure and discharge rate.

During the short drought cycle of 1998–2001 which affected all of the Middle East, drilling of new deep wells under the Food and Agriculture Organisation (FAO) Programme in the northern Iraq region and intensive groundwater use ensured that people and livestock were saved from forced migration resulting from lack of water. In particular, some 500 new wells were drilled during the 3-year period based on results of the conducted hydrogeological survey (Stevanović and Iurkiewicz 2004). The Swarawa village (Qaradagh basin) near the city of Sulaimani is one of the locations where the most successful results were achieved. After conducting short geological and geophysical surveys, the first hole that was ever drilled in the area reached and tapped a rich semi-confined karst-fissured aquifer. Well SW-1 is located relatively near the outcrop contact of impervious Lower Fars Fm. and underlain karstified Pila Spi limestones (Fig. 15.65). The total depth of the well is 126 m. From the depth of 89–126 m, SW-1 is drilled through white, "sugary," coarse crystalline, vuggy, and cavernous limestone of Pila Spi Fm.

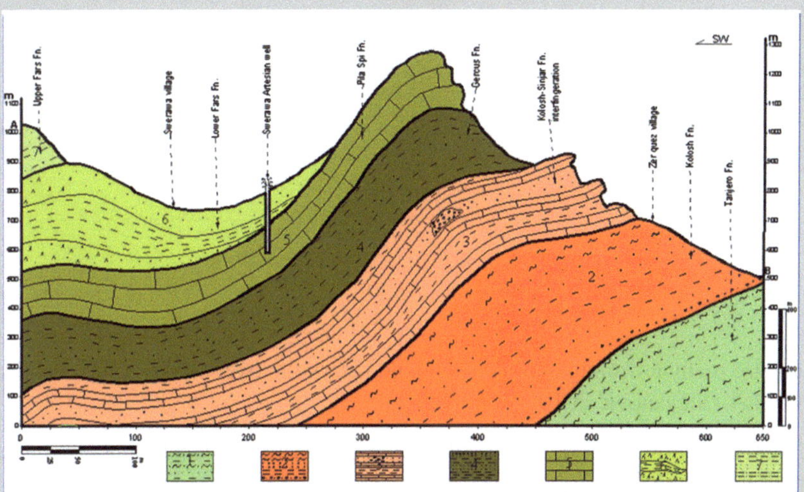

Fig. 15.65 Cross section through drilled well SW-1 (Swarawa). *Legend* (*1–4*) impervious and low permeable upper Cretaceous and Tertiary sediments, (*5*) Pila Spi limestones, (*6–7*) lower and upper Fars Fms. impervious claystone, siltstone, and sandstone (after Stevanović et al. 2005; printed with permission)

Beyond 126 m, further drilling was not possible due to enormous water pressure and eruption, which eventually reached 50 l/s. Despite the fact that a reservation column was installed up to a depth of 10 m (including cementation), groundwater erupted around the hole and the pipes. After final casing, the artesian outflow was reduced to 20 l/s. A planned new irrigation project for the whole Qaradagh basin had to be postponed due to mismanagement of local water resources: The farmers from a nearby village complained that the new well affected their spring which was not in fact the case and refused any solution other than the drilling of a new well for them (Stevanović et al. 2005).

- *Subsurface (underground) dam* is based on retention effect: Increasing the water level at the discharge point results in an increase in the hydraulic head inside the aquifer. Such increase is possible if the discharge point is fully or mostly blocked and artificially controlled. There are two main engineering options which make the creation of an additional reservoir inside the aquifer (above the natural water table) possible:

1. Constructing the dam at the land surface directly in front of the spring. The dam should be properly founded at the bottom but also laterally coupled with impermeable rocks to avoid leakage (Fig. 15.66);
2. Sealing the spring channel either by a concrete plug or by a watertight barrage or carpet (usually built from concrete, and rarely from clay).

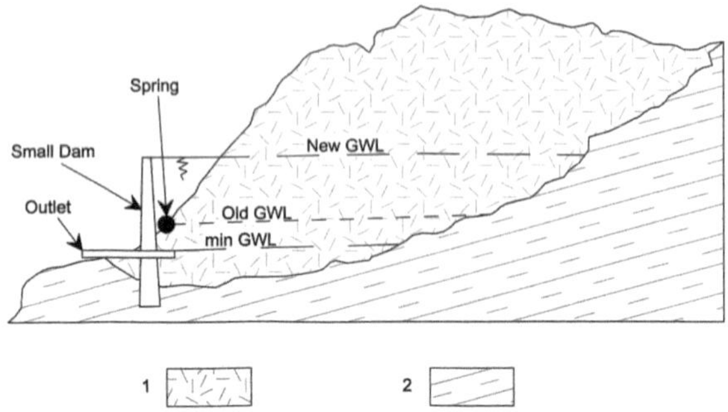

Fig. 15.66 Typical underground dam and storage as a result of dam construction. *Legend 1* karstic aquifer, *2* impervious rocks, GWL groundwater level

There are many advantages of underground reservoirs over surface ones. A few are as follows (Stevanović 2010b):

- No problems with flooding of infrastructure, fertile land, monuments, or compensation to reallocated people;
- No threat of dam collapse and major destruction downstream;
- No immense evaporation rate as in surface reservoirs; and
- No negative impact on water quality as in open reservoirs (eutrophication, sedimentation).

The additional storage can be calculated by use of formulae 6.9–6.11 presented in Chap. 6. The augmented water table is the result of effective storage (aquifer storativity) and thickness of the newly saturated part (difference between new and old groundwater level on Fig. 15.66). For the total volume of new water, the autogenic surface area of the aquifer should be exclusively considered.

Over and above the expensive research and construction, there are two main problems with an underground reservoir: The first is the possible leakage from the reservoir, and the second is that storage space within the aquifer is not fully predictable.

It is a fact that however large the investigation program undertaken may be, there can be no 100 % guarantee against leakage (Milanović 2000b, see also Chap. 13 and Sect. 16.1). The problem can be mitigated by grouting which regularly supports the dam construction. The aim of a grout curtain is to seal bottom and lateral flows. However, sometimes, even a very dense curtain (boreholes at a distance of 1 m) cannot stop the undesirable leakage, particularly when very karstified rocks are present. When the saturated zone is not very deep, grouting the cutoff walls or watertight diaphragm instead could be supportive elements for regulating and directing the flow (Stevanović 2010a).

The issue of uncertain volume of a new reservoir and consequently of benefits which complex dam structure would provide is even more challenging. The main problem is to properly assess effective storativity in a non-homogenous media, the issue broadly discussed in Chap. 3.

The specific type of construction is the subsurface dam built in alluviums. Their construction often takes place in countries with arid and semiarid climatic

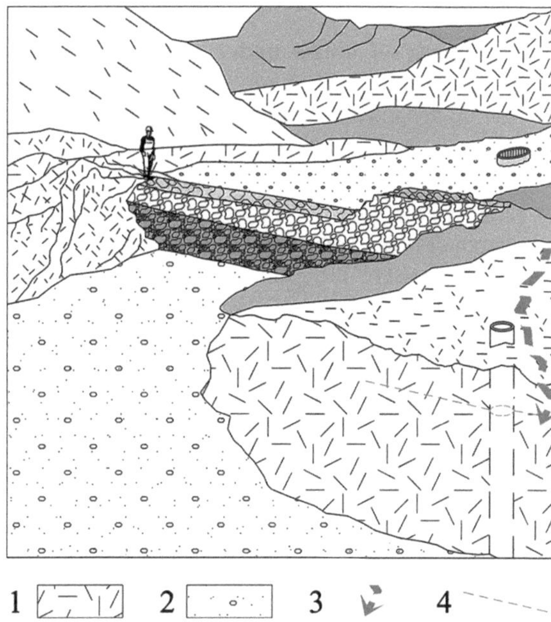

Fig. 15.67 Subsurface dam placed in alluvium with connected karst aquifer. *Legend 1* alluvium, *2* karstic aquifer, *3* groundwater flow direction, *4* groundwater table (after Stevanović 2014)

conditions, and many of these structures have been completed in the Middle East and northern and eastern African countries. But what is the relationship between a dam built in alluvium and karstic aquifer?

The main purpose of the alluvial type of dam is to collect water from temporary streams *(oued, wadi, kouri, togga)* and extend water availability for upstream/downstream consumers throughout the year. Short periods of river flow, unstable regime, and fast water propagation through alluvial deposits are the main reasons why such systems should be applied. In an open reservoir, evaporation losses from the water table are very high, so the best way to store water is to locate the dam in the ground (Stevanović 2014). Additionally, in the case that alluvium is overlaying karstic aquifer, a more intense and longer period of recharge of the connected aquifer takes place (Fig. 15.67).

Box 15.5.6

Concerning the construction of underground dams, there are several designed projects and experiments conducted in Europe and in North America (Perić 1981; Milanović 2000b, 2004), but the main practical experience has been obtained from such project implementation in China (Lu et al. 1973; Lu 1986), Indonesia (KIT Germany), Japan (Ishida et al. 2005), and Iraq (Stevanović 2014).

Case study—Chinese underground dams. In China, some 20 underground reservoirs have been completed for different purposes (water supply, irrigation, hydropower), and they enable water storage ranging from 1×10^5 m^3 to 1×10^7 m^3 in each of them.
Different interventions are applied. However, both earlier described solutions with a dam in front of the spring and with the plugging of the discharge channel were among the most frequently implemented (Fig. 15.68a, b). There were different combinations of underground and small surface reservoirs. An increase in water table and filling the sinking (swallow) hole with water was the result obtained in some cases, while artificially created hydraulic head may provide additional energy benefits (Fig. 15.69a, b).

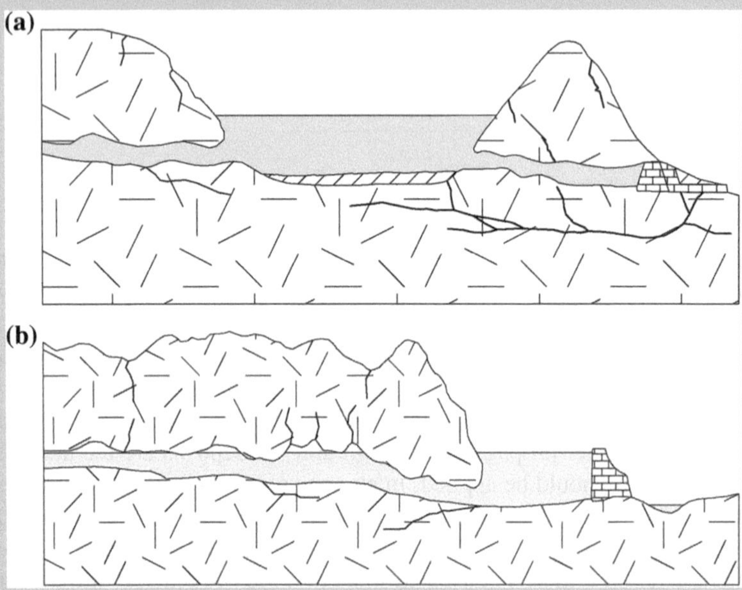

Fig. 15.68 **a, b** Implemented project with surface dam and underground reservoir in China (modified from Lu 1986)

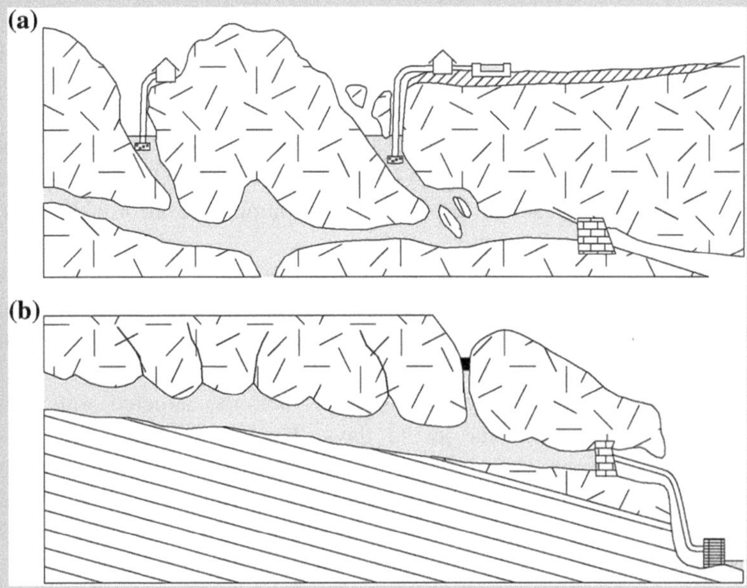

Fig. 15.69 **a** Implemented project with underground dam and pumping of backwater from the deep sinkholes in China (modified from Lu 1986), **b** hydraulic head in underground dam and gravity water transport (modified from Lu 1986)

The underground reservoir and connected hydropower plant are not even small, as expected of such structures (see also Sect. 13.3). The large reservoir Quibei (Yunnan province) is formed behind a 25-m-high underground dam, whereas the head at 109 m generates 25 MW at the exit power station.

Case study—underground dam in Indonesia. Similarly, an artificially created hydraulic head resulted from the German Indonesian pilot project "Accessing and managing underground karst waters in Central Java" which enabled the construction of a small hydropower plant and the supply of some 80,000 people with water and energy (KIT, and personal communication with Dr Peter Oberle).

The project was launched in 2002 by the Institute for Water and River Basin Management (IWG) of the Karlsruhe Institute of Technology (KIT), with the aim of constructing a demonstration hydropower plant in Gunung Sewu, and involved seven institutes from different specialist disciplines as well as industrial partners from the fields of tunneling, pumping, and controlling technology. Following intensive on-site research, the (Gua) Bribin cave has been chosen for the construction work. The cave has storage for some 300,000 m^3 and a subterranean yield of over 1 m^3/s even in dry seasons. The design included a barrage in the cave aiming to ensure storage of water, a part of which is to be transported along a 100 m riser pipe to a small hydropower

plant. Based on results of speleological survey and use of state-of-the-art laser technology, assessed potential water losses, corrosion effects, and the long-term stability of the system, firstly a small diameter exploratory borehole and then 100-m-deep shaft with a diameter of 2.5 m was constructed. Taking into consideration control, repairs, and maintenance as well as expertise of local technical personnel, German experts decided to use inversely operated pumps instead of turbines. These pumps are affordable, highly robust, and easy to maintain.
"Following successful completion of the dam and installation of the first pump module, the first test of the system was performed in August 2008 amid great public interest. Based on saturation processes in the surrounding mountains, it was estimated that it would take one to 2 weeks to fill the man-made reservoir—in actual fact, the targeted water level of 16 m was reached after just 2 days. The first filling of the reservoir was also accompanied by a test of the first pump module, during which the scientists measured a delivery rate of 20 l/s at the end of the 100 m vertical pipe—the results were thus in line with expectations. A further four pump modules were then installed, along with an electric system to control the plant" (http://ressourcewasser.fona.de/reports/bmbf/annual/2010/nb/English/405010/2_5_01-adapted-technology-an-underground-hydro-power-plant-on-java.html?printReport=1).

Case study—underground dams in Japan. Underground dams are constructed for irrigation in the Ryukyu and Amami Islands that consist of high-permeable Quaternary Ryukyu limestone. Some dams (called megaunderground dams) store 1×10^6 m^3 or more of groundwater. Groundwater is tapped by more than 100 tube wells in that area. On Miyakojima Island, a large national irrigation project including an underground dam was also executed and some changes in the groundwater level and flow as well as variation of chemical composition have been observed (Ishida et al. 2005).

Case study—subsurface dam in Iraq. Gali Basera, a subsurface dam near Dohuk, northern Iraq, was constructed in a short period and on a very low budget (Fig. 15.70). The results were successful: The period of stored water in alluvium was much longer, and even more importantly, the duration of recharge of the Pila Spi karstic aquifer of the Neogene age connected to the alluvium was extended, resulting in an increase in the pumping rate of drilled wells during the critical summer months (Stevanović and Iurkiewicz 2004; Stevanović 2014).

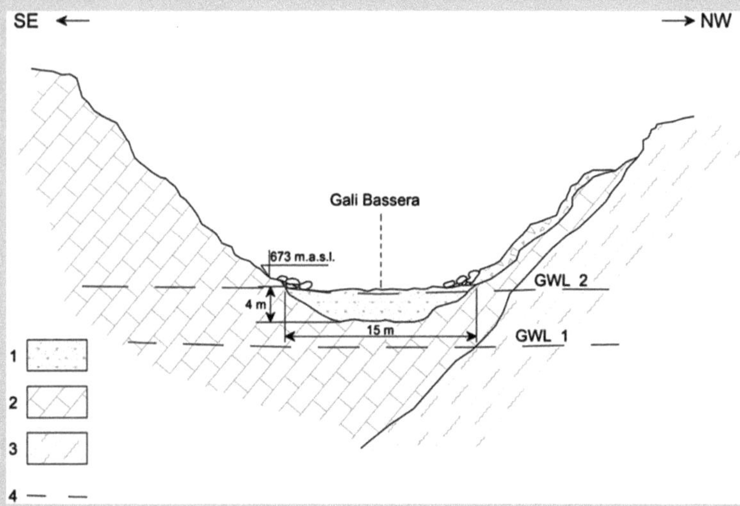

Fig. 15.70 Cross section over Gali Bassera valley. *Legend* (*1*) alluvium, (*2*) Pila Spi karstic aquifer, (*3*) impervious Gercus Fm., (*4*) groundwater table: GWL 1(min), GWL 2 (max)

- *Artificial recharge* is a term often used in water practice to describe any kind of intervention on water resources aiming to expand their volume. This approach is not correct, and here, the term will be used in the sense of its real meaning: adding some new water to the aquifers. Therefore, overpumping and decrease in water table are definitely not *artificial recharge* solutions. The idea to build an underground dam in karst does not necessarily involve "new water" but just focuses on keeping the existing resource inside the system. It is thus again not an *artificial recharge* concept.

Another, more academic matter is where to put *artificial recharge*, as we distinguish the two groups of interventions: those in the discharge zone and those in the entire karst catchment. The logic to stick to one or another depends on where we introduce new water to the aquifer: If near the discharge point, we may consider this as local intervention directly contributing to increase in discharge. However, in an environment such as karst is, the artificial recharge is better managed as far as possible from the discharge area, simply because of fast water movement through the aquifer with regularly limited attenuation capacity.

The water source for recharge is by definition surface water, taken from rivers, canals, or lakes. Even some projects in the Middle East have evaluated recharge of local karstic aquifers by brackish or even seawaters after their initial treatment. But none by definition can expect water purification and quality alteration in karst. It is thus the order of the day to control the quality of any water intentionally used for recharge and prevent deterioration of existing karstic waters. The quality of the introduced water should, though, be superior to the natural waters.

The kinds of non-intentional artificial recharge are briefly noted in Chaps. 3 and 6. They include seepage from built dams or canals, return flow from irrigation, and losses from utility pipeline.

15.5.5 Regulations in Wider Catchment Area

Aquifer regulation does not imply only management in discharge zones, but also all measures in the entire catchment area aiming at changing the conditions under which these waters are formed and circulate within the aquifer, such as riverbed regulation, directing groundwater to other more promising catchment areas, closing or regulating ponors (swallow holes), building impermeable barriers, and similar. Different technical solutions in the catchment area may be applied with respect to the specific circumstances, depending primarily on the effectiveness of such interventions, their cost, and finally the achieved economical and environmental benefits.

15.5.5.1 Building Small Dams (Weirs, Barriers) and Reservoirs

In many karstic basins, non-karstic areas provide allogenic recharge to the aquifer. The recharge zones inside karst terrains are commonly located along the riverbeds in which water from perennial streams start to sink and infiltrate. Such a situation is presented on Fig. 15.71. The infiltration in karst may take place over concentric and large swallow holes, but diffuse percolation over small holes is also frequent.

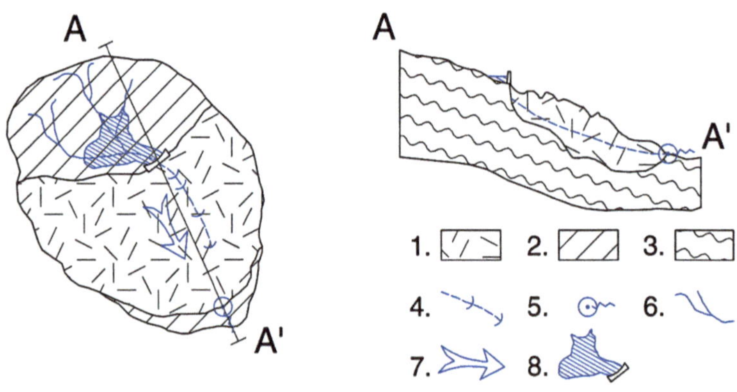

Fig. 15.71 Scheme of one karst basin with karstic and non-karst terrains. The perennial streams in non-karstic terrains become sinking streams after entering into karstic terrains. Dam is built at the contact of two formations. *Legend 1* karst aquifer, *2*, *3* impervious rocks, *4* swallow holes in sinking stream, *5* spring, *6* perennial stream, *7* groundwater flow direction, *8* dam and reservoir

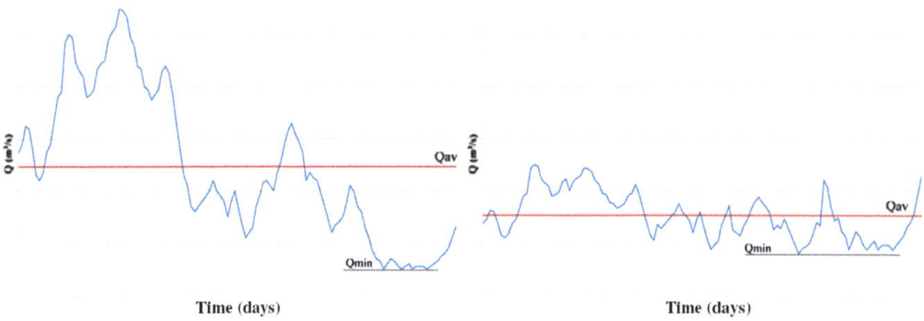

Fig. 15.72 Spring hydrographs, before (*left*) and after dam construction (*right*)

Whatever the case, direct impact on spring hydrograph exists (Fig. 15.72 left). But if we build a dam near the contact with karst and locate a reservoir in a non-karstic area, such a solution enables runoff control, water conservation, and management of downstream streamflow.

By building a dam and reservoir and by controlling the upstream catchment as in the presented case, we also directly influence the discharge regime. The new hydrograph is shown on Fig. 15.72 right side. It is clear that we artificially "stabilize" hydrograph by reducing high peaks and, more importantly, by improving minimal springflow (Q_{min}). One may ask why in the case of river regulation the average riverflow (Q_{av}) becomes lower than in natural conditions. Where does this water disappear to? The answer is in evaporation from the newly created reservoir. In fact, most such reservoirs are small, created by building of small weirs, barrages, or similar barrier structures. From shallow reservoirs, evaporation is regularly more intensive than from deep ones and not surprisingly the loss in evaporation is high especially in arid areas. And so instead of having this water infiltrated in karst, we lose it?! Therefore, feasibility studies which assess advantages and disadvantages of such solutions should provide appropriate answers concerning the logic of such interventions, impacts on the environment, and cost–benefit analysis.

15.5.5.2 Plunging or Regulating Ponors

In the situation presented on Fig. 15.71, plunging of large ponors or building dams around them can also mitigate fast streamflow sinking and reduce the impact of high runoff peaks on spring discharge. Even if not all ponors can be regulated, the blockade of major ones may slow recharge and extend periods of high/mid-water discharge and conversely shorten the low-water period. Such projects are relatively rare in terms of drinking water supply, but frequent in dam and reservoir construction. Such examples from Nikšić polje (Montenegro) are discussed in Chap. 13 and Sect. 16.1.

15.5.5.3 Constructing Grout Curtain

Redirection or prevention of groundwater flows can also be achieved by grouting curtains which are regularly applied and support dam and reservoirs built in karst. This also involves previously discussed underground dams. But a grout curtain can be placed far from the dam site or discharge points and prevent sinking and percolation of surface waters. As earlier commented, grouting does not always provide an impermeable wall and even 1 m distance between boreholes in the curtain can result in considerable water losses (Box 15.5.7).

Box 15.5.7

Case study—Grout curtain in Pljevlja coal basin, Montenegro

In the Pljevlja coal basin in northern Montenegro, the stratigraphical succession includes formations from Carboniferous to Upper Jurassic overlaid by thick Tertiary (Mid-Miocene) lacustrine sediments with coal. Paleorelief and margins of the main coal field consist of Triassic limestone stratified in thick beds. The coal layer is homogenous, 20 m thick. Karstic aquifer spreads over the surface of about 500 km^2. The most important drainage of groundwater into the open coal pit has been done by the Tvrdaš Spring, very close to the zone reached by mining works.

In the 1990s, the primary elevation of Tvrdaš Spring was 773.5 asl, and a one-row grouting curtain, 500 m long, was made in order to prevent water outburst from this zone and to redirect groundwater to laterally flow out of the bottom of the basin and actual mining works. Unfortunately, the curtain did not prove to be the right solution, and very soon completely lost its function. Tvrdaš waters successively lowered down the slope to the bottom of the pit (660 asl). Maximal Tvrdaš discharge into the open pit was 500 l/s, and the average was 125 l/s. A dewatering solution is found in the pumping of gravitationally collected waters at the pit bottom (Fig. 15.73).

Fig. 15.73 Open coal pit in Pljevlja basin. Photograph taken from direction of Tvrdaš Spring and grout curtain along the basin's margin

15.5.5.4 Channeling the Streams

Covering the riverbed by impervious material may prevent water losses or extend the riverbed to the desired location for infiltrating of surface water. Such paved shotcrete was placed along 65 km of riverbed of the Trebišnjica River (Popovo polje, Bosnia and Herzegovina, Fig. 15.74) before this intervention the largest sinking river in all of Europe (Milanović 2000b, 2004).

Fig. 15.74 Regulated Trebišnjica River in Popovo polje

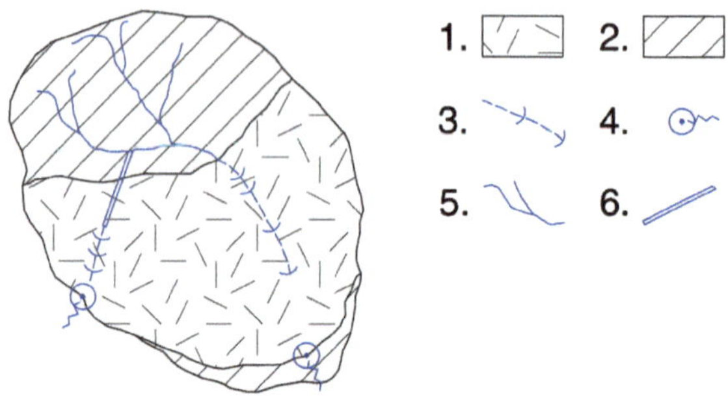

Fig. 15.75 Rearrangement of hydrographic network and water channeling toward another subcatchment. *Legend* (*1*) karst aquifer; (*2*) impervious rocks; (*3*) sinking stream and successive ponors in its riverbed; (*4*) karstic spring; (*5*) perennial stream; (*6*) channel or tube for water transfer

15.5.5.5 Transferring Water (Redirect Surface Streams)

Management of surface waters aiming to make impact on underground system also may include transfer of water from one to another catchment and in this way rearrangement of the hydrographic network and changes in water availability in subcatchments. Such a situation is shown on Fig. 15.75.

15.5.6 Recommended Methods and Programme in Hydrogeology Survey

The basic and general concept of hydrogeological research for the karst aquifer regulation should include but not be limited to the following activities:

- interpretation of data of previous investigations and historical data of groundwater regime
- climatic and hydrological data evaluation
- installation of new stations for observation of climatic, hydrological, and hydrogeological parameters
- remote sensing
- geological and hydrogeological mapping
- speleological and aqua speleological survey
- geophysical investigations in the drainage zones and areas planned for aquifer regulation
- tracing tests
- drilling and testing of exploratory boreholes and wells (or other intakes)
- hydrochemistry
- long-term pumping tests
- stochastic analyses, water budgeting, evaluation of available water resources for regulation (replenishment potential)

- aquifer storativity assessment and calculation of groundwater reserves (incl. waters for dependent ecosystems)
- aquifer vulnerability assessment and hazard of potential pollution
- definition of sanitary protection zones and preventive protection measures
- establishment of permanent monitoring system (groundwater quantity and quality)
- definition of exploitable groundwater reserves, i.e., of optimal flow rates and operational regime (periodical during dry season or continual)
- design of regulation structure (final solution)
- construction/installation of regulation water tapping structures (intakes and wells)

To improve and maintain the capacity of a karstic source during critical periods is very delicate engineering work. Knowledge of aquifer characteristics (the discharge regime and the position of the hydraulic head in the aquifer), the thickness of the saturated zone, and the aquifer storage capacity is of key importance. Sustainable water management of regulated karst aquifer requires sufficient replenishment potential for overpumped waters followed by detailed monitoring of the effects of applied measures and their impact on groundwater quantity and quality.

Apart from a request to increase the minimal discharge of the springs, there is an obligation to prevent aquifer overexploitation. Adequate management of water resources should be based on the following:

- Balance of extracted and available resources in accordance with replenishment potential,
- Prevention of water sources deterioration,
- Protection of water ecosystems, and
- Sustainable use of both ground and surface waters artificially controlled (when appropriate).

References

References to Section 15.1

Ashton K (1966) The analysis of flow data from karst drainage systems. Trans Cave Res Group Great Br 7:161–203

Atkinson TC (1977a) Diffuse flow and conduit flow in limestone terrain in Mendip Hills, Somerset (Great Britain). J Hydrol 35:93–100

Bailly-Comte V, Martin JB, Jourde H, Screaton EJ, Pistre S, Langston A (2010) Water exchange and pressure transfer between conduits and matrix and their influence on hydrodynamics of two karst aquifers with sinking streams. J Hydrol 386:55–66

Bauer S, Liedl R, Sauter M (2005) Modeling the influence of epikarst evolution on karst aquifer genesis: a time-variant recharge boundary condition for joint karst-epikarst development. Water Resour Res 41:W09416. doi:10.1029/2004WR003321

Bonacci O (1993a) Karst spring hydrographs as indicators of karst aquifers. Hydrol Sci J 38: 51–62

Bonacci O (1995) Ground water behaviour in karst: example of the Ombla Spring (Croatia). J Hydrol 165:113–134

Bonacci O, Živaljevic R (1993) Hydrological explanation of the flow in karst: example of the Crnojevica spring. J Hydrol 146:405–419
Brutsaert W (1994) The unit response of ground water outflow from a hill-slope. Water Resour Res 30(10):2759–2763
Celico P (1981) Relazioni tra l'idrodinamica sotterranea e terremoti in Irpinia (Campania) (Hydrodynamical underground relationship in Irpinia (Campania); in Italian). Rend Soc Geol Ital 4:103–108
Celico P, Civita M (1976) Sulla tettonica del massiccio del Cervialto (Campania) e le implicazioni idrogeologiche ad essa connesse (On tectonics of Cervialto massif and related hydrogeological implications (Campania); in Italian). Boll Soc Natur Naples 85:555–580
Celico F, Mattia C (2002) Analisi degli effetti indotti dal sisma del 23/11/1980 sugli equilibri idrogeologici della sorgente Sanità (Campania), mediante simulazione ragionata delle dinamiche di ricarica e esaurimento (Analyses of seismic effects induced by 23 November 1980 earthquake on the hydrogeological behaviour of Sanità spring (Caposele, Campania), simulating the recharge and recession dynamic periods; in Italian). Quaderni di Geologia Applicata 1:5–18
Civita M (1969) Idrogeologia del massiccio del Terminio-Tuoro (Campania) (Hydrogeology of Terminio-Tuoro massif (Campania); in Italian). Memorie e Note Istituto di Geol Appl Univ di Napoli 11:5–102
Cotecchia V, Salvemini A (1981) Correlazione tra eventi sismici e variazioni di portate alle sorgenti di Caposele e Cassano Irpino, con particolare riferimento al sisma del 23 novembre 1980 (Correlation between Caposele and Cassano Irpino spring discharge and earthquakes, with particular reference to 23 November 1980 earthquakes; in Italian). Geol Applicata e Idrogeologia, 16:167–192
Dreybrodt W, Romanov D, Kaufmann G (2010) Evolution of caves in porous limestone by mixing corrosion: a model approach. Geologia Croat 63(2):129–135
Drogue C (1980) Essai d'identification d'un type de structure de magasins carbonatés fissurés: application à l'interprétation de certains aspects du fonctionnement hydrogéologique. Mémoire hors série de la Société Géologique de France 11:101–108
Fiorillo F (2009) Spring hydrographs as indicators of droughts in a karst environment. J Hydrol 373:290–301
Fiorillo F (2011a) Tank-reservoir emptying as a simulation of recession limb of karst spring hydrographs. Hydrogeol J 19:1009–1019
Fiorillo F (2011b) The role of the evapotranspiration in the aquifer recharge processes of Mediterranean areas. Evapotranspiration—from measurements to agricultural and environmental applications. InTech, Croatia, pp 373–388
Fiorillo F (2014) The recession of spring hydrographs focused on karst aquifers. Water Res Manag 28:1781–1805
Fiorillo F, Doglioni A (2010) The relation between karst spring discharge and rainfall by the cross-correlation analysis. Hydrogeol J 18:1881–1895
Fiorillo F, Guadagno FM (2010) Karst spring discharges analysis in relation to drought periods, using the SPI. Water Res Manag 24:1867–1884
Fiorillo F, Guadagno FM (2012) Long karst spring discharge time series and droughts occurrence in Southern Italy. Environ Earth Sci 65(8):2273–2283
Fiorillo F, Pagnozzi M, Ventafridda G (2014) A model to simulate recharge processes of karst massifs. Hydrological Processes. doi:10.1002/hyp.10353
Fiorillo F, Revellino P, Ventafridda G (2012) Karst aquifer drainage during dry periods. J Cave Karst Stud 74(2):148–156
Ford D, Ewers RO (1978) The development of limestone cave systems in the dimensions of length and depth. Can J Earth Sci 15(11):1783–1798
Ford D, Williams P (2007) Karst hydrogeology and geomorphology. Wiley, Chichester
Gabrovšek F, Dreybrodt W (2010) Karstification in unconfined limestone aquifers by mixing of phreatic water with surface water from a local input: a model. J Hydrol 386(1–4):130–141

Gunn J (1986) A conceptual model for conduit flow dominated karst aquifers. In: Günay G, Johnson AI (eds) Karst water resources. Proceedings of Ankara symposium. vol 161 July 1985, IAHS :587–596

Halihan T, Wicks CM, Engeln JF (1998) Physical response of a karst drainage basin to flood pulses: example of the Devil's Icebox Cave system (Missouri, USA). J of Hydrol 204:24–36

Kaufmann G (2003) A model comparison of karst aquifer evolution for different matrix flow formulations. J Hydrol 283(1–4):281–289

Kiraly L (2002) Karstification and groundwater flow. In: Gabrovšek F (ed) Evolution of karst: from prekarst to cessation. Zalozba ZRC, Postojna-Ljubljana, pp 155–190

Kovács A, Perrochet P, Király L, Jeannin YP (2005) A quantitative method for characterisation of karst aquifers based on the spring hydrograph analysis. J Hydrol 303:152–164

Kresic N (2010) Types and classification of springs. In: Kresic N, Stevanović Z (eds) Groundwater hydrology of springs. Engineering, theory, management and sustainability. Elsevier Inc. BH, Amsterdam, pp 31–85

Malík P, Vojtková S (2012) Use of recession-curve analysis for estimation of karstification degree and its application in assessing overflow/underflow conditions in closely spaced karstic springs. Environ Earth Sci 65(8):2245–2257

Martin JB, Dean RW (2001) Exchange of water between conduits and matrix in the Floridan aquifer. Chem Geol 179(1–4):145–165

Mangin A (1975) Contribution à l'étude hydrodynamique des aquiféres karstiques (A contribution to the study of karst aquifer hydrodynamics). 3éme partie, Annales de Spéléogie 30(1):21–124

Muir-Wood R, King GCP (1993) Hydrological signature of earthquake strain. J Geophys Res 98(B12):22035–22068

Rorabaugh MI (1964) Estimating changes in bank storage and ground-water contribution to streamflow. Int Assc Sci Hydrol 63:432–441

Stevanović Z, Milanović S, Ristić V (2010a) Supportive methods for assessing effective porosity and regulating karst aquifers. Acta Carsologica 39(2):301–311

Tallaksen LM, Van Lanen HLJ (eds) (2004) Drought as a natural hazard. In: Hydrological drought: processes and estimation methods for stream flow and groundwater. Elsevier, pp 3–53

Thornthwaite CW, Mather JR (1957) Instructions and tables for computing potential evapotranspiration and the water balance. Publication in climatology 10, Drexel Institute of Technology, Centerton

White WB (1988) Geomorphology and hydrology of karst terrain. Oxford Press University, Oxford p 464

White WB (2002) Karst hydrology: recent developments and open questions. Eng Geol 65:85–105

References to Section 15.2

Eisenlohr L, Kiraly I, Bouzelboudjen M, Rossier I (1997) Numerical versus statistical modeling of natural response of a karst hydrogeological system. J Hydrol 202:244–262

Jemcov I, Ristić V, Prohaska S, Stevanović Z (1998) The use of autocross-regression model for analysis and simulation of karst springflow. J Min Geol Sci 37: 55–642

Jukic D (2005) The role of transfer functions in karst water budgeting and runoff modeling (in Croatian). Ph.D. thesis, University of Split, Faculty of Civil Engineering and Architecture, Split

Kresic N (2010) Modeling. In: Kresic N, Stevanović Z (eds) Groundwater hydrology of springs. Engineering, theory, management and sustainability. Elsevier Inc., BH, Amsterdam, pp 165–230

Kresic N, Stevanović Z (eds) (2010) Groundwater hydrology of springs. Engineering, theory, management and sustainability. Elsevier Inc., BH, Amsterdam

Mangin A (1984a) Pour une meilleure connaissance des systemes hydrologiques a partir des analyses correlatioire et spectrale. J Hydrol 67:25–43

Milanović S, Vasić LJ, Dašić T (2014) Determination of ecological flow rates of karst springs featuring large seasonal fluctuations (in Serbian). In: Proceedings of the 16th congress of Serbian geologists, Donji Milanovac, pp 363–368

Nikolić I, Kocić V, Ristić-Vakanjac V (2012) Groundwater monitoring by the Serbian national monitoring network (in Serbian). In: Proceedings of 14th Serbian symposium on hydrogeology, Zlatibor pp 45–50

Prohaska S (2006) Hydrology, Part II. Jaroslav Černi Institute for the development of water resources and University of Belgrade, Faculty of Mining and Geology, Belgrade, p 578

Prohaska S, Petković T, Simonović S (1977) Application of multiple nonlinear standardized correlation in calculating correlation relations, vol 58. Saopštenja (Papers) of the Jaroslav Černi Institute for the development of water resources, Belgrade, pp 25–34

Prohaska S, Petković T, Simonović S (1979) Mathematical model for spatial transfer and interpolation of hydro-meteorological data. Saopštenja (Papers) of the Jaroslav Černi Institute for the development of water resources, Belgrade, p 64

Prohaska S, Ristić V, Srna P, Marčetić I (1995) The use of mathematical MNC model in defining karst spring flows over the years. In: Proceedings of the 15th congress of the Carpatho-Balkan geological association, vol 4, no 3, Athens, pp 915–919

Prohaska S, Ristić V, Majkić B (2006) A cross-correlation analysis of the effects of atmospheric precipitation and water level in the karst polje of East Herzegovina on the Bregava River flow regime. In: Stevanović Z, Milanović P (eds) Water resources and environmental problems in karst. Proceedings of international conference KARST 2005, University of Belgrade, Institute of Hydrogeology, Belgrade, pp 531–538

Ristić V (2007) Development of simulation model for estimation of karst spring daily discharges (in Serbian). Ph.D. thesis. Faculty of Mining and Geology, University of Belgrade, Belgrade

Ristić Vakanjac V, Polomčić D, Blagojević B, Čokorilo M, Vakanjac B (2012a) Simulation of karst spring daily discharges. In: Proceedings of the conference on water observation and information system for decision support—BALWOIS 2012, Ohrid, pp 1–10

Ristić Vakanjac V, Stevanović Z, Milanović S (2012b) WP4—availability of water resources, In: Stevanović Z, Ristić Vakanjac V, Milanović S (eds) Climate change and impacts on water supply, University of Belgrade—Faculty of Mining and Geology, pp 133–176

Ristić Vakanjac V, Prohaska S, Polomčić D, Blagojević B, Vakanjac B (2013a) Karst aquifer average catchment area assessment through monthly water balance equation with limited meteorological data sets: application to Grza spring in Eastern Serbia. Acta Carsologica 42(1):109–119

Ristić Vakanjac V, Stevanović Z, Polomčić D, Blagojević B, Čokorilo M, Bajić D (2013b) Determination of dynamic storage volume and water budget of the Veliko Vrelo aquifer (South Beljanica) (in Serbian), vol 261–263. Vodoprivreda, pp 97–110

Stevanović Z (1987) Hydrogeological characteristics of karst aquifers in eastern Serbia with special reference to the water supply potential (in Serbian), Ph.D. thesis, Faculty of Mining and Geology, University of Belgrade, Belgrade

Stevanović Z (1992) Regime of karst springflows in Eastern Serbian Carpatho-Balkanides. Geol Ann Balkan Peninsula 56(1):411–436

Stevanović Z, Milanović S, Ristić V (2010b) Supportive methods for assessing effective porosity and regulating karst aquifers. Acta Carsologica 39(2):313–329

Stevanović Z, Ristić Vakanjac V, Milanović S, Vasić L, Petrović B (2011) The importance of groundwater monitoring in Serbian karst (in Serbian). In: Proceedings of the 7th symposium on karst protection, Bela Palanka, pp 21–28

Stevanović Z, Ristić Vakanjac V, Milanović S (2014) On the need to set up a new national groundwater monitoring network in Serbia (in Serbian). In: Proceedings of the 16th congress of Serbian geologists, Donji Milanovac, pp 313–319

Thornthwaite CW (1948) An approach to a rational classification of climates. Geograph Rev 35:55–94

References to Section 15.3

Atkinson TC (1977b) Diffuse flow and conduit flow in limestone terrain in the Mendep Hills, Somerset (GB). J Hydrol 35:93–103

Bonacci O (1987) Karst hydrology with special reference to the Dinaric karst. Springer, Berlin

Bakalowicz M (2005) Karst groundwater: a challenge for new resources. Hydrogeol J 13:148–160

Cornaton F, Perrochet P (2002) Analytical 1D dual-porosity equivalent solutions to 3D discrete single-continuum models. Application to karstic spring hydrograph modelling. J Hydrol 262:165–176

Fan Y, Huo X, Hao Y, Liu Y, Wang T, Liu Y, Yeh JT (2013) An assembled extreme value statistical model of karst spring discharge. J Hydrol 504:57–68

Debieche TH, Guglielmi Y, Mudry J (2002) Modeling the hydraulical behavior of a fissured-karstic aquifer in exploitation conditions. J Hydrol 275:247–255

Fleury P, Plagnes V, Bakalowicz M (2007) Modelling of the functioning of karst aquifers with a reservoir model: application to Fontaine de Vaucluse (South of France). J Hydrol 345:38–49

Fleury P, Ladouche B, Conroux Y, Jourde H, Dörfliger N (2009) Modelling the hydrologic functions of a karst aquifer under active water management—the Lez spring. J Hydrol 365:235–243

Hugman R, Stigter YT, Monteiro JP (2013) The importance of temporal scale when optimising abstraction volumes for sustainable aquifer exploitation: a case study in semi-arid South Portugal. J Hydrol 490:1–10

Jemcov I (2007a) Water supply potential and optimal exploitation capacity of karst aquifer systems. Environ Geol 51(5):767–773

Jemcov I (2014) Estimating potential for exploitation of karst aquifer: case example on two Serbian karst aquifers. Environ Earth Sci 71(2):543–551

Jemcov I, Petrič M (2009) Measured precipitation versus effective infiltration and their influence on the assessment of karst systems based on results of the time series analysis. J Hydrol 379:304–314

Jemcov I, Milanović S, Milanović P, Dašić T (2011) Analysis of the utility and management of karst underground reservoirs: case study of the Perućac karst spring. Carbonates Evaporites 26(1):61–68

Ladouche B, Marechal JC, Dörfliger N (2014) Semi-distributed lumped model of a karst system under active management. J Hydrol 509:215–230

Mangin A (1975b) Contribution a l'étude hydrodynamique des aquifères karstiques. Ann Spéléol 30(1):21–124

Milanović P (1981) Karst hydrogeology. Water Resources Publications, Littleton

Milanović P (1988) Artificial underground reservoirs in the karst experimental and project examples. In: Proceedings of the IAH 21st congress karst hydrogeology and karst environment protection, vol XXI, Part I, Guilin China, pp 76–87

Milanović P (2004) Water resources engineering in karst. CRC Press, Boca Raton

Mohrlok U. Kienie J. Teutsch G. (1997) Parameter identification in double-continuum models applied to karst aquifers. In: Proceedings of the 12th international congress of speleology, vol 2, pp 163–166

Pulido-Bosch A (1999) Karst water exploitation. In: Drew D, Hötzl H (eds) Karst hydrogeology and human activities: impacts, consequences and implications: IAH international contributions to hydrogeology, vol 20, pp 225–256

Stevanović Z (1994) Karst ground waters of Carpatho-Balkanides in Eastern Serbia. In: Stevanović Z, Filipović B (eds) Ground waters in carbonate rocks of the Carpathian-Balkan mountain range, Spec. edn. of CBGA, Allston Hold., Jersey, pp 203–237

White WB (1969) Conceptual models for carbonate aquifers. Ground Water 7(3):15–21

References to Section 15.4

Dragišić V, Milanović S, Špadijer S (2004) An approach of karst investigation for water supply needs, case example Miroc karst massif. In: Proceedings of the symposium: karstology—XXI century: theoretical and practical significance, Perm

Fazeli MA (2007) Construction of grout curtain in karstic environment case study: Salman Farsi Dam. Environ Geol 51(5):791–796

Garašić M (2005) Some new speleological research of caverns in route of the highways in Croatian Karst. In: Proceedings of the 14th international congress of speleology, Athens

Global Underwater Explorers (GUE) web site: https://www.globalunderwaterexplorers.org. Visited on 3 June 2014

Jeannin PY, Groves C, Häuselmann P (2007) Speleological investigations. In: Goldscheider N, Drew D (eds) Methods in karst hydrogeology. International contribution to hydrogeology, IAH, vol 26. Taylor & Francis/Balkema, London, pp 25–44

Jevtić G, Dimkić D, Dimkić M, Josipović J (2005a) Regulation of the Krupac spring outflow regime. In: Stevanović Z, Milanović P (eds) Water resources and environmental problems in karst. Proceedings of international conference KARST 2005, University of Belgrade, Institute of Hydrogeology, Belgrade, pp 321–326

Milanović P (2000a) Geological engineering in karst. Zebra Publishing Ltd., Belgrade

Milanović S (2005) Underground karst morphology for applied hydrogeology purpose. MS thesis, department of hydrogeology, University of Belgrade—Facualty Mining and Geology, Belgrade

Milanović S (2007) Hidrogeological characteristics of some deep siphonal springs in Serbia and Montenegro karst. Environ Geol 51(5):755–759

Milanović S, Stevanović Z, Jemcov I (2010) Water losses risk assessment: an example from Carpathian karst. Environ Earth Sci 60(4):817–827

Potié L, Ricour J, Tardieu B (2005) Port-Miou and Bestouan freshwater submarine springs (cassis–France) investigations and works (1964–1978). In: Stevanović Z, Milanović P (eds) Water resources and environmental problems in karst. Proceedings of international conference KARST 2005, University of Belgrade, Institute of Hydrogeology, Belgrade, pp 283–290

Stevanović Z, Dragišić V, Dokmanović P, Mandić M (1996) Hydrogeology of Miroč karst massif, Eastern Serbia, Yugoslavia. Theor Appl Karstol 9:89–97

Stevanović Z, Dragišić V (2002) Paleohydrogeological reconstruction of Bogovina Cave. Spec. edn. Serb Acad. Sci & Arts, Receuil des rap. du Com. pour le karst et spel., Belgrade VII, vol DCL, pp 49–59

Toulomudjian C (2005) The springs of Montenegro and Dinaric karst. In: Stevanović Z, Milanović P (eds) Water resources and environmental problems in karst. Proceedings of international conference KARST 2005, University of Belgrade, Institute of Hydrogeology, Belgrade, pp 443–450

Zlokolica M, Mandić M, Ljubojević V (1996) Some significant caves at the western rim of the Miroč Karst (Yugoslavia). Theoretical and Applied Karstology, vol 9. Academia Romana, Bucharest

References to Section 15.5

Avias J (1984) Captage des sources karstiques avec pompage en periode d'etiage. L'example de la source du Lez. In: Burger A, Dubertret L (eds) Hydrogeology of karstic terrains. Case histoires. International contributions to hydrogeology, IAH, vol 1. Verlag Heinz Heise, Hannover, pp 117–119

Bonacci O (1993b) Karst spring hydrographs as indicators of karst aquifers. Hydrol Sci 38(1):51–62

Burdon D, Safadi C (1963) Ras-El-Ain: the great karst spring of Mesopotamia. An hydrogeological study. J Hydrol 1:58–95

Burke JJ, Moench HM (2000) Groundwater and society: resources, tensions and opportunities. Spec ed. of DESA and ISET, UN public. st/esa/265, New York, p 170

Custodio E (1992) Hydrogeological and hydrochemical aspects of aquifer overexploitation. Selective paper of IAH, vol 3. Verlag Heinz Heise, pp 3–27

Hole F, Smith R (2004) Arid land agriculture in northeastern Syria—will this be a tragedy of the commons? In: Gutman G et al. (eds) Land change science. Kluwer Academy Publication, pp 209–222

Ishida S, Tsuchihara T, Fazeli MA, Imaizumi M (2005) Evaluation of impact of an irrigation project with a mega-subsurface dam on nitrate concentration in groundwater from the Ryukyu limestone aquifer, Miyako island, Okinawa, Japan. In: Stevanović Z, Milanović P (eds) Water resources and environmental problems in karst. Proceedings of international conference KARST 2005, University of Belgrade, Institute of Hydrogeology, Belgrade, pp 121–126

Jemcov I (2007b) Water supply potential and optimal exploitation capacity of karst aquifer systems. Environ Geol 51(5):767–773

Jevtić G, Dimkić D, Dimkić M, Josipović J (2005b) Regulation of the Krupac spring outflow regime. In: Stevanović Z and Milanović P (eds) Water resources and environmental problems in karst. Proceedings of international conference KARST 2005, University of Belgrade, Institute of Hydrogeology, Belgrade, pp 321–326

Lu Y, Jie XA, Zhang SH (1973) The development of karst in China and some of its hydrogeological and engineering geological conditions. Acta Geol Sinica 1:121–136

Lu Y (1986) Karst in China. Landscapes, types, rules (in Chinese). Special edition of Geological publications house, Beijing, p 288

Karlsruhe Institute of Technology (KIT) Accessing and managing underground karst waters in Central Java. Web site: http://ressourcewasser.fona.de/reports/bmbf/annual/2010/nb/English/405010/2_5_01-adapted-technology-an-underground-hydro-power-plant-on-java.html?printReport=1 visited on 15 May 2014

Kresic N (2009) Groundwater resources: sustainability, management and restoration. McGraw-Hill, New York

Kullman E (1984a) Vyhladávanie a moznosti zachytenia skryte vstupujúcich puklinovo-krasových vod do povrchových tokov (Possibilities of capturing of hidden fissured-karstic waters flowing to surface streams, in Slovakian). In: Proceeddings of VIII State hydrogeology conference on "Puklinové a puklinovo-krasové vody a problémy ich ochrany", Geol. ustav "D. Stúra", Liptov, pp 75–86

Kullman E (1984b) Etude en vie du captage et de l'exploitation les plus favorable de la source karstique de Jergaly (Velika Fatra, Tchecolovaquie). In: Burger A, Dubertret L (eds) Hydrogeology of karstic terrains. Case histoires. International contributions to hydrogeology, IAH, vol 1. Verlag Heinz Heise, Hannover, pp 54–56

Kullman E (1990) Krasovo—puklinové vody (Karst-fissure waters, in Slovakian). Spec. edn. Geol. ustav "D. Stúra", Bratislava, p 184

Kullman E, Hanzel V (1994) Karst—fissure waters in Mesozoic carbonate rocks of West Carpathians (Slovakia). In: Stevanović Z, Filipović B (eds) Ground waters in carbonate rocks of the Carpathian—Balkan mountain range, Spec. edn. of CBGA, Allston, Jersey, pp 113–148

Mangin A (1984b) Pour une meilleure connaisance des systèmes hydrologiques à partir des analyses corrélatoire et spectrale. J Hydrol 67:25–43

Margat J., 1992. The overexploitation of aquifers, Selective paper of IAH, vol 3. Verlag Heinz Heise, pp 29–40

Mediterranean Agronomic Institute from Bari (Italy) http://www.delsyr.ec.europa.eu/en/eu_and_syria_new/projects/24.htm). Visited 15.09.2008 (site is no longer available; some information still available at http://hal.archives-ouvertes.fr/docs/00/52/20/55/PDF/Galli_Sustainable.pdf)

Meinzer OE (1920) Quantitative methods of estimating groundwater supplies. Bull Geol Soc Am 31:329–338

Milanović P (2000b) Geological engineering in karst. Dams, reservoirs, grouting, groundwater protection, water tapping, tunneling. Zebra Publications ltd., Belgrade, p 347
Milanović P (2004) Water resources engineering in karst. CRC Press, Boca Raton
Paloc H, Mijatović B (1984) Captage et utilisation de l'eau des aquiferes karstiques. In: Burger A, Dubertret L (eds) Hydrogeology of karstic terrains. Case histoires. International contributions to hydrogeology, IAH, vol 1. Verlag Heinz Heise, Hannover, pp 101–112
Perić J (1981) Underground reservoirs as hydrogeo-technical structures for economical utilization of groundwater as a for water supply source (in Serbian). Trans Fac Min Geol 21:81–93
Radulović M (2000) Karst hydrogeology of Montenegro. Sep. issue of Geological Bulletin, vol. XVIII. Special edition Geological Survey of Montenegro, Podgorica, p 271
Stevanović Z, Dragišić V, Hajdin B and Miladinović M (1994). Essai de pompage de la source siphonate en vue de la base de definir les decisions de regulation de l'aquifère karstique. Transactions of the Fac. Min. & Geol. Belgrade, vol 32, 33, pp 80–87
Stevanović Z (1995) Identification of subsurface outflow of karst aquifer- basement for regulation of regime. In: Proceedings of XV congress of the Carpatho-Balkan geological Associations, B. vol 4, no 3, Athens, pp 927–930
Stevanović Z, Dragišić V (1995) An example of regulation of karst aquifer. In: Gunay G, Johnson I (eds) Karst waters and environmental impacts, Balkema, pp 19–26
Stevanović Z, Dragišić V (1998) An example of identifying karst groundwater flow. Environ Geol 35(4):241–244
Stevanović Z, Iurkiewicz A (2004) Hydrogeology of Northern Iraq. General hydrogeology and aquifer systems, vol 2. Spec. Edn. TCES, FAO, Rome, p 175
Stevanović Z, Ali S, Iurkiewicz A, Lowa F, Andjelić M, Motasam E. (2005) Tapping and managing a highly productive semi-confined karstic aquifer—Swarawa near Sulaimaniyah (Iraq), In: Stevanović Z, Milanović P (eds) Water resources and environmental problems in karst. Proceedings of international conference KARST 2005, University of Belgrade, Institute of Hydrogeology, Belgrade, pp 327–334
Stevanović Z, Jemcov I, Milanović S (2007) Management of karst aquifers in Serbia for water supply. Environ Geol 51(5):743–748
Stevanović Z (2009) Karst groundwater use in the Carpathian-Balkan region. In: Paliwal B (ed) Global groundwater resources and management. Scientific Publishers, Jodhpur, pp 429–442
Stevanović Z (2010a) Utilization and regulation of springs. In: Kresic N, Stevanović Z (eds) Groundwater hydrology of springs. Engineering, theory, management and sustainability. Elsevier Inc., BH, Amsterdam, p 339–388
Stevanović Z (2010b) Regulacija karstne izdani u okviru regionalnog vodoprivrednog sistema "Bogovina" (Management of karstic aquifer of regional water system "Bogovina", Eastern Serbia). University of Belgrade—Faculty of Mining and Geology, Belgrade, p 247
Stevanović Z (2011) Menadžment podzemnih vodnih resursa (Management of groundwater resources; in Serbian). University of Belgrade—Faculty of Mining and Geology, Belgrade, p 340
Stevanović Z, Milanović S, Dokmanović P, Ristić Vakanjac V, Petrović B, Vasić L (2013) Engineering regulation of karst aquifer as a response to minimal flows in sensitive areas. In: Proceedings of international conference on waters in sensitive and protected areas, 13–15 June 2013, Zagreb, pp 109–112
Stevanović Z (2013) Global trend and negative synergy: Climate changes and groundwater overextraction. In: Proceedings of the International Conference "Climate Change Impact on Water Resources", 17–18 Oct 2013, Institute of Water Management J. Cerni & WSDAC, Belgrade, pp 42–45
Stevanović Z (2014) Subsurface dams as a solution for supplementary recharge and groundwater storage in karst aquifers in arid areas, In: Lollino G, Manconi A, Guzzetti F, Culshaw M, Bobrowsky P, Luino F (eds) Engineering geology for society and territory, vol 5. Proceedings of the XII IAEG congress, Torino 2014, Springer, pp 471–474

Chapter 16
Prevent Leakage and Mixture of Karst Groundwater

Saša Milanović, Veselin Dragišić, Milan M. Radulović and Zoran Stevanović

16.1 Choosing Optimal Dam Sites and Preventing Leakage from Reservoirs

Saša Milanović

Centre for Karst Hydrogeology, Department of Hydrogeology, Faculty of Mining and Geology, University of Belgrade, Belgrade, Serbia

16.1.1 Introduction

With respect to risk factors in the dam and the reservoir construction in karst, particular attention must be paid to choosing optimal dam sites and preventing leakage from reservoirs (Milanović et al. 2010). An appropriate project concept prior to exploration can significantly reduce the risks of water losses or at least minimize them to

S. Milanović (✉) · Z. Stevanović
Centre for Karst Hydrogeology, Department of Hydrogeology,
Faculty of Mining and Geology, University of Belgrade, Belgrade, Serbia
e-mail: sasa.milanovic@rgf.bg.ac.rs

Z. Stevanović
e-mail: zstev_2000@yahoo.co.uk

V. Dragišić
Department of Hydrogeology, Faculty of Mining and Geology,
University of Belgrade, Belgrade, Serbia
e-mail: v.dragisic@rgf.bg.ac.rs

M.M. Radulović
Faculty of Civil Engineering, University of Montenegro,
Cetinjski put bb, 81000 Podgorica, Montenegro
e-mail: radulovicmilan33@yahoo.com

Z. Stevanović (ed.), *Karst Aquifers – Characterization and Engineering*,
Professional Practice in Earth Sciences, DOI 10.1007/978-3-319-12850-4_16

acceptable levels (Therond 1972; Zogović 1980; Milanović 2000a; Bruce 2003; Ford and Williams 2007a; Fazeli 2007), while the absence or reduction of exploratory works can increase them (Sahuquillo 1985). Milanović et al. (2010) stated that many analyses show that once the reservoir is filled up, groundwater flow often reactivates currently unsaturated (fossilized conduits) pathways and form a reverse discharge outside of the reservoir area. In response to these findings, it is of great importance that at an early stage of the research, the main emphasis must be on the establishing of the complex conditions of karst groundwater circulation genesis. As discussed in Chap. 13, some inadequately explored dam sites or reservoirs constructed in karst have never fully filled up with water or have been abandoned after unsuccessful attempts to reduce enormous water losses. Others have had sudden water losses even after years of successful operation or increases during years of operation. However, construction of the dam and reservoir in karst does not automatically result in a leakage problem by choosing optimal dam sites with suitable geological and hydrogeological conditions, leakage problems can be partially or totally avoided (Milanović 2000a).

16.1.2 General Overview of Procedures for Preventing Leakage and Choosing Dam Sites

Choosing optimal dam sites is a very complicated task due to the nature of karst and the insecurity of water storage due to leakage from reservoirs. In this regard, a few preliminary questions of great importance must be raised and answered:

1. How do we select an appropriate location for the dam and reservoir?
2. How do we select appropriate investigation methods?
3. What are the main tasks during investigations?
4. Is our survey sufficient to provide answers about the general feasibility of the project?
5. Will the data expected from the survey be sufficient for a proper assessment of potential water losses?
6. Can we find a technical solution either to avoid leakage or reduce it to an acceptable level?

The answers to the last three of these six questions depend strongly on positive answers to the first three questions. And even though a guarantee of reservoir or grout curtain tightness is unlikely, the risk of large water losses can be minimized (Milanovic 2000). Based on previous approaches to leakage from reservoirs applied, the proposed methodology should be applied in the most probable leakage zones and pathways at the dam site and reservoir and can be highlighted generally as three cases:

- dam sites and reservoirs in karst poljes—leakage in different directions through Quaternary deposits on the bottom of polje (probability of high risk),
- dam sites and reservoirs in river valleys—leakage usually below the dam site, in the direction of the river stream (medium to high risk),
- dam sites in deep and narrow canyons with different morphology than in the reservoir area—leakage through dam site banks (medium risk).

Fig. 16.1 Reinforced concrete cylindrical dam around Opačica ponor (sinkhole) (*left*); non-returned valve constructed on some small sinkhole in the area of Vrtac Reservoir in Nikšić polje (*right*)

A variety of dam sites may be suitable for potential dam construction, but investigation of the relevant geological and hydrogeological factors and issues will reveal which sites will best achieve the aim of water tightness, while still fulfilling the intended purpose of the proposed dam and reservoir.

From the hydrogeological point of view, karst poljes are probably the most complex areas for dam and reservoir construction. Usually, wide and very large zones of concentrated infiltration and discharge are located in these depressions. And usually, when the bottom of the polje consists of impervious strata, some sinkholes, estavelles, and intermittent springs are situated along the foothills. If alluvial deposits cover the bottom, the zones of concentrated infiltration can occur everywhere. As previously indicated, one good example is the reservoir at Vrtac with its enormous losses (27 m^3/s) (Vlahović 2005). However, after considerable work performed on the sinking zones by constructing cylindrical dams (Fig. 16.1) and non-returned valves, losses are not reduced at all.

A dam site can be located in narrow karstic canyons where the riverbeds are usually the deepest erosion base levels. In this case, the karstic features (different kinds of conduits) below the river bottom are usually very rare and only in some cases can be predisposed by hypogene karstification. In such cases, the water table is either connected with or very close to the riverbed, and dam sites are generally reasonably impermeable. Still, the dam site and reservoir banks can be prone to leakage. Solving the problem of the water tightness of banks can sometimes be very complicated. Salman Farsi dam in Iran is a good example of finding karstification as a main water tightness problem on dam site banks and both abutments (Box 16.1.1).

Very often, the spatial position of dam sites and reservoirs is located in karstic river valleys. Choosing optimal dam sites as well as preventing leakage from reservoirs, there needs more attention than in the case of canyon sites. Usually, the base of karstification is below the riverbed, and the presence of huge karst conduits (caverns and channels) can be hundreds of meters below the river or future reservoir bottom (Box 16.1.2). Reactivation of old deep conduits due to huge pressures from reservoir water is often a problem in these cases.

During the dry period of the year, the groundwater level is deep below the surface, while during the high water period, the water table rises abruptly and a considerable part of the bottom is under the influence of a strong uplift. The most common and unpredictable defects found during reservoir operation are karst channels naturally plugged by clay and covered by alluvium and *terra rossa* and reactivated by water pressure, suffusion, or air pressure effect (Milanović 2000a). As a result, collapses, holes, and huge open cracks in the reservoir bottom appear. Many examples of dams and reservoirs (successful constructions or total failures in terms of water tightness) in karst poljes illustrate the complexity of karst hydrogeology in dam and reservoir constructions.

Generally, the investigations should be very serious and decisions on water tightness treatment (during construction or later remediation work during exploitation) should be carefully provided and analyzed. Those investigations are focused on the one hand on defining the type of underground and deep treatment, and on the other hand, on surface treatment (Box 16.1.3).

The provision of good quality geological and hydrogeological data is necessary to establish a crucial base for making correct decisions in the process of choosing optimal dam sites and preventing leakage from reservoirs. Also, the geological, hydrogeological, speleological, and other special investigation procedures should be a permanent activity during the design stage, during the construction of the dam site and filling of the reservoir, as well as during exploitation. Having a good map, database, models, and geological, hydrogeological, and other 2D and 3D layers increases the chances of successful dam site choosing and minimizes the possibilities of further leakage from reservoirs below the dam site and through the reservoir and dam site bankment. The following procedure is necessary for the acquisition of some of the basic information (with characteristic examples) for choosing an optimal dam site and preventing leakage from reservoirs in karst formations:

1. Detailed surface geomorphological analysis of reservoir and dam site area,
2. Speleological and cave diving as analysis of karst interior,
3. Detailed hydrogeological analysis of characteristics of lithological units,
4. Structural analysis of the spatial position of folded structures, especially the hydrogeological role of anticline core,
5. Analysis of characteristics of main faults (elongation, depth, nature of crushing material between tectonic blocks, and disturbance of ruptures by karstification),
6. Analysis of tracer tests data as a base for groundwater flow defining,
7. Detailed analysis of speleogenesis,
8. Detailed analysis of karst conduit distribution,
9. Analysis of groundwater monitoring data of the wider area (fluctuation, defining minimum, and maximum level of water table, etc.),
10. Detailed analysis of the investigation borehole (thermal methods, borehole radar, different geophysical logging methods, core analysis, TV logging of borehole, etc.),

11. Excavation of investigation and grouting galleries and its arrangement are very important. All huge open caverns should be directly observed using access galleries and shafts. The vertical distance between grouting galleries should be less than 50 m (Milanović 2000a),
12. Analysis of water pressure tests,
13. Forming 3D spatial and physical models of the reservoir and dam site area.

Box 16.1.1

Case Study—Salman Farsi dam—Preventing possible leaking problem during construction

Salman Farsi Dam (125 m high) is located on the Ghareh-Agaj River in Fars province, in southern Iran (Fig. 16.2). From the geological and hydrogeological point of view, this dam is one of the most complicated sites in Iran.

Fig. 16.2 Photo of dam site during construction and Salman Farsi Dam in the last stage of building

During the excavation of grouting galleries, some large caves at both abutments were discovered (see Sect. 15.4, Box 15.4.5). The volume of the biggest one (Golshan's Cave) exceeds 150,000 m^3. Due to these conditions, large-scale underground geotechnical treatment was needed to prevent leaking and improve the water tightness of the dam site. The main question was how to prevent leaking through the dam abutments (Fig. 16.3).

The dam site is located in the northern flank of the Changal anticline, whose hydrogeological characteristics played a key role in the choice of the dam site. The Pabdeh Formation downstream of the dam site (consisting of marl and shale layers) is located in the core of the anticline. This formation has acted like a thick and deep impervious barrier against underground water filtration from the upper erosion base levels to the lower levels (Fig. 16.3), which is also the key function of the water tightness of the Salman Farsi grout curtain heading downstream. However, upon finding large caverns in the abutments of the dam site, the question of water tightness and how to prevent leaking from the reservoir. DEM with main faults and the spatial position of the dam, grouting gallery, and Golshan's Cave are shown on Fig. 16.4.

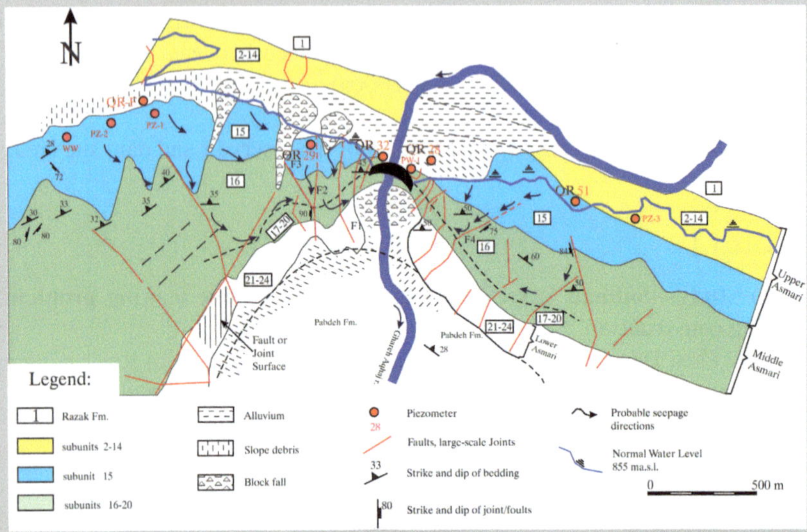

Fig. 16.3 Geological sketch of dam area (Fazeli 2005)

During the folding process, the anticline is disturbed by some fractures, faults, and joint sets. Some of them are distinguished at the dam site. From the point of view of water tightness, two of them are very important: faults with the strike of NE–SW and other sets of faults with the strike of N–S (Fig. 16.3). Both sets are subvertical (dip about 80°). The dip of the bedding planes is 50–60 degrees toward upstream (Fazeli 2005). Well-developed faults sets together with steep joints and bedding planes have made a well-connected network for water filtration.

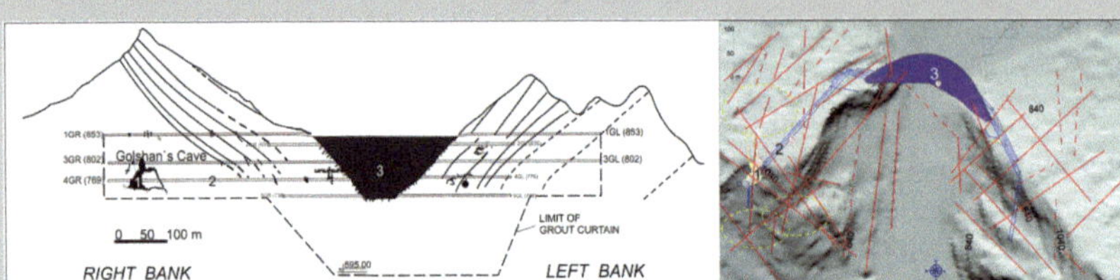

Fig. 16.4 DEM of dam site with spatial position of the main caverns, grouting galleries, and faults. *1* Golshan's Cave, *2* Grouting galleries, and *3* Salman Farsi Dam

Prevent leaking through dam abutments—grouting gallery redesign

The construction of a grout curtain in dam abutments with a highly random distribution of caverns and channels has some uncertainties. After the discovery of caverns during the excavation of grouting galleries (see Sect. 15.4, Box 15.4.5), it was necessary to make some redesigns in order to prevent leakage of water from the reservoir.

Cleaning of the caverns and caves entails removing all kinds of material (clay deposits, blocks, crushed material, and water) from the investigated caverns and caves. Usually, removal of the materials and cleaning will continue until sound rock is reached. It often needs some extra shaft and gallery excavation as well as blasting. Cavern treatment was provided 20 m upstream and 10 m downstream of the curtain. In the Salman Farsi Dam, it took years to clean some big caverns (Fig. 16.5).

Fig. 16.5 *Left*—geological investigation in grouting galleries, *Right*—cleaning of the caverns

Then after, the following measures were also applied:

1. All caverns and caves which are accessible and do not have a big volume were filled with concrete. Also, caverns with a big volume, and where the spatial position is only a few meters upstream and downstream of the curtain, were filled (this is called concrete plug).
2. Final grouting is performed after executing the concrete plug. This grouting is done to fill the empty spaces between the concrete and rock by grouting.
3. After completing the above phases, drilling new grouting holes and connecting the curtain to the plug are recommended.

In response to new data, it was decided that an upstream bypass of Golshan's cave is the best solution. Four galleries (853R, 802R, 769R, and 738R) were shifted upstream of Golshan's cave, and the grout curtain was extended by 1,560 m (Fig. 16.6). The karst treatment took about 1.5 year, but the final results seem satisfactory. The karst treatment was done and it took only 1,020 m^3 to fill the karst with concrete. The old route of the curtain will be used to monitor the curtain performance.

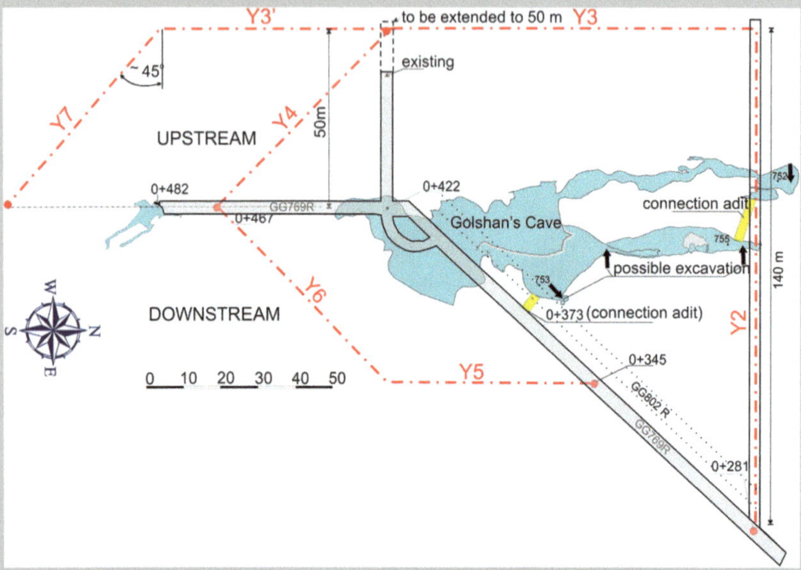

Fig. 16.6 Grout curtain bypass (Layout of Golshan's Cave after D. Vučković, S. Milanović, 2001, Salman Farsi (Ghir) Dam Project, Speleological Investigations, Mission Report), unpublished, (From Stucky-Electrowatt 1996–2004 and Fazeli 2005)

Box 16.1.2

Case example—Problem of karstification (leakage) below the Ourkiss dam site

The Ourkiss dam site is located 5 km from the city Oum el Bouaghi (NE Algeria) at the entrance of Oued Ourkiss gorge. Oued Ourkiss is one of the tributaries of the salty lake Garaet el Guelif. The wide valley inside the future reservoir consists of Miopliocene and Quaternary deposits which overlie the Albian and Aptian well—karstified limestones. A rock-filled dam with clay core is proposed in the gorge cutting the limestone outcrops (Fig. 16.7).

Fig. 16.7 Ourkiss dam site and reservoir area during construction

The designed reservoir at the altitude of 951 m a.s.l. will store the waters of the temporary Oued Ourkiss stream, but over 90 % will flow in as waters pumped from another reservoir, the Oued Athmenia (Stevanović et al. 2010). The dam height is 35 m with a crest length of 400 m, width of 8 m, and width of foundation of 216 m.

This case example summarizes the analysis of important discontinuity systems and their role in karst genesis at the dam site. The Ourkiss dam site was developed in limestone rock, under particular hydraulic and structural conditions. Therefore, these karst systems are composed of many segments (small channels, caverns with speleo forms, shafts, cracks, etc.), each of which is contained entirely by distinct discontinuities, such as bedding planes, joints and fractures, faults, or along interception of those structures. Survey of those structural elements is the initial step in definition of pathways of the groundwater circulation. The most important task for the future prevention of leakage below the dam site, however, is to know exactly what the karst process and level of karstification (depth and spatial position) are.

Results and discussion

The geological setting indicated possible water losses from a new reservoir. Moreover, during one of the floods, a large amount of water sunk into one of the discovered ponors. After an almost complete loss of large amounts of water, excavation was carried out at the dam site and its right side is a registered cavern with a length of 0.5 m (240/60), and which can be visually traced to approximately 20 m deep. Following the discovery of the caverns, further research in the area of the dam has devoted far more attention to possible indicators of intense karstification. Performed by gravimetric measurements on a wider profile dam, a strong anomaly at the dam site was also noted. In order to collect sufficient data for possible protective measures such as grouting and impermeable blanket, (Milanović et al. 2007) an investigation program has been defined.

Generally, some methods can, however, help to lessen the probability of undesired effects. Direct observation of interior karst was the first among different methods applied at the Ourkiss dam site. Recording of karstified pathways by a specially constructed video camera was more effective in this case than some geophysical methods as well as, logging, caliper or tracing (Fig. 16.8). Recording of open caverns and boreholes on both embankments of the Ourkiss Dam very much helped to orient further exploration and to define protective measures which include sealing and the construction of an impermeable clay blanket in the wider area (S. Milanović, 2007, Results of recording of karstic features—Ourkiss dam project. Report. Hidrotehnika, Belgrade, 2012).

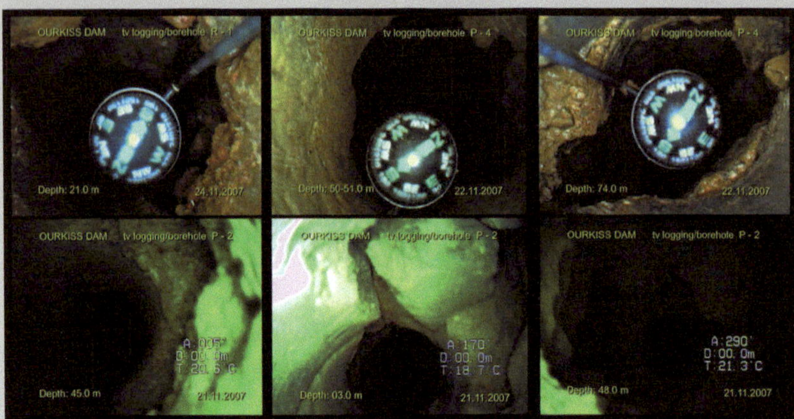

Fig. 16.8 Some of the recorded caverns in the boreholes at the Ourkiss dam site

The inventory covers a large number of cavities, in total 460 in ten surveyed boreholes (Table 16.1). Most of those cavities are currently above maximal groundwater table, but will be fully impounded after the reservoir filling.

Table 16.1 The results of camera recording of karst conduit network at the Ourkiss dam site

Borehole	Elevation (m a.s.l.)	Depth (m)	GWL (m a.s.l.)	No. of karstified intervals	General direction/^{0}	Maximum length (m)	Minimum length (m)	Calculated porosity (%)
P2	919.57	62		28	325–335	9	0.1	12.1
P3	923.62	67	863.62	19	335–360	4	0.3	7.3
P4	939.13	85	854.13	22	300–325	5	0.2	8.2
R1	924.5	74		16	295–318	4	0.25	4.1
R2	913.75	50	863.75	14	310–322	3	0.1	1.9
R3	932.35	60		19	288–303	3	0.1	2.2
R4	932.65	78	857.65	17	310–340	2	0.05	1.5
GD-III-25	921.25	33		9	312–345	3	0.2	1.7
GD-III-30	921.08	37.5		4	300–330	2	0.3	1.8
GD-III-35	920.97	42		5	290–325	3	0.25	2.1

Based on this survey, the sealing of the main shafts and the construction of an impermeable blanket in the wider area are proposed.

The spatially large and deep Ourkiss conduit system would probably not be fully explored by using conventional geophysical or other remote methods for the following reasons:

- The caliper's bars are too short to record the size of the found cavities.
- Tracing tests provide valuable information through recorded velocities or tracer quantitative analyses, but cannot be successfully conducted in unsaturated parts.
- Good information about the size and volume of empty space can be obtained by a water injection test into wells, but there are also many obstacles. For example, a large amount of water can be absorbed even by a small, but well-connected fissure system. This means that water losses registered by the classical test do not provide sufficient data to distinguish enormous from medium-sized or even small cavities.

Lastly, obtained photos or movies recorded by camera are also, at most, relative proof. Although powerful, the lamp can still be insufficient to light the furthest or curved parts. Besides, similar to the above-mentioned methods, the camera evidence will not result in absolute values of storativity, but will always provide better insight into situations and orient further research (Fig. 16.9).

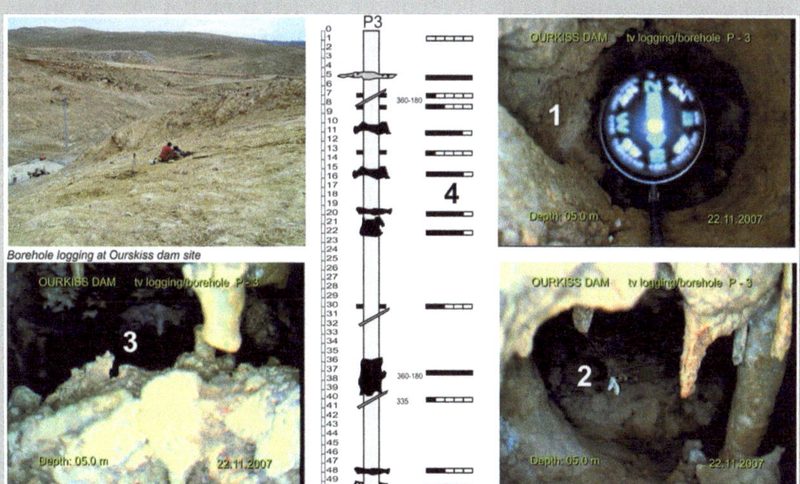

1 Vertical view through borehole 90 °. Manual orientation of foults and badding plans. Diameter of borehole 101 mm. Depth 5 m from surface.
2 Subhorizontal view 40° of cevarn on 5 m depth. Unseen on the record above - with vertical position of borehole camera.
3 Horizontal view 0° of cevarn on 5 m depth. Unseen on the record above - with vertical position of borehole camera.
4 Level of karstification in borehole observation

Fig. 16.9 Detail of borehole P-3 video logging

Data of monitoring boreholes with detailed zones of karstification are shown on the comparative cross sections on Fig. 16.10

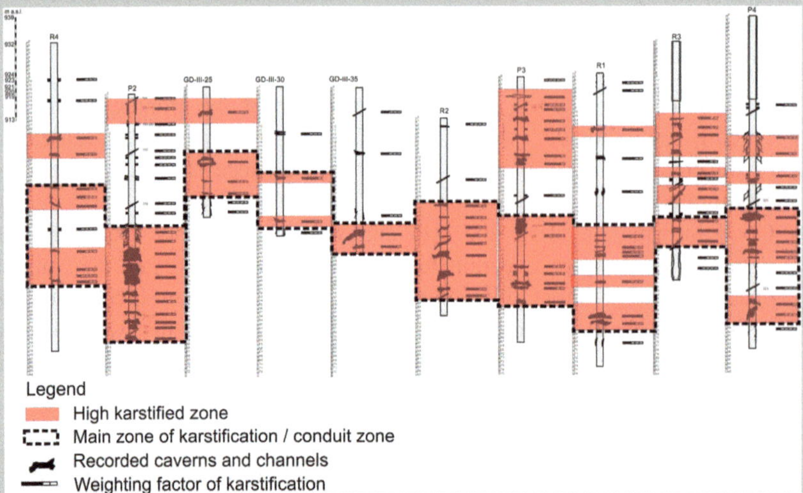

Fig. 16.10 Cross section of boreholes with data of karstified zones

In all boreholes, in parallel with the drilling, experiments concerning test permeability were carried out (Lugeon test). The greatest losses of over 100 Lu were registered in the borehole on the left bank P-4 (45–50 m) and downstream in the river valley, in P2 (45–50, 60–65, and 65–70 m).

For purposes of correlation, data of karstification in the narrow zone of the dam site at Ourkiss, and a detailed layout and cross section of all known karst forms of the borehole were made, as shown on Fig. 16.10. The borehole video logging data and borehole profiles previously derived from zones of karstification in Table 16.1 show the spatial development of the main karstified zone at the dam site at Ourkiss (Fig. 16.11).

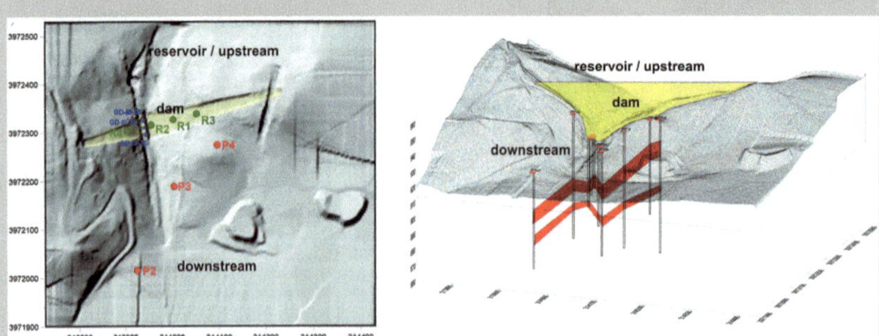

Fig. 16.11 DEM of Ourkiss dam site with investigation borehole position (*left*); 3D model of main karstification distribution (*right*)

On the basis of data obtained during research in the area of the dam site and reservoir area as a solution for the Ourkiss project leakage under the dam body, three solutions were considered.

1. underground sealing—grout curtain,
2. surface sealing (impermeable carpet)—geomembrane and geotextile, and
3. surface sealing—shotcrete.

Upon analysis of all the alternative solutions, it was concluded that the second solution and the use of geomembrane and geotextile will be the most trusted solution for this condition for highly developed karstification and very insecure and troubled grouting. Finally, for preventing leakage under the dam site as well as in the reservoir area, an impermeable carpet was made geomembrane (520,000 m^2) and geotextile (550,000 m^2) (Fig. 16.12).

Fig. 16.12 Photos from the phase of impermeable carpet installation (geomembrane and geotextile) http://www.hidrotehnika.rs/alzir/brana-ourkiss/

Box 16.1.3

Case Study—Višegrad dam—Problem of increasing leakage during exploitation (current investigation)

The Višegrad hydropower plant is situated in Dinaric karst on the River Drina, 2.7 km upstream from the town of Višegrad (Bosnia and Herzegovina). It was built from 1985 to 1989. The dam of the Višegrad HPP is a concrete gravity dam. An integral part of the dam is the 594-m-long grouting curtain (325 m beneath the dam structure and 65 m in the left abutment and 204 m in the right abutment) and 50–130 m deep. Investigation works related to the development of the dam and power plant began in 1976, and the design documentation for the principal objects that constitute the power plant was developed before 1983.

Already in the first year of the exploitation of the Višegrad dam, the occurrence of submerged downstream springs was noticed. From the measurements of water quantities that appear in the springs downstream from the dam, it was established that the quantity had increased from the 1.4 m^3/s (in 1990) to 13.92 m^3/s (in 2008), and 14.68–15.00 m^3/s (in the 2012) (Fig. 16.13). In order to define the positions of karst conduits along which the groundwater circulates under the dam site, in 2009 and again in 2013, special-purpose investigations and remediation works (still in progress) were performed. Geological investigations of the karst setting, required to address leakage beneath the Višegrad dam, were focused on a rather narrow area containing a refill "sinking" zone and a drainage "discharge" zone. The basic problem was how to perform a quality analysis, or how to state the problem whose goal was to reduce or to stop leakage below the dam site. The problem was approached from three parallel directions: a theoretical approach, which initially played a major role and provided guidelines for field activities; detailed and highly complex field investigations and development of a basic input 3D model; and an empirical approach and later also a mathematical approach aimed at producing the final form of the model.

Fig. 16.13 *Left*—Submerged discharge points during tracer test (*downstream, left photo*); Spatial position of main sinkhole (location of tracer injection at a depth of 50 m) (*upstream/reservoir; right photo*)

Overview of results during previous investigations, design construction, and exploitation

Geological investigations were performed during all phases of design and construction of the dam of the Višegrad HPP. During the previous period of dam exploitation, certain multidisciplinary investigations were also performed, primarily with the goal of choosing the optimum dam site as well as definition of possible seepage. All these investigations have yielded a large body of results, a part of which is significant for the solution of the leakage problem under present conditions below the dam site.

For the needs of the development of the main project, within the scope of the development of geological data relevant to dam construction, detailed geological mapping of the dam site (in the 1:50–1:1000 scale) had been performed. Mapping of the catchment area of the Višegrad reservoir and dam site was performed on the scale of 1:10000, while an engineering geological map of the dam site was performed on the scale of 1:500. Engineering geological mapping of the foundation tectonic was performed during the course of excavation of the dam foundation and riverbed downstream. Some of the investigative drilling was completed during the course of the development of the preliminary design, but most of it was performed during the preparation of the basic design. More than 4,470 m of drilling was performed during previous investigation phases together with water pressure tests WPTs. Also, groundwater tracer tests were performed in the phase of choosing the optimal dam site. Systematic monitoring from the stage of previous investigation and design construction up to the exploitation of water seepage beneath the dam has been done from 1991 to the present (Fig. 16.14) (Institute for Development of Water Resources "Jaroslav Černi", 2009, Design on rehabilitation regarding water seepage beneath the dam of the Višegrad hydropower plant, Summary Report on Performed Investigations, Belgrade, unpublished).

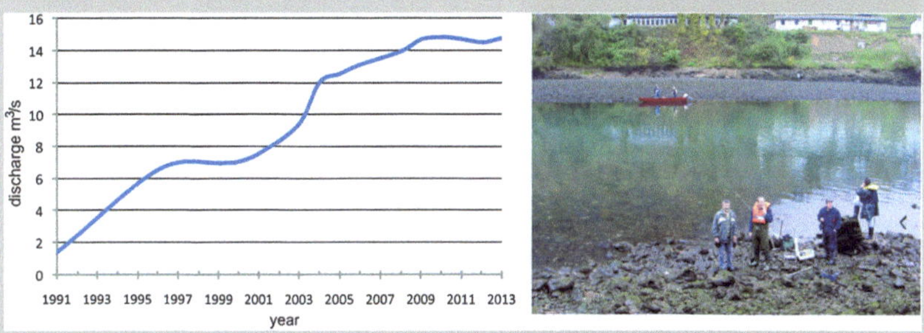

Fig. 16.14 Diagram of water discharge (seepage) through time (*left*). Measuring discharge from springs (*right*)

Groundwater levels were also monitored in the piezometers located in dam abutments and abutment injection galleries during the course of trial filling and then on a continuous basis through the years 1991, 1994, and 1995. At present, the groundwater level regime is monitored twice a month in 58 piezometers.

Outline of results of special-purpose investigations

In 2009, special-purpose investigations were performed in order to define the conduits along which the water leakage occurs below the dam site. Since the study and systematization of the results of previous investigations generally shown above, the investigations have started with the determination of the geological structure of the wider area of the Višegrad reservoir that encompassed an area of cca. 6 km^2. The first step was remote sensing investigations, based on the analysis of satellite images and aerophoto images from the period before the construction of the reservoir. Along with that, the geological mapping of the terrain was performed, with the corresponding petrologic investigations and detailed ruptural investigations. Detailed investigations of the narrower area around the dam site started with the geodetical survey of the dam and the appurtenant structures on the scale of 1:1000 (Institute for Development of Water Resources "Jaroslav Černi", 2009, Design on rehabilitation regarding water seepage beneath the dam of the Višegrad hydropower plant, Summary Report on Performed Investigations, Belgrade, unpublished). Geological investigations of the dam site started with new, high-quality detailed geological mapping of the terrain (Fig. 16.15). Parallel with these investigations, cross-hole geoelectric scanning in the left and right dam abutment and over the reservoir was performed, as well as reflective seismic investigations in dam galleries.

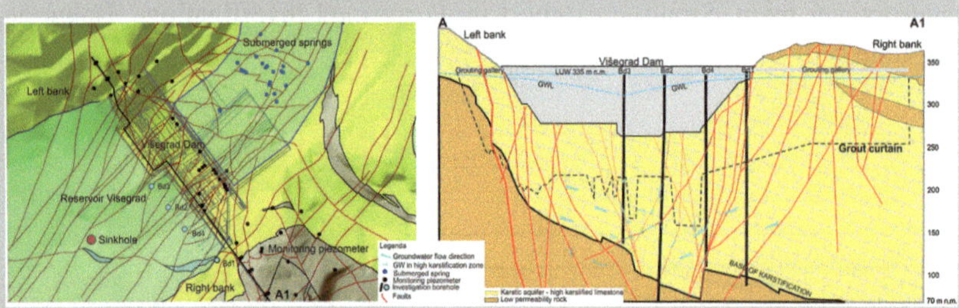

Fig. 16.15 Hydrogeological map and cross section of Višegrad dam site with position of grout curtain and investigation boreholes

The bathymetric recording of the bottom of the part of the storage near the dam and measurement of self-potential were performed simultaneously in the storage zone. These investigations in the storage zone led to the detection of anomalous locations that indicated possible water percolation. At locations of detected anomalies, detailed diving investigations and underwater video camera recordings were also performed. These investigations revealed the existence of a sinkhole of great size and capacity, which had an impact on the adjustment and re-direction of the subsequent investigations. Immediately after the detection of the "large sinkhole," its measurement was taken and the determination of the water inlet velocities made. The detected sinkhole was also used for the test of spatial detection of groundwater flows upstream from the dam by the "misse- a- la- masse" method. Downstream from the dam, in the part of the riverbed near the dam, the bathymetric recordings of the bottom were performed, as well as the diving investigations and underwater video camera recording, along with the measurements of water outflow at the locations of the detected springs. In the meantime, the investigative drilling and corresponding investigations in the boreholes (video endoscopy, carotage, etc.) (Fig. 16.16) were performed. Four investigation boreholes were drilled. The locations of these boreholes were determined in succession, depending on the results of all previously performed investigations.

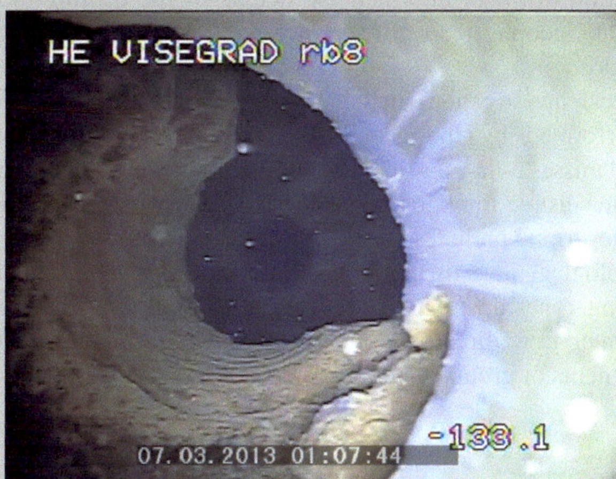

Fig. 16.16 Entrance to cavern at the depth of 133 m from an altitude 337 m a.s.l (video endoscopy)

In accordance with the progress of the investigation works, a large number of sodium fluorescein tracer tests were conducted (used more than 60 kg), during which the tracer was injected in the existing piezometers upstream from the dam, then in detected sinkholes, as well as into the newly drilled boreholes (Fig. 16.16). Also, very important are the tracer tests repeated several times in the large sinkhole.

The salt tracing test was also performed, as part of the system for real-time monitoring. A large quantity of the tracer was injected into the "large sinkhole". Investigations were also performed in all 4 special-purpose boreholes, existing piezometers, and the springs downstream from the dam. Finally, after drilling of all boreholes and the completion of the corresponding investigations in all 4 boreholes (Fig. 16.17), cross-hole geoelectric and seismic tomography were conducted between the boreholes.

The results of all special-purpose investigations were necessary to perform such investigations as these:

- geodetic surveys in which the results were used for geodetic surveys of the contours of the dam, terrain, and structures in the surrounding area, as well as the survey of the storage bottom and riverbed bottom downstream from the dam in the zone of springs and all other spatial data,
- geological investigations in which the geological data were used for gathered from all available results of performed investigations (surface mapping, core mapping, analysis of tectonic, etc.),
- investigation drillings in which the details of the drilling method, geological data, geophysical carotage investigations, WPT investigations and video endoscopy,
- geophysical investigations in which the results were used for investigations performed by the geoelectrical scanning method, self-potential method, "misse- a- la- masse" method, cross-hole, geoelectric tomography, cross-hole seismic tomography, and measurements of electrical resistance on samples from boreholes, and
- hydrogeological investigations in which the results of investigations in the sinking zones upstream from the dam were used for determination of groundwater flow directions, measurements of discharge velocities, and measurements of groundwater levels.

All those data and results can be presented schematically as on Fig. 16.17.

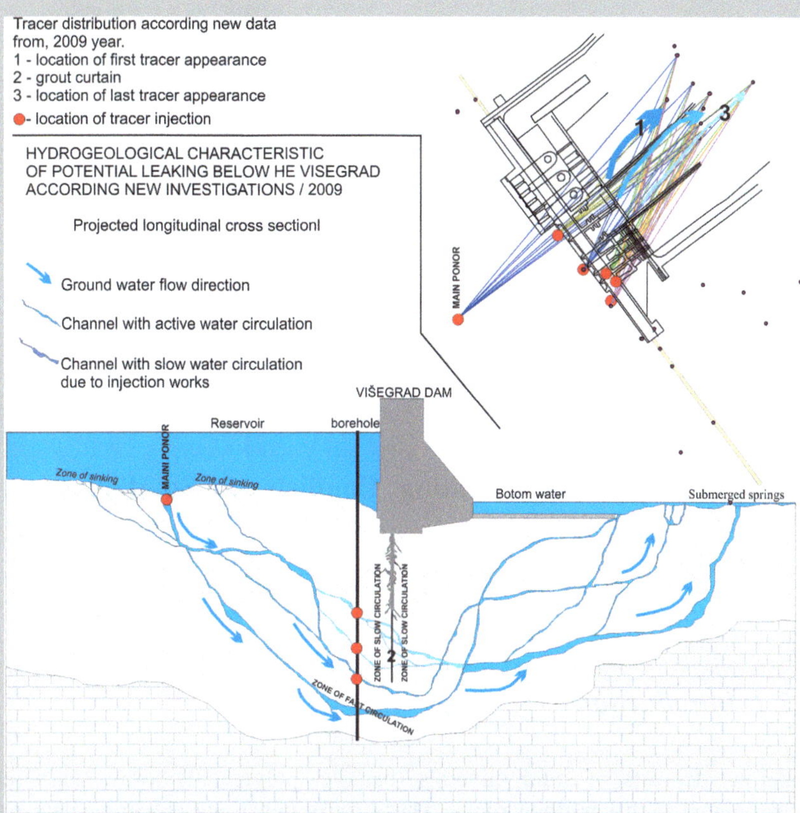

Fig. 16.17 Schematic longitudinal cross sections with conduit distribution. Layout tracer test results (S. Milanović, 2009 Report on special investigation on dam site Višegrad, Inst. for Develop. of Water Resources "Jaroslav Černi", Belgrade, unpublished)

Based on the above problem statement, it was safe to assume either that interactive work or integrated use of known 2D and partly defined 3D parameters is sufficient to produce an output of a three-dimensional conduit defined as 3D physical model. This case example presents certain investigations performed especially for the needs of the possible establishment of karst conduit genesis and spatial position below dam site (Fig. 16.18). The new data collected during grouting works give us evidence that the method which was applied for the construction of a 3D geometrical (or physical) model and parametric model of karst aquifer, aided by an incomplete data series, is feasible.

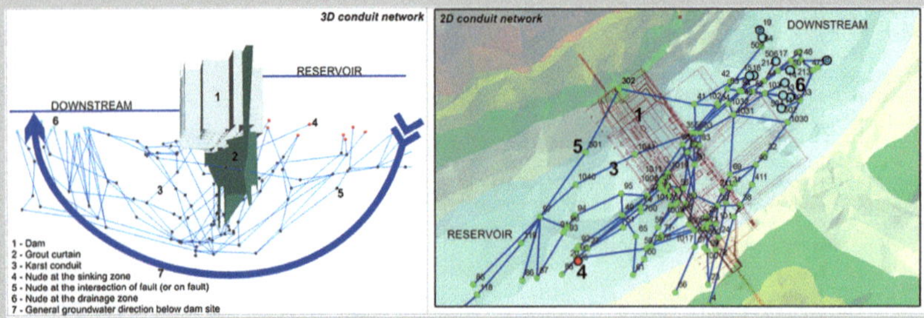

Fig. 16.18 2D conduit network of possible leakage pathways below dam site (model result; *right*); 3D conduit network of possible leakage pathways below dam site (*left*)

16.2 Karst Aquifers and Mining: Conflicts and Solutions

Veselin Dragišić

Department of Hydrogeology, Faculty of Mining and Geology, University of Belgrade, Belgrade, Serbia

16.2.1 Introduction

Numerous deposits of solid minerals (bauxite, polymetallic ores, skarn, etc.) are associated with karstified rocks. These rocks constitute the immediate overburden or basement, or both, of many ore deposits. In certain cases, the karstified rocks themselves are the mineral resources that are being mined. Groundwater from karst aquifers can be a major nuisance, threatening the mining operations with which they come into contact. This is especially true of underground mining, where ore is extracted from far below the water table. In advertent, intersecting of major karst conduits, due to a lack of knowledge about the hydrogeological conditions prevailing in the immediate environment, often leads to inrushes of groundwater into mining operations. A mine can be flooded over a very short time, causing human casualties and considerable loss. Additionally, intensive drainage of a karst aquifer can lead to subsidence and caving of the ground surface.

It is often difficult to predict water inflows to mining operations due to nonhomogeneous karstic character. In addition to water wells on the ground surface, mine drainage is provided by a combination of drilled drains leading out of the mining operations, drainage galleries, and drainage shafts.

One of the issues that mines in a karst environment have to cope with is the transformation of karst groundwater quality. Specifically, in carbonate karst, high-quality low-TDS groundwater transforms into highly acidic mine water, whose uncontrolled flow affects the quality of the immediate environment.

Operations involving evaporite extraction and drainage are faced with special problems. Extremely rapid karstification can lead to sudden groundwater discharges, often with immeasurable consequences.

16.2.2 Hydrogeological Types of Ore Deposits in a Karst Environment

More than 40 different types of ore deposits have been associated with karst. Also, the karstified rocks themselves can constitute mineral resources, such as limestones, marbles, calcareous tuff, sulfate rocks (gypsum-anhydrite), chloride rocks (halite and sylvite), and other rocks prone to karstification (Dublyansky and Nazarova 2004). The diverse ore deposits found in karst include bauxite, non-ferrous metals (Ni, Sb, Hg, Zn, and Cu), manganese, iron, and oil shales (Lunev et al. 2004). Uranium deposits have also been associated with carbonate rocks; in Uzbekistan and the USA, there are large deposits of this radioactive element in caves, caverns, enlarged conduits, and fractures (Bell 1963). Ore deposits in karst occur as mineral ore accumulations in contemporary karst depressions (ravines, sinkholes, and poljes), at the points of contact between carbonate rocks and igneous intrusives. The third type includes ore deposits overlain by younger sediments (Ford and Williams 2007b).

The positions of the ore deposits and mining operations relative to the water table are of special significance for the magnitude of inflow and the preferred drainage method. These deposits can be found in the vadose zone or the saturation zone in karst. Additionally, some ore deposits in karstified rocks, under the influence of a karst aquifer, are covered with younger sediments. Apart from these, ore deposits in coastal areas are distinguished because of the special way in which the inflow is formed. Evaporite deposits constitute a separate group due to a number of specific features.

16.2.2.1 Ore Deposits in the Vadose Zone

In general, mining operations above a karst aquifer are characterized by low and sporadic water enrichment. Intermittent pit water inflows occur solely after heavy rainfall. Dewatering of these mines does not pose a significant problem. One of the specific features of this type of ore deposits is the formation of perched aquifers, especially in bauxites (Box 16.2.1).

Box 16.2.1

Case Study—Bauxite deposits watering

Bauxite deposits near the City of Nikšić (Montenegro) are a typical example of ore deposits in the vadose zone. The bauxites are inter-stratified within carbonate formations (between Jurassic and Upper Cretaceous or between Lower and

Upper Cretaceous strata). Water inflows occur solely after heavy rainfall. High karstification of underlying limestones and a considerable depth to groundwater relative to the ore deposits enable direct infiltration of atmospheric precipitation in transit to the deeper reaches of the carbonate formation (Vasiljević et al. 1988). However, the bauxites in this part of Montenegro have a significant hydrogeological function. Being semipermeable rocks, they separate karstified overlying rocks from karstified underlying carbonate rocks. They form small (and the only) groundwater reservoirs in this part of the Montenegrin karst (perched aquifers). In certain localities, the aquifers are drained via perched springs, whose minimum discharge in places is as high as 1.0 l/s (Fig. 16.19). The significance of such springs is considerable, given that they occur in a typical karst environment, where springs are very rare while the demand for groundwater is high (Korać and Kecojević 1988; Radulović 2000a).

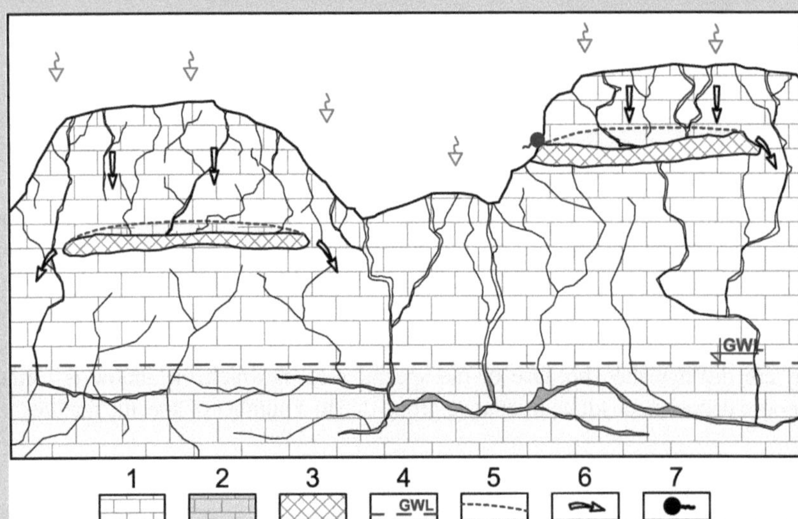

Fig. 16.19 Ore deposits in the vadose zone. *Legend 1* karstified limestone (vadose zone); *2* karstified limestone (saturated zone); *3* ore body (bauxite); *4* general groundwater level; *5* perched aquifer groundwater level; *6* groundwater direction; and *7* perched spring

16.2.2.2 Ore Deposits in the Saturated Zone

A large number of ore deposits are found below the water table in karst. They feature high groundwater inflow rates, which often hinder ore extraction, particularly underground mining. The ore is deposited within carbonate rocks, where the karst aquifer is unconfined (Fig. 16.20). The aquifer is recharged by infiltration of precipitation and water from surface streams. Bauxite deposits in the Urals, Russia, and lead and zinc deposits in Mirgalimsai, Kazakhstan, are found within a several hundred meters thick formation of karstified carbonate rocks (Ershov et al. 1989).

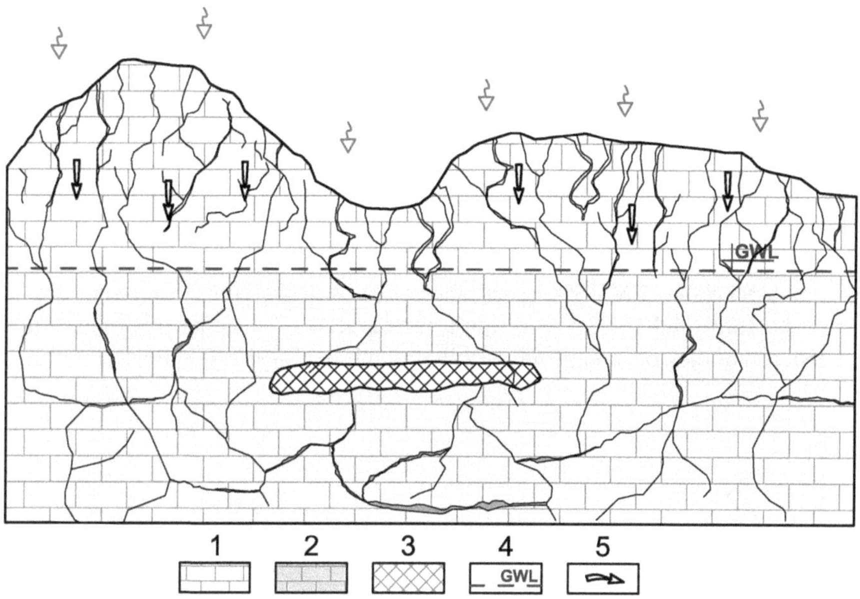

Fig. 16.20 Ore deposits in the saturated zone. *Legend 1* karstified limestone (vadose zone); *2* karstified limestone (saturated zone); *3* ore body (bauxite); *4* groundwater level; and *5* groundwater direction

They feature extremely high inflow rates: about 9 m^3/s in the Urals (Kleiman 1982; Plotnikov and Roginec 1987) and up to 3.3 m^3/s at Mirgalimsai (Kleiman 1982). Apart from underground operations, high inflow rates have also been recorded in opencast mines, such as in the Estonian oil shale deposits between karstified Ordovician dolomitic limestones, during periods of snowmelt and heavy rainfall, when they were as high as 3.0 m^3/s (Abramov and Skirgello 1968).

16.2.2.3 Ore Deposits on a Karst Bedrock Covered with Younger Sediments

Many ore deposits worldwide occur on a karst bedrock, which subsequent sedimentation processes have covered with younger, generally semipermeable strata. The groundwater is for the most part pressurized, and the inflow rates are extremely high. For example, the inflow rates to the Fan Gezhuang coal mine in China, from underlying karstified Ordovician limestones, are as enormous as 34 m^3/s (Gongyu and Wanfang 2006).

The bauxite deposits in Hungary north of Balaton, in the Halimba and Nyrad districts, directly overlie water-bearing Triassic and Jurassic limestones and dolomites, which are 500–600 m thick. Upper Cretaceous or Eocene and Pleistocene non-carbonate semipermeable formations lie discordantly over the bauxite deposits. The groundwater in most of the deposits is pressurized. The amounts of water pumped from the karst aquifer were rather impressive in the early 1980s, about 5 m^3/s, with a drawdown of 120 m (Alliquander 1982).

Box 16.2.2

Case Study—Confined karstic aquifer

The Maoče Coal Mine in the Pljevlja Coal Basin, Montenegro, is a good example of how pressurized groundwater can affect the inflow rate. The coal seam either overlies Middle Triassic limestones directly, or there is a transition marked by a thin bed of clay sediments (Nikolić and Dimitrijević 1990). A confined karst aquifer underlies the coal seam. Immediately after drilling, a borehole (BM-159) issued about 120 l/s at the point of artesian flow (Figs. 16.21 and 16.22). As a result, the capacity of some karst springs was reduced or they ran completely dry (Radulović et al. 1987; Radulović 2000b).

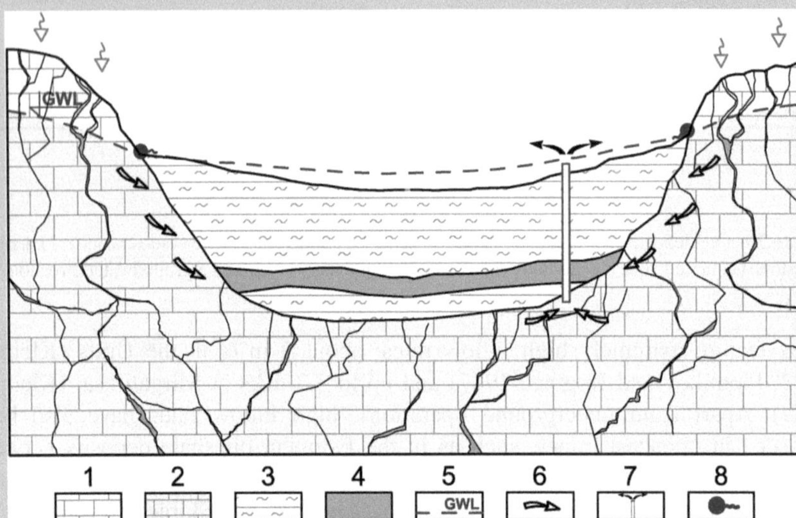

Fig. 16.21 Ore deposits over a karst bedrock, covered with younger sediments (Pljevlja Coal Basin, Montenegro). *Legend 1* karstified limestone (vadose zone); *2* karstified limestone (saturated zone); *3* Neogene sediments; *4* coal; *5* groundwater level; *6* groundwater direction; *7* exploration well; and *8* spring

Fig. 16.22 Artesian flow from exploration well BM-159, limestones in the background (Courtesy of Mićko Radulović)

16.2.2.4 Ore Deposits in Coastal Areas

Inflow rates to ore deposits in coastal areas can be boosted by seawater intrusion. Examples of this include the Raša Coal Mine (Croatia) and lead and zinc mines in Sardinia (Italy). The coal seams at Raša directly overlie Upper Cretaceous limestones. Drainage of the mining operations removed about 0.3 m^3/s of groundwater on average. Seawater intruded into the mining operations, which were as deep as 250 m below sea level (Šarin and Tomašič 1991).

At the Iglesias Lead and Zinc Mine, the ores are extracted from deposits located within a karst aquifer, 100 m below sea level. The karst groundwater is recharged by both atmospheric precipitation and seawater, such that high pumping rates caused seawater intrusion into the mining operations (Carta et al. 1982).

16.2.2.5 Evaporite Deposits

Evaporite rocks (salt and gypsum-anhydrite) are highly soluble and often form karst features that are generally found in limestone and dolomite deposits. A specific trait of evaporite karst, compared to carbonate karst, is that karst features are created rapidly, within several days, weeks, or years, while those in carbonate rocks need years, decades, or centuries (Johnson 2004).

Salt rocks are particularly prone to karstification under the influence of groundwater that is not saturated with chlorides. Low-TDS groundwater from other aquifers flows to the salt rock deposits, dissolving them and creating diverse karst features. The springs that naturally drain the karst aquifer remove a large amount of salt and deplete the deposits (Korotkevich 1970). Groundwater discharges through the newly created features in the ore deposits (caverns and caves) have often flooded mining operations (Abramov and Skirgello 1968).

Mining of evaporite deposits, primarily salt deposits, has produced numerous examples of subsidence and caving. There are two well-known cases of subsidence as a result of salt extraction: the Bereznikovsky Salt Mine in the Perm region of Russia and the salt mines in Cheshire, England (Ford and Williams 2007b).

Caving can also occur when not evaporites, but other deposits in their vicinity are mined. One example is the Gays River Mine in Canada. Lead and zinc ores are deposited in limestones covered with gypsum-anhydrite rocks, overlain by fluvio-glacial sediments. The drawdown within the ore deposits, due to the high solubility of gypsum-anhydrite rocks, resulted in subsidence and caving of the ground surface, as well as elevated rates of groundwater inflow to the mining operations. Despite numerous attempts to prevent the inflow, the rates increased from 100 to 250 l/s. This rendered mining unprofitable, and the mine was eventually shut down (McKee and Hannon 1985).

Box 16.2.3

Case Study—Collapse in salt mine area

In 1986, the potassium salt mine at Bereznikovsky in the Perm region of Russia experienced caving of gigantic proportions. It was caused by freshwater from 350-m-thick clastic sediments above the salt deposits, which reached the deposits and proceeded to destroy them. The inflow rates to the mining operations began to increase in January, when they were about 15 l/s, and reached more than 500 l/s on the night of 8/9 March. The mine was flooded. Within a few months, the mining operations of about 15 million m^3 became submerged. A large cavity was created above the mine, whose volume was in excess of 1 million m^3. The ground surface collapsed on the night of 25/26 July. The sink was ellipsoid on the surface, with 40×80 m sides, and about 170 m deep. Subsequent caving increased the sidewalls to

220 × 150 m (Fig. 16.23). The water table of the newly created pond (sink) was initially at about 100 m, but later rise to 60–70 m below the ground surface (Andreychuk 1996).

Fig. 16.23 Bereznikovsky Sink (Perm region, Russia)

16.2.3 Groundwater Inrush into Mining Operations

Groundwater inrush is frequent in mining of ore deposits in a karst aquifer environment. They are generally a result of insufficient hydrogeological exploration and a lack of preventative drainage measures. The results are very rapid flooding of the mining operations, substantial losses and, at times, human casualties. Such occurrences can be a consequence of different relationships between the ore deposits and the karst aquifer. Contrary to the carbonate karst, inrushes into evaporate karst are associated with intensive dissolution of the deposits under the influence of freshwater inflow, generally from overlying formations.

The brown coal mine at Vrdnik on Fruška Gora Mt. (Serbia) is a good example of how a lack of knowledge about hydrogeological circumstances can be devastating. Drilling of an exploratory pit in 1929 caused an inrush of some 500 l/s (water temperature 34 °C), resulting in flooding and abandoning of the mining operations. The sudden inflow occurred when mining operations passed through semi-permeable Tertiary sediments and entered underlying karstified Triassic limestones (Luković 1939).

Intensive drainage of a karst aquifer during the course of mining, without prior dewatering, has been one of the most frequent causes of groundwater inrushes in the past. For instance, at the bauxite mines in the Urals, at about 100–150 m below the ground surface, 93 inrushes were registered in 25 years. The highest measured inflow rate was 1.2 m^3/s (Abramov and Skirgello 1968).

Sudden intrusion of karst groundwater into mining operations can also be a result of heavy precipitation and rapid infiltration into the karst. For instance, the underground bauxite mine at Trobukva in West Herzegovina (Bosnia and Herzegovina), experienced several inrushes (0.18 m^3/s in 1982 and 0.5 m^3/s in 1987), caused by heavy rainfall (Slišković 1984; Bilopavlović 1988).

A hydraulic contact between surface water and karst groundwater is another potential cause of inrushes into mining operations. Operations at a dolomite quarry (in Pennsylvania, USA) cut through a karst conduit at a depth of 40 m, which was connected with surface water. Some 0.6 m^3/s of water from the conduit intruded into the mining operations (Lolcama 2005).

Box 16.2.4

Case Study—Sudden inrush from isolated aquifer

An atypical inrush of karst groundwater occurred in 1980 in the brown coal mine at Lipov Deo, belonging to the Senj-Resava Mines (Serbia). The inflow was caused by exploratory adit N-9 at an level of 327 m, penetrating the roof of a coal seam comprised of red Permian sandstones, which were only 2 m thick, and entering water rich Jurassic limestones (Fig. 16.24). The limestones, along with the Permian sandstones, were tectonically positioned so that they formed the roof of the coal seam. The initial inflow rates were 50 l/s, but after 6 days of continuous discharge, the mining facilities were flooded. Maximum inflow rates were about 170 l/s, but they later decreased to 20 l/s and were about 1.5 l/s in the next 1981 (Fig. 16.25). This pattern suggested an isolated karst aquifer whose recharge was very slow (Miladinović and Dragišić 1998; Miladinović 2000).

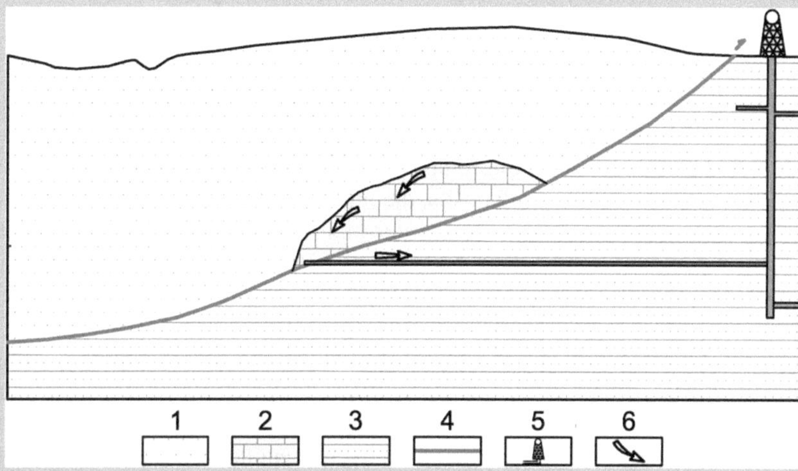

Fig. 16.24 Section of exploratory adit N-9 at the Lipov Deo Coal Mine (Serbia). *Legend* *1* Permian sandstones; *2* Jurassic karstified limestone (saturated zone); *3* Neogene sediments with coal; *4* overthrust; *5* shaft; and *6* groundwater direction

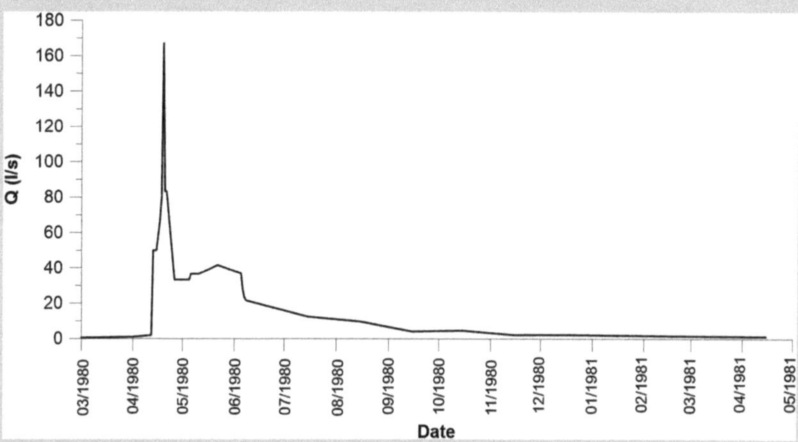

Fig. 16.25 Karst groundwater discharge hydrograph at adit N-9 of the Lipov Deo Coal Mine (Serbia)

16.2.4 Dewatering of Ore Deposits in a Karst Aquifer Environment

Safe mining in a karst aquifer environment requires efficient and timely drainage, or the prevention of pit water inflow (Table 16.2). In practice, a series of measures are implemented to dewater mining operations in a karst environment, depending on the type and size of the ore deposits, the inflow rate, the extent of karstification, hydraulic connection between groundwater and surface water, economic drivers, and similar parameters (Plotnikov and Roginec 1987). Drainage methods generally include the following: drainage wells on the ground surface; drilled drains leading out of the mining operations; drainage galleries and shafts; grouting of karst conduits; grout curtains; re-alignment of surface streams; tamping of ponors and river channels; and the like (Kleiman 1982). Due to the highly diverse nature of groundwater flow, lowering of the water table inside the deposits by means of a special configuration of wells is often not effective enough, so it tends to be combined with underground drainage works.

Table 16.2 Extreme karst groundwater inrushes into a number of mines worldwide (historical data)

Country	Mine	Mine product	Type of operation	Water inflow (m^3/s)		Reference
				Reg.	Max.	
China	Fangzhuang	Coal	Underground		34.2	Wenyong et al. 1991
					33.0	Gongyu and Wanfang (2006)
Hungary	Transdanubian region	Coal and bauxite	Underground and open pit	12.5		Kleiman (1982)
Hungary	Transdanubian region	Bauxite	Underground	5.0		Alliquander (1982)
Russia	Northern ural	Bauxite	Underground		9.1	Kleiman (1982)
						Plotnikov and Roginec (1987)
France	Region var	Bauxite	Underground		4.2	Tilmat (1973)
USA	Friendsville mine (Pemsilvania)	Zn	Underground	1.7	3.8	Kleiman (1982)
Kazakhstan	Mirgalimsai	Pb–Zn	Underground	2.7–3.3		Kleiman (1982)
Estonia	Estonian mines	Oil shale	Open pit		3.0	Abramov and Skrigello (1968)
Poland	Olkusz mine (3 mines)	Pb-Zn	Underground		7.3	Motyka and Czop (2010)
Poland	Lubin-Glogow	Cu	Underground	1.0		Bochenska et al. (1995)

Prior dewatering plays an important role because it ensures safe access to the ore deposits. It is commonly undertaken by means of drainage wells (water table lowering boreholes), located in water rich zones, in combination with drainage galleries and other works. One example of effective prior dewatering is the Olkusz Lead and Zinc Mine (Poland), which is one of the European mines that feature the highest inflow rates. The ore bodies are found in paleokarst dolomite cavities, at a depth of 200–300 m. The dolomites are overlain by a thick sequence of water-bearing Quaternary sands that store large amounts of groundwater and are hydraulically linked with a karst aquifer. Prior to extraction, the water table was lowered by means of wells and drainage galleries under each horizon (Ford and Williams 2007b).

Bauxite and coal deposits in Hungary, in an environment of Mesozoic water-bearing limestones and dolomites, are drained by galleries and large-diameter wells (shafts). In the early 1980s, the pumping rate was 5 m^3/s and the drawdown was 120 m (Alliquander 1982). The diameters of the drainage wells were 1.35–2.95 m, and each well was equipped with three submersible pumps (Tóth 1982).

Mine drains are generally drilled in cases where the karst aquifer to be drained overlies or is located to the side of the mining operations. At the Velenje Coal Mine in Slovenia, for example, such a system drains a karst aquifer in Triassic dolomites and limestones, which had caused several groundwater inrushes in the past, two of which were disastrous in 1918 and 1973 (Mramor 1984).

One of the measures undertaken to reduce groundwater inflow to mining operations is grouting of fractures and caverns, which constitute the greatest barriers in underground mining. Karst conduits are often exposed to high water pressures, destructive turbulent flows, and enormous inflow rates. Groundwater inflow is usually very fast and rates often measure hundreds or even thousands of l/s (Milanović 2000b).

This method can be effective but surprises are not rare. A typical example of ineffective grouting is the previously mentioned Trobukva Bauxite Mine in West Herzegovina (Bosnia and Herzegovina). Following a groundwater inrush and flooding of the pit in 1982, the karst conduits, which constituted the main groundwater pathways, were grouted between 90 and 240 m, section of decline. Water inflow rates in underground mining works decreased significantly, to amount of 2.5–12.0 l/s during the year. This water inflow prevention method had been cited as an effective underground bauxite mine drainage solution in the Dinaric karst of Herzegovina (Slišković 1984). However, a sudden inrush of more than 500 l/s in 1987 contradicted previous claims (Bilopavlović 1988).

The dolomite quarry in West Virginia is another example of ineffective grouting. The largest bitumen grout curtain in the North American karst was emplaced there to prevent river water inflow through karst conduits. However, a new karst conduit developed parallel to the stratification planes and river water continued to intrude into the mining operations (Lolcama 2005).

High drainage rates sometimes result in large-scale regional drawdown, as registered in the Lubin-Glogow copper mining region at depths of 600–1,200 m, where inflow rates from a karst aquifer in limestones and dolomites measured about 1.0 m^3/s. The drawdown extended over a surface area of 2,500 km^2 (Bochenska et al. 1995).

When evaporite deposits (primarily those of halite and sylvine) are dewatered, most of the groundwater needs to be evacuated before it gets into contact with the highly soluble ore deposits. For example, at the Solotvyno Salt Mine (Ukraine), "fresh" groundwater flows from an alluvial aquifer through fractures, destroying the salt rocks and creating karst features. This in turn boosts groundwater inflow and floods the mining operations. The task in such a mine is to undertake prior dewatering and prevent freshwater from reaching the ore deposits. Storm water also needs to be evacuated away from the deposits in good time (Abramov and Skirgello 1968).

As mining developed, mining methods were optimized and conventional mining gave way to solution mining, where ore deposits are dissolved by injected freshwater (Ford and Williams 2007b).

Ore extraction and mine drainage in a karst environment involve many risks. Apart from inrushes and contamination of karst groundwater, the most frequent adverse effects are subsidence and the creation of sinks on the ground surface and landslides in opencast mines.

High ore deposit/mine drainage rates result in dramatic drawdowns and subsidence. Possibly, the worst incident occurred in 1962 in South Africa, where intensive drainage of a karst aquifer in dolomites and limestones overlying gold deposits caused an ore preparation facility at the Dreifonte in Gold Mine to collapse. Twenty-nine people perished (Ford and Williams 2007b).

As a result of bauxite mine drainage in the northern Urals, a 500–600 km^2 cone of depression was formed on the ground surface. Karst-suffusion processes triggered by high drainage rates led to the development of more than 1,000 sinkholes, considerably enhancing infiltration of precipitation (Plotnikov 1989).

Box 16.2.5

Case Study—Landslides in contact zone

Depending on its position relative to the ore deposits, karst groundwater can have an adverse effect on rock stability of the ore deposits and thus be a threat to mining safety. The copper mine at Veliki Krivelj in eastern Serbia is an example of how karst groundwater causes landslides in opencast mining operations. The copper deposits are of the porphyry type, created in hydrothermally altered igneous (kaolinated and chloritized) rocks. These rocks, along with the ore deposits, are in tectonic contact with the Jurassic limestones of Veliki Krš Mt. (Fig. 16.26). Pressurized karst groundwater intrudes into the surrounding rocks, as corroborated during the course of drilling of exploratory adits within the deposits and surrounding rocks, as well as by chemical analyses of the groundwater (Dragišić and Stevanović 1984; Stevanović et al. 1991; Dragišić 1992; Stevanović and Dragišić 1995). A lack of timely drainage caused large-scale landslides (Fig. 16.27).

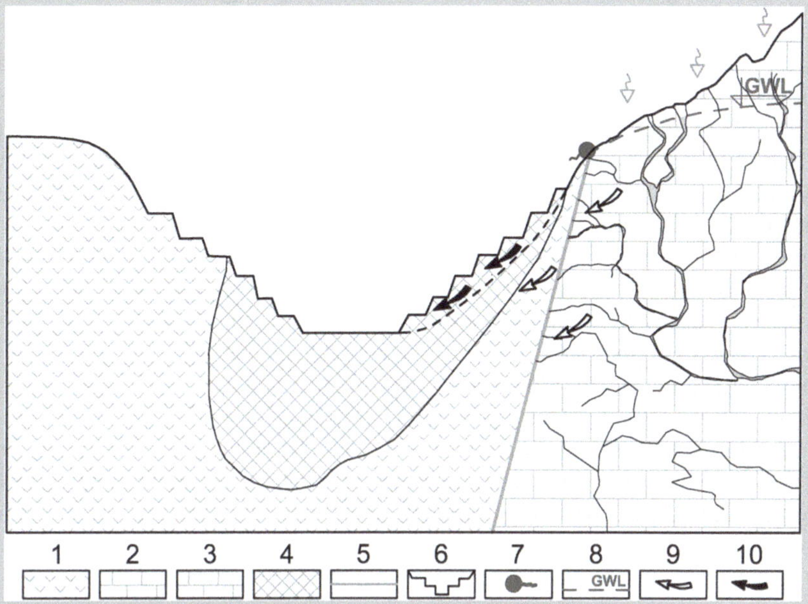

Fig. 16.26 Hydrogeological section through porphyry copper deposits at Veliki Krivelj (Serbia). *Legend 1* andesite (hydrothermal alteration); *2* karstified limestone (vadose zone); *3* karstified limestone (saturated zone); *4* copper deposit; *5* fault; *6* open pit; *7* spring; *8* groundwater level; *9* groundwater direction; *10* landslide

Fig. 16.27 Landslide at the opencast mine of Veliki Krivelj (Serbia)

16.2.5 Transformation of Karst Groundwater Quality

Ore extraction and high mine drainage rates in a karst environment often alter the quality of karst groundwater. This especially applies to non-ferrous ore deposits (such as those of Cu, Pb, Zn, Sb, and Hg), but is associated with other ores as well.

Prior to reaching mining operations, the quality of karst groundwater is generally good and such water is often used for drinking water supply, irrigation, and other needs. Since the discharge rates are frequently high (several hundreds or even thousands of liters per second), the significance of such groundwater cannot be overstated. Consequently, in order to use karst groundwater for practical purposes, it needs to be tapped before it comes into contact with mining operations. There are numerous examples of karst groundwater use. The quality of the groundwater pumped from the bauxite deposits in Hungary is very high, such that this water is used for drinking water supply (Alliquander 1982).

An 8-km-long drainage gallery in the lead mine at Homesford (USA) taps some 870 l/s, of which 460–580 l/s is used for drinking water supply (James 1997). Several hundred liters per second of high-quality karst groundwater, tapped before it reaches lead and zinc mining operations at Mirgalimsai in Kazakhstan, is used for drinking water supply and irrigation of farmland (Plotnikov and Roginec 1987).

As karst groundwater flows into mining operations, it comes into close contact with the ore and other minerals that make up the ore body. As a result of complex geochemical processes, generally high-quality karst groundwater of the HCO_3-Ca type, with a pH level of 7–7.5 and TDS < 1,000 mg/l, becomes transformed into the SO_4-Ca type, with low pH and high TDS levels and elevated concentrations of Fe, Al, and other elements. A typical example of such transformation are the copper mines at Veliki Krivelj and Majdanpek in eastern Serbia, where the inflow is formed from karst aquifer discharge or at locations where tailings and waste rock dumps have been formed on karst (Dragišić 1992; Dragišić 1994; Stevanović and Dragišić 1995).

A special type of karst groundwater transformation is encountered under the influence of high-TDS seawater in coastal areas. Namely, high pumping rates at mining operations in such areas lead to over-pumping of low-TDS groundwater and intrusion of high-TDS seawater. Typical examples are the lead and zinc mines in Sardinia (Carta et al. 1982) and the Raša Coal Mine in Croatia (Šarin and Tomašič 1991).

Box 16.2.6

Case Study—Groundwater quality transformation
The Kizelkovsky Coal Basin in the Perm region of Russia is a classic example of karst groundwater quality transformation. The stone coal seam is Lower Cretaceous, emplaced between carbonate rocks. Accompanying minerals include pyrite, calcite, siderite, and gypsum. Underground coal extraction began more than 200 years ago, reaching depths up to 1,000 m below

the ground surface. The mine has been abandoned because of uncontrolled water discharges from 12 pits, which created a major environmental issue. During the course of coal extraction, underground mining formed a drain for karst groundwater discharge from the carbonate cover. The mine pits in this region were among those with highest water inflows in Russia. The pits were abandoned and flooded in 1990. However, groundwater of transformed quality is still being discharged from the karst aquifer overlying the coal seam (Fig. 16.28). The transformed groundwater is typical acidic mine water (Bankovskaya and Krasavin 2004; Maksimovich 2004).

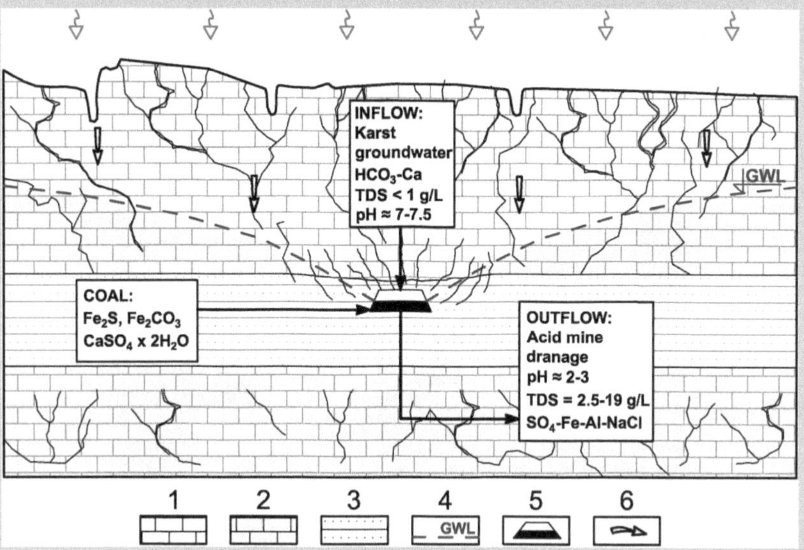

Fig. 16.28 Schematic representation of the formation of acidic mine water at the Kizelkovsky Coal Mines. *Legend 1* karstified limestone (vadose zone); *2* karstified limestone (saturated zone); *3* clastic sediments with coal; *4* groundwater level; *5* mining works with mine water; and *6* groundwater direction

The transformation process can be summarized as follows: There is a karst aquifer in the roof made up of non-carbonate rocks with coal, in highly fractured and karstified limestones, whose thickness in places is more than 1,000 m. The limestones are partially exposed and heavily recharged by infiltration of atmospheric precipitation. The quality of the karst aquifer groundwater is excellent. The aquifer provides drinking water supply to the extended area of the mine (Bankovskaya and Krasavin 2004). After the mining operations were abandoned, the pits that served as karst groundwater drains became flooded. The geochemical processes that take place in coal mining operations in the presence of water and accompanying minerals (pyrite, siderite, gypsum, etc.) tend to create acidic mine water (Lovell 1983; Lottermoser 2007; Dimitrijević 2013).

The mine water is typical acidic water of the SO_4–Fe–Al–Na–Ca type. TDS generally measures 2.5–19.0 g/l, but in certain cases as much as 35 g/l. The pH levels are up to 2–3. Compared to low-TDS water, such mine water features a number of times higher concentrations of lead, copper, zinc, silver, nickel, cobalt, and other minerals (Maksimov 2004).

When the morphological conditions are favorable, the mine water is gravitationally discharged untreated, from the pits to the ground surface. One of the pits is the Kalinin Mine from which acidic mine water flows to a nearby river (Figs. 16.29 and 16.30).

Fig. 16.29 Gravitational discharge of acidic mine water from the Kalinin Mine

Fig. 16.30 Flow of acidic mine water to the river (Kalinin Mine)

16.3 Remote Techniques for the Delineation of Highly Karstified Zones

Milan M. Radulović

Faculty of Civil Engineering, University of Montenegro, Cetinjski put bb, 81000 Podgorica, Montenegro

16.3.1 Introduction

The majority of karst terrains are characterized by a high degree of heterogeneity. The results obtained by applying methods for assessment of local karstification (e.g., borehole tests) often cannot be reliable to extrapolate to a wider area. The use of remote sensing provides the opportunity to assess the spatial distribution of karstification in the subregional scale. Analysis of satellite and aerial images allows identification of geomorphological and tectonic forms that may indicate the highly karstified zones. There are several factors that indicate the karstification, and which can be mapped by remote sensing, and this contribution is focused on two of them: surface karstification (K_{sf}) and density of faults (T_f). By overlapping maps of these two factors using geographical information system (GIS) techniques, the final map expressed through a *KARST* index is obtained. The application of this approach provides an image of the spatial distribution of karstification, even for areas that are inaccessible for direct field research. The obtained map can be used as a basis for solving some of engineering problems in karst that are related to the regulation of water, extraction of groundwater, and protection of karst aquifers from contamination.

16.3.2 The Complexity and Categorization of Karst Terrains

The complexity of karst terrains, and often inaccessibility for direct field observations, induce a need for application of remote sensing, as one of the additional methods for karst research. The observation of karst terrains from a distance, except obtaining a more general image of study area, provides an opportunity for indirect delineation of zones with different degree of karstification, which is the main objective of the approach described in this section. Application of the mapping approach is shown on the example of catchment area of Karuč springs (Montenegro, Dinarides; Fig. 16.31).

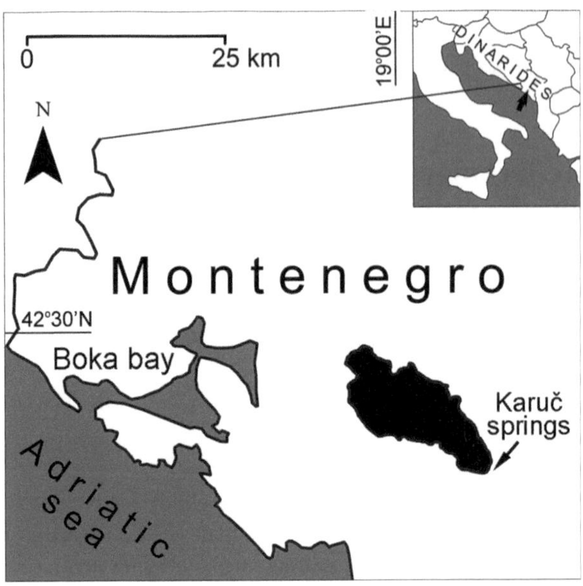

Fig. 16.31 Location of Karuč springs catchment area

For the purposes of categorization of karst terrains, different criteria have been used by different authors. In most cases, when considering the regional scale, the categories have been delineated according to the distribution of karst landforms such as sinkholes and caves (*number of sinkholes*/km^2, *area of sinkholes*/km^2, *volume of sinkholes*/km^2, *number of caves*/km^2, *length of caves*/km^2, and *volume of caves channels*/km^2). There are also some examples of satisfactory mapping of karstification through a so-called sinkhole index, which is based on the mean spacing of closed contours in a given area (Gregory et al. 2001).

The results of various tests (pumping test, slug test, Lugeon test, and packer test) and geophysical loggings (borehole televiewer, caliper logging, electromagnetic induction logging) give excellent data about local porosity and karstification, but the problem is the limitation of the results only on a narrow area around the tested borehole. A number of field geophysical methods can also provide an image of karstification, but for the analysis of a complete catchment area that are often partly impassable, these methods also appear to be impractical. In recent years, airborne electromagnetic (AEM) techniques for remote mapping of karstification have been developed (Smith et al. 2005; Supper et al. 2009; Gondwe et al. 2012). These techniques are especially suitable for wider areas. Their application gives an image which shows electrical conductivity anomalies that are mainly related to surface karst landforms (Smith et al. 2005). However, these methods are still under development, but they will likely play an important role in the future to address this issue. Also, it is interesting to note airborne and satellite thermal imaging sensors which have been used for detecting temperature anomalies. Temperature anomalies could occur in the opening of the caves due to the difference in outside air temperature and the temperature of air that outflow from the caves (Zurbuchen and Kellenberger 2008; Wynnea et al. 2008). Furthermore, underwater karstification, precisely the locations of vruljas, where colder groundwater discharge below the level of sea or lake (Figs. 16.32 and 16.35), could be detected by using thermal imaging sensors.

The approach presented in this section consists of analysis of satellite and aerial images in order to identify geomorphological and tectonic forms that are often indicators of highly karstified zones. The procedure of mapping is extracted from the more complex KARSTLOP method which is used for assessing the spatial distribution of recharge of karst aquifers (Radulović et al. 2012). Two subfactors, that can be mapped using remote sensing, are extracted from the aforementioned method. These subfactors are the surface karstification (K_{sf}) and the density of faults (T_{f}). Through them, by using GIS techniques, the final map of the karstification assessed by remote sensing techniques (*KARST*) index is obtained. The *KARST* index represents the index that indirectly reflects the degree of karstification.

The possibilities of application and some limitations of the presented approach are also discussed in this section.

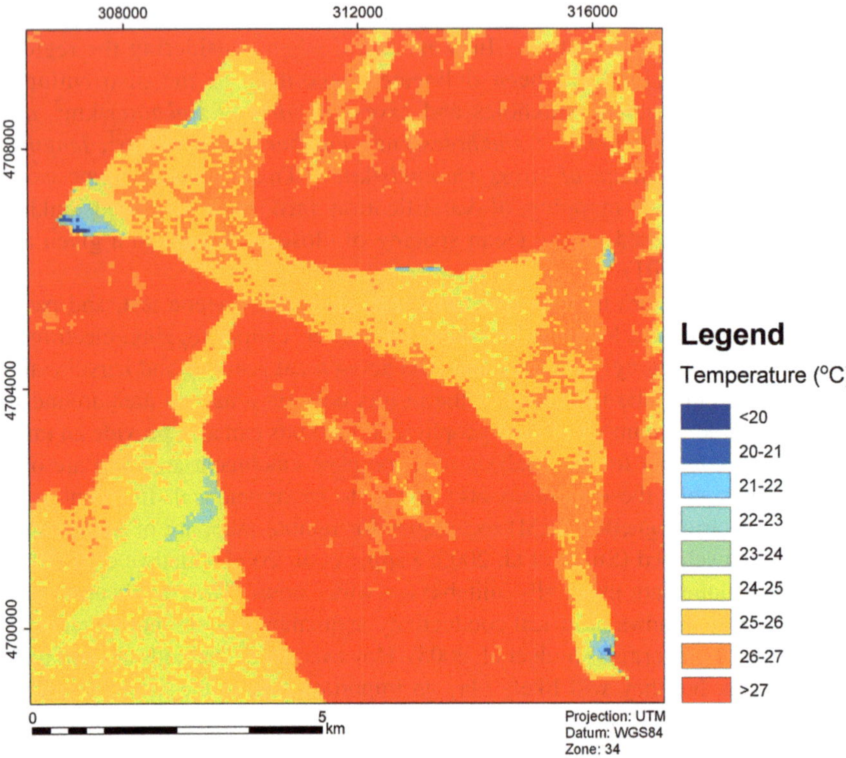

Fig. 16.32 The map of sea temperature in the Boka bay (Montenegro) which obtained by satellite image LANDSAT 7 ETM+ (thermal infrared band–Band 6; resolution 60 m; date of capturing: June 23, 2002). The temperature anomalies (*blue tones*) indicate the locations of the submarine springs where colder groundwater discharges below the sea level

16.3.3 The Concept of Mapping of Karstification by Using Remote Sensing and GIS

This subsection presents the approach for obtaining the map of karstification, which has been developed and applied to highly karstified terrains of External Dinarides. There are two factors that indicate the degree of karstification, and which can be mapped from satellite and aerial images. Those are factor K_{sf} (surface karstification) and factor T_f (the density of faults).

16.3.3.1 Map of Surface Karstification

Special attention, when analyzing the degree of surface karstification, should focus on the analysis of the distribution of surface karst landforms. The surface karst landforms are recognizable on aerial and detailed satellite images, such as Quick

Bird and SPOT (resolution about 2.5 m). Also, consideration should be given to an appropriate scale in which the map of surface karstification will be created (preferably between 1:25000 and 1:100000).

A coverage of karst terrains with sinkholes and uvalas can sometimes reflect the degree of karstification, since they are often elongated along the faults which were additionally extended by corrosive actions of water. A distribution of sinkholes is, from many authors, accepted as a criterion for categorization of karst terrains (Gregory et al. 2001; Angel et al. 2004). The problem is that sinkholes are in large numbers mainly formed on the karst plateau, but they rarely occur on steep slopes which were formed by later erosion. For this reason, the use of density of sinkholes as the only criterion does not give a complete image of the degree of karstification. There are karst terrains where sinkholes are totally absent, while other indicators (karren fields, caves, and fluctuation of discharge) suggest a high degree of karstification.

Since the karren fields are often the only karst landforms on the slopes, it is reasonable that one criterion for mapping the surface karstification is an area of karren fields (or degraded zones) per 1 km^2 and that the second criterion is an area of karst depressions (sinkholes, uvalas, poljes, and dry valleys) also per 1 km^2. By overlapping input maps created by these two criterions, a map that will satisfactorily show the degree of surface karstification can be obtained. According to this conception, the highly karstified terrains would be areas with uvalas and sinkholes, whose slopes are scarred by karren.

So there are two subfactors, K_{sf1} (area of degraded zones/km^2) and K_{sf2} (area of karst depressions/km^2), according to which the categorization of karst terrains is performed (Table 16.3). Limit values between the categories are defined by measurements on a number of sites along the External Dinarides.

Factor of surface karstification (K_{sf}) represents the average of subfactors K_{sf1} and K_{sf2} (Table 16.3).

Figure 16.33 shows some of the karst terrains that could be the subject of mapping according to the described concept. The first image (Fig. 16.33a) shows the karst terrain with poorly expressed surface karst landforms, from which karst depressions and karren fields are almost totally absent. The second image

Table 16.3 The categorization of karst terrains according to the subfactors K_{sf1}, K_{sf2}, and according to the factor K_{sf}

Area of degraded zone (karren fields, ruin-like relief, etc.) per unit square (10^3 m^2/km^2)	K_{sf1}	Area of karst depressions per unit square (10^3 m^2/km^2)	K_{sf2}	$K_{sf} = (K_{sf1} + K_{sf2})/2$
<60	1	<25	1	1
60–120	2	25–50	2	>1–2
120–180	3	50–75	3	>2–3
180–240	4	75–100	4	>3–4
>240	5	>100	5	>4–5

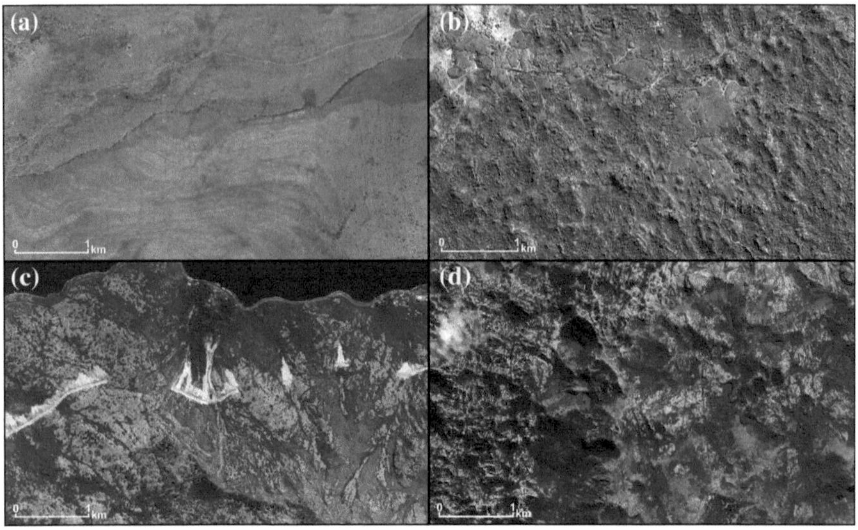

Fig. 16.33 Satellite images of karst terrains **a** terrain with poorly expressed surface karst landforms; **b** terrain with karst depressions but without degraded zones; **c** karst slope where the karren fields are well developed; **d** terrain where karst depressions and karren fields are together

(Fig. 16.33b) shows the karst terrain on which karst depressions (sinkholes and uvalas) are very common, while degraded zones are poorly represented. The third image (Fig. 16.33c) shows a karst slope where the karren fields are well developed, while sinkholes and other karst depressions are completely absent. Finally, the fourth image shows (Fig. 16.33d) the terrain with a high degree of karstification, where there are both karst depressions (sinkholes and uvalas) and karren fields.

16.3.3.2 Map of Fault Density

Karst landforms, especially karst channels, are often elongated along the faults. Intersections of two or more faults represent zones with a greater chance for the existence of caves. For this reason, it is considered that is reasonable to introduce a map of fault density as one of input maps for assessing the degree of karstification. Map of fault density (T_f) sows contours which are categorized according to the length of faults per unit of surface (km/km^2). Before the creation of this map, it is necessary to create a digital map of fault lines. The Landsat 7 ETM+ satellite images (resolution of 30 $\times$ 30 m), after appropriate processing, can be a good basis for creating the map of fault lines. Karstic terrains of Dinarides are generally photogenic for the analysis and interpretation of images (Pavlović et al. 2001), so that the degree of subjectivity in identifying faults is significantly lower than in the analysis of other terrains.

Table 16.4 The categorization of terrains according to the density of faults

Density of faults (km/km^2)	T_f
0–1	1
1–2	2
2–3	3
3–4	4
>4	5

Table 16.5 The categorization of karst terrains according to the value of the *KARST* index

Karstification	*KARST* index
Very low	2
Low	>2–4
Moderate	>4–6
High	>6–8
Very high	>8–10

Production of a map of fault density is greatly facilitated by the use of appropriate software for spatial analysis. The contours of the map of fault density are obtained based on the categorization shown in the Table 16.4.

16.3.3.3 Karstification Map

The *KARST* index is obtained through the two previously described factors (K_{sf} and T_f). It represents the degree of karstification which is assessed by geomorphological and tectonic analysis of satellite and aerial images. The *KARST* index is calculated by adding the factors K_{sf} and T_f (*KARST* index $= K_{sf} + T_f$).

A *KARST* map is obtained by overlapping the map of surface karstification (K_{sf} map) and the map of fault density (T_f map), using appropriate GIS software. The result is a map that shows the spatial distribution of the *KARST* index, which ranges from 2 to 10. Higher values of the *KARST* index should correspond to highly karstified terrains, while lower values should be related to less karstified terrains built of less permeable carbonate rocks. The Table 16.5 shows the categorization of karst terrains according to the value of the *KARST* index.

Box 16.3.1

Case Study—Application to the catchment area of Karuč springs
The study area is located in the southeastern part of the Dinaric Alps, in the territory of Montenegro (Fig. 16.31). That is the catchment area of Karuč springs which occupies an area of about 116 km^2. The complete catchment area is built of carbonate rocks (limestones and dolomites). The soil is very

thin or completely absent. The catchment area of Karuč springs belongs to the geomorphological unit named "Old Montenegrin karst plateau," which is characterized by a large number of surface karst landforms (sinkholes, uvalas, poljes, and dry valleys).

The recharge of karst aquifer occurs primarily by precipitation (mean annual precipitation is about 2,700 mm). Hydraulic conductivity of the carbonate aquifer is relatively high. Groundwater flows mainly through privileged directions which are marked with faults and joints. Fictive velocity of groundwater flow, measured by artificial tracers, is from 0.65 to 5.40 cm/s (Radulović 2010). Discharge of karst aquifer occurs through sublacustrine springs (vruljas) that occur along the coastal part of the Skadar Lake (Figs. 16.34 and 16.35). The mean annual discharge of Karuč springs is 7 m^3/s.

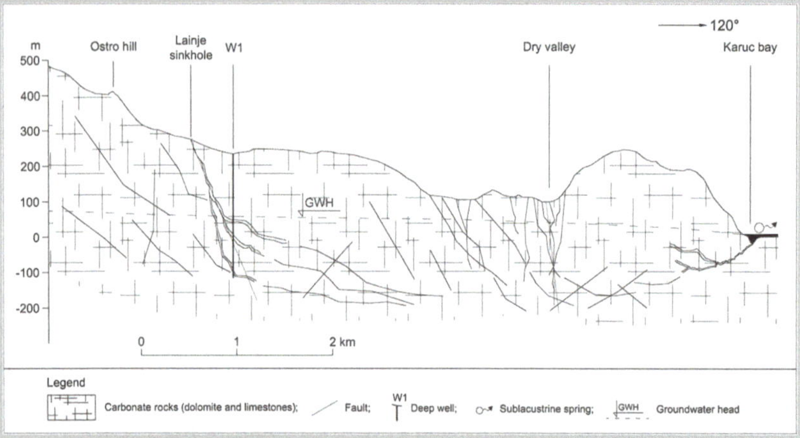

Fig. 16.34 Hydrogeological section of Ostro hill–Karuč bay (Skadar Lake)

Fig. 16.35 **a** Photo of Karuč bay captured in visible spectrum; **b** photo of Karuč bay captured in thermal infrared spectrum, from which it can be seen the temperature anomaly (*purple tones*), i.e., the main location of groundwater discharge

***K_{sf}* map of Karuč springs catchment area**

As a basis for mapping the surface karstification, orthorectified satellite images (Quick Bird and SPOT images with a resolution of about 2.5 m), and aerial images (captured by photogrammetric camera WILD RC30 in the scale of about 1:5000) were used.

One of the first steps in the process of mapping the surface karstification has been the identification and mapping of surface karst landforms, so separately the map of karren fields and separately the map of the karst depressions have been drawn (Fig. 16.36a, b). From these two maps, the K_{sf1} and K_{sf2} maps have been obtained (Fig. 16.36c, d), and by their overlapping, the K_{sf} map has been created (Fig. 16.36e, f).

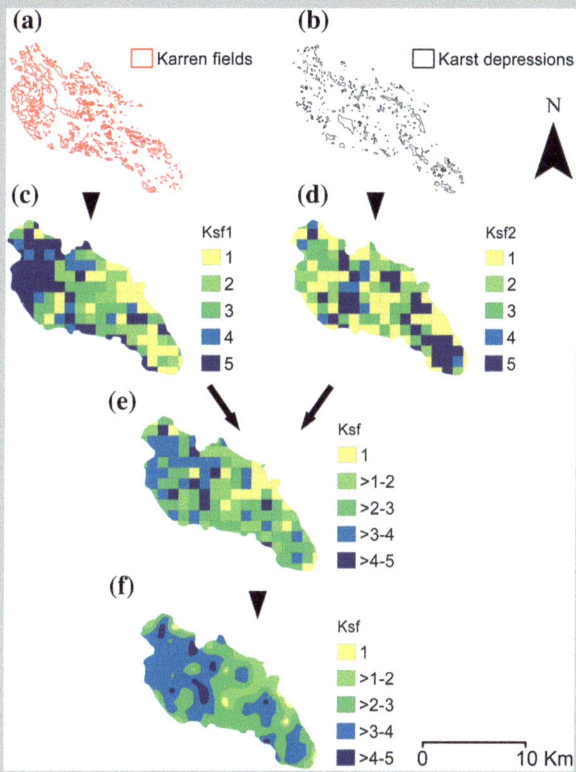

Fig. 16.36 Flow chart of procedure for derivation K_{sf} map. **a** Map of karren fields; **b** map of karst depressions; **c** K_{sf1} matrix map; **d** K_{sf2} matrix map; **e** K_{sf} matrix map; and **f** K_{sf} contour map (modified from Radulović et al. 2012). Springer and the Environmental Earth Sciences, 65(8), 2012, 2221–2230, A new approach in assessing recharge of highly karstified terrains–Montenegro case studies, Radulović MM, Stevanović Z, Radulović M, Fig. 2, Copyright 2012. With kind permission from Springer Science and Business Media

T_f map of Karuč springs catchment area

The map of fault lines has been obtained by the interpretation of Landsat ETM+ color composite image created from bands 4, 5, and 7, which is considered as one of the most suitable for geological research (Won-In and Charusiri 2003). The images have been previously processed in order to enhance their quality. Enhancing contrast of raw images has been performed by selective linear transformations of the original pixel values.

Based on the digital map of fault lines (Fig. 16.37), by applying previously presented categorization (Table 16.4) and using ArcGIS 10.1 software (Line Density tool), map of fault density of Karuč springs catchment area has been obtained (Fig. 16.38).

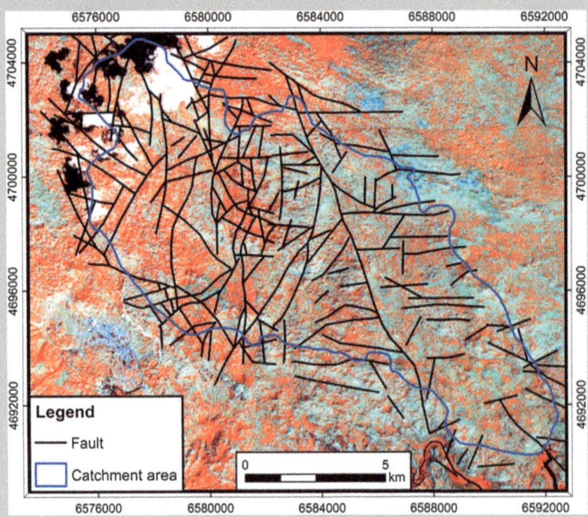

Fig. 16.37 The map of fault lines of Karuč springs catchment area (the background image is LANDSAT ETM+ color composite created by bands 4, 5, and 7)

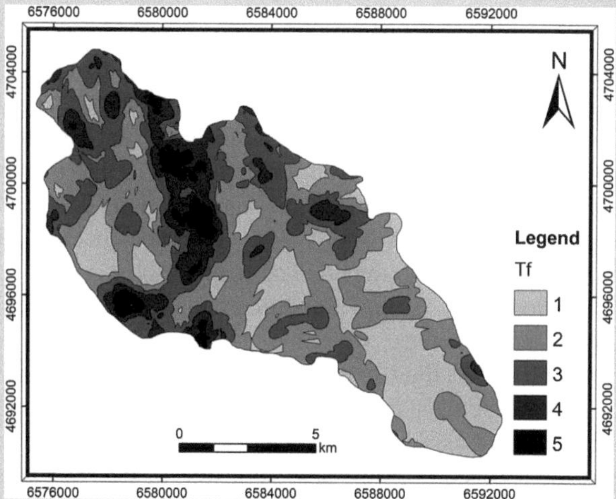

Fig. 16.38 The map of fault density of Karuč springs catchment area

The map of the KARST index

The map of the *KARST* index is created by overlapping K_{sf} and T_f maps, adding the values of these two factors and categorizing obtained contours according to the Table 16.5.

From the map (Fig. 16.39), it can be seen that the highly karstified terrains, i.e., areas with an increased *KARST* index, are represented in the northwestern part of the study area. Also, from the map, it can be seen that the terrains that are assessed as less karstified are represented in the central and southeastern parts of study area. Terrains that have been classified in the category of moderate KARST index occupy the largest surface on the catchment area of Karuč springs.

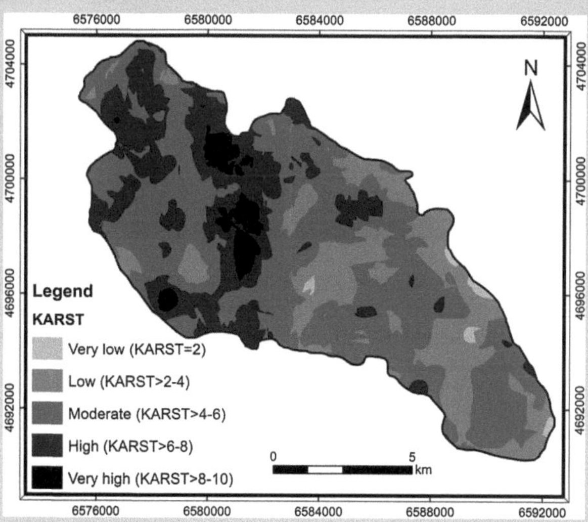

Fig. 16.39 The map of the *KARST* index of Karuč springs catchment area

16.3.4 Discussion

For the first time, the mapping approach has been applied to the catchment area of Karuč springs (Montenegro). The proposed approach allows simple mapping of karst terrains according to the degree of karstification, but it is only the assessed image that can significantly deviate from the real state. Large discrepancies could be expected especially in the case of deeper karstification. By this approach, it is only possible to identify the areas where caves could be expected, but the real axis of cave channels cannot be determined. The map of the *KARST* index more reliably indicates the degree of shallow karstification, which has been developed to the depth of a few tens of meters.

After preparation of the final map, in order to verify the assessed degree of karstification, visits of some parts of the mapped area have been done. On the most visited sites, the assessed degree of karstification by using remote sensing mainly matches the subjective assessment from the field (Fig. 16.40). Therefore, assessed degree of karstification by using remote sensing mainly matches to the field assessment of shallow karstification.

Maps obtained by the presented concept are the most desirable to create in the early stages of research. Thus, the research can focus on potential areas with increased karstification, which would be checked by applying geophysical methods and exploratory drilling on the field.

The presented approach has its advantages and disadvantages. Some of which are shown in the Table 16.6.

Fig. 16.40 **a** Terrains of the southeastern part of the mapped area which have been assessed (by remote sensing) as low to moderate karstified; **b** terrains of northwestern part of the mapped area which have generally been assessed (by remote sensing) as highly karstified terrains

Table 16.6 The advantages and disadvantages of the presented approach

Disadvantages	Advantages
There may be certain errors (interpolation errors, errors in analysis and interpretation of satellite images)	Maps are produced without major investments of funds
Obtained maps, due to the limited details, are not especially reliable for the analysis in the local scale	The result is spatial distribution of karstification that is not possible to obtain by applying most of the other methods
Maps are not especially reliable for the assessing the karstification of deeper zones, but are mainly intended to assess shallow karstification	The results are also obtained for inaccessible karst areas, where it is not possible to perform the conventional on-site investigations
It is not possible to obtain quality maps for the areas covered by vegetation	The maps are obtained based on easily available data (aerial and satellite images)

16.3.5 Conclusion

Preparation of the map of karstification by using remote sensing represents a relatively easy way to delineate highly karstified zones in the regional scale.

The map can be used as an additional base for solving some of the engineering problems in karst. For example, it could be used for assessing suitability of the terrain for the construction of reservoirs (in terms of permeability), for identifying potential sites for the abstraction of groundwater, for identifying sites for artificial aquifer recharge, as one of basis for the modeling of groundwater flow in the karst, as one of the input maps for mapping groundwater vulnerability, for a first assessment of suitability of the terrain for the construction of underground facilities, and as a basis for planning research in order for rationalization.

The approach needs to be tested on a greater number of karst areas in order to achieve additional improvements. The concept of mapping is adapted for highly karstified terrains, so it is important in the following stages to highlight the possible shortcomings and take eventual modification for the terrains with lower degree of karstification.

In the future, it is necessary to devote additional attention to this issue, considering the importance of the assessment of karstification for more complete research and protection of karst aquifers.

16.4 Combat Mixture of Groundwater and Surface Waters in Karst

Zoran Stevanović

Centre for Karst Hydrogeology, Department of Hydrogeology, Faculty of Mining and Geology, University of Belgrade, Belgrade, Serbia

16.4.1 Introduction

The mixture of fresh groundwater and surface water is a frequent problem in karst, and most problematic for the sustainable use of fresh groundwater. This problem results mostly from the high permeability and low attenuation capacity of karst aquifers, particularly those formed in open (unconfined) structures. The existence of such a mixture makes it complicated to distinguish and separate fresh groundwater from surface flows or water reservoirs such as a lake or sea. Even more problematic is the restoration of the natural quality of groundwater polluted by surface water flows or seepage, as discussed under the topic of remediation in Sect. 17.4 of this book.

Concerning the kinds of surface water and their negative impact on adjacent aquifers, surface water can generally be distinguished as follows:

1. Waste surface waters (already contaminated and untreated or partially treated, or leakage from various sources such as landfills, industry, communal systems; pollutant liquids may also be flowing through canals or be being stored in ponds in direct or indirect contact with karst),
2. Sea and brackish waters (saline to various degrees),
3. Lake waters (potentially polluted),
4. River waters (potentially polluted).

Regardless of which is present, it is always better to prevent their contact with karstic waters. Interventions may be very different and should be adapted to the concrete circumstances. Some of the remedial measures that may be applied in karst are as follows:

- making an impermeable seal along canal bottoms or riverbeds,
- building small dams or weirs,
- constructing grouting curtains,
- diverting surface waters to other directions and catchments,
- plunging ponors,
- making so-called reactive barriers by pumping additional freshwater into aquifer.

How to tap fresh groundwater in coastal karst and avoid the intrusion of surface, particularly saline, waters is the practical problem on which this discussion is further focused.

16.4.2 Historical Experience

Tapping coastal aquifers and distinguishing freshwater from seawaters are regularly a very difficult task. The Phoenicians constructed special structures such as collective boats with iron funnels and leather pipes to force freshwaters to flow upwards to the surface to supply their citizens. Freshwater was also tapped by tubes or specially constructed bells driven down spring outlets or amphorae turned upside down to catch the flow (Bakalowicz et al. 2003a).

Many cities in the Mediterranean basin, one of the World's cradles of civilizations, were built near large sources of freshwater (Stevanović and Eftimi 2010; Stevanović 2010a). An important role in this case was played by the position of impermeable barriers which dictate the drainage point to be below sea level or above it. Therefore, where this barrier is well above sea level, the chances to utilize freshwaters are increased and this was well known by the skilled Greeks and Romans who defined the position of many famous historical towns. For instance, on the eastern Adriatic shoreline (Dalmatian coast, Croatia), the town of Split (lat. *Spalatum*) was built in close proximity to the major karstic spring in the Jadro area which is around 30 m above sea level (see Fig. 2.4), and the town of Dubrovnik (lat. *Ragusa*) was built near the huge spring of Ombla which discharged at only 3 m a.s.l. Some other springs such as the Bistrica group of springs in Albania (including the famous Syri Kalter—Blue Eye, see Fig. 3.41) are envisaged as one of the potential sources even for transboundary distribution to southern Italy (Puglia Region) due to their huge minimal discharge and secure discharge points, some 50 m a.s.l. (Eftimi 2003). In the Adriatic basin, the main karstic aquifer system is from the Mesozoic age while the main barrier is represented by younger Eocene flysch. The deep and regional flow from inland is thus adapted to the regional erosion base as is the sea level (Fig. 16.41).

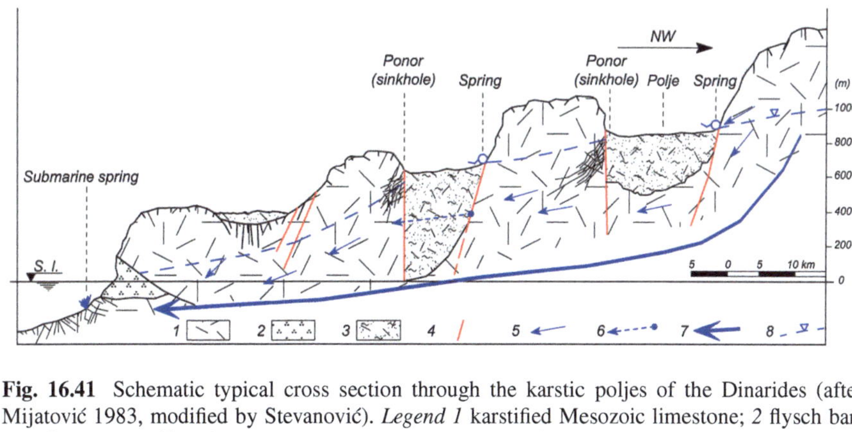

Fig. 16.41 Schematic typical cross section through the karstic poljes of the Dinarides (after Mijatović 1983, modified by Stevanović). *Legend 1* karstified Mesozoic limestone; *2* flysch barrier; *3* porous aquifer of polje; *4* fault; *5* groundwater flow; *6* direction of flow around the barrier; *7* regional flow; *8* groundwater table

During the geological history in Neogene and the Messinian salinity crisis, the connection between the Mediterranean Sea and the Atlantic Ocean was closed and the sea level in the Mediterranean basin declined by several hundred meters. As a result of intensive karstification and first a specific arid and then a glacial climate, many springs along the paleocoast were also opened at a lower position but today, with the sea level significantly raised, they are functioning as submerged springs. Therefore, the hydrogeological evolution of the Mediterranean littoral karst was decisively affected by the rise in level of about 100 m during the last interglacial state (Mijatović 2007).

In contrast to drainage points above sea level, the submarine springs (vruljas) occur in places where the impermeable barrier is missing or is much below the sea. Bakalowicz et al. (2003a), Fleury et al. (2007) and Dörfliger et al. (2010) emphasized that the location of many of these springs in the Mediterranean basin is relatively well known and documented. An overview of the largest submarine springs in the Mediterranean basin, but also in other regions with large submarine discharge (Florida, Caribbean basin, Black Sea, Persian Gulf, and Pacific islands), has been provided in an unpublished report by M. Bakalowicz (2014, Karst submarine and related coastal springs of the world. Report on WOKAM project, unpublished) for the ongoing World Karst Map project (WOKAM, in press).

16.4.3 Hydraulic Mechanism and Methods to Identify Submerged Flows

The classical Ghyben–Herzberg formula which defines the relationship and interface between fresh and salty water is generally valid for karst aquifer (Stringfield and LeGrand 1971; Bonacci 1987; Arfib et al. 2005), but its application, as in the case of Darcy law, should be used with caution (Fig. 16.42) because direct contact of two fluids established in saturated karstic channels and cavities and transitional zone of brackish water could be very much extended (Fig. 16.43).

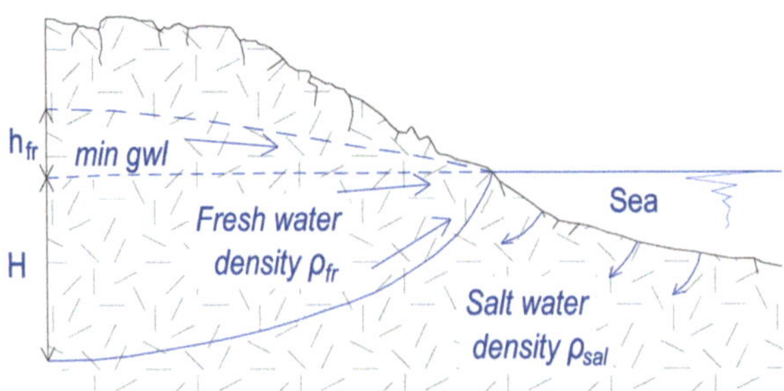

Fig. 16.42 Sketch for Ghyben–Herzberg hydraulic relationship

Fig. 16.43 Open structure of highly karstified rocks in direct contact with the sea. Porto Vromi (Zakhintos, Greece)

The hydraulic relationships are driven by the difference in water density between fresh and seawater, but the more important factor in this relation is the hydraulic head in the aquifer system (h_{sr}). The freshwater is of lower density (ρ_{fr}) and overlies saltwater (ρ_{sal}) as a lens. Consequently, when head (lens' thickness) is larger, the chances for mixture are reducing. Although the regime of karst aquifer regularly varies throughout the year, the same aquifer may discharge fresh or brackish, or even totally saline water and spring discharge is activated permanently or seasonally.

The Ghyben–Herzberg formula for depth of fresh—saline water interface can be calculated as follows:

$$H = \frac{\rho_{fr}}{\rho_{sal} - \rho_{fr}} h_{fr} \tag{16.1}$$

where,

H depth to interface fresh—saline water
ρ_{fr} density of freshwater
ρ_{sal} density of saltwater
h_{fr} freshwater head above sea (and minimal groundwater level)

Although the density of freshwater is 1.0 and the density of saltwater is 1.025, under hydrostatic equilibrium, the depth to the interface is 40 times the height of the water level above the sea. Consequently, one meter of drawdown of freshwater results in an intrusion of saltwater by 40 m and more. And more intense pumping will cause fast salinization of the aquifer.

The problem with the Ghyben–Herzberg hydraulic law is the assumption that there is no flow under hydrostatic equilibrium and that the head of freshwater

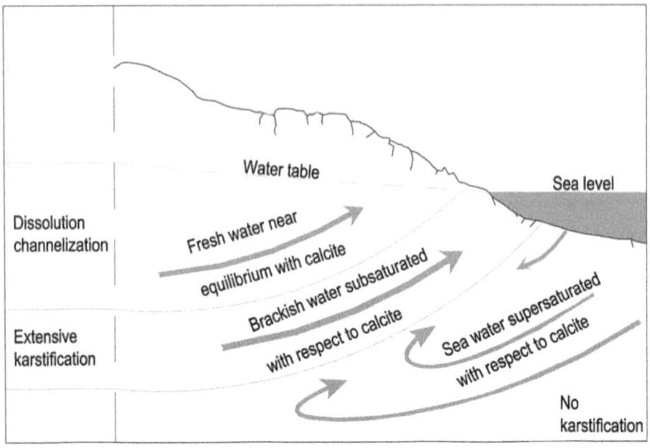

Fig. 16.44 Schematic model of different salinity zones in karst of Yucatan (After Beck, 1992)

comes to zero at the point of contact with the sea. Because the movement and discharge of fresh groundwater have to be taken into account and the interface is not sharp but transitional, many new formulae have been developed to overcome these deficiencies since the law was imposed in the early 1900s (Arfib et al. 2005).

The dynamic of coastal aquifer discharge and the problem of interface salty–freshwaters have been studied for many years and by many authors (Zektser et al. 1973; Mijatović 1983; Drogue and Bidaux 1986; Drogue 1996; Bakalowicz et al. 2003a; Fleury et al. 2007). A simple scheme is also presented by Back (1992) who studied the famous Yucatan karstic aquifer in Mexico (Fig. 16.44).

Box 16.4.1

Yucatan Peninsula (Mexico)—Open and vulnerable karst system

The Yucatan Peninsula aquifer, a huge open system at low altitude (average 30 m a.s.l.), consists of highly permeable limestones and rapid infiltration of rainfall but also discharges into the sea. However, when in some areas such as the city of Merida in the NW or the tourist center of Cancun, pumping of groundwater becomes intensive, the attained drawdown results in the intrusion of salty water inland as well as its movement upward. Back (1992) identified large hydraulic conductivity of karst and lack of freshwater head as reasons why a saltwater body underlies a freshwater body and is extended over almost one third of the peninsula's surface. The mixing zone between fresh and seawater is marked by brackish water whose salinity increases downward. Back (1992) states that a geochemical zone with increased salinity also stimulated the karstification process and this is why many cenotes such as karstic inlets (shaft-like sinkholes with extended saturated deep and long canals), lagoons, coves, and crescent bays are well developed particularly along the east cost of Yucatan (Fig. 16.45).

Fig. 16.45 Cenote in the middle of Yucatan, near Valladolid

Many cases of deterioration of littoral aquifers and deep intrusion of salty water have been noticed worldwide. Where there is absolutely no problem of salt intrusion, there is almost no littoral karst, at least during droughts or low water seasons. Particularly, problematic might be the situation in the islands where freshwater bodies are completely surrounded by seawaters.

Different methods are used in hydrogeological practice to identify the presence of deep groundwater flows in coastal zones in order to capture them before they reach the shoreline or start to mix with salty waters.

The reconnaissance or initial stage of survey of coastal aquifers includes remote sensing methods. A satellite image represents a digital signature of reflected electromagnetic energy. Images from a multispectral scanner include bands from different spectral areas: blue, red, green, infrared, etc. Analysis of the lithology, tectonic pattern, vegetation index, and moisture vegetation index can also be supported by careful observation of the shoreline and possible changes in the color or intensity of the reflected signal in surface water near the banks.

Besides optical locating of the sublacustrine springs, infrared thermography techniques, which are based on the detection of temperature anomalies, could also be applied. The temperature anomalies usually appear at locations where there are discharges of colder groundwater below the level of the warmer sea/lake water, i.e., at locations of submerged springs. But, the reverse situation of warmer groundwater discharging into colder seawater is also possible in some regions particularly in the Northern Hemisphere. Examples of identification of sublacustrine springs along the shoreline of Skadar Lake in Montenegro are explained in Sect. 16.3. The same section also presents the use of terrestrial recording by thermal infrared camera which may help in identifying submerged springs (see Fig. 16.35).

Different geophysical methods, including geoelectrical resistivity, spontaneous potential, mise à la masse, electrical tomography, very low frequency, microseismics, gravimetry, and geophysical logging, can be applied for defining groundwater flows (Arandjelović 1976). One very specialized method is induced polarization which was successfully applied for the first time in the Adriatic islands in the 1970s. This method helps to distinguish further caverns filled with water from caverns filled with clay, as both provide similar values of resistivity when an electrical resistivity method is applied.

Along with classical tracing tests which enable visualization of colored waters, different types of robots or underwater vehicles are constructed for identification and sampling of water. For example, the autonomous underwater vehicle, Taipan 1 and 2, has been developed in LIRMM laboratory (France) for measurements of conductivity and temperature values of water at certain depths as well as GPS positioning of sampling sites (Lapierre et al. 2008). It is a torpedo-shaped vehicle with a single propeller (very similar to a current meter), and relatively small in size (1.80 m length, 0.2 m diameter, 60 kg).

Thanks to this vehicle, specialized divers can enter saturated karst channels of large dimensions and collect valuable information on their extension as well as do water sampling (see Box 3.13). Mapping the karstic deep channels may significantly support engineering projects on aquifer regulation.

Finally, the exploratory drilling, the subsequent pumping test, and the hydrochemistry analyses provide the most precise data on groundwater flow, optimal discharge, and water quality which are essential elements when evaluating the relationship between fresh and salty water.

16.4.4 Sustainable Tapping and Use of Fresh Karstic Waters

According to Mijatović (1983), one of the main tasks of hydrogeological research is to distinguish the following:

- aquifers opened toward the sea,
- aquifers having an incomplete barrier (resulting in the presence of a wide range of salinity from brackish waters), and
- aquifers discharging over complete barriers.

The last are "safer" from seawater intrusion and provide the most convenient ambience for tapping and engineering regulation of aquifer. As already stated, in the case of the Mediterranean coast, the freshwater supply greatly depends on the barrier extension and local position.

During the last half century, many successful projects have been implemented in the Adriatic basin which solved the problem of increased water demands and pressures on already vulnerable coastal aquifers (Alfirević 1963; Mijatović 1984; Komatina 1984; Biondić and Goatti 1984; Milanović 2000c; Eftimi 2003; Biondić and Biondić 2003; Al Charideh 2007). Some of these are explained in *Hydrogeology of karstic terrains, case histories*, which is edited by Burger and Dubertret, and which is Volume 1 of the IAH series International Contributions to Hydrogeology (1984).

Similarly, the book *Groundwater management of coastal karstic aquifer* resulted from the project COST action 621 (2005) and contains discussions and results from 62 selected observation and studied sites from 8 Mediterranean countries. The aquifers are displayed in the four groups, class A (40 % of the total studied aquifers) being most vulnerable to salt intrusion due to unconfined character and high permeability. Classes A and B comprise almost 80 % of all studied aquifers "where salt intrusion is a real problem and over-exploitation is the common way to manage the aquifer."

Many attempts have been made worldwide to find an appropriate solution to tap freshwater from submerged springs. Some of them are based on previously described principles first applied by the Phoenicians. For instance, Bakalowicz et al. (2003b) described the function and result of testing of a tapping device constructed by the French company Nymphea water in the 1970s and improved in 2003. The secure cylinder enclosing the spring was designed to follow the seabed contour. The chimney is topped with a half sphere and pipe conveying water upward. The system relies on differences in density between spring water and seawater, and freshwaters are driven to the surface by the denser, more brackish water below.

Although any attempt to capture submerged freshwater flow is difficult and the final result is uncertain, the most appropriate solution is to tap fresh groundwater flow as far as possible from the sea and direct discharge zone. In many cases, the horizontal galleries placed directly into the conduit which enables gravity flow are a better option than pumping from drilled wells or shafts. But whatever intake is applied, systematic monitoring of discharges or pumping rate versus water salinity should ensure sustainable aquifer development and prevent the effects of water salinization or quality deterioration (Fig. 16.46).

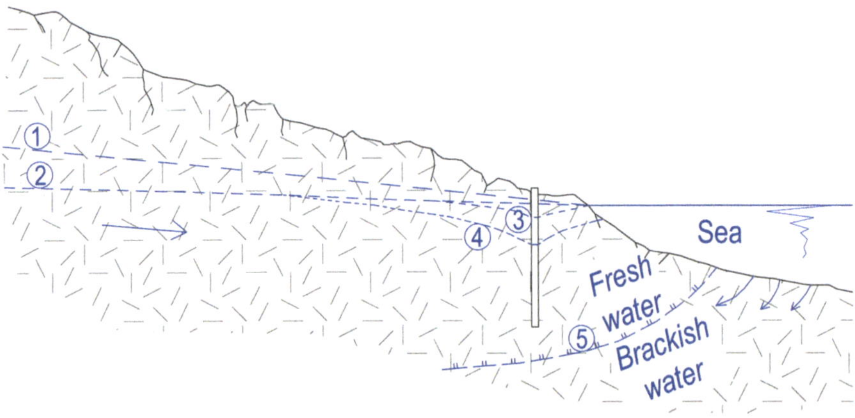

Fig. 16.46 Typical cross section in littoral karstic aquifer where pumping rate from drilled well has to be accommodated to actual hydraulic head. *Legend 1* Static water table in high water period, large hydraulic head prevent mixture of fresh with salty water; *2* Static water table in low water period, declined hydraulic head does not prevent mixture of fresh with salty water and a counterbalance is established; *3* A small pumping rate and radius of well are still feasible and slight increase in mineralization of tapped water is only possible. *4* A large pumping rate disturbs interface of fresh and salty water and intrusion of the latter is very probable. *5* Approximated zone delineates brackish and farther away salty water body from freshwater body. It is movable in accordance with status of hydraulic head (pressure in aquifer system)

Box 16.4.2

Ain Zeina, Lybia—saline water intrusion due to over-pumping

In order to improve the water supply of Benghazi (Libya), a proposal to tap Ain Zeina brackish spring water far from the coastal zone was provided by experts of the Geological Survey of Serbia (B. Mijatović, 2006, Geological and hydrogeological framework of integrated water resources management in Libya. Report. Geological Survey of Serbia, Belgrade, unpublished). Research undertaken in the Jebel Akhdar area resulted in 10 exploratory wells drilled in well-karstified numulitic Eocene limestones in the most prosperous zone of Sidi Mansur which is 11 km from the coast (Fig. 16.47). The yield of these wells ranged from just a few l/s to 20 l/s. More importantly, the Cl ion content was regularly lower than 400 ppm.

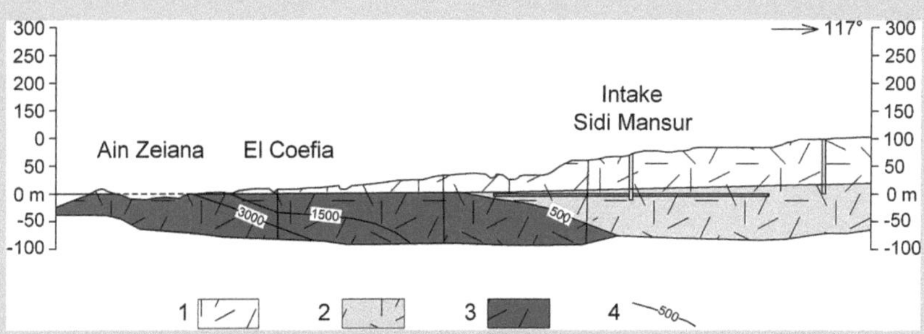

Fig. 16.47 Hydrogeological cross section of the Sidi Mansur (Jebel Akhdar)—Ain Zeina (modified from B. Mijatović, 2006, Geological and hydrogeological framework of integrated water resources management in Libya. Report. Geological Survey of Serbia, Belgrade, unpublished). *Legend 1* Unsaturated Eocene limestones; *2* Karstic Eocene aquifer with freshwater; *3* Karstic Eocene aquifer with brackish water; *4* Salinity iso lines, values in ppm

Based on these results, one vertical shaft with a long horizontal galley was designed and constructed (Fig. 16.48). It was the first horizontal gallery executed in Africa in modern times (B. Mijatović, 2006, Geological and hydrogeological framework of integrated water resources management in Libya. Report. Geological Survey of Serbia, Belgrade, unpublished).[1] The vertical shaft is excavated to a depth of 82 m. The bottom is at 1.9 m a.s.l. while the average groundwater table is at 3.5 m a.s.l. From the shaft bottom, the 500-m-long gallery is excavated perpendicular to the groundwater flow (2 × 250 m from the center of shaft). While the Ain Zeina spring discharged an average yield of 1,000 l/s of brackish water, the Sidi Mansur tapping structure enabled the pumping of 400 l/s (with a Cl content of 300–350 ppm).

Unfortunately, with no established proper monitoring and control over the extensive pumping from the shaft, the salinity front intruded further inland and the groundwater became brackish even in the Sidi Mansur. After 15 years of operation, the average groundwater salinity reached a value of 1,500 ppm.

[1] Collecting galleries were used for the water supply of ancient Cyrene (Cyrenaica) in Roman times.

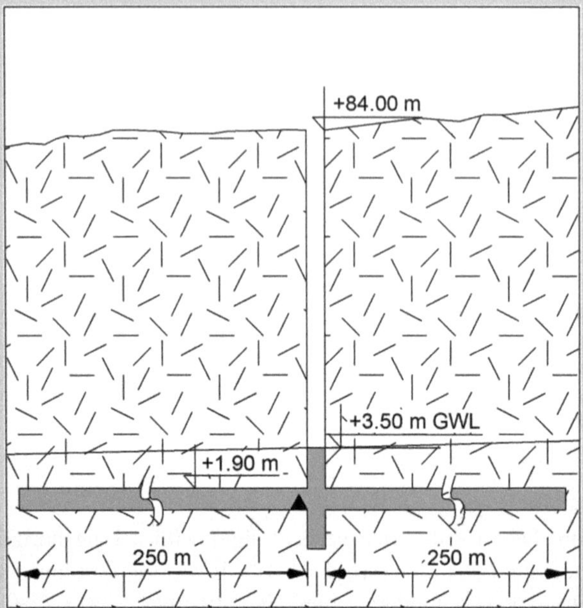

Fig. 16.48 Sidi Mansur tapping structure—vertical shaft and gallery (redrawn based on personal explanation of B. Mijatović)

Box 16.4.3

Port-Miou (France)—Regulation of brackish springs by underground dam

One of the world's most known projects in coastal karst is the intake and underground dam constructed for the city of Marseille with the aim to distinguish fresh and seawater flows. The two large springs, Port-Miou and Bestouan, are in fact underground drowned karst rivers issuing from the Calanques range of hills on the coastline between Marseilles and Cassis. They are discharging much more groundwater (3 m^3/s) than calculations from conventional formulae allow for and an apparent surface catchment of maximal 150 km^2. It has been assessed that only infiltration from a close alluvial plain fed by water from irrigation canals together with inflows from more distant limestone formations can cause such a huge discharge (Potié et al. 2005).

In order to find an appropriate response to the fast-growing water demands of the city of Marseilles, in 1964 two French institutions, the Geological Survey of France Bureau de Recherches Géologiques et Minières (BRGM) and a local water company from Marseille (SEM), conducted an extensive geological, geophysical, topographic survey including tracing tests and diving exploration (described also in Box 15.4.4 of Sect. 15.4). The survey of Port-Miou resulted in the construction of the first underground dam in 1972 and a complete closure of the gallery with a second underground dam (Potié et al. 2005). The objective of the phase-1 dam was to prevent saltwater intrusion without modifying the aquifer's pressure. The principle of a "chicane" dam was adopted. This involved building a pair of dams 530 m from the sea outlet: an upstream chicane dam on the gallery floor and a downstream chicane dam on the gallery roof (Fig. 16.49).

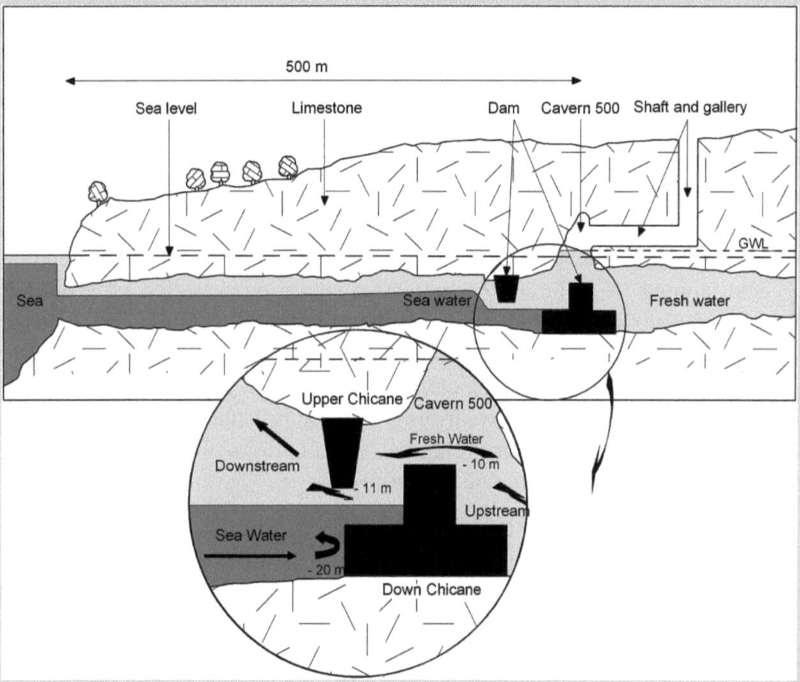

Fig. 16.49 Port Miou underground dams (after Potié 2005). The inverted dam forces freshwater downwards, while the dam on the gallery floor prevents saltwater from penetrating farther upstream into the exploitation gallery

Construction of chicane dams did not entirely stop salinization of groundwater, but it had a significant positive effect: Salinity of the water dropped from 4–5 to 2–3 g/l at the surface, and upstream of the dam at a depth of 20 m below sea level from 18–20 to 3–4 g/l. However, complete prevention of salinization was not possible.

There are also many sublacustrine springs or spring/seepage registered along the riverbeds which are not easy to tap and separate from surface waters. An example of very intense underground drainage along lake shores is the Skadar/Shköder Lake, the largest in the Balkans, shared between Montenegro and Albania. Only on the Montenegrian shoreline has the average groundwater flux from the submerged springs been estimated at more than 10 m^3/s. From this, one of the largest reservoirs of freshwaters in all of Europe, one recent project successfully conducted, has ensured the water supply of all Montenegrian ports and touristic cities on the coast (Stevanović 2010b). Experience collected from this project (Box 16.4.4) and a sophisticated solution to distinguishing lake and groundwater waters can also be accordingly applied at many other submerged springs along the lakes and Adriatic coast.

Box 16.4.4

Bolje sestre intake (Montegero)—Separated lake water and groundwater
The new Montenegro Regional Water Supply System (RWSS) covers the water supply of all municipalities along the coast. The RWSS is designed on a maximal capacity of 1.5 m^3/s; the total pipeline length is 140 km, of which the continental part from Skadar Lake to the coast comprises 33 km. This new system, functioning since 2010, has solved the acute problem of water shortage due to the small discharge of previously tapped coastal springs in summer months when the tourist season dictates maximal water demands.

The tapping of Bolje sestre sublacustrine spring was first proposed by Radulović (2000b) because of its shallow depths of submerged discharge points and its assessed sufficient and stable flow throughout the year. After conducting complex hydrological, hydrochemical, hydrogeological surveys, tracing tests, exploratory drilling, and identification of the position of the main groundwater transient channel, three options of tapping freshwater flow were evaluated (Stevanović 2010b):

1. by enclosing the area of the source discharge, and using a small coffer dam, the discharging water can be captured and pumped into the scheme;
2. custom-built metallic pipes can be jacked into the ground of the lake, and anchored by underwater cementation, creating intake shafts for pumping to the scheme,
3. a shaft of large diameter located on land, approx 5–6 m above the lake level, cased off with steel casing, enclosed in a suitable pump—control house.

Although option 3 guarantees the capture of groundwater flow before it reaches the lake, the problem of possible large drawdown and backflow of lake water, along with the energy required for pumping, oriented further study toward solution 1.

Following the request to avoid a mixture of ground and surface waters, the concrete elliptical structure—coffer dam has been designed and constructed. It covers a surface of some 300 m^2. Considering the importance and influence of lake water fluctuations on the intake, the special removable spillway section (rubber gate) has been designed and installed on the dam crest. The idea is to allow the non-captured spring water to overflow easily into the lake and to manipulate the height of the water level inside the coffer dam when the lake level rises (annual amplitude can be 5–6 m). The two automatic and remotely operated compressors are used to direct the air into a rubber tube and activate the gate when necessary (Fig. 16.50).

Fig. 16.50 Intake Bolje sestre and three stages of its construction

To date, the system is functioning without interruptions and failure. The minimal discharge of 2.3 m^3/s is far above actual demands. More importantly, the design and implementation of this excellent intake solution are between just 1 and 2 percent of the total investment in this complex water supply project. However, the question is, be proud of this fact or consider it an underestimation of intellectual property?

References

References to Section 16.1

Bruce DA (2003) Sealing of massive water inflows through karst by grouting: principles and practice. In: Back B (ed) Sinkholes and the engineering and environmental impacts of karst. ASCE Geotech. Spec. publ. no. 122

Fazeli MA (2005) Construction of grout curtain in karstic environment case study: Salman Farsi dam. In: Stevanović Z, Milanović P (eds) Water resources and environmental problems in karst, Proceedings of international conference KARST 2005. University of Belgrade, Institute of Hydrogeology, Belgrade, pp 659–666

Fazeli MA (2007) Construction of grout curtain in karstic environment case study: Salman Farsi Dam. Environ Geol 51(5):791–796

Ford D, Williams P (2007a) Karst hydrogeology and geomorphology. Willey, Chichester

Hidrotehnika (2012), Belgrade. Web site: http://www.hidrotehnika.rs/alzir/brana-ourkiss/

Milanović P (2000a) Geological engineering in karst. Monograph, Zebra Publ. Ltd, Belgrade

Milanović P, Stevanović Z, Beličević V (2007) Barrage Hammam Grouz, Saf Saf, Ourkiss. Raport d'Expertise, ANBT, Algeria

Milanović S, Stevanović Z, Jemcov I (2010) Water losses risk assessment: an example from Carpathian karst. Environ Earth Sci 60(4):817–827

Sahuquillo A (1985) Spanish experience in karst water resources, vol 161. IAHS Publication, Ankara

Stevanović Z, Milanović S, Ristić V (2010) Supportive methods for assessing effective porosity and regulating karst aquifers. Acta Carsologica 39(2):313–329

Stucky-Electrowatt Joint Venture (1996–2004) Salman Farsi Dam—Reports on the design of the grout curtain. Zürich

Therond R (1972) Recherche sur l'etancheite des lacs de barrage en pays karstique. Eyrolles, Paris

Vlahović M (2005) Surface reservoir in the karst of Nikšićko polje and problems of their maintenance. In: Stevanović Z, Milanović P (eds) Water resources and environmental problems in karst, Proceedings of international conference KARST 2005. University of Belgrade, Institute of Hydrogeology, Belgrade, pp 671–678

Zogović D (1980) Some methodological aspects of hydrogeological analysis related to dam and reservoir construction in karst. In: Proceedings of 6th yugoslav conference for hydrogeology and engineering-geology. Portorož, pp 283–291

References to Section 16.2

Abramov KS, Skirgelo BO (1968) Osušenie šahtnih ikarernih polej. Nedra, Moscow, p 255

Alliquander E (1982) Experience and ideas of development on the control of mining under karstic water hazard. In: Proceedings of the first international mine water congress (IMWA), Vol. B. Budapest, pp 7–18

Andreychuk VN (1996) Bereznikovsky Sink. Urals Branch of Russian Academy of Sciences, Perm, p 133

Bankovskaja MV, Krasavin PA (2004) Protection of underground waters against pollution at liquidation of mines in karst region of Kizel Coal Pool. In: Proceedings of the international symposium karstology-XXI century: theoretical and practical significance. Perm, pp 303–306

Bell KG (1963) Uranium in carbonate rocks. Shorter Contributions to General Geology. Geological Survey. Professional Paper 474-A, A1–A20

Bilopavlović V (1988) Utvrđivanje pravca kretanja podzemnih voda u području jame Trobukva-Studena vrela (Definition of groundwater flow direction in vicinity of Trobukva-Studena pithole; in Serbian). In: Proceedings of the VI Yugoslav symposium on bauxite research. Herceg Novi, pp 161–165

Bochenska T, Limisiewicz P, Poprawski L (1995) Long-term changes in the shallow water table in mining area: the Lubin-Glogow copper region, southwestern Poland. Hydrogeol J 3(3):41–52

Carta M, Ghiani M, Rossi G (1982) Uticaj rudarske industrije na okolinu: mere zaštite sa ciljem sprečavanja ili otklanjanja štete (Impact of mining industry on environment: protection measures—prevention and sanitation). In: Proceedings of the 11th world mining congress, Vol. B. Belgrade, pp 452–466

Dimitrijević DM (2013) Oksidacija pirita i kisele rudničke vode (Pyrite oxidation and acidic mining waters; in Serbian). University of Belgrade—Technical Faculty in Bor, p 187

Dragišić V, Stevanović Z (1984) O mogućnostima podzemnog isticanja dela karstnih izdanskih voda po obodu Timočke eruptivne oblasti (Possibility of subterranean inflow along the edge of Timok Eruptive Zone; in Serbian). In: Proceedings of the VIII Yugoslav Symposium on Hydrogeology and Engineering-Geology, Vol 1. Budva, pp 95–101

Dragišić V (1992) Hidrogeologija ležišta bakra istočne Srbije (Hydrogeology of copper deposits of eastern Serbia; in Serbian). University of Belgrade—Faculty of Mining and Geology, Belgrade

Dragišić V (1994) Hidrohemijske karakteristike ležišta bakra "Veliki Krivelj" (Hydrochemical characteristics of copper mine "Veliki Krivelj"; in Serbian). In: Proceedings of the XXVI October conference of miners and metallurgists. Donji Milanovac, pp 42–45

Dublyanskiy VN, Nazarova VU (2004) Karst rocks as mineral resources. In: Proceedings of the international symposium karstology-XXI century: theoretical and practical significance. Perm, pp 279–284

Ershov VV, Eremin VU, Popova BG, Tihomirov ME (1989) Geologija i razvedka mestoroždenij poleznih iskopaemih (Geology and exploration of mineral resource deposits; in Russian). Nedra, Moscow, p 399

Ford D, Williams P (2007b) Karst hydrogeology and geomorphology. Wiley, Chichester

Gongyu L, Wanfang Z (2006) Impact of karst water on coal mining in North China. Environ Geol 49:449–457

James R (1997) Mine drainage and water resources. Bull Peak District Min Hist Soc 13(4):74–80

Johnson KS (2004) Evaporite-Karst processes, distribution, and problems in the United States. In: Proceedings of the international symposium karstology-XXI century: theoretical and practical significance. Perm, pp 14–19

Kleiman BD (1982) Mine drainage in karst. In: Proceedings of the first international mine water congress (IMWA), Vol B. Budapest, pp 501–516

Korać M, Kecojević V (1988) Rezultati analize podataka geomehaničkih ispitivanja radne sredine na primjeru ležišta „Štitovo" (Nikšić) (Results of geomechanical analyses of working environment on example of "Štitovo" mine (Nikšić); in Serbian). In: Proceedings of the VI Yugoslav symposium on bauxite research. Herceg Novi, pp 191–206

Korotkevich GV (1970) Soljanoj karst (Salty karst; in Russian). Nedra, Moscow, p 255

Lolcama J (2005) Case studies of massive flow conduits in karst limestone. 10th multidisciplinary conference on sinkholes and the engineering and environmental impacts of karst, Proceedings sinkholes engineering and environmental impacts on karst Texas, United States, pp 57–65

Lovell HL (1983) Coal mine drainage in the United States—an overview. Water Sci Tech 15:1–25

Lottermoser B (2007) Mine wastes characterization, treatment, environmental impact, 2nd edn. Springer, New York

Luković M (1939) Novija promatranja u rudniku Vrdniku u vezi sa pojavom termalne vode u potkopima (Novelties on Vrdnik Mine in connection with thermal water occurrences in the galleries; in Serbian). Compte rendu des société Serbe de géologie 1938, Belgrade, pp 13–16

Lunev BS, Naumova BO, Naumov VA (2004) Karst mineral resources. In: Proceedings of the international symposium karstology-XXI century: theoretical and practical significance. Perm, pp 23–29

Maksimovich GN (2004) Ways of the decision of environmental problems connected with development of the karst in coal-mining areas. In: Proceedings of the international symposium karstology-XXI century: theoretical and practical significance. Perm, pp 306–311

McKee DM, Hannon PJ (1985) The hydrogeological environment at the Gays River mine. Int J Mine Water 4:13–34

Miladinović B, Dragišić V (1998) Prognoza rudničkih voda u rudarskim radovima nekih ležišta uglja Senjsko-resavskog basena (Forecasting water inflow in mine works of some Senj-Resava coal basin; in Serbian), vol 8. Podzemni radovi (Ground Works) University of Belgrade—Faculty of Mining and Geology, Belgrade, pp 53–60

Miladinović B (2000) Hydrogeology of REMBAS. Zadužbina Andrejević p 110

Milanović P (2000b) Geological engineering in karst. Zebra Publishing Ltd., Belgrade

Motyka J, Czop M (2010) Influence of karst phenomena on water inflow to Zn-Pb mines in the Olkusz District (S Poland). In: Andreo B et al (eds) Advances in research in karst media. Springer, Berlin, Heidelberg, pp 449–454

Mramor J (1984) Problematika odvodnjavanja triade v rudniku lignita Velenje (Problems in dewatering of Velenje mine field; in Slovenian). In: Proceedings of the VIII Yugoslav symposium on hydrogeology and engineering-geology, Vol 1. Budva, pp 417–425

Nikolić P, Dimitrijević D (1990) Coal of Yugoslavia. Pronalazaštvo, Belgrade, p 462

Plotnikov NI, Roginec II (1987) Gidrogeologija rudnih mestoroždenij. Nedra, Moscow, p 1987

Plotnikov NI (1989) Tehnogennie izmenenia gidrogeologičeskih uslovij (Technogenetic modifications of hydrogeological conditions; in Russian). Nedra, Moscow, p 268

Radulović M, Popović Z, Damjanović M (1987) Osvrt na hidrogeološke uslove ovodnjenosti ugljonosnog basena Maoče kod Pljevalja (On hydrogeological conditions of watering of coal basin Maoče near Pljevlja; in Serbian). In: Proceedings of the IX Yugoslav symposium on hydrogeology and engineering geology, Vol 1. Priština, pp 277–288

Radulović M (2000a) Karst hydrogeology of Montenegro, vol XVIII. The Institute for Geological Explorations of Montenegro. Separate Issues of Geological Bulletin, Podgorica, p 271

Slišković I (1984) Geološko-hidrogeološke karakteristike ležišta boksitne rude Trobukva-Studena vrela i mogućnost smanjenja dotoka vode u jamske prostorije (Geology and hydrogeology of the bauxite deposits Trobukva-Studena vrela and opportunities to reduce inflows into mine pit; in Croatian). In: Proceedings of the VIII Yugoslav symposium on hydrogeology and engineering geology, Vol 1. Budva, pp 559–571

Stevanović Z, Dragišić V, Filipović B (1991) The influence of the karst aquifer on ore deposits in east Serbia, Yugoslavia. Int J Mine Water 1:114–119

Stevanović Z, Dragišić V (1995) Some cases of accidental karst water pollution in the Serbian Carpathians. Theor Appl Karstology 8:137–144

Šarin A, Tomašič M (1991) Hydrogeological review of the Raša Coal Mine in the coastal karst of Croatia. In: Proceedings of the 4th congress Intern. mine water association, Vol. 1. Ljubljana (Slovenia)-Pörtscach (Austria), pp 125–131

Tilmat G (1973) Hydrologie des gisements karstiques Francais. In: Yugoslav II (ed) Symposium on resource and exploration of Bauxites, vol A-XVII. Tuzla, pp 1–22

Tóth B (1982) Shaft drilling activity for the water protection of bauxite mines in Nyrád. In: Proceedings of the 1st congress of the Intern. mine water association (IMWA). B, Budapest, pp 244–258

Vasiljević S, Koprivica V, Nikolić R, Raonić M, Todorović M, Kilibarda S (1988) Tehnologija podzemne eksploatacije u složenim rudarsko-geološkim uslovima rudnika boksita "Nikšić" sa osvrtom na mogućnost poboljšanja i tendencije razvoja (Technology of underground exploitation in complex mining-geological conditions of Nikšić Mine—development tendencies and improvement; in Serbian). In: Proceedings of the VI Yugoslav symposium on bauxite research, Herceg Novi, pp 305–316

Wenyong P, Zhizhong L, Hongze G (1991) Ordovician limestone water control in north China coalfield, 4th Intern. mine water association congress, Ljubljana (Slovenia) – Portschach (Austria), pp 207–211

References to Section 16.3

Angel JC, Nelson DO, Panno SV (2004) A comparison of manual and GIS-based methods for determining sinkhole distribution and density: an example from Illinois' sinkhole plain. J Cave Karst Stud 66(1):9–17

Gondwe BRN, Ottowitz D, Supper R, Motschka K, Merediz-Alonso G, Bauer-Gottwein P (2012) Regional-scale airborne electromagnetic surveying of the Yucatan karst aquifer (Mexico): geological and hydrogeological interpretation. Hydrogeol J 20:1407–1425

Gregory AS, Hugh HM, Jason ED (2001) A simple map index of karstification and its relationship to sinkhole and cave distribution in Tennessee. J Cave Karst Stud 63(2):67–75

Pavlović R, Čupković T, Marković M (2001) Daljinska detekcija (Remote sensing; in Serbian). Faculty of Mining and Geology, University of Belgrade, Belgrade

Radulović M (2010) Groundwaters of Montenegro (in Serbian). In: Djordjević B, Sekulić G, Radulović M, Šaranović M, Jaćimović M (eds) Water potentials of Montenegro (in Serbian). Montenegrin academy of sciences and arts, Podgorica, pp 62–105

Radulović MM, Stevanović Z, Radulović M (2012) A new approach in assessing recharge of highly karstified terrains–Montenegro case studies. Environ Earth Sci 65(8):2221–2230

Smith BD, Smith DV, Paine JG, Abraham JD (2005) Airborne and ground electrical surveys of the Edwards and Trinity aquifers, Medina, Uvalde, and Bexar Counties, Texas. In: US Geological Survey Karst interest group proceedings, USGS, Rapid City, 12–15 Sept 2005

Supper R, Motschka K, Ahl A, Bauer-Gottwein P, Gondwe B, Alonso GM, Romer A, Ottowitz D, Kinzelbach W (2009) Spatial mapping of submerged cave systems by means of airborne electromagnetics: an emerging technology to support protection of endangered karst aquifers. Near Surf Geophys 7(5–6):613–627

Won-In K, Charusiri P (2003) Enhancement of thematic mapper satellite images for geological mapping of the Cho Dien area, Northern Vietnam. Int J Appl Earth Obs Geoinf 4(3):183–193

Wynnea JJ, Timothy NT, Diazd GC (2008) On developing thermal cave detection techniques for Earth, the Moon and Mars. Earth Planet Sci Lett 272(1–2):240–250

Zurbuchen L, Kellenberger T (2008) Detection of cave entrances with airborne thermal imaging. EGU General Assembly, University of Zurich, Vienna, pp 13–18

References to Section 16.4

Al Charideh AR (2007) Environmental isotopic and hydrochemical study of water in the karst aquifer and submarine springs of the Syrian coast. Hydrogeol J 15:351–364

Alfirević S (1963) Hydrogeological investigations of submarine springs in the Adriatic. AIH Publ. Réunion de Belgrade. Comité National Yougoslave pour la Géologie de Génie Civil et l'Hydrogéologie, Belgrade, pp 255–264

Arandjelović D (1976) Geofizika na karstu (Geophysics in karst). Spec. ed, Geozavod, Belgrade

Arfib B, Dörfliger N, Wittwer C (2005) General overview of seawater intrusion in coastal aquifers. In: Cost Environment Action COST 621, Tulipano L, Fidelibus MD, Panagopoulos A (eds) Groundwater management of coastal karstic aquifer. COST Office, Luxemburg, pp 76–80

Back W (1992) Coastal karst formed by ground-water discharge. Yucatan, Mexico. In: Paloc H, Back W (eds) Hydrogeology of selected karst regions. International contributions to hydrogeology, IAH, vol 13. Heise, Hannover, pp 461–466

Bakalowicz M, Fleury P, Dörfliger N, Seidel JL (2003a) Coastal karst aquifers in Mediterranean regions. A valuable ground water resource in complex aquifers. vol 2, no 8. IGME Publ., Technologia de la Intrusion de Agua de Mar en Acuiferos Costeros: Paises Mediterraneos (TIAC). Hydrogeologia y aguas subterraneas, Alicante, pp 125–140

Bakalowicz M, Fleury P, Jouvencel B, Prome JJ, Becker P, Carlin T, Dörfliger N, Seidel JL, Sergent P (2003b). Coastal karst aquifers in Mediterranean regions. A methodology for exploring, exploiting and monitoring submarine springs. Technologia de la Intrusion de Agua de Mar en Acuiferos Costeros: Paises Mediterraneos, IGME, Madrid, pp 673–680

Biondić B, Goatti V (1984) La galerie souterraine "Zvir II" a Rijeka (Yougoslavie). In: Burger A, Dubertret L (eds) Hydrogeology of karstic terrains. Case histories. International contributions to hydrogeology, IAH, vol 1, Verlag Heinz Heise, Hannover, pp 150–151

Biondić B, Biondić R (2003) State of seawater intrusion of the Croatian coast vol 2. IGME Publ., TIAC, Tecnologia de la Intrusion de Agua de Mar un Acuiferos Costeros: Paises Mediterraneos, Alicante, pp 225–238

Bonacci O (1987) Karst hydrology with special reference to the Dinaric karst. Springer, Berlin

Burger A, Dubertret L (eds) (1984) Hydrogeology of karstic terrains. Case histories. International contributions to hydrogeology, IAH, vol 1. Verlag Heinz Heise, Hannover

Cost Environment Action COST 621 (2005) Tulipano L, Fidelibus MD, Panagopoulos A (eds) Groundwater management of coastal karstic aquifer. COST Office, Luxemburg

Dörfliger N, Fleury P, Bakalowicz M, El-Hajj H, Al Charideh A, Ekmekci M (2010) Specificities of coastal karst aquifers with the hydrogeological characterisation of submarine springs—overview of various examples in the Mediterranean basin. In: Bonacci O (ed) Sustainability of the karst environment—Dinaric karst and other karst regions. IHP series on Groundwater-VII/2010/GW-2, UNESCO, Paris, pp 41–48

Drogue C and Bidaux P (1986) Simultaneous outflow of fresh water and inflow of sea in a coastal spring. Nature 322/6077:361–363

Drogue C (1996) Groundwater discharge and freshwater-saline water exchange in karstic coastal zones. In: Proceedings of international symposium groundwater discharge in the coastal zone, vol 8. LOICZ-IGBP, LOICZ reports and studies, Moscow, pp 37–43

Eftimi R (2003) Some considerations on seawater-freshwater relationship in Albanian coastal area vol 2. In: IGME Publ., TIAC, Tecnologia de la Intrusion de Agua de mar un acuiferos Costeros: paises mediterraneos, Alicante, pp 239–250

Fleury P, Bakalowicz M, de Marsily G (2007) Submarine springs and coastal karst aquifers: a review. Hydrol J 339:79–92

Komatina M (1984) Control of underground flow in the littoral karst, Orebic, Yugoslavia. In: Burger A, Dubertret L (eds) Hydrogeology of karstic terrains. Case histories. International contributions to hydrogeology, IAH, vol 1. Verlag Heinz Heise, Hannover, pp 156–159

Lapierre L, Creuse V, Jouvencel B (2008) Robust diving control of an AUV. Ocean Eng 36(1):92–104
Mijatović B (1983) Problems of sea water intrusion into aquifers of the coastal Dinaric karst. In: Mijatovic B (ed) Hydrogeology of the Dinaric Karst, Field trip to the Dinaric karst, Yugoslavia. "Geozavod" and SITRGMJ, Belgrade, pp 89–108 15–28 May 1983
Mijatović B (1984) Captage par galerie dans un aquifere karstique de la cote Dalmate: Rimski bunar, Trogir (Yougoslavie). In: Burger A, Dubertret L (eds) Hydrogeology of karstic terrains. Case histories. International contributions to hydrogeology, IAH, vol 1. Verlag Heinz Heise, Hannover, pp 152–155
Mijatović B (2007) The groundwater discharge in the Mediterranean karst coastal zones and freshwater tapping: set problems and adopted solutions. Case studies. Environ Geol 51:737–742
Milanović P (2000c) Geological engineering in karst. Dams, reservoirs, grouting, groundwater protection, water tapping, tunnelling. Zebra Publishing Ltd, Belgrade, p 347
Potié L, Ricour J, Tardieu B (2005) Port-Miou and Bestouan freshwater submarine springs (Cassis—France) investigations and works (1964–1978). In: Stevanović Z, Milanović P. Water resources and environmental problems in Karst. Proceedings of the international conference KARST 2005, University of Belgrade, Institute of Hydrogeology, Belgrade, pp 249–257
Radulović M (2000b) Karst hydrogeology of Montenegro. Special issue of geological bulletin, vol. XVIII. Spec. Ed. Geol. Survey of Montenegro, Podgorica, p 271
Stevanović Z (2010a) Major springs of southeastern Europe and their utilization. In: Kresic N, Stevanović Z (eds) Groundwater hydrology of springs. Engineering, theory, management and sustainability. Elsevier Inc. BH, Amsterdam, pp 389–410
Stevanović Z (2010b) Intake of the Bolje Sestre karst spring for the regional water supply of the Montenegro. In: Kresic N, Stevanović Z (eds) Groundwater hydrology of springs. Engineering, theory, management and sustainability. Elsevier Inc. BH, Amsterdam, pp 457–478
Stevanović Z, Eftimi R (2010) Karstic sources of water supply for large consumers in southeastern Europe—sustainability, disputes and advantages. Geol Croat 63(2):179–186
Stringfield VT, LeGrand HE (1971) Effects of karst features on circulation of water in carbonate rocks in coastal areas. Hydrol J. 14:139–157
Zektzer IS, Ivanov VA, Meskheteli AV (1973) The problem of direct groundwater discharge to the seas. Hydrol J 20:1–36

Chapter 17
Hazards in Karst and Managing Water Resources Quality

Mario Parise, Nataša Ravbar, Vladimir Živanović, Alex Mikszewski, Neven Kresic, Judit Mádl-Szőnyi and Neno Kukurić

17.1 Hazards in Karst Environment and Mitigation Measures

Mario Parise

National Research Council, IRPI, Bari, Italy

17.1.1 Peculiarity of Karst

The karst environment is among the most fragile and vulnerable on Earth, as an effect of its main hydrological, geomorphological, and hydrogeological features (White 1988; Ford and Williams 2007a; Parise and Gunn 2007). In karst, water

M. Parise (✉)
National Research Council, IRPI, Bari, Italy
e-mail: m.parise@ba.irpi.cnr.it

N. Ravbar
Karst Research Institute, Research Centre of the Slovenian Academy of Sciences and Arts, Postojna and Urban Planning Institute of the Republic of Slovenia, Ljubljana, Slovenia
e-mail: natasa.ravbar@zrc-sazu.si

V. Živanović
Department of Hydrogeology, Faculty of Mining and Geology, University of Belgrade, Belgrade, Serbia
e-mail: v.zivanovic@rgf.bg.ac.rs

A. Mikszewski · N. Kresic
AMEC Environment and Infrastructure, Inc., Kennesaw, GA, USA
e-mail: alex.mikszewski@amec.com

Z. Stevanović (ed.), *Karst Aquifers – Characterization and Engineering*, Professional Practice in Earth Sciences, DOI 10.1007/978-3-319-12850-4_17

typically does not flow at the surface, but infiltrates underground through the complex network of conduits and caves that typically characterize the carbonate rocks. Enlargement of the voids in the rock mass by solution and mechanical erosion produces the genesis of caves that later may become beautifully decorated by a large variety of speleothems, locally becoming a source of attraction to tourists.

Besides the lack or scarcity of water at the surface, definition of the hydrological boundaries of catchments is a very delicate matter in karst, due to its very peculiar hydrological functioning (Gunn 2007; Palmer 2007). Not necessarily there is correspondence between surface and underground watersheds: water can infiltrate at a certain site and, due to complexity of the subterranean systems, be transported in another, nearby watershed. Thus, the only way to get information about the course of water in karst is following it underground, when possible (by means of caving explorations and surveys), or using dye tracers (Goldscheider and Drew 2007a) to mark its path whenever man is not able to enter narrow or flooded passages.

The peculiarities of karst environment make it highly vulnerable to a number of geohazards (De Waele et al. 2011; Gutierrez et al. 2014): As concerns the natural hazards, the main categories are sinkholes, slope movements, and floods. To these, anthropogenic hazards have to be added, as pollution events, land use changes resulting in loss of karst landscape, destruction of karst landforms, etc. Even carrying out engineering works without taking into the due consideration, the peculiar aspects of karst can be extremely dangerous and cause risk to the natural environment, as well as to the man-made infrastructures and buildings.

In the second half of last century, man has definitely become one of the most powerful factors that can cause changes in the karst environment, produce direct damage, predispose the territory to threatening events, and increase with mismanagement actions the negative effects deriving from natural hazards (Milanovic 2002).

In this chapter, the main hazards in karst are briefly described with particular focus on the natural hazards and with the help of some case studies.

Box 17.1.1

Exercise—Identification of likely sinkholes in karst landscape

Karst landscapes have as the most typical surface feature the presence of sinkholes (or dolines, following the term more consolidated in Europe; see Sauro 2003). When sinkholes are deep and well marked with respect to the surrounding terrain, their identification is generally straightforward and

J. Mádl-Szőnyi
Faculty of Science, Eötvös Loránd University, Pázmány P. Stny.1/C, Budapest 1117, Hungary
e-mail: madlszonyi.judit@gmail.com

N. Kukurić
IGRAC, International Groundwater Resources Assessment Centre, Delft, The Netherlands
e-mail: neno.kukuric@un-igrac.org

immediate. However, in many karst settings, the low topography, as well as the vegetational cover, makes their identification more difficult. In order to recognize these landforms and represent them accordingly in a karst geomorphological map, a morphological analysis of the terrains must be carried out. This may start from the analysis of the topographic maps, even at different scales. Availability of digital elevation models (DEM) may greatly help in this sense. Further, in vegetated areas, Light Detection and Ranging (LIDAR) may represent the best technique to outline landforms masked by forests and vegetation (Weishampel et al. 2011).

Use of aerial photographs, and their analysis and interpretation in stereoscopic view, allows to map (in function of the scale of the air photos) topographically depressed areas and sinkholes (Brinkmann et al. 2008). In particular in low topography, however, the features identified through air photos need to be carefully checked in the field, since some may be produced by other processes or by human activities.

Along coastal areas, in particular in low coasts, occurrence of sinkholes in the proximity of the coastline can be widespread (Margiotta et al. 2012; Basso et al. 2013). In some cases, the formation and the later evolution of sinkholes may strongly control the pattern of the coastline, by creating bays and inlets produced by nearby, coalescing sinkholes (Bruno et al. 2008).

Analysis of the surface hydrology is also of great importance for sinkhole identification: Observing water lines, and ranking the visible streams according to Strahler's classification, provides the first indications about the surface hydrology. This will be then integrated by the identification of existing swallow holes and the collection of data about caves (presence, density, development, and depth).

Box 17.1.2

Case study—Surface water in karst? A view at lakes of karst origin

Karst terrains have as one of the main features the scarce presence of water at the surface: Water typically infiltrates rapidly underground, through fissures and other types of discontinuities in the soluble rock mass, and begins the action of dissolving rocks and the formation of karst caves. However, when some conditions are satisfied, water can remain at the surface even in karst: For instance, the presence of topographically depressed areas, where the residual deposits from the karst process (i.e., terra rossa, clays) may accumulate, creates the ideal situation for keeping water at, or in the direct proximity of, the surface.

Conversano is a village located in the Low Murge (Apulia, southern Italy) where the above conditions are satisfied (Lopez et al. 2009). Around the village, at least 10 lakes of karst origin are located (Fig. 17.1), which have played a very important role in the historical development of human settlements in the area. In fact, the first settlements were established in the surroundings of two of these lakes, which testify that the function of the sites to keep some amount of water in the dry season was already known at that time. Other archaeological evidence states the occupation of the sites during the Middle Age, while in more recent times in all the 10 lakes several deep shafts, coated with carbonate rocks, were built in order to work as tanks during the long and dry summer season.

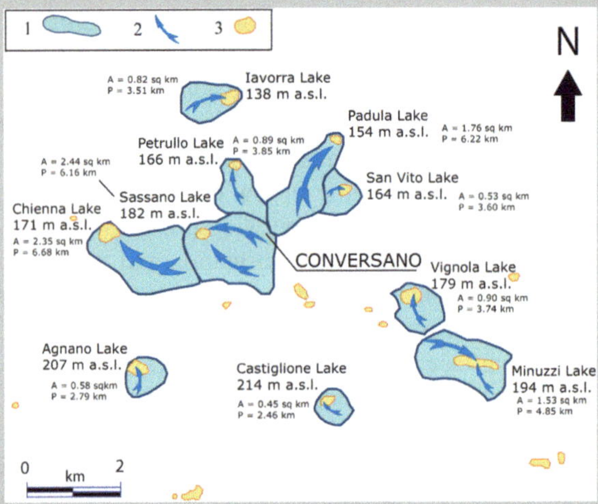

Fig. 17.1 The 10 lakes of karst origin in the Conversano area (simplified after Lopez et al. 2009). *A* catchment area; *P* catchment perimeter

Not all the lakes share the same origin: The most topographically pronounced have been produced by likely collapses (collapse sinkhole), while others, showing a less clear morphological evidence, were formed as solution sinkholes. Further, size of the lake, as well as of the likely catchment feeding it from the surface, is quite different.

17.1.2 Sinkholes

Sinkholes are probably the most typical hazards in karst areas (Waltham et al. 2005; Gutierrez et al. 2014). In a broader sense, sinkholes are a form of instability and can be considered as a category of mass movements, even though they generally affect very low-gradient, or sub-horizontal, slopes in soluble rocks.

Being related to presence of underground caves, they are considered as the main feature of karst, at the same time often representing the only way to access and

Fig. 17.2 Natural sinkhole (*left*) and collapse sinkhole (*right*), the latter being related to presence of an underground quarry

explore the underground voids. In addition to natural caves, sinkholes can also be linked to man-made (artificial) cavities (Parise et al. 2013) that are often at the origin of collapses of the ground with consequent damage to the built-up area above. This may occur even in non-karstic areas (for instance, in settings with volcanic rocks).

In terms of the hazard that sinkholes pose to man, velocity of the process is extremely important, especially as regards the catastrophic phase, that is, the collapse (Fig. 17.2). In this term, the most dangerous typologies are those characterized by high velocity, namely collapse and cover-collapse sinkholes (Tharp 1995). Lack of premonitory signs and the rapid evolution in the catastrophic phase of the event are still elements worth to be further investigated by scientists, in the attempt to improve our understanding of the phenomena and move toward mitigation of the risk deriving from sinkholes.

In the case of anthropogenic sinkholes, man and his activities play a crucial role in the development of sinkholes: The main problem is that underground cavities, once used by man for several functions, are later on "forgotten" or abandoned and may become part of the built-up areas, once these expand above the zones where the cavities are located. Knowledge of the precise position of these voids, of their characteristics (depth, width, length, etc.), and of their stability conditions as well is therefore fundamental to avoid the possibility of causing damage in the above roads or buildings (Table 17.1).

Table 17.1 Geotechnical properties of rock materials at the underground quarries of Cutrofiano (modified after Parise and Lollino 2011)

	γ (kN/m^3)	E′ (kPa)	ν'	c′ (kPa)	φ' (°)	σ_t (kPa)	σ_c (kPa)
Sand	18	70,000	0.3	0	28	0	–
Clay	20	40,000	0.25	15	20	0	–
Mazzaro	17.5	180,000	0.3	360	33	300	2,400
Soft calcarenite	15.5	100,000	0.3	160	30	160	1,400

Key to symbols γ unit weight; E' modulus of elasticity; ν' Poisson ratio; c' cohesion; φ' friction angle; σ_t tensile strength; σ_c unconfined compressive strength

Box 17.1.3

Case study—Stability conditions of underground caves and prediction of the likely occurrence of a sinkhole (Cutrofiano, Italy)

The presence of caves (natural or of anthropogenic origin) affected by instability features and occurrence of underground failures in the rock mass may potentially cause the development of a sinkhole at the ground surface (Parise and Lollino 2011; Lollino et al. 2013). In order to assess the possibility of such a failure (that may directly damage the human structures and buildings), the following steps are needed:

1. knowledge of the spatial and altimetric development of the cave (topographic or speleological survey);
2. mapping of instability features;
3. description of the local stratigraphy;
4. definition of the physical–mechanical properties of involved rocks;
5. ascertaining the presence of water;
6. ascertaining the effects of weathering processes in the rock mass;
7. performing a geological–mechanical characterization of the rock mass;
8. implementation of numerical codes.

Step 1: The pattern of the cave and its spatial development (in terms of length, direction of development, height, and width of the rooms) need to be known. This requires a speleological survey, using reliable techniques and tools in order to have a detailed cartographic outcome, in function of the required degree of precision. In particular, the access to the cave must be carefully located by means of Global Positioning System (GPS) survey; this is very important, because the cave plan will be then superimposed on the built-up areas, to check any possible interferences.

The use of old speleological surveys, carried out with low degree of precision due to the tools utilized at that time, must be discouraged, since it would inevitably result in errors and misuse of the surveys (Martimucci and Parise 2012).

Step 2: The cave map (in plan view) needs to be integrated by observations aimed at the recognition of the different mechanisms of failure (partial or total falls from the vault, failures from the walls, deformation at the edge of pillars or along the walls, etc.; Parise 2013). The already occurred failures must be differentiated from those incipient that may be identified by means of a number of premonitory signs such as swelling in the walls and wedge extrusions.

Step 3: The local stratigraphy of the overburden (Fig. 17.3) must be reconstructed in detail, using available data, performing new field observations, and, whenever possible, carrying out boreholes.

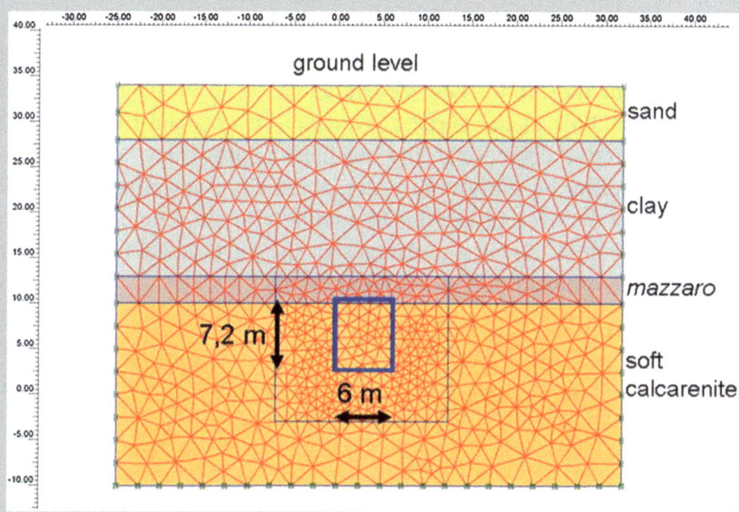

Fig. 17.3 Geological model in the Cutrofiano area. The *blue rectangle* indicates the location of a typical quarry and the average size. The term *mazzaro* indicates the uppermost portion of the calcarenite formation, made of stiff deposits that were used as the vault of the underground quarries

Step 4: The geotechnical characteristics of the involved rock materials have to be assessed. Ideally, this should be done by sampling the different rocks and performing laboratory tests; however, this is rarely possible due to lack of funds. Therefore, the geotechnical properties may be, in first approximation, extracted from the scientific literature.

Step 5: During the phases of the steps 1 and 2, a particular focus may be given to observing the presence of water in the cave and any other feature that may indicate the infiltration of water from the surface. All the sites where this is observed must be carefully mapped.

Step 6: Directly linked with the previous step, the presence of weathered materials must be ascertained. This may be done through simple tests in the cave (by hand, with the geologist hammer, by means of simple tools, etc.) and provides a first assessment of the degree of weathering in the rock mass. Weathered materials are generally favored by water coming in the cave (continuously or occasionally) and by the local micro-climate conditions (temperature and humidity).

Step 7: In case of presence of discontinuities in the rock mass (bedding planes, joints, and faults), a geo-mechanical characterization must be performed, by individuation of the main families of discontinuities, and of their features, following the international standards (ISRM 1978).

Step 8: Once the above steps are done, implementation of specific codes may allow to identify, starting from the features observed underground, the more likely development of shear surfaces bounding the sinkhole at the surface. This can be obtained by simulating a progressive decrease in the mechanical properties, as an effect of weathering by infiltrating water, and/or by a lowering in the thickness of the overburden, due to continuous detachment from the cave roof (Fig. 17.4), or by changes in the cave geometry as a consequence of lateral failures from the walls.

Different methods have to be implemented, depending upon the presence of a material that can be treated as a continuum (due to absence or scarce importance of discontinuities) or as a non-continuum medium. Accordingly, finite element method or distinct element method will be applied.

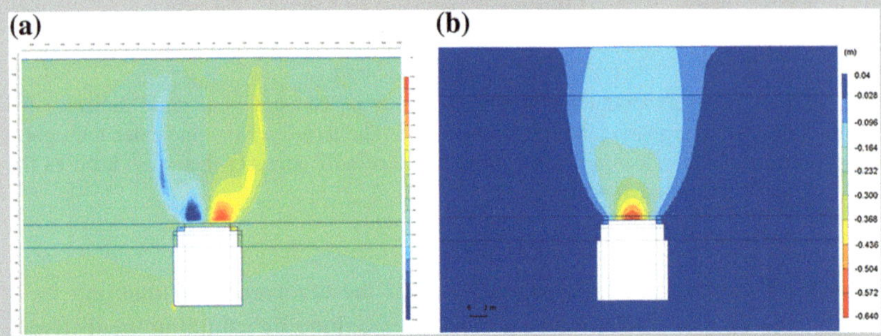

Fig. 17.4 Shear strains (**a**) and vertical displacements (**b**) due to increased size of the quarry (simplified, after Lollino et al. 2013)

17.1.3 Mass Movements

Slope movements are frequent in karst settings, where the typical conditions favoring the occurrence of such phenomena (high relief, fractured rock mass, and low values in the geotechnical properties of affected materials) combine to further discontinuities produced by karst (Waltham 2002; Santo et al. 2007; Parise 2008; Palma et al. 2012): from those of larger size (caves accessible to man), to phreatic conduits and small fissures, all these features are additional paths available to water, where to develop chemical (solution) and physical (erosion) actions. The result is an increase in the overall weakness of the rock mass, potentially leading to failures.

However, establishing the connection between karst processes and the occurrence of mass movements is not an easy task, and the wide literature on landslides does not offer many examples where the role of karst in the development of slope failures has been clearly defined, or even taken into account.

Karst, on the other hand, offers a unique opportunity for a double view on gravitational processes, adding to the usual view from the surface the possibility to look at instability from underground, by entering the caves and observing the failures therein occurring. It is well known that breakdown processes, mostly occurring through progressive failures from the vault, represent the main type of evolution of caves, once they have been left by water (White and White 1969). Through such failures, the cave evolves toward a stable configuration (typically, in the shape of an arch) or, depending upon depth of the underground voids and the geo-mechanical properties of the rock mass, may continue to produce progressive detachment, eventually reaching the surface, thus producing a sinkhole (Culshaw and Waltham 1987; Waltham and Lu 2007; Parise 2008).

Analysis of breakdown deposits within cave systems, in relation to geology, and to the above existing man-made structures at the surface are therefore of extreme importance for hazard assessment, by adding a view from the inside to the common analysis carried out at the surface (Klimchouk and Andrejchuk 2002; Iovine et al. 2010).

To evaluate the sectors within cave systems mostly prone to further failures, the weathering processes in the rock mass must also be carefully examined (Fookes and Hawkins 1988; Hajna 2003; Lollino et al. 2004): these often cause strong reductions in the mechanical properties of the rocks, thus contributing to its overall decrease in strength, and facilitating its proneness to failures.

17.1.4 Floods

Given the content of the present book, floods in karst and pollution events are not discussed in great detail here, being the specific topic of several other chapters. Nevertheless, some key points are delineated, as floods in karst represent a quite frequent hazard, and are at the origin of many damages and economic losses in many karst areas worldwide.

Water generally infiltrates underground in karst through the complex network of karst conduits, caves, and fissures, until it is transferred to the outflow zones. There is therefore a very limited amount of surface runoff; when the swallow holes become, for some reasons, partially or totally clogged, water is not able to enters the ground, and thus it accumulates at the surface, creating floods and inundating areas (White and White 1984; Bonacci et al. 2006; Delle Rose and Parise 2010; Farfan et al. 2010), especially in the case of very flat, polje-like valleys. Floods in karst may be frequent and destructive. Many examples worldwide show that, unfortunately, man tends to loose the memory of past events and keeps planning and building in karst areas without taking into account that in the past those areas had been affected by flooding and inundation (Parise 2003). As a consequence of such behavior, we have to repeatedly face great damage on the occasion of floods, and probably it is easy to forecast further, increasing damage in the next future.

Pollution events in karst are often related to abandoned quarries, or some types of anthropogenic cavities, where solid and liquid wastes can be discharged, with severe impact on the quality of the karst groundwater resources (Zwahlen 2004). In particular social situations, such as in post-conflict scenarios, the situation becomes even worst and severe (Calò and Parise 2009). Natural caves, too, are not immune by pollution and are frequently interested by dumping of different types of wastes (even toxic and highly dangerous).

17.1.5 Loss of Karst Landscape

Land use changes in karst are typically related to the search for new pieces of land to agriculture: At this aim, terracing slopes, filling the depressions, and clearing the fields from the rocky stones are among the main activities carried out in karst terrains. In particular when topography is very flat, it is extremely easy to cancel some of the karst landforms. The loss of karst landscape has undoubtedly strongly increased during the last decades: As a matter of fact, in ancient times, stone clearing was performed by hand, and the collected stones were used to build the drystone walls, a typical element in the rural architecture of many Mediterranean karst areas (Nicod 1972; Parise 2012a). In the last decades, thanks to the extensive use of machinery to remove and destroy the rocks, stone clearing has completely changed the natural setting, destroying the epikarst (Williams 2008a), diverting the surface runoff, causing erosion even on the occasion of small amount of rainfall, until determining in many areas a strong tendency toward desertification. As a further consequence, many rocks taken out from the field are dumped into caves or accumulated at their entrances, in both cases determining a serious danger to cavers (Parise and Pascali 2003a).

One of the most destructive activities by man for karst areas is quarrying (Gunn 1993, 2004; Parise 2010): quarries may cause partial or total destruction of caves, degradation of the landscape, and heavy changes in the natural hydrography. Once the activity stops, these sites often become illegal landfills, inevitably causing pollution to karst ecosystems. Beside surface quarries, quarrying may also be developed underground, with a long and difficult work to realize subterranean galleries, and bring at the surface the excavated material. Underground quarrying activity may be at the origin of the development of sinkholes at the surface (Parise 2012b), because of instability problems underground, moving progressively upward (see Box 17.1.3).

17.1.6 Mitigating Hazards in Karst

Karst is extremely fragile, probably among the most delicate environments on Earth. Understanding the complex geological and anthropogenic hazards that may occur in karst and mitigating the related risks require specific approaches and techniques, given the peculiarities of this environment, especially as regards its hydraulic and hydrogeological behavior. The main priority should be protection of karst terrains and safeguard of its natural resources, first and foremost groundwater. At this aims, issuing of regulations and laws and strict implementation of the existing rules are important and necessary actions. More important, however, is their real enforcement: A law can be written in an excellent way, but if it remains only on paper, and no technical standards for actual implementation are released, the effects are practically null.

And, more powerful than any prohibition or law issue, education and direct involvement of the local inhabitants living in karst are fundamental. People should be educated to understand the reasons why they should not carry out certain actions, by explaining them the consequences of their activities, and that the damage they are producing does not only affect the environment, but also their own life, that of their children, the next generations, the water they drink, the food they crop and eat.

Scientists and cavers should produce stronger efforts in this direction: Educating people, talking about karst to those living in karst lands, carrying their knowledge outside the too often much closed world of karst science and speleology. We have to take not for granted that everybody knows what karst is, and what problems may affect it (Fig. 17.5).

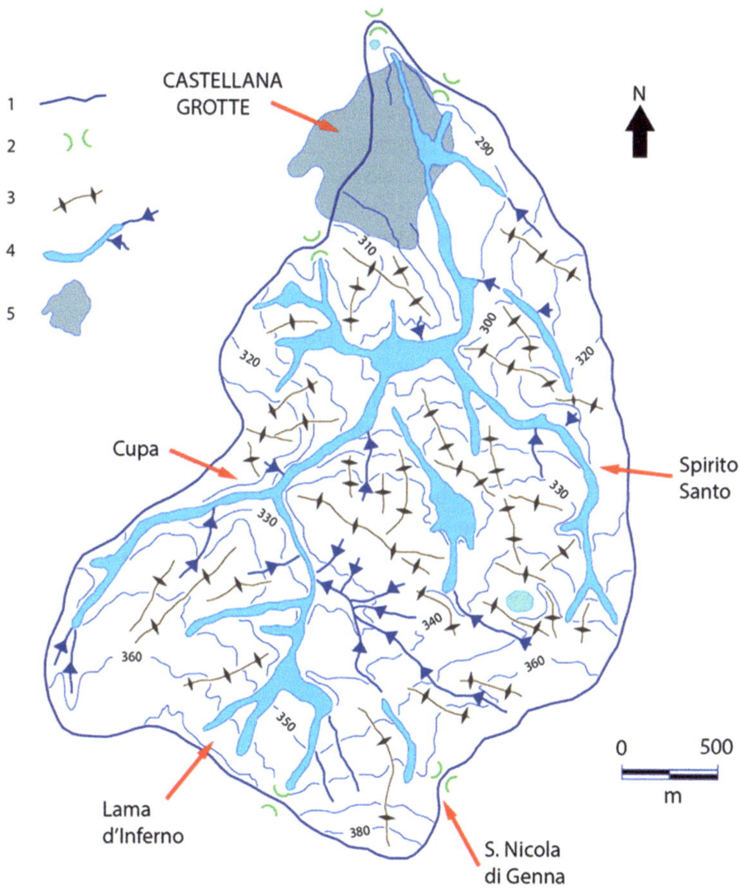

Fig. 17.5 The karst valley at Castellana-Grotte, S Italy. Key: *1* valley boundary; *2* saddle; *3* water divide; *4* sinkhole and lame (ephemeral streams); *5* town

Box 17.1.4

Case study—Evaluation of the disturbance induced by man to karst environments at the local and regional scale

The Karst Disturbance Index (KDI, proposed by van Beynen and Townsend 2005, and modified by North et al. 2009) is a way to determine the impact deriving from human activities to the karst environment. It has been applied to different karst settings worldwide (Calò and Parise 2006; De Waele 2009; North et al. 2009; Day et al. 2011). Taking into account a number of indicators, subdivided into five different categories the disturbance to the karst environment deriving by anthropogenic actions can be evaluated, to provide some insights about how the karst landscape has negatively been modified by man.

KDI consists of 31 environmental indicators contained within five categories; namely, (1) geomorphology, (2) hydrology, (3) atmosphere, (4) biota, and (5) cultural. Within these broad categories, many components of karst systems are taken into account and assessed by taking advantage of past experiences, studies, and researches. Such environmental index approach to assess karst terrains enables scientists to measure, compare, and contrast the condition of karst environments spatially and temporally and to guide the future mitigation action aimed at minimizing the disturbance to the natural environment. In order to apply the index, a thorough research of the available data must be pursued, and the obtained dataset must be carefully analyzed in terms of quality, reliability, and updating of the collected information.

Apulia, in southern Italy, is a region almost entirely made of carbonate rocks, where therefore karst is the main agent sculpturing the landscape, both at the surface and underground. In its territory, over 2,100 natural caves have been explored and surveyed; further, due to the long human history in the area, over 1,000 artificial caves, excavated by man in different epochs and for different uses, have to be counted. In such a setting, all the activities performed by man have been evaluated, from original data or the examination of available documents and works. The 31 indicators of the KDI have been assessed, and, as shown in Table 17.2, the disturbance caused by man to the Apulian karst is calculated. It has a value of 0.58, which corresponds to the class of *significant disturbance*, in the KDI classification modified by North et al. (2009).

Table 17.2 Indicator and total disturbance scores for Apulia (Italy) (After North et al. 2009)

Category	Attribute	Indicator	Apulia (Italy)
Geomorphology	Surface landforms	Quarrying/mining	3
		Human-induced hydrological change	1
		Stormwater flow into sinks	2
		Infilling of sinkholes	3
		Dumping refuse into sinks	3
	Soils	Soil erosion	1
		Soil compaction	1
	Subsurface karst	Flooding of subsurface karst	2
		Decoration removal/vandalism	2
		Mineral or sediment removal	2
		Floor sediment compaction	1
Atmosphere	Air quality	Desiccation	1
		Condensation corrosion	1
Hydrology	Water quality from surface practices	Pesticide and herbicide use	2
		Industrial/petroleum spills or dumping	2
		Leakage from underground petroleum storage tanks	–
	Spring water quality	Concentration of harmful chemical constituents in springs	1
	Water quantity	Changes in water table	2
		Changes in cave drip waters	1
Biota	Vegetation disturbance	Vegetation removal	3
	Subsurface cave biota	Species richness of cave biota	1
		Population density of cave biota	1
	Subsurface groundwater biota	Species richness of groundwater biota	1
		Population density of groundwater biota	1
Cultural	Human artifacts	Destruction/removal of historical artifacts	2
	Stewardship of a karst region	Regulatory protection	2
		Enforcement of regulations	3
		Public education	2
	Building infrastructure	Building of roads	2
		Building over karst features	2
		Construction within caves	1
Total indicator score (A)			52
Total possible score (B)			90
Disturbance level (A/B)			0.58

17.2 Advanced Strategies in Managing and Sustaining Karst Water Quality

Nataša Ravbar

Karst Research Institute, Research Centre of the Slovenian Academy of Sciences and Arts, Postojna and Urban Planning Institute of the Republic of Slovenia, Ljubljana, Slovenia

17.2.1 Karst Groundwater Environmental Issues and Protection

In the past, karst landscapes have, with the exception of those under warm and humid climates, been considered as less suitable for human settlement and with rather poor prospects for agriculture. Main reasons were unfavorable natural conditions, such as variegated relief, shallow soils, and shortage of available water sources. In the past few decades, karst areas gradually became economically very important (Kranjc et al. 1999; Ford and Williams 2007b).

Besides traditional sources, such as minerals and stones, karst areas have, with the technological and economic development, grown to be among the most important drinking water, oil, and gas reservoirs and have experienced agriculture intensification (Gunn and Bailey 1993; Nicod et al. 1997; Moore 2001; Bakalowicz 2005). Furthermore, particularly numerous caves and geomorphologically remarkable scenic spots became attractions and provoked development of tourism (Hamilton-Smith 2007; Williams 2008b). An increasing demand for clean energy resources calls for intensive use of geothermal energy from karst systems (Liedl and Sauter 1998; Goldscheider et al. 2010a).

Due to consequential variety of activities and accompanying urban development, an increased pressure on karst environment, especially by intensive and unsustainable spread of settlement, infrastructure and industry have occurred (Ravbar and Kovačič 2013). Besides the possible future effects of climate change, these impacts may cause varying groundwater-related problems, connected to different types of contamination (Fig. 17.6), overexploitation, and ecological damage that can further lead to other adverse consequences such as depletion of water resources, sea water intrusions, and land subsidence (Kundzewicz et al. 2008; Parise and Pascali 2003b; Bonacci et al. 2009; Guo et al. 2012). Some examples of long-term human degradation of karst groundwater have already been highlighted by Pulido-Bosch et al. (1995), Han (1998), Drew and Hötzl (1999), Kaçaroğlu (1999), Loáiciga et al. (2000), Mamon et al. (2002), Vesper and White (2004), Kovačič and Ravbar (2005), Pronk et al. (2006), Kresic (2009a), and others.

Accordingly, contamination prevention has been a major concern of local, regional, and state agencies worldwide, and urban karstology has been confronted with a challenge, how to maintain the equilibrium between proper protection and management of karst water resources and a growing demand of different interest groups.

Fig. 17.6 The Krupa karst spring in SE Slovenia represents the most important potential source of drinking water in the region (for about 27.000 inhabitants), but has been contaminated with the PCBs since 1985 due to illegal dumping in its catchment

17.2.2 The Concept of Vulnerability and Contamination Risk Assessment

Along the past half of the century, much effort has been dedicated to the development of the aquifer vulnerability and contamination risk mapping concept. Main principles, definitions, and terminology have been introduced (Vrba and Zaporozec 1994) and numerous assessment methods have been developed. The purpose of this concept is to identify the most vulnerable areas of the aquifer that need the highest protection and to optimize the land use in the catchment areas of the captured water sources. Protection of most vulnerable areas can minimize contamination of groundwater.

The concept of groundwater vulnerability to contamination is based on quantifying the neutralizing capacities of a hydrological system against contamination. **Intrinsic vulnerability** denotes the intrinsic characteristics of the hydrological system. It is founded on the assessment of basic factors that determine the

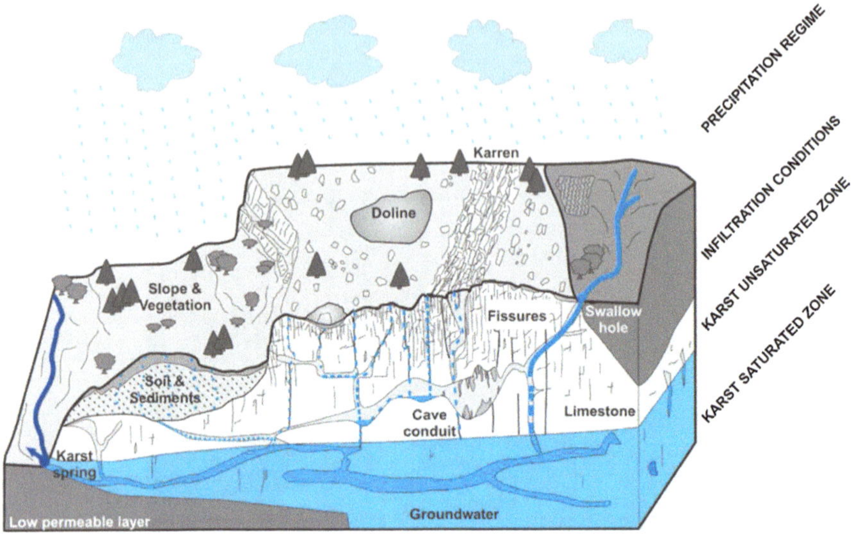

Fig. 17.7 Conceptual model of a karst aquifer and factors influencing the vulnerability of water resource and source (modified after Ravbar 2007a)

degree of natural protection to reduce negative influences of contamination and to re-establish the equilibrium of the environment (Fig. 17.7). Its assessment bases on taking into account geological, geographical, hydrological, and hydrogeological characteristics of the area, but is independent of the nature of the contaminant. **Specific vulnerability** denotes vulnerability of groundwater to a particular contaminant and additionally considers contaminant's properties and its interaction with the hydrological system (Vrba and Zaporozec 1994; Daly et al. 2002).

There are two general approaches to groundwater protection and vulnerability mapping: **Resource protection** aims to protect the entire aquifer, and the resource vulnerability map consequently considers the vertical percolation of water and contaminants from the land surface through the unsaturated zone toward the groundwater table; **source vulnerability** assessment focuses on a particular spring or well and additionally considers the lateral pathway within the saturated zone (Daly et al. 2002).

With regard to possible damage of groundwater, the term **contamination risk** is used for the probability of a specific adverse consequence occurring. It takes into account the interaction between the natural characteristics of an aquifer, i.e., the vulnerability of the aquifer, and the infiltrating contaminant load, pointing out the consequences for the groundwater if a hazardous event occurs (De Ketelaere et al. 2004).

17.2.3 Commonly Applied Methods in Karst

Classical vulnerability assessment methods proved not appropriate for application in karst aquifers because they do not consider the specific nature of water flow, such as heterogeneity and duality of infiltration processes. In particular, they disregard the existence of allogenic recharge and infiltration into swallow holes. Consequently, their application to karst tends to categorize karst terrains in the highest vulnerability classes with no differences on the aquifer level (Gogu et al. 2003; Goldscheider 2005).

As karst aquifers host some of the most prolific aquifers in the world, aims at defining guidelines to vulnerability and risk mapping for the protection of karst aquifers have been penetrated through an international project COST Action 620 (Daly et al. 2002; Zwahlen 2004b; Foster et al. 2013, see also Chap. 8). The resulting European Approach based on the EPIK Method (Dörfliger and Zwahlen 1998). The EPIK Method was the first that considered the specific properties of water flow in karst aquifers. Later, European Approach stemmed from a number of vulnerability assessment methods, like for example the COP Method (Andreo et al. 2009). A complete implementation of the European Approach that anticipates vulnerability, hazard, and risk assessment was developed by Ravbar and Goldscheider (2007a). Assessment principle of a very detailed Slovene Approach and an example of its application is shown in Figs. 17.8 and 17.9. Furthermore, the PaPRIKa Method (Dörfliger and Plagnes 2009) demonstrates a slightly different approach to vulnerability assessment. The recently developed hydrogeological conceptualization of karst systems, the KARSYS Method, can among others also be used as a base for groundwater protection purposes (Jeannin et al. 2012). Main differences between these approaches are in regard to various complex parameters, their interaction and weighting systems.

Box 17.2.1

Case study—Usefulness of vulnerability and contamination risk maps
Exercise:
The criteria for protecting water sources are usually based primarily on the distance from the water source or the velocity of the groundwater flow. Why are these criteria unsuitable for the mitigation of karst water sources contamination? Why has the concept of vulnerability and contamination risk assessment been recognized to be appropriate alternative for source protection zoning in karst? What is the usefulness of vulnerability and contamination risk maps? Compare the case of the Korentan karst spring that drains about 6 km^2 large binary karst system—see Figs. 17.9b and 17.10.

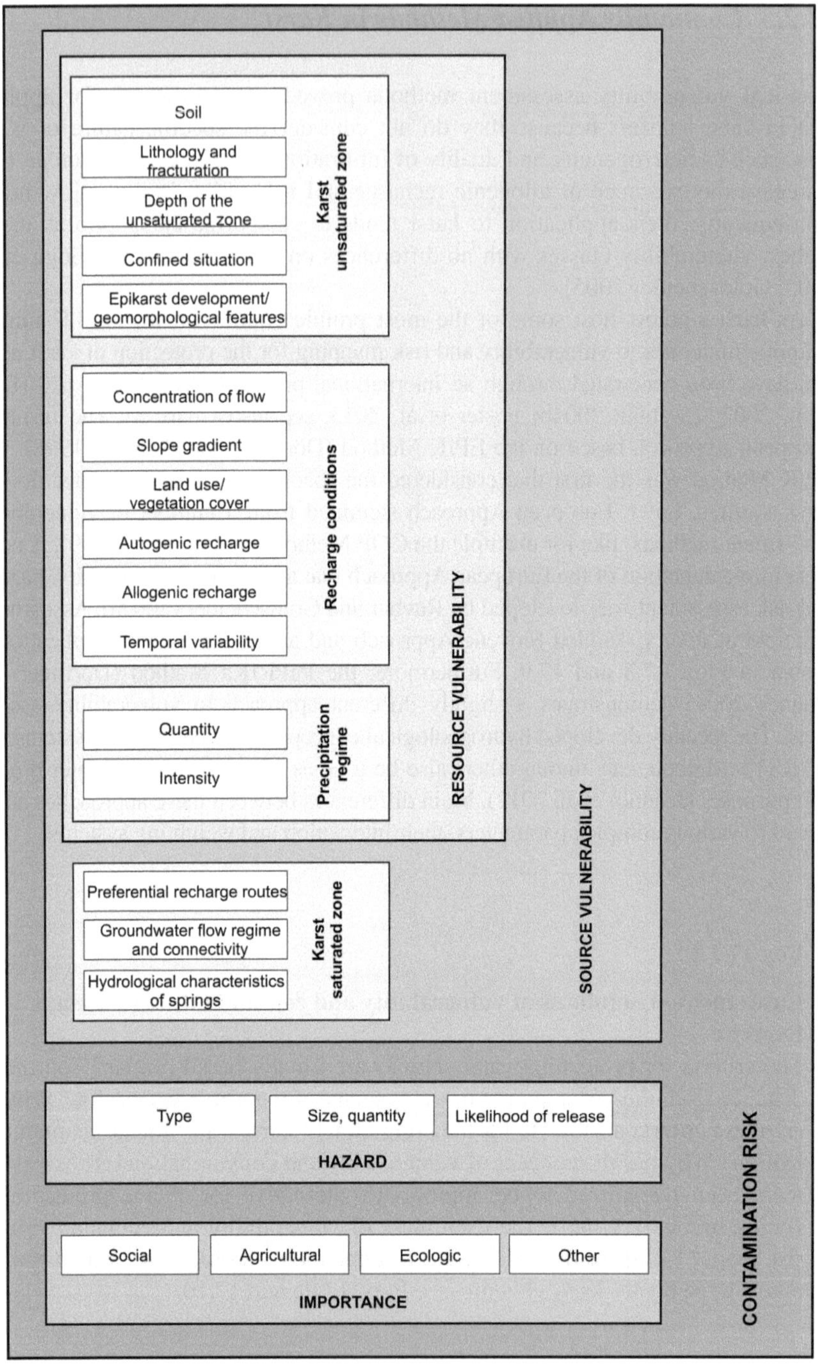

Fig. 17.8 Flowchart diagram showing factors and data required by the Slovene Approach for the groundwater vulnerability and risk to contamination assessment

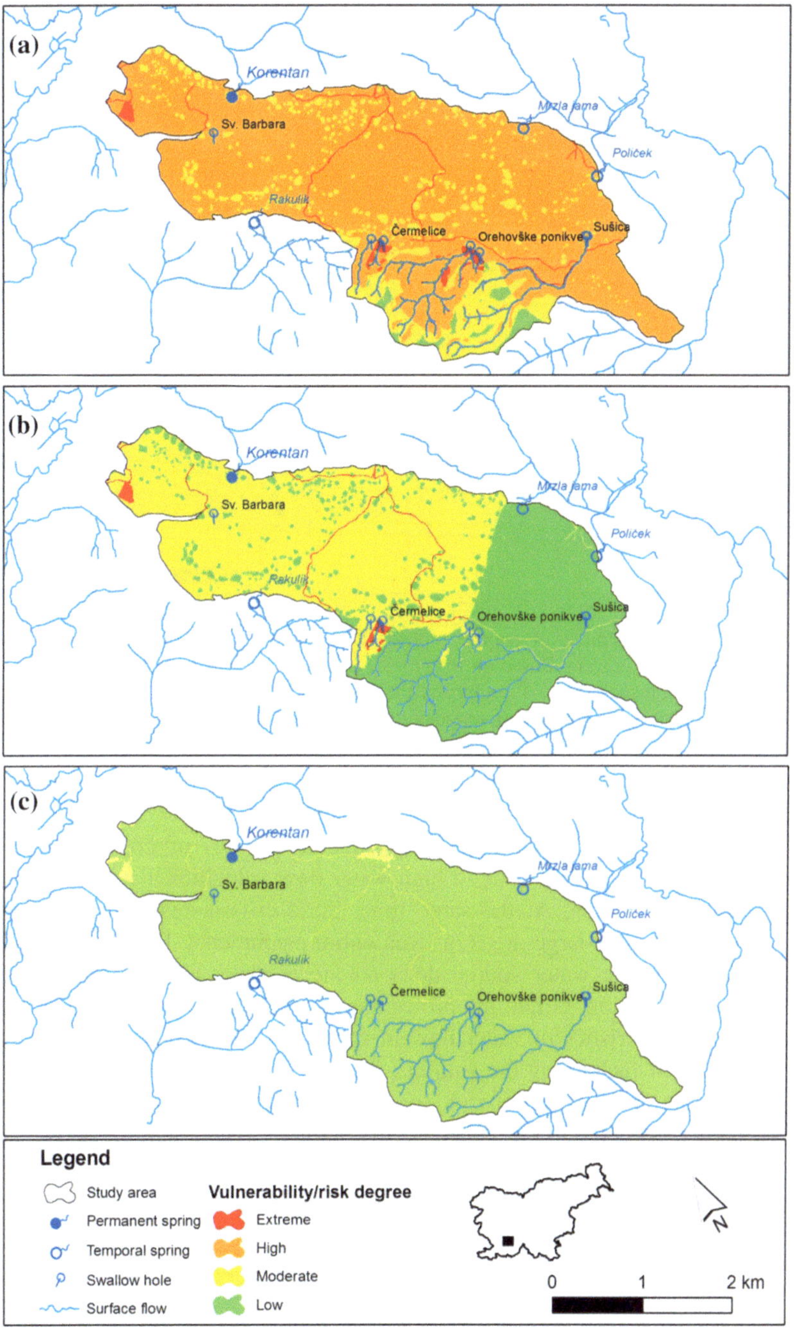

Fig. 17.9 Examples of groundwater **a** resource and **b** source vulnerability maps, and **c** contamination risk map assessed by the Slovene Approach (modified after Ravbar et al. 2013)

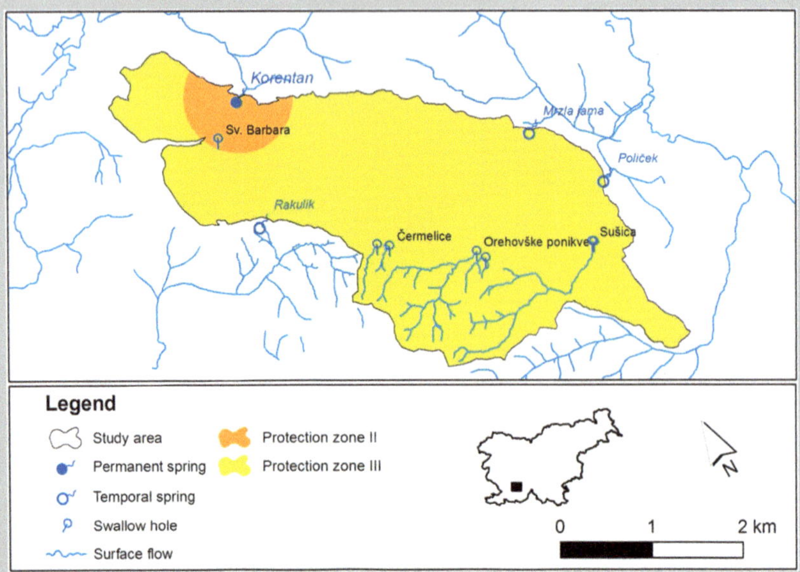

Fig. 17.10 Arrangement of source protection zones set up on the velocity of the groundwater flow criteria. Based on tracer test, groundwater flow velocity was estimated to 25 m/h. Protection Zone I comprises the area immediately surrounding the spring, Zone II comprises the area within one day isochrones, and Zone III comprises the rest of the aquifer

Answer:
Consideration of distance criteria and retention times of groundwater in karst alone is insufficient due to high water flow velocities that hinder contaminant degradation. At the same time, these criteria result in declaring high protection for large areas or even entire catchments. However, greater distance from the water source does not necessarily mean greater safety from contamination. Often, the protection requirements do not consider the special characteristics of water flow in karst such as the heterogeneity and complexity of recharging the aquifers. The role of protective layers and the development of the karst network are usually not taken into account as well. These are considered in karst specific vulnerability assessment methods.

The concept of assessing vulnerability and contamination risk upgrades protection strategies in karst by integrating the specific nature of recharge and groundwater flow. It offers a balance between protection on the one hand and spatial planning and economic interests on the other. The final result are maps showing various levels of vulnerability to pollution of karst waters in different colors that can possibly be converted into water protection zones or showing areas of the highest contamination risk level that call for immediate action and rehabilitation. The identification of the most vulnerable areas and areas with the highest risk of contamination allows the optimization of water protection zones, appropriate and prudent management of water sources, and a foundation for planning the monitoring of water quality.

These maps are very useful since those responsible for making decisions about land use can employ them to identify areas inside individual catchment areas suitable for certain human activities, areas in need of protection, and the level of their protection, which can result in banning certain activities. It is also possible to envisage rehabilitation measures for actual sources of pollution and determine a schedule for their implementation relative to the level of risk. Maps of natural vulnerability and the pollution risk to groundwater therefore provide a useful decision making basis for national and local authorities responsible for spatial planning and land use.

Box 17.2.2

Case study—Importance of temporal hydrological variations for the karst groundwater vulnerability assessment
Exercise:
Some karst systems, for example Dinaric karst, or karst areas under subtropical/monsoon climates often show fast and strong hydrological variations in response to precipitation events or snowmelt, including groundwater level variations (reaching order of tens or hundreds of meters), change from open channel to pressurized flow, change from laminar to turbulent flow, changing flow directions, and velocities (velocities during high-flow conditions can be ten or even more times higher than during low-flow conditions). Consequently, shifting groundwater divides and different types of surface–groundwater interaction, such as activating temporarily active streams and swallow holes, may occur (Fig. 17.11). Why and how these variations are relevant with respect to contaminant transport and groundwater vulnerability?

Fig. 17.11 One of the largest intermittent karst lakes is Cerkniško Jezero of the Slovene Dinaric karst. Several springs, estavelles, surface streams, and direct precipitation recharge the lake. It can extend over 26 km^2 and contain more than 82 million m^3 of water. The lake usually fills twice per year but remains empty in dry years; in wet years, it occurs several times per year and/or does not dry up entirely

Answer:
Many common methods of karst vulnerability assessment take into account the thickness of the unsaturated zone. However, large water level fluctuations result in a variable thickness of this zone. Rising water levels means decreasing unsaturated zone thickness, decreasing protectiveness of the overlying layers, and increasing vulnerability.

Furthermore, temporal variations may induce different surface–groundwater interactions and preferential flow to occur, which permit contaminants to bypass overlying layers that may otherwise attenuate them. Most existing methods classify swallow holes, sinking streams, and their catchment areas as zones of high or extreme vulnerability. Some swallow holes are only frequently active and may operate only during exceptional hydrological events, sometimes less than once per year. However, only in case of permanently or frequently active swallow holes, a contaminant release would always and rapidly reach the groundwater without significant attenuation and can thus be considered as highly vulnerable.

Variations in groundwater flow velocities and variability of transport processes may also have implication for vulnerability, as higher water flow velocities reduce underground retention. Some contaminants may be largely immobile during low-water conditions, as they might be stored in the unsaturated zone, in cave sediments or at the bottom of water-filled cavities. During high-flow events, these contaminants may be remobilized and cause sanitary problems.

Variations of flow directions can result in contributions from different parts of the aquifer to a particular spring, and thus in variable catchment boundaries which are crucial for delineation of the spring catchment area and for source protection. The areas that always directly and fully contribute to the spring should be differentiated from the areas that are only temporally or indirectly connected to the spring, areas where a connection is not sure and very remote parts of the aquifer system.

Box 17.2.3

Case study—Vulnerability and contamination risk assessment reliability
Exercise:
Groundwater vulnerability and contamination risk maps are increasingly used for protection zoning and management. Since groundwater vulnerability cannot be measured or directly obtained in the field, vulnerability and risk maps are simplifications of natural conditions and their uncertainty is inherent in all methods. These are liable predominantly to subjectiveness and unreliability in assessment procedure. The results are thus often influenced by diverse aspects (e.g., selection and evaluation of different parameters, quality and accuracy of data, and their availability and interpretation). Because water protection provisions require limitations to urban development and activities, the assessment of vulnerability and risk to contamination must be objective and guarantee credibility. Which methods and techniques are most appropriate for the independent validation of the vulnerability and contamination risk maps?

Answer:
To avoid bias in research and to ensure reliable interpretation of vulnerability indices, the validation of vulnerability maps should be done by default. However, until now, validation has not become a standard practice and there is no commonly accepted method. The following criteria are the most relevant in validating the obtained vulnerability results: travel time of possible contaminant, its peak concentration, and duration of contamination. Various conventional hydrogeological methods and techniques can be used to obtain these aspects: hydrograph, chemograph and bacteriological analyses, tracer techniques, water balances, numerical simulations, and analog studies.

The validation should follow main concepts of vulnerability and should at the same time be performed independently from the map making processes.

Tracer techniques using artificial tracers are the most straightforward and most commonly used validation methods (Fig. 17.12). However, cost-effective artificial tracing can only be applied in limited areas or over a small surface (limited number of injection points). Consequently, the spatial distribution of aquifer vulnerability cannot be deduced. Reliable groundwater flow modeling, such as neural networks and fuzzy logic techniques, and statistical verification, together with geographic information system (GIS) can provide reliable and cost-effective nonpoint verification at a regional scale.

Fig. 17.12 Validation of vulnerability results by execution of an artificial tracer test. By spreading a tracer on the land surface, a simulation of contamination can be done. By monitoring of the tracer appearance at the spring(s), the level of "contamination" can be achieved

17.3 Delineation of Karst Groundwater Protection Zones

Vladimir Živanović

Department of Hydrogeology, Faculty of Mining and Geology, University of Belgrade, Belgrade, Serbia

17.3.1 Introduction

One of very important tasks of modern hydrogeological engineering is the protection of groundwater against pollution. As a result of human activity, groundwater is often exposed to slow and long-lasting pollution. The residence time of groundwater in a different rock formation can be as long as several 1,000 years, such that pollutants remain in the groundwater for an extended period. Remediation of polluted groundwater also requires a long time, as well as enormous money spending. In karst environments, however, the contaminant travel time can be very short, even less than one day, causing such groundwater sources to be classified as highly vulnerable. However, both slow and fast contamination requires a preventative approach to conserve groundwater quality.

In particular, sensitive are drinking water supply sources, where protection generally involves the establishment of various sanitary protection zones, in areas where groundwater flows toward intake structures (tapped springs, wells, galleries, and the like). Restrictive policies are implemented within such sanitary protection zones, which restrict or prohibit activities that might threaten existing or future groundwater sources. Preventative measures aimed at groundwater protection have considerable socioeconomic impacts; however, as a rule, these impacts are of a much smaller magnitude than those of groundwater pollution.

Sanitary protection zones are delineated on the ground surface, generally in concentric spheres around the tapping structure. The number of such zones generally depends on local legislation and prevailing hydrogeological conditions (Fig. 17.13). In practice, the following sanitary protection zones are commonly delineated (Chave et al. 2006):

- *Immediate* sanitary protection zone which is the area adjacent to a spring or well, whose function is to prevent direct input of pollutants through the intake structure;
- *Inner* sanitary protection zone which is usually determined by the time needed for the pathogen count to be reduced to acceptable levels;
- *Outer* sanitary protection zone delineated with respect to the expected time of degradation or reduction in pollutant concentrations. In certain cases, this zone also reflects the time needed to implement remediation measures and remove pollutants whose concentrations are not reduced by filtration through a porous medium; and
- Further, much larger protection zone which encompasses the *entire catchment area* drained via the protected intake structure. This zone is expected to ensure long-term sustainability of groundwater quality.

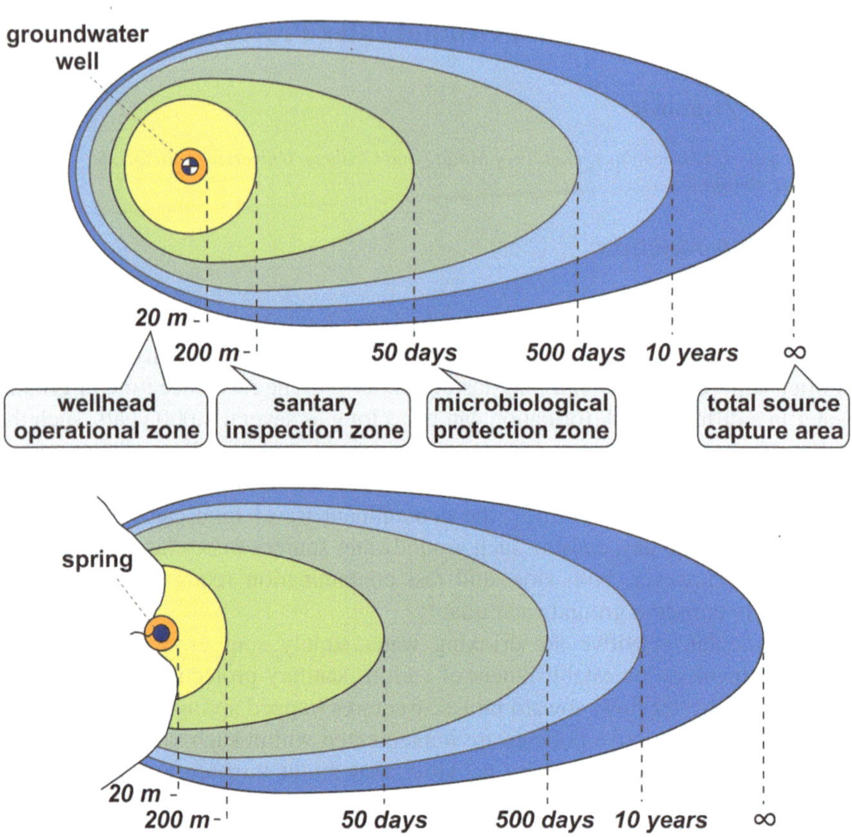

Fig. 17.13 Idealized scheme of groundwater capture areas and transit time perimeters around a water well and springhead (modified after Foster et al. 2002, Published with kind permission of the World Bank. All rights reserved); *above* for unconfined well source, *below* for unconfined spring source

17.3.2 Current Approaches to Sanitary Protection Zoning

The delineation of sanitary protection zones is an inherently uncertain process because the drivers that determine the size of such a zone are often not sufficiently known, and they also vary over space and sometimes over time as well. For that reason, the development of a hydrogeological study of groundwater source protection is a comprehensive and multidisciplinary task. Although the hydrogeologist plays the most important role in the delineation of sanitary protection zones, the task also needs to involve specialists in related disciplines, such as chemistry, microbiology, spatial planning, and administration (Waegeningh 1985). The final outcome of the research activity needs to result in the establishment of sanitary protection zones that constitute a practicable and reasonable solution. The sanitary protection zones should not be too large, so as not to interfere with socioeconomic interests. On the other hand, narrowed-down zones require detailed knowledge of groundwater flow and discharge conditions, which in turn necessitates additional spending.

The sanitary protection zone sizes and shapes depend on several factors (European Commission 2007):

- Hydrogeological characteristics of the aquifer;
- Size of the groundwater source (rate of groundwater extraction);
- Types of pollutant(s) and pollution source(s);
- Vulnerability of the aquifer; and
- Zoning approach.

Sanitary protection zone delineation methods generally fall under three categories: non-hydrogeological, quasi-hydrogeological, and hydrogeological (Kresic 2007, 2009b, 2013). The first group includes approaches where the sanitary protection zone is determined by an arbitrary radius around the source, with no scientific basis. The quasi-hydrogeological group is comprised of methods that include estimation of the well's radius of influence, assessments of the effect of the well on a uniform flow field, and delineation of sanitary protection zones based on hydrogeological modeling not founded upon the outcomes of actual hydrogeological mapping. Additionally, this group includes methods that address maximum well drawdown during pumping, assimilative capacity methods (based on attenuation in unsaturated and saturated zones), and flow boundary methods where the zone boundaries are determined by demarcation of recharge zones and other hydrological features that control groundwater flow (Chave et al. 2006). Although these methods are commonly used for zoning purposes, the outcomes are often debatable because such methods do not consider many significant parameters, such as natural groundwater flow, aquifer heterogeneity and anisotropy, and the safeguard role of the unsaturated zone. The third group includes methods based on hydrogeological models and the results of detailed hydrogeological exploration, where the hydrogeological characteristics of the rocks and the hydrodynamic conditions of the groundwater flow are known rather well.

A general description of commonly used protection zone delineation methods is provided below, with special reference to karst groundwater sources.

17.3.2.1 Specific Hydrogeological Features of Karst in Sanitary Protection Zoning

Infiltration and groundwater flow in karst are often very rapid, such that no geological setting is deemed to be as prone to pollution as a karst aquifer. As such, the delineation of sanitary protection zones in karst requires much more detailed exploration and far better knowledge of the entire system. The zoning focus should be on aquifer heterogeneity and anisotropy. In hydrogeological engineering practice, however, karst systems are often seen as equivalents to porous media and that is why modeling may result in erroneous zoning.

In order to better understand the specific nature of karst aquifer protection, special consideration needs to be given to several facts (Milanović 2000):

- Aquifer recharge is rapid, especially if it takes place through zones of concentrated infiltration;
- Enormous amounts of water tend to flow through karst conduits;

- The groundwater flow velocity is high, generally greater than 2 cm/s, but in some cases it is as high as 20–50 cm/s. Compared to other aquifer types, the residence time in karst is very short;
- The hydrogeological anisotropy of karst is highly pronounced; and
- The self-purification capacity of karst is limited.

Pollution of the water supply source in Walkerton, Ontario, Canada, is a good example of the high vulnerability of karst aquifers and how human lives can be threatened by rapid transport of pollutants (Worthington 2003, 2011; Goldscheider and Drew 2007b; Kresic 2013). Walkerton received its water supply from three wells that tapped a 70 m-thick Paleozoic karst aquifer, overlain by 3–30 m-thick clay sediments. The aquifer was contaminated after heavy rainfall in May 2000. Investigations revealed that cow manure was the source of pathogens. The consequences were seven fatalities and 2,300 people becoming ill. Subsequent research (38 exploratory boreholes, surface and downhole geophysics (logging), pumping tests, and chemical and bacteriological analyses of the water) served as a basis for the development of a numerical model of groundwater flow using MODFLOW and modeling the aquifer as an Equivalent Porous Medium with an effective porosity of 5 %. Modeling of a 30-day residence time in the aquifer produced zones extending 150–290 m around the production wells. Given that the original hydrodynamic model did not provide a suitable solution, supplemental research was undertaken with dye-tracing tests conducted by karst experts (Worthington 2011). The tests revealed that the actual velocities were several hundred times higher. This suggested that the pollution source could have been far more distant than the previous research indicated.

17.3.3 Delineation of Sanitary Protection Zones Based on Fixed Radius and Travel Time

The most commonly used approach to the delineation of sanitary protection zones is a combination of a fixed radius and the pollutant travel time to the tapping structure. The fixed radius method is by far the simplest, where delineation of sanitary protection zones is not based too much on local hydrogeological conditions. The zones are generally represented by polygons or circles around the water-tapping object, with a predefined radius. This approach is followed in many countries (Table 17.3), particularly for the delineation of the **first zone**, i.e., immediate or wellhead/spring protection zone (Fig. 17.14). It usually covers an area of at least 10 m around the tapping structure, regardless of the prevailing hydrogeological conditions. Only a few countries designate a larger area for taping structures in karst (e.g., 30 m in Germany, 20 m in Slovenia and 100 m in Turkey).

The generally applied delineation method for the **second zone** (the inner protection zone) is the time approach. In practice, hydrodynamic modeling is used to delineate the zones in accordance with relevant hydrogeological parameters. A minimum radius is an additional criterion in some countries. In the case of karst aquifers, a 50-day travel time through the aquifer often encompasses the entire catchment area

Table 17.3 Examples of sanitary protection zoning in several European countries, with special reference to karst groundwater sources (modified after COST Action 65 1995, Margane 2003; Chave et al. 2006; Carey et al. 2009)

Country	Wellhead/spring protection zone (Zone I)	Inner protection zone (Zone II)	Outer protection zone (Zone III)	Comment
Croatia	IA: "Immediate spring" IB: "Immediate catchment area"	II: 24 h = "Zone of strict limitation"	IV: 10–50 days = "zone of limited protection"	"Specially protected zone water reserve area" for whole catchment
		III: 1–10 days = "Zone of limitation and control"		
Germany	10 m well	50 days	Whole catchment (subdivided for large catchments, based on radius of 2 km)	II and III zone can be reduced if impermeable overlying layers are present
	20 m spring	Whole catchment for uncovered karst aquifers where $t < 50$ days		
	30 m well in karst aquifers		Whole catchment for uncovered karst aquifers	Zone II not necessary for deep karst aquifers
Portugal	20–60 m	50 days or 40–280 m depending on aquifer type	3,500 days or 350–2,400 m depending on aquifer type	
Serbia	3–10 m	50 days (min 50 m)	200 days (min 500 m)	II and III zone can be reduced if impermeable overlying layers are present
		1 day (min 500 m) for karst aquifers	Whole catchment for karst aquifers	
Slovenia	10 m	50 days (min 50 m)	Whole catchment	Middle zone
	20 m for karst aquifers	<12 h for karst aquifers		400 days (min 25 % of recharge catchment)
				>12 h for karst aquifers
Spain	24 h	50 days (porous)	1 year	
		100 days (karst aquifers)		
Switzerland	10 m	10 days (min 100 m from Zone I)	Double size of the middle zone	
		Based on vulnerability for karst aquifers	Based on vulnerability for karst aquifers	
Turkey	50 m (porous)	50–250 m (porous)	Recharge area	
	100 m (karst)	100–500 m (karst)		

Fig. 17.14 Designation of immediate protection zone for the karst spring of Gaura Mare in Eastern Serbia

(e.g., in Germany), such that in many countries a much shorter travel time is used for the second protection zone (e.g., 12 h in Slovenia, 1 day in Croatia and Serbia, and 10 days in Switzerland). Certain countries use the groundwater vulnerability concept to delineate the second and third zones (e.g., Switzerland), as below discussed.

The **third zone** (outer protection zone) covers the space of 200 days, 1 year or 10 years, or encompasses an entire catchment area, depending on local legislation. In the case of karst groundwater sources, the third zone usually extends across the entire drainage area of the source. This delineation approach may result in a large outer protection zone, such that many countries limit the zone or divide it into two parts where different restrictions apply (e.g., in Germany, the division is based on a 2 km radius).

Despite considerable differences between the numbers of sanitary protection zones and the delineation methods, the travel time generally determines zoning in karst. However, the specific features of karst certainly need to be considered, such that in addition to travel time Milanović (2000) proposes the inclusion of flow velocity, any points of concentrated infiltration, and the like. Based on these criteria, the same author proposes the delineation of four protection zones (Fig. 17.15):

- Zone 1 (protection area of a spring or drinking water intake structure): 50 m upstream from the tapping structure.
- Zone 2 (immediate protection area—zone requiring stringent protection and restrictions): Portion of the catchment area where groundwater travels 24 h or more to the tapping structure, particularly if groundwater conduit flows are faster than 5 cm/s. Should also include Zone 3 ponors from which water travels to the tapping structure in less than 24 h (or more).
- Zone 3 (protection area): needs to include the portion of the catchment area where groundwater travels to the water-tapping facility for 10 days or more. Should also include all Zone 4 permanent or intermittent ponors directly connected to the protected spring or well.
- Zone 4 (external protection area): the remainder of the catchment area. This zone is not directly connected with the intake structure, and flow velocities are less than 1 cm/s.

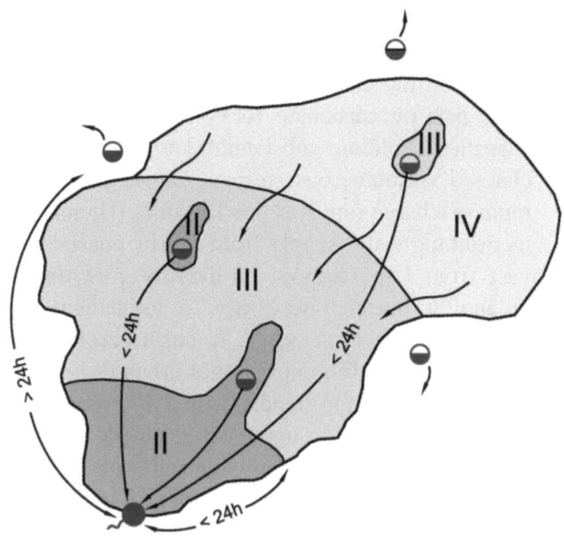

Fig. 17.15 Schematic of the zone of protection against groundwater pollution in karst (Modified after Milanovic 2000)

Box 17.3.1

Case study—Protection of karst groundwater sources based on horizontal travel tame: Rijeka region, Croatia

Protection of karst groundwater worldwide is generally based on the designation of sanitary protection zones with regard to the horizontal travel of groundwater. Various time limits are used for delineation purposes (from 24 h to 50 days for the inner protection zone and from 10 days to the entire catchment for the outer protection zone). As a result of specific issues associated with the protection of karst groundwater sources, certain countries have introduced more than three sanitary protection zones to implement different restrictions and facilitate a trade-off between land use and protection in the drainage area of the source. One such example is Croatia, where five sanitary protection zones are designated for karst groundwater sources (Biondic 2000). The first zone ("strict regime zone") encompasses the area around the water-tapping facility (Zone 1A), as well as the immediate (upland) drainage area of the spring (1B). The second zone ("strict limitation zone") includes the area where groundwater travels for less than 24 h, or parts of the catchment where groundwater flow velocities are >3 cm/s. The third zone ("limitation and control zone") encompasses terrains where the travel time is 1–10 days and the flow velocity 1–3 cm/s. The fourth zone ("limited protection zone") is comprised of parts of the catchment area from which groundwater flows for 50 days and where the flow velocities are less than 1 cm/s. The fifth zone ("specially protected zone") is delineated in hilly and mountains terrains and needs to include the main recharge zones in the catchment area.

This approach to water supply source protection zoning has been implemented in the City of Rijeka, Croatia. The city and its environs lie for the most part on carbonate rocks of the Dinaric karst. The rocks are highly karstified, holding substantial karst groundwater reservoirs that are discharged via numerous springs. The largest is Rječina Spring, whose maximum discharge rate is about 150 m^3/s (Biondic et al. 1998). The city obtains its drinking water supply from nearby coastal springs, whose discharge rates vary from 1 to 15 m^3/s. As the city grew and spread across the carbonate upland, it became necessary to implement comprehensive water supply source protection measures. To ensure groundwater quality in the long term, and at the same time not restrict urban development, five sanitary protection zones were introduced, each with a different level of restrictions (Fig. 17.16).

Even though this approach reconciled protection zone restrictions and urban development, certain shortfalls were identified, primarily those related to the fact that assessments did not take into account the safeguard function of the unsaturated zone. As a result, more recent research (Biondic et al. 1998; Biondic 2000) proposes groundwater vulnerability assessment as an additional criterion for introducing restrictions, particularly in the second zone.

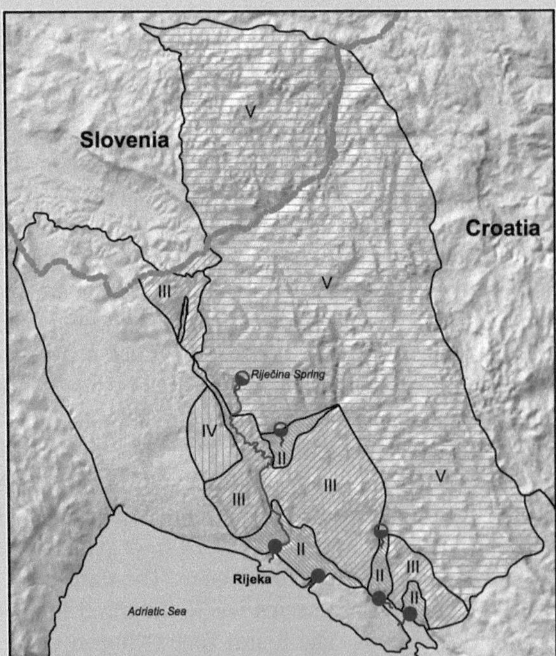

Fig. 17.16 Sanitary protection zones within the territory of the City of Rijeka (modified after Biondic et al. 1998)

17.3.4 Delineation of Sanitary Protection Zones with Vulnerability Assessment

In some countries (such as Ireland, the UK, Switzerland, and Serbia), the preferred sanitary protection zone delineation approach involves the use of groundwater vulnerability maps. Groundwater vulnerability assessment methods generally increase delineation accuracy and improve safety. Such methods assess the safeguard function of the unsaturated zone, which, in combination with other methods (well contribution zone assessment and the like) tend to optimize protection zone delineation, particularly in karst, where this approach is becoming increasingly important.

One example of the use of vulnerability maps to delineate sanitary protection zones is discussed in the so-called Irish approach to groundwater vulnerability assessment (Department of Environment and Local Government, Environmental Protection Agency, and Geological Survey of Ireland 1999). A map of sanitary protection zones is generated by overlaying a map of source protection areas (showing the zone of contribution) and a groundwater vulnerability map. As a result, four levels of protection can be identified in the inner protection zone (SI/E, SI/H, SI/M, and SI/L—extreme, high, moderate, and low vulnerability of the inner protection zone). The same principle applies to the outer protection zone (SO/E, SO/H, SO/M, and SO/L). Different restrictions apply in each zone.

One of the first applications of the groundwater vulnerability approach to delineate karst groundwater source sanitary protection zones was the EPIK method (Doerfliger and Zwahlen 1997). According to the authors, the vulnerability of a karst aquifer is a direct function of epikarst development, soil composition and thickness, infiltration conditions, and the extent of the karst network. Sanitary protection zones are directly related to the level of groundwater vulnerability.

17.3.5 Delineation of Sanitary Protection Zones Using a Combined Approach

In recent years, groundwater vulnerability maps have increasingly been used in karst groundwater source protection. The list of methods for creation of vulnerability maps is presented in Sect. 17.2. For the sanitary protection zoning often is used the “European approach” (Zwahlen 2004c) which combines vulnerability map with the fixed radius or travel time. For the karst groundwater source protection, the identification of four factors (O, C, P, and K) is required. The first three are related to the vulnerability of the resource, while K is used for source vulnerability assessment.

One of the methods that is fully aligned with the conceptual model of the “European approach” is the Slovene approach (Ravbar and Goldscheider 2007b, see Sect. 17.2), developed for the purposes of protecting karst aquifers and springs in Slovenia (Box 17.3.2). Another method, called PaPRIKa (Kavouri et al. 2011), for assessing resource and source vulnerability, has been developed in France to define protection zones as accurately as possible. Source vulnerability is defined by means of an I_{source} map, where vulnerability is not only a function of infiltration conditions, but also of the groundwater travel time. In addition, four isochrones (12, 24, 36, and 48 h) are proposed, resulting in four source vulnerability maps showing source vulnerability as a function of time.

Box 17.3.2

Case study—Use of vulnerability assessment maps in groundwater source protection: Podstenjšek Springs (Slovenia)

As proposed by the "European approach," in order to determine source vulnerability, a map of factor K needs to be developed in addition to resource vulnerability maps. The purpose of the K factor map is to better define karst groundwater flow in the aquifer zone. One of the first implementations is the Slovene approach to the protection of Podstenjšek Springs in southwestern Slovenia (Ravbar and Goldscheider 2007b).

Factor K was defined on the basis of three sub-factors: t, n, and r. The first sub-factor involves information on travel time in the saturated zone. According to Slovene legislation, three classes were proposed: <1 day, 1–10 days, and >10 days. The results of dye-tracing tests conducted under relatively high water conditions were needed to determine sub-factor t. Sub-factor n indicated the presence (or absence) of an active conduit network. According to this factor, high vulnerability encompassed parts of the terrain above these conduits. The third sub-factor, r, needed to describe the degree of connection and contribution of the parts of the karst aquifer that were permanently or intermittently connected with the considered groundwater source. When these three sub-factors were superimposed, a map of factor K was produced which, in combination with the resource vulnerability map, resulted in a vulnerability map of Podstenjšek Springs (Fig. 17.17). This map can be used to produce sanitary protection zone maps of the springs by simple transformation of different vulnerability classes into different protection zones (Ravbar 2007b).

Despite the fact that the use of vulnerability maps considerably increases the level of detail in sanitary protection zoning, this approach has certain lacks, primarily those related to the merit points subjectively given to certain parameters used in the assessment and insufficient reliance on travel time.

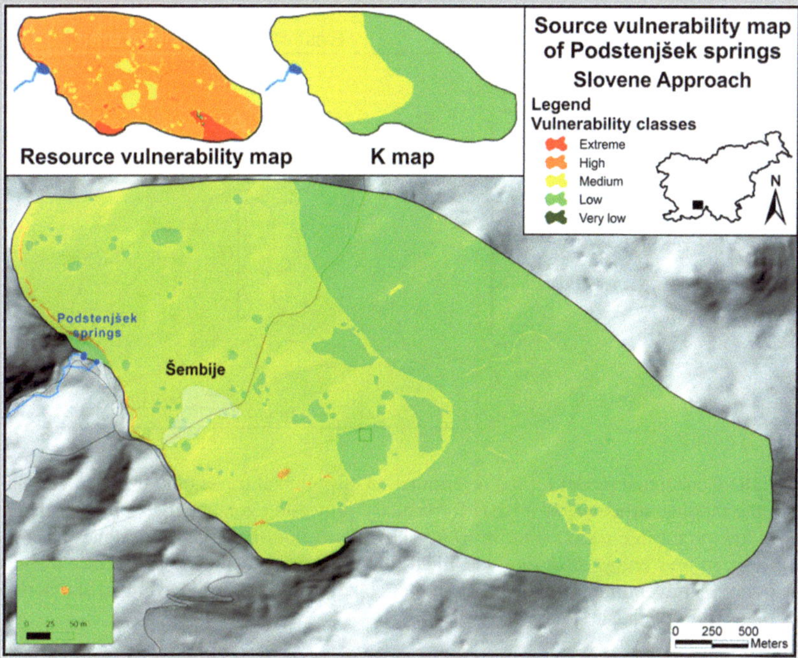

Fig. 17.17 Source vulnerability map of the Podstenjšek springs catchment area assessed by the Slovene approach (Ravbar 2007)

One of the main disadvantages of the groundwater vulnerability approach to protection zone delineation is that source vulnerability is not solely a function of time, especially when surface water travel to the ponor and travel through the unsaturated zone are evaluated. This increases subjectivity and renders the validation of results rather difficult. The travel time through the above-aquifer zone has been incorporated in the vulnerability of karst groundwater by the Time-Input Method (Kralik and Keimel 2003). This approach can be optimized further and used to delineate sanitary protection zones. Namely, since the Time-Input Method produces a map where each point represents the time needed for a pollutant to reach the aquifer, the time can be combined with the estimated travel time of groundwater from that point to the tapping structure (Fig. 17.18, point A). Consequently, each point on the map represents the total travel time of the pollutant to the spring or well (Total Karst Travel Time). Given that, it is extremely important to predict the travel of a potential pollutant in the drainage area of a ponor or sinking stream in karst, this method can be considerably enhanced by analysis of surface water velocity and travel time to the ponor (Fig. 17.18, points B and C).

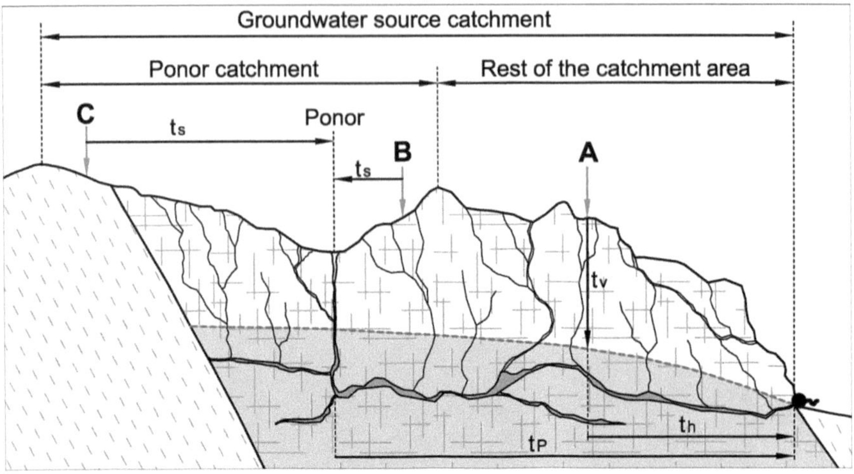

Fig. 17.18 Conceptual model for the determination of pollutant travel time from the ground surface to the tapping structure, as a basis for sanitary protection zone delineation (Fm. first from the *left*) low-permeable rocks; (Fm. in *white* above *dash line*) karstified limestone (unsaturated zone); (Fm. in *grey* below *dash line*) karstified limestone (saturated zone); t_s surface travel time to the ponor; t_P travel time from the ponor to the intake structure; t_v and t_h vertical and horizontal travel times from point *C* to the intake structure

The total travel time of the water (Karst Total Travel Time), and likewise of the pollutant, from the ground surface to the water-tapping facility can be estimated as follows:

$t_{tot} = t_v + t_h$ (for parts of the karst beyond the drainage area of the ponor or sinking stream, for example point A)

$t_{tot} = t_s + t_P$ (for parts of the karst within the drainage area of the ponor or sinking stream, for example, points B and C)

where:

t_{tot} total travel time;
t_v vertical travel time from the ground surface to the water table;
t_h horizontal travel time to the water-tapping facility;
t_s surface travel time within the ponor drainage area; and
t_P travel time from the ponor to the water-tapping facility.

Certain components used in the estimation of the total travel time can be calculated directly (from dye-tracing tests) or indirectly (based on knowledge about the hydrogeological system).

The surface travel time in the drainage area of a ponor (t_s) is the time a surface runoff particle needs to go from a point in the ponor drainage area to the ponor. The factors that affect the travel time are the length of the flow, the terrain slope and roughness, and the intensity of precipitation (Conservation Engineering Division 1986). The surface water travel time is usually computed as the sum of the times of several segments of the surface stream (Fig. 17.19). Sheet flow

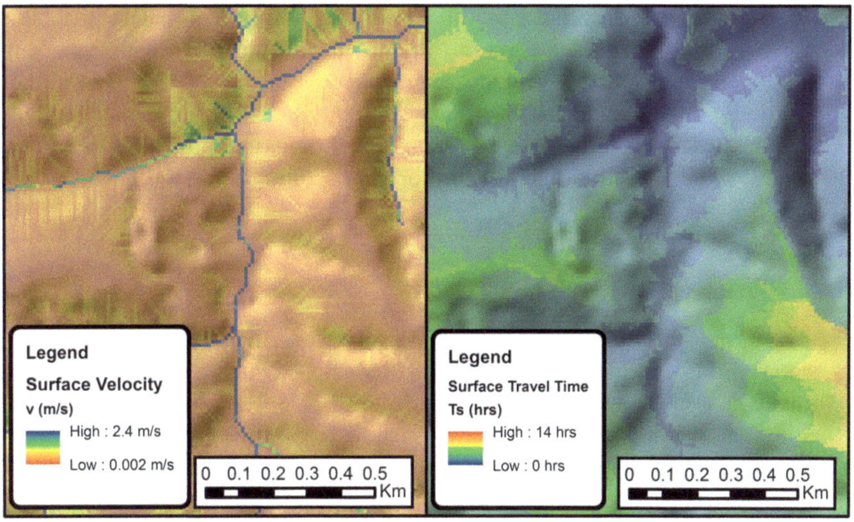

Fig. 17.19 Estimation of surface water velocity (*left*) and travel time (*right*) based on terrain slope and roughness (CORINE layer) and precipitation intensity

is generally found in the highest parts of the catchment area. Such flow occurs over short distances, usually up to 100 m, after which sheet flow becomes shallow concentrated flow. When surface water reaches the channel of an intermittent or perennial surface stream, shallow concentrated flow becomes open channel flow. Each of these components can be calculated applying Manning's equation for sheet flow or open channel flow. Dye-tracing tests constitute the basis for estimating travel time from the ponor to the tapping structure (t_P).

For parts of the catchment area of the water supply source outside the drainage area of the ponor, the travel time of polluted water from the ground surface to the groundwater source can be derived directly, based on the travel time of an injected or natural dye-tracer (Pronk et al. 2009; Worthington 2007), or indirectly by evaluating the vertical t_v and horizontal t_h travel times (Fig. 17.18). The first component can be determined directly using the Time-Input Method, while speleological surveys, borehole dye-tracing, and the like facilitate calculations of horizontal travel time. Indirectly, the travel time to the water-tapping facility, or the residence time along the way, can be assessed by spring hydrograph analysis. Based on a discharge study of more than 200 wells in Italy, Civita provided an example (2008) of how groundwater flow velocity can be assessed using the annual Maximum Discharge Half-Time (MDHT), a parameter that represents the time needed for the maximum recorded discharge of a spring at the beginning of a recession period to drop to Qmax/2.

A groundwater source vulnerability map is produced by adding up the travel times from each point in the catchment area of the source to the intake structure. Given that vulnerability is a function of time, sanitary protection zones can then easily be delineated as required by local legislation.

One of the advantages of this approach to source vulnerability assessment is the fact that vulnerability is a function of the travel time of a pollutant, which makes it easier for the hydrogeologist to validate results. Given that the water table in karst can vary by as much as 300 m (Milanović 2000) and that within the same system the groundwater flow velocity can vary by a factor of 5–10 or even more (Göppert and Goldscheider 2008; Milanović 2000), by assessing the vulnerability of a groundwater source based on the total travel time, it is possible to determine the needed protection for different hydrological conditions.

Box 17.3.3

Case study—Delineation of sanitary protection zones based on surface water and groundwater travel times: Blederija Spring (Serbia)
Blederija is a karst spring whose average discharge is 250 l/s (Prohaska et al. 2001). It drains the Miroč karst massif in eastern Serbia. Most of the groundwater is stored in a karst aquifer formed in fractured and highly karstified massive Upper Jurassic limestones. A significant portion of the spring catchment area is made up of non-carbonate formations, especially in its Western part where a network of surface streams is formed, which deplete as they pass through the karstic parts of the catchment area. At times of high water flows, most of the surface streams sink via the Cvetanovac ponor, which is directly connected with Blederija Spring (Fig. 17.20).

The Karst Total Travel Time Model (Fig. 17.18) was used to delineate sanitary protection zones. The following calculations were included:

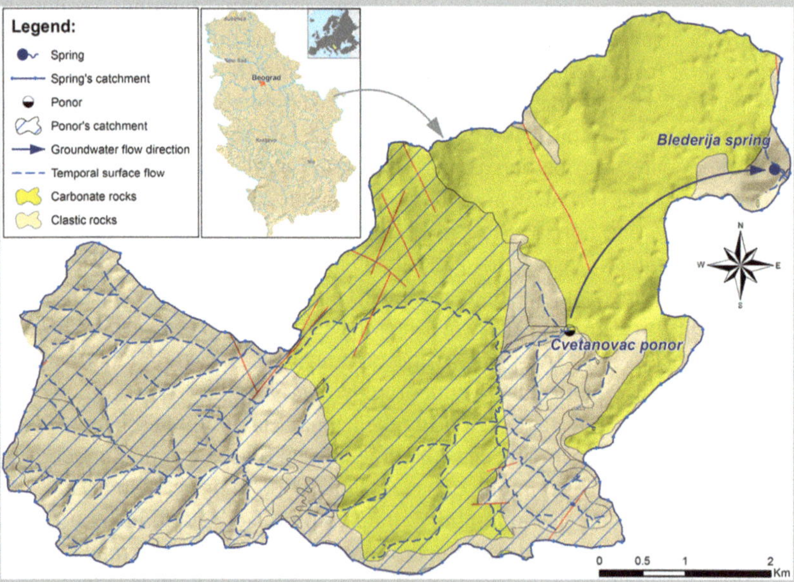

Fig. 17.20 Map showing carbonate outcrops and location of the Cvetanovac ponor in the catchment area of Blederija Spring

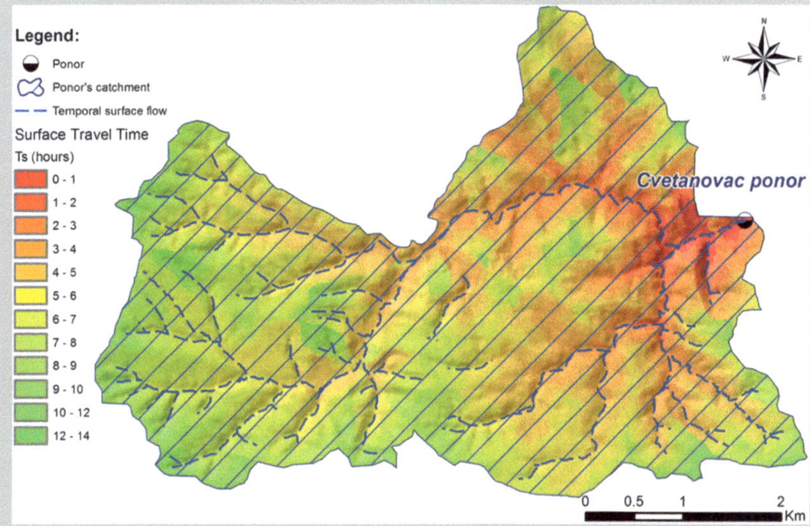

Fig. 17.21 Surface travel time t_S (in hours) in the ponor's catchment area

(a) t_S—Surface water travel times from each point in the drainage area of the Cvetanovac ponor to the ponor itself. The time was the sum of the times of sheet flow, shallow concentrated flow, and open channel flow. A map was produced (Fig. 17.21), showing the time needed for the surface water, and thus the contaminant as well, to reach the ponor zone. When the map of total travel time in the catchment area of Blederija Spring was generated, $t_P = 2.5$ days were added to each point in the catchment area, which relate to the time a dye-tracer took to travel from the ponor to the spring.

(b) t_V—The travel time from the ground surface to the saturated zone was estimated for each point beyond the drainage area of the ponor, applying the Time-Input Method (Kralik and Keimel 2003). First, the thickness and hydraulic conductivity of the soil and rock strata in the above-aquifer zone were assessed at each point, and then the travel time through this zone (the time factor) was estimated. The time was additionally adjusted for the infiltration rate at each point (input factor). The produced map (Fig. 17.22, left) shows that in most of the area less than eight days were needed for the water from the surface to be infiltrated into the saturation zone. The only exceptions were the parts of the karst covered with semi-permeable rocks, in the south-western and eastern sections outside from the ponor's drainage area.

(c) t_h—The horizontal travel time to the spring was also assessed for each point beyond the ponor's catchment area. Dye-tracer test results were the basis for these calculations. The mean flow velocity through the

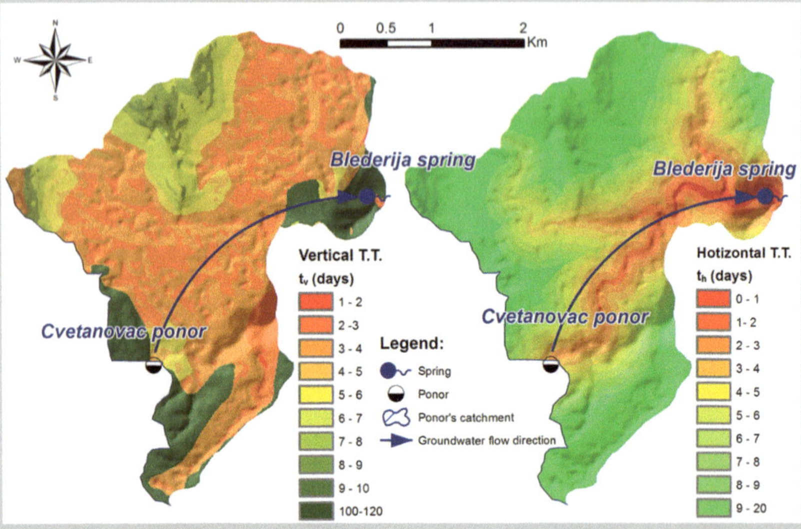

Fig. 17.22 Vertical t_{v} (*left*) and horizontal t_{h} (*right*) travel times (in days) in the remainder of the catchment area of Blederija Spring

main conduits, whose routing was assumed from the configuration of the surface karst features, was calculated using the tracing time. The lateral groundwater travel time toward the main conduits was calculated using significantly lower velocities. The produced map of horizontal travel time (Fig. 17.22, right) shows that considerably more time was needed in the parts of the terrain away from the main conduits, for the water to reach the spring.

The data that represented different components of the travel time were overlaid in an ArcGIS environment (ESRI 2011), and a source vulnerability map was produced where each point in the catchment area showed the estimated travel time to the spring (Fig. 17.23). Such a map can easily be converted into a map of sanitary protection zones, as required by applicable legislation. The map of the total water travel time demonstrates how important it is to analyze travel time in the drainage area of a ponor. In such drainage areas, water (or consequently a pollutant) needs much less time to reach a water supply source, compared to some parts of the drainage area where diffuse infiltration is predominant.

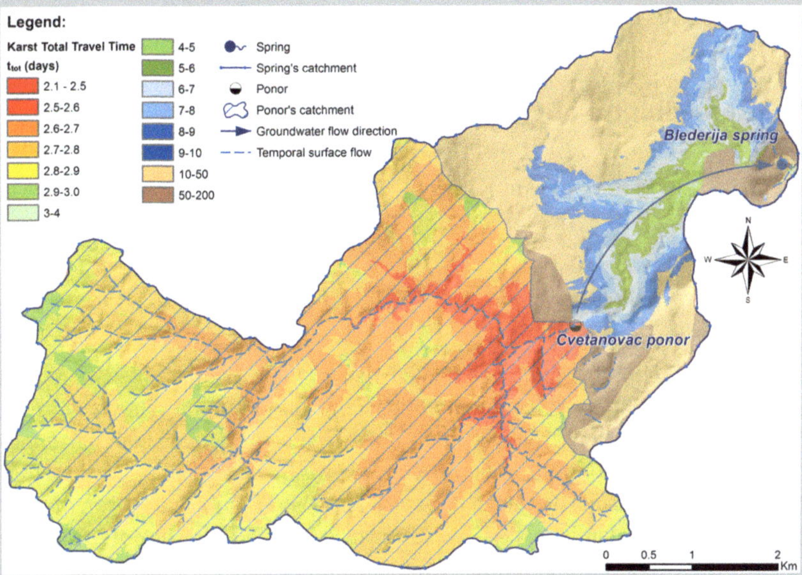

Fig. 17.23 Map of total water travel time to Blederija Spring

17.3.6 Delineation of Sanitary Protection Zones Using Groundwater Models

In the past, hydrodynamic models have rarely been used to delineate sanitary protection zones of water supply sources in karst, due to the need for a considerable amount of data which were not available in many cases. The use of this model can result in significant errors, as demonstrated in the Walkerton case. However, great progress was made in the functional simulation of karst systems with the advent of MODFLOW CFP (Reimann and Hill 2009) and MODFLOW-USG (Panday et al. 2013). However, these software packages are still under development and the modules needed to delineate sanitary protection zones, such as fate and transport modeling and particle tracking modules, are still not available (Kresic 2013).

17.3.7 Monitoring in Support of Groundwater Source Protection

The implementation of sanitary protection zones is only one segment of karst groundwater source protection. Another equally import segment is the establishment of a robust surface water and groundwater quality monitoring network. The best locations for surface water monitoring are reservoir entry and exit points, while in the case of groundwater, the preferred points are ponors, springs, piezometers, and wells connected to a karst conduit network (Milanović 2000, see also Chap. 12).

Acknowledgments I express my gratitude to Veselin Dragišić and Igor Jemcov for all valuable advices and to Nataša Ravbar for providing several maps.

17.4 Remediation of Groundwater in Karst

Alex Mikszewski and Neven Kresic

Amec Foster Wheeler, Kennesaw, GA, USA

17.4.1 Introduction

At the vast majority of contaminated karst sites in the USA, the involved regulatory agencies and consultants alike attempt their remediation in the same way as in any other, non-karstic setting, by applying either a prescribed, cookie-cutter approach or the most popular current country-wide approach. Karst sites are therefore no different than all other sites in that, as we learn more and more over time about the behavior of contaminants in the subsurface and the efficacy of various technologies, the preferred technology from regulatory and commercial perspectives changes frequently. Changing trends in the preferred remedial technology selected at Superfund sites are presented on Figs. 17.24 and 17.25, from USEPA (2013).

One complicating factor at many karst sites is that the karstified rock is covered by an uneven layer of residuum which may be tens of feet thick. Since, in most cases, the contamination of the underlying karst aquifer had to come from the land surface and through the residuum, the unconsolidated residuum sediments are part of the problem as well. It is for this reason that various remedial technologies which may work for residuum (e.g., SVE or injection-based technologies

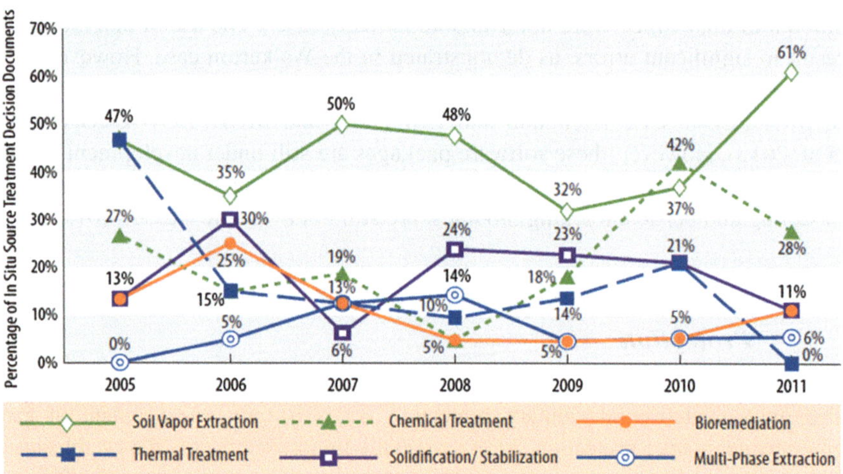

Fig. 17.24 Percentage of preferred in situ remedial technologies for source zone treatment selected at superfund sites (USEPA 2013)

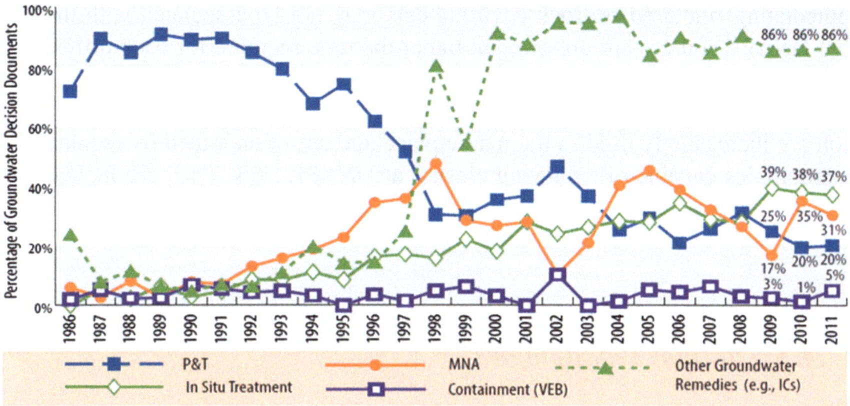

Fig. 17.25 Percentage of preferred remedial technologies for dissolved phase (groundwater plume) treatment selected at superfund sites (USEPA 2013)

such as ISCO or bioremediation) do not work effectively in karst. The presence of very high permeable conduits of unknown extent makes remediation in karst more challenging than in any other hydrogeological environment.

17.4.2 In Situ Treatment Technologies

Remediation at individual sites can cost anywhere from several hundred thousand dollars to tens of millions of dollars (for example, see McDade et al. 2005), and it is rarely (if ever) possible to remove all of the contaminant. The benefit of source remediation efforts is that by removing source mass, it tends to reduce the mass discharge to the plume (Rao et al. 2001; Falta et al. 2005a; Jawitz et al. 2005). The reduced plume loading following source remediation may or may not be sufficient to allow natural attenuation processes to keep the plume concentrations within acceptable limits (Falta et al. 2005a, b).

As emphasized by the USEPA, a reasonable strategy for many sites would be some combination of source and plume remediation. Selection of the optimal remedy for a site, in terms of the degree of remediation, must consider the inherent coupling of the source remediation to the plume remediation (Falta et al. 2007).

As demonstrated by Lipson et al. (2005), back diffusion from the rock matrix to the open fractures can occur for very long periods of time following the removal of DNAPL sources in fractured bedrock. The back diffusion process takes a longer period of time to remove a certain amount of contaminant mass from the rock matrix than the forward diffusion process takes to initially place the mass into the rock matrix. This stems from the fact that the forward diffusion process is unidirectional, whereas the back diffusion is associated with local diffusion back into the open fractures while some forward diffusion is still taking place further in the rock matrix. The implication of this result is that the timescales associated with remediating fractured bedrock environments may not be associated with the time that DNAPL sources are present, but rather the time required for back diffusion of aqueous phase contamination from the rock matrix.

In situ thermal treatment, in situ chemical oxidation, and in situ bioremediation are increasingly used in the marketplace and are often touted by regulators as technologies capable of achieving closure at DNAPL sites. They are increasingly being applied at fractured rock and karst sites. A discussion of these technologies is presented in the following section.

17.4.3 Thermal Technologies

Thermal remediation technology is derived from the petroleum industry, where for many years subsurface heating and steam injection have been used to enhance oil recovery from high-gravity deposits and oil sands and oil shales. The application of thermal technology to groundwater remediation projects is similar, as in many cases contamination is caused by NAPLs such as oils and chlorinated solvents. In situ thermal treatment is an enhanced form of SVE, as volatilized contaminants must be captured and treated in the vapor phase. Thermal treatment also typically involves multi-phase extraction wells. Thermal heating of contaminated zones greatly enhances mass removal relative to ambient temperature SVE by (from USACE 2009):

- Increasing contaminant vapor pressures;
- Decreasing NAPL viscosities;
- Increasing contaminant solubilities and diffusivities; and
- Increasing biological activities.

In addition, the thermal conductivity of soil is generally less variable than conventional remediation parameters such as soil permeability (hydraulic conductivity). Therefore, in situ thermal treatment can address low-permeability materials such as silt and clay better than injection-based technologies. The three primary commercially available forms of in situ thermal treatment are as follows:

- Electrical Resistivity Heating (ERH);
- Steam-enhanced Extraction (SEE); and
- Conductive Heating.

For a detailed description of these technologies, the reader is referred to USACE (2006, 2009), USEPA (1998a, 2004), Davis (1997, 1998), Powell et al. (2007), Kresic (2009c), and Kresic and Mikszewski (2013).

The predominant design concerns for ERH and conductive heating systems are moisture loss and high groundwater flows (Kingston et al. 2009). High groundwater seepage velocities in excess of several feet per day can cause significant heat losses as warm water is continuously flushed out of the treatment area and replaced with cool water. This limitation is typical of karst aquifers with preferential flowpaths. A management system consisting of up-gradient pumping wells

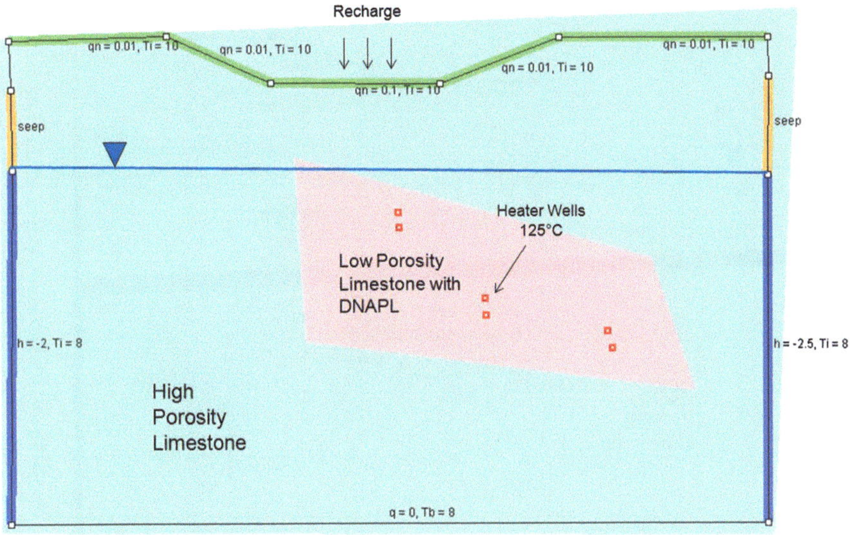

Fig. 17.26 Setup of a heat transfer numeric model for testing remedial concept of thermal treatment

and/or down-gradient injection wells can help reduce excessive groundwater flux and associated heat losses (Kingston et al. 2009).

The influence of high groundwater flux preferential flowpaths within karst can be seen using thermal conduction modeling in VS2DHI, a two-dimensional unsaturated–saturated heat transfer model produced by the USGS. This example illustrates the value of modeling for remedial feasibility assessment and design.

The objective of the thermal remediation is to heat a low porosity, low-permeability matrix within a shallow limestone formation that has retained chlorinated solvents DNAPL. The goal of the thermal treatment is to heat this matrix to 80 °C, a temperature determined during bench testing at which most target contaminants volatilize. The setup of the remedial concept and model is presented in Fig. 17.26. The temperature distribution after approximately 2 months of heating is presented in Fig. 17.27. While the remedial goal has not been met across the low-porosity unit, heat transfer is occurring between the heater wells.

Fig. 17.27 The temperature distribution after approximately 2 months of simulated heating with the model setup shown in Fig. 17.26

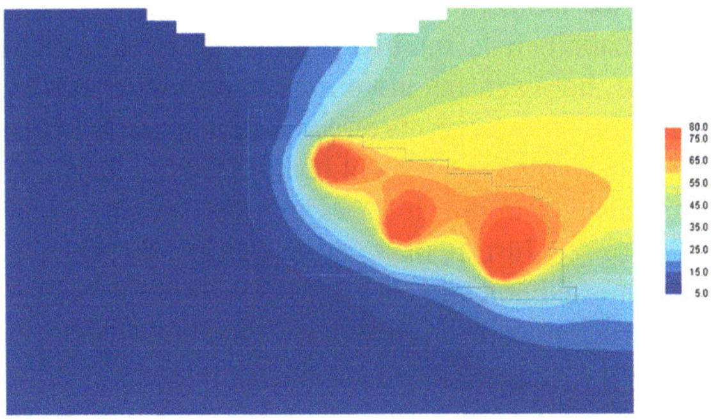

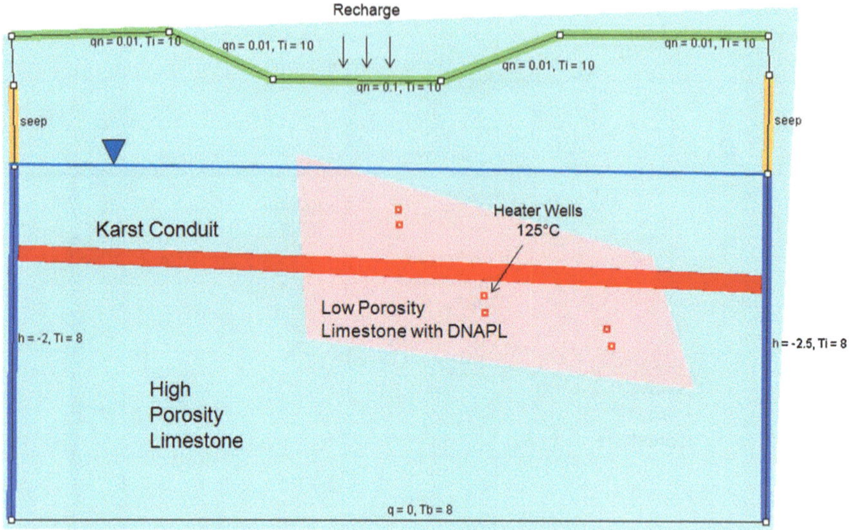

Fig. 17.28 Numeric model setup for simulating a high transmissivity karst conduit intersecting the low-permeability matrix

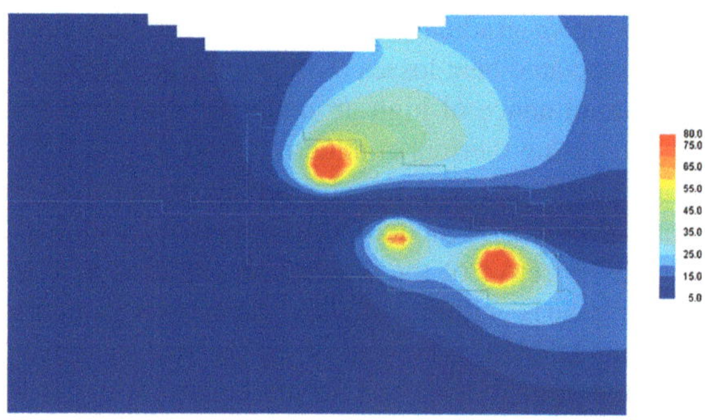

Fig. 17.29 The temperature distribution after approximately 2 months of simulated heating with the model setup shown in Fig. 17.28

Figure 17.28 presents an alternative situation where a high transmissivity karst conduit intersects the low-permeability matrix. This preferential flow pathway will convey a large flux of groundwater through the treatment zone. Figure 17.29 presents the temperature distribution after approximately 2 months of heating with the conduit. It is intuitive that water in the conduit itself cannot practically be heated; however, the more important implication is that the conduit acts as a heat sink. This sink prevents effective heat transfer between the heater wells and results in a much smaller radius of heating around the wells.

Preferential flow features in karst present a challenge to any in situ remedial technology. Just as the conduit quickly disperses heat, it would quickly disperse any injected amendment for ISCO or bioremediation, preventing effective contact with contaminants in the matrix. One must locate conduits and account for them in remedial design to avoid wasting resources.

SEE for hazardous waste remediation involves the use of steam injection wells to create a pressure gradient for recovery of NAPLs and to heat the subsurface to volatilize and extract contaminants in the vapor phase. Similar to ERH, SEE is capable of heating the subsurface to a maximum temperature equal to the boiling point of water or the steam distillation temperature at approximately 100 °C. After NAPL has been displaced and recovered in the liquid phase, pressure cycling may be used to enhance mass removal in the vapor phase. Pressure cycling promotes contaminant volatilization by creating thermodynamically unstable conditions within soil pores (USACE 2006).

As emphasized by Gudbjerg (2003) in the case of non-karstic granular porous media, SEE technology is expensive and difficult to implement and should only be used in highly contaminated source zones where rapid remediation is required. Steam injected in the saturated zone is strongly influenced by buoyancy, which causes the steam zone to rise to the unsaturated zone, depending on the anisotropy of the porous media. Only in simple cases, it is possible to predict steam zone development in the saturated zone without doing a full-scale complex numerical simulation. Downward migration of contaminant may also be a potential problem when remediating with steam. Although there are no widely reported cases or analyses of the SEE applications in karst, the above complexities are sure to be accentuated.

17.4.4 In Situ Chemical Oxidation (ISCO)

In situ chemical oxidation (ISCO) involves the introduction of a chemical oxidant into the subsurface for transforming groundwater or soil contaminants into less harmful chemical species. There are several different forms of oxidants that have been used for ISCO; however, the following four are the most commonly used:

1. permanganate (MnO_4^-),
2. hydrogen peroxide (H_2O_2) and iron (Fe) (Fenton-driven or H_2O_2-derived oxidation),
3. persulfate ($S_2O_8^{2-}$), and
4. ozone (O_3).

The type and physical form of the oxidant indicates the general materials handling and injection requirements. As discussed by Huling and Pivetz (2006), the persistence of the oxidant in the subsurface is important since this affects the contact time for advective and diffusive transport and ultimately the delivery of oxidant to targeted zones in the subsurface. For example, permanganate persists for long periods of time, and diffusion into low-permeability materials and greater transport distances through porous media is possible. H_2O_2 has been reported to persist in soil and aquifer material for minutes to hours, and the diffusive and advective transport distances will be relatively limited. Radical intermediates formed using hydrogen peroxide, persulfate, and ozone that are largely responsible for various contaminant transformations react very quickly and persist for very short periods of time (<1 s).

Four major factors play a role in determining whether an oxidant will react with a certain contaminant in the field:

1. kinetics,
2. thermodynamics,
3. stoichiometry, and
4. delivery of oxidants.

On a microscale, kinetics or reaction rates are perhaps the most important. In fact, reactions that would be considered thermodynamically favorable based on E_0 values may be impractical under field conditions. The rates of oxidation reactions are dependent on many variables that must be considered simultaneously, including temperature, pH, concentration of the reactants, catalysts, reaction by-products, and system impurities such as natural organic matter (NOM) and oxidant scavengers (ITRC 2005). The oxidant chemicals react with the contaminants, producing innocuous substances such as carbon dioxide, water, and—in the case of chlorinated compounds—inorganic chloride. However, there may be many chemical reaction steps required to reach those end points, and some reaction intermediates, as in the case of polyaromatic hydrocarbons and organic pesticides, that are not fully identified at this time.

The remediation of groundwater contamination using ISCO involves injecting oxidants and potentially amendments directly into the source zone and downgradient plume. If the contamination is relatively shallow, in the residuum zone and immediately below in the karst aquifer, it may sometimes be possible to overdose the residuum with the oxidant in an attempt to deliver it to the underlying karst aquifer over wider areas. Typical delivery methods include injection through temporary or permanent wells, gravity infiltration through trenches, and direct mixing or tilling into soil. For deeper contamination in bedrock, it will be necessary to install a dense array of bedrock wells at a much higher cost. In any case, the main challenge with ISCO in karst, similar to all other fluid-based remediation technologies, is the feasibility of delivering fluids to the target area, assuming such area is well defined, at the right dosage and residence time. In order to oxidize contaminants diffused into the rock matrix and/or moving with slow advective transport, the oxidant has to persist in such portions of the aquifer long enough. On the other hand, oxidant injected directly into or entering preferential flowpaths (conduits) where groundwater and contaminant flow is rapid will obviously be quickly diluted and washed away from the target zone. For these reasons, ISCO applications in karst in many cases will not be feasible.

In general, the major benefits of ISCO are that minimal remedial waste material is generated and that treatment can sometimes be completed in a relatively short timeframe (i.e., weeks or months versus years). However, in case of non-aqueous phase liquids (NAPLs), oxidants that are in a water-based solution will only be able to react with the dissolved phase of the contaminant, since the two will not mix. As a result, treatment kinetics may become desorption limited, and the destruction of NAPLs, although possible, may require a cost-prohibitive oxidant dose and multiple applications (ITRC 2005).

Contaminants amenable to treatment by ISCO include benzene, toluene, ethylbenzene, and xylenes (BTEX); methyl *tert*-butyl ether (MTBE); total petroleum hydrocarbons (TPH); chlorinated solvents (ethenes and ethanes); polyaromatic hydrocarbons (PAHs); polychlorinated biphenyls (PCBs); chlorinated benzenes (CBs); phenols; organic pesticides (insecticides and herbicides); and munitions constituents (RDX, TNT, HMX, etc.).

A recent advance in ISCO technology is the development of solid-phase oxidants that can be placed in the subsurface to dissolve passively over several years. For example, RemOx® SR is a slow-release form of potassium permanganate that has been dispersed in a solid, paraffin wax cylinder. When placed in the subsurface, potassium permanganate dissolves and diffuses from the wax matrix into surrounding groundwater. While this technology cannot practically deliver the thousands of pounds of oxidant necessary for DNAPL source zone remediation, it has great promise in the treatment of residual, low-level contamination in low-permeability strata, or as part of an ISCO barrier to plume migration (analogous to a permeable reactive barrier).

This technology is not applicable for karst conduits with high groundwater flow rates, as the cylinders would dissolve in a very short timeframe. However, they may be useful in treating contamination in residuum or in the limestone matrix. The longevity of permanganate cylinders in the subsurface can be evaluated at sites using a dissolution model provided by Carus Corporation. Model input parameters and example output are displayed in Figs. 17.30 and 17.31.

17.4.5 Bioremediation

In situ enhanced bioremediation of halogenated volatile organic compounds (VOCs), and in particular of chlorinated aliphatic hydrocarbons (CAHs), is the most rapidly growing groundwater remediation technology. This is not surprising since VOCs are the most frequently occurring type of contaminant in soil and groundwater at Superfund and other hazardous waste sites in the USA. Innovative technologies, such as in situ bioremediation, are being developed and implemented in an effort to reduce the cost and time required to clean up those sites. In situ bioremediation is increasingly being selected to remediate hazardous waste sites because, when compared to other technologies, it is usually less expensive, does not require waste extraction or excavation, and is more publicly acceptable as it relies on natural processes to treat contaminants.

Engineered bioremediation is a technology of adjusting the concentration of various electron acceptors, electron donors (substrates), and nutrients in groundwater in order to stimulate biodegradation of contaminants by native (indigenous) microorganisms. This process is commonly referred to as biostimulation. Bioremediation also includes bioaugmentation or the addition of microbes to the subsurface where organisms able to degrade specific contaminants are deficient. Such microbes may be "seeded" from populations already present at a site and grown in aboveground

Oxidant Release Parameters	Symbol	Value	Cell Type
Oxidant	$KMnO_4$	permang	
Cylinder Diameter (inches)	d	1.35	
Cylinder Diameter (cm)	d	3.43	af
Initial Cylinder Radius (cm)	ro	1.715	af
Oxidant solubility (g/cm^3)	Cs	0.03	given
Mass of Cylinder (g)		815	af
Mass of MnO_4 (g)		652	af
% MnO_4 available		0.9	given
Mass of MnO_4 available (g)		587	af
Effective diffusion coefficient (cm^2s^{-1})	De	4.000E-07	given
Amount of available oxidant (g/cm^3)	A	1.39	af
Cylinder Height (cm)	h	45.7	given
Cylinder Volume (cm^3)	V	422.2	af
Cylinder Density (g/cm^3)	ρ	1.9	given
Molecular Weight of MnO_4 (g/mol or mg/mmol)	mw	119	given

Time and distance of Interest	Symbol	Value	Cell Type
Time of Interest (days)	t	171	sp
Time Step (days)	Δt	2	given
Downgradient Distance of interest (ft)	x	20	sp

Site Characteristics	Symbol	Value	Cell Type
2nd order NOD rate (L/mmol-day)	K_2	0.0245	sp
2nd order NOD rate (L/mg-day)	K_2	2.06E-04	af
Hydraulic conductivity (cm/s)	K	3.70E-04	sp
Hydraulic conductivity (cm/d)	K	32.0	af
Hydraulic gradient (dh/dl)	dh/dl	0.010	sp
Porosity	n	0.02	sp
Groundwater Velocity (cm/day)	v	15.98	af
Cross Sectional Area of Flow (cm^2)	CSA	246	af
Groundwater Flow (L/day)	Qf	3.9	af

Fig. 17.30 Input parameters for a dissolution model evaluating longevity of permanganate cylinders in the subsurface, courtesy of Carus Corporation

reactors, or they can be specially cultivated non-indigenous (exogenous) strains of bacteria having known capabilities to degrade specific contaminants.

An electron acceptor is a compound capable of accepting electrons during oxidation–reduction reactions. Microorganisms obtain energy by transferring electrons from electron donors, such as organic compounds (or sometimes reduced inorganic compounds, such as sulfide), to an electron acceptor. Electron acceptors are compounds that are reduced during the process and include oxygen, nitrate, iron (III), manganese (IV), sulfate, carbon dioxide, and, in some cases, chlorinated aliphatic hydrocarbons, such as carbon tetrachloride (CT), PCE, TCE, DCE, and VC. Electron donors are compounds that are oxidized during the process and include fuel hydrocarbons and native organic carbon (USEPA 2000). Hydrogen is the preferred electron donor for reductive dechlorination

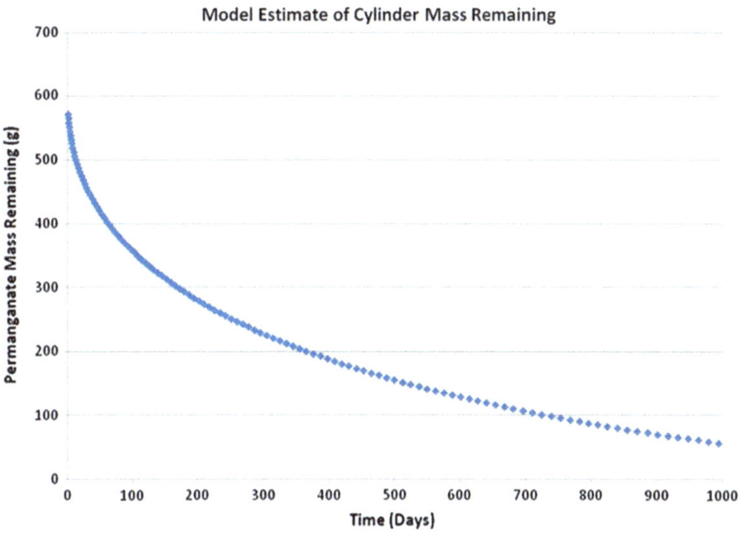

Fig. 17.31 Result of the permanganate dissolution model run using parameters shown in Fig. 17.30

of chlorinated solvent PCE by the process of bacterial growth using PCE as the terminal electron acceptor (TEA). This process is termed dehalorespiration (Magnuson et al. 1998). The identification of *Dehalococcoides* (Dhc) was a watershed moment in the advancement of in situ bioremediation technologies and led to the proliferation of bioremediation technologies in the 2000s.

Nutrients are elements required for microbial growth such as carbon, hydrogen, oxygen, nitrogen, and phosphorus. A substrate is a source of energy or molecular building block used by a microorganism to carry out biological processes and reproduce. Substrates include various forms of solid and liquid organic carbon such as carbohydrates.

The goal of engineered biodegradation is to promote growth and stimulate the activity of those groups of microorganisms best capable of degrading certain contaminants. For example, biostimulation may be successful in degrading PCE and TCE, but it may result in a buildup of cis-DCE and VC, which cannot be as successfully degraded by the present bacteria. Another type of biostimulation or bioaugmentation would then have to be used downgradient from the first treatment zone to create conditions that promote biodegradation of cis-DCE and VC.

Enhanced anaerobic bioremediation will generally not be effective in the following conditions:

- Sites with impacted receptors, or with short travel time or distance to potential receptors;
- The contaminant cannot be anaerobically degraded;
- Strongly reducing conditions cannot be generated;
- A microbial community capable of driving the process is not present or cannot be introduced to the subsurface;

- A fermentable carbon source cannot be successfully distributed throughout the subsurface treatment zone;
- There are unknown or inaccessible DNAPL sources;
- Difficult hydrogeological characteristics that preclude cost-effective delivery of amendments, such as low permeability or a high degree of aquifer heterogeneity;
- Geochemical factors (e.g., unusually low or high pH) that inhibit the growth and development of dechlorinating bacteria.

Settings with the extremes of very high and very low rates of groundwater flow, which is typical of karst aquifers, impose significant limits to applying bioremediation. It may be impractical to maintain reducing conditions in high-flow settings, due to the magnitude of groundwater and native electron acceptor flux. On the other hand, it may be difficult to inject substrates into tight formations including low-porosity rock matrix. Under low-flow settings, mixing of substrate with groundwater due to advection and dispersion may be limited (Parsons 2004).

17.4.6 Groundwater Containment: Pump and Treat

The most serious conceptual error when applying containment technologies to karst sites is to treat the underlying karst aquifer as an equivalent porous medium. Unfortunately, this is also the most frequent error at most karst sites in the USA. Remedial systems are designed to extract groundwater from the porous media as if the flow of groundwater and transport of contaminants dissolved in were taking place relatively uniformly "everywhere" within the karst aquifer. If likely influence of any known or inferred preferential flow paths, in three dimensions and in transient conditions (baseflow and storm flow), is not evaluated appropriately or such analysis is disregarded altogether, there is high likelihood that any remediation system in karst will be ineffective. Unfortunately, in addition to various difficulties and high costs associated with defining preferential flowpath in karst aquifers, the problem is compounded by the lack of applicable fate and transport numeric programs for coupled continuum conduit flow (CCCF) models.

Although plume containment is often required from a regulatory perspective, it still is important to address both its feasibility and the rationale in every site-specific karst settings. It is the experience of the authors and many colleagues working in karst that containment of contaminant plumes at any reasonably complex karst site is highly uncertain. The only true evidence that such containment is achieved would be contaminant concentrations less than maximum contaminant level (MCL) at all points of compliance monitoring and, most importantly, at all groundwater receptor locations because of which the pump-and-treat system has been installed in the first place. In other words, any other indirect "evidence" such as the hydraulic heads in the aquifer during pumping and the related potentiometric maps, however created, would be pointless as long as the contaminant concentrations at key locations, including springs and water supply wells, remain above MCLs.

As discussed throughout this book, groundwater flow in karst aquifers is very complex and takes place through both the rock matrix and the preferential

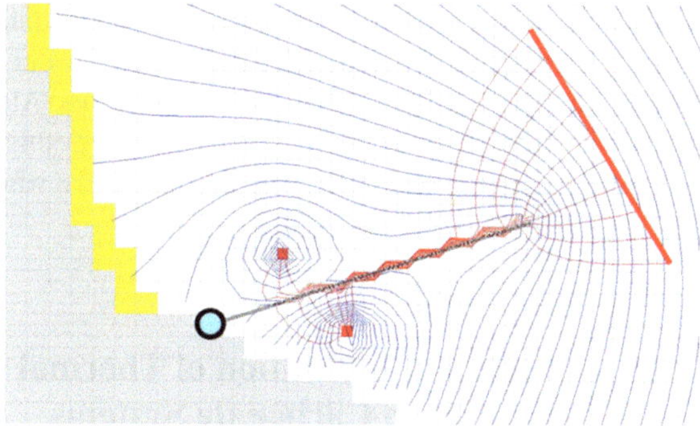

Fig. 17.32 Simulation of a pump-and-treat system for plume containment using EPM approach. Tracks of all particles (*red lines*) enter the EPM conduit and are eventually captured by the wells. (From Mikszewski and Kresic 2014; printed with permission)

flowpaths which can be very deep and are often non-definable at practical field (site) scales. Using a model based on an equivalent porous media (EPM) formulation that neglects turbulent, high-velocity flow in conduits could result in a failed pump-and-treat design. This is illustrated in Figs. 17.32 and 17.33.

Figure 17.32 shows a pump-and-treat model using a high permeability, low-porosity matrix to simulate a conduit zone (i.e., the EPM, "fake conduit" approach). The model shows that all particles emanating from the source zone will be captured by the pump-and-treat wells. In contrast, Fig. 17.33 shows the same situation modeled with the conduit flow process (CFP) to simulate more realistic turbulent flow in conduits. In this model, all particles enter the conduit and discharge directly to a spring, rather than being captured by the pumping wells.

In conclusion, the main difficulties with achieving feasible plume containment in karst with pump-and-treat systems are as follows: (1) finding all preferential flow paths that facilitate transport of the dissolved contaminant to the points of compliance including receptors and (2) hydraulically stopping this transport by pumping both contaminated and clean water from the aquifer. Assuming that, after a sufficient expenditure of resources, the plume containment can be achieved and can be fully verifiable; the following main question still remains: Is it all necessary from a risk perspective and who exactly will benefit from it? This includes comparison with logical alternatives such as groundwater treatment at the receptor locations (springs, water supply wells) and institutional controls of groundwater use. The overall cost of preventing contaminant migration to the receptors using

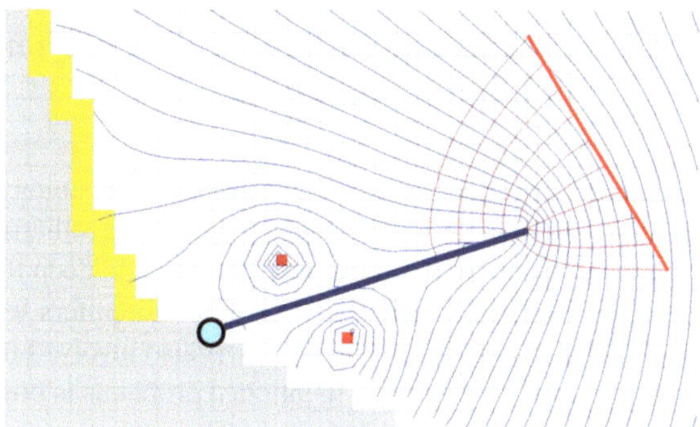

Fig. 17.33 Simulation of a pump-and-treat system for plume containment with CFP. Tracks of all particles (*red lines*) enter the CFP conduit and flow to the spring (*blue circle*) without being captured. (From Mikszewski and Kresic 2014; printed with permission)

the pump-and-treat system should be fully accounted for and compared with the cost of alternatives. This includes the value of water that will be extracted from the aquifer prior to reaching downgradient receptors. Although in highly litigious societies with inflexible regulations and regulators this comparison may not achieve much, it is highly recommended in situations where sustainable approaches to groundwater restoration prevail.

17.5 Genesis and Utilization of Thermal Flow in Deep Carbonate Systems

Judit Mádl-Szőnyi

Faculty of Science, Eötvös Loránd University, Pázmány P. Stny.1/C, 1117 Budapest, Hungary

17.5.1 Introduction

Deep carbonate systems commonly involve thermal flow which can be utilized in different ways. Practical concerns related to karst hydrogeology generally appear on a regional scale such as sustainable water management and geothermal utilization. Contrarily, there are local effects related to deep flow, such as stability issues of near-surface caves passages and contamination of aquifers which also can be interpreted in broader context. The understanding of the scale effect and the regional and hydraulically connected nature of these carbonate systems is a key issue in resolving practical problems. The hydraulic continuity and the relatively high hydraulic diffusivity of karstified carbonates can be the cause of the effects of natural or artificial stresses on the hydraulic head of groundwater propagating greater distances and depths than in siliciclastic sedimentary basins.

The goals of this section are as follows: firstly, to summarize the basic knowledge regarding deep flow in regional flow context; secondly, to illustrate the practical consequences of it based on case studies from Tata, Budapest (Transdanubian Range, Hungary); and thirdly, to display the air drilling techniques in deep karsts.

17.5.2 Problem of Scales and Type of Flows in Karst Research

The practical concerns related to karst hydrogeology are often on a regional scale such as sustainable water management and geothermal utilization of karst terrains. However, there are also local effects related to deep flow, such as geotechnical issues of caves, sinkholes, and contamination of aquifers which appears on local scale. At regional discharge zones, deep flow is also involved in near-surface processes, so the interpretation of the above-mentioned problems is possible only in a broader context.

Water engineers working on solving practical karst problems often have to face different types of groundwater flow associated with different types of permeability of the rock volume (intergranular flow, fissured or enlarged fracture flow, and cross-formational flow). The different types of flow occur in different rock volumes and operate on different timescales.

As a consequence, one of the key issues in finding an appropriate solution for a particular karst problem is to understand the flow field around it on the required space and timescales (Tóth 2013). Due to the lack of information, it is a difficult task to recognize the regional aspects of seemingly local scale problems. We can state that an appropriate solution for deep flow-related karst problems supposes a general understanding of the gravity-driven groundwater flow (Tóth 2009).

This goal is supported by the fact that there is a sharp contrast between the number of published papers in karst hydrogeology related to fissure, fracture scale studies, and the negligibility of those involving the concept of gravity-driven groundwater flow on regional scale. Among the few, the first review paper concerning the subject was published by Goldscheider et al. (2010b). The gravity-driven flow concept helps to understand the common genesis of thermal flow and provides a basis for planning the utilization of lukewarm and thermal waters.

17.5.3 Genesis of Thermal Flow in Deep Carbonates and Consequences for Utilization

The regionally extended carbonate rock volume can be handled simply as a special type of a hydrogeological environment (Tóth 1971, 1984) where the rock matrix basically consists of soluble formations with relatively high hydraulic conductivities. These formations can be fully or partially covered by siliciclastic sediments (Ford and Williams 2007c). Nevertheless, the driving forces acting in the lithosphere can be similar, independent from the nature of the rock matrix (Klimchouk 2007; Goldscheider et al. 2010b). The principle of fluid potential (Hubbert 1940) has to be valid in the karstic environment and its effect on fluids also must be ubiquitous. We can also presume the existence of a continuous water table in deep and thick carbonate sequences, which determines the effect of the topographic driving forces on the fluids of the system. Contrary to siliciclastic regional systems, in regional carbonates, fractures and faults more strongly control the preferred groundwater flow paths. High-permeability faults can act as conduits and are crucial in the discharge of deep karstic flow systems. Hydraulic continuity exists on regional scale in these systems (Klimchouk 2007; Goldscheider et al. 2010b). The hydraulic continuity is defined as the ratio of the induced change in hydraulic head (or pore pressure) to an inducing change of head (pressure) (Tóth 1995). As a consequence, the subsurface rock body is considered hydraulically continuous on a given timescale, if a change in hydraulic head at any of its points causes a head change at any other points, within a time interval that is measurable on the specified time interval (Tóth 1995). The hydraulic continuity and the relatively high hydraulic diffusivity of karstified carbonates can cause the effects of natural or artificial stresses on the hydraulic head of groundwater to propagate greater distances and depths than in siliciclastic sedimentary environment.

The hydrogeological environment in karst systems regularly consists of intergranular micro-scale flow, which exists in a matrix of blocks divided by bedding planes, fractures, and solution channels. In spite of these heterogeneities, the groundwater movement on a regional scale has to be handled as cross-formational flow which incorporates all of the above-mentioned flow types (Ford and Williams 2007c). Consequently, flow is not restricted only to wider fractures of the rock. Therefore, on a regional scale, a hydraulically continuous flow evolves in deep continental carbonates induced by fluid potential differences at the water table (Tóth 1963; Klimchouk 2007; Goldscheider et al. 2010b) (Fig. 17.34).

Figure 17.34 is suitable for summarizing our knowledge of the nature and genesis of deep flow in carbonate systems. It displays the schematic illustration of groundwater flow and karstification processes in a deep and mostly hypogenic inland carbonate system. The flow system is basically gravity-driven, caused by topographic gradients; moreover, sedimentary compaction, tectonic compression, and density differences act as additional driving forces; thermal convection can occur near discharge zones. The arrows in the figure indicate flow directions; blue, orange, and red colors detach cold, lukewarm, and hot waters, respectively. The different position of springs in flow systems can be derived from the figure. The thermal and lukewarm springs at discharge regions, connected to the deeper regional and intermediate flow, can also be seen in the figure. The processes of

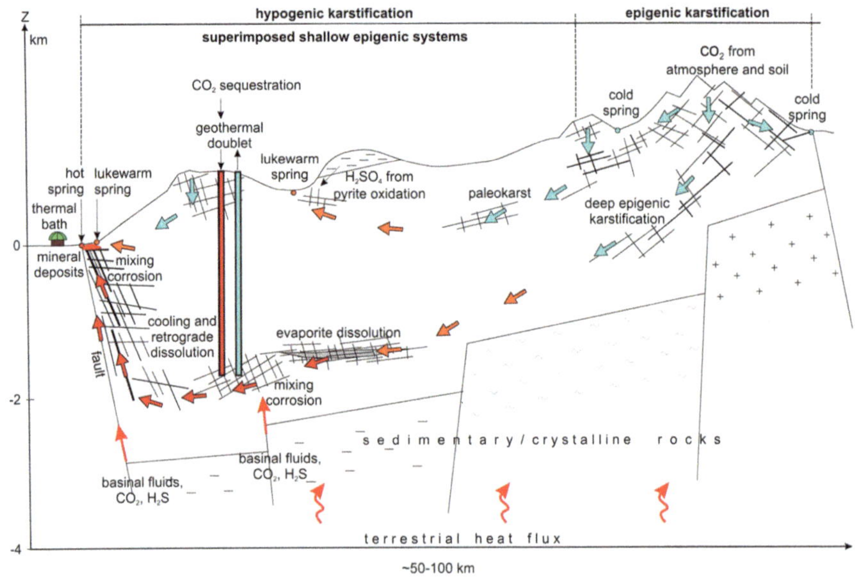

Fig. 17.34 Schematic illustration of groundwater flow and karstification processes in a deep carbonate system (Goldscheider et al. 2010b)

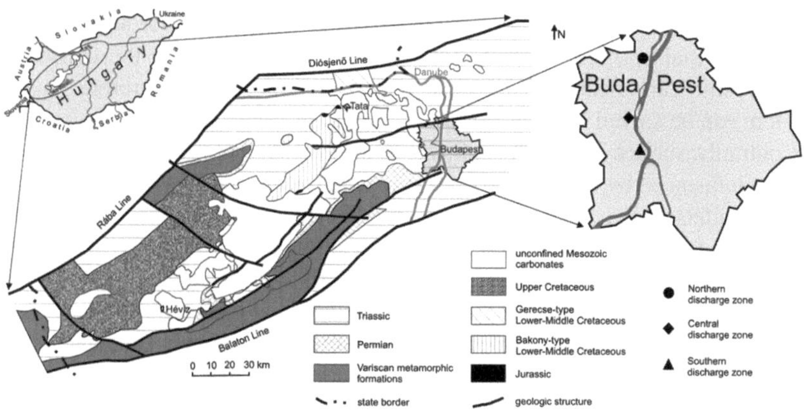

Fig. 17.35 The Transdanubian range unit: distribution of Paleozoic-Mesozoic formations (after Fülöp in Haas 2001, Fig. 20)

epigenic karstification: meteoric and soil origin CO_2 and hypogenic karstification: mixing corrosion, evaporite dissolution, basinal origin H_2S and CO_2 etc. can also be seen (Goldscheider et al. 2010b, Fig. 17.35).

Genesis of deep thermal flow is in the focus of this paper which can be easily explained by its regional discharge position. It can be said that thermal springs are one of the most characteristic features of regional discharge zones in carbonate systems. The temperature of the lukewarm and thermal springs in carbonates depends on the local heat flow and the penetration depth of intermediate and regional flow systems. The heat transport by flowing groundwater is an efficient process therefore upward flow of thermal groundwater increases near-surface geothermal gradients and can cause positive heat anomaly (Bredehoeft and Papadopulos 1965; Sass 2007). Not only the lukewarm and thermal springs but the hypogenic caves also are characteristic in the surroundings of discharge zones (Mádl-Szőnyi and Erőss 2013).

The first step in solving practical problems in hydrogeology is the recognition of the appropriate scale of the problem. The difficulties of this issue in regional carbonate regions are obvious from Király's considerations (1975) who displayed the increase of hydraulic conductivity by ten orders of magnitude from a laboratory sample of a matrix toward a whole carbonate basin.

The thick and deep carbonate systems are important as drinking water, mineral water, and thermal water resources. These systems are also suitable for geothermal installations and the heat content of the effluent water can also be used with heat pumps. Due to different flow systems in carbonates, the sustainable water production for a whole basin has to be planned not simply with the conventional methods. The amount of exploitable water from a basin depends on the weighted average permeabilities of the basin wide representative elementary volume (REV), or even on the available recharge for the whole basin. It has to be highlighted that the S (storativity) and T (transmissivity) calculated on laboratory or well's scale are not appropriate for use on regional scale numerical modeling.

However, if we found stability problems above unknown caves in the surroundings of thermal springs, this local effect can be interpreted as the consequence of hypogenic karstification connected to deep regional flow. Therefore, this local problem can be solved if we are aware of its regional background.

Contrarily, surface origin contaminants in regional discharge zones of deep carbonates are influenced by the positive hydraulic effect of upwelling water. It can prevent deeper aquifers from contamination because it forces advective contaminant transport laterally. Water abstraction at discharge zones can obviously disturb water upwelling. Therefore, these phenomena can be handled properly only in a broader context.

Confined deep carbonates are prosperous places for CO_2 sequestration because CO_2 increases transmissivity therefore increasing the economic efficiency of such installations. The CO_2 sequestration can be coupled with geothermal utilization (Goldscheider et al. 2010b).

Box 17.5.1

Case study—Tata and Budapest discharge regions, Transdanubian Range, Hungary

The Transdanubian Range is part of the central region of the Pannonian Basin. Geological boundaries of the Range are tectonic lines: Rába Line in the NW, Diósjenő Line in the N and Balaton Line in the South. The system is made up of slightly metamorphosed Variscan rocks and overlain by Alpine sequences. The Mesozoic carbonate sequences are made up in 90 % of Main dolomite and Dachstein limestone. During Middle Cretaceous tectogenesis, large NE–SW synforms and antiforms determined the structural setting of the range (Haas 2001). The unit came into its position during the Late Cretaceous-early Paleogene (Fig. 17.35). The Pannonian Basin is a Neogene structure formed as a result of extension-related attenuation of the lithosphere in late Early to Late Miocene times (Royden and Horváth 1988). Inversion of the basin began in the latest Miocene and contributed to the uplift of the Transdanubian Range. The whole Pannonian Basin is characterized by elevated heat flux (~100 mW/m^2) due to the thinned lithosphere (Lenkey et al. 2002).

The Transdanubian Range with its >10,000 km^2 area is the biggest karstified carbonate aquifer system in Hungary (Csepregi 2007 in Alföldi and Kapolyi 2007). The natural discharge of the system is manifested through cold, lukewarm and thermal springs and wetlands, creeks and rivers (Fig. 17.36). The groundwater of the system is used mainly for public water supply and as mineral waters, the lukewarm and thermal water are extracted for healing and heating purposes. The region was affected by coal and bauxite mining dewatering from the 1950s to 1990s. The whole water production from the system was >700 m^3/min between 1965 and 1990, but in some years it reached 800 m^3/min. This amount exceeded the natural recharge, 485 m^3/min, by ~60 % (Csepregi 2007 in Alföldi and Kapolyi 2007). This huge artificial intervention caused regional hydraulic head decrease in the reservoir.

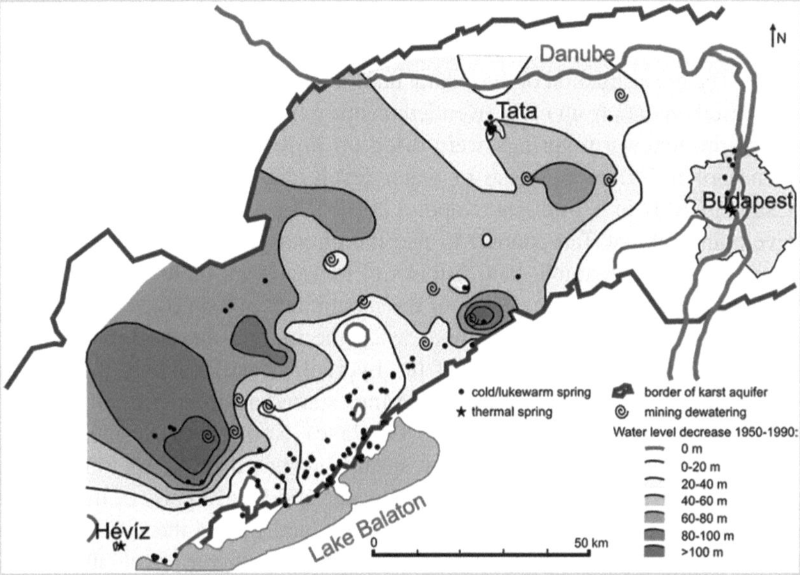

Fig. 17.36 The water level decrease in the Transdanubian range (1950–1990) due to coal and bauxite mining dewatering (based on the data of Csepregi 2007 in Alföldi and Kapolyi 2007)

The fact of regional hydraulic continuity of the Transdanubian Range, based on the definition of Tóth (1995), was evidenced by the long-lasting regional dewatering activity related to underground mines in the area. The places of active dewatering of mines and the hydraulic head (water level) drop between 1950 and 1990 can be seen in Fig. 17.36 (based on Lorberer 1986; Csepregi 2007 in Alföldi and Kapolyi 2007). The inducing hydraulic head changes by intense dewatering caused hydraulic head changes, the drying up of most of the springs in the whole system.

The environmental impacts of the dewatering on the hydraulically continuous system give a warning example from those periods when environmental impact assessment was not an issue (Alföldi and Kapolyi 2007). The regeneration of the system is also a long-lasting process, the numerical modeling presages quasi natural conditions only for 2030 (Csepregi 2007 in Alföldi and Kapolyi 2007). However, this example can demonstrate the hydraulic reaction of a hydraulically continuous regional carbonate system for stresses induced by over-exploitation.

Environmental impact of long-term water production and recovery: Tata lukewarm springs

The environmental impact of mining dewatering, on a part of a regional system, is demonstrated in the example of Tata, which is named "the town of water" in the NE part of the Transdanubian Range (Figs. 17.35 and 17.36). This area is characterized by lukewarm springs and wetlands. The springs discharge at cross-cutting faults. The original volume of springs was 1.33 m/s (Maller and Hajnal 2013). The reconstruction of the natural undisturbed water level (118–140 m asl) was based on data from early twentieth century (Horusitzky 1923).

All the lukewarm springs were dried up in the early 1970s as a consequence of a ~40 m decrease of water levels due to dewatering of mines (Csepregi 2007 in Alföldi and Kapolyi 2007) (Fig. 17.36). During 1970s, the developing city of Tata started to use the already dried areas, which were previously used as arable land. Blocks of houses were built in the formerly wetland areas and now the 30 % of the inhabitants of Tata live here (Kovács and Szőcs 2014). In the meantime, when the mining dewatering had been stopped in the 1990s, recovery of the hydraulic head began in the system (Tóth et al. 1999). The discharge of springs started again in 2002 in the lowest elevation part of the region (Fig. 17.37a–c). Not only the former springs have been reactivated, but also new springs appeared in the region (Maller and Hajnal 2013). The reappearance of water in different parts of the former wetland area requires the re-examination of land use of the city (Ballabás 2004). It is supposed that the highest original discharge level of springs, around 140 m asl, will be reached around year of 2018 based on the forecast

Fig. 17.37 **a** The recovery of dried up lukewarm springs at lowest elevation, in Tata, Hungary (photo was taken by the author in 2006); **b** discharge features around the recovered spring lake (photo was taken by Malmos in 2013), **c** underwater discharge in the spring lake (photo courtesy of diver Siska, taken in 2013)

using permanent infiltration and water abstraction circumstances (Kovács and Szőcs 2014). The recovery of the springs causes significant geotechnical and water quality problems in the area.

Sustainable utilization of thermal water at the perennial discharge region in Budapest

The uprising thermal water in the discharge region of regional flow systems in carbonates can be used for geothermal purpose as wells. The Transdanubian Range has two distinct regional discharge zones with thermal springs (>30 °C) (temperature limit based on Hungarian regulation). One of them is situated in the surroundings of Hévíz Lake and the other in Budapest (Figs. 17.35 and 17.36). The Buda part is partly unconfined while the carbonate reservoir is deep seated and confined under the Pest part of the city (Fig. 17.35).

The discharge zone of the system is concentrated in the surroundings of the Danube, which is the base level of erosion, and situated within the city center, historically in a fully urbanized environment (Mindszenty 2013) (Fig. 17.35). In the northern discharge zone, there are lukewarm springs (18–22 °C, TDS: <1,000 mg/l), in the central zone, lukewarm (21−27 °C, TDS: <1,000 mg/l) and thermal springs (50−63 °C, TDS: >1,000 mg/l), while in the southern part of the system, thermal springs have discharged (33−45 °C, TDS: 1,500–2,000 mg/l) (Papp 1942; Alföldi et al. 1968). Because of sanitation reasons, the natural springs have been substituted gradually by safety deep wells, from the second part of the nineteenth century.

Due to regional flow systems, the Budapest perennial discharge region is in hydraulic connection with the whole Transdanubian Range system toward the W-SW. The hydraulic connectivity was directly proven by mining dewatering which affected the discharge zone as well. The effect of mining dewatering is superimposed on the effect of local thermal water withdrawal from the 1970s. This coupled effect led to temperature and volume decreases in the thermal and lukewarm waters of the discharge region. With termination of mining dewatering, the system has recovered (Csepregi 2007 in Alföldi and Kapolyi 2007).

Nowadays, there is strict restriction for thermal water production, and no additional water utilization from deep carbonate reservoir is allowed by the authorities. The famous thermal Spas in Budapest (Fig. 17.38b) are supplied roughly by deep wells. Therefore, the water of natural springs flows directly to the Danube mainly without any utilization. These unused effluent spring waters transport significant amount of heat with the water to their receiver.

The restricted thermal water production brings up the recognition of the possibility of utilization of the heat content of the effluent lukewarm water by heat pumps. Owing to the regional discharge zone position of these springs (Fig. 17.35), their volume and temperature is independent from the individual precipitation events, and at the same time the transient effect of the Danube can be established (Alföldi et al. 1968; Mádl-Szőnyi et al. 2001).

These statements were proven by recent continuous measurements in Central discharge region (Fig. 17.35), with the ascertainment of the

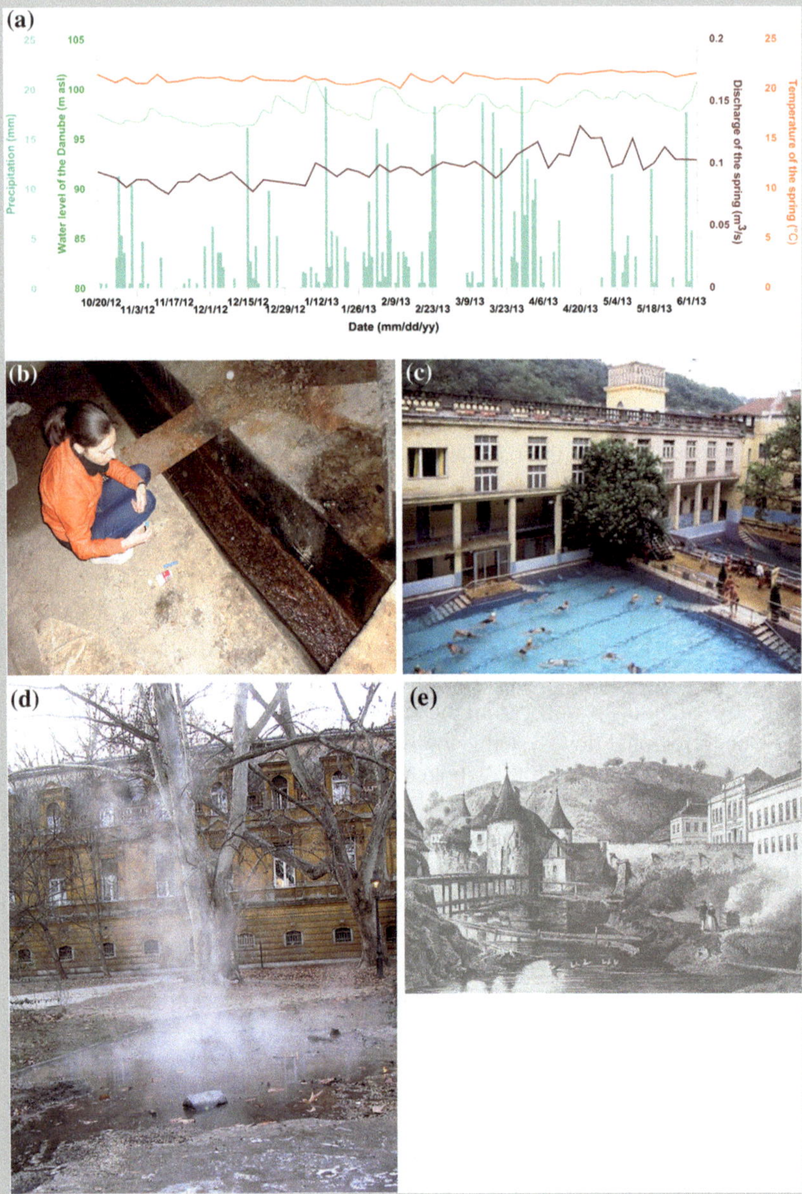

Fig. 17.38 Central discharge region of Budapest; **a** recent measurement data for effluent lukewarm spring water: discharge volume, temperature, supplemented by the precipitation, and water level data of the Danube (based on the measuring data of Lovrity and Bodor 2014); **b** measurements regarding effluent spring water (photo courtesy of Kátay, taken in 2013); **c** utilization of lukewarm water in the Spa; **d** underground cavities and thermal water revealed by near-surface geoworks; **e** View of the discharge region in the mid 19th century (engraving of Rohbock)

minimum volume discharge: 7,000 m^3/day and the temperature, 22.5 °C (Lovrity and Bodor 2014), which can provide a basis for planning heat utilization of the spring water by heat pumps (Fig. 17.38a, b).

Similarly to springs, the thermal Spa (Fig. 17.38c), which uses thermal water from fully or partly drilled wells, also produces wastewater. Simple volumetric calculation was restricted to only one spring in the Central discharge region and the waste water from the four biggest spas has proven the importance of this issue. These effluent waters carry 782 TJ/y (25 MWh) of waste heat (Erőss et al. 2013).

The local scale specialities connecting to regional discharge zones were well demonstrated by geoworks and were achieved around the Spa buildings in the Central discharge zone at the end of 2008. They revealed near-surface cavities with thermal water infilling (>40 °C), proving the existence of hypogenic cavities and coexistence of distinct thermal and lukewarm water components in the discharge zone (Fig. 17.38d). This local effect can be interpreted only in broader flow context.

Nevertheless, not only the effluent spring water but also wastewater of spas can be used for heating purposes in terminal discharge zones of carbonates. The heat content of water in deep carbonate reservoirs is also exploitable (Goldscheider et al. 2010b). The potential for installation of geothermal heating systems in Budapest was evaluated based on the analysis of temperature conditions and the depth and reconnaissance of the deep carbonate reservoir (Fig. 17.39). The difference in hydrogeological position (unconfined or semi-unconfined and confined) between Buda and Pest is reflected on the map. The NW part of Buda is not appropriate for revealing thermal water. The SW part of the city, in spite of better temperature conditions, is not as prosperous

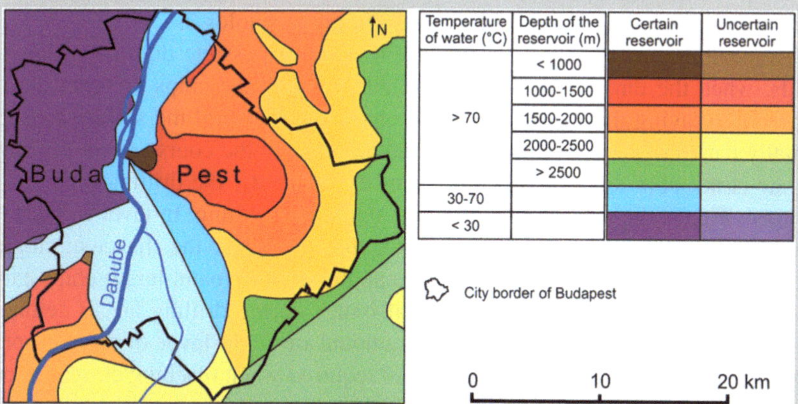

Fig. 17.39 Simplified geothermal potential map for the utilization of heat content of thermal water of the carbonate reservoir, Budapest, Hungary (modified based on the map in Mádl-Szo˝nyi et al. 2009)

because of the uncertainties in the lithology of the reservoir. However, under the Pest part of the city, the thermal water is theoretically well exploitable. The depth of the reservoir and the potential installation increases toward the East. It is clear from the geothermal map of the region that the area is promising for geothermal installation for heating purposes.

However, the hydraulic and temperature reaction of the system as a consequence of hydraulic connectivity of the whole Transdanubian Range provides the opportunity to use the heat content of the deep carbonate reservoir only with geothermal doublets. These are highlighted by the requirements of local water authorities as well. The use of geothermal doublets (production and injection wells) can be the only basis for sustainable thermal water utilization in the area.

17.5.4 Drilling in Deep Karstified Formations

The method of air drilling is principally applied in execution of thermal wells in karstic formations (Szabó 2006). In the case of such type of drilling, the bit rotates and hammers on the bottom of wellbore and grinds the rock (Fig. 17.40a). Drilling literature often terms air drilling "dusty," because the chips are usually a very fine powder-like material. Bigger chips stay at the bottom until those are ground to an adequate size to be carried away (Pratt 1989; Johnson 1995).

The circulation of a large amount of air in the drilling tools carries drilling ground to the surface. An advantage of the method is that when the drilling reaches fractures, the presence of water in the circulating air can be immediately observed; and the accidental pollution of layers occurring during drilling with circulating fluid (up to the loss of all of the drilling fluid) can be avoided. In the case of karstic formations, air drilling is one of the most efficient technologies.

The drilling process can very quickly become impossible during drilling with fluids, when the circulating fluid is not able to carry the chips up from the bottom. After leaving the wellbore bottom, chips travel toward the surface but occasionally can get stuck in the fractures and caverns. In the course of air drilling, if the air is not able to circulate chips from the wellbore, the drilling pipe can get clogged. This is the most important threat during air drilling. In favor of increased efficiency of the method dosing foam is required: therefore the chance of clogging becomes minimal. Nevertheless, with the application of foam, larger chips can be taken to the surface. The most important issues during drilling are to define the velocity of circulation and quantify the amount of air in favor of an appropriate flushing of the well. Applying foam is also required when reaching closed caverns. In this case, the grinding and/or flushing of chips can become insolvable. To handle this problem, foam must be dosed to increase viscosity and density.

Regarding the velocity of circulation, the main question is how the air could carry the chips to the surface. It was found that when the chips are coarser, then it is possible to move them only at higher air velocity. Chip transporting is strongly

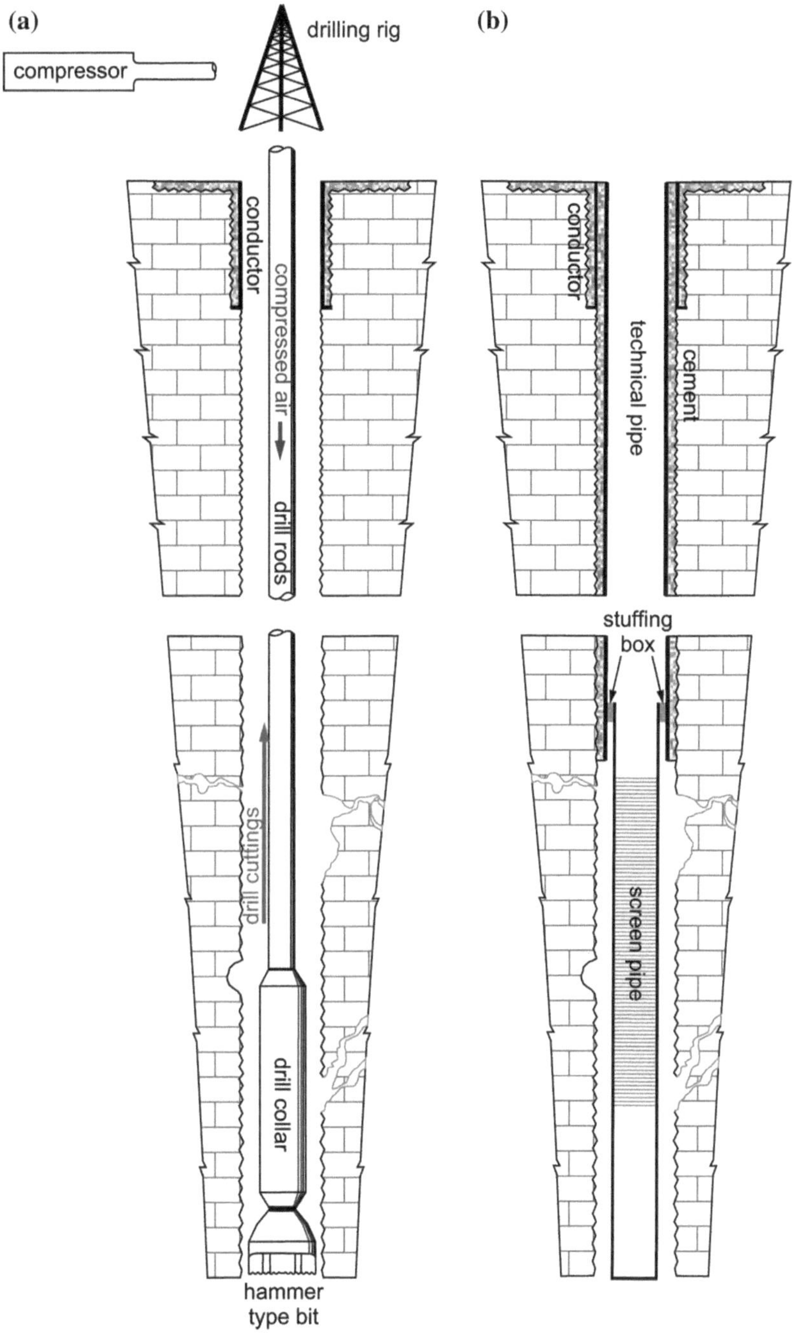

Fig. 17.40 **a** The technological sketch of air drilling in karst (modified after the manuscript of Bede, Aquaplus Ltd. 2014); **b** thermal well structure in karst (modified after the manuscript of Bede, Aquaplus Ltd. 2014)

influenced by compressibility of air. Pressure-loss of air heading upward in the annulus raises the bottom-pressure of air.

The quantity of chips and air, just like the air density, is directly proportional to the air pressure. As a consequence—supposing that the geometry of the well does not alter—the velocity of air is smaller at deeper levels. So, a constantly increasing air circulation rate is needed to drill deeper. Rotating the drilling pipe helps to carry chips by stirring up the settled chips back into the stream (Angel 1958).

The geometry of the annulus also influences the motion of the chips. According to practices, at the lower section of wellbore, mainly around drill collar usually there are no problems, because the annulus is narrow here, so the air velocity and pulling forces are the greatest in the section. On the other hand, a dangerous situation can come into existence at the upper end of the drill collar, where diameter of the annulus increases suddenly and air velocity drops. If circulation velocity is not high enough to carry chips, they will settle at this location. This is obvious, because velocity is proportional to the square of annulus-diameter, and a slightly wider annulus (e.g., at washouts) can lead to significant velocity-drop, destroying the efficiency of chip carrying. Significant amount of air entering the wellbore can really increase circulation velocity, also increasing the pressure-loss of circulation from the kicking level to surface, and increasing air pressure below kicking point. This reduces air velocity below kicking point and also reduces the efficiency of chip carrying. As a conclusion, to manage a successful air drilling in karst, it is necessary to accurately define the pressure and quantity of air, taking the diameter of wellbore into account.

The depth capacity of the air drilling method depends on the performance of the operating compressor. Applying hydraulic hammer-type bit, air drilling can be managed in depths of up to 3,000 m. Thermal wells deepened into karst do not have any specification except not having gravel mantle around the screen pipe (Fig. 17.40b). After building the well structure, during the well completion, zones intended to screen must be submitted to intense compressing and acidifying to clean fractures containing water in order to enhance water movements. As regards the well structure, there is not any difference between air-drilled and mud-drilled wells.

If the devices required for air drilling are not available, drilling in karst can be managed with mud as well. In this case, a special plugging material, i.e., lost circulation material, is needed to prevent the huge mud loss. This, after plugging the open fractures, must be drilled off, and thus the mud loss can be minimized. To remove plugging material from the screening sections, intense compressing is required during the course of well completion.

17.5.5 Summary and Conclusion

Practical concerns related to karst hydrogeology are often on a regional scale such as sustainable water management and geothermal utilization. Therefore, it is of high importance to understand the regional and hydraulically connected nature of these carbonate systems. It can provide a basis for finding an appropriate solution to these particular problems. The importance of the gravity-driven flow concept is that it helps to understand the common genesis of thermal flow and therefore can lay the foundation for the sustainable utilization of thermal water.

Understanding of the scale effect is highlighted to resolve practical problems. An important consequence of hydraulic continuity and relatively higher hydraulic diffusivity in karst is that the effects of natural or artificial stresses on the groundwater level can propagate greater distances and depths than in siliciclastic sedimentary basins.

The Transdanubian Range and the Tata and Budapest case studies (can) provide examples for environmental intervention by long-lasting mining dewatering. However, as a scientifically positive effect, it produced an "in situ example" for the operation of hydraulic continuity as well.

We can adapt the practical consequences of this to similar regions. These can be used not only in water management but for better understanding local scale contamination and geotechnical problems and also during the planning of geothermal doublet systems at the regional discharge zones of the system. Moreover, the possibility of over-exploitation is drawing the attention to the importance of the use of effluent lukewarm and thermal springs and the heat utilization of wastewater of spas in discharge zones of thermal water.

The unused lukewarm and thermal springs and the wastewater of spa pools in regional discharge zones of thermal springs do not require further thermal water production. Therefore, their utilization can contribute to the sustainable thermal water and energy consumption.

In deep karstic formations, mainly the method of air drilling is applied as deep as 3000 m. While managing a successful air drilling in karst, it is necessary to accurately define the pressure and quantity of air, taking the diameter of wellbore into account. Thermal wells of deep karsts do not have any specification except not having gravel mantle around the screen pipe.

17.6 Transboundary Aquifers in Karst

Neno Kukurić

IGRAC, International Groundwater Resources Assessment Centre, Delft, The Netherlands

17.6.1 Introduction

Many aquifers cross political boundaries or rather the other way around: Many national borders cross aquifers (because aquifers were there prior the borders). Inadequate management of transboundary aquifers (TBAs) can lead to various groundwater quality (changes in groundwater flow, levels, and volumes) and quantity (dissolved substances) problems. These problems are more difficult to prevent, mitigate, and solve in an international context than in the case of national aquifers. International cooperation is necessary to ensure an appropriate assessment, monitoring, and management of transboundary groundwater resources. International agreements are made to prevent potential conflicts and to improve the overall benefit from groundwater. In practice, agreements, to be made and respected, require a sufficient knowledge on the resource, its current state, and the trends. This is often a challenge for invisible groundwater and especially in a complex hydrogeological environment like karst. Aquifers in karst are very vulnerable as well, asking for an additional attention of national and international water authorities.

This chapter describes DIKTAS, a case study of transboundary aquifers in the Dinaric karst region; it addresses motivation for international water cooperation, methodological approach, achieved results, and current efforts. The experience obtained in this region can be replicated to the similar karst regions in other parts of the world. The chapter starts with a brief review of global efforts to improve assessment, monitoring, and management of transboundary groundwater resources. After the DIKTAS case study is presented, the chapter is rounded off with some concluding comments about TBAs in general and in karst particularly.

17.6.2 Transboundary Aquifers of the World

The end of twentieth century marked first systematic efforts to make an overview of internationally shared groundwaters. In 1999, UNECE Task Force on Monitoring and Assessment conducted an inventory of transboundary groundwaters in Europe. This pioneering work pointed out the challenges of international groundwater data availability and harmonization. In 2007, the first Assessment of transboundary rivers, lakes, and groundwaters in UNECE region was carried out, followed by the second one in 2011. The first assessment included a separate overview of TBAs in the regions of South-Eastern Europe and of Caucasus and Central Asia. In the second assessment, however, aquifers were assessed as a part of river drainage basis.

In 2000, in recognition of the importance of transboundary aquifer systems as a source of freshwater in certain regions of the world, UNESCO decided to launch an initiative to promote studies on transboundary aquifers. In cooperation with IAH, International Association of Hydrogeologist (that already took a similar initiative), the Internationally Shared Aquifer Resources Management (ISARM) programme was established. ISARM operates as an umbrella programme, (co)organizing regional activities, and promoting international groundwater cooperation. The main technical support an input to the program is provided by International Groundwater Resources Assessment Centre (IGRAC) that is UNESCO Global Groundwater Centre. While the first map of Transboundary Aquifers of the world was produced (in 2006) in the framework of WHYMAP (another UNESCO programme), the updates of the map (2009, 2012, and 2014) are prepared under the umbrella of ISARM. Additionally, ISARM produced the Atlas of Transboundary Aquifers, organized a very-well received international conference on transboundary aquifers in 2010, and supported the work of the

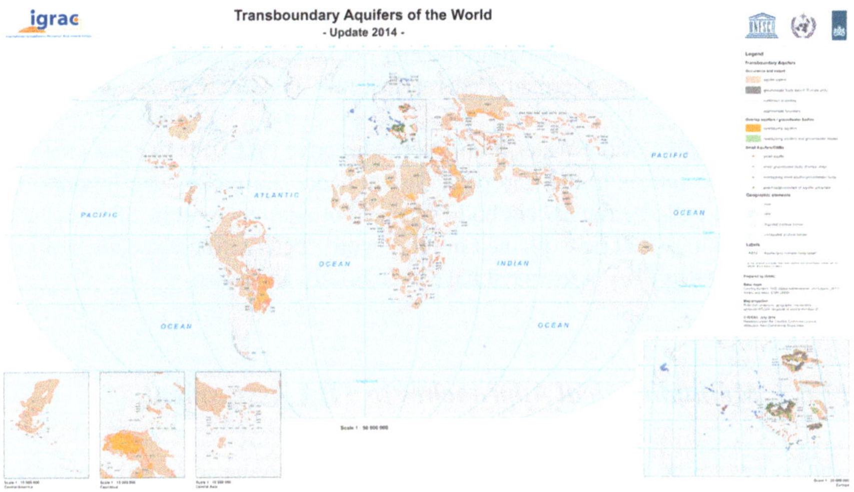

Fig. 17.41 Transboundary Aquifers of the World map (IGRAC 2014)

International Law Commission in formulating so-called Articles on the Law of Transboundary Aquifers. More information on ISARM and on TBAs in general can be found via the ISARM portal www.isarm.org (Fig. 17.41).

Begin of new millennium also marked increased attention to groundwater in the International Water portfolio of GEF, Global Environment Facility. Within a several years, GEF initiated and co-funded a several projects dedicated to internationally shared groundwaters, such as in Southern Africa (Limpopo), South America (Guarani), North Africa (Nubian), and Europe (DIKTAS). The full name of DIKTAS project is Protection and Sustainable Use of the Dinaric Karst Transboundary Aquifer System. The project experiences, results, and current activities of DIKTAS will be elaborated further in this chapter as the case study on TBAs in karst.

The most relevant global assessment of internationally shared waters is ongoing in the framework of the GEF Transboundary Water Assessment Programme (TWAP). The TWAP (www.geftwap.org) aims to provide a baseline assessment to identify and evaluate changes in transboundary water systems. The TWAP full-sized project is the first truly global comparative assessment for transboundary water system categories (aquifers, lakes, rivers and large marine ecosystems, and the open ocean). The TWAP groundwater component (www.twap.isarm.org) is assessing approximately 170 transboundary aquifers and 43 groundwater systems of small island developing states. Data are being collected through networks of regional and national experts. The TWAP groundwater is executed by UNESCO in close cooperation with IGRAC.

In recent years, there has been an increase of initiatives undertaken to better assess regional water resources. These initiatives, often financially supported by European and North American development agencies, are taking place almost exclusively in Africa and Asia. The most comprehensive assessment recently accomplished in the Inventory of Shared Water Resources in Western Asia, conducted by the United Nations Economic and Social Commission for Western Asia (UN-ESCWA) and the German Federal Institute for Geosciences and Natural Resources (BGR).

One of the tasks of the Water Framework Directive (WFD) is to assess the state of the water resources in the European Union, hence across 28 state borders. The WFD uses so-called groundwater bodies instead of aquifers as units of assessment. This methodology will be explained in more words below, where specifics of various approaches to TBA assessment have been briefly addressed.

17.6.3 Methodological Approaches to TBA Assessment

Approaches to the TBA assessment vary, mostly due to the scale and the purpose of the study. In any case, the assessment needs to include all relevant aspects of transboundary groundwaters. ISARM Framework Document (UNESCO 2001) distinguishes five aspects of the transboundary aquifers, namely hydrogeological, legal, socioeconomical, institutional, and environmental. Hydrogeological assessment starts with delineation and description (i.e., collection and interpretation of hydrogeological information) and is followed by classification, diagnostic analysis, and possible zoning (yielding information necessary for decision making). Data harmonization and information management are the third task of the hydrogeological assessment and the more demanding at an international (than at a national) level due to differences in language, classifications, terminologies, formats, reference systems, etc. (Kukuric et al. 2008).

In the UNECE inventories (UNECE 2007, 2011), so-called Driving Forces-Pressures-State-Impact-Responses (DPSIR) framework was used, resulting in a regional overview of pressures to groundwater, of current groundwater status and of future trends and prospects. However, the implemented DPSIR provided very few quantitative (or quantifiable) data on individual aquifers (Lipponen and Kukuric 2010).

Currently under implementation, the TWAP assessment methodology aims at assigning indicator values for each national segment of the identified TBAs. Ten core indicators, encompassing the hydrogeological, environmental, socioeconomic, and governance dimensions of the systems, have been identified (UNESCO-IHP 2011). For some of the core indicators, projections will be developed for the years 2030 and 2050 using the global hydrological model. Although carried out at the global scale, TWAP assessment yields considerable quantitative information, required to calculate the indicators. This info is often geo-referenced and will be made available via IGRAC global groundwater information system (GGIS) already this year.

While GEF TWAP is the global assessment, the most of GEF groundwater projects address one aquifer, allowing more in-depth analysis. These projects implement a transboundary diagnostic analysis (TDA) that is a GEF-prescribed assessment procedure. The GEF prescribes procedural and organisational steps, including the milestones and reporting; the content of the assessment is defined by the project parties and anchored in a Project Document. GEF Groundwater Portfolio Analysis pointed out that projects include hydrogeological, ecological, socioeconomical, organisational, legal, and political facts relevant for actual/potential disputes and conflicts; hence, very much in the same line with the ISARM Framework document. Implementation of the GEF TDA will be presented in the DIKTAS Case Study further in this chapter.

European Water Framework Directive uses groundwater bodies (GWBs) as a basis assessment unit. Unlike aquifers, GWBs are (not necessary hydrogeological but) managerial units. The guidelines for interpretation of GWBs (European Commission 2003) are quite general and their interpretation varies substantially from country to country. According to the submission/compilation status and evaluation report of European Environmental Agency from 2011, out of 7019 GWBs in total holding specifications related to this attribute, only 124 transboundary GWBs have been reported. A simple GIS operation performed by IGRAC while preparing the update of the TBA map showed that the number of GWBs along national borders is several times larger than reported. Since new EU members, candidates and EU neighboring countries are adjusting water management approaches according to WFD. In karst, WFD needs a further specification, as the assessment of Dinaric Karst aquifers also showed. For example, what would be a purpose of marking "non-productive groundwater bodies" in karst?

In recent years, several documents are published addressing transboundary assessment methodology, such as "Handbook for Integrated Water Resources Management in Basins" (INBO and GWP 2012) and "Toward a joint management of transboundary aquifer systems" (Machard de Gramont et al. 2010). None of them, however, deals with specifics of assessment of transboundary aquifers in karst.

17.6.4 International Agreements on Transboundary Aquifers

When countries share an aquifer that is exploited or threaten by pollutants, they need an agreement about the aquifer management and protection. Sometimes, bilateral agreements concentrate only on a specific, existing problem, often including additional monitoring and assessment activities. Regional or global guidelines, protocols, conventions, and laws are in general very instrumental to the process of making bilateral agreements, either providing guidelines or being a binding document to the parties.

The only global agreement that relates (also) to groundwater is the Convention on the Law of the Non-Navigational Uses of International Watercourses. The convention entered in force just recently (17 August 2014) after 35 countries ratified it. Since the preparation of the convention took a several years, about two decennia in total were needed to have it ratified by the minimal prescribed number of countries. Obviously, most of the countries are not eager to sign these kinds of international obligations, especially countries located upstream in river catchment areas. Anyway, the Convention was prepared for international surface waters and it addresses groundwater only when is related to a surface water body and when flows to the same terminus, which is a limited view on aquifers.

Therefore, UN International Law Commission added to its program the topic of shared natural resources and in 2008 completed preparation of nineteen draft articles on the law of transboundary aquifers (Stephan 2011). The draft articles were then deferred to the UN General Assembly, which adopted Resolution A/RES/63/124 including the draft articles in annex. In the Resolution, the UN GA "encourages the States concerned to make appropriate bilateral or regional arrangements for the proper management of their transboundary aquifers, taking into account the provisions of these draft articles."

The most influential agreement at the regional level is certainly the Convention on the Protection and Use of Transboundary Watercourses and International Lakes from 1992 (also known as "Helsinki Convention" or "UNECE Convention"). The convention appeared to be extremely helpful for execution of the assessments in the UNECE region. Initially negotiated as an instrument within the UNECE region, the Convention was amended to allow accession by all the United Nations Member States. The groundwater part of the Convention is recently amended, using the draft articles on the law of transboundary aquifers and the main inspiration source. This, together with new (global) applicability, makes the UNECE Convention a very useful tool while agreeing about management of individual transboundary aquifers.

A number of bilateral groundwater dedicated agreements are rapidly increasing in last years; the most exemplary are as follows:

- Convention on the protection, utilization, recharge, and monitoring of the Franco-Swiss Genovese aquifer,
- Guarani Aquifer Agreement (Argentina, Brazil, Paraguay, and Uruguay),
- Constitution of the Joint Authority for the Nubian Sandstone Aquifer Waters (Chad, Egypt, Libya, and Sudan).

Some other international agreements address groundwater as well, such as Convention on Wetlands of International Importance (Ramsar Convention). Naturally, it is not a question of choosing between one or the other agreement, but all those that can contribute to agreements at aquifer level.

Box 17.6.1

Case study—DIKTAS: The karst transboundary aquifer

Protection and Sustainable Use of the Dinaric Karst Transboundary Aquifer System (DIKTAS) is a regional project aimed at improving the management of karst groundwaters in the Dinaric Karst shared by several countries in South-Eastern Europe. As such, the project is the first ever attempt to globally introduce integrated management principles in a transboundary karst freshwater aquifer system of such magnitude.

Groundwaters of the Dinaric Karst System in South-Eastern Europe form some of the World's largest and most prolific karst aquifers which host numerous first magnitude springs. The system extends from north-eastern Italy through Slovenia, Croatia, Bosnia and Herzegovina, Serbia, Montenegro, Albania to FYR Macedonia, and Greece. For the most part, this region is characterized by still pristine environments and a variety of unique geomorphological landforms. It also hosts numerous karst underground species, many of which are endemic such that some of the Dinaric karst localities are recognized as the most biodiverse worldwide.

The DIKTAS Project started in 2010 and will continue until the end of 2014. The project was initiated by the aquifer-sharing states and is a full-size global environment facility (GEF) regional project, implemented by United Nations Development Programme (UNDP) and executed by UNESCO. The activities of the project focus on Albania, Bosnia and Herzegovina, Croatia, and Montenegro. Several other countries and international organizations have also joined this challenging project and are providing valuable contributions to the realization of its objectives.

At the global level, the project aims to focus the attention of the international community on the abundant but vulnerable water resources contained in karst aquifers. The project's main outputs include the transboundary diagnostic analysis (TDA), the establishment of cooperation mechanisms at national and regional level, and the adoption of a regional Strategic Action Program (SAP).

DIKTAS—Transboundary diagnostic analysis, approach, and outcomes
The TDA was conducted in the period 2011–2013 by the DIKTAS Project Team in accordance with the GEF guidelines provided in the TDA/SAP Training Manual. The TDA was based on a substantial regional analysis that was required in order to fully understand the context of transboundary issues. The regional analysis was particularly important given the complexity of the karst environment and regime and the interconnectivity of karst aquifers. The regional analysis also enabled a delineation of transboundary aquifers shared by the project countries. The project team was organized in four working groups, reflecting the main issues of the regional analysis: (1) environmental and socioeconomic analysis; (2) hydrogeology of the Dinaric karst; (3) legal and institutional framework and policy; and (4) stakeholder analysis.

The regional analysis was followed by an in-depth analysis of the transboundary aquifer areas that focused on observed and potential issues of transboundary concern. The analysis was carried out in a systematic way that included climate, hydrology, hydrogeology, groundwater reserves and their utilization, groundwater quality, and water resources protection. For each aquifer, the major issues of concern were determined and priority actions proposed (Fig. 17.42).

The TDA showed that the state of groundwater in the DIKTAS project region is generally good in terms of both quantity and quality with a few exceptions and with a number of serious potential threats. The main threat

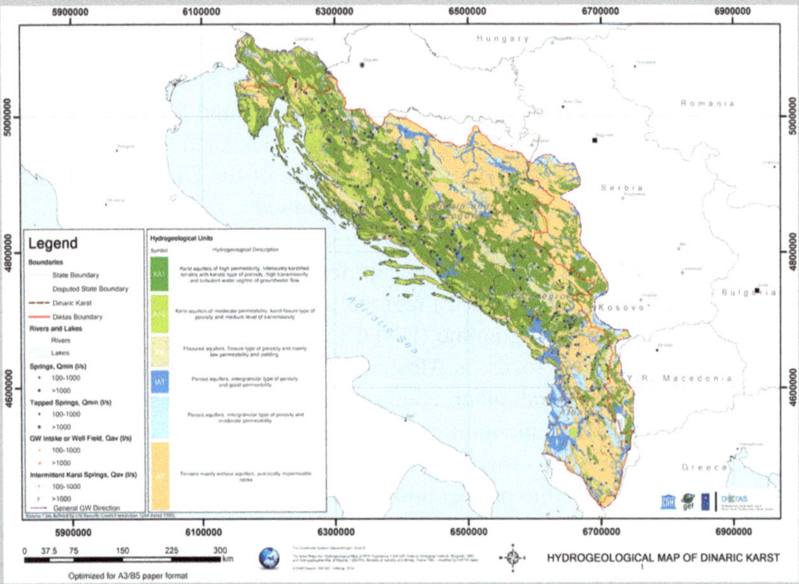

Fig. 17.42 Hydrogeological map of the Dinaric karst with the DIKTAS project area and state boundaries

to the overall groundwater quality in the DIKTAS region is solid waste and wastewater disposal. There are hundreds of unregulated landfills and illegal dumping sites in the four project countries. The number of wastewater treatment plants is insufficient, with about half of the population not connected to this service. For the vulnerable karst environment of the Dinaric region, which has a very limited auto-purification capacity, this is the most serious current as well as (potential) future problem. To a lesser degree, karst groundwater resources in the region are also being contaminated by agricultural and industrial activities.

Currently, no common legal framework and no common criteria exist for (a) the delineation of water source sanitary protection zones and (b) setting cost-efficient measures for groundwater protection in the Dinaric Karst region. This was identified as the main issue of concern in the TBAs with centralized public water supply systems: Trebišnjica, Neretva, Cetina, and Una.

Stakeholder analysis revealed a pressing need for transparent, public sharing of knowledge, information, and scientific data on the many unique characteristics of karst aquifers in the DIKTAS region. Stakeholders view DIKTAS as an opportunity for cooperation, networking and communication between government authorities, agencies, non-governmental organizations (NGOs), and other actors at transboundary level and, most importantly, for the harmonization of legal and karst aquifers management frameworks among the countries. Opportunities for participation in the decision-making process are also among the most widely anticipated outcomes of DIKTAS.

A major added value of the TDA was the collection and harmonization of a large amount of data and information relevant for the assessment and management of karst groundwater resources in the region. This gathered information was not always complete, and in some cases, there were still significant information gaps. Nevertheless, the DIKTAS TDA was the first thorough regional groundwater analysis that covers Albania, Montenegro, Bosnia and Herzegovina, and Croatia. Outputs of the TDA, including GIS materials such as thematic maps and databases, and quantitative hydrogeological analyses, form the basis for developing groundwater resources management models at both regional and local scales.

While the TDA has produced a fair assessment of groundwater resources in the region, it also revealed limitations of knowledge on their actual state and trends in terms of quality and quantity. The main obstacle for this was a lack of monitoring data at both regional and local scales, such as in the vicinity of solid waste and wastewater disposal (treatment) sites, mines, intensive agriculture areas, and industrial facilities handling and generating hazardous materials. Therefore, a strong message resulting from the TDA was a request for improvement of the groundwater monitoring network throughout the region and the need to intensify capacity building in the public sector.

DIKTAS—Strategic action plan

Within the DIKTAS project, two mechanisms have been created to improve consultation and exchange of information between the governmental entities involved in water resources management. The mechanisms are the National Inter-ministerial Committees (NICs) in each of the project countries, and the Consultation and Information Exchange (CIE) body at the regional level.

The NICs are aimed at:

- involving all relevant governmental institutions in an effort to implement integrated land and water resources management and harmonize existing policy frameworks at the national level, and
- contributing to the preparation, review, and adoption of key DIKTAS outputs.

In a case of that a governmental body with similar inter-ministerial setup and mandate would already existed in project countries, the DIKTAS Steering Committee recommended to use/extend these existing structures. However, only one country had a similar body (appreciating an external incentive). In meantime, the NICs were established in all project countries and had two meetings, one at the national level and other, together at the regional level. These meetings were very useful for the finalization of the TDA report and the of ongoing SAP preparation process. A TDA is a fact-finding exercise carried out by the national experts nominated by project countries. Therefore, procedurally the TDA report required neither evaluation nor an official approval by NICs. Nevertheless, by involving NICs, many useful comments and pieces of information were obtained, leading to improvement of the report. Besides, the report has consent of about all ministries in the region that are related to water.

Unlike TDA, a SAP is a negotiated policy process meant to identify policy, legal and institutional reforms and investments needed to address the priority transboundary problems. Endorsed at the highest level, it establishes clear priorities for action to resolve the priority problems which were identified in the TDA. The preparation of a SAP is a cooperative process among the countries of the region. NICs have been involved in formulation of SAP, starting from overall long-term vision on transboundary groundwater resources: To achieve joint sustainable and equitable use and protection of Dinaric karst aquifer system. The vision is elaborated in the main water resources objectives and those further in outcomes, outputs and finally in priority actions: The main objectives are as follows:

- *Groundwater Quantity*: Ensure sufficient groundwater availability in dry periods, especially for water supply and to support environmental flow.
- *Groundwater Quality*: To maintain and improve (where required) quality of karst groundwater in the Dinaric region.
- *Protection of Groundwater-Dependent Ecosystems* (GDE): To ensure protection of GDE, specific features and their ecosystem services for the future
- *Equitable Use*: Support equitable use of Groundwater resources
- *Capacity Building*: Raise awareness and capacity building related to karst water and dependant ecosystem

At this moment, the SAP draft document is prepared awaiting translation in five project languages before being discussed by NICs again. Once the NICs comments and suggestions are collected and processed as possible, the draft will be submitted to the CIE for the final negotiation and agreements. Unlike NICc that are made up of water-related specialists, the CIE is composed of high-level political representatives of the project countries. Once the SAP is approved and funded secured, the specified priority actions can be undertaken.

The CIE is seen as a possible core of long-term international groundwater cooperation in the region, also after the project termination. The CIE is open to other countries sharing the Dinaric karst groundwater upon their request. The final form of the CIE after the project completion will be decided by the member countries; it might join some existing international cooperation mechanism or have its own secretariat. If latter, the secretariat would continue some of "project" activities, particularly related to capacity building and public awareness.

17.6.5 Concluding Remarks

There has been a significant increase of attention for transboundary aquifers in the last 15 years, also due to increased awareness of climate variability and water security. Assessment and management of TBAs is more complex than of aquifers within national borders due to technical, legal, organisational, and political differences. Often, extensive harmonization of data and information is required because of difference in languages, classifications, reference systems, etc. used in aquifer-sharing countries. International cooperation mechanisms and agreements are necessary to ensure sustainable use and protection of internationally shared aquifers.

A fair overview of transboundary aquifers has gradually been obtained through various projects and programs and regularly compiled in the worldwide map by IGRAC. The ongoing TWAP programme will provide substantial quantitative and qualitative overview on TBAs around the world, including a global overview of Transboundary aquifers in karst. Karst nature of some TBAs is already known but TWAP country experts will provide official reading that is very relevant due to a sensitive nature of the TBAs.

The DIKTAS case study delivered a significant insight in nature and state of groundwaters in the project area. In particular at the regional level, the project activities resulted in a major consolidation and update of knowledge, including all relevant aspects, namely hydrogeological, legal, socioeconomical, environmental, legal, and institutional. Substantial regional analysis was required in order to fully understand the context of transboundary issues. From the hydrogeological point of view, aquifers in the Dinaric karst region do not form a single aquifer system (there are rather a few distinct systems). Nevertheless, from the management point of view (hence including all the TBA aspects), groundwater of Dinaric Karst should be considered as a unique system. Although crossed by several state boundaries, Dinaric Karst is quite a unique region, in terms of water resources, but also environment and socioeconomic development. This uniqueness requires intensive cooperation among the states in the region. It seems that cooperation mechanism established in the DIKTAS project provides a technical–political interface that will ensure wide participation and joint commitment to sustainable and equitable use and protection of transboundary groundwaters of the Dinaric Karst.

References

References to Section 17.1

Basso A, Bruno E, Parise M, Pepe M (2013) Morphometric analysis of sinkholes in a karst coastal area of southern Apulia (Italy). Environ Earth Sc 70(6):2545–2559

Bonacci O, Ljubenkov I, Roje-Bonacci T (2006) Karst flash floods: an example from the Dinaric karst Croatia. Nat Hazards Earth Sys Sci 6:195–203

Brinkmann R, Parise M, Dye D (2008) Sinkhole distribution in a rapidly developing urban environment: Hillsborough County, Tampa bay area, Florida. Eng Geol 99:169–184

Bruno E, Calcaterra D, Parise M (2008) Development and morphometry of sinkholes in coastal plains of Apulia, southern Italy. Preliminary sinkhole susceptibility assessment. Eng Geol 99:198–209

Calò F, Parise M (2006) Evaluating the human disturbance to karst environments in southern Italy. Acta Carsologica 35(2):47–56

Calò F, Parise M (2009) Waste management and problems of groundwater pollution in karst environments in the context of a post-conflict scenario: the case of Mostar (Bosnia Herzegovina). Habitat Int 33:63–72

Culshaw MG, Waltham AC (1987) Natural and artificial cavities as ground engineering hazards. Quart J Eng Geol 20:139–150

Day M, Halfen A, Chenoweth S (2011) The Cockpit country, Jamaica: boundary issues in assessing disturbance and using a karst disturbance index in protected areas planning. In: Van Beynen PE (ed) Karst management. Springer, Dordretch, pp 399–414

Delle Rose M, Parise M (2010) Water management in the karst of Apulia, southern Italy. In: Proceedings of international interdisciplinary scientific conference on sustainability of the karst environment. Dinaric karst and other karst regions, vol 2, Plitvice Lakes. IHP-UNESCO, series on groundwater, pp 33–40

De Waele J (2009) Evaluating disturbance on Mediterranean karst areas: the example of Sardinia (Italy). Environ Geol 58(2):239–255

De Waele J, Gutierrez F, Parise M, Plan L (2011) Geomorphology and natural hazards in karst areas: a review. Geomorph 134(1–2):1–8

Farfan H, Dias C, Parise M, Aldana C (2010) Scenarios of groundwater pollution in a karst watershed: a case study in the Pinar del Rio province at Cuba. In: Carrasco F, La Moreaux JW, Duran Valsero JJ, Andreo B (eds) Advances in research in karst media. Springer, Berlin, pp 287–292

Fookes PG, Hawkins AB (1988) Limestone weathering: its engineering significance and a proposed classification scheme. Quart J Eng Geol 21:7–31

Ford DC, Williams P (2007a) Karst hydrogeology and geomorphology. Wiley, Chichester

Goldscheider N, Drew D (eds) (2007) Methods in karst hydrogeology. International contributions to hydrogeology 26. Int Ass Hydrogeol. Taylor & Francis, London

Gunn J (1993) The geomorphological impacts of limestone quarrying. Catena 25:187–198

Gunn J (2004) Quarrying of limestones. In: Gunn J (ed) Encyclopedia of cave and karst science. Routledge, London, pp 608–611

Gunn J (2007) Contributory area definition for groundwater source protection and hazard mitigation in carbonate aquifers. In: Parise M, Gunn J (eds) Natural and anthropogenic hazards in karst areas: recognition, analysis and mitigation. Geological Society of London, London, pp 97–109 (sp publ 279)

Gutierrez F, Parise M, De Waele J, Jourde H (2014) A review on natural and human-induced geohazards and impacts in karst. Earth Sci Rev 138:61–88

Hajna NZ (2003) Incomplete solution: weathering of cave walls and the production, transport and deposition of carbonate fines. Carsologica, Postojna-Ljubljana

International Society for Rock Mechanics (1978) Suggested methods for the quantitative description of discontinuities in rock masses. Int J Rock Mech Min Sci Geomech Abs 15:319–368

Iovine G, Parise M, Trocino A (2010) Breakdown mechanisms in gypsum caves of southern Italy, and the related effects at the surface. ZeitGeomorph 54(suppl 2):153–178

Klimchouk A, Andrejchuk V (2002) Karst breakdown mechanisms from observations in the gypsum caves of the Western Ukraine: implications for subsidence hazard assessment. Int J Speleol 31(1/4):55–88

Lollino P, Parise M, Reina A (2004) Numerical analysis of the behavior of a karst cavern at Castellana-Grotte, Italy. In: Proceedings of 1st international UDEC/3DEC symposium, Bochum, 29 Sept–1 Oct 2004, pp 49–55

Lollino P, Martimucci V, Parise M (2013) Geological survey and numerical modeling of the potential failure mechanisms of underground caves. Geosys Eng 16(1):100–112

Lopez N, Spizzico V, Parise M (2009) Geomorphological, pedological, and hydrological characteristics of karst lakes at Conversano (Apulia, southern Italy) as a basis for environmental protection. Environ Geol 58(2):327–337
Margiotta S, Negri S, Parise M, Valloni R (2012) Mapping the susceptibility to sinkholes in coastal areas, based on stratigraphy, geomorphology and geophysics. Nat Hazards 62(2):657–676
Martimucci V, Parise M (2012) Cave surveys, the representation of underground karst landforms, and their possible use and misuse. In: 20th international karst school "karst forms and processes", Postojna, 18–21 June 2012, Guide Book and Abstracts, pp 69–70
Milanovic P (2002) The environmental impacts of human activities and engineering constructions in karst regions. Episodes 25:13–21
Nicod J (1972) Pays et paysages du calcaire. Presses Universitaires de France, Paris
North LA, van Beynen PE, Parise M (2009) Interregional comparison of karst disturbance: West-Central Florida and Southeast Italy. J Environ Manag 90:1770–1781
Palma B, Parise M, Reichenbach P, Guzzetti F (2012) Rock-fall hazard assessment along a road in the Sorrento Peninsula, Campania, southern Italy. Nat Hazards 61(1):187–201
Palmer AN (2007) Cave geology. Cave Books
Parise M (2003) Flood history in the karst environment of Castellana-Grotte (Apulia, Southern Italy). Nat Hazards Earth Syst Sc 3(6):593–604
Parise M (2008) Rock failures in karst. In: Cheng Z, Zhang J, Li Z, Wu F, Ho K (eds) Landslides and engineered slopes. Proceedings of 10th international symposium on landslides, vol 1, Xi'an, pp 275–280
Parise M (2010) The impacts of quarrying in the Apulian karst. In: Andreo B, Carrasco F, Duran JJ, La Moreaux JW (eds) Advances in research in karst media. Springer, Berlin, pp 441–447
Parise M (2012a) A present risk from past activities: sinkhole occurrence above underground quarries. Carbonates and Evaporites 27(2):109–118
Parise M (2012b) Management of water resources in karst environments, and negative effects of land use changes in the Murge area (Apulia). Karst Devel 2(1):16–20
Parise M (2013) Recognition of instability features in artificial cavities. In: Proceedings of 16th international congress speleology, Brno, vol 2, 21–28 July 2013, pp 224–229
Parise M, Gunn J (eds) (2007) Natural and anthropogenic hazards in karst areas: recognition, analysis and mitigation. Geological Society of London, London (sp publ 279)
Parise M, Lollino P (2011) A preliminary analysis of failure mechanisms in karst and man-made underground caves in southern Italy. Geomorph 134(1–2):132–143
Parise M, Pascali V (2003a) Surface and subsurface environmental degradation in the karst of Apulia (southern Italy). Environ Geol 44:247–256
Parise M, Galeazzi C, Bixio R, Dixon M (2013) Classification of artificial cavities: a first contribution by the UIS Commission. In: Filippi M, Bosak P (eds) Proceedings of 16th international congress speleology, Brno, vol 2, 21–28 July 2013, pp 230–235
Santo A, Del Prete S, Di Crescenzo G, Rotella M (2007) Karst processes and slope instability: some investigations in the carbonate Apennine of Campania (southern Italy). In: Parise M, Gunn J (eds) Natural and anthropogenic hazards in karst areas: recognition, analysis and mitigation. Geological Society of London, London, pp 59–72 (sp publ 279)
Sauro U (2003) Dolines and sinkholes: aspects of evolution and problems of classification. Acta Carsologica 32(2):41–52
Tharp TM (1995) Mechanics of upward propagation of cover-collapse sinkholes. Eng Geol 52:23–33
van Beynen PE, Townsend K (2005) A disturbance index for karst environments. Environ Manag 36:101–116
Waltham AC (2002) The engineering classification of karst with respect to the role and influence of caves. Int J Speleol 31(1/4):19–35
Waltham T, Lu Z (2007) Natural and anthropogenic rock collapse over open caves. In: Parise M, Gunn J (eds) Natural and anthropogenic hazards in karst areas: recognition, analysis and mitigation. Geological of Society London, London, pp 13–21 (sp publ 279)

Waltham T, Bell F, Culshaw M (2005) Sinkholes and subsidence: karst and cavernous rocks in engineering and construction. Springer, Berlin
Weishampel JF, Hightower JN, Chase AF, Chase DZ, Patrick RA (2011) Detection and morphologic analysis of potential below-canopy cave openings in the karst landscape around the maya polity of Caracol using airborne lidar. J Cave Karst Stud 73(3):187–196
White WB (1988) Geomorphology and hydrology of karst terrains. Oxford University Press, New York
White E, White W (1969) Processes of cavern breakdown. Bull Natl Speleol Soc 31(4):83–96
White EL, White WB (1984) Flood hazards in karst terrains: lessons from the Hurricane Agnes storm. In: Burger A, Dubertret L (eds) Hydrogeology of karst terrains, vol 1, pp 261–264
Williams PW (2008a) The role of the epikarst in karst and cave hydrogeology: a review. Int J Speleol 37:1–10
Zwahlen F (ed) (2004) Vulnerability and risk mapping for the protection of carbonate (karstic) aquifers. Final report COST action 620. European Commission, Brüssel

References to Section 17.2

Andreo B, Ravbar N, Vías JM (2009) Source vulnerability mapping in carbonate (karst) aquifers by extension of the COP method: application to pilot sites. Hydrogeol J 17(3):749–758
Bakalowicz M (2005) Karst groundwater: a challenge for new resources. Hydrogeol J 13(1):148–160
Bonacci O, Pipan T, Culver DC (2009) A framework for karst ecohydrology. Environ Geol 56:891–900
Daly D, Dassargues A, Drew D, Dunne S, Goldscheider N, Neale S, Popescu IC, Zwahlen F (2002) Main concepts of the "European approach" to karst-groundwater-vulnerability assessment and mapping. Hydrogeol J 10:340–345
De Ketelaere D, Hötzl H, Neukum C, Civita M, Sappa G (2004) Hazard analysis and mapping. In: Zwahlen F (ed) COST action 620. Vulnerability and risk mapping for the protection of carbonate (Karstic) aquifers. Final report COST action 620. European Commission, Directorate-General for Research, Brüssel, Luxemburg, pp 106–107
Dörfliger N, Zwahlen F (1998) Practical guide. Groundwater vulnerability mapping in karstic regions (EPIK). Swiss Agency for the Environment, Forests and Landscape, Bern
Dörfliger N, Plagnes V (2009) Cartographie de la vulnérabilité des aquifères karstiques guide méthodologique de la méthode PaPRIKa. Rapport BRGM RP-57527-FR
Drew D, Hötzl H (eds) (1999) Karst hydrology and human activities. International contributions to hydrogeology. IAH, vol 20. Taylor & Francis/Balkema, London
Ford DC, Williams PW (2007b) Karst hydrogeology and geomorphology. Wiley, Chichester
Foster S, Hirata R, Andreo B (2013) The aquifer pollution vulnerability concept: aid or impediment in promoting groundwater protection? Hydrogeol J 21:1389–1392
Gogu RC, Hallet V, Dassargues A (2003) Comparison of aquifer vulnerability assessment techniques. Application to the Neblon river basin (Belgium). Environ Geol 44(8):881–892
Goldscheider N (2005) Karst groundwater vulnerability mapping—application of a new method in the Swabian Alb, Germany. Hydrogeol J 13:555–564
Goldscheider N, Madl-Szonyi J, Eross A, Schill E (2010a) Review: thermal water resources in carbonate rock aquifers. Hydrogeol J 18(6):1303–1318
Gunn J, Bailey D (1993) Limestone quarrying and quarry reclamation in Britain. Environ Geol 21(3):167–172
Guo F, Jiang GH, Yuan DX, Polk JS (2012) Evolution of major environmental geological problems in karst areas of southwestern China. Environ Earth Sci 69(7):2427–2435
Hamilton-Smith E (2007) Karst and world heritage status. Acta Carsologica 36:291–302
Han ZS (1998) Groundwater for urban water supplies in northern China—an overview. Hydrogeol J 6:416–420
Jeannin PY, Eichenberger U, Sinreich M, Vouillamoz J, Malard A, Weber E (2012) KARSYS: a pragmatic approach to karst hydrogeological system conceptualisation. Assessment of groundwater reserves and resources in Switzerland. Environ Earth Sci 69:999–1013

Kaçaroğlu F (1999) Review of groundwater pollution and protection in karst areas. Water Air Soil Pollut 113:337–356

Kovačič G, Ravbar N (2005) A review of the potential and actual sources of pollution to groundwater in selected karst areas in Slovenia. Nat Hazards Earth Syst Sci 5(2):225–233

Kranjc A, Likar V, Huzjan Ž (eds) (1999) Karst: landscape, life, people. ZRC Publishing, ZRC SAZU, Ljubljana

Kresic N (2009a) Groundwater resources: sustainability, management, and restoration. McGraw-Hill, New York

Kundzewicz ZW, Mata LJ, Arnell NW, Doll P, Jimenez B, Miller K, Oki T, Sen Z, Shiklomanov I (2008) The implications of projected climate change for freshwater resources and their management. Hydrol Sci J 53:3–10

Liedl R, Sauter M (1998) Modelling of aquifer genesis and heat transport in karst systems. Bull d`Hydrogéologie 16:185–200

Loáiciga HA, Maidment DR, Valdes JB (2000) Climate-change impacts in a regional karst aquifer, Texas, USA. J Hydrol 227:173–194

Mamon BA, Azmeh MM, Pitts MW (2002) The environmental hazards of locating wastewater impoundments in karst terrains. Environ Geol 65:169–177

Moore CH (2001) Carbonate reservoirs porosity evolution and diagenesis in a sequence stratigraphic framework. Elsevier, Amsterdam

Nicod J, Julian M, Anthony E (1997) A historical review of man-karst relationships: miscellaneous uses of karst and their impact. Rivista di Geografia Italiana 103:289–338

Parise M, Pascali V (2003b) Surface and subsurface environmental degradation in the karst of Apulia (southern Italy). Environ Geol 44:247–256

Pronk M, Goldscheider N, Zopfi J (2006) Dynamics and interaction of organic carbon, turbidity and bacteria in a karst aquifer system. Hydrogeol J 14(4):473–484

Pulido-Bosch A, Morell I, Andreu JM (1995) Hydrogeochemical effects of groundwater mining of the Sierra de Crevillente Aquifer (Alicante, Spain). Environ Geol 26(4):232–239

Ravbar N, Goldscheider N (2007a) Proposed methodology of vulnerability and contamination risk mapping for the protection of karst aquifers in Slovenia. Acta Carsologica 36(3):397–411

Ravbar N (2007a) The protection of karst waters. ZRC Publishing, ZRC SAZU, Postojna

Ravbar N, Kovačič G (2013) Analysis of human induced changes in a karst landscape—the filling of dolines in the Kras plateau, Slovenia. Sci Total Environ 447:143–151

Ravbar N, Kovačič G, Marin AI (2013) Abandoned water resources as potential sources of drinking water—a proposal for management of the Korentan karst spring near Postojna. Acta Geogr Lovan 53(2):295–316

Vesper DJ, White WB (2004) Spring and conduit sediments as storage reservoirs for heavy metals in karst aquifers. Environ Geol 45(4):481–493

Vrba J, Zaporozec A (eds) (1994) Guidebook on mapping groundwater vulnerability. International contributions to hydrogeology, IAH, vol 16. Verlag Heinz Heise, Hannover

Williams PW (2008) World heritage caves and karst: a thematic study. IUCN world heritage studies, no. 2. Gland, Switzerland

Zwahlen F (ed) (2004) Vulnerability and risk mapping for the protection of carbonate (karstic) aquifers. Final report COST action 620. European Commission, Directorate-General for Research, Brüssel

References to Section 17.3

Biondić B, Biondić R, Dukarić F (1998) Protection of karst aquifers in the Dinarides in Croatia. Environ Geol 34(4):309–319

Biondić B (2000) Karst groundwater protection: the case of the Rijeka region, Croatia. Acta Carsologica 29/1, 2:33–46

Carey M, Hayes P, Renner A (2009) Groundwater source protection zones. Review of methods, integrated catchment science programme, science report: SC070004/SR1, Environment Agency, Bristol

Chave P, Howard G, Schijven J, Appleyard S, Fladerer F, Schimon W (2006) Groundwater protection zones. In: Schmoll O, Howard G, Chilton J, Chorus I (eds) Protecting groundwater for health, managing the quality of drinking-water sources. IWA Publishing, London

Conservation Engineering Division (1986) Urban hydrology for small watersheds, National Resources Conservation Service, U.S. Dep. Agric., Tech. Rel. No. 55 (1975)

COST Action 65 (1995) Final report on hydrogeological aspects of groundwater protection in karstic areas. European Commission, Luxembourg

Department of Environment and Local Government, Environmental Protection Agency and Geological Survey of Ireland (1999) Groundwater protection schemes. Department of Environment and Local Government, Environmental Protection Agency and Geological Survey of Ireland, Dublin

Dörfliger N, Zwahlen F (1997) EPIK: a new method for outlining of protection areas in karstic environment. In: Günay G, Johnson I (eds) Karst waters and environmental impacts. Balkema, Rotterdam, pp 117–123

European Commission (2007) Common implementation strategy for the water framework directive (2000/60/EC). Guidance document no. 16. Guidance on groundwater in drinking water protected areas. Luxembourg

ESRI (2011) ArcGIS desktop: release 10. Environmental Systems Research Institute, Redlands, CA

Foster S, Hirata R, Gomes D, D'Elia M, Parise M (2002) Groundwater quality protection, a guide for water utilities, municipal authorities, and environment agencies. The International Bank for Reconstruction and Development/The World Bank, Washington

Goldscheider N, Drew D (eds) (2007b) Methods in karst hydrogeology. International contribution to hydrogeology, IAH. Taylor & Francis/Balkema, London

Göppert N, Goldscheider N (2008) Solute and colloid transport in karst conduits under low- and high-flow conditions. Ground Water 46(1):61–68

Kavouri K, Plagnes V, Tremoulet J, Dörfliger N, Reijiba F, Marchet P (2011) PaPRIKA: a method for estimating karst resource and source vulnerability—application to the Ouysse karst system (southwest France). Hydrogeol J 19:339–353

Kralik M, Keimel T (2003) Time-input, an innovative groundwater-vulnerability assessment scheme: application to an alpine test site. Environ Geol 44:679–686

Kresic N (2007) Hydrology and groundwater modelling, 2nd edn. CRC Press, Taylor & Francis Group, Boca Raton

Kresic N (2009b) Groundwater resources: sustainability, management and restoration. McGraw-Hill, New York

Kresic N (2013) Water in karst: management, vulnerability, and restoration. McGraw-Hill, New York

Margane A (2003) Guideline for the delineation of groundwater protection zones. Technical cooperation project management, protection and sustainable use of groundwater and soil resources in the Arab region, technical reports, vol 5, prepared by BGR & ACSAD, BGR archive no. 122917:5, Damascus, p 329

Milanović P (2000) Geological engineering in karst, Dams, reservoirs, grouting, groundwater protection, water tapping, tunneling. Zebra Publishing Ltd., Belgrade

Panday S, Langevin CD, Niswonger RG, Ibaraki M, Hughes JD (2013) MODFLOW-USG version 1: an unstructured grid version of MODFLOW for simulating groundwater flow and tightly coupled processes using a control volume finite-difference formulation: U.S. Geological survey techniques and methods, book 6, chap. A45, p 66

Prohaska S, Ristić V, Dragišić V (2001) Proračun bilansa i dinamičkih rezervi podzemnih voda karstnog masiva Miroč (Groundwater budget and dynamical reserves estimation of the Miroč karst massif; in Serbian). Vodoprivreda, 33/189–194:35–40

Ravbar N (2007) Vulnerability and risk mapping for the protection of karst waters in Slovenia—application to the catchment of the Podstenjšek springs. PhD thesis, University of Nova Gorica Graduate School, Nova Gorica

Ravbar N, Goldscheider N (2007b) Proposed methodology of vulnerability and contamination risk mapping for the protection of karst aquifers in Slovenia. Acta Carsologica 36(3):397–411

Reimann T, Hill ME (2009) MODFLOW-CFP: a new conduit flow process for MODFLOW–2005. Ground Water 47(3):321–325

Pronk M, Goldscheider N, Zopfi J, Zwahlen F (2009) Percolation and particle transport in the unsaturated zone of a karst aquifer. Ground Water 47(3):361–369

Van Waegeningh HG (1985) Overview of the protection of groundwater quality. In: Matthess G, Foster SSD, Skinner ACH (eds) Theoretical background, hydrogeology and practice of groundwater protection zones. International contributions to hydrogeology, IAH, vol 6. Heise, Hanover, pp 156–159

Worthington SRH (2003) The Walkerton karst aquifer. Can Caver 60:42–43

Worthington SRH (2007) Groundwater residence times in unconfined carbonate aquifers. J Cave Karst Stud 69(1):94–102

Worthington SRH (2011) Management of carbonate aquifers. In: van Beynen PE (ed) Karst management. Springer, Berlin

Zwahlen F (ed) (2004) Vulnerability and risk mapping for the protection of carbonate (karstic) aquifers. Final report COST action 620. European Commission, Directorate-General for Research, Brüssel

References to Section 17.4

Davis E (1997) Ground water issue: how heat can enhance in-situ soil and aquifer remediation: important chemical properties and guidance on choosing the appropriate technique, EPA 540/S-97/502. U.S. Environmental Protection Agency, Office of Research and Development, Ada, Oklahoma, p 18

Davis EL (1998) Steam injection for soil and aquifer remediation. EPA/540/S-97/505, U.S. Environmental Protection Agency, Office of Research and Development, Ada, Oklahoma, p 16

Falta RW, Rao PS, Basu N (2005a) Assessing the impacts of partial mass depletion in DNAPL source zones: I. Analytical modeling of source strength functions and plume response. J Contam Hydrol 78(4):259–280

Falta RW, Basu N, Rao PS (2005b) Assessing the impacts of partial mass depletion in DNAPL source zones: II. Coupling source strength functions to plume evolution. J Contam Hydrol 79(1):45–66

Falta RW, Stacy MB, Ahsanuzzaman ANM, Wang M, Earle RC (2007) REMChlor, remediation evaluation model for chlorinated solvents; User's manual, version 1.0. U.S. Environmental protection agency, Center for subsurface modeling support, National Risk Management Research Laboratory, Ada, Oklahoma, p 79

Gudbjerg J (2003) Remediation by steam injection. PhD thesis, environment and resources DTU, Technical University of Denmark, p 137

Huling SG, Pivetz BE (2006) In-situ chemical oxidation. Engineering issue, EPA/600/R-06/072, U.S. Environmental Protection Agency, Office of Research and Development, National Risk Management Research Laboratory, Cincinnati, OH, p 58

ITRC (2005) Technical and regulatory guidance for in situ chemical oxidation of contaminated soil and groundwater, 2nd edn. In Situ Chemical Oxidation Team, Interstate Technology and Regulatory Council, Washington, p 71 + appendices

Jawitz JW, Fure AD, Demmy GG, Berglund S, Rao PS (2005) Groundwater contaminant flux reduction resulting from non-aqueous phase liquid mass reduction. Wat Resour Res 41(10):10408–10423

Kingston JT, Dahlen PR, Johnson PC, Foote E, Williams S (2009) State-of-the-practice overview: critical evaluation of state-of-the-art in situ thermal treatment technologies for DNAPL source zone treatment. ESTCP project ER-0314. Available at http://cluin.org/techfocus/default.focus/sec/Thermal_Treatment%3A_In_Situ/cat/Guidance/

Kresic N (2009c) Hydrogeology and groundwater modeling, 2nd edn. CRC/Taylor & Francis, Boca Raton

Kresic N, Mikszewski A (2013) Hydrogeological conceptual site models: data analysis and visualization. CRC/Taylor & Francis Group, Boca Raton
Lipson DS, Kueper BH, Gefell MJ (2005) Matrix diffusion-derived plume attenuation in fractured bedrock. Ground Water 43(1):30–39
Magnuson JK, Stern RV, Gossett JM, Zinder SH, Burris DR (1998) Reductive dechlorination of tetrachloroethene to ethene by a two-component enzyme pathway. Appl Environ Microbiol 64:1270–1275
McDade JM, McGuire TM, Newell CJ (2005) Analysis of DNAPL source-depletion costs at 36 field sites. Remediat J 15(2):9–18
Mikszewski A, Kresic N (2014) Numeric modeling of well capture zones in karst aquifers. In: Kukuric N, Stevanović Z, Kresic N (eds) Proceedings of international conference and field seminar karst without boundaries, 11–15 June 2014, Trebinje, Bosnia and Herzegovina, Dubrovnik, Croatia, DIKTAS, pp 31–38
Parsons (Parsons Corporation) (2004) Principles and practices of enhanced anaerobic bioremediation of chlorinated solvents. Air Force Center for Environmental Excellence (AFCEE), Brooks City-Base, Texas; Naval Facilities Engineering Service Center Port Hueneme, California; Environmental Security Technology Certification Program, Arlington, Virginia, various paging
Powell T, Smith G, Sturza J, Lynch K, Truex M (2007) New advancements for in-situ treatment using electrical resistance heating. Remediat J 17:51–70
Rao PS, Jawitz JW, Enfield CG, Falta RW, Annable MD, Wood AL (2001) Technology integration for contaminated site remediation: cleanup goals and performance criteria. In: Groundwater quality: natural and enhanced restoration of groundwater pollution. Publication no. 275, IAHS, Wallingford, United Kingdom, pp 571–578
USACE (U.S. Army Corps of Engineers) (2006) Design: in situ thermal remediation. UFC 3-280-05. Unified Facilities Criteria (UFC). U.S. Army Corps of Engineers, Naval Facilities Engineering Command (NAVFAC), Air Force Civil Engineer Support Agency (AFCESA)
USACE (2009) Design: in-situ thermal remediation. Manual 1110-1-401536, p 226. Available at http://www.usace.army.mil/inet/usace-docs/
USEPA (1998a) Steam injection for soil and aquifer remediation. EPA/540/S-97/505, Office of solid waste and emergency response, U.S. Environmental Protection Agency, Washington, p 16
USEPA (1998b) Permeable reactive barrier technologies for contaminant remediation. EPA/600/R-98/125, Office of solid waste and emergency response, U.S. Environmental Protection Agency, Washington, p 94
USEPA (2000) Engineered approaches to in situ bioremediation of chlorinated solvents: fundamentals and field applications. EPA 542-R-00-008. Available at http://cluin.org/download/remed/engappinsitbio.pdf. Accessed 12 Aug 2011
USEPA (2004) In situ thermal treatment of chlorinated solvents; fundamentals and field applications. EPA 542/R-04/010. Office of solid waste and emergency response, U.S. Environmental Protection Agency, Washington, various paging
USEPA (United States Environmental Protection Agency) (2013) Superfund remedy report, 14th edn. Office of solid waste and emergency response, EPA-542-R-13-016, p 22+appendices

References to Section 17.5

Alföldi L, Bélteky L. Böcker T, Horváth J, Korim K, Rémi R (eds) (1968) Budapest hévizei (Thermal waters of Budapest: in Hungarian). VITUKI (Institute for water resources research), Budapest, p 365
Alföldi L, Kapolyi L (eds) (2007) Bányászati karsztvízszintsüllyesztés a Dunántúli-középhegységben (Mining dewatering in the transdanubian range; in Hungarian). MTA Földrajztudományi Kutatóintézet (Geography Institute of Hungarian academy of sciences) 138
Angel RR (1958) Volume requirements for air and gas drilling. Gulf Publishing Co., Houston

Ballabás G (2004) Visszatérő karsztforrásokkal kapcsolatos településfejlesztési és környezetvédelmi lehetőségek és veszélyek Tata város példáján (Land use and environmental possibilities and dangers regarding the re-operating of karst springs in the example of Tata; in Hungarian). VIII national conference for Geographer PhD students Szeged. CD 11 http://geogr.elte.hu/TGF/TGF_Cikkek/ballabas2.pdf

Bredehoeft JD, Papadopulos IS (1965) Rates of vertical groundwater movement estimated from the earth's thermal profile. Water Resour Res 1:325–328

Csepregi A (2007) A karsztvíztermelés hatása a Dunántúli-középhegység vízháztartására (The effect of water withdrawal on the water balance of the Transdanubian Range; In Hungarian). 77–112. In: Alföldi L, Kapolyi L (eds) Bányászati karsztvízszintsüllyesztés a Dunántúli-középhegységben (Mining dewatering in the transdanubian range; in Hungarian) MTA Földrajztudományi Kutatóintézet (Geography Institute of Hungarian Academy of Sciences), p 138

Erőss A, Zsemle F, Pataki L, Csordás J, Zsuppán K, Pulay E (2013) Heat potential evaluation of effluent and used thermal waters in Budapest, Hungary. In: Szőcs T, Fórizs I (eds) '013) Proceedings of the IAH Central European groundwater conference, Mórahalom, Hungary 08–10.05.2013. Szeged University Press, Szeged, pp 98–99

Ford DC, Williams PW (2007c) Karst hydrogeology and geomorphology. Wiley, Chichester

Goldscheider N, Mádl-Szőnyi J, Erőss A, Schill E (2010b) Review: thermal water resources in carbonate rock aquifers. Hydrogeol J 18(6):1303–1318

Haas J (ed) (2001) Geology of Hungary. Eötvös University Press Budapest, p 317

Horusitzky H (1923) Tata és Tóváros hévforrásainak hidrogeológiája és közgazdasági jövője (Hydrogeology of Tata, the town of lakes and its economic future; in Hungarian). A Magyar Királyi Földtani Intézet Évkönyve (Yearbook of the Hungarian Royal Geological Institute) XXV. köt. 3. Budapest, pp 37–83

Hubbert MK (1940) The theory of ground-water motion. J Geol XLVIII 8(1):785–944

Johnson PW (1995) Design techniques in air and gas drilling: cleaning criteria and minimum flowing pressure gradients. J Can Pet Tech (May)

Király L (1975) Rapport sur l'état actuel des connaissances dans le domaine des caractéres physiques des roches karstiques. In: Burger A, Dubertret L (eds) Hydrogeology of karstic terrains. IAH, International Union of Geological Sciences, Series B, 3, pp 53–67

Klimchouk AB (2007) Hypogene speleogenesis: hydrogeological and morphogenetic perspective. Special paper no.1, National Cave and Karst Research Institute, Carlsbad, 106

Kovács A, Szőcs T (2014) Prediction of karst water recovery following regional mine depressurization in the tata area, Hungary. In: Kukurić N, Stevanović Z, Krešic N (eds) Proceedings of the DIKTAS conference: "Karst without boundaries", Trebinje, 11–15 June 2014, pp 165–170

Lenkey L, Dövényi P, Horváth F, Cloething SAPL (2002) Geothermics of the Pannonian basin and its bearing on the neotectonics. EUG Stephan Mueller special publication series, 3:29–40

Lorberer Á (1986) A Dunántúli-középhegység karsztvízföldtani és vízgazdálkodási helyzetfelmérése és döntés előkészítő értékelése (evaluation and outline of the karst hydrogeology and water management of the Transdanubian range; in Hungarian) VITUKI Témajelentés Kézirat (manuscript for the Institute for water resources research)

Lovrity V, Bodor P (2014) A Boltív-forrás vízhozamának és fizikai, kémiai paramétereinek változása a csapadékesemények és a Duna vízállás függvényében. Értékelés archív adatok és recens mérések alapján (the changing of discharge volume, physical and chemical parameters of Boltív Spring in comparison with the precipitation and the level of the Danube. Evaluation based on archive data and recent measurements; in Hungarian). Young student research thesis. ELTE Physical and Applied Geology Department 80

Maller M and Hajnal G (2013) A tatai források hidrogeológiai vizsgálata Hydrogeological investigations related to the springs of Tata (in Hungarian) 7–18. In: Török Á, Görög P, Vásárhelyi B (eds) (2013) Mérnökgeológia-Kőzetmechanika (Engineering geology and rock mechanics) http://mernokgeologia.bme.hu/ocs/index.php/konferencia/2013/paper/viewFile/7/7

Mádl-Szőnyi J, Leél-Őssy Sz, Kádár M, Angelus B, Zsemle F, Erőss A, Kalinovits S, Segesdi J, Müller I (2001) In: Mindszenty A (ed) A Budai Termálkarszt-rendszer hidrodinamikájának vizsgálata nyomjelzéssel (evaluation of hydrodynamics of Buda thermal Karst by tracing experiments; in Hungarian). Manuscript ELTE, Physical and Applied Geology Department 456

Mádl-Szőnyi J, Zsemle F, Lenkey L, Virág M (2009) Termálvízalapú geotermikus fűtési rendszerek potenciáltérképe Budapesten. Kétkutas rendszerek telepítési terve (potential map of implementation of geothermal doublets in Budapest; in Hungarian), vol 1–2, Manuscript ELTE Erdélyi Mihály Foundation
Mádl-Szőnyi J and Erőss A (2013) Effects of regional groundwater flow on deep carbonate systems focusing on discharge zones. In: Proceedings of the international symposium on regional groundwater flow: theory, applications and future development, 21–23 June Xi'an, China. China Geological Survey, Commission of Regional Groundwater Flow, IAH, pp 71–75
Mindszenty A (ed) (2013) Budapest: földtani értékek és az ember. Városgeológiai tanulmányok („In urbe et pro urbe") (Budapest, geological values and the man. Urban geological studies; in Hungarian) ELTE Eötvös Kiadó, Budapest, p 311
Papp F (1942) Budapest meleg gyógyforrásai (thermal medicinal springs of Budapest; in Hungarian). A Budapesti Központi Gyógy- és Üdülőhelyi Bizottság Rheuma és Fürdőkutató Intézet kiadványa, Budapest, p 252
Pratt CA (1989) Modifications to and experience with air-percussion drilling, SPE Drilling Engineering, December
Royden LH, Horváth F (eds) (1988) The Pannonian basin—a study in basin evolution. Amer Assoc Petrol Geol Memoir 45, Tulsa, 394
Sass I (2007) Geothermie und Grundwasser (geothermics and groundwater). Grundwasser 12(2):93
Szabó T (2006) Alulegyensúlyozott fúrási technológia folyadékainak vizsgálata (Examination of technological liquids of below balanced drilling technology). PhD thesis, University of Miskolc 106
Tóth J (1963) A theoretical analysis of groundwater flow in small drainage basins. J Geophys Res 68:4795–4812
Tóth J (1971) Groundwater discharge: a common generator of diverse geologic and morphologic phenomena. IASH Bull 16(1–3):7–24
Tóth J (1984) The role of regional gravity flow in the chemical and thermal evolution of ground water. In: Hitchon, B, Wallick, EI (eds) Proceedings of Ist Canadian/American conference on hydrogeology, practical applications of ground water geochemistry. National Water Well Association and Alberta Research Council, Worthington
Tóth J (1995) Hydraulic continuity in large sedimentary basins. Hydrogeol J 3(4):4–16
Tóth J (1999) Groundwater as a geologic agent: an overview of the causes, processes, and manifestations. Hydrogeol J 7:1–14
Tóth J (2009) Gravitational systems of groundwater flow theory, evaluation, utilization. Cambridge University Press, Hardback 294
Tóth J (2013) Groundwater flow systems: analysis, characterization and agency in karst genesis. A1. REGFLOW and MANKARST training course. International symposium on hierarchical flow systems on karst regions. 2–3 Sept 2013 Budapest, Hungary, Eötvös Loránd University. Supplementary notes on session 2 1–14 http://www.karstflow2013.org/?nic=training-course
Tóth M, Dorn F, Fürst Á, Lorberer Á, Sárváry I (1999) A tatai források visszatérésével kapcsolatos vizsgálatok és cselekvési program, Tata (tasks and activities regarding the re-operation of springs at tata; in Hungarian). Manuscript hydrosys Ltd, Monumentum Ltd, Equilibrium Ltd, Municipality of Tata

References to Section 17.6

IGRAC (2014) Transboundary aquifers of the World map, update 2014. www.un-igrac.org
INBO and GWP (2012) Handbook for integrated water resources management in basins, Paris, Sweden
DIKTAS Project Team (2013) Transboundary diagnostic analysis, prepared in the framework of the 'protection and sustainable use of the Dinaric karst transboundary aquifer system' (DIKTAS) project http://diktas.iwlearn.org

Duscher K (2011) Groundwater GIS reference layer, submission/compilation status and evaluation, EEA/NSV/10/002—ETC/ICM, European Environmental Agency
Machard de Gramont H et al (2010) Toward a joint management of transboundary aquifer systems, AFC, French Development Agency
European Commission (2003) Working Group on Water Bodies, Common implementation strategy for the water framework directive (2000/60/EC), Identification of water bodies, guiding document no 2
Kukuric N, Gun van der J, Vasak S (2008) Towards a methodology for the assessment of internationally shared ground-waters. In: Proceedings of 4th international symposium on transboundary waters management, Thessaloniki
Lipponen A, Kukuric N (2010) Assessment of transboundary aquifers in the context of the second assessment of transboundary waters in the United Nations Economic Commission for Europe. International conference transboundary aquifers: challenges and new directions
Stephan RM (2011) The draft articles on the law of transboundary aquifers: the process at the UN ILC, 13 Int'l Comm L Rev 223
UNECE Task Force on Monitoring and Assessment (1999) Inventory of transboundary groundwaters UNECE, Geneva, Switzerland
UNECE (2007) The first assessment of transboundary rivers. Lakes and groundwaters UNECE, Geneva, Switzerland
UNECE (2011) The second assessment of transboundary rivers. Lakes and groundwaters UNECE, Geneva
UNESCO (2001) ISARM framework document www.isarm.org
UNESCO-IHP (2011) Methodology for the GEF transboundary waters assessment programme. vol 2, methodology for the assessment of transboundary aquifers, UNEP, vi + 113 pp
UNESCO (2012) GEF Groundwater portfolio analysis (working draft)
UN-ESCWA and BGR (2013) Inventory of shared water resources in Western Asia. Beirut, Lebanon http://waterinventory.org

List of Keywords

Z. Stevanović (ed.), *Karst Aquifers – Characterization and Engineering*,
Professional Practice in Earth Sciences, DOI 10.1007/978-3-319-12850-4

MIX
Papier aus verantwortungsvollen Quellen
Paper from responsible sources
FSC® C105338

If you have any concerns about our products,
you can contact us on
ProductSafety@springernature.com

In case Publisher is established outside the EU,
the EU authorized representative is:
Springer Nature Customer Service Center GmbH
Europaplatz 3, 69115 Heidelberg, Germany

Printed by Libri Plureos GmbH
in Hamburg, Germany